Sparse Solutions of Underdetermined Linear Systems and Their Applications

Sparse Solutions of Underdetermined Linear Systems and Their Applications

Ming-Jun Lai

University of Georgia
Athens, Georgia

Yang Wang

Hong Kong University of Science and Technology
Hong Kong, China

Society for Industrial and Applied Mathematics
Philadelphia

Publications Director	Kivmars H. Bowling
Executive Editor	Elizabeth Greenspan
Developmental Editor	Mellisa Pascale
Managing Editor	Kelly Thomas
Production Editor	Louis R. Primus
Copy Editor	Julia Cochrane
Production Manager	Donna Witzleben
Production Coordinator	Cally A. Shrader
Compositor	Cheryl Hufnagle
Graphic Designer	Doug Smock

Library of Congress Cataloging-in-Publication Data
Names: Lai, Ming-Jun, author. | Wang, Yang, 1963- author.
Title: Sparse solutions of underdetermined linear systems and their
    applications / Ming-Jun Lai, Yang Wang.
Description: Philadelphia : Society for Industrial and Applied Mathematics,
    [2021] | Includes bibliographical references and index.
Identifiers: LCCN 2021002903 (print) | LCCN 2021002904 (ebook) | ISBN
    9781611976502 (paperback) | ISBN 9781611976519 (ebook)
Subjects: LCSH: Sparse matrices. | Linear systems.
Classification: LCC QA188 .L35 2021 (print) | LCC QA188 (ebook) | DDC
    512.9/434--dc23
LC record available at *https://lccn.loc.gov/2021002903*
LC ebook record available at *https://lccn.loc.gov/2021002904*

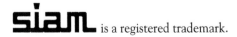 is a registered trademark.

Contents

List of Figures

List of Tables

List of Algorithms

Preface

A linear system of equations consists of a known matrix A and a known vector $\mathbf{b}$ such that $A\mathbf{x} = \mathbf{b}$ for an unknown vector $\mathbf{x}$, where A is of size $m \times n$. When $m = n$ and $m > n$, methods of solving for $\mathbf{x}$ are well known and are discussed in standard numerical analysis textbooks. However, when $m < n$, $A\mathbf{x} = \mathbf{b}$ is called an underdetermined linear system. Such a linear system has become a subject of research in the last 15 years as part of an extremely active study in the community of compressive sensing. Mathematically, an underdetermined linear system has many solutions. The aim of the study is to find a sparse solution in the sense that the number of nonzero entries of the solution is the smallest. Although there exists a sparse solution, the difficulty is that it may take years to find such an exact solution using our current computer power. Therefore, we must look for an approximation of the sparse solution which can be found within reasonable time and tolerance. Compressed sensing and sparse solutions of underdetermined linear system have been actively studied in many branches of mathematics, e.g., approximation theory, Banach space theory, combinatorics, discrete and continuous optimization, numerical analysis, and probability, as well as in many other sciences, e.g., computer science, electrical engineering, medical sciences, optimization, and statistics.

Many approaches, including constrained and unconstrained optimization methods, nonconvex minimization approaches, approximation theory based on greedy ideas, iterative thresholding techniques, and alternating projections, have been rigorously developed to approximate sparse solutions. Hence, many aspects of the sparse solution of underdetermined linear systems are now very well understood. Many algorithms are available to compute sparse solutions together with convergence theory and mathematical heuristics. They are ready for application problems. Therefore, it is time to summarize various parts of sparse solutions in book format and make them available to more people, including math/applied math graduate students, computer scientists, statisticians, and engineers.

The authors are active researchers in this subject and have done their best to digest a certain amount of the papers, make sure that the results are correct, and then rewrite them in a single book. In addition, many results are added in the format of exercises after each chapter to provide the reader with a deeper understanding of the material. This book is among the first few to specialize in the sparse solution of underdetermined linear systems. Such a book is necessary if one wants to understand linear systems completely and to be able to solve linear systems of any kind (square, tall, and narrow as well as flat and long rectangular systems).

The book consists of 11 chapters. The reader is advised to read Chapter 1 first and then decide which other chapters to read next. The first chapter is the most important of the book. It contains a definition of sparse solutions, why one needs to find them, how to solve them by various methods and approaches without much detail, and which chapter contains which type of method. Since the set of sparse vectors does not form a subspace or a convex set, we must introduce brand-new concepts and definitions, such as the restricted isometry property (RIP), the null space property (NSP), and the mutual coherence property, to deal with the mathematical analysis of convergence of various methods. These will appear in Chapter 2. Chapter 3 explains several greedy ideas and various greedy algorithms to find sparse solutions. Starting with an orthogonal matching pursuit (OMP) algorithm, we explain several generalized versions, e.g., economical OMP to improve the effectiveness and efficiency of OMP in finding sparse solutions. Under the RIP conditions or the mutual coherence conditions, we show that these algorithms are convergent. In Chapter 4, we develop other approaches, such as iterative hard threshold pursuit and the alternating projection algorithm, to compute sparse solutions. In Chapters 5 and 6, we explain various methods based on constrained and unconstrained optimization theory to find sparse solutions. Chapter 7 contains a lot of sufficient conditions to ensure that the solution of the ℓ_1 convex minimization is the sparse solution. In addition, we explain some sufficient conditions for the solution to the ℓ_q quasi-norm minimization to be a sparse solution. In Chapter 8, we explain that many random matrices satisfy the RIP. We will show that the graph Laplacian matrices satisfy the RIP in Chapter 10. The last three chapters, Chapters 9, 10, and 11, are extensions and applications of the sparse solution of underdetermined linear systems. For instance, matrix completion and low rank matrix recovery are typical extensions of the sparse solution. The classic graph clustering problem can be converted into a sparse solution problem, which leads to an efficient algorithm. For another example, the sparse phase retrieval problem is to find a solution of an overdetermined system of quadratic equations. If one knows the sign of the square root of each quadratic equation, one can solve the corresponding linear system. Because the choices of signs increase exponentially as the size of the problem increases, it is impossible to exhaust all the possibilities of linear systems within a reasonable time. Thus, how to find approximations is similar to the study of sparse solutions of underdetermined linear systems. There are so many approaches to finding sparse solutions. Each approach has its own merits and advantages as well as disadvantages. The authors include several algorithms, e.g., Wirtinger flow, Gauss–Newton, and DC-based algorithms, for the reader to choose from. Certainly, it is impossible to summarize all the results in this book. The authors would like to apologize in advance to the many researchers in the community of compressed sensing and other areas of study for having not included their favorable results and/or even those results which are better than those in this book. One of the useful features of this book is that it contains 64 algorithms, some of them already in MATLAB format, which can be applied to practical problems immediately or after a variation. These 64 algorithms are strongly supported by the mathematical theories provided in this book. They are explained in a self-contained format and an easy-to-understand fashion.

This book arose from our studies, so some theorems are named according to the papers we found them in. Some of the results may earlier have appeared in other papers. We apologize to those researchers whose papers we had not read before the publication of this book. Also, several parts of the book are based on colloquium talks at various places, including Auburn University, Beihang University, Brown University, Case Western Reserve University, Chinese University of Science and Technology, Georgia Institute of Technology, Georgia Southern University, Harvard University, Hong Kong University of Science and Technology, Purdue University, Rice University, Southwest Jiaotong University, Suzhou University, Tufts University, University of Alabama at Tuscaloosa, University of Calgary, University of California at Los Angeles, Universität Heidelberg, and Zhejiang University, as well as Oak Ridge National Labs and Institute of Computational Mathematics, Academy of Mathematical and Systems Science, and Chinese Academy of Sciences. The authors added more material to these talks to expand them into the current book format. Hopefully, more people will be interested in this book, leading them to do research on these topics and compute sparse solutions of underdetermined linear systems for practical problems. We strongly recommend that the reader do more literature review beyond the references in the book.

The authors would like to take this opportunity to thank their colleagues and collaborators—Kenneth Allen, Simon Foucart, Johnny Guzman, Xiaozhe Hu, Meng Huang, Jinsil Lee, Wenyuan

Liao, Song Li, Louis Y. Liu, Zhaosong Lu, Daniel Mckenzie, Xiaoli Meng, Eric Perkerson, Alexander Petukhov, Zhaiming Shen, Abraham Varghese, Jie Wang, Zheng Wang, Jinmin Wen, Jiaxin Xie, Yangyang Xu, Zhiqiang Xu, Jieping Ye, Wotao Yin, Guannan Zhang, and Rui Zhang—for stimulating discussions on the study of sparse solutions of underdetermined linear systems, various parts of the theory on compressed sensing, phase retrieval, matrix completion, graph clustering, matrix cross approximation, matrix sparsification, optimization, etc. In particular, both authors would like to thank Dr. Rui Zhang for his patience in reading the book thoroughly and checking the correctness of some chapters. In addition, both authors would like to take this opportunity to thank Dr. Charles K. Chui for his encouragement and support for many years.

The first author would like to thank the National Science Foundation for its constant support of his research on multivariate splines (both polynomial splines over triangulation and GBC splines over polygonal partitions) and their applications to constructing multivariate wavelets and wavelet frames and reconstructing smooth curves and surfaces from given data points, and for numerical solution of linear and nonlinear partial differential equations, e.g., Navier–Stokes equations, etc.

Finally, both authors are very grateful to their wives, Lingyun Ma and Weixin Xu, for their understanding and support.

Ming-Jun Lai and Yang Wang

University of Georgia,
Athens, GA
U.S.A.

and

Hong Kong University of Science and Technology
Clear Water Bay
Kowloon, Hong Kong, China

Chapter 1

Introduction

1.1 • Sparse Solutions

Let m, n be two positive integers, Φ be a matrix of size $m \times n$, and $\mathbf{b}$ be a vector of size $m \times 1$. We are interested in numerically solving or approximating the solution of the linear system

$$\Phi\mathbf{x} = \mathbf{b}. \tag{1.1}$$

In practice, we also deal with $\Phi\mathbf{x} \approx \mathbf{b}$ because there may be some noise in $\mathbf{b}$ and/or Φ. When $m = n$ and Φ is invertible, the solution $\mathbf{x}$ is unique, and we often use Gaussian elimination or an iterative method such as Gauss–Jacobi, Gauss–Seidel, steepest descent, or conjugated gradient iterative algorithms to solve the linear system (1.1). When $m > n$, the linear system is called overdetermined. In this situation, we solve $A^\top A\mathbf{x} = A^\top \mathbf{b}$ or find a least squares solution to $\min_{\mathbf{x} \in \mathbb{R}^n} \|\Phi\mathbf{x} - \mathbf{b}\|_2$, where $\|\mathbf{x}\|_2$ is the standard Euclidean norm of vector $\mathbf{x}$. Even when Φ is not of full rank, we look for the least squares solution which has the minimum norm, i.e., the pseudo-inverse solution. However, when $m < n$, $\Phi\mathbf{x} = \mathbf{b}$ is an underdetermined linear system with many possible solutions. We naturally look for a solution $\mathbf{x}$ with the least number of nonzero entries. Such a solution is called a sparse solution to the underdetermined linear system $\Phi\mathbf{x} = \mathbf{b}$. This leads to the problem of finding sparse solutions of underdetermined linear systems, which is the central topic of this book. When $m \ll n$, i.e., m is much smaller than n, finding the sparse solution is more complicated and thus more challenging and more interesting to study in research and in practice. Note that these underdetermined linear systems arise from many engineering and applied sciences problems, including those in the biological and social sciences. Indeed, often there are many contributing factors to a system of interest, but some factors are more important than others. Also, measurements of the system are usually hard to observe or difficult to carry out. Instead of obtaining all the necessary observations to form a linear system with $m = n$ or $m > n$, we may simply solve an underdetermined linear system for a sparse solution to find major contributing factors.

Let us state the central problem more precisely. Throughout this book, we use $\|\mathbf{x}\|_0$ to denote the number of nonzero entries of $\mathbf{x}$. Let $s > 0$ be an integer, and let

$$\mathcal{R}_s := \mathcal{R}_s(\mathbb{R}^n) = \{\Phi\mathbf{x}, \mathbf{x} \in \mathbb{R}^n, \|\mathbf{x}\|_0 \le s\} \tag{1.2}$$

be the range of Φ of all the s-sparse vectors $\|\mathbf{x}\|_0 \le s$. Furthermore, let

$$\mathcal{D}_s := \mathcal{D}_s(\mathbb{R}^n) = \{\mathbf{x} \in \mathbb{R}^n, \quad \|\mathbf{x}\|_0 \le s\} \tag{1.3}$$

1

be the collection of all s-sparse vectors. Note that $\mathcal{R}_s$ or $\mathcal{D}_s$ is neither a linear space nor a convex set. Fix an integer $s > 0$ called the sparsity. Given $\mathbf{b} \in \mathcal{R}_s$, we would like to find a vector $\mathbf{x}$ in $\mathcal{D}_s \subset \mathbb{R}^n$ which solves the minimization problem

$$\min\{\|\mathbf{x}\|_0, \quad \mathbf{x} \in \mathbb{R}^n, \Phi\mathbf{x} = \mathbf{b}\}. \tag{1.4}$$

Under the assumption that Φ is of full rank, the above minimization will have a solution $\mathbf{x} \in \mathbb{R}^n$ for any given vector $\mathbf{b}$, since the feasible set $\{\mathbf{x} \in \mathbb{R}^n, \Phi\mathbf{x} = \mathbf{b}\}$ is not empty. The problem is how to find $\mathbf{x} \in \mathcal{D}_s$ which solves (1.4) under two settings: (1) the sparsity s is not precisely given but can be assumed to exist, and (2) s is given.

As mentioned above, the solution of the minimization problem (1.4) is called the sparse solution of the underdetermined linear system $\Phi\mathbf{x} = \mathbf{b}$. Φ is often called a sensing matrix, and $\mathbf{b}$ is called a measurement vector. Note that this optimization problem is not a traditional one in the sense that the minimizing function $\|\mathbf{x}\|_0$ is not a continuous function of $\mathbf{x}$ and the feasible set is not convex.

Write $\Phi = [\phi_1, \phi_2, \ldots, \phi_n]$ with column vector $\phi_i, i = 1, \ldots, n$. We can choose $A = [\phi_{i_1}, \ldots, \phi_{i_m}]$ from Φ. If A is nonsingular, we can find a solution $\mathbf{z}$. Otherwise, we choose another A. By exhausting all $m \times m$ nonsingular submatrices A from Φ and solving these sublinear systems of equations, we can see which solution has the smallest number of nonzero entries. Although this is a completely correct mathematical idea, it may not be a realizable approach. Note that for a sensing matrix of size $m \times n$, there could be $\binom{n}{m}$ such nonsingular linear systems from $\Phi\mathbf{x} = \mathbf{b}$ to be solved. Consider a rectangular matrix $\Phi = [(x_j)^i]_{i=0,\ldots,m, j=1,\ldots,n}$ for distinct real numbers x_i and any $m \times m$ submatrix from Φ of full rank. The combinatoric number $\binom{n}{m} = \frac{n!}{(n-m)!m!}$ grows exponentially fast as m and n go to ∞. For example, when $n = 2m$, we use Stirling's formula $n! = \sqrt{2\pi n}\left(\frac{n}{e}\right)^n e^{\theta/(12n)}$ with $\theta \in (0, 1)$ to get

$$\binom{n}{m} = \frac{(2m)!}{(m!)^2} = \frac{\sqrt{2}}{\sqrt{2\pi m}} 2^{2m} e^{\theta_1/(24m) - \theta_2/6m},$$

where θ_1, θ_2 are real positive numbers in $(0, 1)$. For the common case $m = 512$ and $n = 1024$, we need to solve at least 2^{512} linear systems of size 512×512. Let us be more precise:

2^{512} = 13407807929942597099574024998205846127479365820592393377723561443721
       76403007354697680187429816690342769003185818648605085375388281194656
       9946433649006084096.

This is impossible to do within a reasonable time. It is a nonpolynomial time (NP) problem (cf. [50]). That is, although the idea is correct, the approach cannot solve the minimization problem (1.4) in real time. Hence, we have to seek a more effective, more efficient approach in order to find the sparse solution or an approximation of the sparse solution.

1.2 ▪ Motivations

Let us present several examples to motivate the need to find such sparse solutions.

Example 1.1 (Free-Knot Linear Splines). It is well known that for any continuous function h, one can use a piecewise linear interpolant L_h (called a linear spline) of h to approximate h. A linear spline L_h can approximate h very well over any given bounded interval as long as the number of knots (the locations of corners of the linear spline) is sufficiently large and the knots are evenly spread out over the given interval. See Figure 1.1 (left). However, for some functions h, one does not need so many knots to approximate h well, as shown in Figure 1.1 (right). We

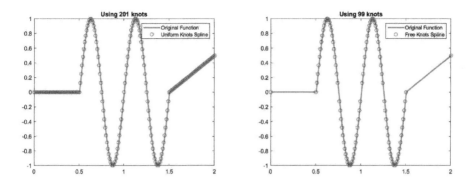

Figure 1.1. *The uniform knot linear spline and a free-knot spline (only values at knots, indicated in circles, are shown).*

are interested in finding such an economical approximation S_h of h with a minimal number of knots. The spline S_h is called the free-knot spline problem (cf. Chapter 7 in [53]).

More precisely, given a continuous function h defined on $[a, b]$ and given an tolerance ϵ, we need to find a continuous piecewise linear function called the linear spline S_h based on a knot sequence $\mathcal{P}_n = \{a = y_0 < y_1 < \cdots < y_n < y_{n+1} = b\}$ with minimal number n such that $|h(t) - S_h(t)|_\infty \leq \epsilon$ for all $t \in [a, b]$. Then S_h will be the most economical representation of h. However, the difficulty is in finding such a sequence $\mathcal{P}_n$. Without loss of generality, we may assume that the interval $[a, b] = [0, 1]$. Intuitively, such an optimal sequence exists and there may be more than one such sequence with the minimal number n^* in general. Let us write $\mathcal{P}^*$ to be an optimal sequence with $\mathcal{P}^* = \{0 < y_1^* < y_2^* < \cdots < y_n^* < 1\}$. Assume that we will be happy with our computational solution $y_k, k = 1, \ldots, n^*$, with a given tolerance $\delta > 0$ in the sense that $|y_k - y_k^*| \leq \delta, k = 1, \ldots, n^*$. One way to attack the problem is to choose a sufficiently large integer n and set $y_k = k/(n+1), k = 0, \ldots, n+1$, to be a sequence $\mathcal{P}_n$ of the knots such that $|y_k - y_{k+1}| \leq \delta$ and $\|h(x) - S_h(x, \mathcal{P}_n)\|_\infty \leq \epsilon$, where $S_h(x, \mathcal{P}_n) := \sum_{k=0}^{n} c_k (x - k/(n+1))_+$, where $(x - b)_+$ is the so-called ReLU function or activated function or truncated power function of linear order. Next we should remove some y_k's to choose the least number of knots to approximate h. This can be done by making some coefficients $c_k = 0$ in S_h while adjusting other coefficients in S_h. That is, we introduce a regularization term $\|\mathbf{c}\|_0$, where $\mathbf{c} = (c_0, c_1, \ldots, c_n)^\top \in \mathbb{R}^{n+1}$ and $\|\mathbf{c}\|_0$ is the number of nonzero entries in $\mathbf{c}$. Hence, we solve the following minimization problem:

$$\min_{\mathbf{c} \in \mathbb{R}^{n+1}} \{\|\mathbf{c}\|_0 : \|h(x) - S_h(x, \mathcal{P}_n)\|_\infty \leq \epsilon\}. \tag{1.5}$$

Next we further assume that $h(x)$ is known at a sequence of locations x_j for $j = 1, \ldots, m$, which may not be the same sequence of knots, and we will be happy if $|h(x_j) - S_h(x_j, \mathcal{P}_n)| \leq \epsilon, j = 1, \ldots, m$. That is,

$$\min_{\mathbf{c} \in \mathbb{R}^{n+1}} \{\|\mathbf{c}\|_0 : |S_h(x_j, \mathcal{P}_n) - h(x_j)| \leq \epsilon, j = 1, \ldots, m\}. \tag{1.6}$$

This can be formulated as follows. Let

$$\Phi = [(x_i - k/(n+1))_+]_{1 \leq i \leq m, 1 \leq k \leq n}$$

be the sensing matrix. Write $\Phi\mathbf{c} = \mathbf{h}$ with $\mathbf{h} = [h(x_1), \ldots, h(x_m)]^\top$. We consider

$$\min_{\mathbf{c} \in \mathbb{R}^{n+1}} \{\|\mathbf{c}\|_0 : \|\Phi\mathbf{c} - \mathbf{h}\|_2 \leq \epsilon, j = 1, \ldots, m\}. \tag{1.7}$$

This problem can now be solved by using the techniques in this book. We use one of our solution methods to find a desired solution, as shown in Figure 1.1, and the maximum errors are $1e-14$ for both the uniform knot spline and the free-knot spline.

Example 1.2 (Variable Selection). A popular data fitting problem is the least squares regression. That is, for a given set of data $\{(x_i, y_i), i = 1, \ldots, n\}$, one chooses an appropriate basis function such as $\{1, x\}$ to fit the data values to see the pattern of the data. Sometimes, if the data exhibits a periodic property or an exponential decay property, one uses $\sin(x), \cos(x)$, or $\exp(-x)$ to model the given data set. This is a very common approach in statistical analysis. In general, one does not know which basis functions are most appropriate. Furthermore, one wishes to find the best basis functions without a human interface. To do so, one approach is to use all the basis functions or at least as many as possible. Let Φ be the matrix based on the evaluation of the basis functions at the given data locations and $\mathbf{b}$ be the vector of the given data values. Let $\mathbf{x}$ be the coefficient vector of the basis functions. Then one solves

$$\min_{\mathbf{x} \in \mathbb{R}^n} \{\|\mathbf{b} - \Phi\mathbf{x}\|_2, \quad \|\mathbf{x}\|_0 \leq s\} \tag{1.8}$$

for an appropriate sparsity s. The residual $\|\mathbf{b} - \Phi\mathbf{x}\|_2$ for various s will help us decide which s is the most appropriate. Note that Φ may be possible for an overdetermined linear system.

To solve the minimization in (1.8), we may minimize

$$\min_{\mathbf{x} \in \mathbb{R}^n} \{\|\mathbf{b} - \Phi\mathbf{x}\|_2^2 + 2\lambda\|\mathbf{x}\|_0\} \tag{1.9}$$

by adjusting $\lambda > 0$. However, it is difficult to deal with the $\|\mathbf{x}\|_0$ in the above minimization. One approach is to replace $\|\mathbf{x}\|_0$ by $\|\mathbf{x}\|_q^q$ with $q \in [0, 2]$. This leads to

$$\min_{\mathbf{x} \in \mathbb{R}^n} \{\|\mathbf{b} - \Phi\mathbf{x}\|_2^2 + 2\lambda\|\mathbf{x}\|_q^q\}. \tag{1.10}$$

Any global or local minimizer of the minimization (1.10) is called a bridge estimator in the statistical literature (cf. [33]). In particular, when $q = 1$, it is called the lasso estimator in [54]. A good method of approximating the solution of (1.10) is to consider the minimization

$$\min_{\mathbf{x} \in \mathbb{R}^n} \{\|\mathbf{b} - \Phi\mathbf{x}\|_2^2 + \lambda\|\mathbf{x}\|_1 + \nu\|\mathbf{x}\|_2^2\}, \tag{1.11}$$

which is called the naive elastic net approach (cf. [58]). We refer the reader to [37] for the statistical approach to sparsity analysis.

Example 1.3 (Error Correcting Code). Let $\mathbf{z}$ be a vector encoded for $\mathbf{x}$ by a redundant linear system A of size $m \times n$ with $m > n$. Then the code $\mathbf{z} = A\mathbf{x}$ is transmitted through a noisy channel. The channel corrupts some random entries of $\mathbf{z}$, resulting in a new vector $\mathbf{w} = \mathbf{z} + \mathbf{v}$. Finding the vector $\mathbf{v}$ is equivalent to correcting the errors.

To do so, we extend A to a square matrix B of size $m \times m$ by adding $A^\perp$, i.e., $B = [A; A^\perp]$. Assume that A satisfies $A^\top A = I_n$, where I_n is the identity matrix of n. Then we can choose $A^\perp$ such that $BB^\top = I_m$. Now

$$B^\top \mathbf{w} = B^\top \mathbf{z} + B^\top \mathbf{v} = \begin{bmatrix} \mathbf{x} \\ 0 \end{bmatrix} + \begin{bmatrix} A^\top \mathbf{v} \\ (A^\perp)^\top \mathbf{v} \end{bmatrix}.$$

Let $\mathbf{y} = (A^\perp)^\top \mathbf{v}$, which is the last $m - n$ entries of $B^\top \mathbf{w}$. Note that the size of $\mathbf{v}$ is $m \times 1$ and the size of $(A^\perp)^\top$ is $(m - n) \times m$. If the receiver can solve the minimization problem (1.4) with $\Phi = (A^\perp)^\top$, i.e., he can find $\mathbf{v}$ such that $\mathbf{y} = (A^\perp)^\top \mathbf{v}$, then he gets the correct $\mathbf{z}$. Thus, this error correcting problem is equivalent to the sparse solution problem (1.4). See [16] and [17] for more detail.

Example 1.4 (Best s-Term Approximation). Let $\mathbf{D} = \{\phi_1, \ldots, \phi_n\}$ be a collection of vectors in $\mathbb{R}^m$ with $n \gg m$. Given a vector $\mathbf{b} \in \mathbb{R}^m$, we look for the best s column vectors $\phi_{i_1}, \ldots, \phi_{i_s}$ in the collection $\mathbf{D}$ to approximate $\mathbf{b}$. That is,

$$\left\| \mathbf{b} - \sum_{k=1}^{s} c_k \phi_{i_k} \right\|_2 = \min\{\|\mathbf{b} - \Phi\mathbf{x}\|_2, \|\mathbf{x}\|_0 \le s\},$$

where $\Phi = [\phi_1, \phi_2, \ldots, \phi_n]$. When s is large enough, say $s \ge m$, the minimal value above is zero. One problem is to find the smallest integer s such that the minimal value is zero. One way to solve this problem is to solve (1.4). In this way, we find s as well as the solution $(c_1, \ldots, c_s)$ and their indices $i_1, \ldots, i_s$ to express $\mathbf{b}$. See [20] for more explanation. In general, let $F = \{f_1, \ldots, f_n\}$ be a sequence of functions defined on a bounded domain Ω, say F is a collection of orthonormal polynomials over $[-1, 1]$. Suppose a function g has a sparse representation in terms of F. That is, $g = \sum_{k=1}^{s} c_k f_{i_k}$ with $s \ll n$, where c_k is the inner product of g with f_{i_k}. To find these nonzero coefficients c_k, we have to compute all the inner products $\langle g, f_i \rangle$. Note that these inner products have to be computed by using numerical quadratures, which get more and more difficult when n is large. Instead, we choose a set of discrete points $-1 \le t_1 < t_2 < \cdots < t_m = 1$ of $[-1, 1]$ and form vectors $\phi_i = [f_i(t_j), j = 1, \ldots, m]^\top$ as well as $\mathbf{b} = [g(t_1), \ldots, g(t_m)]^\top$ and solve (1.4) to obtain $c_1, \ldots, c_s$ to express $g = \sum_{k=1}^{s} c_k f_{i_k}$. See [29] for an example.

Example 1.5 (Data Compression). It is known that a tight wavelet frame Φ of size $m \times n$ with $m < n$ is usually an excellent tool to represent a signal or an image $\mathbf{b}$. See [44] and [35] for a general method of constructing tight wavelet frames using multivariate box splines and image edge analysis. To compress an image $\mathbf{b}$, we look for a sparse approximation $\mathbf{x}$ satisfying

$$\min\{\|\mathbf{x}\|_0, \quad \mathbf{x} \in \mathbb{R}^n, \|\Phi\mathbf{x} - \mathbf{b}\| \le \theta\}, \tag{1.12}$$

where $\theta > 0$ is a tolerance. In particular, for lossless compression, i.e., $\theta = 0$, (1.12) is the central problem (1.4) of this book. Similarly, when $\mathbf{b} = \mathbf{y} + \mathbf{e}$ with noise in vector $\mathbf{e}$, we may use (1.12) to remove the noise and recover the original image $\Phi\mathbf{x}$. Because m is usually very large, e.g., an image of size 512×512 pixel values, the right-hand side $\mathbf{b}$ is of size $512^2 \times 1$. Since $n \gg m$, solving (1.4) with such a pair of m, n is still a challenge to the best of the authors' knowledge. Certainly, in practice, we cut a 512×512 image into many 32×32 blocks and use a dictionary Φ of size $m \times n$ with $m = 32^2$ and $n \gg m$ to compress each block of the image.

Example 1.6 (Cryptography). Suppose that we have a class of matrices Φ of size $m \times n$ with $m < n$ which admit a computationally efficient algorithm for solving the minimization (1.4). Let Ψ be an invertible random matrix of size $m \times m$ and $A = \Psi\Phi$.

Suppose that a sender (e.g., an online shopper) needs to send their credit card number and expiration date through the internet. The receiver (the online store) sends to the sender the matrix A in a public channel. After receiving A, the sender computes $\mathbf{z} = A\mathbf{x}$ and sends $\mathbf{z}$ back to the receiver in a public channel. As mentioned above, the solution of the problem of finding the sparsest solution $\mathbf{x}$ from $\mathbf{z}$ takes nonpolynomial time. With overwhelming probability, this $\mathbf{x}$ cannot be found by other parties.

However, the receiver is able to get $\mathbf{x}$. Indeed, the receiver first computes $\mathbf{y} = \Psi^{-1}\mathbf{z} = \Phi\mathbf{x}$ and then uses the supposed known solution method to find $\mathbf{x}$, which is the central problem (1.4). By changing Ψ frequently enough, the receiver is able to get secured data every time by preventing hackers from developing a sparse solver for any particular system.

Figure 1.2. *A given image with wires in the sky (left) and an in-painted image.*

Example 1.7 (Lost Data Recovery or Image In-painting). The next motivation is to recover lost data. For example, an image has an undesired part which needs to be removed or replaced, so the image needs a repair. See Figure 1.2 for an image with wires (left). We need to replace these wires with an appropriate fill-in, as shown on the right of Figure 1.2. More precisely, let $\mathbf{z}$ be an image and $\widetilde{\mathbf{z}}$ a partial image of $\mathbf{z}$. That is, $\mathbf{z}$ loses some data to become $\widetilde{\mathbf{z}}$. Suppose that we know or we can find the location where the data are lost. We would like to recover the original image $\mathbf{z}$ from the partial known image $\widetilde{\mathbf{z}}$. Let Φ be a tight wavelet frame or a dictionary of size $m \times n$ with $n \gg m$, as mentioned in previous examples, such that $\Phi\mathbf{x} = \mathbf{z}$. Suppose that $\mathbf{z}$ has a sparse representation with a sparse solution $\mathbf{x}$ under this tight wavelet frame or dictionary Φ. Let Ψ be the residual matrix from Φ by dropping off the rows corresponding to the unavailable entries, i.e., the missing data locations. Say the number of known entries in $\mathbf{z}$ is ℓ. Note that $\Phi\Phi^\top = I_m$ and, hence, $\Psi\Psi^\top = I_\ell$ with $\ell < m$. It follows that $\Psi\mathbf{x} = \Psi\Phi^\top\mathbf{z} = \widetilde{\mathbf{z}}$ by the orthonormality of the rows of Φ. Thus, we need to find the sparse solution $\mathbf{x}$ from the given $\widetilde{\mathbf{z}}$ such that $\widetilde{\mathbf{z}} = \Psi\mathbf{x}$, which is exactly the same problem as the central problem (1.4) which will be studied in this book. Once we have $\mathbf{x}$, we can find $\mathbf{z}$, which is $\mathbf{z} = \Phi\mathbf{x}$. (See Chapter 9, where the image in-painting problem in this example is formulated as a matrix completion problem.)

Example 1.8 (Economical Representation of a PDE Solution). Consider a standard boundary value problem of the Poisson equation:

$$\left\{ -\Delta u = f, \quad \Omega \subset \mathbb{R}^2 u = 0, \quad \text{on } \partial\Omega \right\}, \tag{1.13}$$

where Ω is a bounded domain with Lipschitz boundary $\partial\Omega$. Let $S_n, n \geq 1$, be a sequence of nested finite-dimensional subspaces of the Sobolev space $H_0^1(\Omega)$, i.e., $S_1 \subset S_2 \subset \cdots \subset S_n$. For example, we can choose a nested triangulation $\triangle_n$ of Ω and let $S_n = S_d^1(\triangle_n)$ be the bivariate spline space of degree d and smoothness one over triangulation $\triangle_n$. See [42] for theory and [2] for computational detail. Write $S_n = \text{span}\{\phi_{n,1}, \ldots, \phi_{n,N_n}\}$, where N_n is the dimension of S_n. In the weak formulation, we look for a solution $u_n \in S_n$ such that

$$\langle \nabla u_n, \nabla \phi_{n,j} \rangle = \langle f, \phi_{n,j} \rangle, \quad j = 1, \ldots, N_n,$$

where $\langle f, g \rangle$ stands for the inner product of the Hilbert space $L_2(\Omega)$. Since the solution u may change rapidly over one subregion and slowly over other regions of Ω, it is not very economical to use a space of high dimensions, or more dimensions than necessary, to approximate the solution. Traditionally, in order to achieve high accuracy, one has to refine a triangulation many times to find an approximating solution. Hence, the dimension of the spline space is increasing significantly and one needs to use a lot of coefficients to represent an approximation. However, the solution may not need many coefficients over the regions when the solution changes slowly, e.g., the solution behaves like a constant or a linear polynomial. Thus, we let

$$\Phi_j = [a(\phi_{n,i}, \phi_{j,1}), a(\phi_{n,i}, \phi_{j,2}), \ldots, a(\phi_{n,i}, \phi_{j,N_j})]_{i=1,\ldots,N_n}, \quad j = 1, \ldots, n,$$

where $a(\phi_{j,1}, \phi_{n,i}) = \langle \nabla \phi_{j,1}, \nabla \phi_{n,i} \rangle$ for all $j = 1, \ldots, n, i = 1, \ldots, N_n$; $\Phi = [\Phi_1, \Phi_2, \ldots, \Phi_n]$; and $\mathbf{b} = [\langle f, \phi_{n,1} \rangle, \ldots, \langle f, \phi_{n,N_n} \rangle]^\top$. We look for a solution $\mathbf{x} \in \mathbb{R}^{N_1 + \cdots + N_n}$ such that

$$\min\{\|\mathbf{x}\|_0, \quad \Phi \mathbf{x} = \mathbf{b}\}.$$

Let $\mathbf{x_b}$ be the sparse solution with $\|\mathbf{x_b}\|_0 < N_n$. Write

$$\Psi = [\phi_{1,1}, \ldots, \phi_{1,N_1}, \phi_{2,1}, \ldots, \phi_{2,N_2}, \ldots, \phi_{n,1}, \ldots, \phi_{n,N_n}]$$

and let $u^* = \Psi \mathbf{x_b}$. Then $u^* \in S_n$ and satisfies

$$\langle \nabla u^*, \nabla \phi_{n,j} \rangle = \langle f, \phi_{n,j} \rangle \quad \forall j = 1, \ldots, N_n.$$

By the uniqueness of the weak solution, u^* is the weak solution in S_n for (1.13). However, the number of nonzero coefficients is the smallest. That is, it is the most economical representation of the weak solution in the nested subspace sequence $\{S_1, S_2, \ldots, S_n\}$. We can show that the economical representation of the Poisson equation achieves the same approximation error as the standard spline weak solution. This approach may be compared with the standard adaptive algorithms for numerical solution of the PDE. See [38] for more detail.

Example 1.9 (Portfolio Selection). In 1952, Markowitz introduced a mathematical framework for assembling a portfolio of assets and proposed a mean variance model for portfolio selection (see [48]). This won him a Nobel Prize in Economics in 1990. The Markowitz model can be described as follows:

$$\min_{\mathbf{x}} \mathbf{x}^\top C \mathbf{x}, \text{ such that } \mathbf{x}^\top \mathbf{p} = \rho, \sum_{i=1}^n \mathbf{x}_i = 1, \tag{1.14}$$

where $\mathbf{x} = (x_1, \ldots, x_n)^\top$ is the portfolio vector with each x_i associated with the ith asset, C is the covariance matrix of returns which can be obtained from historical data, $\mathbf{p}$ is a vector of expected return rates from all assets, and ρ is the total expected return.

Clearly, we cannot pay attention to all assets, so we need to limit the number of positions to monitor, buy, or sell. Thus, we need a model to promote the sparsity from all positions in $\mathbf{x}$. It is reasonable to select a small number of assets. Let us rewrite the Markowitz model (1.14) first. Note that $\mathbf{x}^\top C \mathbf{x} = \mathcal{E}(\mathbf{x}^\top (\mathbf{p}_t - \mathbf{p})(\mathbf{p}_t - \mathbf{p})\mathbf{x}) = \mathcal{E}(|\mathbf{p}_t^\top \mathbf{x} - \rho|^2)$, where $\mathcal{E}$ stands for the expectation. For the empirical implementation, we replace expectations by sample averages due to the law of large numbers. We can recast the Markowitz model as follows:

$$\min_{\mathbf{x}} \frac{1}{N} \sum_{t=1}^N (\mathbf{p}_t^\top \mathbf{x} - \rho)^2, \text{ subject to } \mathbf{x}^\top \hat{\mathbf{p}} = \rho, \sum_{i=1}^n \mathbf{x}_i = 1. \tag{1.15}$$

Let us use $\mathbf{1} = (1, 1, \ldots, 1) \in \mathbb{R}^n$ and $\Phi = [\mathbf{p}_1^\top \cdots \mathbf{p}_n^\top]$. Then we can formulate the following:

$$\min_{\mathbf{x}} \|\Phi \mathbf{x} - \rho \mathbf{1}\|^2 + \lambda \|\mathbf{x}\|_1, \text{ subject to } \mathbf{x}^\top \hat{\mathbf{p}} = \rho, \sum_{i=1}^n \mathbf{x}_i = 1. \tag{1.16}$$

When $\lambda = 0$, the above is our sparse solution problem. See [8] for $\lambda > 0$ for more detail.

Example 1.10 (Compressed Sensing). Let $\mathbf{y} \in \mathbb{R}^n$ be a vector, and suppose that $\mathbf{y}$ is a sparse vector such that $\|\mathbf{y}\|_0 \leq s < m \ll n$. We would like to use a set of measurements, say vector inner products, to find the entries of $\mathbf{y}$. That is, we look for a sensing matrix Φ of size $m \times n$ with $m > s$ and a computational algorithm called $\triangle$ such that $\triangle \Phi \mathbf{y} = \mathbf{y}$. Of course, if Φ is the identity matrix of size $n \times n$, we can find the entries of $\mathbf{y}$ easily. The key points are (1) to design a sensing matrix Φ of size $m \times n$ with $m \ll n$ and (2) to design a computational algorithm $\triangle$ to find $\mathbf{y} = \mathbf{x_b}$ from $\mathbf{b}$ by first measuring $\mathbf{x_b}$ by the rectangular matrix Φ to form $\mathbf{b} = \Phi \mathbf{x_b}$ and then computing $\triangle(\mathbf{b})$ to obtain $\mathbf{x_b} = \triangle(\mathbf{b})$ for any sparse vector $\mathbf{x_b} \in \mathcal{D}_s \subset \mathbb{R}^n$. That is, we find the matrix Φ of size $m \times n$ and $\mathbf{x_b} \in \mathbb{R}^n$ with $\|\mathbf{x_b}\|_0 \leq s$ such that

$$[\Phi, \mathbf{x_b}] := \arg \min_{\Phi \in \mathbb{R}^{m \times n}, \mathbf{x} \in \mathbb{R}^n} \{\|\Phi \mathbf{x} - \mathbf{b}\|_2^2, \|\mathbf{x}\|_0 \leq s\} \tag{1.17}$$

for a class of observation vectors $\mathbf{b}$. Thus, if Φ is known, letting $\mathbf{b} = \Phi \mathbf{x_b} \in \mathcal{R}_s$ for any $\mathbf{x_b} \in \mathcal{D}_s$ in (1.4), the solution of (1.4) will find the entries of the vector $\mathbf{x_b}$. See [12], [25], [3], [9], and references therein for more explanation of this research topic. We refer the reader to [56] for discussion of constructing various sensing matrices.

1.3 ▪ Basic Approaches

In order to fully understand the computation of (1.4), we next introduce many approaches. One is to solve the minimization problem

$$\min\{\|\mathbf{x}\|_0, \quad \mathbf{x} \in \mathbb{R}^n, \|\Phi \mathbf{x} - \mathbf{b}\|_2 \leq \delta\}, \tag{1.18}$$

where $\mathbf{b} = \Phi \mathbf{x_b} + \mathbf{z}$, with $\|\mathbf{z}\|_2 \leq \epsilon$ and $\|\mathbf{x_b}\|_0 \leq s$. That is, $\mathbf{b}$ has some measurement error and we compute a solution $\mathbf{x}$ within accuracy δ. This is a δ variance of the (1.4) problem. It is called a *noisy recovery problem*. The original problem (1.4) is referred to as the *noiseless sparse recovery problem*.

In addition, the sensing or measurement matrix Φ may have some noise or errors. We consider a totally perturbed compressive sensing problem: Suppose that $\mathbf{b} = \Phi \mathbf{x}^* + \mathbf{e}$, and let $\widetilde{\Phi} = \Phi + E$, where E is a small perturbation. Defining

$$\mathbf{x}^{\#} := \arg\min\{\|\mathbf{x}\|_0 : \quad \mathbf{x} \in \mathbb{R}^n, \|\widetilde{\Phi} \mathbf{x} - \mathbf{b}\|_2 \leq \delta\}, \tag{1.19}$$

we see that $\mathbf{x}^{\#}$ is a good approximation to $\mathbf{x}^*$ if $\|\mathbf{x}^*\|_2 \in \mathcal{D}_s$.

There are many approaches to numerically approximating the sparse solution in (1.4), (1.18), and (1.19). The study of these methods is necessary since the situation of sparse solutions is more complicated than the solution of the standard linear system $\Phi \mathbf{x} = \mathbf{b}$ with square matrix Φ. Indeed, for an underdetermined linear system Φ of size $m \times n$, the case $n = m + 1, \ldots, m + k$ for $k = o(m)$ is different from the case $n = 2m, \ldots, kn$ for $k = o(m)$. A method that works for $n = 2m$ may not work very well for $n = 10m$. When the sparsity s is small (relative to m), many methods work well. However, when the sparsity s is large, e.g., $s = m/2$ or $s \geq m/2$ but less than m, many methods in this book may fail to find a sparse solution. These hint that more research needs to be done beyond this book.

The methods discussed in this book can be summarized in three categories: constrained minimization approaches; unconstrained minimization approaches; and the other approaches, e.g., greedy approaches. There are two kinds of constrained minimization approaches: convex minimization and nonconvex minimization. Each convex minimization has a dual problem associated with it. There are many methods for solving the dual problem, e.g., the primal method and the dual method. Similarly, for the unconstrained minimization approach, the methods can be divided into two kinds: convex and nonconvex. Typical methods in the third approach are greedy

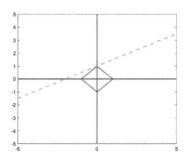 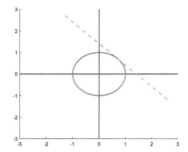

Figure 1.3. *The ℓ_1 norm will find a sparse solution of $\Phi\mathbf{x} = \mathbf{b}$, shown in light blue (left), and the ℓ_2 norm will not find a sparse solution (right).*

algorithms, iterative hard thresholding algorithms, screen rules, convex-concave programming (CCCP), alternating projection algorithms, etc. There are many methods in between these three approaches. We shall give a little more detail in this section.

1.3.1 ▪ Constrained Minimization Approaches

The first method to solve the problem in (1.4) is to use the constrained minimization

$$\min\{\|\mathbf{x}\|_1, \quad \Phi\mathbf{x} = \mathbf{b}, \mathbf{x} \in \mathbb{R}^n\}, \tag{1.20}$$

where $\|\mathbf{x}\|_1 = \sum_{i=1}^n |x_i|$ for $\mathbf{x} = (x_1, x_2, \ldots, x_n)^\top$. This is called the ℓ_1 minimization for sparse recovery.

The geometric meaning of the minimization (1.20) in $\mathbb{R}^n$ can be explained as follows. It is known that $\|\mathbf{x}\|_1 = c$ for $\mathbf{x} \in \mathbb{R}^n$ is a diamond-shaped function, as shown in the left graph of Figure 1.3. Note that the solution of $\Phi\mathbf{x} = \mathbf{b}$ forms an affine subspace or hyperplane. As shown in Figure 1.3, $\Phi\mathbf{x} = \mathbf{b}$ is only one linear equation with two variables (the light blue dashed line). The norm $\|\mathbf{x}\|_1$ may touch the hyperplane at a vertex which is a sparse solution. When the ℓ_2 norm $\|\mathbf{x}\|_2$ is used, we consider the minimization $\min\{\|\mathbf{x}\|_2, \Phi\mathbf{x} = \mathbf{b}\}$, and the solution is the point tangent to the hyperplane, which has two nonzero components unless the hyperplane is parallel to one of the coordinate planes. So the convex minimization (1.20) is a good approach for (1.4) to find a sparse solution.

Since the ℓ_1 minimization problem is equivalent to linear programming (to be explained in Chapter 6), this converts the problem into a convex minimization problem, which is a tractable computational problem. The ℓ_1 minimization problem is called *basis pursuit*. We shall explain for what kind of matrix Φ the minimization in (1.20) has a unique sparse solution and that the minimizer is the sparse solution we are looking for. To this end, we shall introduce the null space property (NSP), the spherical section property (SSP), the mutual coherence property, and the restricted isometry property (RIP) of rectangular matrices. These properties will be explained in Chapter 2. In addition, we shall give some sufficient conditions on Φ to ensure that the solution of (1.20) is a solution of (1.4). These will be discussed in later chapters.

Besides the standard interior point method or simplex method to numerically solve the minimization (1.20), there are also several other approaches. Because $\|\mathbf{x}\|_1$ is not a differentiable function, one approach is to regularize $\|\mathbf{x}\|_1$ by using $\|\mathbf{x}\|_{1,\epsilon} := \sum_{j=1}^n (x_j^2 + \epsilon^2)^{1/2}$ for another parameter $\epsilon > 0$ and considering

$$\min\{\|\mathbf{x}\|_{1,\epsilon}, \quad \Phi\mathbf{x} = \mathbf{b}, \mathbf{x} \in \mathbb{R}^n\}. \tag{1.21}$$

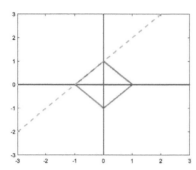 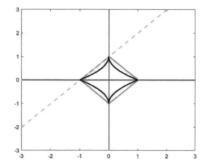

Figure 1.4. *The ℓ_1 norm will not find a sparse solution of $\Phi x = b$, the line shown in light blue (left graph), but the ℓ_q norm (the black curves in the right graph) minimization will find two sparse solutions.*

Another approach is to add $\beta \|x\|_2^2$ to regularize $\|x\|_1$ with $\beta > 0$ small enough. That is, we consider

$$\min \left\{ \|x\|_1 + \frac{\beta}{2} \|x\|_2^2, \quad \Phi x = b, x \in \mathbb{R}^n \right\}. \tag{1.22}$$

In the noisy recovery setting, we consider the minimization

$$\min\{\|x\|_1, \quad x \in \mathbb{R}^n, \|\Phi x - b\|_2 \le \delta\}. \tag{1.23}$$

Therefore, we have

$$\min\{\|x\|_{1,\epsilon}, \quad \|\Phi x - b\|_2 \le \delta, x \in \mathbb{R}^n\} \tag{1.24}$$

and

$$\min \left\{ \|x\|_1 + \frac{\beta}{2} \|x\|_2^2, \quad \|\Phi x - b\|_2 \le \delta, x \in \mathbb{R}^n \right\}. \tag{1.25}$$

Next let us introduce the ℓ^q quasi-norm for $x \in \mathbb{R}^n$. That is, for $0 < q < 1$, we define

$$\|x\|_q = \left(\sum_{i=1}^n |x_i|^q \right)^{1/q}$$

to be the standard ℓ^q quasi-norm for $x \in \mathbb{R}^n$. The following result is easy to see.

Lemma 1.11. *We have*

$$\lim_{q \to 0+} \|x\|_q^q = \|x\|_0 \tag{1.26}$$

for any nonzero vector $x \in \mathbb{R}^n$.

Proof. The proof is easy, since for any number a, $|a|^q \to 1$ if $a \ne 0$ and $|a|^q = 0$ if $a = 0$. The details are left to the interested reader. □

We thus use $\|x\|_q^q$ to approximate $\|x\|_0$, which leads to the constraint minimization

$$\min\{\|x\|_q^q, \quad x \in \mathbb{R}^n, \Phi x = b\} \tag{1.27}$$

for $0 < q \le 1$ as an approximation of the original research problem (1.4) (cf. [19], [31]). Figure 1.4 shows that (1.27) can find sparse solutions. In fact, we have the following result.

Theorem 1.12. *There exists a $q^* \in (0, 1)$ such that, for any $q \in (0, q^*)$, each minimizer of (1.27) is a minimizer of (1.4).*

Proof. If the result is not true, there exists a sequence $q_k \to 0$ when $k \to \infty$ and $\mathbf{x}_k \in \mathbb{R}^n$ which is a minimizer of (1.27) with $q = q_k$ but $\mathbf{x}_k$ is not a minimizer of (1.4). That is, $\|\mathbf{x}_k\|_0 > \|\mathbf{x}_b\|_0$ for a minimizer $\mathbf{x}_b$ of (1.4). Next we use the result in Theorem 1.13 below that the union of all minimizers of (1.27) with $q \in (0, 1)$ is finite. Consequently, there is a subsequence x_{k_j} which repeats the same minimizer $\widehat{\mathbf{x}}$ and $\|\widehat{\mathbf{x}}\|_0 > \|\mathbf{x}_b\|_0$. Now taking the minimizer $\mathbf{x}_b$ of (1.4), since it is not the minimizer of (1.27), we have

$$\|\mathbf{x}_0\|_0 = \lim_{q_{k_j} \to 0} \|\mathbf{x}_0\|_{q_{k_j}}^{q_{k_j}} \geq \lim_{q_{k_j} \to 0} \|\mathbf{x}_{k_j}\|_{q_{k_j}}^{q_{k_j}} = \lim_{q_{k_j} \to 0} \|\widehat{\mathbf{x}}\|_{q_{k_j}}^{q_{k_j}} = \|\widehat{\mathbf{x}}\|_0 > \|\mathbf{x}_b\|_0.$$

That is a contradiction. $\square$

In the proof above, we have used a fact which is established as follows.

Theorem 1.13. *Let Λ_q be the set of all minimizers of (1.27) for a fixed $q \in (0, 1)$. Then Λ_q is a finite set. Furthermore, let $\Lambda(\Phi, \mathbf{b}) = \bigcup_{0 < q < 1} \Lambda_q$. Then $\Lambda(\Phi, \mathbf{b})$ is a finite set.*

Proof. Clearly, we can divide the Euclidean space $\mathbb{R}^n$ into 2^n octants by using coordinate hyperplanes; e.g., $\mathbb{R}_+^n = \{(x_1, x_2, \ldots, x_n), x_i \geq 0, i = 1, \ldots, n\}$ is one of them. Each minimizer $\mathbf{x}$ of (1.27) for a fixed $q \in (0, 1)$ is in one of these octants, say $\mathbf{x} \in \mathcal{O}_i$ for an $i \in \{1, \ldots, 2^n\}$. The minimizer is also on the hyperplane $H = \{\mathbf{x} : \Phi\mathbf{x} = \mathbf{b}\}$. Thus $\mathbf{x}$ is on the intersection of the hyperplane H and $\mathcal{O}_i$, which is a convex polyhedron P. We now claim that $\mathbf{x}$ is a vertex of the polyhedron. Otherwise, there would be two points $\mathbf{u}, \mathbf{v}$ in $P \subset \Lambda_q$ and $\alpha \in (0, 1)$ such that $\mathbf{x} = \alpha\mathbf{u} + (1 - \alpha)\mathbf{v}$. It follows from the fact that $\mathbf{u}, \mathbf{v}, \mathbf{x}$ are in the same $\mathcal{O}_i$ that

$$\|\mathbf{x}\|_q^q = \|\alpha\mathbf{u} + (1 - \alpha)\mathbf{v}\|_q^q = \sum_{i=1}^n |\alpha u_i + (1 - \alpha)v_i|^q$$

$$= \sum_{i=1}^n (\alpha|u_i| + (1 - \alpha)|v_i|)^q > \alpha\|\mathbf{u}\|_q^q + (1 - \alpha)\|\mathbf{v}\|_q^q \geq \|\mathbf{x}\|_q^q$$

because $\mathbf{u}, \mathbf{v} \in \Lambda_q$ have the same minimal value. This is a contradiction, and hence $\mathbf{x}$ is a vertex of P. It is easy to see that the number of vertices of P is finite. Thus, Λ_q is a finite set for any $q \in (0, 1)$ because it is contained in the union of the vertices of the polyhedron $H \cap \mathcal{O}_i$ for all 2^n octants $\mathcal{O}_i$. Furthermore, $\Lambda(\Phi, \mathbf{b})$ is a finite set. $\square$

In order to compute such minimizers of (1.27), we may regularize the minimizing functional in the following sense: Let

$$\|\mathbf{x}\|_{q,\epsilon}^q = \sum_{i=1}^n (|x_i|^2 + \epsilon)^{q/2}$$

for a parameter $\epsilon > 0$ and study the computation of

$$\min\{\|\mathbf{x}\|_{q,\epsilon}^q, \quad \mathbf{x} \in \mathbb{R}^n, \Phi\mathbf{x} = \mathbf{b}\} \tag{1.28}$$

for $0 < q \leq 1$ (cf., e.g., [45] and [46]).

Furthermore, we can reformulate (1.20) in the format

$$\min_{\substack{\mathbf{x} \in \mathbb{R}^n \\ \mathbf{z} \in \mathbb{R}^m}} \lambda\|\mathbf{x}\|_1 + \frac{1}{2}\|\mathbf{z}\|_2^2, \text{ such that } \mathbf{z} = \Phi\mathbf{x} - \mathbf{b}. \tag{1.29}$$

Then this is a typical problem of minimization over two independent variables. It can be solved by using the alternating direction method of multipliers (ADMM). It is known that ADMM is convergent. We can use the ADMM algorithm to find the sparse solutions.

1.3.2 ▪ Unconstrained Minimization Approximation

An unconstrained version of minimizations (1.20) and (1.23) is

$$\min\left\{\|\mathbf{x}\|_1 + \frac{1}{2\lambda}\|\Phi\mathbf{x} - \mathbf{b}\|_2^2, \quad \mathbf{x} \in \mathbb{R}^n\right\}. \tag{1.30}$$

For $\lambda > 0$ small enough, the solution of (1.30) approximates the solution in (1.20) and (1.23). This statement will be made more precise in Chapter 5. Intuitively, let $\mathbf{x_b}$ be a minimizer of (1.20) and $\mathbf{x}_{\mathbf{b},\lambda}$ a minimizer of (1.30). Then it follows from (1.30) that

$$\|\mathbf{x}_{\mathbf{b},\lambda}\|_1 + \frac{1}{2\lambda}\|\Phi\mathbf{x}_{\mathbf{b},\lambda} - \mathbf{b}\|_2^2 \le \|\mathbf{x_b}\|_1.$$

That is, $\|\mathbf{x}_{\mathbf{b},\lambda}\|_1 \le \|\mathbf{x_b}\|_1$ and $\|\Phi\mathbf{x}_{\mathbf{b},\lambda} - \mathbf{b}\|_2^2 \le 2\lambda\|\mathbf{x_b}\|_1$. Obviously, if $\lambda = 0$, then $\mathbf{x}_{\mathbf{b},\lambda}$ is a minimizer of (1.20). In general, we cannot let $\lambda = 0$. If $\lambda > 0$ is so small that $\sqrt{2\lambda\|\mathbf{x_b}\|_1} \le \delta$, then $\mathbf{x}_{\mathbf{b},\lambda}$ is an excellent candidate for a minimizer of (1.23). Thus, the unconstrained minimization has the advantage of being able to combine the noisy recovery and noiseless recovery problems. Thus, we shall spend a great deal of effort studying the solution of (1.30). In particular, we shall explain how to compute such a solution numerically.

The unconstrained minimization (1.30) is also a typical lasso problem, as explained before. Many schemes for lasso can be used. Because $\|\mathbf{x}\|_1$ is not Lipschitz differentiable, we use the proximal gradient method, which is the same as the fixed point method in this case (see [36]). This scheme can be sped up by using the Nesterov acceleration, as seen in [5]. These will be studied in Chapter 5.

The following regularized versions may be studied:

$$\min\left\{\|\mathbf{x}\|_{1,\epsilon} + \frac{1}{2\lambda}\|\Phi\mathbf{x} - \mathbf{b}\|_2^2, \quad \mathbf{x} \in \mathbb{R}^n\right\} \tag{1.31}$$

and

$$\min\left\{\|\mathbf{x}\|_1 + \beta\|\mathbf{x}\|_2^2 + \frac{1}{2\lambda}\|\Phi\mathbf{x} - \mathbf{b}\|_2, \mathbf{x} \in \mathbb{R}^n\right\}. \tag{1.32}$$

The convergence of various numerical schemes for these regularized minimizations will be discussed below.

Similar to (1.27), the unconstrained version is

$$\min\left\{\|\mathbf{x}\|_q^q + \frac{1}{2\lambda}\|\Phi\mathbf{x} - \mathbf{b}\|_2^2, \quad \mathbf{x} \in \mathbb{R}^n\right\}. \tag{1.33}$$

For $\lambda > 0$ small enough, the minimizer approximates the solution in (1.27). These are nonconvex unconstrained minimization problems. They are not well understood so far. We will make an effort to understand them better in later chapters. Similarly, the regularized version is

$$\min\left\{\|\mathbf{x}\|_{q,\epsilon}^q + \frac{1}{2\lambda}\|\Phi\mathbf{x} - \mathbf{b}\|_2^2, \quad \mathbf{x} \in \mathbb{R}^n\right\}. \tag{1.34}$$

We shall study the solution of the above minimization and derive a computational scheme based on the minimization. We shall show that the computational scheme converges and local minimizers can be found. The local minimizers are close to a minimizer of the associated constrained minimization by an amount $C\lambda$. Similar to the discussion above, they are close to a sparse solution of (1.4) by an amount $C\lambda$.

In general, we shall consider the general regularized minimizing functional of $\mathbf{x} \in \mathbb{R}^n$, i.e.,

$$L(\mathbf{x}, \lambda) = \sum_{j=1}^{n} f(x_j) + \frac{1}{2\lambda} \|\Phi\mathbf{x} - \mathbf{b}\|^2, \tag{1.35}$$

and find solutions of the minimization

$$\min_{\mathbf{x} \in \mathbb{R}^n} L(\mathbf{x}, \lambda), \tag{1.36}$$

where the function $f(x)$ is a single-variable C^1 real function satisfying the following properties:

(P1) $f(x)$ is nondecreasing for $x > 0$.

(P2) $f(x) = f(|x|)$.

(P3) $f'(x)/x$ is nonincreasing for $x > 0$.

Note that (P2) implies

$$\frac{f'(x)}{x} = \frac{f'(|x|)}{|x|} \tag{1.37}$$

by differentiating $f(x) = f(|x|)$. Functions satisfying the above properties include $\ln(x^2 + \epsilon)$, $(x^2 + \epsilon)^{q/2}$ with $q \in (0, 1]$, etc. For the functions $\ln(|x| + \epsilon)$ and $(|x| + \epsilon)^q$ for $0 < q < 1$, only (P1) and (P2) are satisfied.

Our (1.36) is an approximation of the constrained minimization

$$\min_{\mathbf{x} \in \mathbb{R}^n} \sum_j f(\mathbf{x}_j), \quad \text{subject to } \Phi\mathbf{x} = \mathbf{b}. \tag{1.38}$$

The linearization of the nonlinear equations leads to iteratively reweighted ℓ_1 or ℓ_q algorithms (cf. [18], [39], [23], [46]). We shall study this general minimization in Chapter 6.

Even more generally, we can study the $\ell_q + \ell_p$ minimization problem

$$\min_{\mathbf{x} \in \mathbb{R}^n} \|\mathbf{x}\|_q^q + \frac{1}{p\lambda} \|\Phi\mathbf{x} - \mathbf{b}\|_p^p \tag{1.39}$$

for $q \in (0, 1]$ and $p \geq 1$ or

$$\min \left\{ \|\mathbf{x}\|_{q,\epsilon}^q + \frac{1}{p\lambda} \|\Phi\mathbf{x} - \mathbf{b}\|_p^p, \quad \mathbf{x} \in \mathbb{R}^n \right\} \tag{1.40}$$

with $\lambda > 0$ and $\epsilon > 0$. We shall leave this to the interested reader.

1.3.3 ▪ The Duality Approach

For each convex minimization problem, there is an associated dual problem. That is, letting $J(\mathbf{x})$ be a convex functional, the following problem is called a primal problem:

$$\min_{\mathbf{x} \in \mathbb{R}^n} \{J(\mathbf{x}), \quad \Phi\mathbf{x} = \mathbf{b}\}. \tag{1.41}$$

We let $f(\lambda) = \min_{\mathbf{x} \in \mathbb{R}^n} J(\mathbf{x}) + \lambda^\top (\Phi\mathbf{x} - \mathbf{b})$ and consider

$$\max_{\lambda \in \mathbb{R}^m} f(\lambda), \tag{1.42}$$

which is called the Lagrangian dual problem. It is easy to see that for any feasible solution $\mathbf{x}$ with $\Phi\mathbf{x} = \mathbf{b}$, we have

$$f(\lambda) \leq J(\mathbf{x}) + \lambda^\top(\Phi\mathbf{x} - \mathbf{b}) = J(\mathbf{x})$$

for any λ. It follows that $\max_\lambda f(\lambda) \leq J(\mathbf{x})$. If we have a real number λ such that $f(\lambda) = J(\mathbf{x})$ for a feasible solution $\mathbf{x}$, then $\mathbf{x}$ is the minimizer. See [52] for more properties of duality.

For the minimization problem (1.20), the Lagrangian dual problem is

$$\max_{\mathbf{y} \in \mathbb{R}^m} \{-\mathbf{b}^\top\mathbf{y}, \quad \|\Phi^\top\mathbf{y}\|_\infty \leq 1\}. \tag{1.43}$$

Indeed, in this case, we have $J(\mathbf{x}) = \|\mathbf{x}\|_1$. Note that $\|\mathbf{x}\|_1 = \max\{\mathbf{x}^\top\mathbf{y}, \mathbf{y} \in \mathbb{R}^n, \|\mathbf{y}\|_\infty \leq 1\}$. We have

$$\begin{aligned}
f(\lambda) = \|\mathbf{x}\|_1 + \lambda^\top(\Phi\mathbf{x} - \mathbf{b}) &= \min_{\mathbf{x}} \max_{\|\mathbf{y}\|_\infty \leq 1} \mathbf{x}^\top\mathbf{y} + \lambda^\top\Phi\mathbf{x} - \lambda^\top\mathbf{b} \\
&= \max_{\|\mathbf{y}\|_\infty \leq 1} \min_{\mathbf{x}} \mathbf{x}^\top\mathbf{y} + \lambda^\top\Phi\mathbf{x} - \lambda^\top\mathbf{b} \\
&= \max_{\|\mathbf{y}\|_\infty \leq 1} -\lambda^\top\mathbf{b} \quad \text{when } \mathbf{y}^\top + \lambda^\top\Phi = 0 \\
&= \max\{-\lambda^\top\mathbf{b}, \|\Phi^\top\lambda\|_\infty \leq 1\}.
\end{aligned}$$

For the minimization problem (1.22), the dual problem is

$$\max_{\mathbf{y},\mathbf{z}} \left\{-\mathbf{b}^\top\mathbf{y} + \frac{1}{2\beta}\|\Phi^\top\mathbf{y} - \mathbf{z}\|_2^2, \quad \mathbf{z} \in [-1, 1]^n\right\}. \tag{1.44}$$

We leave the verification to Exercise 10.

1.3.4 ▪ The Other Approaches

In addition to the approaches above, there are many other approaches to solving problems (1.4), (1.21), and (1.19). We list them as follows.

(1) **Iterative Hard Thresholding Methods.** If we have a guesstimate on the sparsity s, we may consider

$$\min\left\{\frac{1}{2}\|\Phi\mathbf{x} - \mathbf{b}\|_2, \quad \|\mathbf{x}\|_0 \leq s\right\}. \tag{1.45}$$

Since $\|\mathbf{x}\|_0$ is hard to enforce as a constraint, we may replace it by fixing $p \in [0, 1]$ and finding

$$\min\left\{\frac{1}{2}\|\Phi\mathbf{x} - \mathbf{b}\|_2, \quad \|\mathbf{x}\|_p \leq c\right\}, \tag{1.46}$$

with $c = \|\mathbf{x}^*\|_p$, where $\mathbf{x}^*$ is the sparse solution we are looking for. The unconstrained version of (1.46) will be (1.30) when $p = 1$ and (1.33) for $p \in (0, 1)$. One interesting approach to solving (1.45) when $p = 1$ is the iterative hard thresholding algorithm (cf. [6] and [7]): Guess a sparsity $s > 1$ and let $H_s(\mathbf{x}) = \mathbf{x}_T$ be the s-thresholding operator, where T is the index set of the s largest in magnitude entries of $\mathbf{x}$ and $\mathbf{x}_T$ is zero everywhere except for the entries in T which are the corresponding entries in $\mathbf{x}$. Starting from $\mathbf{x}^{(k)}$, we find

$$\mathbf{x}^{(k+1)} = H_s(\mathbf{x}^{(k)} - \nu\Phi^\top(\Phi\mathbf{x}^{(k)} - \mathbf{b})), \quad k \geq 1,$$

where $\nu > 0$ is a step size.

This type of algorithm has several variations. For example, we can choose $\nu < 1$ and ν dependent on iterative number k. In addition, to ensure that $\mathbf{x}^{(k)}$ satisfies $\Phi\mathbf{x}^{(k)} = \mathbf{b}$, one approach is to find the sparse locations

$$S_{k+1} = \text{support } (H_s(\mathbf{x}^{(k)} - \nu\Phi^\top(\Phi\mathbf{x}^{(k)} - \mathbf{b})))$$

and then minimize the routine least squares problem

$$\mathbf{x}^{(k+1)} := \arg\min\{\|\Phi\mathbf{z} - \mathbf{b}\|_2, \quad \mathbf{z} \in \mathbb{R}^n, \text{support}(\mathbf{z}) \subset S_{k+1}\}. \tag{1.47}$$

We shall present a convergent analysis of these algorithms in Chapter 4. See also [32]. In fact, the convergence rate of these algorithms can be determined and the number of iterations can be estimated to achieve a certain accuracy. When $p \in (0,1)$, we shall describe a projected gradient method to solve (1.45) and discuss the convergence of the algorithm associated with this method.

(2) **Various Greedy Algorithms.** Next we can recast the minimization problem (1.4) in the format

$$\min_{s,\mathbf{x}\in\mathbb{R}^n} \left\{\frac{1}{2}\|\Phi\mathbf{x} - \mathbf{b}\|^2, \|x\|_0 = s\right\}. \tag{1.48}$$

One greedy approach is to choose $s = 1$ and find the least squares solution

$$\min_{i,c} \|\mathbf{b} - c\phi_i\|, \tag{1.49}$$

where $\Phi = [\phi_1, \phi_2, \dots, \phi_n]$. Let c_1 and ϕ_{n_1} be the minimizer pair. Then replace $\mathbf{b}$ by $\mathbf{b} - c_1\phi_{n_1}$ and repeat the above step. This leads to the orthogonal matching pursuit (OMP) algorithm or the weak orthogonal greedy (WOG) algorithm (cf. [26], [55], [20]) and its variations. Instead of minimizing one term in (1.49), we can minimize two terms. A variant is called the quasi-orthogonal matching pursuit (QOMP) algorithm. We find a few columns of Φ which have a very large inner product value with residual vector $\mathbf{b}$ and compute the least squares approximation of $\mathbf{b}$. Then we replace $\mathbf{b}$ by its approximating residual $\widetilde{\mathbf{b}} = \mathbf{b} - \Phi\mathbf{x}^{(1)}$ and repeat the above greedy technique, where $\mathbf{x}^{(1)}$ is a linear combination of these columns. Fewer iterations will be needed to obtain an approximate sparse solution. We shall present a convergent analysis of several greedy algorithms in Chapter 3.

(3) **Screen Rules.** Another approach to dealing with linear systems of large size is the screen rule approach. Indeed, the aim of a screening scheme is to screen all components of $\mathbf{x}$ and remove all inactive ones or set them to zero. That is, the method first reduces the size of $\Phi\mathbf{x} = \mathbf{b}$ by removing all inactive features and then applying any good convex minimization approach to find the approximation of a minimizer of (1.4). Although there are many screen rules, it seems that the screen rules developed in [57] are among the best.

(4) **GIST Algorithms.** Another approach is the generalized iterative shrinkage and thresholding (GIST) algorithms (cf. [34]), which can be viewed as an extension of FISTA (cf. [5]). These algorithms use a difference of convex (DC) functions as a regularizer and then use the ideas of the proximal gradient descent method and the monotonic line search technique from standard optimization. We refer to [34] for details. The GIST algorithm includes many concave and differentiable functions as regularizers. For example, we can use the difference of the ℓ_1 and ℓ_2 norms as a minimizing function. For another example, we can use the capped ℓ_1 norm as a minimizing function. The regularized ℓ_q norm minimization for $q < 1$ studied in [45] is a

special case. In this setting, the GIST extends the study of classic convex-concave programming (CCCP).

(5) **Iteratively Reweighted Methods.** Let us mention here some iterative reweighted methods, such as the iterative reweighted ℓ_1 algorithm (cf. [18]), the iterative reweighted least squares algorithm (cf. [23]), the L_1 greedy algorithm (cf. [39]), and the iterative reweighted ℓ_q algorithm with $q < 1$ (cf. [45] and [46]). We shall address the convergence of these algorithms in Chapters 3 and 6.

(6) **Jacobi–Proximal ADMM Method.** To deal with underdetermined linear systems of very large size, we propose using a parallel computational scheme called Jacobi–Proximal ADMM to deal with (1.20) (see [24]). That is, we minimize

$$\min_{\mathbf{x}_1,\ldots,\mathbf{x}_n} \sum_{i=1}^{n} f_i(\mathbf{x}_i), \qquad \sum_{i=1}^{n} \Phi_i \mathbf{x}_i = \mathbf{b}. \tag{1.50}$$

Several linear systems of size $150,000 \times 300,000$ with sparsity 3000 were tested for sparse solutions on an Amazon EC2 cluster.

The progressive variation-reduced proximal stochastic gradient method, coordinate descent algorithms, and the randomized descent algorithm, as well as accelerated randomized algorithms, have recently been studied to deal with sparse solutions of huge systems.

(7) **The Alternating Projection Method.** Finally, constrained minimization, i.e., (1.4), can be approached directly by using the alternating projection method. Define two projection operators as follows. Recall that $\mathcal{D}_s$ is the set of s-sparse vectors in the Euclidean space $\mathbb{R}^n$. For any vector $\mathbf{x} \in \mathbb{R}^n$, we let $\mathcal{P}_s$ be the projection from $\mathbb{R}^n$ to $\mathcal{D}_s$ by

$$\mathcal{P}_s(\mathbf{x}) := \arg \min_{\mathbf{y} \in \mathbb{R}^n} \|\mathbf{x} - \mathbf{y}\|_2 \tag{1.51}$$

in the ℓ_2 norm. Note that $\mathcal{P}_s$ is the same as H_s used before. We can easily check that $\mathcal{P}_s(\mathbf{x})$ is the s largest entries in magnitude of $\mathbf{x}$. For a sensing matrix Φ and a measurement vector $\mathbf{b}$, let

$$\mathcal{A} := \mathcal{A}(\Phi, \mathbf{b}) = \{\mathbf{x} \in \mathbb{R}^n : \Phi\mathbf{x} = \mathbf{b}\} \tag{1.52}$$

be the affine space of all $\mathbf{x} \in \mathbb{R}^n$ satisfying the underdetermined linear system $\Phi\mathbf{x} = \mathbf{b}$. Now we define the projection $\mathcal{P}_\Phi$ mapping $\mathbb{R}^n$ to $\mathcal{A}$ by

$$\mathcal{P}_\Phi(\mathbf{x}) := \min_{\mathbf{y} \in \mathcal{A}} \|\mathbf{x} - \mathbf{y}\|_2. \tag{1.53}$$

The alternating projection method to solve (1.4) is to start with an initial guess $\mathbf{x}^{(0)}$; project it to $\mathcal{D}_s$, i.e., $\mathbf{y}^{(1)} = \mathcal{P}_s(\mathbf{x}^{(0)})$, and then project it to $A(\Phi, \mathbf{b})$ by $\mathbf{x}^{(1)} = \mathcal{P}_\Phi(\mathbf{y}^{(1)})$. We repeat these two projection procedures to obtain a sequence $\{\mathbf{x}^{(k)}, k \geq 1\}$. We will show that the sequence is convergent and the limit is the sparse solution $\mathbf{x}^* = \lim_{k \to \infty} \mathbf{x}^{(k)}$. It would be interesting to know when the sequence has a convergent limit. See the discussion in Chapter 4.

1.4 ▪ Some Extensions

The ideas of finding a sparse solution can be extended to deal with other applied problems. Let us mention a few of them.

1.4.1 ▪ The Block Measurement Setting

An extension of the sparse solution problem is called the block sparse solution problem. It was introduced and studied in [27], [4], which have many practical applications, such as DNA microarrays [51], multiband signals [49], and magnetoencephalography [28]. In recovering the sparse solution $\mathbf{x}$ from $\Phi\mathbf{x} = \mathbf{b}$, the entries of $\mathbf{x}$ are grouped into blocks. That is, $\mathbf{x} = (\mathbf{x}_{t_1}, \mathbf{x}_{t_2}, \dots, \mathbf{x}_{t_\ell})^\top$, with $\mathbf{x}_{t_i}$ being a block of entries for each i. We look for the least number of nonzero blocks $\mathbf{x}_{t_i}$ such that $\Phi\mathbf{x} = \mathbf{b}$. Letting

$$\||\mathbf{x}\||_{2,q} = \left(\sum_{i=1}^{\ell} \|\mathbf{x}_{t_i}\|_2^q \right)^{1/q}$$

be a mixed norm with $\|\mathbf{x}_{t_i}\|_2$ being the standard ℓ_2 norm of vector $\mathbf{x}_{t_i}$, we find the block sparse solution

$$\min\{\||\mathbf{x}\||_{2,q}, \quad \Phi\mathbf{x} = \mathbf{b}\}$$

(cf., e.g., [27] for $q = 1$). The basic properties and computational algorithms studied in this book can be extended to the block measurement setting.

1.4.2 ▪ The Multiple Measurement Setting

The problem of computing sparse solution recovery of underdetermined linear systems can be extended to the sparse solution vectors for multiple measurement vectors. That is, letting Φ be a sensing matrix of size $m \times n$ with $m \ll n$ and given multiple measurement vectors $\mathbf{b}^{(k)}, k = 1, \dots, r$, we are looking for solution vectors $\mathbf{x}^{(k)}, 1, \dots, r$, such that

$$\Phi\mathbf{x}^{(k)} = \mathbf{b}^{(k)}, \quad k = 1, \dots, r, \tag{1.54}$$

and the vectors $\mathbf{x}^{(k)}, k = 1, \dots, r$, are jointly sparse, i.e., have nonzero entries at the same locations and have as few nonzero entries as possible. Such problems arise in source localization (cf. [47]), neuromagnetic imaging (cf. [21]), and equalization of sparse communication channels (cf. [22], [30]).

A popular approach to this new problem for multiple measurement vectors is to solve the following optimization:

$$\text{minimize} \left\{ \left(\sum_{j=1}^{n} \|(x_{1,j}, \dots, x_{r,j})\|_q^p \right)^{1/p} : \mathbf{x}^{(k)} \in \mathbb{R}^n, k = 1, \dots, r, \right.$$

$$\left. \text{subject to } \Phi\mathbf{x}^{(k)} = \mathbf{b}^{(k)}, k = 1, \dots, r \right\}, \tag{1.55}$$

where $\mathbf{x}^{(k)} = (x_{k,1}, \dots, x_{k,n})^\top$ for all $k = 1, \dots, r$ and $\| \cdot \|_q$ is the standard ℓ_q norm for $q \geq 1$ and $p \geq 1$. Clearly, it is a generalization of the standard ℓ_1 approach for the sparse solution. That is, when $r = 1, q = p = 1$, we find the sparse solution $\mathbf{x}$ solving the minimization problem (1.20). Again, this approach can be extended to the ℓ_q setting. Let us consider a joint recovery from multiple measurement vectors via

$$\text{minimize} \sum_{i=1}^{N} \left(\sqrt{x_{1,j}^2 + \cdots + x_{r,j}^2} \right)^p, \text{ subject to } \Phi\mathbf{x}^{(1)} = \mathbf{b}^{(1)}, \dots, \Phi\mathbf{x}^{(r)} = \mathbf{b}^{(r)}, \tag{1.56}$$

for all $0 < p \leq 1$, where $\mathbf{x}^{(k)} = (x_{k,1}, \ldots, x_{k,n})^\top \in \mathbb{R}^n$ for all $k = 1, \ldots, r$. We refer the reader to [40]. Note that when $p \to 0_+$, we have $\left(\sqrt{x_{1,j}^2 + \cdots + x_{r,j}^2}\right)^p \to 1$ if any of $x_{1,j}, \ldots, x_{r,j}$ is nonzero, and hence

$$\sum_{i=1}^{n} \left(\sqrt{x_{1,j}^2 + \cdots + x_{r,j}^2}\right)^p \to s,$$

which is the joint sparsity of the solution vectors $\mathbf{x}^{(k)}, k = 1, \ldots, r$. Thus, the minimization in (1.56) makes sense. The study in Chapters 2–6 can be extended to dealing with this problem. We let interested readers explore this on their own.

1.4.3 ▪ Matrix Completion

Given partial information about a matrix M of size $m \times n$, e.g., entries of a subset Ω from the index set $\{(i,j), 1 \leq i \leq m, 1 \leq j \leq n\}$, we look for a matrix X such that $M_{ij}, (i,j) \in \Omega$, match the given entries. For example, Netflix has a giant collection of movies and a huge list of registered customers. Only some movies have been rated by some customers. In order to recommend each customer movies that best match the customer's preferences, Netflix needs a whole rating matrix called a recommendation system. Note that two customers having similar taste means that the two rating vectors associated with these two customers are linearly dependent on each other. Many matrices match the given partial information. To have a meaningful solution, we look for the matrix X of smallest rank, with $X_{ij} = M_{ij}, (i,j) \in \Omega$. That is,

$$\min\{\operatorname{rank}(X), \quad X \in \mathbb{R}^{m \times n}, X_{ij} = M_{ij}, (i,j) \in \Omega\}. \tag{1.57}$$

This minimization can be approximated by

$$\min\{\|X\|_*, \quad X \in \mathbb{R}^{m \times n}, X_{ij} = M_{ij}, (i,j) \in \Omega\}, \tag{1.58}$$

where $\|X\|_* = \sum_{j \geq 1} \sigma_j(X)$ and $\sigma_j(X)$ stands for the jth singular value of X. Here $\|X\|_*$ is called the nuclear norm of X. All the approaches for the sparse solution can be extended to study this matrix completion problem. See, e.g., [14] and [11]. We leave the discussion of the existence and uniqueness of low rank matrix completion, including computational algorithms, to Chapter 9. A typical application of low rank matrix completion is to complete a recommendation matrix, e.g., movie rating matrix, a joke rating matrix, or an image in-painting. In the following example, we show the effectiveness of one of these matrix completion methods.

Example 1.14. Let us start with a common image called "bank" and use SVD to convert it into a low rank matrix of rank 100 (see Figure 1.5). We randomly sample 100 rows and 100 columns to form a matrix M_Ω with the known entries from the 100 rows and 100 columns shown on the left of Figure 1.6. Then we apply one of the algorithms to recover the original image. The result is excellent, seeing that the maximal error of the completed image and the rank 100 approximation of the original image is less than or equal to $1e{-}6$. See the image on the right of Figure 1.6. That is, the algorithm works very well. For another example, the images on the front cover of the book are calculated based on this approach.

1.4.4 ▪ Graph Clustering

Let V be a given set of vertices in high-dimensional space $\mathbb{R}^n$, and let E be a given list of edges among these vertices. Let $G = (V, E)$, which is called a graph of vertices V and edges E. For

The Original

Rank 100 Matrix

Figure 1.5. *Original image and low rank reduction.*

To Be Completed

After Allen-Lai Method

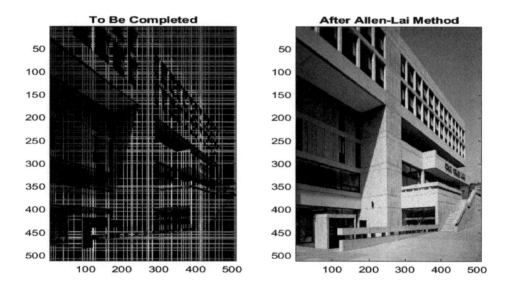

Figure 1.6. *The recovered image agrees with the rank reduced image with error* 1e−6.

example, friendship relations among 10 people form a graph G with $V = \{p_1, \ldots, p_{10}\}$ and $E = \{e_{ij}, i, j = 1, \ldots, 10\}$, where $e_{ij} = 1$ if person i and person j are friends and $e_{ij} = 0$ if they are not friends. We can write E in an adjacency matrix A with entries $a_{ij} = 1$ if $e_{ij} = 1$ and $a_{ij} = 0$ if $e_{ij} = 0$. For the given graph $G = (V, E)$, the graph clustering problem is to divide the vertex set V into subsets $V = C_1 \cup \cdots \cup C_k$ such that there are many intracluster edges (edges between vertices in the same cluster) and few intercluster edges (edges between vertices in different clusters) in E. Using the example of friendships among 10 people, we can divide this group into a few friendship clusters. For another example, in Figure 1.7, 10 faces in 10 postures are shown. One interesting problem is to sort these face images automatically. This is a standard

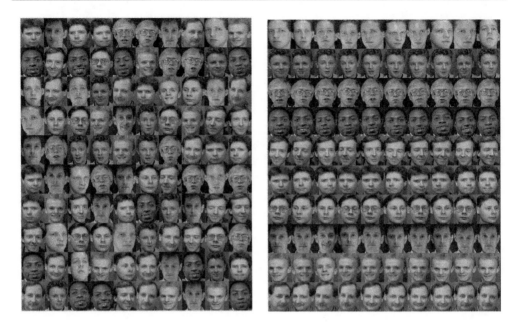

Figure 1.7. *An image of randomized human faces (left) and an image of sorted human faces by a sparse solution method discussed in Chapter 10. (From AT&T Laboratories Cambridge.)*

computational problem in graph clustering. Classically, one solves the graph clustering problem by a spectral clustering method, i.e., by determining the eigenvectors $\mathbf{v}_1, \ldots, \mathbf{v}_k$ associated with the k smallest eigenvalues of the adjacency matrix A associated with G. Suppose that $|V| = n$, and let $\mathbf{1}_{C_i} \in \mathbb{R}^n$ denote the indicator vector of the ith cluster. For convenience, we introduce

$$\mathcal{L}_G = I_n - D^{-1}A, \tag{1.59}$$

the graph Laplacian $\mathcal{L}_G$, where I_n is the identity matrix of size $n \times n$; $A = [e_{ij}]$ is the adjacency matrix; and $D = \text{diag}(d_1, \ldots, d_n)$, with d_i being the summation of the ith row of A. Note that for human face images, we can compute the similarities among these faces to build an adjacency matrix A with weighted entries. Consider the subspace spanned by them, $\text{span}\{\mathbf{v}_1, \ldots, \mathbf{v}_k\} \subset \mathbb{R}^n$. We can show that, under certain conditions, this subspace is "close" to $\text{span}\{\mathbf{1}_{C_1}, \ldots, \mathbf{1}_{C_k}\}$, and hence we can use some graph cutting strategies to find the support patterns of the basis $\{\mathbf{1}_{C_1}, \ldots, \mathbf{1}_{C_k}\}$ and thus determine the clusters $C_1, \ldots, C_k$. We refer the reader to [1] and [10], for example. Note that $\mathcal{L}_G \mathbf{1}_{C_i} = 0$ and the indicator vector $\mathbf{1}_{C_i}$ will be *sparse*. We can adapt sparse recovery algorithms for (1.4) to solve the problem

$$\arg\min \|\mathcal{L}_G \mathbf{x}\|_2, \quad \text{subject to } x_j = 1 \text{ and } \|\mathbf{x}\|_0 \leq n_0 \tag{1.60}$$

to determine directly an approximation to the indicator vector of the cluster containing the vertex j, where n_0 is a guesstimate of the size of a cluster which contains the index j. We can then recover the cluster by considering the support of this vector. Repeating this procedure, we can iterate the same algorithm to find all the clusters of G.

Example 1.15. Let us present an example of using a sparse solution technique to sort human faces. Suppose that we are given the set of human faces shown in the left image in Figure 1.7. A method in this book is used to sort the faces as shown in the image on the right of Figure 1.7. We shall explain this computation in detail in Chapter 10. More examples can be found there or in [41] and [43].

1.4.5 ▪ Phase Retrieval

The phase retrieval problem has been extensively studied due to its numerous applications, such as X-ray diffraction and other imaging applications, communication, and waveform. Mathematically, we would like to recover an unknown vector $\mathbf{x} \in \mathbb{R}^n$ through a set of quadratic measurements:

$$y_1 = |\langle \mathbf{a}_1, \mathbf{x} \rangle|^2, \ldots, y_m = |\langle \mathbf{a}_m, \mathbf{x} \rangle|^2. \tag{1.61}$$

That is, we need to solve the system of these quadratic equations. Because the system is very large for any real-life application, using a computational method from algebraic geometry, e.g., a Gröbner basis, is not realistic. We shall discuss how to use numerical methods to find an approximation to (1.61). It is clear that if $\mathbf{x}$ satisfies the quadratic conditions (1.61), so does $-\mathbf{x}$. Thus, we look for a solution $\mathbf{x}$ up to a phase factor. This problem has a counterpart in the complex value setting. That is, $\mathbf{x} \in \mathbb{C}^n$ and $\mathbf{a}_i \in \mathbb{C}^n$. Then if $\mathbf{x}$ is a solution, letting $c = e^{i\theta}$, $c\mathbf{x}$ is also a solution for any angle variable θ, where $\mathbf{i} = \sqrt{-1}$. For simplicity, let us consider the real value setting first. Note that if we know the signs of the equations in (1.61), e.g., suppose we know

$$\langle \mathbf{a}_i, \mathbf{x} \rangle = \sqrt{y_i}, \quad i = 1, \ldots, m,$$

then we simply solve the linear system to obtain $\mathbf{x}$. The difficulty is that we do not know the signs. Since the choices of signs for $\sqrt{y_i}$ grow exponentially as m increases, we cannot exhaust all the choices of signs and solve each linear system within a reasonable time.

Besides the difficulty of nonlinearity, the number m of measurements is usually not large compared to n, e.g., $m = O(n \log(n))$ or $m \geq 3n$, due to the expenses and/or difficulties of obtaining measurements. We need to find out what the smallest m for each n is to have a solution, assuming the measurement vectors $\mathbf{a}_i, i = 1, \ldots, m$, are in a general position. We shall present this kind of result in Chapter 11. In addition, there are many approaches to attacking this nonlinear problem under various assumptions such as $\mathbf{x}$ being sparse. One approach is to linearize the problem by letting $X = \mathbf{x}\mathbf{x}^\top$ and $A_j = \mathbf{a}_j\mathbf{a}_j^\top$, $j = 1, \ldots, m$. In this case, the constraints in (1.61) can be rewritten as

$$y_j = |\mathbf{a}_j^\top \mathbf{x}|^2 = \mathbf{x}^\top \mathbf{a}_j \mathbf{a}_j^\top \mathbf{x} = \mathbf{x}^\top A_j \mathbf{x} \tag{1.62}$$

$$= \operatorname{tr}(\mathbf{x}^\top A_j \mathbf{x}) = \operatorname{tr}(A_j \mathbf{x}\mathbf{x}^\top) = \operatorname{tr}(A_j X) \tag{1.63}$$

for $j = 1, \ldots, m$, where tr is the trace operator, i.e., $\operatorname{tr}(A) = \sum_{i=1}^n A_{ii}$ if $A = [A_{ij}]_{1 \leq i,j \leq n}$. So the phase retrieval problem can be recast as a matrix recovery problem (cf. [13]): Find $X \in M$ satisfying the linear measurements $\operatorname{tr}(A_j X) = y_j, j = 1, \ldots, m$, where

$$M = \{X \in \mathbb{R}^{n \times n} : X \succeq 0; \operatorname{rank}(X) \leq 1\}. \tag{1.64}$$

In other words, we solve the minimization

$$\min\{\operatorname{rank}(X) : \operatorname{tr}(A_j X) = y_j, j = 1, \ldots, m, X \succeq 0\}. \tag{1.65}$$

This problem is more difficult than the low rank matrix recovery problem due to the nonlinear constraint $X \succeq 0$. One may replace $\operatorname{rank}(X)$ by other minimal functionals. See Chapter 9 and also [15].

One standard numerical approach to solving (1.61) is to solve the following minimization:

$$\min_{\mathbf{x} \in \mathbb{R}^n \text{ or } \mathbb{C}^n} \sum_{i=1}^m (|\langle \mathbf{a}_i, \mathbf{x} \rangle|^2 - y_i)^2. \tag{1.66}$$

Naturally, a gradient descent method (called Wirtinger flow in the complex value setting) and the Gauss–Newton method will be developed for the numerical solution of (1.66). In addition, we shall develop a DC-based algorithm for (1.66). All the details can be found in Chapter 11.

1.5 ▪ Exercises

Exercise 1. Let $\mathbf{x_b} \in \mathbb{R}^n$ be a sparsest solution of underdetermined linear system $\Phi\mathbf{x} = \mathbf{b}$ with sparsity $\|\mathbf{x_b}\|_0 = s < m$, where Φ is of size $m \times n$ and $m < n$. Let $x_{i_1}, \ldots, x_{i,s}$ be nonzero components of $\mathbf{x_b}$. Writing $\Phi = [\phi_1, \ldots, \phi_n]$, show that $\phi_{i_1}, \ldots, \phi_{i_s}$ are linearly independent.

Exercise 2. Let $\mathbf{x_b} \in \mathbb{R}^n$ be a sparse solution of underdetermined linear system $\Phi\mathbf{x} = \mathbf{b}$ with sparsity $\|\mathbf{x_b}\|_0 = s < m$, where Φ is of size $m \times n$ and $m < n$. Let $\mathbf{y}$ be another sparse solution to $\Phi\mathbf{x} = \mathbf{b}$ with the same sparsity. Show that if $x_{i_j} y_{i_j} \neq 0, j = 1, 2, \ldots, s$, then $\mathbf{y} = \mathbf{x_b}$.

Exercise 3. Given an underdetermined linear system $\Phi\mathbf{x} = \mathbf{b}$ with Φ of size $m \times n$ and $m < n$, let $S = \{\mathbf{x} : \Phi\mathbf{x} = \mathbf{b}, \|\mathbf{x}\|_0 < m\}$ be the collection of all sparse solutions. Suppose that for $0 < s < m$, $A_s = \{\mathbf{x} \in S, \|\mathbf{x}\|_0 = s\}$ is nonempty. Then show that the cardinality of A_s is less than or equal to $C_s^n = n!/(s!(n-s)!)$.

Exercise 4. Let $\mathbf{x} = (x_1, \ldots, x_n)^\top \in \mathbb{R}^n$ be is divided into k blocks, i.e., $\mathbf{x} = (\mathbf{x}_1, \ldots, \mathbf{x}_k)^\top$. For example, for $n = 100$ and $k = 10$, $\mathbf{x}_i = (x_{10(i-1)+j}, j = 1, \ldots, 10)$ for $i = 1, \ldots, 10$. Suppose that the division is fixed. For any real numbers $p \geq 0, q \geq 0$, define

$$\|\mathbf{x}\|_{p,q} = \left(\sum_{i=1}^k \|\mathbf{x}_i\|_p^q \right)^{1/q}.$$

Show that when $q \geq 1$, $\|\mathbf{x}\|_{p,q}$ is a norm for any $p > 0$.

Exercise 5. Let $\mathbf{x} = (x_1, \ldots, x_n)^\top \in \mathbb{R}^n$. Define $\|\mathbf{x}\|_p = (\sum_{i=1}^n |x_i|^p)^{1/p}$ for $p > 0$. Show that when $p \in (0, 1)$, $\|\mathbf{x}\|_p$ is not a norm. Furthermore, show that $\|\mathbf{x} + \mathbf{y}\|_p^p \leq \|\mathbf{x}\|_p^p + \|\mathbf{y}\|_p^p$ for all $\mathbf{x}, \mathbf{y} \in \mathbb{R}^n$.

Exercise 6. Prove Lemma 1.11.

Exercise 7. Show that any unconstrained ℓ_q minimization problem

$$\min_{\mathbf{x} \in \mathbb{R}^n} \|\Phi\mathbf{x} - \mathbf{b}\|_2^2 + \frac{1}{2\lambda} \|\mathbf{x}\|_q^q$$

with $\lambda > 0$ can be converted to the standard minimization problem

$$\min_{\mathbf{z} \in \mathbb{R}^n} \|\widetilde{\Phi}\mathbf{z} - \widetilde{\mathbf{b}}\|_2^2 + \frac{1}{2} \|\mathbf{z}\|_q^q$$

for another sensing matrix $\widetilde{\Phi}$ and measurement vector $\widetilde{\mathbf{b}}$ for any $q \in (0, 1]$.

Exercise 8. Consider the following constrained minimization problem:

$$\min_{\mathbf{x} \in \mathbb{R}^n} \|\mathbf{x}\|_q^q, \quad \text{subject to } \|\Phi\mathbf{x} - \mathbf{b}\|_2 \leq \epsilon. \tag{1.67}$$

Suppose that $\|\mathbf{b}\|_2 > \epsilon$. Show that for $0 < q \leq 1$, if $\mathbf{x_b}$ is a minimizer of (1.67), then $\|\Phi\mathbf{x_b} - \mathbf{b}\|_2 = \epsilon$.

Exercise 9. Let $\mathbf{x_b}$ be a minimizer of (1.67). Show that

$$\|\mathbf{x_b}\|_2 \leq \frac{\epsilon\sqrt{m} + \|\mathbf{b}\|_2}{\min_{\sigma_i(\Phi) > 0} \sigma_i(\Phi)}, \tag{1.68}$$

where $\sigma_i(\Phi)$ stands for the ith singular value of Φ.

Exercise 10. Prove (1.44).

Exercise 11. Formulate the matrix completion problem in (1.57) in the noisy setting using an unconstrained minimization.

Exercise 12. For a matrix $A = [a_{ij}]_{1 \leq i \leq m, 1 \leq j \leq n}$ of size $m \times n$, the matrix norm is defined by

$$\|A\|_p = \max_{\mathbf{x} \neq 0} \frac{\|A\mathbf{x}\|_p}{\|\mathbf{x}\|_p}$$

for all $p > 0$. Show that

$$\|A\|_1 = \max_{1 \leq j \leq n} \sum_{i=1}^{m} |a_{ij}| \quad \text{and} \quad \|A\|_\infty = \max_{1 \leq i \leq m} \sum_{j=1}^{n} |a_{ij}|.$$

Show that $\|A\|_2 = \sqrt{\lambda_{\max}(A^\top A)}$ and $\|A\|_2^2 \leq \|A\|_1 \|A\|_\infty$.

Exercise 13. Let $\mathrm{tr}(A) = \sum_{i=1}^{n} A_{ii}$ for matrix $A = [A_{ij}]_{1 \leq i,j \leq n}$. tr is called the trace operator. Show that $\mathrm{tr}(AB) = \mathrm{tr}(BA)$ for any square matrices A and B.

Exercise 14. Let A be a symmetric nonnegative definite matrix of size $n \times n$. Show that for any vectors $\mathbf{a}, \mathbf{b} \in \mathbb{R}^n$,

$$|\langle \mathbf{a}, A\mathbf{b} \rangle| \leq (\langle \mathbf{a}, A\mathbf{a} \rangle)^{1/2} (\langle \mathbf{b}, A\mathbf{b} \rangle)^{1/2}, \tag{1.69}$$

which is called the Cauchy–Schwarz inequality in matrix format.

Exercise 15. Let A be a symmetric positive definite matrix of size $n \times n$. Show that for any vectors $\mathbf{a}, \mathbf{b} \in \mathbb{R}^n$,

$$|\langle \mathbf{a}, \mathbf{b} \rangle| \leq (\langle \mathbf{a}, A\mathbf{a} \rangle)^{1/2} (\langle \mathbf{b}, A^{-1}\mathbf{b} \rangle)^{1/2}, \tag{1.70}$$

which is called the generalized Cauchy–Schwarz inequality in matrix format.

Exercise 16. For a matrix $A = [a_{ij}]_{1 \leq i \leq m, 1 \leq j \leq n}$ of size $m \times n$, let

$$\|A\|_F = \left(\sum_{1 \leq i \leq m, 1 \leq j \leq n} |a_{ij}|^2 \right)^{1/2}.$$

Show that $\|A\|_F$ is a norm. Also, show that $\|AB\|_F \leq \|A\|_F \|B\|_F$ whenever the product AB is defined.

Exercise 17. For a matrix $A = [a_{ij}]_{1 \leq i \leq m, 1 \leq j \leq n}$ of size $m \times n$, for any real numbers $p \geq 0, q \geq 0$, define

$$\|A\|_{p,q} = \max_{\mathbf{x} \neq 0} \frac{\|A\mathbf{x}_i\|_p}{\|\mathbf{x}\|_q}.$$

Show that when $q \geq 1$, $\|A\|_{p,q}$ is a norm for any $p > 0$.

Exercise 18. Use various approaches in section 1.3 to formulate computational methods for matrix completion. In particular, use the alternating projection method to formulate a low rank matrix completion problem.

Exercise 19 (Matrix Sparsification). Recall that any matrix $M = [m_{ij}]_{1 \leq i,j \leq n}$ can be written as

$$M = \sum_{i,j=1}^{n} m_{ij} \mathbf{e}_i \mathbf{e}_j^\top,$$

where $\mathbf{e}_i \in \mathbb{R}^n$ is the unit vector with one in the ith entry and zero otherwise. In general,

$$M = \sum_{i=1}^{N} m_i \mathbf{u}_i \mathbf{v}_i^{\top} \tag{1.71}$$

for $i = 1, \ldots, N \gg n$, where $\mathbf{u}_i, \mathbf{v}_i \in \mathbb{R}^n$ are vectors. For example, when $\mathbf{v}_i = \mathbf{u}_i$ and $m_i > 0$, M is a positive definite matrix. Because $N \gg n$, we are interested in how to simplify M by matrix sparsification. For example, find

$$M_s = \sum_{i=1}^{N} x_i \mathbf{u}_i \mathbf{v}_i^{\top} \tag{1.72}$$

such that $\|M - M_s\|_F \le \epsilon$ and $\|\mathbf{x}\|_0 \le s \ll N$. Formulate (1.72) as a compressive sensing problem.

Exercise 20. Suppose

$$M = \sum_{i=1}^{N} m_i \mathbf{u}_i \mathbf{u}_i^{\top}, \tag{1.73}$$

where $\mathbf{u}_i \in \mathbb{R}^n$ are vectors and $m_i > 0$. So M is a positive definite matrix. Because $N \gg n$, we are interested in simplifying M by sparse subset selection. That is, find

$$M_s = \sum_{i=1}^{N} x_i \mathbf{u}_i \mathbf{u}_i^{\top}, \quad x_i \ge 0, \tag{1.74}$$

such that $\|M - M_s\|_F \le \epsilon$ and $\|\mathbf{x}\|_0 \le s \ll N$. Formulate (1.74) as a constrained sparse solution problem.

Bibliography

[1] J. M. Aldous and R. J. Wilson, Graphs and Applications: An Introductory Approach, Springer Verlag, 2000. (Cited on p. 20)

[2] G. Awanou, M.-J. Lai, and P. Wenston, The multivariate spline method for scattered data fitting and numerical solution of partial differential equations, in Wavelets and Splines: Athens 2005, Nashboro Press, Brentwood, TN, 2006, 24–74. (Cited on p. 6)

[3] R. Baraniuk, Compressive sensing, IEEE Signal Process. Mag., 24 (2007), 118–121. (Cited on p. 8)

[4] R. G. Baraniuk, V. Cevher, M. F. Duarte, and C. Hegde, Model-based compressive sensing, IEEE Trans. Inform. Theory, 56 (2010), 1982–2001. (Cited on p. 17)

[5] A. Beck and M. Teboulle, A fast iterative shrinkage-thresholding algorithm for linear inverse problems, SIAM J. Imaging Sci., 2 (2009), 183–202. (Cited on pp. 12, 15)

[6] T. Blumensath and M. E. Davies, Iterative hard thresholding for compressed sensing, Appl. Comput. Harmon. Anal., 27 (2009), 265–274. (Cited on p. 14)

[7] T. Blumensath and M. E. Davies, Normalized iterative hard thresholding: Guaranteed stability and performance, IEEE J. Sel. Top. Signal Process., 4 (2010), 298–309. (Cited on p. 14)

[8] J. Brodie, I. Daubechies, C. De Molc, D. Giannoned, and I. Lorisc, Sparse and stable Markowitz portfolios, Proc. Nat. Acad. Sci. USA, 106 (2009), 12267–12272. (Cited on p. 7)

[9] A. M. Bruckstein, D. L. Donoho, and M. Elad, From sparse solutions of systems of equations to sparse modeling of signals and images, SIAM Rev., 51 (2009), 34–81. (Cited on p. 8)

[10] F. R. K. Chung, Spectral Graph Theory, AMS, Providence, RI, 1997. (Cited on p. 20)

[11] J. Cai, E. Candès, and Z. Shen, A singular value thresholding algorithm for matrix completion, SIAM J. Optim., 20 (2010), 1956–1982. (Cited on p. 18)

[12] E. J. Candès, Compressive Sampling, International Congress of Mathematicians, Vol. III, European Mathematical Society, Zürich, 2006, pp. 1433–1452. (Cited on p. 8)

[13] E. J. Candès, Y. Eldar, T. Strohmer, and V. Voroninski, Phase retrieval via matrix completion, SIAM J. Imaging Sci., 6(1) (2013), 199–225. (Cited on p. 21)

[14] E. Candés and B. Recht, Exact matrix completion via convex optimization, Found. Comput. Math., 9 (2009), 717. (Cited on p. 18)

[15] E. J. Candès, T. Strohmer, and V. Voroninski, PhaseLift: Exact and stable signal recovery from magnitude measurements via convex programming, Comm. Pure Appl. Math., 66(8) (2013), 1241–1274. (Cited on p. 21)

[16] E. J. Candès and T. Tao, Decoding by linear programming, IEEE Trans. Inform. Theory, 51 (2005), 4203–4215. (Cited on p. 4)

[17] E. J. Candès and T. Tao, Near-optimal signal recovery from random projections: Universal encoding strategies, IEEE Trans. Inform. Theory, 52 (2006), 5406–5425. (Cited on p. 4)

[18] E. J. Candès, M. B. Wakin, and S. Boyd, Enhancing sparsity by reweighted ℓ_1 minimization, J. Fourier Anal. Appl., 14 (2008), 877–905. (Cited on pp. 13, 16)

[19] R. Chartrand, Exact reconstruction of sparse signals via nonconvex minimization, IEEE Signal Process. Lett., 14 (2007), 707–710. (Cited on p. 10)

[20] A. Cohen, W. Dahmen, and R. DeVore, Compressed sensing and best k-term approximation, J. Amer. Math. Soc., 22 (2009), 211–231. (Cited on pp. 5, 15)

[21] S. F. Cotter and B. D. Rao, Sparse channel estimation via matching pursuit with application to equalization. IEEE Trans. Commun., 50(3) (2002), 374–377. (Cited on p. 17)

[22] S. F. Cotter, B. D. Rao, K. Engang, and K. Kreutz-Delgado, Sparse solutions to linear inverse problems with multiple measurement vectors, IEEE Trans. Signal Process., 53 (2005), 2477–2488. (Cited on p. 17)

[23] I. Daubechies, R. DeVore, M. Fornasier, and C. S. Guntuk, Iteratively reweighted least squares minimization for sparse recovery. Commun. Pure Appl. Math., 63(1) (2010), 1–38. (Cited on pp. 13, 16)

[24] W. Deng, M. J. Lai, Z. Peng, and W. T. Yin, Parallel multi-block ADMM with $o(1/k)$ convergence, J. Sci. Comput., 71 (2017), 712–736. (Cited on p. 16)

[25] D. L. Donoho, Compressed sensing, IEEE Trans. Inform. Theory, 52 (2006), 1289–1306. (Cited on p. 8)

[26] R. A. DeVore and V. N. Temlyakov, Some remarks on greedy algorithms, Adv. Comput. Math. 5 (1996), 173–187. (Cited on p. 15)

[27] Y. C. Eldar and M. Mishali, Robust recovery of signals from a structured union of subspaces, IEEE Trans. Inform. Theory, 55 (2009), 5302–5316. (Cited on p. 17)

[28] Y. C. Eldar, P. Kuppinger, and H. Bolcskei, Block-sparse signals: Uncertainty relations and efficient recovery, IEEE Trans. Signal Process., 58 (2010), 3042–3054. (Cited on p. 17)

[29] R. Feng, A. Huang, M.-J. Lai, and Z. Shen, Reconstruction of Sparse Polynomials via Quasi-Orthogonal Matching Pursuit Method, submitted for publication, 2020. (Cited on p. 5)

[30] I. J. Fevrier, S. B. Gelfand, and M. P. Fitz, Reduced complexity decision feedback equalization for multi-path channels with large delay spreads, IEEE Trans. Commun., 47(6) (1999), 927–937. (Cited on p. 17)

[31] S. Foucart and M.-J. Lai, Sparsest solutions of underdetermined linear systems via ℓ_q-minimization for $0 \leq q \leq 1$, Appl. Comput. Harmon. Anal., 26 (2009), 395–407. (Cited on p. 10)

[32] S. Foucart and H. Rauhut, A Mathematical Introduction to Compressive Sensing, Birkhäuser Verlag, 2013. (Cited on p. 15)

[33] I. E. Frank and J. H. Freidman, A statistical view of some chemometrics regression tools (with discussion), Technometrics, 35 (1993), 109–148. (Cited on p. 4)

[34] P. Gong, C. Zhang, Z. Lu, J. Huang, and J. Ye, A general iterative shrinkage and thresholding algorithm for non-convex regularized optimization problems, in Proceedings of the 30th International Conference on Machine Learning, Volume 28, 2013, 37–45. (Cited on p. 15)

[35] W. H. Guo and M.-J. Lai, Box spline wavelet frames for image edge analysis, SIAM J. Imaging Sci., 6 (2013), 1553–1578. (Cited on p. 5)

[36] E. T. Hale, W. Yin, and Y. Zhang, Fixed-point continuation for ℓ_1-minimization: Methodology and convergence, SIAM J. Optim., 19 (2008), 1107–1130. (Cited on p. 12)

[37] T. Hastie, R. Tibshirani, and M. Weinwrite, Statistical Learning with Sparsity, CPC Press, 2015. (Cited on p. 4)

[38] H. M. Kang, M.-J. Lai, and X. Li, An economical representation of PDE solution by using compressive sensing approach, J. Comput. Aided Design, 115 (2019), 78–86. (Cited on p. 7)

[39] I. Kozlov and A. Petukhov, Sparse solution of underdetermined linear systems, in Handbook of Geomathematics, W. Freeden, M. Z. Nashed, and T. Sonar (Eds.), Springer, 2010, 1243–1259. (Cited on pp. 13, 16)

[40] M.-J. Lai and Louis Y. Liu, The null space property for sparse recovery from multiple measurement vectors, Appl. Comput. Harmon. Anal., 30 (2011), 402–406. (Cited on p. 18)

[41] M.-J. Lai and D. Mckenzie, Compressive sensing for cut improvement and local clustering, SIAM J. Math. Data Sci., 2 (2020), 368–395. (Cited on p. 20)

[42] M.-J. Lai and L. L. Schumaker, Spline Functions over Triangulations, Cambridge University Press, 2007. (Cited on p. 6)

[43] M.-J. Lai and Z. Shen, An Effective Approach to Semi-supervised Cluster Extraction, submitted, 2020. (Cited on p. 20)

[44] M.-J. Lai and J. Stoeckler, Construction of multivariate compactly supported tight wavelet frames, Appl. Comput. Harmon. Anal., 21 (2006), 324–348. (Cited on p. 5)

[45] M.-J. Lai and J. Wang, An unconstrained ℓ_q minimization for sparse solution of underdetermined linear systems, SIAM J. Optim., 21 (2011), 82–101. (Cited on pp. 11, 15, 16)

[46] M.-J. Lai, Y. Y. Xu, and W. T. Yin, Improved iteratively reweighted least squares for unconstrained smoothed ℓ_q minimization, SIAM J. Numer. Anal., 51 (2013), 927–957. (Cited on pp. 11, 13, 16)

[47] D. Malioutov, M. Cetin, and A. S. Willsky, A sparse signal reconstruction perspective for source localization with sensor arrays, IEEE Trans. Signal Process., 53 (2005), 3010–3022. (Cited on p. 17)

[48] H. M. Markowitz, Portfolio selection, J. Finance, 7 (1952), 77–91. (Cited on p. 7)

[49] M. Mishali and Y. C. Eldar, Blind multi-band signal reconstruction: Compressed sensing for analog signals, IEEE Trans. Signal Process., 57 (2009), 993–1009. (Cited on p. 17)

[50] B. K. Natarajan, Sparse approximate solutions to linear systems, SIAM J. Comput., 24 (1995), 227–234. (Cited on p. 2)

[51] F. Parvaresh, H. Vikalo, S. Misra, and B. Hassibi, Recovering sparse signals using sparse measurement matrices in compressed DNA microarrays, IEEE J. Sel. Top. Signal Process., 2 (2008), 275–285. (Cited on p. 17)

[52] R. T. Rockafellar, Convex Analysis, Princeton University Press, 1970. (Cited on p. 14)

[53] L. L. Schumaker, Spline Functions, Cambridge University Press, 2007. (Cited on p. 3)

[54] R. Tibshirani, Regression shrinkage and selection via the lasso, J. R. Statist. Soc. B, 58 (1996), 267–288. (Cited on p. 4)

[55] J. A. Tropp, Greed is good: Algorithmic results for sparse approximation, IEEE Trans. Inform. Theory, 50 (2004), 2231–2242. (Cited on p. 15)

[56] M. Vidyasagar, An Introduction to Compressed Sensing, SIAM, 2019. (Cited on p. 8)

[57] J. Wang and J. Ye, Lasso screening rules via dual polytope projection, J. Mach. Learn. Res., 16 (2015), 1063–1101. (Cited on p. 15)

[58] H. Zou and T. Hastie, Regularization and variable selection via the elastic net, J. R. Stat. Soc. Ser. B, 67 (2005), 301–320. (Cited on p. 4)

Chapter 2

Basic Concepts and Properties

In this chapter we shall present some basic concepts and properties of sparse solutions and sensing matrices. In particular, we shall introduce null space properties, mutual coherence, the restricted isometry property, and completely full rankness of rectangular matrices. In addition, we shall discuss some properties of noisy sensing matrices.

2.1 ▪ Null Space Properties

Let $\mathcal{N}(\Phi) = \{\mathbf{x} \in \mathbb{R}^n : \Phi\mathbf{x} = 0\}$ be the null space of Φ. Recall that $\mathcal{R}_s = \{\Phi\mathbf{x} : \mathbf{x} \in \mathbb{R}^n, \|\mathbf{x}\|_0 \leq s\}$ is the range of Φ of all s-sparse vectors $\mathbf{x}$ with $\|\mathbf{x}\|_0 \leq s$. We begin with the following result to characterize the solution (1.20).

Theorem 2.1. $\mathbf{x}^* = (x_1^*, x_2^*, \ldots, x_n^*)^\top \in \mathbb{R}^n$ *solves the minimization problem* (1.20) *if and only if for any vector* $\eta \in \mathcal{N}(\Phi)$,

$$\left| \sum_{x_i^* \neq 0} \operatorname{sign}(x_i^*)\eta_i \right| \leq \sum_{x_i^* = 0} |\eta_i|. \tag{2.1}$$

Moreover, $\mathbf{x}^*$ *is a unique solution if and only if*

$$\left| \sum_{x_i^* \neq 0} \operatorname{sign}(x_i^*)\eta_i \right| < \sum_{x_i^* = 0} |\eta_i| \tag{2.2}$$

for all $\eta \in \mathcal{N}(\Phi)\backslash\{0\}$.

Proof. For any $\eta \in \mathcal{N}(\Phi)$, we have $\Phi(\mathbf{x}^* + t\eta) = \mathbf{b}$ for all real numbers t. For all real numbers t,

$$\|\mathbf{x}^* + t\eta\|_1 \geq \|\mathbf{x}^*\|_1.$$

Let us be precise. For t sufficiently small,

$$\|\mathbf{x}^* + t\eta\|_1 = \sum_{x_i^* \neq 0} \operatorname{sign}(x_i^*)(x_i^* + t\eta_i) + \sum_{x_i^* = 0} |t||\eta_i|$$

$$= \sum_{x_i^* \neq 0} |x_i^*| + t \sum_{x_i^* \neq 0} \operatorname{sign}(x_i^*)\eta_i + |t| \sum_{x_i^* = 0} |\eta_i| \geq \|\mathbf{x}^*\|_1.$$

It follows that

$$t \sum_{x_i^* \neq 0} \operatorname{sign}(x_i^*)\eta_i + |t| \sum_{x_i^*=0} |\eta_i| \geq 0$$

for any sufficiently small $t \in (-1, 1)$. Thus, we have (2.1).

On the other hand, by (2.1),

$$\|\mathbf{x}^*\|_1 = \sum_{x_i^* \neq 0} \operatorname{sign}(x_i^*)(x_i^* + \eta_i) - \sum_{x_i^* \neq 0} \operatorname{sign}(x_i^*)(\eta_i)$$

$$\leq \sum_{x_i^* \neq 0} |x_i^* + \eta_i| + \sum_{x_i^*=0} |\eta_i| = \|\mathbf{x}^* + \eta\|_1.$$

Since $\mathbf{x}^* + \eta$ is feasible for all $\eta \in \mathcal{N}(\Phi)$, $\mathbf{x}^*$ is a minimizer of (1.20).

When $\mathbf{x}^*$ is unique, the inequalities discussed above have to be strict for all nonzero $\eta \in \mathcal{N}(\Phi)$. This completes the proof. $\square$

When $\mathbf{x}^* = (x_1^*, \dots, x_n^*)$ with nonzero entries, we have the following characterization theorem.

Theorem 2.2. *The results in Theorem 2.1 hold for overdetermined linear systems, i.e., when $m \geq n$. If $\mathbf{x}^* = (x_1^*, x_2^*, \dots, x_n^*)^\top \in \mathbb{R}^n$ with $x_j^* \neq 0, j = 1, \dots, n$, then $\mathbf{x}^*$ solves the minimization problem (1.20) if and only if*

$$\sum_{j=1}^{n} \operatorname{sign}(x_j^*)\eta_j = 0$$

for any vector $\eta \in \mathcal{N}(\Phi)$.

Proof. We leave the proof to Exercise 1. $\square$

Similarly, we can characterize the solution of the unconstrained ℓ_1 minimization (1.30). We have the following result.

Theorem 2.3. $\mathbf{x}^* = (x_1^*, x_2^*, \dots, x_n^*)^\top \in \mathbb{R}^n$ *solves the minimization problem (1.30) if and only if for any vector $\eta \in \mathcal{N}(\Phi)$, the null space of Φ satisfies (2.1). Moreover, $\mathbf{x}^*$ is a unique solution if and only if the strict inequality (2.2) is satisfied for all $\eta \in \mathcal{N}(\Phi)\backslash\{0\}$.*

Proof. We leave the proof to Exercise 2. $\square$

See [22] for more results on L_1 approximation. It is clear that the discussion can be generalized to characterize the solution of (1.27) and (1.33). We have the following theorem.

Theorem 2.4. *Fix $0 < q \leq 1$. Suppose that $\mathbf{x_b}$ is a sparse solution of $\Phi\mathbf{x_b} = \mathbf{b}$ with sparsity s. Let $\mathbf{x}^*$ solve the minimization problem (1.27). Let $S \subset \{1, \dots, n\}$ be the index set of the nonzero entries of the sparse solution $\mathbf{x_b}$. Then $\mathbf{x}^*$ is the unique sparse solution and hence is equal to $\mathbf{x_b}$ if any nonzero vector $\mathbf{v} \in \mathcal{N}(\Phi)$ satisfies*

$$\sum_{i \in S} |v_i|^q < \sum_{i \notin S} |v_i|^q, \tag{2.3}$$

where $\mathbf{v} = (v_1, v_2, \dots, v_n)^\top \in \mathbb{R}^n$.

Proof. Let S^c be the complement of S in $\{1, 2, \ldots, n\}$. Let $\mathbf{v}$ be specified as $\mathbf{v} := \mathbf{x_b} - \mathbf{x}^*$. Clearly, $\mathbf{v} \in \mathcal{N}(\Phi)$. Because $\mathbf{x}^*$ is a minimizer of (1.27), we have

$$\|\mathbf{x_b}\|_q^q \geq \|\mathbf{x}^*\|_q^q, \qquad \text{i.e.,} \qquad \|(\mathbf{x_b})_S\|_q^q + \|(\mathbf{x_b})_{S^c}\|_q^q \geq \|(\mathbf{x}^*)_S\|_q^q + \|(\mathbf{x}^*)_{S^c}\|_q^q.$$

By the standard triangle inequality for quasi-norm $\|\mathbf{x}\|_q^q$, i.e., $\|\mathbf{x} + \mathbf{y}\|_q^q \leq \|\mathbf{x}\|_q^q + \|\mathbf{y}\|_q^q$, we have

$$\|(\mathbf{x_b})_S\|_q^q + \|(\mathbf{x_b})_{S^c}\|_q^q \geq \|(\mathbf{x}^*)_S\|_q^q + \|(\mathbf{x}^*)_{S^c}\|_q^q \geq \|(\mathbf{x_b})_S\|_q^q - \|\mathbf{v}_S\|_q^q + \|\mathbf{v}_{S^c}\|_q^q - \|(\mathbf{x_b})_{S^c}\|_q^q.$$

Rearranging the latter yields the inequality

$$\|\mathbf{v}_{S^c}\|_q^q \leq 2 \|(\mathbf{x_b})_{S^c}\|_q^q + \|\mathbf{v}_S\|_q^q = \|\mathbf{v}_S\|_q^q \tag{2.4}$$

since $(\mathbf{x_b})_{S^c} = 0$. That is, $\|\mathbf{v}_{S^c}\|_q^q \leq \|\mathbf{v}_S\|_q^q$ since the exact solution $\mathbf{x_b}$ is s-sparse and S is the index set of the nonzero entries of $\mathbf{x_b}$. If we had (2.3) and if $\mathbf{v}$ were nonzero, we would have a contradiction: $\|\mathbf{v}_{S^c}\|_q^q \leq \|\mathbf{v}_S\|_q^q < \|\mathbf{v}_{S^c}\|_q^q$. □

Now we are ready to characterize the minimizer of (1.27).

Theorem 2.5. *Fix $0 < q \leq 1$. The minimization problem* (1.27) *finds the unique sparse solution for any* $\mathbf{b} \in \mathcal{R}_s$ *if and only if any nonzero vector* $\mathbf{v} \in \mathcal{N}(\Phi)$ *satisfies*

$$\sum_{i \in S} |v_i|^q < \sum_{i \notin S} |v_i|^q \qquad \forall S \subset \{1, \ldots, n\} \text{ with cardinality } \#(S) = s. \tag{2.5}$$

Proof. The "if" part can be seen by Theorem 2.4 when S is the special index set of all nonzero entries of the solution $\mathbf{x_b}$.

On the other hand, suppose that there is an index set S and a nonzero vector $\mathbf{v} \in \mathcal{N}(\Phi)$ such that (2.5) does not hold while the minimization problem (1.27) can find the unique solution $\mathbf{x}_q$. That is, for this special set S and this special vector $\mathbf{v}$, we have

$$\sum_{i \in S} |v_i|^q \geq \sum_{i \notin S} |v_i|^q.$$

Let $\mathbf{x_b} \in \mathbb{R}^n$ be the vector whose entries are $\mathbf{v}_S$ if they are in S and zero otherwise. Then letting $\mathbf{b} = \Phi \mathbf{x_b} = \Phi \mathbf{x_b} - \Phi \mathbf{v} = \Phi(\mathbf{x_b} - \mathbf{v})$, we have

$$\|\mathbf{x_b}\|_q^q = \sum_{i \in S} |x_i|^q = \sum_{i \in S} |v_i|^q \geq \sum_{i \notin S} |v_i|^q$$
$$= \|\mathbf{x_b} - \mathbf{v}\|_q^q.$$

For this special $\mathbf{b}$, the ℓ_q minimization finds the unique sparse solution $\mathbf{x}_q$ with $\|\mathbf{x}_q\|_0 \leq s$. If $\mathbf{x}_q \neq \mathbf{x_b}$, then the minimizer $\mathbf{x}_q$ satisfies $\|\mathbf{x}_q\|_q \leq \|\mathbf{x_b} - \mathbf{v}\|_q \leq \|\mathbf{x_b}\|_q$, which is a contradiction to $\mathbf{x_b}$ being the unique solution to the minimization. □

The condition in (2.5) is called the null space property (NSP) of Φ in the ℓ_q quasi-norm with sparsity s. One immediate consequence is that the sparsity $s < n/2$. Otherwise, (2.5) cannot hold for an S dependent on $\mathbf{v} \in \mathcal{N}(\Phi)$ if $s \geq n/2$. Indeed, take S to be the index set of the $s \geq n/2$ largest entries of $\mathbf{v}$ in absolute value, and then $\sum_{i \in S} |v_i|^q \geq \sum_{i \in S^c} |v_i|^q$, which violates (2.5).

Due to the finite dimensionality of $\{\mathbf{z}, \mathbf{z} \in \mathcal{N}(\Phi), \|\mathbf{z}\|_q = 1\}$, there exists a constant $\gamma \in (0, 1)$ such that

$$\sum_{i \in S} |v_i|^q \leq \gamma \sum_{i \notin S} |v_i|^q \qquad \forall S \subset \{1, \ldots, n\} \text{ with cardinality } \#(S) = s \tag{2.6}$$

when we have (2.5). Thus, the condition (2.6) is equivalent to (2.5).

With this property, we have the following lemma.

Lemma 2.6 (Daubechies, DeVore, Fornasier, and Güntuk, 2010 [7]). *Suppose that Φ satisfies the NSP with $\gamma < 1$ for sparsity s in the ℓ_q (quasi-)norm. Then for any vector $\mathbf{v} \in \mathcal{N}(\Phi)$, writing $\mathbf{v} = \mathbf{x} - \mathbf{y}$,*

$$\|\mathbf{x} - \mathbf{y}\|_q^q \leq \frac{1+\gamma}{1-\gamma} \left(\|\mathbf{x}\|_q^q - \|\mathbf{y}\|_q^q + 2\|\mathbf{y}_{S^c}\|_q^q \right)$$

for any index set $S \subset \{1, 2, \ldots, n\}$ of size s, where S^c is the complement of S in $\{1, 2, \ldots, n\}$.

Proof. Let S be any subset of $\{1, 2, \ldots, n\}$ of size s. Then

$$\begin{aligned}
\|(\mathbf{x} - \mathbf{y})_{S^c}\|_q^q &\leq \|(\mathbf{x})_{S^c}\|_q^q + \|(\mathbf{y})_{S^c}\|_q^q = \|\mathbf{x}\|_q^q - \|\mathbf{x}_S\|_q^q + \|(\mathbf{y})_{S^c}\|_q^q \\
&= \|\mathbf{x}\|_q^q - \|\mathbf{y}\|_q^q + \|\mathbf{y}_S\|_q^q - \|\mathbf{x}_S\|_q^q + 2\|(\mathbf{y})_{S^c}\|_q^q \\
&\leq \|\mathbf{x}\|_q^q - \|\mathbf{y}\|_q^q + \|(\mathbf{y} - \mathbf{x})_S\|_q^q + 2\|(\mathbf{y})_{S^c}\|_q^q \\
&\leq \|\mathbf{x}\|_q^q - \|\mathbf{y}\|_q^q + \gamma\|(\mathbf{y} - \mathbf{x})_{S^c}\|_q^q + 2\|(\mathbf{y})_{S^c}\|_q^q,
\end{aligned}$$

where we have used the NSP (2.6). It follows that

$$(1 - \gamma)\|(\mathbf{x} - \mathbf{y})_{S^c}\|_q^q \leq \|\mathbf{x}\|_q^q - \|\mathbf{y}\|_q^q + 2\|(\mathbf{y})_{S^c}\|_q^q$$

and

$$\begin{aligned}
\|\mathbf{x} - \mathbf{y}\|_q^q = \|(\mathbf{x} - \mathbf{y})_S\|_q^q + \|(\mathbf{x} - \mathbf{y})_{S^c}\|_q^q &\leq (1 + \gamma)\|(\mathbf{x} - \mathbf{y})_{S^c}\|_q^q \\
&\leq \frac{1+\gamma}{1-\gamma} \left(\|\mathbf{x}\|_q^q - \|\mathbf{y}\|_q^q + 2\|(\mathbf{y})_{S^c}\|_q^q \right).
\end{aligned}$$

This completes the proof. □

With Lemma 2.6, we can give a second proof that any minimizer of (1.27) is the sparse solution. That is, suppose that $\mathbf{b} = \Phi\mathbf{y}$ with another sparse solution $\mathbf{y}$ of sparsity s. Let $\mathbf{x}$ be a solution of the minimization (1.27). Then we have $\|\mathbf{x}\|_q^q \leq \|\mathbf{y}\|_q^q$ and $\mathbf{x} - \mathbf{y} \in \mathcal{N}(\Phi)$. When Φ has the NSP of sparsity s in the ℓ_q quasi-norm, Lemma 2.6 shows that $\|\mathbf{x} - \mathbf{y}\|_q^q \leq 0$ or $\mathbf{x} = \mathbf{y}$.

Next it is easy to see that we can rewrite (2.5) as

$$2\sum_{i \in S} |v_i|^q < \|\mathbf{v}\|_q^q \quad \text{or} \quad \frac{\sum_{i \in S} |v_i|^q}{\|\mathbf{v}\|_q^q} < \frac{1}{2}$$

for all $S \subset \{1, \ldots, n\}$ with $\#(S) \leq s$. That is,

$$\max_{\substack{S \subset \{1, \ldots, n\} \\ \#(S) \leq s}} \frac{\sum_{i \in S} |v_i|^q}{\|\mathbf{v}\|_q^q} < \frac{1}{2}. \tag{2.7}$$

This enables us to prove the following thereom.

Theorem 2.7. *If Φ satisfies the NSP in the ℓ_q quasi-norm with $q \leq 1$ with sparsity s, then it satisfies the NSP in all ℓ_p quasi-norms for $p \in (0, q]$ with sparsity s. Equivalently, if the minimization problem (1.27) for $q = q_0 \in (0, 1]$ finds the unique sparse solution for any $\mathbf{b} \in \mathcal{R}_s$, then so does the minimization problem (1.27) for all $q \in (0, q_0]$.*

Proof. Without loss of generality, we may assume that Φ satisfies the NSP in the ℓ_1 norm with sparsity s. Let us show that Φ satisfies the NSP in the ℓ_q norm for $0 < q \leq 1$ with sparsity s.

Fix S to be an index set with cardinality s. For any $\mathbf{v} \in \mathcal{N}(\Phi)$, we may write $\eta = |\mathbf{v}|$, i.e., $\eta = (\eta_1, \ldots, \eta_n)^\top \in \mathbb{R}^n$ with $\eta_i = |v_i| \geq 0, i = 1, \ldots, n$, just for convenience. Also, we may assume that $\eta_i, i = 1, \ldots, n$, are decreasing, and hence $S = \{1, 2, \ldots, s\}$ in this case. For any $\eta_i \geq \eta_{s+j}$ with $i \in \{1, \ldots, s\}$ and $j \geq 1$, we have

$$\frac{\eta_i^q}{\eta_{s+j}^q} \leq \frac{\eta_i}{\eta_{s+j}} \quad \text{or} \quad \eta_i^q \eta_{s+j} \leq \eta_i \eta_{s+j}^q.$$

Thus, $\sum_{i=1}^{s} \eta_i^q \sum_{j \geq 1} \eta_{s+j} \leq \sum_{i=1}^{s} \eta_i \sum_{j \geq 1} \eta_{s+j}^q$ or

$$\frac{\sum_{j \geq 1} \eta_{s+j}}{\sum_{i=1}^{s} \eta_i} \leq \frac{\sum_{j \geq 1} \eta_{s+j}^q}{\sum_{i=1}^{s} \eta_i^q}.$$

It follows that

$$\frac{\sum_{j \geq 1} \eta_{s+j} + \sum_{i=1}^{s} \eta_i}{\sum_{i=1}^{s} \eta_i} \leq \frac{\sum_{j \geq 1} \eta_{s+j}^q + \sum_{i=1}^{s} \eta_i^q}{\sum_{i=1}^{s} \eta_i^q}$$

or

$$\frac{\sum_{i=1}^{s} \eta_i}{\|\eta\|_1} \geq \frac{\sum_{i=1}^{s} \eta_i^q}{\|\eta\|_q^q}.$$

Since $\frac{\sum_{i=1}^{s} \eta_i}{\|\eta\|_1} < 1/2$ by the NSP in the ℓ_1 norm based on (2.7), we see that

$$\frac{\sum_{i=1}^{s} \eta_i^q}{\|\eta\|_q^q} < \frac{1}{2}.$$

Thus, Φ satisfies the NSP in the ℓ_q norm with sparsity s for all $q \in (0, 1]$ by using (2.7).

By using the same proof as the one for Theorem 2.5, we can show that the minimizer $\mathbf{x}^*$ of (1.27) is unique and hence is $\mathbf{x_b}$. $\square$

At the end of this section, we discuss a sufficient condition to ensure the NSP.

Definition 2.8 (Δ-SSP). *Let m and n be two integers with $m > 0$ and $m < n$. An $(n-m)$-dimensional subspace $V \subset \mathbb{R}^n$ has the Δ-spherical section property (SSP) if*

$$\frac{\|h\|_1}{\|h\|_2} \geq \sqrt{\frac{m}{\Delta}} \tag{2.8}$$

holds for all nonzero $h \in V$, where $\Delta > 0$ is a fixed number.

One of the significances of (2.8) is the following.

Lemma 2.9 (Zhang, 2008 [28]). *Suppose that the null space $\mathcal{N}(\Phi)$ satisfies the SSP in the following sense:*

$$\frac{\|h\|_1}{\|h\|_2} \geq 2\sqrt{s} \quad \forall h \in \mathcal{N}(\Phi) \setminus \{0\}. \tag{2.9}$$

Then Φ satisfies the NSP in the ℓ_1 norm.

Proof. Let $\mathbf{x}, \mathbf{y} \in \mathbb{R}^n$ and $S = \{i, \mathbf{x}_i \neq 0\}$ be the support index set of $\mathbf{x}$. Suppose $s = \#(S)$ is the cardinality of S. We first claim that $\|\mathbf{x}\|_1 < \|\mathbf{y}\|_1$ if

$$\frac{\|\mathbf{x} - \mathbf{y}\|_1}{\|\mathbf{x} - \mathbf{y}\|_2} \geq 2\sqrt{s}.$$

Indeed, write $\mathbf{y} = \mathbf{x} + h$ and we have

$$\|\mathbf{y}\|_1 = \|\mathbf{x}_S + h_S\|_1 + \|h_{S^c}\|_1 = \|\mathbf{x}\|_1 + \|h_{S^c}\|_1 - \|h_S\|_1 + \|\mathbf{x}_S + h_S\|_1 - \|\mathbf{x}_S\|_1 + \|h_S\|_1$$
$$\geq \|\mathbf{x}\|_1 + \|h_{S^c}\|_1 - \|h_S\|_1.$$

Thus, $\|\mathbf{y}\|_1 > \|\mathbf{x}\|_1$ if $\|h_{S^c}\|_1 - \|h_S\|_1 > 0$ or $\|h\|_1 > 2\|h_S\|_1$. Since $\|h_S\|_1 \leq \sqrt{s}\|h_S\|_2$ by the Cauchy–Schwarz inequality, we will have $\|h_{S^c}\|_1 > \|h_S\|_1$ if $\|h\|_1 \geq 2\sqrt{s}\|h\|_2 \geq 2\sqrt{s}\|h_S\|_2 \geq 2\|h_S\|_1$.

Consequently, for any sparse solution $\mathbf{x}$ with $\Phi\mathbf{x} = \mathbf{b}$, for any $h \in \mathcal{N}(\Phi)$ with $h \neq 0$, if (2.9) holds, we will have $\|h\|_1 > 2\|h_S\|_1$ or $\|h_{S^c}\|_1 > \|h_S\|_1$, and hence $\|\mathbf{x} + h\|_1 > \|\mathbf{x}\|_1$. That is, $\mathbf{x}$ is the unique sparse solution. Hence, Φ satisfies the NSP in the ℓ_1 norm.   □

It is easy to extend the result in Lemma 2.9 to the ℓ_q setting with $q \in (0, 1)$ (cf. [24]).

Lemma 2.10. *Fix any $q \in (0, 1)$. Suppose that the null space $\mathcal{N}(\Phi)$ satisfies the SSP in the following sense:*

$$\frac{\|h\|_q^q}{\|h\|_1^q} \geq 2s^{1-q} \quad \forall h \in \mathcal{N}(\Phi)\backslash\{0\}. \tag{2.10}$$

Then Φ satisfies the NSP in the ℓ_q quasi-norm.

We leave the proof to Exercise 21. It would be interesting to know whether a Φ with $\mathcal{N}(\Phi)$ exists satisfying the above property.

2.2 ▪ Matrix Spark Estimates

In this section, we will prove bounds on the sparsity of the sparsest solution of the linear system $\Phi\mathbf{x} = \mathbf{b}$, We will assume $\mathbf{b} \neq 0$. Without loss of generality, we will assume that $\|\mathbf{b}\| = 1$. Let us begin by defining a *spark* (cf. [9]).

Definition 2.11. *The* spark *of a matrix A, denoted by $\sigma(A)$, is the smallest number k such that there exists a set of k columns in A which are linearly dependent. That is,*

$$\mathrm{spark}(A) = \min_{\mathbf{x} \neq 0} \|\mathbf{x}\|_0 \quad \textit{subject to } \mathbf{x} \in \mathrm{Null}(A),$$

where $\|\mathbf{x}\|_0$ denotes the number of nonzero components of the vector $\mathbf{x}$.

It is clear that $\sigma(A) \leq \mathrm{rank}(A) + 1$. A full rank rectangular matrix A may have $\sigma(A)$ smaller than the rank. For example, let B be a nonsingular square matrix of size $m \times m$ with $m > 2$ and $A = [B\ B]$. Then $\sigma(A) = 2$, but rank(A) is m. Now we define a new matrix $\tilde{A}$ as follows: For each $i = 1, 2, \ldots, n$, the ith column $\tilde{A}_i$ of the matrix $\tilde{A}$ is obtained from ith column A_i of the matrix A by orthogonalizing it with respect to the vector $\mathbf{b}$. In other words,

$$\tilde{A}_i = A_i - \langle A_i, \mathbf{b}\rangle \mathbf{b}$$

and

$$\tilde{A} = [\tilde{A}_1 \mid \tilde{A}_2 \mid \cdots \mid \tilde{A}_n].$$

More concisely, after normalizing the system $A\mathbf{x} = \mathbf{b}$ so that $\|\mathbf{b}\| = 1$, $\tilde{A}$ can be described as

$$\tilde{A} = (I_m - \mathbf{b}\mathbf{b}^\top)A,$$

where I_m is the $m \times m$ identity matrix. Our main result is as follows.

Theorem 2.12. *If* $\mathrm{spark}(\tilde{A}) \neq \mathrm{spark}(A)$, *then* $\mathrm{spark}(\tilde{A}) < \mathrm{spark}(A)$ *and* $\|\mathbf{x}^\star\|_0 = \mathrm{spark}(\tilde{A})$, *where* $\mathbf{x}^\star$ *is the sparsest solution satisfying* $A\mathbf{x}^\star = \mathbf{b}$.

To prove this result, we need some preparatory results. The first one is the lower bound.

Proposition 2.13. *We have*

$$\{\mathbf{x} \mid A\mathbf{x} = \mathbf{b}\} \subseteq \{\mathbf{x} \mid \tilde{A}\mathbf{x} = 0\}. \tag{2.11}$$

In particular,

$$\mathrm{spark}(\tilde{A}) \leq \|\mathbf{x}^\star\|_0. \tag{2.12}$$

Proof. If $\mathbf{x} = (x_1, \ldots, x_n) \in \{\mathbf{x} \mid A\mathbf{x} = \mathbf{b}\}$, then

$$\begin{aligned}
\tilde{A}\mathbf{x} = x_1 \tilde{A}_1 + \cdots + x_n \tilde{A}_n &= x_1(A_1 - \langle A_1, \mathbf{b}\rangle \mathbf{b}) + \cdots + x_n(A_n - \langle A_n, \mathbf{b}\rangle \mathbf{b}) \\
&= x_1 A_1 + \cdots + x_n A_n - \langle x_1 A_1 + \cdots + x_n A_n, \mathbf{b}\rangle \mathbf{b} \\
&= \mathbf{b} - \langle \mathbf{b}, \mathbf{b}\rangle \mathbf{b} = 0.
\end{aligned}$$

The last step follows from the assumption that $\|\mathbf{b}\| = 1$. Thus, we have (2.11).

To prove (2.12), we use (2.11) to obtain

$$\mathrm{spark}(\tilde{A}) = \min_{\{\mathbf{x} \mid \tilde{A}\mathbf{x}=0\}} \|\mathbf{x}\|_0 \leq \min_{\mathbf{x} \in \{\mathbf{x} \mid A\mathbf{x}=\mathbf{b}\}} \|\mathbf{x}\|_0 = \|\mathbf{x}^\star\|_0.$$

This completes the proof. □

Next we will need the following proposition.

Proposition 2.14. *If* $\mathbf{x}$ *is a vector such that* $\tilde{A}\mathbf{x} = 0$, *then exactly one of the following two conclusions holds true:*

- $A\mathbf{x} = 0$.

- *There exists a nonzero scalar* λ *such that* $A(\frac{\mathbf{x}}{\lambda}) = \mathbf{b}$.

Proof. Assume that $\mathbf{x} = (x_1, \ldots, x_n)^\top$ is a vector such that $\tilde{A}\mathbf{x} = 0$. Using the definition of $\tilde{A}$, we get

$$\begin{aligned}
0 = \tilde{A}\mathbf{x} &= x_1 \tilde{A}_1 + \cdots + x_n \tilde{A}_n \\
&= x_1 A_1 + \cdots + x_n A_n - \langle x_1 A_1 + \cdots + x_n A_n, \mathbf{b}\rangle \mathbf{b}. \tag{2.13}
\end{aligned}$$

Now we have two cases:

Case 1: $\langle x_1 A_1 + \cdots + x_n A_n, \mathbf{b}\rangle = 0$. In this case, using equation (2.13), we get $x_1 A_1 + \cdots + x_n A_n = 0$. That is, $A\mathbf{x} = 0$.

Case 2: $\langle x_1 A_1 + \cdots + x_n A_n, \mathbf{b}\rangle \neq 0$. For this case, set $\lambda := \langle x_1 A_1 + \cdots + x_n A_n, \mathbf{b}\rangle$. Then we get $A(\frac{\mathbf{x}}{\lambda}) = \mathbf{b}$.

It is clear that the two cases cannot hold together. Indeed, if $A\mathbf{x} = 0$ and $A(\frac{\mathbf{x}}{\lambda}) = \mathbf{b}$, then $\mathbf{b} = 0$, which contradicts our assumption that $\|\mathbf{b}\| = 1$. Thus the proof is complete. $\square$

Proposition 2.15. *Let $\tilde{A}$ be the matrix defined above. Then*

$$\mathrm{spark}(\tilde{A}) \leq \mathrm{spark}(A). \tag{2.14}$$

Proof. From equation (2.13), we have $\{\mathbf{x} \mid A\mathbf{x} = 0\} \subseteq \{\mathbf{x} \mid \tilde{A}\mathbf{x} = 0\}$. Hence,

$$\mathrm{spark}(\tilde{A}) := \min_{\{\mathbf{x} \mid \tilde{A}\mathbf{x}=0\}} \|\mathbf{x}\|_0 \leq \min_{\mathbf{x} \in \{\mathbf{x} \mid A\mathbf{x}=0\}} \|\mathbf{x}\|_0 =: \mathrm{spark}(A),$$

which is (2.14). $\square$

Proposition 2.16. *Let $\mathbf{x}^\star \in \mathcal{D}_s(\mathbb{R}^n)$ be a sparse solution of $A\mathbf{x} = \mathbf{b}$ for a positive integer s. Then either of the following conclusions holds true:*

1. $\mathrm{spark}(A) = \mathrm{spark}(\tilde{A}) \leq \|\mathbf{x}^\star\|_0.$

2. $\|\mathbf{x}^\star\|_0 = \mathrm{spark}(\tilde{A}) \leq \mathrm{spark}(A).$

Proof. Let $\tilde{\mathbf{x}} = \arg\min_{\mathbf{x} \text{ s.t. } \tilde{A}\mathbf{x}=0} \|\mathbf{x}\|_0$. That is, $\|\tilde{\mathbf{x}}\|_0 = \mathrm{spark}(\tilde{A})$. Now using Proposition 2.14, either $A\tilde{\mathbf{x}} = 0$ or there exists a nonzero scalar λ such that $A(\frac{\tilde{\mathbf{x}}}{\lambda}) = \mathbf{b}$.

Case 1: $A\tilde{\mathbf{x}} = 0$. In this case, from the definition of spark, we get $\mathrm{spark}(A) \leq \|\tilde{\mathbf{x}}\|_0 = \mathrm{spark}(\tilde{A})$. Now the first conclusion follows by using Propositions 2.13 and 2.15.

Case 2: $A\tilde{\mathbf{x}} \neq 0$. We have a nonzero scalar λ such that $A(\frac{\tilde{\mathbf{x}}}{\lambda}) = \mathbf{b}$. Since $\mathbf{x}^\star$ is the sparsest solution to the system $A\mathbf{x} = \mathbf{b}$, we have

$$\|\mathbf{x}^\star\|_0 \leq \left\| \frac{\tilde{\mathbf{x}}}{\lambda} \right\|_0 = \|\tilde{\mathbf{x}}\|_0 = \mathrm{spark}(\tilde{A}).$$

Combining the above equation with Proposition 2.13, we obtain $\|\mathbf{x}^\star\|_0 = \mathrm{spark}(\tilde{A})$. Now, using (2.15), we obtain the second conclusion. $\square$

Furthermore, let us derive another lower bound for the sparsity. Define the matrix A_{aug} as

$$A_{\mathrm{aug}} := [A \mid \mathbf{b}].$$

That is, A_{aug} is the augmented matrix obtained by concatenating the vector $\mathbf{b}$ as a new column to matrix A. Then we have the following lower bound on sparsity.

Proposition 2.17.

$$\mathrm{spark}(A_{\mathrm{aug}}) - 1 \leq \|\mathbf{x}^\star\|_0. \tag{2.15}$$

Proof. Since $A\mathbf{x} = \mathbf{b}$, we have $A_{\mathrm{aug}} \begin{bmatrix} \mathbf{x}^\star \\ -1 \end{bmatrix} = 0$. Hence,

$$\|\mathbf{x}^\star\|_0 + 1 = \left\| \begin{bmatrix} \mathbf{x}^\star \\ -1 \end{bmatrix} \right\|_0 \geq \mathrm{spark}(A_{\mathrm{aug}}),$$

which is (2.15). $\square$

Finally, we end with the following theorem.

Theorem 2.18 (Donoho and Elad, 2003 [9]). *The underdetermined linear system* $\mathbf{b} = \Phi\mathbf{x_b}$ *has a unique sparse solution if* $\|\mathbf{x_b}\|_0 < spark(\Phi)/2$.

Proof. Suppose that there are two sparsest solutions, $\mathbf{x}^1$ and $\mathbf{x}^2$, with $\|\mathbf{x}^1\|_0 \leq s$ and $\|\mathbf{x}^2\|_0 \leq s$, solving $\mathbf{y} = \Phi\mathbf{x}$. Then $\Phi(\mathbf{x}^1 - \mathbf{x}^2) = 0$. So $\|(\mathbf{x}^1 - \mathbf{x}^2)\|_0 \leq 2s$, but $\|(\mathbf{x}^1 - \mathbf{x}^2)\|_0 \geq spark(\Phi)$. It follows that $s \geq spark(\Phi)/2$. Hence, when $s < spark(\Phi)/2$, the sparsest solution is unique. That is, if we find a solution $\mathbf{x_b}$ of $\Phi\mathbf{x_b} = \mathbf{b}$ with $\|\mathbf{x_b}\|_0 < spark(\Phi)/2$, then $\mathbf{x_b}$ is the sparsest solution. $\square$

To conclude this section, we refer the reader to [11] for more properties of the concept of the matrix spark.

2.3 ▪ Mutual Coherence

Next we introduce the concept of the mutual coherence of matrix Φ. Assume that each column of Φ is normalized. Let $G = \Phi^\top\Phi$, which is a square matrix of size $n \times n$. Writing $G = (g_{ij})_{1 \leq i,j \leq n}$, the mutual coherence of Φ is

$$M = M(\Phi) = \max_{\substack{1 \leq i,j \leq n \\ i \neq j}} |g_{ij}|.$$

Clearly, $M \leq 1$. We would like to have matrix Φ such that its mutual coherence M is as small as possible. However, $M(\Phi)$ cannot be too small, as shown in the following lemma.

Lemma 2.19. *Assume that* $\Phi \in \mathbb{R}^{m \times n}$ *is of full rank. Then* $M(\Phi) \geq \sqrt{\frac{n-m}{m(n-1)}}$. *In particular, if* $n \geq 2m$, $M(\Phi) \geq (2m)^{-1/2}$.

Proof. Let $\lambda_i, i = 1, \dots, n$, be eigenvalues of G. Since G is positive semidefinite, all $\lambda_i \geq 0$. Since the rank of G is equal to m, only m eigenvalues λ_i are nonzero. Since $\sum_i \lambda_i$ is equal to the trace of G, which is n since $g_{ii} = 1$, we have

$$n = \sum_i \lambda_i \leq \sqrt{m}\sqrt{\sum_i \lambda_i^2}. \tag{2.16}$$

On the other hand, using a property of the Frobenius norm of G, we have

$$\sum_i \lambda_i^2 = \|G\|_F^2 = \sum_{1 \leq i,j \leq n} (g_{ij})^2. \tag{2.17}$$

It follows from (2.16) and (2.17) that

$$(n^2 - n)M(\Phi)^2 + n \geq \sum_{1 \leq i,j \leq n} (g_{ij})^2 \geq \frac{n^2}{m}.$$

That is, $M(\Phi) \geq \sqrt{\frac{n-m}{m(n-1)}} \geq \sqrt{\frac{n-m}{mn}}$. In particular, when $n \geq 2m$, we have $M(\Phi) \geq (2m)^{-1/2}$. That is, $M(\Phi) \in ((2m)^{-1/2}, 1]$. $\square$

Note that the lower bound $M(\Phi) \geq \sqrt{\frac{n-m}{m(n-1)}}$ is called Welch's bound (cf. [26]).

Example 2.20. When $n = \alpha m$ with $\alpha > 1$, $M(\Phi) \geq \sqrt{\frac{(\alpha-1)}{\alpha m}}$. In particular, for $\alpha = 1.01$, $M(\Phi) \geq \sqrt{\frac{1}{101m}}$, and for $\alpha = 101$, $M(\Phi) \geq \sqrt{\frac{100}{101m}}$. This shows that the rectangular size of Φ will have an effect on the mutual coherence.

To explore more properties of mutual coherence, we need the following lemma.

Lemma 2.21. *Let $s < 1/M + 1$. For any $S \subset \{1, \ldots, n\}$ with $\#(S) \leq s$ and Φ_S the matrix consisting of the s columns of Φ with column indices in S, the singular values of Φ_S are bounded below by $(1 - M(s-1))^{1/2}$ and above by $(1 + M(s-1))^{1/2}$.*

Proof. For any vector $\mathbf{v} \in \mathbb{R}^n$ with support on S, we have

$$\mathbf{v}^\top G \mathbf{v} = \mathbf{v}_S^\top \Phi_S^\top \Phi_S \mathbf{v}_S = \|\mathbf{v}\|^2 + \sum_{\substack{i \neq j \\ i,j \in S}} v_i g_{ij} v_j.$$

Since

$$\left| \sum_{\substack{i \neq j \\ i,j \in S}} v_i g_{ij} v_j \right| \leq M \sum_{\substack{i \neq j \\ i,j \in S}} |v_i v_j| \leq M \left(\sum_{i,j \in S} |v_i v_j| - \|\mathbf{v}\|_2^2 \right) \leq M \|\mathbf{v}\|_2^2 (s-1),$$

we have

$$\mathbf{v}_S^\top \Phi_S^\top \Phi_S \mathbf{v}_S \geq \|\mathbf{v}\|_2^2 - M(s-1)\|\mathbf{v}\|_2^2 = (1 - M(s-1))\|\mathbf{v}\|_2^2.$$

Similarly,

$$\mathbf{v}^\top G \mathbf{v} = \mathbf{v}_S^\top \Phi_S^\top \Phi_S \mathbf{v}_S = \|\mathbf{v}\|^2 + \sum_{\substack{i \neq j \\ i,j \in S}} v_i g_{ij} v_j \leq \|\mathbf{v}\|_2^2 + M(s-1)\|\mathbf{v}\|_2^2 = (1 + M(s-1))\|\mathbf{v}\|_2^2.$$

This completes the proof. $\square$

We first show that if $s < (1 + 1/M)/2$ and for any $\mathbf{b} \in \mathcal{R}_s$, the sparse solution of (1.4) will be unique, as explained in the following.

Theorem 2.22. *Suppose $s < (1 + 1/M)/2$. For any $\mathbf{b} \in \mathcal{R}_s$, the sparse solution of (1.4) is unique.*

Proof. Let $\mathbf{b} = \Phi \mathbf{x_b} \in \mathcal{R}_s$. Let $\mathbf{x}_1$ be a solution of (1.4) with $\|\mathbf{x}_1\|_0 \leq s$. Then $\|\mathbf{x_b} - \mathbf{x}_1\|_0 \leq 2s$. By Lemma 2.21, we have

$$(1 - M(2s-1))\|\mathbf{x_b} - \mathbf{x}_1\|_2^2 \leq \|\Phi(\mathbf{x_b} - \mathbf{x}_1)\|_2^2 = 0.$$

Since $1 - M(2s-1) \neq 0$, it follows that $\mathbf{x_b} = \mathbf{x}_1$. $\square$

For noiseless recovery of the sparse solution, the ℓ_1 minimization can find the unique solution.

Theorem 2.23 (Gribonval and Nielson, 2003 [13]). *Let M be the mutual coherence of Φ. For any $\mathbf{x_b}$ with $\|\mathbf{x_b}\|_0 \leq s$, the minimizer of (1.20) is the sparsest solution under the sufficient condition*

$$s < (1/M + 1)/2.$$

Proof. Let $\mathbf{x}^*$ be a minimizer of (1.20) and $h = \mathbf{x}^* - \mathbf{x_b}$. Then $h \in \mathcal{N}(\Phi)$. Suppose that $h \neq 0$. Then we claim $\|h\|_1 \le 2\|h_S\|_1$, where S is the index set of the support of $\mathbf{x_b}$. Otherwise, we have $\|h\|_1 = \|h_S\|_1 + \|h_{S^c}\|_1 > 2\|h_S\|_1$ or $\|h_{S^c}\|_1 > \|h_S\|_1$. Since $h_{S^c} = \mathbf{x}^*_{S^c}$, we have

$$\|h_{S^c}\|_1 = \|\mathbf{x}^*_{S^c}\|_1 > \|h_S\|_1 = \|\mathbf{x}^*_S - (\mathbf{x_b})_S\|_1 \ge \|\mathbf{x_b}\|_1 - \|\mathbf{x}^*_S\|_1.$$

In other words, $\|\mathbf{x}^*\|_1 = \|\mathbf{x}^*_{S^c}\|_1 + \|\mathbf{x}^*_S\|_1 > \|\mathbf{x_b}\|_1$, which contradicts the property of the minimizer $\mathbf{x}^*$.

Next, since $\Phi^\top \Phi h = 0$, for each $i = 1, \dots, n$, we have

$$\phi_i^\top \Phi h = 0 \quad \text{or} \quad \sum_{j=1}^n \phi_i^\top \phi_j h_j = 0,$$

where $\Phi = [\phi_1, \phi_2, \dots, \phi_n]$, with column vectors $\phi_i, i = 1, \dots, n$. It follows that $h_i(1 + M) = -\sum_{j \neq i} \phi^\top \phi_j h_j + M h_i$. That is,

$$(1 + M)|h_i| \le M\|h\|_1 \text{ or } |h_i| \le (1 + M)^{-1} M\|h\|_1.$$

Thus, $\|h_S\|_1 \le (1 + M)^{-1} M s \|h\|_1 < (1 + M)^{-1} M (1/M + 1)/2 \|h\|_1 \le \|h_S\|_1$ by the claim. That is, we obtained a contradiction. Thus, $h \equiv 0$. □

For noisy recovery of the sparse solution using the ℓ_1 minimization, we have the following theorem.

Theorem 2.24 (Donoho, Elad, and Temlyakov, 2006 [10]). *Let M be the mutual coherence of Φ. Suppose that*

$$s < (1/M + 1)/4.$$

For any $\mathbf{x_b}$ with $\|\mathbf{x_b}\|_0 \le s$, let $\widehat{\mathbf{x}}_{\epsilon, \delta}$ be the solution of (1.23). Suppose that $\|\widehat{\mathbf{x}}_{\epsilon, \delta}\|_0 \le s$. Then

$$\|\widehat{\mathbf{x}}_{\epsilon, \delta} - \mathbf{x_b}\|_2^2 \le \frac{(\epsilon + \delta)^2}{1 - M(4s - 1)}.$$

Proof. Write $w = \widehat{\mathbf{x}}_{\epsilon, \delta} - \mathbf{x_b}$. Clearly, $\|\widehat{\mathbf{x}}_{\epsilon, \delta}\|_1 = \|w + \mathbf{x_b}\|_1 \le \|\mathbf{x_b}\|_1$ when computing the ℓ_1 minimization. Let $S \subset \{1, 2, \dots, n\}$ be the index set where $\mathbf{x_b}$ is supported. Since $\|w + \mathbf{x_b}\|_1 \ge \|\mathbf{x_b}\|_1 - \sum_{i \in S} |w_i| + \sum_{i \in S^c} |w_i|$, we have $\sum_{i \in S^c} |w_i| \le \sum_{i \in S} |w_i|$ or

$$\|w\|_1 \le 2 \sum_{i \in S} |w_i| \le 2\sqrt{s}\|w\|_2, \tag{2.18}$$

where S^c denotes the complement set of S in $\{1, 2, \dots, n\}$.

On the other hand, $\|\Phi \widehat{\mathbf{x}}_{\epsilon, \delta} - \mathbf{y}\|_2 \le \delta$ and $y = \Phi \mathbf{x_b} + z$ imply that $\|\Phi w + z\|_2 \le \delta$. That is, $\|\Phi w\|_2 \le \|\Phi w + z\|_2 + \epsilon \le \delta + \epsilon$.

Finally, $\|\Phi w\|_2^2 = \|w\|_2^2 + w^\top (G - I) w \ge \|w\|_2^2 - M(\|w\|_1^2 - \|w\|_2^2) \ge (1 + M)\|w\|_2^2 - M 4s \|w\|_2^2$ by the estimate (2.18) above. It follows that

$$\|w\|_2^2 \le \frac{1}{1 + M - 4Ms} \|\Phi w\|_2^2 \le \frac{(\epsilon + \delta)^2}{1 - M(4s - 1)}$$

by the estimate in the previous paragraph. This concludes the proof of this theorem. □

However, for noisy recovery of the sparse solution using the ℓ_0 approach, we can prove the following theorem.

Theorem 2.25 (Donoho, Elad, and Temlyakov, 2006 [10]). *Let M be the mutual coherence of Φ. Suppose that*

$$s < (1/M + 1)/2.$$

For any $\mathbf{x_b}$ with $\|\mathbf{x_b}\|_0 \leq s$, let $\widetilde{\mathbf{x}}_{\epsilon,\delta}$ be the solution of (1.18). Suppose that $\|\widetilde{\mathbf{x}}_{\epsilon,\delta}\|_0 \leq s$. Then

$$\|\widetilde{\mathbf{x}}_{\epsilon,\delta} - \mathbf{x_b}\|_2^2 \leq \frac{(\epsilon + \delta)^2}{1 - M(2s - 1)}.$$

Proof. By Lemma 2.21, we have

$$\|\widetilde{\mathbf{x}}_{\epsilon,\delta} - \mathbf{x_b}\|_2^2 \leq \frac{1}{1 - M(2s - 1)} \|\Phi(\widetilde{\mathbf{x}}_{\epsilon,\delta} - \mathbf{x_b})\|_2^2$$

$$= \frac{1}{1 - M(2s - 1)} \|\Phi\widetilde{\mathbf{x}}_{\epsilon,\delta} - \mathbf{y} + \mathbf{z}\|_2^2 \leq \frac{(\epsilon + \delta)^2}{1 - M(2s - 1)}.$$

This completes the proof.   □

We remark that the difference of the results in the above two theorems explains a difference between ℓ_0 minimization and ℓ_1 minimization.

Next we show that the mutual coherence M of the sensing matrix Φ with Gaussian random entries will go to zero when $m, n \to \infty$.

Theorem 2.26 (Lai and Shen, 2020 [19]). *Let $\Phi \in \mathbb{R}^{m \times n}$ be a Gaussian sensing matrix, i.e., each entry ϕ_{ji} of the sensing matrix $\Phi = [\phi_{ji}]_{1 \leq j \leq m, 1 \leq i \leq n}$ is a Gaussian random variable with zero mean and variance one, i.e., $\phi_{ji} \sim \mathcal{N}(0, 1)$. Then for large m, the mutual coherence*

$$M(\Phi) = \max_{1 \leq i,j \leq n, i \neq j} \frac{|\phi_i^\top \phi_j|}{\|\phi_i\| \cdot \|\phi_j\|}$$

satisfies $M(\Phi) \leq \frac{1}{f(m)}$ for some function f, where $f(m) = o(\sqrt{m})$ (e.g., $f(m) = \sqrt{m}/\log(m)$) as $m \to \infty$, with high probability.

Proof. First, we observe that when m gets large, the norm $\|\phi_i\|$ of each column vector $\phi_i = [\phi_{ji}]_{1 \leq j \leq m}$ of Φ is comparable to $\sqrt{m}$. Indeed, for each ϕ_i, we have $\|\phi_i\|^2 = \sum_{j=1}^m \phi_{ji}^2$. Since each $\phi_{ji} \sim \mathcal{N}(0, 1)$, we have $\mathbb{E}(\phi_{ji}^2) = 1$ for all $j = 1, 2, \ldots, m$ by the well-known strong law of large numbers, $\frac{1}{m}\|\phi_i\|^2 = \frac{1}{m}\sum_{j=1}^m \phi_{ji}^2 \to 1$ almost surely. Hence $\|\phi_i\| \to \sqrt{m}$ almost surely as $m \to \infty$.

Now we have $M(\Phi) = \max_{1 \leq i,j \leq n, i \neq j} \frac{|\phi_i^\top \phi_j|}{\|\phi_i\| \cdot \|\phi_j\|} \approx \max_{1 \leq i,j \leq n, i \neq j} \frac{|\phi_i^\top \phi_j|}{m}$. Note that $|\phi_i^\top \phi_j| = |\sum_{k=1}^m \phi_{ki}\phi_{kj}|$, and by letting $X_k = \phi_{ki}\phi_{kj}$, we have $\frac{1}{m}|\phi_i^\top \phi_j| = \frac{1}{m}|\sum_{k=1}^m X_k|$. By the independence of ϕ_{ki} and ϕ_{kj}, the expectation of each X_k satisfies

$$\mathbb{E}(X_k) = \mathbb{E}(\phi_{ki}\phi_{kj}) = \mathbb{E}(\phi_{ki})\mathbb{E}(\phi_{kj}) = 0,$$

and the variance of each X_k satisfies

$$\text{Var}(X_k) = \text{Var}(\phi_{ki}\phi_{kj}) = (\mathbb{E}(\phi_{ki}))^2\text{Var}(\phi_{kj}) + (\mathbb{E}(\phi_{kj}))^2\text{Var}(\phi_{ki}) + \text{Var}(\phi_{ki})\text{Var}(\phi_{kj}) = 1,$$

which implies that $\mathbb{E}(|X_k|^2) = \mathbb{E}(X_k^2) = \text{Var}(X_k) + \mathbb{E}(X_k))^2 = 1$. By the strong law of large numbers again, we have

$$\frac{1}{m}\sum_{k=1}^m X_k = 0, \quad \text{almost surely}$$

as $m \to \infty$. Hence, $\frac{1}{m}|\phi_i^\top \phi_j| \to 0$ as $m \to \infty$ for all $i, j = 1, \ldots, n$ with $i \neq j$. Therefore, the mutual coherence $M(\Phi) \to 0$ almost surely.

Furthermore, since the measure of a probability space is always one, which is bounded, we have $L^p(X_k) \subset L^2(X_k)$ since $p < 2$ for all k. Therefore, $\mathbb{E}(|X_k|^p) < \infty$ for $p < 2$ and for all k.

Now, applying Lemma 2.27 (below) to the sequence of random variables X_k, $k = 1, 2, \ldots, n$, we have

$$m^{-1/p} \left(\sum_{k=1}^{m} X_k \right) = m^{-1/p} \left(\sum_{k=1}^{m} \phi_{ki} \phi_{kj} \right) = m^{-1/p} (\phi_i^\top \phi_j) \to 0 \qquad (2.19)$$

almost surely for all $p \in (0, 2)$. Since the limit is zero, we have

$$m^{-1/p} \left| \sum_{k=1}^{m} X_k \right| = m^{-1/p} \left| \sum_{k=1}^{m} \phi_{ki} \phi_{kj} \right| = m^{-1/p} |\phi_i^\top \phi_j| \to 0. \qquad (2.20)$$

Note that for any $f(m) = o(\sqrt{m})$, there is some $p < 2$ such that $f(m) \cdot m^{\frac{1}{p}-1}$ converges to zero almost surely. Indeed, we can take $\frac{1}{p} = 3/4 - 1/2 \cdot \log_m f(m)$. Since $f(m) = o(\sqrt{m})$, we will have $\log_m f(m) < \log_m \sqrt{m} = 1/2$ for large m, so $\frac{1}{p} = 3/4 - 1/2 \cdot \log_m f(m) > 3/4 - 1/2 \cdot \log_m \sqrt{m} = 1/2$ or $p < 2$. By plugging p into $f(m) \cdot m^{\frac{1}{p}-1}$, we have $f(m) \cdot m^{\frac{1}{p}-1} = (\frac{f(m)}{\sqrt{m}})^{1/2} \to 0$ as $m \to \infty$. Hence we have

$$m^{-1} \cdot |\phi_i^\top \phi_j| \cdot f(m) = m^{-1/p} \cdot |\phi_i^\top \phi_j| \cdot f(m) \cdot m^{\frac{1}{p}-1} \to 0 \qquad (2.21)$$

almost surely as $m \to \infty$. Therefore, by taking the supremum over all n, we have

$$M(\phi) \cdot f(m) = \sup_{1 \leq i,j \leq n, i \neq j} \frac{|\Phi_i^\top \phi_j|}{m} \cdot f(m) \to 0 \qquad (2.22)$$

almost surely as $m \to \infty$. Hence, with high probability, we have $M(\Phi) \cdot f(m) \leq 1$ for large m, and the result is proved. $\quad\square$

In the proof above, we have used the following standard result from probability.

Lemma 2.27 (Strong Law of Large Numbers in the Kolmogorov, Marcinkiewicz, and Zygmund Sense). *Let $\xi, \xi_1, \xi_2, \ldots$ be independent and identically distributed (i.i.d.) random variables, and fix any $p \in (0, 2)$. Then $n^{-1/p} \sum_{k \leq n} \xi_k$ converges almost surely if and only if $\mathbb{E}(|\xi|^p) < \infty$ and either $p \leq 1$ or $\mathbb{E}(\xi) = 0$. In that case, the limit equals $\mathbb{E}(\xi)$ for $p = 1$ and is 0 otherwise.*

Proof. We refer the reader to Theorem 3.23 in [17] for a proof. $\quad\square$

The result in Theorem 2.26 demonstrates that the conditions $s < (1 + 1/M(\Phi))/2$ and $s < (1 + 1/M(\Phi))/4$ in the previous theorems can indeed be satisfied, e.g., $s < (1 + f(m))/2$. By Lemma 2.19, the mutual coherence $M(\Phi)$ of a sensing matrix with Gaussian random variables as entries satisfies $O(\frac{1}{\sqrt{m}}) \leq M(\Phi) \leq \frac{1}{o(\sqrt{m})}$. We may be interested in looking for a sensing matrix Φ whose mutual coherence achieves or is close to the Welch bound. One approach can be found in the following example.

Example 2.28. Let $\Phi = [I_m\ F_m]$ be a sensing matrix with

$$F_m = [F_{jk}]_{1 \le j,k \le m},$$

where $F_{jk} = \frac{1}{\sqrt{m}} \exp(2\pi i(j-1)(k-1)/N)$ with $m \le N$ and $m > N - \sqrt{m}$, and I_m is the identity matrix of size $m \times m$. One can show that $M(\Phi) = 1/\sqrt{m}$. We leave the verification to Exercise 14.

More examples, as well as many other sensing matrices, can be found in [25]. In particular, we refer the reader to [27] for construction of a partial Fourier matrix for a sensing matrix with the mutual coherence close to the Welch bound. In practice, we may not have much freedom to choose a sensing matrix Φ.

Finally, we remark on equal-angle frames. For convenience, let $\Phi = \{\phi_1, \dots, \phi_n\}$ be the collection of columns $\phi_i, i = 1, \dots, n$. Assume that each ϕ_i is of norm 1 for $i = 1, \dots, n$. Suppose that the span of Φ is $\mathbb{R}^m$. Φ is called a dictionary or a frame for $\mathbb{R}^m$. The equal-angle frame is the sensing matrix Φ whose angles $\langle \phi_i, \phi_j \rangle$ are the same for $i \ne j$. For example, when $m = n$, any orthonormal matrix Φ forms an equal-angle frame. When $m < n$, construction of such equal-angle frames is difficult for any given m and n. We leave the problem to the interested reader.

2.4 ▪ The Restricted Isometry Property

Another way to characterize the solution of (1.30) is to use the so-called restricted isometry property (RIP) of the matrix Φ (cf. [3]).

Definition 2.29. *Letting $0 < s < m$ be an integer and Φ_S be a submatrix of Φ which consists of columns of Φ whose column indices are in $S \subset \{1, 2, \dots, n\}$, the s restricted isometry constant (RIC) δ_s of Φ is the smallest quantity such that the two inequalities*

$$(1 - \delta_s)\|\mathbf{x}_S\|_2^2 \le \|\Phi_S \mathbf{x}\|_2^2 \le (1 + \delta_s)\|\mathbf{x}_S\|_2^2 \tag{2.23}$$

hold for all subsets S with cardinality $\#(S) \le s$. If a matrix Φ has such a constant $\delta_s > 0$ for some s, Φ is said to possess the RIP of order s.

With this concept, it is easy to see that if $\delta_{2s} < 1$, then the solution of (1.4) is unique. Indeed, if there were two solutions $\mathbf{x}^{(1)}$ and $\mathbf{x}^{(2)}$ such that

$$\Phi(\mathbf{x}^{(1)} - \mathbf{x}^{(2)}) = 0,$$

then we would choose the index set S which contains the indices of the nonzero entries of $\mathbf{x}^{(1)} - \mathbf{x}^{(2)}$ and see that $\#(S) \le 2s$, which implies

$$(1 - \delta_{2s})\|\mathbf{x}^{(1)} - \mathbf{x}^{(2)}\|_2^2 \le \|\Phi_T(\mathbf{x}^{(1)} - \mathbf{x}^{(2)})\|_2^2 = 0.$$

It follows that $\|\mathbf{x}^{(1)} - \mathbf{x}^{(2)}\|_2 = 0$ when $\delta_{2s} < 1$. That is, the solution is unique.

An alternative definition of the RIC reads as follows.

Lemma 2.30 (Foucart, 2010 [12]).

$$\delta_s = \max_{\#(T) \le s} \|I_m - \Phi_T^\top \Phi_T\|_2, \tag{2.24}$$

where I_m is the identity matrix of size $m \times m$ and T is any index subset of $\{1, 2, \dots, n\}$ with cardinality $\#(T)$.

Proof. Fix any index set $S \subset \{1, 2, \ldots, n\}$ with $\#(S) \leq s$. We note that $I_m - \Phi_S^\top \Phi_S$ is a symmetric matrix and that $\|I_m - \Phi_S^\top \Phi_S\|$ is associated with its largest eigenvalue $L(S)$ in absolute value. It is known that

$$L(S) = \max_{\substack{\mathbf{x} \in \mathbb{R}^{\#(S)} \\ \|\mathbf{x}\|_2 = 1}} \mathbf{x}^\top (I_m - \Phi_S^\top \Phi_S) \mathbf{x}.$$

Using (2.23),

$$\mathbf{x}^\top (I_m - \Phi_S^\top \Phi_S) \mathbf{x} = \|\mathbf{x}\|_2^2 - \|\Phi_S \mathbf{x}\|_2^2 \leq \|\mathbf{x}\|_2^2 - (1 - \delta_s) \|\mathbf{x}\|_2^2 = \delta_s \|\mathbf{x}\|_2^2.$$

That is, $L(S) \leq \delta_s$ for all S with $\#(S) \leq s$. Since δ_s is the smallest real number satisfying the RIP, we know that $\max L(S) = \delta_s$. This completes the proof. □

The following is another simple property of δ_s.

Lemma 2.31 (Candés, 2006 [2]). *Suppose that Φ possesses RIP of order $k + k'$ for two positive integers k and k'. For any $\mathbf{x}, \mathbf{y} \in \mathbb{R}^n$ with $\|\mathbf{x}\|_0 \leq k$ and $\|\mathbf{y}\|_0 \leq k'$, if the indices of the nonzero entries of $\mathbf{x}$ do not overlap with those of $\mathbf{y}$, then*

$$|\langle \Phi \mathbf{x}, \Phi \mathbf{y} \rangle| \leq \delta_{k+k'} \|\mathbf{x}\|_2 \|\mathbf{y}\|_2. \tag{2.25}$$

Proof. Suppose that $\mathbf{x}$ and $\mathbf{y}$ are unit vectors with disjoint supports, with $\|\mathbf{x}\|_0 = k$ and $\|\mathbf{y}\|_0 = k'$. Then for $+$ or $-$, we have

$$2(1 - \delta_{k+k'}) = (1 - \delta_{k+k'}) \|\mathbf{x} \pm \mathbf{y}\|_2^2 \leq \|\Phi(\mathbf{x} \pm \mathbf{y})\|_2^2 \leq (1 + \delta_{k+k'}) \|\mathbf{x} \pm \mathbf{y}\|_2^2 = 2(1 + \delta_{k+k'}).$$

Using the parallelogram identity, we get

$$|\langle \Phi \mathbf{x}, \Phi \mathbf{y} \rangle| = \frac{1}{4} \left| \|\Phi(\mathbf{x} + \mathbf{y})\|_2^2 - \|\Phi(\mathbf{x} - \mathbf{y})\|_2^2 \right| \leq \delta_{k+k'}.$$

This completes the proof. □

It is easy to see that δ_s is an increasing function of $s \geq 1$. More precisely, we have the following lemma.

Lemma 2.32 (Foucart, 2010 [12]). *For integers $t \geq s \geq 1$, we have*

$$\delta_s \leq \delta_t \leq \frac{t - d}{s} \delta_{2s} + \frac{d}{s} \delta_s, \quad \text{where } d := \gcd(s, t). \tag{2.26}$$

Proof. We prove only the second inequality. Write $s = jd$ and $t = kd$ with $j \leq k$ for a common divisor d. Given a vector $\mathbf{x} \in \mathbb{R}^n$ with sparsity t, let T be the index set of the support of $\mathbf{x}$. We split the indices in T into k subsets $S_1, \ldots, S_k$ of length s in such a way that each index in T belongs to exactly j sets. Thus, we have

$$\mathbf{x} = \frac{1}{j} \sum_{i=1}^{k} \mathbf{x}_{S_i} \quad \text{and} \quad \|\mathbf{x}\|_2^2 = \frac{1}{j} \sum_{i=1}^{k} \|\mathbf{x}_{S_i}\|_2^2. \tag{2.27}$$

Now we use Lemmas 2.31 and 2.30 to obtain

$$
\begin{aligned}
\left| \|\Phi\mathbf{x}\|_2^2 - \|\mathbf{x}\|_2^2 \right| &= \left| \langle (\Phi^\top\Phi - \mathrm{Id})\mathbf{x}, \mathbf{x} \rangle \right| \\
&\leq \frac{1}{j^2} \sum_{1\leq i\leq k} \sum_{1\leq \ell\leq k} \left| \langle (\Phi^\top\Phi - \mathrm{Id})\mathbf{x}_{S_i}, \mathbf{x}_{S_\ell} \rangle \right| \\
&\leq \frac{1}{j^2} \left(\sum_{1\leq i\neq \ell\leq k} \delta_{2s} \|\mathbf{x}_{S_i}\|_2 \|\mathbf{x}_{S_\ell}\|_2 + \sum_{1\leq i\leq k} \delta_s \|\mathbf{x}_{S_i}\|_2^2 \right) \\
&= \frac{\delta_{2s}}{j^2} \left(\sum_{1\leq i\leq k} \|\mathbf{x}_{S_i}\|_2 \right)^2 - \frac{\delta_{2s}-\delta_s}{j^2} \sum_{1\leq i\leq k} \|\mathbf{x}_{S_i}\|_2^2 \\
&\leq \left(\frac{\delta_{2s}k}{j^2} - \frac{\delta_{2s}-\delta_s}{j^2} \right) \sum_{1\leq i\leq k} \|\mathbf{x}_{S_i}\|_2^2 \\
&= \left(\frac{\delta_{2s}k}{j} - \frac{\delta_{2s}-\delta_s}{j} \right) \|\mathbf{x}\|_2^2 = \left(\frac{t}{s}\delta_{2s} - \frac{d}{s}(\delta_{2s}-\delta_s) \right) \|\mathbf{x}\|_2^2
\end{aligned}
$$

by using (2.27). The constant $\left(\frac{t}{s}\delta_{2s} - \frac{d}{s}(\delta_{2s}-\delta_s) \right)$ can be as small as possible if we choose $d := \gcd(s,t)$. This completes the proof. □

For example, letting $t = 3s$, we have $\delta_{3s} \leq 2\delta_{2s} + \delta_s \leq 3\delta_{2s}$ by using Lemma 2.32. In the next section, we shall present some matrices which satisfy the RIP with $\delta_s > 0$. In Chapter 8, we shall explain how random matrices subject to Gaussian distribution or other distributions will satisfy a RIP condition with high probability when their size m, n goes to ∞. In addition, we will show that the graph Laplacian matrix will satisfy some RIP condition with high probability for some random graphs in Chapter 10.

Next we introduce another constant of interest.

Definition 2.33. *Let* $\theta_{k,k'}$ *be the smallest number such that*

$$
|\langle \Phi\mathbf{x}, \Phi\mathbf{y} \rangle| \leq \theta_{k,k'} \|\mathbf{x}\|_2 \|\mathbf{y}\|_2
$$

for any k-*sparse vector* $\mathbf{x} \in \mathbb{R}^n$ *and* k'-*sparse vector* $\mathbf{y} \in \mathbb{R}^n$. $\theta_{k,k'}$ *is called the* k, k' *restricted orthogonality constant.*

By the definition, it is easy to see that $\theta_{k,k'}$ is an increasing function of k and k'. Note that by Lemma 2.31, we have $\theta_{k,k'} \leq \delta_{k+k'}$. This gives one connection to the RIC. The other connection is the following, whose proof is left to Exercise 3.

Lemma 2.34. *For integers* k, k', $\theta_{k,k'} \leq \delta_{k+k'}$. *On the other hand,* $\delta_{k+k'} \leq \theta_{k,k'} + \max\{\delta_k, \delta_{k'}\}$.

The following is another useful property of $\theta_{k,k'}$.

Lemma 2.35. *Suppose that* $a \geq 1$. *For any integers* k *and* $ak' < n$,

$$
\theta_{k,ak'} \leq \sqrt{a}\,\theta_{k,k'}.
$$

Proof. For any $\mathbf{x} \in \mathbb{R}^n$ with $\|\mathbf{x}\|_0 \leq k$ and $\mathbf{y} \in \mathbb{R}^n$ with $\|\mathbf{y}\|_0 \leq ak'$, without loss of generality, we assume that $\mathbf{y} = (y_1, y_2, \ldots, y_n)^\top$ with $y_i \neq 0$ for $i = 1, \ldots, ak'$ and $y_i = 0$ for $i \geq ak'+1$.

Let

$$
\mathbf{z}_j = \begin{cases}
(\underbrace{0,\ldots,0}_{j-1}, y_j,\ldots,y_{j+k'-1},0,\ldots,0), j = 1,\ldots,ak'-k'+1, \\
(y_1,\ldots,y_{i-1},\underbrace{0,\ldots,0}_{k'}, y_{(a-1)k'-1+i}, y_{(a-1)k'+i},\ldots,y_{ak'},0,\ldots,0), \\
j = (a-1)k'+i, i = 2,\ldots,k'.
\end{cases}
$$

For example, when $a = 2$, $k' = 3$, and $\mathbf{y} = (y_1, y_2,\ldots,y_6,0,\ldots,0)$, we let

$$
\begin{aligned}
\mathbf{z}_1 &= (y_1, y_2, y_3, 0, 0, 0, 0,\ldots,0), \\
\mathbf{z}_2 &= (0, y_2, y_3, y_4, 0, 0, 0,\ldots,0), \\
\mathbf{z}_3 &= (0, 0, y_3, y_4, y_5, 0, 0,\ldots,0), \\
\mathbf{z}_4 &= (0, 0, 0, y_4, y_5, y_6, 0,\ldots,0), \\
\mathbf{z}_5 &= (y_1, 0, 0, 0, y_5, y_6, 0,\ldots,0), \\
\mathbf{z}_6 &= (y_1, y_2, 0, 0, 0, y_6, 0,\ldots,0).
\end{aligned}
$$

We have $\mathbf{y} = \sum_{j=1}^{ak'} \mathbf{z}_j/k'$. It now follows that

$$
\begin{aligned}
|\langle \Phi\mathbf{x}, \Phi\mathbf{y}\rangle| &= \left| \left\langle \Phi\mathbf{x}, \frac{1}{k'}\sum_{j=1}^{ak'}\Phi\mathbf{z}_j \right\rangle \right| \\
&\leq \frac{1}{k'}\theta_{k,k'}\sum_{j=1}^{ak'}\|\mathbf{x}\|_2\,\|\mathbf{z}_j\|_2 \\
&\leq \frac{1}{k'}\theta_{k,k'}\|\mathbf{x}\|_2 \sqrt{ak'\sum_{j=1}^{ak'}\|\mathbf{z}_j\|_2^2} \\
&= \sqrt{a}\theta_{k,k'}\|\mathbf{x}\|_2\|\mathbf{y}\|_2.
\end{aligned}
$$

This completes the proof.   ☐

We now discuss the relations among the eigenvalues and the RIC.

Lemma 2.36. *Suppose that the sensing matrix Φ satisfies the RIP with RIC $\delta_s < 1$. Then for any index set T with $\#(T) \leq s$, the symmetric matrix $(\Phi_T)^\top \Phi_T$ is positive definite and largest eigenvalue $\lambda_1 \leq 1 + \delta_s$ and the smallest eigenvalue $\lambda_s \geq 1 - \delta_s$.*

Proof. Let $\mathbf{y} \in \mathbb{R}^s$ be an eigenvector associated with the smallest eigenvalue λ_s. We claim that $\lambda_s > 0$. Indeed, since $(\Phi_T)^\top \Phi_T \mathbf{y} = \lambda_s \mathbf{y}$, we have

$$
\|\Phi_T\mathbf{y}\|_2^2 = \mathbf{y}^\top (\Phi_T)^\top \Phi_T \mathbf{y} = \lambda_s \mathbf{y}^\top \mathbf{y} = \lambda_s \|\mathbf{y}\|_2^2.
$$

Extend $\mathbf{y}$ to be a vector in $\mathbb{R}^n$ whose entries in the index set T are $\mathbf{y}$. For convenience, we still use $\mathbf{y}$ to denote this. The above equality can be rewritten as

$$
\lambda_s \|\mathbf{y}_T\|_2^2 = \|\Phi_T\mathbf{y}_T\|_2^2 \geq (1-\delta_s)\|\mathbf{y}_T\|_2^2
$$

by using the RIP. Thus, $\lambda_s \geq 1 - \delta_s > 0$. The proof for the largest eigenvalue is similar.   ☐

Next we present a result which explains a relationship between the RIP and the NSP.

Theorem 2.37 (Cohen, Dahman, and DeVore, 2009 [6]). *If Φ is a matrix satisfying the RIP of order $(j + k)s$ with $\delta_{(j+k)s} < 1$, then Φ satisfies the NSP in the ℓ_1 norm of length js with constant $\gamma = \frac{\sqrt{j(1+\delta_{(j+k)s})}}{\sqrt{k(1-\delta_{(j+k)s})}}$. When $k > j$ is large enough, then Φ satisfies the NSP with $\gamma < 1$.*

Proof. For any $\eta \in \mathcal{N}(\Phi)$, let T be the set of indices of the largest js entries in magnitude of η. Let T_1 denote the set of indices of the next ks largest entries of η and T_2 the next ks largest, and so on.

We define $\eta_0 = \eta_T + \eta_{T_1}$. Since $\eta \in \mathcal{N}(\Phi)$, we have $\Phi\eta_0 = -\Phi(\sum_{i \geq 2} \eta_{T_i})$, and hence

$$\|\eta_T\|_2 \leq \|\eta_0\|_2 \leq \frac{\|\Phi\eta_0\|_2}{\sqrt{1 - \delta_{(j+k)s}}} = \frac{1}{\sqrt{1 - \delta_{(j+k)s}}}$$

$$= \frac{1}{\sqrt{1 - \delta_{(j+k)s}}} \sum_{i \geq 2} \|\Phi\eta_{T_i}\|_2 \leq \frac{\sqrt{1 + \delta_{(j+k)s}}}{\sqrt{1 - \delta_{(j+k)s}}} \sum_{i \geq 2} \|\eta_{T_i}\|_2,$$

where we have used both bounds of the RIP. Since for any $\ell \in T_{i+1}$ and $\ell' \in T_i$, we have $|\eta_\ell| \leq |\eta_{\ell'}|$ and $|\eta_\ell| \leq \|\eta_{T_i}\|_1/(ks)$,

$$\|\eta_{T_{i+1}}\|_2 \leq \|\eta_{T_i}\|_1/\sqrt{ks} \quad \forall i \geq 2.$$

That is, we have

$$\|\eta_T\|_2 \leq \frac{\sqrt{1 + \delta_{(j+k)s}}}{\sqrt{1 - \delta_{(j+k)s}}} \|\eta_{T^c}\|_1/\sqrt{ks}.$$

Therefore, by the Cauchy–Schwarz inequality,

$$\|\eta_T\|_1 \leq \sqrt{js}\|\eta_T\|_2 \leq \frac{\sqrt{j(1 + \delta_{(j+k)s})}}{\sqrt{k(1 - \delta_{(j+k)s})}} \|\eta_{T^c}\|_1.$$

This completes the proof.   □

The concept of the RIP can be extended to the quasi-norm setting (cf. [4], [5]), which will be discussed more in detail in Chapter 8.

Definition 2.38. *Fix $q \in (0,2)$. Letting $0 < s < m$ be an integer and Φ_S be a submatrix of Φ which consists of columns of Φ whose column indices are in $S \subset \{1, 2, \ldots, n\}$, the s-RIC $\delta_{s,q}$ of Φ is the smallest quantity such that*

$$(1 - \delta_{s,q})\|\mathbf{x}_S\|_2^q \leq \|\Phi_S\mathbf{x}\|_q^q \leq (1 + \delta_{s,q})\|\mathbf{x}_S\|_2^q \tag{2.28}$$

for all subsets S with cardinality $\#(S) \leq s$. The constant $\delta_{s,q}$ is called the q-RIC. If a matrix Φ has such a constant $\delta_{s,q} > 0$ for some s, Φ is said to possess q-RIP of order s.

It is easy to see that $\delta_{s,q}$ is an increasing function of $s \geq 1$. We leave the interested reader to develop more properties of $\delta_{s,q}$ for $q \in (0,2)$.

To conclude this section we extend the concept of the mutual coherence a little bit. Recall that $G = \Phi^\top\Phi$ is a square matrix of size $n \times n$ and write $G = (g_{ij})_{1 \leq i,j \leq n}$. The 1-mutual coherence function $M_1(s)$ of Φ is defined by

$$M_1(s) = \max_{\substack{1 \leq j \leq n \\ S \subset \{1,\ldots,n\}, \#(S) \leq s}} \sum_{i \in S} |g_{ij}|. \tag{2.29}$$

Then we can show the relations between the standard mutual coherence and the RIC δ_s.

Lemma 2.39 (Rauhut, 2010 [21]). *For any sensing matrix* Φ,

$$\delta_s \le M_1(s) \le (s-1)M(\Phi),$$

where δ_s *is the RIC.*

Proof. For any subset S of $\{1, 2, \ldots, n\}$ with $\#(S) \le s$, we use

$$\|\Phi_S^\top \Phi_S - I\|_{1 \to 1}$$

to denote the norm of the operator $\Phi_S^\top \Phi_S - I$ mapping $\mathbb{R}^s$ to itself in the ℓ_1 norm. Then by the ℓ_1 norm for matrices, we have

$$\|\Phi_S^\top \Phi_S - I\|_{1 \to 1} = \max_{k \in S} \sum_{j \in S} |(\Phi_S^\top \Phi_S - I)_{jk}| = \max_{\substack{k \in S}} \sum_{\substack{j \in S \\ j \ne k}} |g_{jk}| \le M_1 \le M(\Phi)(s-1).$$

Next we note, by Lemma 2.30, that

$$
\begin{aligned}
\delta_s &= \max_{\substack{S \subset \{1,2,\ldots,n\} \\ \#(S) \le s}} \|\Phi_S^\top \Phi_S - I\|_{2 \to 2} \\
&\le \max_{\substack{S \subset \{1,2,\ldots,n\} \\ \#(S) \le s}} \max\{\|\Phi_S^\top \Phi_S - I\|_{1 \to 1}, \|\Phi_S^\top \Phi_S - I\|_{\infty \to \infty}\} \\
&= \max_{\substack{S \subset \{1,2,\ldots,n\} \\ \#(S) \le s}} \|\Phi_S^\top \Phi_S - I\|_{1 \to 1},
\end{aligned}
$$

where we have used the Riesz–Thorin interpolation theorem for the inequality above.   □

Example 2.40. Let $\Phi = [I_m \; F_m]$ be a sensing matrix with

$$F_m = [F_{jk}]_{1 \le j, k \le m},$$

with $F_{jk} = \frac{1}{\sqrt{m}} \exp(2\pi i(j-1)(k-1)/m)$. One can see that $M_1(s) = s/\sqrt{m}$ for $s \ll m-1$. By Lemma 2.39, $\delta_s \le s/\sqrt{m}$.

2.5 ▪ Completely Full Rankness

Finally, let us introduce the concept of completely full rank for matrices as in [20].

Definition 2.41. *We say a matrix* A *of size* $m \times n$ *with* $m < n$ *is of completely full rank if any submatrices of* A *of size* $m \times m$ *are of full rank.*

A matrix of completely full rank is sometimes called full spark in the literature. For example, $A = [(x_j)^{i-1}]_{1 \le i \le m, 1 \le j \le n}$ with x_j distinct is a matrix of completely full rank due to the property of a Vandermonde determinant. For another example, let A, with

$$A^\top = [1, \cos(x_j), \sin(x_j), \ldots, \cos(mx_j), \sin(mx_j)]_{j=1,\ldots,n}$$

for all $x_j \in [0, 2\pi), j = 1, \ldots, n$, be a matrix of size $(2m+1) \times n$. Then A is of completely full rank since A is a Chebyshev system (cf. [16]). For another example, let $\mathbf{v}_i = (x_i, y_i), i =$

$1, \ldots, n$, be distinct points in $\mathbb{R}^2$. Assume that no three points from $V = \{\mathbf{v}_i, i = 1, \ldots, n\}$ are collinear. Then

$$
A = \begin{bmatrix} 1 & 1 & 1 & \cdots & \cdots & 1 \\ x_1 & x_2 & x_3 & \cdots & \cdots & x_n \\ y_1 & y_2 & y_3 & \cdots & \cdots & y_n \end{bmatrix}
$$

is a matrix of completely full rank. Similarly, we can use points in $\mathbb{R}^{m-1}$ to form a completely full rank matrix as long as points $\mathbf{v}$ satisfy a necessary and sufficient condition: none of the m points are on the same hyperplane.

In addition, let $\mathcal{D} = \{\phi_1, \ldots, \phi_n\}$ be the collection of all columns from Φ. Then $\mathcal{D}$ is called a dictionary if $\|\phi_i\|_2 = 1$ for all $i = 1, \ldots, n$. We say $\mathcal{D}$ is full spark if any m distinct elements from $\mathcal{D}$ span the space $\mathbb{R}^m$. This concept is introduced in [9]. We can see that Φ being of completely full rank is the same as the dictionary $\mathcal{D}$ being full spark.

Theorem 2.42 (Lai and Wang, 2011 [20]). *Any matrix A of completely full rank can be normalized to have the RIP of order m. On the other hand, if A has the RIP of order m, then A is of completely full rank.*

Proof. Let $I = \{T \subset \{1, \ldots, n\}, \#(T) = m\}$ be the collection of all index sets, each of which is a subset of $\{1, \ldots, n\}$ with cardinality m, i.e., $T = \{i_1, \ldots, i_m\}$ with $1 \leq i_1 < i_2 < \cdots < i_m \leq n$ for all choices of $i_1, \ldots, i_m$. Let α_T and β_T be the smallest and largest singular values of A_T, which consists of columns with indices in T. Since A is of completely full rank, there exist α and β such that $\alpha_T \geq \alpha > 0$ and $\beta_T \leq \beta < \infty$ for all $T \in I$. If we normalize A by $\hat{A} := A/\beta$, then we have

$$
\frac{\alpha^2}{\beta^2}\|\mathbf{x}_T\|_2^2 \leq \|\hat{A}_T\mathbf{x}_T\|_2^2 \leq \|\mathbf{x}\|_2^2 \quad \forall T \in I.
$$

By letting $1 - \delta_m = \frac{\alpha^2}{\beta^2}$, we know that $\hat{A}$ possesses an RIP of order m.

When A has an RIP of order m, i.e., $\delta_m < 1$, then the definition $\|A_T\mathbf{x}_T\|^2 \geq (1-\delta_m)\|\mathbf{x}_T\|^2$ for all T with $\#(T) = m$ implies that A_T is of full rank for all T with $\#(T) = m$. That is, A is of completely full rank. $\square$

If A is of completely full rank, so is αA for any scalar α. This gives us a method to check if a matrix A has the RIP of order m or can be scaled to have an RIP of order m. We can use this property to construct a matrix satisfying an RIP for any given $s < m$. Indeed, take a matrix $\tilde{A}$ of size $s \times n$ which is of completely full rank and add $m - s$ rows of any real numbers of length n to form a matrix B of size $m \times n$. Then by using the method in the proof above, B can be rescaled to A so that A possesses the RIP of order s.

On the other hand, many random matrices satisfy an RIP with overwhelming probability. We shall explain this point in Chapter 8.

To conclude this section, let us consider the $N \times N$ discrete Fourier transform (DFT) matrix scaled to have entries of unit modulus:

$$
\begin{bmatrix} 1 & 1 & 1 & \cdots & 1 \\ 1 & \omega & \omega^2 & \cdots & \omega^{N-1} \\ 1 & \omega^2 & \omega^4 & \cdots & \omega^{2(N-1)} \\ 1 & \omega^3 & \omega^6 & \cdots & \omega^{3(N-1)} \\ \vdots & \vdots & \vdots & \cdots & \vdots \\ 1 & \omega^{N-1} & \omega^{2(N-1)} & \cdots & \omega^{(N-1)(N-1)} \end{bmatrix}, \tag{2.30}
$$

where $\omega = \exp(i2\pi/N)$. It is easy to see that the first $m \leq N/2$ rows of the above matrix form a Vandermonde matrix which is of completely full rank. However, randomly choosing n rows may not yield a matrix of completely full rank. For example, consider $N = 4$. Then the DFT matrix is

$$\begin{bmatrix} 1 & 1 & 1 & 1 \\ 1 & i & -1 & -i \\ 1 & -1 & 1 & -1 \\ 1 & -i & -1 & i \end{bmatrix}.$$

Then the first and third rows do not form a matrix of completely full rank. Nevertheless, we have the following result.

Theorem 2.43 (Chebotarëv, 1996). *Let N be a prime integer. Then every square submatrix of the $N \times N$ discrete Fourier transform matrix in (2.30) is invertible.*

Proof. We refer the reader to [23] for a proof.   □

That is, any $0 < m < N$ rows from (2.30) will be of completely full rank. More designs of matrices of completely full rank from DFT can be found in [1].

2.6 ▪ Noisy Sensing Matrices

In this section, we mainly explain the relations between the RIC constants δ_s of Φ and $\tilde{\delta}_s$ of $\tilde{\Phi} = \Phi + E$ and, similarly, the relations between the mutual coherences $\nu(s)$ of Φ and $\tilde{\nu}(s)$ of $\tilde{\Phi}$, where E is a perturbed matrix. To do so, we shall use the following norm ($p \geq 1$) and quasi-norm ($p < 1$) for matrices $B \in \mathbb{R}^{m \times n}$:

$$\|B\|_{p,s} := \max_{\substack{S \subset [1,\ldots,n] \\ |S|=s}} \|B_S\|_p. \tag{2.31}$$

When $p = 2$, we have $\|B\|_{2,s} = \max_{\substack{S \subset [1,\ldots,n] \\ |S|=s}} \sigma_{max}(B_S)$, where σ_{max} is the largest singular value of B_S, the columns with indices in B.

Theorem 2.44 (Herman and Strohmer, 2010 [15]). *Suppose that $\tilde{\Phi} = \Phi + E$. Let δ_s and $\tilde{\delta}_s$ denote the s RICs of Φ and $\tilde{\Phi}$, respectively. Define $\epsilon_{s,\Phi} := \|E\|_{2,s}/\|\Phi\|_{2,s}$. Then*

$$\tilde{\delta}_s \leq (1 + \delta_s)(1 + \epsilon_{s,\Phi})^2 - 1. \tag{2.32}$$

Proof. We mainly use Lemma 2.30 to obtain

$$\tilde{\delta}_s = \max_{\substack{S \subset \{1,\ldots,n\} \\ |S| \leq s}} \|I - (\Phi_S + E_S)^\top (\Phi_S + E_S)\|_2$$

$$= \max_{\substack{S \subset \{1,\ldots,n\} \\ |S| \leq s}} \|I - (I + \Phi_S^{-1} E_S)^\top \Phi_S^\top \Phi_S (I + \Phi_S^{-1} E_S)\|$$

$$= \max_{\substack{S \subset \{1,\ldots,n\} \\ |S| \leq s}} \|(I + \Phi_S^{-1} E_S)^\top \Phi_S^\top \Phi_S (I + \Phi_S^{-1} E_S)\| - 1$$

$$\leq (1 + \epsilon_{s,\Phi})^2 \max_{\substack{S \subset \{1,\ldots,n\} \\ |S| \leq s}} (\|\Phi_S^\top \Phi_S - I\| + \|I\|) - 1$$

$$= (1 + \epsilon_{s,\Phi})^2 (1 + \delta_s) - 1,$$

where we have used Lemma 2.30 again. This completes the proof.   □

The following is the standard stable analysis for a linear system of equations.

Lemma 2.45. *Let A be an invertible matrix and $\tilde{A}$ be a perturbation of A satisfying $\|A^{-1}\|\|A - \tilde{A}\| < 1$. Suppose that $\mathbf{x}$ and $\tilde{\mathbf{x}}$ are the exact solutions of $A\mathbf{x} = \mathbf{b}$ and $\tilde{A}\tilde{\mathbf{x}} = \mathbf{b}$, respectively. Then*

$$\frac{\|\mathbf{x} - \tilde{\mathbf{x}}\|}{\|\mathbf{x}\|} \leq \frac{\kappa(A)}{1 - \kappa(A)\frac{\|A-\tilde{A}\|}{\|A\|}} \left[\frac{\|A - \tilde{A}\|}{\|A\|} + \frac{\|\mathbf{b} - \tilde{\mathbf{b}}\|}{\|\mathbf{b}\|} \right],$$

where $\kappa(A)$ denotes the condition number of matrix A, i.e., $\kappa(A) = \|A\|\|A^{-1}\|$.

Proof. The proof, which can be found in any standard numerical analysis book, is left to the interested reader. □

For a sensing matrix Φ of size $m \times n$, we can define $s < m$ as the condition number of Φ as follows. Suppose that $\mathbf{x}_\mathbf{b}$ is a sparse solution of $\Phi\mathbf{x} = \mathbf{b}$ with $S = \text{support}(\mathbf{x}_\mathbf{b})$. It is easy to see that $\mathbf{x}_\mathbf{b}$ is the least squares solution to $\Phi_S\mathbf{x} = \mathbf{b}$ since $s < m$. That is, $\mathbf{x}_\mathbf{b}$ is the solution of

$$(\Phi_S)^\top \Phi_S \mathbf{x} = (\Phi_S)^\top \mathbf{b}. \tag{2.33}$$

We are now ready to define the s-condition number of Φ by

$$\kappa_s(\Phi) = \max_{\substack{S \subset \{1,\dots,n\} \\ |S| \leq s}} \kappa(\Phi_S^\top \Phi_S), \tag{2.34}$$

where $1 \leq s < m$.

When Φ has an RIP with RIC $\delta_s \in (0, 1)$, we can see that

$$\kappa_s(\Phi) \leq \frac{1 + \delta_s}{1 - \delta_s}. \tag{2.35}$$

We now study the stability of sparse solution $\mathbf{x}_\mathbf{b}$.

Theorem 2.46. *Suppose that Φ has an RIP with $\delta_s < 1$. Suppose that $\widetilde{\Phi} = \Phi + E$ with small perturbed matrix E such that $\kappa_s(\Phi)\|E\| < \sqrt{1 - \delta_s}$. Let $\mathbf{x}$ and $\tilde{\mathbf{x}}$ be sparse solutions of $\Phi\mathbf{x} = \mathbf{b}$ and $\widetilde{\Phi}\tilde{\mathbf{x}} = \tilde{\mathbf{b}}$. Without loss of generality, we may assume that the supports of $\mathbf{x}$ and $\tilde{\mathbf{x}}$ are contained in the same index set S with $s = |S|$. Then*

$$\frac{\|\mathbf{x} - \tilde{\mathbf{x}}\|}{\|\mathbf{x}\|} \leq \frac{\kappa_s(\Phi)}{1 - \kappa_s(\Phi)\|E\|/\sqrt{1 - \delta_s}} \left(\frac{\|E\|}{\sqrt{1 - \delta_s}} + \frac{\|\mathbf{b} - \tilde{\mathbf{b}}\|}{\|\mathbf{b}\|} \right). \tag{2.36}$$

Proof. We simply apply Lemma 2.45 to (2.33) and conclude this result. Indeed, we know that $\Phi_S\mathbf{x}_S = \mathbf{b}$ and $\widetilde{\Phi}_S\tilde{\mathbf{x}}_S = \tilde{\mathbf{b}}$. We only consider Φ_S and $\widetilde{\Phi}_S$ and use Lemma 2.45. Note that $\|E_S\| \leq \|E\|$ and $\|\mathbf{x} - \tilde{\mathbf{x}}\| = \|\mathbf{x}_S - \tilde{\mathbf{x}}_S\|$. By the RIP, we have $1 - \delta_s \leq \|\Phi_S\|^2 \leq 1 + \delta_s$ by dividing by $\|\mathbf{x}_S\|$ on both sides. It follows that $1/\|\Phi_S\| \leq 1/\sqrt{1 - \delta_s}$. Details are left to the interested reader. □

2.7 ▪ Remarks

Remark 2.7.1. *The concept of the NSP can be slightly extended in the following sense. We shall say that Φ has the generalized null space property (GNSP) of order L for $\beta > 0$ if*

$$\|\eta_S\|_q \leq \beta\|\eta\|_q \tag{2.37}$$

for any index set S with cardinality $\#(S) \leq L$ and for all $\eta \in \mathcal{N}(\Phi)$. Here β may be bigger than or equal to $1/2$. Once Φ satisfies the GNSP of order L for a $\beta > 0$, there exists an $s \leq L$ such that Φ satisfies the NSP of order s with $\beta < 1/2$. See Theorem 2.37.

Remark 2.7.2. *We may use a general nonnegative function f to replace the ℓ_q norm. More precisely, let f be a nondecreasing function on $[0, \infty) \mapsto [0, \infty)$ with $f(0) = 0$ and not identically zero such that $f(t)/t$ is nonincreasing on $(0, \infty)$. Let $\mathcal{M}$ be a collection of all such functions f. Then it is easy to see that $f \circ g \in \mathcal{M}$ if both f and $g \in \mathcal{M}$, $\min\{f, g\} \in \mathcal{M}$, and $\max\{f, g\} \in \mathcal{M}$. Define*

$$\|\mathbf{x}\|_f := \sum_{i=1}^{n} f(|\mathbf{x}_i|).$$

We can formulate the minimization problem

$$\min\{\|\mathbf{x}\|_f, \mathbf{x} \in \mathbb{R}^n, \Phi\mathbf{x} = \mathbf{b}\}. \tag{2.38}$$

We can find a similar NSP in the f norm. We leave the discussion to the interested reader. See [14] for more details.

Remark 2.7.3. *The RIC δ_s is not invariant under scaling of the matrix. That is, if Φ is replaced by $c\Phi$ for a nonzero constant c, the RIC of $c\Phi$ is significantly different from δ_s. One possible modification is to define*

$$\alpha_s \|\mathbf{x}_T\|_2^2 \leq \|\Phi_T \mathbf{x}_T\|_2^2 \leq \beta_s \|\mathbf{x}_T\|_2^2$$

for all $\mathbf{x} \in \mathbb{R}^n$ and all $T \subset \{1, 2, \ldots, n\}$ with $\#(T) \leq s$. Then

$$\frac{1 - \delta_s}{1 + \delta_s} = \frac{\alpha_s}{\beta_s} \quad or \quad \delta_s = \frac{1 - \frac{\alpha_s}{\beta_s}}{1 + \frac{\alpha_s}{\beta_s}}.$$

If we define δ_s to be the rightmost quantity of the equation above, then such a δ_s will be scale invariant.

Remark 2.7.4. *Both Theorems 2.24 and 2.25 were proved in [8]. However, the proof of Theorem 2.25 is much simpler than the one in [8]. It is easy to see that there is a gap between the requirements of k. That is, one requirement is that $s < (1 + 1/M)/4$ by any ℓ_1 approach (1.20), and the other is that $s < (1 + 1/M)/2$ by the ℓ_0 (combinatorial) method. Thus, the ℓ_1 minimization method is not optimal yet. It would be interesting to know how we could increase the sparsity s when using the ℓ_1 minimization (1.20).*

Remark 2.7.5. *Many results in this chapter can be extended to the setting of multiple measurements by solving the minimization in (1.56). Let us present the corresponding NSP.*

Theorem 2.47. *Let Φ be a real matrix of size $m \times n$ and $S \subset \{1, 2, \ldots, n\}$ be a fixed index set. Fix $p \in (0, 1]$ and $r \geq 1$. Then the following conditions are equivalent:*
(a) All $\mathbf{x}^{(k)}$ with support in S for $k = 1, \ldots, r$ can be uniquely recovered using (1.56).
(b) For all vectors $\left(\mathbf{u}^{(1)}, \ldots, \mathbf{u}^{(r)}\right) \in (\mathcal{N}(\Phi))^r \backslash \{(0, 0, \ldots, 0)\}$,

$$\sum_{j \in S} \left(\sqrt{u_{1,j}^2 + \cdots + u_{r,j}^2}\right)^p < \sum_{j \in S^c} \left(\sqrt{u_{1,j}^2 + \cdots + u_{r,j}^2}\right)^p. \tag{2.39}$$

(c) *For all vectors $\mathbf{z} \in \mathcal{N}(\Phi)$ with $\mathbf{z} \neq 0$,*

$$\sum_{j \in S} |z_j|^p < \sum_{j \in S^c} |z_j|^p, \tag{2.40}$$

where $\mathbf{z} = (z_1, \ldots, z_n)^T \in \mathbb{R}^n$.

Proof. We leave the proof to the interested reader; see also [18] for a proof. □

2.8 ▪ Exercises

Exercise 1. Prove Theorem 2.2.

Exercise 2. Prove Theorem 2.3.

Exercise 3. Prove Lemma 2.34.

Exercise 4. Suppose that $h \in \mathcal{N}(\Phi) \setminus \{0\}$ does not satisfy the NSP of sparsity s in the ℓ_1 norm. Show that we can choose index set T with $\|T\|_0 \leq s$ such that not only do we have $\|h_{T^c}\|_1 \leq \|h_T\|_1$ but we also have $\|h_{T^c}\|_\infty \leq \|h_T\|_1/s$.

Exercise 5. Prove the following result: If an RIC of Φ of size $m \times n$ satisfies

$$\delta_{as} + b\delta_{(a+1)s} < b - 1$$

with $a = b^{2/(2-p)}$ such that as is an integer, then the minimizer of (1.27) is the sparse solution $\mathbf{x_b}$ with sparsity $\|\mathbf{x_b}\|_0 \leq s$.

Exercise 6. Let $\tilde{\delta}_s := \dfrac{1 - \frac{\alpha_s}{\beta_s}}{1 + \frac{\alpha_s}{\beta_s}}$. Show a property similar to Lemma 2.31.

Exercise 7. Suppose that Φ has RIC $\delta_s < 1$. Suppose that T is a subset of $\{1, 2, \ldots, n\}$ with $\#(T) \leq s$. Show that

$$\|\Phi_T^\dagger \mathbf{x}\|_2 \leq \frac{1}{\sqrt{1 - \delta_s}} \|\mathbf{x}\|, \tag{2.41}$$

where $\Phi_T^\dagger$ is the pseudo-inverse of Φ_T.

Exercise 8. Suppose that Φ has RIC $\delta_s < 1$. Suppose that S, T are subsets of $\{1, 2, \ldots, n\}$ with $\|T\|_0 + \|S\|_0 \leq s$ and $S \cap T = \emptyset$. Show that

$$\|\Phi_S^\top \Phi_T\|_2 \leq \delta_s.$$

Exercise 9. Suppose that Φ has RIC $\delta_s < 1$. Suppose that T is a subset of $\{1, 2, \ldots, n\}$ with $\|T\|_0 < s$. Show that for any $\mathbf{x}$ with $\|\mathbf{x}\|_0 = s$,

$$\|\Phi_T^\top \Phi \mathbf{x}_{T^c}\|_2 \leq \delta_s \|\mathbf{x}_{T^c}\|_2.$$

Exercise 10. Suppose that Φ has RIC $\delta_s < 1$. Suppose that T is a subset of $\{1, 2, \ldots, n\}$ with $\|T\|_0 \leq s$. Show that

$$(1 - \delta_s)\|\mathbf{x}_T\|_2 \leq \|\Phi_T^\top \Phi_T \mathbf{x}_T\|_2 \leq (1 + \delta_T)\|\mathbf{x}_T\|_2$$

and

$$\frac{1}{(1 + \delta_s)}\|\mathbf{x}_T\|_2 \leq \|(\Phi_T^\top \Phi_T)^{-1}\mathbf{x}_T\|_2 \leq \frac{1}{(1 - \delta_T)}\|\mathbf{x}_T\|_2.$$

Exercise 11. Suppose that Φ has RIC $\delta_s < 1$. Show that for any $\mathbf{x} \in \mathbb{R}^n$,

$$\|\Phi\mathbf{x}\|_2 \leq \sqrt{1+\delta_s} \left(\|\mathbf{x}\|_2 + \frac{1}{\sqrt{s}}\|\mathbf{x}\|_2 \right).$$

Exercise 12. Suppose that Φ is a matrix satisfying an RIP of order $(j+k)s$ with $\delta_{(j+k)s} < 1$. Then show that Φ satisfies the NSP in the ℓ_q norm of length js if $k > j$ is large enough, where $q \in (0, 1)$ for any fixed q.

Exercise 13. Show that the function defined in (2.31) is a norm if $p \geq 1$ and a quasi-norm if $p \in (0, 1)$ for matrices.

Exercise 14. Show that the mutual coherence $M(\Phi) = 1/\sqrt{m}$ in Example 2.28.

Exercise 15. Prove Lemma 2.36.

Exercise 16. Prove the estimate given in (2.35).

Exercise 17. Prove Lemma 2.45.

Exercise 18. Study the NSP related to the minimization in (2.38).

Exercise 19. Give a detailed proof of Theorem 2.46.

Exercise 20. Prove Theorem 2.47.

Exercise 21. Prove Lemma 2.10.

Bibliography

[1] B. Alexeev, J. Canhill, and D. Mixon, Full spark frames, J. Fourier Anal. Appl., 18 (2012), 1167–1194. (Cited on p. 49)

[2] E. J. Candés, Compressive sampling, in International Congress of Mathematicians, Vol. III, European Mathematical Society, Zürich, 2006, 1433–1452. (Cited on p. 43)

[3] E. J. Candés and T. Tao, Decoding by linear programming, IEEE Trans. Inform. Theory, 51 (2005), 4203–4215. (Cited on p. 42)

[4] R. Chartrand, Exact reconstruction of sparse signals via nonconvex minimization, IEEE Signal Process. Lett., 14 (2007), 707–710. (Cited on p. 46)

[5] R. Chartrand and V. Steneva, Restricted isometry properties and nonconvex compressive sensing, Inverse Problems, 24 (2008), 035020. (Cited on p. 46)

[6] A. Cohen, W. Dahmen, and R. DeVore, Compressed sensing and best k-term approximation, J. Amer. Math. Soc., 22 (2009), 211–231. (Cited on p. 46)

[7] I. Daubechies, R. DeVore, M. Fornasier, and C. S. Güntuk, Iteratively reweighted least squares minimization for sparse recovery, Commun. Pure Appl. Math., 63(1) (2010), 1–38. (Cited on p. 32)

[8] D. L. Donoho, Compressed sensing, IEEE Trans. Inform. Theory, 52 (2006), 1289–1306. (Cited on p. 51)

[9] D. L. Donoho and M. Elad, Optimally sparse representation in general (nonorthogonal) dictionaries via ℓ_1 minimization, Proc. Natl. Acad. Sci. USA, 100 (2003), 2197–2202. (Cited on pp. 34, 37, 48)

[10] D. L. Donoho, M. Elad, and V. N. Temlyakov, Stable recovery of sparse overcomplete representations in the presence of noise, IEEE Trans. Inform. Theory, 52 (2006), 6–18. (Cited on pp. 39, 40)

[11] M. Elad, Sparse and Redundant Representations, Springer, 2010. (Cited on p. 37)

[12] S. Foucart, A note on guaranteed sparse recovery via ℓ_1-minimization. Appl. Comput. Harmon. Anal.,
      29 (2010), 97–103. (Cited on pp. 42, 43)

[13] R. Gribonval and M. Nielsen, Sparse representations in unions of bases, IEEE Trans. Inform. Theory,
      49 (2003), 3320–3325. (Cited on p. 38)

[14] R. Gribonval and M. Nielsen, Highly sparse representations from dictionaries are unique and inde-
      pendent of the sparseness measure, Appl. Comput. Harmon. Anal., 22 (2007), 335–355. (Cited on
      p. 51)

[15] M. A. Herman and T. Strohmer, General deviants: An analysis of perturbations in compressed sensing.
      IEEE J. Sel. Top. Signal Process., 4 (2010), 342–349. (Cited on p. 49)

[16] S. Karlin, Total Positivity, Stanford University Press, 1968. (Cited on p. 47)

[17] O. Kallenberg, Foundations of modern probability, Probab. Appl., Springer-Verlag, New York, 1997.
      (Cited on p. 41)

[18] M.-J. Lai and L. Y. Liu, The null space property for sparse recovery from multiple measurement
      vectors, Appl. Comput. Harmon. Anal., 30 (2011), 402–406. (Cited on p. 52)

[19] M.-J. Lai and Z. Shen, A Quasi-Orthogonal Matching Pursuit Algorithm for Compressive Sensing,
      preprint arXiv:2007.09534, 2020. (Cited on p. 40)

[20] M.-J. Lai and J. Wang, An unconstrained ℓ_q minimization for sparse solution of underdetermined
      linear systems, SIAM J. Optim., 21 (2011), 82–101. (Cited on pp. 47, 48)

[21] H. Rauhut, Compressive sensing and structured random matrices, in Theoretical Foundations and
      Numerical Methods for Sparse Recovery, Radon Ser. Comput. Appl. Math. 9, Walter de Gruyter,
      Berlin, 2010, 1–92. (Cited on p. 47)

[22] A. Pinkus, On L^1-Approximation, Cambridge University Press, 1989. (Cited on p. 30)

[23] P. Stevenhagen and H. W. Lenstra, Chebotarëv and his density theorem, Math. Intelligencer, 18 (1996),
      26–37. (Cited on p. 49)

[24] S. A. Vavasis, Derivation of Compressive Sensing Theorems from the Spherical Section Property,
      on-line, 2009. (Cited on p. 34)

[25] M. Vidyasagar, An Introduction to Compressed Sensing, SIAM, 2019. (Cited on p. 42)

[26] L. R. Welch, Lower bounds on the maximum cross-correlation of signals, IEEE Trans. Inform. Theory,
      20 (1974), 397–399. (Cited on p. 37)

[27] G. Xu and Z. Xu, Compressed sensing matrices from Fourier matrices, IEEE Trans. Inform. Theory,
      61 (2015), 469–478. (Cited on p. 42)

[28] Y. Zhang, Theory of Compressive Sensing via ℓ_1-Minimization: A Non-RIP Analysis and Extensions,
      CAAM Technical Report TR08–11, Rice University, 2008. (Cited on p. 33)

Chapter 3

Various Greedy Algorithms

Recall that Φ is a sensing matrix of size $m \times n$ with $m < n$ or $m \ll n$ and $\mathbf{b}$ is a vector which is compressible in the sense that there exists a vector $\mathbf{x_b}$ with $\|\mathbf{x_b}\|_0 < m$ such that $\mathbf{b} = \Phi\mathbf{x_b}$. Typically, Φ consists of tight wavelet framelets which can be useful for image compression. Indeed, we find the sparse solution $\mathbf{x_b}$ for a compressible image $\mathbf{b}$ using the minimization

$$\min_{\mathbf{x} \in \mathbb{R}^n} \{\|\mathbf{x}\|_0, \quad \Phi\mathbf{x} = \mathbf{b}\}. \tag{3.1}$$

Note that the size of a standard image is 512×512 and hence it is of size $512^2 \times 1$ after converting it to vector $\mathbf{b}$. The size of the sensing matrix Φ is $512^2 \times n$ with $n \gg 512^2$, where n is the size of the dictionary consisting of various patterns of image blocks. When the sparsity s is relatively small, a greedy algorithm is definitely preferred over many other sparse solution approaches discussed in other chapters.

3.1 ▪ What Are Greedy Algorithms?

A greedy algorithm can be explained as follows. Without loss of generality, we may assume that each column in $\Phi = [\phi_1, \ldots, \phi_n]$ has been normalized. Since we do not know the sparsity s beforehand, the best possible guess is 1. That is, let us first assume that the sparsity is 1, i.e., $\|\mathbf{x}\|_0 = 1$. Certainly, we may not have a sparse solution $\mathbf{x}$ with sparsity 1 to satisfy $\Phi\mathbf{x} = \mathbf{b}$. So we find the least squares solution

$$\min_{\substack{\mathbf{x} \in \mathbb{R}^n \\ \|\mathbf{x}\|_0=1}} \{\|\Phi\mathbf{x} - \mathbf{b}\|^2\}. \tag{3.2}$$

It is easy to solve the above least squares problem. Let $\mathbf{x}_1$ be the solution, where $\mathbf{x}_1 = \langle \mathbf{b}, \phi_{i_1} \rangle \phi_{i_1}$ and $i_1 \in \{1, \ldots, n\}$ is the index such that $|\langle \phi_{i_1}, \mathbf{b} \rangle|$ is the largest among all $|\langle \phi_i, \mathbf{b} \rangle|$, $i = 1, \ldots, n$. Indeed, it is easy to see that the residual $\|\Phi\mathbf{x}_1 - \mathbf{b}\|^2 = \|\mathbf{b}\|^2 - |\langle \phi_{i_1}, \mathbf{b} \rangle|^2$, and the choice i_1 makes the residual the smallest. That is, we choose the index i_1 such that the inner product $|\langle \phi_{i_1}, \mathbf{b} \rangle|$ is largest.

If the residual is not zero, we repeat the above step. That is, let $\mathbf{b}^{(1)} = \mathbf{b} - \mathbf{x}_1$ and Φ_{-i_1} be the resulting dictionary from Φ by dropping off ϕ_{i_1}. We choose ϕ_{i_2} such that $|\langle \phi_{i_2}, \mathbf{b}^{(1)} \rangle|$ is largest. Let us formulate this procedure in the following algorithm.

ALGORITHM 3.1

Pure Greedy Algorithm [16]

1: Let $\Phi = \{\phi_1, \ldots, \phi_n\}$ be a collection of vectors in $\mathbb{R}^n$, and assume that $\|\phi\|_2 = 1, i = 1, \ldots, n$.

2: Let $\mathbf{b} \in \mathbb{R}^n$ be a given vector and $S_0 = \emptyset$. Start with the residual vector $R^0 = \mathbf{b}$ and greedy solution $\mathbf{x}_0 = 0$.

3: **while do** for $k \geq 1$, we find an index i_k such that

$$|\langle R^{k-1}, \phi_{i_k}\rangle| = \max_{1 \leq i \leq n} |\langle R^{k-1}, \phi_i\rangle| \tag{3.3}$$

and form $S_k = S_{k-1} \cup \{i_k\}$,

$$R^k = R^{k-1} - c_{i_k}\phi_{i_k}, \text{ and } \mathbf{x}_k = \mathbf{x}_{k-1} + c_{i_k}\mathbf{e}_{i_k}, \tag{3.4}$$

where $c_{i_k} := \langle R^{k-1}, \phi_{i_k}\rangle$ and $\mathbf{e}_i, i = 1, \ldots, n$, are standard unit vectors with 1 as the ith entry and zero otherwise.

4: **end while**

5: If R^k is not small enough, we advance k and do the above again. Otherwise, if $R^k = 0$ or less than a given tolerance for an integer $k \geq 1$, then $\mathbf{x_b} = \mathbf{x}_k$ is an approximate sparse solution and $\mathbf{b} \approx \Phi\mathbf{x}_k$. S_k is the support set for $\mathbf{x}_k$.

Let us explain why this algorithm works. Recall that our sparse solution can be recast as the best s-term approximation problem: Given $\mathbf{b}$, find the $\mathbf{x} \in \mathbb{R}^n$ with $\|\mathbf{x}\|_0 \leq s$ such that

$$\|\mathbf{b} - \Phi\mathbf{x}\|_2 = \inf_{\substack{c_1, \ldots, c_s \\ n_1, \ldots, n_s \in \{1, \ldots, n\}}} \left\| \mathbf{b} - \sum_{i=1}^{s} c_i\phi_{n_i} \right\|_2. \tag{3.5}$$

Assume that $n = m$ and Φ consists of orthonormal basis vectors. Then the pure greedy Algorithm 3.1 will give the solution to the best s-term approximation (3.5). Indeed, let us use $s = 1$ to illustrate this point. For $R^0 = \mathbf{b}$, from Algorithm 3.1, we know that

$$R^1 = \mathbf{b} - c_{n_1}\phi_{n_1}.$$

It is easy to see that

$$\|R^1\|_2^2 = \|\mathbf{b}\|^2 - |c_{n_1}|^2 = \|\mathbf{b}\|_2^2 - |\langle \mathbf{b}, \phi_{n_1}\rangle|^2 \leq \|\mathbf{b}\|_2^2 - |\langle \mathbf{b}, \phi_i\rangle|_2^2 = \|\mathbf{b} - c_i\phi_i\|_2^2 \tag{3.6}$$

for any $i = 1, \ldots, n$, where $c_i = \langle \mathbf{b}, \phi_i\rangle$. That is, $c_{n_1}\phi_{n_1}$ is the best 1-term approximation of $\mathbf{b}$. By induction, we can show that R^s is the best s-term approximation of $\mathbf{b}$ when $\phi_i, i = 1, \ldots, n$, are orthonormal. Indeed, since $|c_{n_s}| \geq |c_i|$ for $i \notin \{n_1, \ldots, n_s\}$ and by using induction,

$$\|R^s\|_2^2 = \|R^{s-1}\|_2^2 - |c_{n_s}|^2 \leq \|R^{s-1}\|_2^2 - |c_i|^2$$

$$= \left\| \mathbf{b} - \sum_{j=1}^{s-1} c_{n_j}\phi_{n_j} \right\|_2^2 - |c_i|^2 = \left\| \mathbf{b} - \sum_{j=1}^{s-1} c_{n_j}\phi_{n_j} - c_i\phi_i \right\|_2^2$$

for any $i \notin \{n_1, \ldots, n_{s-1}\}$ and any $s - 1$ terms $n_j, j = 1, \ldots, s - 1$, where we have used the orthonormal conditions $\phi_j, j = 1, \ldots, n$.

Obviously, this greedy algorithm does not generate the best s-term approximation of $\mathbf{b}$ for a given sensing matrix. However, it motivates us to find better greedy-type algorithms, e.g., the orthogonal greedy algorithm (OGA), which is also called the orthogonal matching pursuit algorithm (OMP), and the weak orthogonal greedy algorithm (WOGA) (cf. [14]). Many studies of these greedy algorithms are summarized in [15] and [16]. We shall present several greedy algorithms in the next section. Furthermore, we shall provide some sufficient conditions on a sensing matrix Φ such that the greedy algorithm, e.g., the orthogonal matching pursuit (OMP) algorithm (cf. [17]), finds the sparse solution. Earlier references on greedy algorithms can be found in [6], [1], [9], [12], [3], [7], [5], etc.

3.2 ▪ Various Greedy Algorithms

Let us start with some notation to define orthogonal greedy algorithms. For simplicity, we consider $H_n = \mathrm{span}\{\phi_1, \dots, \phi_n\}$, which is the Euclidean space $\mathbb{R}^m$ with the standard inner product $\langle f, g \rangle$ for any two vectors f and g and the norm $\|f\| = (\langle f, f \rangle)^{1/2}$ for any $f \in \mathbb{R}^m$. For convenience, we assume that $\|\phi_i\| = 1$ for all $i = 1, \dots, n$. Then $\mathcal{D} = \{\phi_1, \dots, \phi_n\}$ is called a dictionary, while matrix $\Phi = [\phi_1, \dots, \phi_n]$ is a sensing matrix. Let S be an index set which is a subset of $\{1, 2, \dots, n\}$ and S^c be the complement of S in $\{1, 2, \dots, n\}$. Let Φ_S be the submatrix with the columns from Φ whose indices are in S. Also, $\Phi_{-S} = \Phi_{S^c}$. Because we do not have orthonormality among the columns in Φ, we have to use the least squares approximation in Algorithm 3.1. Let us start with the well-known orthogonal matching pursuit (OMP) algorithm (cf. [17]).

ALGORITHM 3.2

Orthogonal Matching Pursuit Algorithm [17]

1: **Input:** Measurement vector $\mathbf{b}$, sensing matrix $\Phi = [\phi_1, \phi_2, \dots, \phi_n]$, and sparsity s or $s = m/2$.

2: Initialize: $k = 0, R^0 = \mathbf{b}, S_0 = \emptyset$.

3: Iterate the five steps below until the stopping criterion is met.

Step 1: Let $k = k + 1$.

Step 2: Find $s^k = \arg\max_{1 \le i \le n} |\phi_i^\top R^{k-1}| = \arg\max_{1 \le i \le n} |\langle \phi_i, R^{k-1} \rangle|$.

Step 3: Let $S_k = S_{k-1} \bigcup \{s^k\}$.

Step 4: Find $\hat{\mathbf{x}}_k = \arg\min_{\mathbf{x}} \|\mathbf{b} - \Phi_{S_k}\mathbf{x}\|_2$ or $\hat{\mathbf{x}}_k = (\Phi_{S_k}^\top \Phi_{S_k})^{-1}\Phi_{S_k}^\top \mathbf{b}$.

Step 5: Define $R^k = \mathbf{b} - L_k$ with $L_k = \Phi_{S_k}\hat{\mathbf{x}}_k$.

4: **Output:** Let s be the iterative number and $\hat{\mathbf{x}}_s$ be an approximation of the sparse solution.

Note that $(\Phi_{S_k}^\top \Phi_{S_k})^{-1}\Phi_{S_k}^\top$ is the pseudo-inverse of Φ_{S_k}. We shall use the standard notation $\Phi_{S_k}^\dagger$ to denote the pseudo-inverse from now on. This algorithm is also called the orthogonal greedy algorithm (OGA) (cf. [14]).

There are many variants of Algorithm 3.2. Mainly, the greedy step of picking the best index may not lead to the best performance overall. Let us list a few more versions. The first one is called the grouped orthogonal matching pursuit (GOMP) or generalized orthogonal matching pursuit algorithm. That is, instead of choosing one term in Algorithm 3.2, we may choose more, e.g., $g = 2$ terms from each iteration. See [18].

ALGORITHM 3.3

Generalized Orthogonal Matching Pursuit Algorithm

1: **Input:** Measurement vector $\mathbf{b}$, sensing matrix $\Phi = [\phi_1, \phi_2, \dots, \phi_n]$, and sparsity s or $s = m/2$. Also, a number $g > 1$ of terms to choose in each iteration.

2: Initialize: $k = 0, R^0 = \mathbf{b}, S_0 = \emptyset$.

3: Iterate the five steps below until the stopping criterion is met.

Step 1: Let $k = k + 1$.

Step 2: Sort $|\langle \phi_i, R^{k-1} \rangle|, i = 1, \ldots, n$, in increasing order and choose the indices associated with the largest g terms to form a subset s_k.

Step 3: Let $S_k = S_{k-1} \bigcup \{s_k\}$.

Step 4: Find $\hat{\mathbf{x}}_k = \arg\min_{\mathbf{x}} \|\mathbf{b} - \Phi_{S_k}\mathbf{x}\|_2$ or $\hat{\mathbf{x}}_k = (\Phi_{S_k}^\top \Phi_{S_k})^{-1}\Phi_{S_k}^\top \mathbf{b}$.

Step 5: Define $R^k = \mathbf{b} - L_k$ with $L_k = \Phi_{S_k}\hat{\mathbf{x}}_k$.

4: **Output:** Let s be the iterative number and $\hat{\mathbf{x}}_s$ be an approximation of the sparse solution.

Another variation is called the weak OGA (WOGA). See Algorithm 3.4, where $\{t_k \in (0, 1]$, $k = 1, \ldots\}$ is a sequence of real positive numbers. Clearly, when $t_k = 1$, WOGA becomes the original Algorithm 3.2. As soon as we have one index n_k such that (3.7) holds, we immediately compute the best approximation instead of finding more indices to satisfy (3.7).

ALGORITHM 3.4
Weak Orthogonal Greedy Algorithm (WOGA)

1: **Input:** Fix a thresholding sequence $\{t_k, k = 1, 2, \ldots\}$ with $t_k \in (0, 1]$, e.g., $t_k = 0.95$. We are given a sensing matrix Φ and the right-hand side $\mathbf{b} \in H_n$, a tolerance, and a maximum number of iterations.

2: **Initialize:** Let $R^0(\mathbf{b}) = \mathbf{b}$, $L_0(\mathbf{b}) = 0$, $S_0 = \emptyset$.

3: For each $k \geq 1$, find $\phi_{n_k} \in H_n$ with $n_k \in S_{k-1}^c$ such that

$$|\langle R^{k-1}(\mathbf{b}), \phi_{n_k} \rangle| \geq t_k \max\{|\langle R^{k-1}(\mathbf{b}), \phi_i \rangle|, i \in S_{k-1}^c\} \qquad (3.7)$$

and compute the best approximation $L_k(\mathbf{b})$ of $R^{k-1}(\mathbf{b})$ in $H_k = \operatorname{span}\{\phi_{n_1}, \ldots, \phi_{n_k}\}$.

4: Then let $S_k = S_{k-1} \cup \{n_k\}$ and iteratively define

$$R^k(\mathbf{b}) = \mathbf{b} - L_k(\mathbf{b}), \qquad (3.8)$$

where $L_k(\mathbf{b}) = \Phi_{S_k}\Phi_{S_k}^\dagger \mathbf{b}$ and $\Phi_{S_k}^\dagger$ stands for the pseudo-inverse of Φ_{S_k}.

5: Stop the iteration if the norm of $R^k(\mathbf{b})$ is within the given tolerance or the maximum number of iterations reaches a fixed number.

6: **Output:** The support index set S_k and the coefficients $\Phi_{S_k}^\dagger \mathbf{b}$.

Another algorithm is called the modified WOGA (MWOGA). Instead of picking one term in each iteration, we may pick up more dependent on $t_1, t_2, \ldots$. This idea leads to Algorithm 3.5.

ALGORITHM 3.5
Modified Weak Orthogonal Greedy Algorithm (MWOGA)

1: **Input:** A sensing matrix Φ and the right-hand side $\mathbf{b}$. Choose a thresholding sequence $\{t_1, t_2, \ldots\}$ with all $t_k \in (0, 1]$. Give a tolerance and a maximum number of iterations.

2: **Initialize:** Let $S_0 = \emptyset$, $\widehat{R^0}(\mathbf{b}) = \mathbf{b}$, $\widehat{G_0}(\mathbf{b}) = 0$.

3: **Repeat the following steps:**

Step 1: For $k \geq 1$, find $M_k = \max_{i \notin S_{k-1}} |\langle \widehat{R^{k-1}}(\mathbf{b}), \phi_i \rangle|$.

Step 2: Let $S_k = S_{k-1} \cup \{i, |\langle \widehat{R^{k-1}}(\mathbf{b}), \phi_i \rangle| \geq t_k M_k\}$.

Step 3: Let $L_{S_k}(\mathbf{b})$ be the best approximation (least squares approximation) of $\widehat{R^{k-1}}(\mathbf{b})$ in subspace $H_k = \operatorname{span}\{\phi_i, i \in S_k\}$.

Step 4: Update $\widehat{R^k}(\mathbf{b}) = \mathbf{b} - L_{S_k}(\mathbf{b})$.

Step 5: If the norm of $\|\widehat{R^k}(\mathbf{b})\|$ is less than the given tolerance, stop the algorithm. Otherwise, advance k to $k + 1$ and go to Step 1.

4: **Output:** The subspace H_k and coefficient vector of $L_{S_k}(\mathbf{b})$.

For convenience, we will call the algorithm α-MWOGA if $t_k = \alpha \in (0, 1]$ for all $k \geq 1$. Note that this algorithm is slightly different from the WOGA because it chooses variable terms in each iterative step instead of only one term in WOGA. This difference sometimes greatly increases the computational efficiency. Furthermore, if we know the sparsity s or a good guess for s, we may choose the s largest terms in each iteration. This will lead to the subspace pursuit algorithm.

In addition to picking a variable number of vectors from H_n dependent on t_k at the kth iteration, the greedy algorithm introduced below (cf. [13]) computes the best approximation by using an iterative method which reduces the computational time when the size of the associated linear system is large. Let us first introduce an iterative algorithm to approximate the solution using the least squares method.

For a subset S of $\{1, 2, \ldots, n\}$, let P_S be the diagonal matrix of size $n \times n$ with entries 1 if the index is in S and 0 otherwise. We first introduce a computationally efficient algorithm for finding the best approximation of a given vector $\mathbf{b}$ from a subspace H_S spanned by the vectors in Φ_S for a fixed index set S. That is, we find the coefficients $c_i, i \in S$, in the least squares approximation $L_S(\mathbf{b}) = \sum_{i \in S} c_i \phi_i$ such that $\|\mathbf{b} - L_S(\mathbf{b})\|_2 = \min\{\|\mathbf{b} - g\|_2, g \in H_S\}$, where $\mathcal{D}_S$ is the span of vectors $\phi_i, i \in S$. In general, $L_S(\mathbf{b})$ can be computed directly by inverting a Gramian matrix $[\langle \phi_i, \phi_j \rangle]_{i,j \in S}$. When the size of S is large, it is more efficient to use the following iterative algorithm to find an approximation of $L_S(\mathbf{b})$ than the direct method.

ALGORITHM 3.6
Least Squares Approximation

1: We start with $R^0(\mathbf{b}) = \mathbf{b}, G_0(\mathbf{b}) = 0$ and $L_{0,\Lambda}(\mathbf{b}) = 0$.

Step 1: For $k \geq 1$, compute

$$R^k(\mathbf{b}) = R^{k-1}(\mathbf{b}) - \Phi_{S_k} P_{S_k} \Phi^\top R^{k-1}(\mathbf{b}) \text{ and } L_k(\mathbf{b}) = L_{k-1}(\mathbf{b}) + \Phi_{S_k} P_{S_k} \Phi^\top R^{k-1}(\mathbf{b}).$$

Step 2: If $\|L_k(\mathbf{b}) - L_{k-1}(\mathbf{b})\|_2$ is small enough, we stop the iteration. Otherwise, we go to Step 1.

We write $L_{\ell_k, S_k}(\mathbf{b})$ to be the output of the above algorithm after enough iterations, say ℓ_k. Then it is known that $L_{k,S_k}(\mathbf{b})$ is a good approximation of $L_S(\mathbf{b})$, as explained in section 3.6. Now we are ready to state the iterative least squares orthogonal greedy algorithm (ILSOGA) (cf. [13] with all $t_k = r \leq 1$) as follows.

ALGORITHM 3.7
Iterative Least Squares Orthogonal Greedy Algorithm (ILSOGA) [13]

1: **Input:** Φ and $\mathbf{b}$. Let $\epsilon > 0$ be a tolerance. Choose a sequence of thresholds $\{t_1, t_2, \ldots\}$ with $t_l \in (0, 1]$.

2: **Initialize:** We start with $S_0 = \emptyset, R^0(\mathbf{b}) = \mathbf{b}, G_0(\mathbf{b}) = 0$.

3: **Repeat:**

Step 1: For $k \geq 1$, find $M_k = \max_{i \notin S_{k-1}} |\langle R^{k-1}(\mathbf{b}), \phi_i \rangle|$.

Step 2: Let $S_k = S_{k-1} \cup \{i | \langle R^{k-1}(\mathbf{b}), \phi_i \rangle \geq t_k M_k\}$.

Step 3: Apply Algorithm 3.6 above to find the best approximation of $R^{k-1}(\mathbf{b})$ within the given precision ϵ. Let us say that the least squares approximation iterative steps are applied ℓ_k times to obtain an approximation $L_{\ell_k, S_k}(\mathbf{b})$.

Step 4: Update $G_k(\mathbf{b}) = G_{k-1}(\mathbf{b}) + L_{\ell_k, S_k}(\mathbf{b})$ and $R^k(\mathbf{b}) = R^{k-1}(\mathbf{b}) - L_{\ell_k, S_k}(\mathbf{b})$.

Step 5: If $\|L_{\ell_k, S_k}(\mathbf{b})\|_2$ is small enough, we stop the algorithm. Otherwise we advance k to $k+1$ and go to Step 1.

4: **Output:** The coefficient vector of $L_{\ell_k, S_k}(\mathbf{b})$.

This ILSOGA was implemented, its convergence was tested numerically, and its efficiency and effectiveness for image compression were demonstrated in [13] for large problems.

Next we present an economical version of the OMP algorithm which performs much faster than Algorithm 3.2 and is even more efficient than Algorithm 3.7. In addition, this new algorithm has the same convergence as Algorithm 3.2. That is, in section 3.4 we shall show some properties of Algorithm 3.2, including its convergence. Similar properties also hold for Algorithm 3.8. The main difference between Algorithms 3.8 and 3.2 is the computation in Step 4. Algorithm 3.8 solves a 2×2 linear system instead of the $k \times k$ linear system in Algorithm 3.2. When s is large, Algorithm 3.8 is more efficient. In fact, it is the most efficient among all the greedy algorithms discussed in this chapter.

ALGORITHM 3.8
Economical OMP Algorithm

1: **Input:** Measurement vector $\mathbf{b}$, sensing matrix Φ, and sparsity s or $s = m/2$.
2: **Initialize:** $k = 0, R^0 = \mathbf{b}, L_0 = 0, S_0 = \emptyset$.
3: **Repeat:** Iterate the five steps below until the stopping criterion is met.

Step 1: Let $k = k + 1$.

Step 2: Find $s^k = \arg\max_{1 \le i \le n} |\phi_i^\top R^{k-1}| = \arg\max_{1 \le i \le n} |\langle \phi_i, R^{k-1} \rangle|$.

Step 3: Let $S_k = S_{k-1} \bigcup \{s^k\}$.

Step 4: Find $[\alpha_1^*, \alpha_2^*] = \arg\min_{\alpha_1, \alpha_2} \|\mathbf{b} - \alpha_1 L_{k-1} - \phi_{s^k}\alpha_2\|_2$.

Step 5: Define $R^k = \mathbf{b} - L_k$ with $L_k = \alpha_1^* L_{k-1} + \alpha_2^* \phi_{s^k}$.

4: **Output:** Let s be the iterative number and $\hat{\mathbf{x}} = \arg\min_{\mathbf{x}:\text{supp}(\mathbf{x})\subset S_s} \|\mathbf{b} - \Phi\mathbf{x}\|_2$.

Finally, we introduce another type of greedy algorithms: a quasi-greedy algorithm. The motivation is as follows. Instead of choosing $s = 1$ each time in (3.6) section 3.1, we are greedy enough to choose two terms as sparse solutions, since most applications have a sparsity of more than 2. We have to solve the best approximations

$$\min_{b_1, b_2} \|b_1\phi_i + b_2\phi_j - \mathbf{b}\| \tag{3.9}$$

for all $i \ne j, i, j = 1, \ldots, n$, to find the residuals. That is, we choose the best index pair, (i_1, j_1), such that the residual is the smallest:

$$\min_{\substack{i,j,b_1,b_2 \\ i \ne j}} \|b_1\phi_i + b_2\phi_j - \mathbf{b}\| = \min_{b_1, b_2} \|\mathbf{b} - b_1\phi_{i_1} - b_2\phi_{j_1}\|. \tag{3.10}$$

Once we find (i_1, j_1) to solve (3.10), we let $\mathbf{r}_1 = \mathbf{b} - b_{i_1}\phi_{i_1} - b_{j_1}\phi_{j_1}$ and repeat the procedure. This leads to the QOMP algorithm (cf. [8]).

ALGORITHM 3.9
Economical Quasi-orthogonal Matching Pursuit (QOMP) Algorithm

1: **Input:** A sensing matrix $\Phi_{m \times n}$, $\mathbf{b}_{n \times 1}$, a sparsity $s \ge 2$, a maximum number $k_{\max}$ ($k_{\max} \le m/2$) of iterations, and a tolerance ϵ.
2: **Initialize:** $S_0 = \emptyset, L_0 = 0, \mathbf{r}_0 = \mathbf{b}, k = 0, \Psi_{m \times n} = \Phi_{m \times n}$.

3: **Repeat:** While $k < k_{\max}$ and $|\mathbf{r_k}| > \epsilon$, do the following 5 steps:

Step 1: $k = k + 1$.

Step 2: Let $\mathbf{Res}_{(i,j)}(\mathbf{r}_{k-1}) = \min_{u,v \in \mathbb{R}}\{\|\phi_i u + \phi_j v - \mathbf{r_{k-1}}\|_2\}$ and compute

$$(i_k, j_k) = \arg \min_{1 \le i \le n, 1 \le j \le n}\{\mathbf{Res}_{(i,j)}(\mathbf{r}_{k-1})\}.$$

Step 3: $\mathcal{S}_k = \mathcal{S}_{k-1} \cup \{i_k, j_k\}$.

Step 4:

$$\{\alpha_k, \beta_k, \gamma_k\} := \arg \min \|\mathbf{b} - \alpha L_{k-1} - \beta\phi_{i_k} - \gamma\phi_{j_k}\|.$$

Step 5: Define $L_k = \alpha_k L_{k-1} - \beta_k\phi_{i_k} - \gamma_k\phi_{j_k}$ and $\mathbf{r}_k = \mathbf{b} - L_k$.

4: **Output:** $\mathcal{S} = \mathcal{S}_k$, $\mathbf{x}_S = \Phi_S^\dagger \mathbf{b}$, and $\mathbf{x}_{S^c} = \mathbf{0}$.

The major problem of this algorithm is the computation in Step 2, which is slow when n is large. One way to overcome this problem is to use a parallel computation technique and/or a computer with GPU processes.

3.3 ▪ Sparse Solutions by Algorithm 3.2

In this section we explain how the OMP algorithm can be used to recover any sparse solution. We begin with a simple sufficient condition. Let $\Lambda_{opt} \subset \{1, \ldots, n\}$ be the supporting index set for a compressible vector $\mathbf{b}$ such that there is a vector $\mathbf{x}$ with support on Λ_{opt} such that $\mathbf{b} = \Phi\mathbf{x}$. We have the following theorem.

Theorem 3.1 (Tropp, 2004 [17]). *Suppose that there exists a vector $\mathbf{x_b}$ with support on $\Lambda_{opt} \subset \{1, \ldots, n\}$ such that $\mathbf{b} = \Phi\mathbf{x_b}$. Suppose $\#(\Lambda_{opt}) \le s$. Let $\Psi_{opt} = [\phi_i, i \in \{1, 2, \ldots, n\}\backslash\Lambda_{opt}]$ be the complement set of $\Phi_{opt} := [\phi_i, i \in \Lambda_{opt}]$. Then Algorithm 3.2 will find $\mathbf{x_b}$ in at most s iterations if*

$$\max_{\psi \in \Psi_{opt}} \|\Phi_{opt}^\dagger \psi\|_1 < 1, \tag{3.11}$$

where $\Phi_{opt}^\dagger$ stands for the pseudo-inverse of Φ_{opt}.

Proof. After the first k iterations in Algorithm 3.2, let $\mathbf{x}_k$ be the solution which is a linear combination of k columns with indices listed in Λ_{opt}. We would like to know a sufficient condition which can guarantee that the next iteration is also optimal.

Observe that the vector $\Phi_{opt}^\top R^k$ contains the inner products between the kth residual R^k and the columns in Φ_{opt}. Similarly, $\Psi_{opt}^\top R^k$ is the vector containing all inner products of the residual vector R^k with columns not in the optimal set Φ_{opt}. Thus, we claim

$$\rho(R^k) = \frac{\|\Psi_{opt}^\top R^k\|_\infty}{\|\Phi_{opt}^\top R^k\|_\infty} < 1.$$

Because the residual vector R^k is supported on Λ_{opt}, we have $(\Phi_{opt}^\top)^\dagger \Phi_{opt}^\top R^k = R^k$. Thus,

$$\rho(R^k) = \frac{\|\Psi_{opt}^\top R^k\|_\infty}{\|\Phi_{opt}^\top R^k\|_\infty} = \frac{\|\Psi_{opt}^\top(\Phi_{opt}^\top)^\dagger \Phi_{opt}^\top R^k\|_\infty}{\|\Phi_{opt}^\top R^k\|_\infty} \le \|\Psi_{opt}^\top(\Phi_{opt}^\dagger)^\top\|_\infty = \|\Phi_{opt}^\dagger \Psi_{opt}\|_1,$$

where we have used the fact that $\|A\|_\infty = \|A^\top\|_1$ for any matrix A of $n \times m$. The condition in (3.11) is sufficient to ensure that $\rho(R^k) < 1$, and hence Algorithm 3.2 will pick up an optimal index for Λ_{k+1}. Also, Algorithm 3.11 will never pick up the same index twice, so it follows that s iterations at most will find all the optimal indices for $\mathbf{x_b}$. $\square$

The above study also explains the convergence of MWOGA.

Theorem 3.2. *A sufficient condition for α-MWOGA to find $\mathbf{x_b}$ in m iterations is*

$$\max_{\psi \in \Psi_{opt}} \|\Phi_{opt}^\dagger \psi\|_1 < \alpha. \tag{3.12}$$

Under an RIP condition, we can show that Algorithm 3.2 will find the sparse solution. We can even tell how many iterations in Algorithm 3.2 are needed. Without loss of generality, we assume that the sparse solution $\mathbf{x_b} = (x_1, \ldots, x_s, 0, \ldots, 0)^\top \in \mathbb{R}^n$. We write

$$S_i = |\langle \phi_i, \mathbf{b} \rangle| = |\langle \phi_i, \Phi\mathbf{x_b} \rangle|, \quad i = 1, \ldots, n.$$

Let $S_0 = \max |S_i|, i = 1, \ldots, s$. We first need the following lemma.

Lemma 3.3. *Suppose that the restricted isometry constant (RIC) δ_{s+1} of a sensing matrix $\Phi = [\phi_1, \ldots, \phi_n]$ with $\|\phi_i\|_2 = 1$ satisfies*

$$\delta_{s+1} < \frac{1}{\sqrt{s} + 1}.$$

Then $S_0 > |S_i|$ for all $i > s$.

Proof. By Lemma 2.31, we have

$$|S_i| = |\langle \Phi e_i, \Phi\mathbf{x_b} \rangle| \le \delta_{s+1}\|\mathbf{x_b}\|_2 \quad \forall i > s. \tag{3.13}$$

Using the assumption, $|S_i| < \frac{1}{(\sqrt{s}+1)}\|\mathbf{x_b}\|_2$ or $\|\mathbf{x_b}\|_2 \ge (1 + \sqrt{s})|S_i|$.

Next we have

$$\langle \Phi\mathbf{x_b}, \Phi\mathbf{x_b} \rangle = \left\langle \Phi \sum_{i=1}^{s} x_i e_i, \Phi\mathbf{x_b} \right\rangle = \sum_{i=1}^{s} x_i \langle \Phi e_i, \Phi\mathbf{x_b} \rangle$$

$$= \sum_{i=1}^{s} x_i S_i \le S_0\|\mathbf{x_b}\|_1 \le S_0\sqrt{s}\|\mathbf{x_b}\|_2.$$

By using the RIP, we have

$$(1 - \delta_{s+1})\|\mathbf{x_b}\|_2^2 \le (1 - \delta_s)\|\mathbf{x_b}\|_2^2 \le \|\Phi\mathbf{x_b}\|_2^2 \le S_0\sqrt{s}\|\mathbf{x_b}\|_2.$$

That is, $S_0 \ge (1 - \delta_{s+1})\|\mathbf{x_b}\|_2/\sqrt{s}$. It follows by (3.13) that

$$S_0 > (1 - \delta_{s+1})(1 + \sqrt{s})|S_i|/\sqrt{s} \ge \left(1 - \frac{1}{\sqrt{s}+1}\right)(1 + \sqrt{s})|S_i|/\sqrt{s} = |S_i|$$

for $i > s$. $\square$

We now prove another main result in this section.

Theorem 3.4 (Mo and Shen, 2011 [11]). *Suppose that Φ satisfies the RIP with $\delta_{s+1} < 1/(1 + \sqrt{s})$. Then for any s-sparse vector $\mathbf{x_b}$, Algorithm 3.2 will recover $\mathbf{x_b}$ from $\mathbf{b} = \Phi\mathbf{x_b}$ in s iterations.*

Proof. Consider the first iteration. By Lemma 3.3, $\delta_{s+1} < 1/(1 + \sqrt{s})$ ensures that $S_0 > S_i$ for $i > s$. Thus, the first selection is a successful one. Note that the same analysis of Lemma 3.3 works for the residual vector R^k instead of $\mathbf{b}$ and $\mathbf{x_b} - \hat{\mathbf{x}}_k$ for $\mathbf{x_b}$ since $\|\mathbf{x_b} - \hat{\mathbf{x}}_k\|_0 \leq s$. Since Algorithm 3.2 makes an orthogonal projection in each iteration, we can see that Algorithm 3.2 selects a different index from $\{1, 2, \ldots, s\}$ in each iteration. This finishes the proof. $\qquad\square$

The condition in Theorem 3.4 is slightly improved in [10].

Theorem 3.5 (Mo, 2015 [10]). *If $\delta_{s+1} < 1/\sqrt{s+1}$, the exact recovery of the s-sparse signal $\mathbf{x}$ can be guaranteed by using Algorithm 3.2 in s iterations.*

To establish the result in Theorem 3.5 including the noise case, we need more notation. Let $\Omega = \text{supp}(\mathbf{x_b})$ be the support set of sparse signal $\mathbf{x_b}$. As usual, $\mathbf{x}_S$ is the subvector of $\mathbf{x_b}$ that contains only the entries indexed by S. For full column rank matrix Φ_S, let $\Phi_S^\dagger = (\Phi_S^\top \Phi_S)^{-1}\Phi_S^\top$ and $\Phi_S^\perp = I - \Phi_S \Phi_S^\dagger$ denote the pseudo-inverse operator and orthogonal complement projector on the column space of Φ_S, respectively. In particular, we have

$$R^k(\mathbf{y}) = \mathbf{y} - \Phi_{S_k}\hat{\mathbf{x}}_{S_k} = \mathbf{y} - \Phi_{S_k}\Phi_{S_k}^\dagger \mathbf{y} = \Phi_{S_k}^\perp \mathbf{y}. \tag{3.14}$$

We mainly follows the ideas in [19]. Let us start with a few useful lemmas.

Lemma 3.6. *Let S_1, S_2 be two subsets of $\{1, \ldots, n\}$ satisfying $|S_2 \setminus S_1| \geq 1$, and let matrix Φ satisfy the RIP of order $|S_1 \cup S_2|$. Then for any vector $\mathbf{x} \in \mathbb{R}^{|S_2 \setminus S_1|}$,*

$$(1 - \delta_{|S_1 \cup S_2|})\|\mathbf{x}\|_2^2 \leq \|\Phi_{S_1}^\perp \Phi_{S_2 \setminus S_1}\mathbf{x}\|_2^2 \leq (1 + \delta_{|S_1 \cup S_2|})\|\mathbf{x}\|_2^2.$$

Proof. The proof is left to Exercise 24. $\qquad\square$

Lemma 3.7. *Suppose that Φ satisfies the RIP of order $s + 1$ with $\delta_{s+1} \in [0, 1)$. Let $S \subset \Omega$ be a subset of Ω with $|S| < |\Omega| \neq 0$. If $\mathbf{x}$ satisfies*

$$\|\mathbf{x}_{\Omega \setminus S}\|_1 \leq \sqrt{\nu}\|\mathbf{x}_{\Omega \setminus S}\|_2 \tag{3.15}$$

for some $\nu \geq 1$, then, letting $\mathbf{z} = \Phi_{\Omega \setminus S}\mathbf{x}_{\Omega \setminus S}$,

$$\|\Phi_{\Omega \setminus S}^\top \Phi_S^\perp \mathbf{z}\|_\infty - \|\Phi_{\Omega^c}^\top \Phi_S^\perp \mathbf{z}\|_\infty \geq \frac{(1 - \sqrt{\nu + 1}\delta_{|\Omega|+1})\|\mathbf{x}_{\Omega \setminus S}\|_2}{\sqrt{\nu}}. \tag{3.16}$$

Proof. Since $S \subseteq \Omega$ with $|S| < |\Omega| \neq 0$, $\|\mathbf{x}_{\Omega \setminus S}\|_1 \neq 0$. We first claim that

$$\|\Phi_S^\perp \mathbf{z}\|_2^2 \leq \sqrt{\nu}\|\mathbf{x}_{\Omega \setminus S}\|_2 \|\Phi_{\Omega \setminus S}^\top \Phi_S^\perp \mathbf{z}\|_\infty. \tag{3.17}$$

Indeed, by using (3.15), it is easy to check that

$$\sqrt{\nu}\|\mathbf{x}_{\Omega \setminus S}\|_2 \|\Phi_{\Omega \setminus S}^\top \Phi_S^\perp \mathbf{z}\|_\infty \geq \|\mathbf{x}_{\Omega \setminus S}\|_1 \|\Phi_{\Omega \setminus S}^\top \Phi_S^\perp \mathbf{z}\|_\infty$$

$$= \left(\sum_{l \in \Omega \setminus S} |\mathbf{x}_l|\right) \|\Phi_{\Omega \setminus S}^\top \Phi_S^\perp \mathbf{z}\|_\infty \geq \|\mathbf{x}_{\Omega \setminus S}^\top \Phi_{\Omega \setminus S}^\top \Phi_S^\perp \mathbf{z}\|_1$$

$$= \|\mathbf{x}_{\Omega \setminus S}^\top \Phi_{\Omega \setminus S}^\top (\Phi_S^\perp)^\top \Phi_S^\perp \mathbf{z}\|_1 = \|\Phi_S^\perp \mathbf{z}\|_2^2,$$

where we have used the property of the projection $\Phi_S^\perp$:

$$(\Phi_S^\perp)^\top \Phi_S^\perp = \Phi_S^\perp \Phi_S^\perp = \Phi_S^\perp. \tag{3.18}$$

Thus, the claim (3.17) holds.

Next we define $\mathbf{h} \in \mathbb{R}^d$ by

$$\mathbf{h} = \frac{\Phi_{\Omega^c}^\top \Phi_S^\perp \mathbf{z}}{\|\Phi_{\Omega^c}^\top \Phi_S^\perp \mathbf{z}\|_2},$$

where Ω^c is the complement of Ω. Then it is easy to see that

$$\mathbf{h}^\top \Phi_{\Omega^c}^\top \Phi_S^\perp \mathbf{z} = \|\Phi_{\Omega^c}^\top \Phi_S^\perp \mathbf{z}\|_2 \tag{3.19}$$

and $\|\mathbf{h}\|_2 = 1$. With $\mathbf{h}$, we define the two sparse vectors

$$\mathbf{u} = \begin{bmatrix} \mathbf{x}_{\Omega \setminus S}^\top & \mathbf{0}^\top \end{bmatrix}^\top \in \mathbb{R}^{(|\Omega \setminus S| + d)} \text{ and } \mathbf{w} = \begin{bmatrix} \mathbf{0}^\top & \mu \|\mathbf{x}_{\Omega \setminus S}\| \mathbf{h} \end{bmatrix}^\top \in \mathbb{R}^{(|\Omega \setminus S| + d)},$$

where $\mu = -\frac{\sqrt{\nu+1}-1}{\sqrt{\nu}}$. By a simple calculation, we have

$$\frac{2\mu}{1-\mu^2} = -\sqrt{\nu}, \quad \frac{1+\mu^2}{1-\mu^2} = \sqrt{\nu+1}. \tag{3.20}$$

Then we have

$$\|\mathbf{u} + \mathbf{w}\|_2^2 = (1 + \mu^2)\|\mathbf{x}_{\Omega \setminus S}\|_2^2, \tag{3.21}$$

$$\|\mu^2 \mathbf{u} - \mathbf{w}\|_2^2 = \mu^2(1 + \mu^2)\|\mathbf{x}_{\Omega \setminus S}\|_2^2. \tag{3.22}$$

Moreover, we let

$$\mathbf{B} = \Phi_S^\perp \Phi_{(\Omega \setminus S) \cup \Omega^c} = \Phi_S^\perp \Phi_{(\Omega \cup \Omega^c) \setminus S}, \tag{3.23}$$

where the second equality is because $S \subset \Omega$. Then

$$\mathbf{B}\mathbf{u} = \Phi_S^\perp \mathbf{z} \tag{3.24}$$

and

$$\mathbf{w}^\top \mathbf{B}^\top \mathbf{B}\mathbf{u} = \mathbf{w}^\top \Phi_{(\Omega \setminus S) \cup \Omega^c}^\top (\Phi_S^\perp)^\top \Phi_S^\perp \mathbf{z} = \mathbf{w}^\top \Phi_{(\Omega \setminus S) \cup \Omega^c}^\top \Phi_S^\perp \mathbf{z}$$
$$= \mu \|\mathbf{x}_{\Omega \setminus S}\|_2 \mathbf{h}^\top \Phi_{\Omega^c}^\top \Phi_S^\perp \mathbf{z} = \mu \|\mathbf{x}_{\Omega \setminus S}\|_2 \|\Phi_{\Omega^c}^\top \Phi_S^\perp \mathbf{z}\|_2,$$

where we have used (3.23), (3.24), (3.18), and (3.19). Therefore, we have

$$\|\mathbf{B}(\mathbf{u} + \mathbf{w})\|_2^2 = \|\mathbf{B}\mathbf{u}\|_2^2 + \|\mathbf{B}\mathbf{w}\|_2^2 + 2\mathbf{w}^\top \mathbf{B}^\top \mathbf{B}\mathbf{u}$$
$$= \|\mathbf{B}\mathbf{u}\|_2^2 + \|\mathbf{B}\mathbf{w}\|_2^2 + 2\mu \|\mathbf{x}_{\Omega \setminus S}\|_2 \|\Phi_{\Omega^c}^\top \Phi_S^\perp \mathbf{z}\|_2$$

and, similarly,

$$\|\mathbf{B}(\mu^2 \mathbf{u} - \mathbf{w})\|_2^2 = \mu^4 \|\mathbf{B}\mathbf{u}\|_2^2 + \|\mathbf{B}\mathbf{w}\|_2^2 - 2\mu^3 \|\mathbf{x}_{\Omega \setminus S}\|_2 \|\Phi_{\Omega^c}^\top \Phi_S^\perp \mathbf{z}\|_2.$$

Combining the two equations above, we have

$$\|\mathbf{B}(\mathbf{u} + \mathbf{w})\|_2^2 - \|\mathbf{B}(\mu^2 \mathbf{u} - \mathbf{w})\|_2^2$$
$$= (1 - \mu^4)\|\mathbf{B}\mathbf{u}\|_2^2 + 2\mu(1 + \mu^2)\|\mathbf{x}_{\Omega \setminus S}\|_2 \|\Phi_{\Omega^c}^\top \Phi_S^\perp \mathbf{z}\|_2$$
$$= (1 - \mu^4) \left(\|\mathbf{B}\mathbf{u}\|_2^2 + \frac{2\mu}{1 - \mu^2}\|\mathbf{x}_{\Omega \setminus S}\|_2 \|\Phi_{\Omega^c}^\top \Phi_S^\perp \mathbf{z}\|_2 \right)$$
$$= (1 - \mu^4)(\|\mathbf{B}\mathbf{u}\|_2^2 - \sqrt{\nu}\|\mathbf{x}_{\Omega \setminus S}\|_2 \|\Phi_{\Omega^c}^\top \Phi_S^\perp \mathbf{z}\|_2), \tag{3.25}$$

where the last equality follows from the first equality in (3.20).

By using (3.23) and Lemma 3.6, we have

$$
\begin{aligned}
&\|\mathbf{B}(\mathbf{u}+\mathbf{w})\|_2^2 - \|\mathbf{B}(\mu^2\mathbf{u}-\mathbf{w})\|_2^2 \\
&\geq (1-\delta_{|\Omega|+1})\|(\mathbf{u}+\mathbf{w})\|_2^2 - (1+\delta_{|\Omega|+1})\|(\mu^2\mathbf{u}-\mathbf{w})\|_2^2 \\
&= (1-\delta_{|\Omega|+1})(1+\mu^2)\|\mathbf{x}_{\Omega\setminus S}\|_2^2 - (1+\delta_{|\Omega|+1})\mu^2(1+\mu^2)\|\mathbf{x}_{\Omega\setminus S}\|_2^2 \\
&= (1+\mu^2)\|\mathbf{x}_{\Omega\setminus S}\|_2^2\big((1-\delta_{|\Omega|+1}) - (1+\delta_{|\Omega|+1})\mu^2\big) \\
&= (1+\mu^2)\|\mathbf{x}_{\Omega\setminus S}\|_2^2\big((1-\mu^2) - \delta_{|\Omega|+1}(1+\mu^2)\big) \\
&= (1-\mu^4)\|\mathbf{x}_{\Omega\setminus S}\|_2^2\left(1 - \frac{1+\mu^2}{1-\mu^2}\delta_{|\Omega|+1}\right) \\
&= (1-\mu^4)\|\mathbf{x}_{\Omega\setminus S}\|_2^2\big(1 - \sqrt{\nu+1}\delta_{|\Omega|+1}\big),
\end{aligned}
\tag{3.26}
$$

where we have used (3.18), (3.23), (3.21), (3.22), and (3.20).

By (3.24), (3.25), (3.26), and the fact that $1-\mu^4 > 0$, we have

$$
\begin{aligned}
\|\Phi_S^{\perp}\mathbf{z}\|_2^2 - \sqrt{\nu}\|\mathbf{x}_{\Omega\setminus S}\|_2\|\Phi_{\Omega^c}^{\top}\Phi_S^{\perp}\mathbf{z}\|_2 &= \|\mathbf{B}\mathbf{u}\|_2^2 - \sqrt{\nu}\|\mathbf{x}_{\Omega\setminus S}\|_2\|\Phi_{\Omega^c}^{\top}\Phi_S^{\perp}\mathbf{z}\|_2 \\
&\geq \|\mathbf{x}_{\Omega\setminus S}\|_2^2\big(1 - \sqrt{\nu+1}\delta_{|\Omega|+1}\big).
\end{aligned}
$$

Thus, combining this with (3.17), we obtain

$$
\sqrt{\nu}\|\mathbf{x}_{\Omega\setminus S}\|_2\big(\|\Phi_{\Omega\setminus S}^{\top}\Phi_S^{\perp}\mathbf{z}\|_\infty - \|\Phi_j^{\top}\Phi_S^{\perp}\mathbf{z}\|_2\big) \geq \|\mathbf{x}_{\Omega\setminus S}\|_2^2\big(1 - \sqrt{\nu+1}\delta_{|\Omega|+1}\big).
$$

Therefore, because $\|\Phi_j^{\top}\Phi_S^{\perp}\mathbf{z}\|_2 \geq \|\Phi_j^{\top}\Phi_S^{\perp}\mathbf{z}\|_\infty$, (3.16) follows.   □

By Lemma 3.7, we obtain the following sufficient condition for the recovery of signals of sparsity s with Algorithm 3.2 from $\mathbf{y} = \Phi\mathbf{x_b} + \mathbf{v}$ with noise vector $\mathbf{v}$.

Theorem 3.8 (Wen, Zhou, Liu, Lai, and Tang, 2019 [19]). *Suppose that the noisy vector $\mathbf{v}$ satisfies $\|\mathbf{v}\|_2 \leq \epsilon$ and Φ satisfies the RIP with δ_{s+1} satisfying*

$$
\delta_{s+1} < \frac{1}{\sqrt{s+1}}.
\tag{3.27}
$$

Then Algorithm 3.2 with stopping criterion $\|R^k\| \leq \epsilon$ exactly recovers the support of the sparse signal $\mathbf{x_b} = [x_1, \ldots, x_n]^\top$ in s iterations provided that

$$
\min_{x_i\neq 0}\{|x_i|\} > \frac{2\epsilon}{1 - \sqrt{s+1}\delta_{s+1}}.
\tag{3.28}
$$

Moreover, the recovery error can be bounded by

$$
\|\mathbf{x_b} - \widehat{\mathbf{x}}_s\|_2 \leq \epsilon,
\tag{3.29}
$$

where $\widehat{\mathbf{x}}_s$ is the output from Algorithm 3.2.

Proof. Let Ω be the support set of the sparsest solution $\mathbf{x_b}$ and $\mathbf{y} = \Phi\mathbf{x_b} + \mathbf{v}$ with noise vector $\mathbf{v}$. The proof consists of two parts. In the first part, we show that Algorithm 3.2 selects correct indices in all iterations. In the second part, we prove that it performs exactly $s + 1$ iterations.

We prove the first step by induction. Suppose that Algorithm 3.2 selects correct indices in the first $k-1$ iterations, i.e., $S_{k-1} \subseteq \Omega$ for $1 \leq k \leq |\Omega|$. Then we need to show that Algorithm 3.2

selects the correct index in the kth iteration, i.e., by Algorithm 3.2. That is, we need to show that $s^k \in \Omega$. Thus, by line 2 of Algorithm 3.2, to show $s^k \in \Omega$, it suffices to show

$$\|\Phi_{\Omega \setminus S_{k-1}}^\top R^{k-1}\|_\infty > \|\Phi_{\Omega^c}^\top R^{k-1}\|_\infty. \tag{3.30}$$

By line 4 of Algorithm 3.2, we have

$$\hat{\mathbf{x}}_{S_{k-1}} = (\Phi_{S_{k-1}}^\top \Phi_{S_{k-1}})^{-1} \Phi_{S_{k-1}}^\top \mathbf{y}. \tag{3.31}$$

Thus, by line 5 of Algorithm 3.2 and (3.31), we recall that $\Phi_S^\perp = I - \Phi_S(\Phi_S^\top \Phi_S)^{-1}\Phi_S^\top$ to obtain

$$
\begin{aligned}
R^{k-1} &= \mathbf{y} - \Phi_{S_{k-1}}\hat{\mathbf{x}}_{S_{k-1}} \\
&= \big(I - \Phi_{S_{k-1}}(\Phi_{S_{k-1}}^\top \Phi_{S_{k-1}})^{-1}\Phi_{S_{k-1}}^\top\big)\mathbf{y} \\
&= \Phi_{S_{k-1}}^\perp(\Phi \mathbf{x_b} + \mathbf{v}) = \Phi_{S_{k-1}}^\perp(\Phi_\Omega \mathbf{x}_\Omega + \mathbf{v}) \\
&= \Phi_{S_{k-1}}^\perp(\Phi_{S_{k-1}} \mathbf{x}_{S_{k-1}} + \Phi_{\Omega \setminus S_{k-1}} \mathbf{x}_{\Omega \setminus S_{k-1}} + \mathbf{v}) \\
&= \Phi_{S_{k-1}}^\perp \Phi_{\Omega \setminus S_{k-1}} \mathbf{x}_{\Omega \setminus S_{k-1}} + \Phi_{S_{k-1}}^\perp \mathbf{v},
\end{aligned} \tag{3.32}
$$

where we have used the fact that $\Omega = \operatorname{supp}(\mathbf{x_b})$, the induction assumption $S_{k-1} \subseteq \Omega$, and $\Phi_{S_{k-1}}^\perp \Phi_{S_{k-1}} = 0$, respectively. Thus, by (3.32), we obtain

$$\|\Phi_{\Omega \setminus S_{k-1}}^\top R^{k-1}\|_\infty \geq \|\Phi_{\Omega \setminus S_{k-1}}^\top \Phi_{S_{k-1}}^\perp \Phi_{\Omega \setminus S_{k-1}} \mathbf{x}_{\Omega \setminus S_{k-1}}\|_\infty - \|\Phi_{\Omega \setminus S_{k-1}}^\top \Phi_{S_{k-1}}^\perp \mathbf{v}\|_\infty, \tag{3.33}$$

which is an estimate for the left-hand side of (3.30). For the right-hand side of (3.30), we have

$$\|\Phi_{\Omega^c}^\top R^{k-1}\|_\infty \leq \|\Phi_{\Omega^c}^\top \Phi_{S_{k-1}}^\perp \Phi_{\Omega \setminus S_{k-1}} \mathbf{x}_{\Omega \setminus S_{k-1}}\|_\infty + \|\Phi_{\Omega^c}^\top \Phi_{S_{k-1}}^\perp \mathbf{v}\|_\infty. \tag{3.34}$$

Therefore, by (3.33) and (3.34), in order to show (3.30), we need to show

$$
\begin{aligned}
&\|\Phi_{\Omega \setminus S_{k-1}}^\top \Phi_{S_{k-1}}^\perp \Phi_{\Omega \setminus S_{k-1}} \mathbf{x}_{\Omega \setminus S_{k-1}}\|_\infty - \|\Phi_{\Omega^c}^\top \Phi_{S_{k-1}}^\perp \Phi_{\Omega \setminus S_{k-1}} \mathbf{x}_{\Omega \setminus S_{k-1}}\|_\infty \\
&> \|\Phi_{\Omega \setminus S_{k-1}}^\top \Phi_{S_{k-1}}^\perp \mathbf{v}\|_\infty + \|\Phi_{\Omega^c}^\top \Phi_{S_{k-1}}^\perp \mathbf{v}\|_\infty.
\end{aligned} \tag{3.35}
$$

By the induction assumption $S_{k-1} \subseteq \Omega$, we have

$$|\operatorname{supp}(\mathbf{x}_{\Omega \setminus S_{k-1}})| = |\Omega| + 1 - k.$$

Thus,

$$\|\mathbf{x}_{\Omega \setminus S_{k-1}}\|_2 \geq \sqrt{|\Omega| + 1 - k} \min_{i \in \Omega \setminus S_{k-1}} |x_i| \geq \sqrt{|\Omega| + 1 - k} \min_{i \in \Omega} |x_i|. \tag{3.36}$$

In the following, we shall give a lower bound on the left-hand side of (3.35) and an upper bound on the right-hand side of (3.35). To bound the left-hand side of (3.35), we use Lemma 3.7 and (3.15) to obtain, since $S_{k-1} \subseteq \Omega$ and $|S_{k-1}| = k - 1$,

$$
\begin{aligned}
&\|\Phi_{\Omega \setminus S_{k-1}}^\top \Phi_{S_{k-1}}^\perp \Phi_{\Omega \setminus S_{k-1}} \mathbf{x}_{\Omega \setminus S_{k-1}}\|_\infty - \|\Phi_{\Omega^c}^\top \Phi_{S_{k-1}}^\perp \Phi_{\Omega \setminus S_{k-1}} \mathbf{x}_{\Omega \setminus S_{k-1}}\|_\infty \\
&\geq \frac{(1 - \sqrt{|\Omega| + 1 - k + 1}\delta_{|\Omega|+1})\|\mathbf{x}_{\Omega \setminus S_{k-1}}\|_2}{\sqrt{|\Omega| + 1 - k}} \\
&\geq \frac{(1 - \sqrt{s+1}\delta_{|\Omega|+1})\|\mathbf{x}_{\Omega \setminus S_{k-1}}\|_2}{\sqrt{|\Omega| + 1 - k}} \\
&\geq \frac{(1 - \sqrt{s+1}\delta_{s+1})\|\mathbf{x}_{\Omega \setminus S_{k-1}}\|_2}{\sqrt{|\Omega| + 1 - k}} \geq (1 - \sqrt{s+1}\delta_{s+1}) \min_{i \in \Omega} |x_i|, \tag{3.37}
\end{aligned}
$$

where we have used the fact that $k \geq 1$ and $\mathbf{x_b}$ is s-sparse (i.e., $|\Omega| \leq s$), taking into account the monotone property of δ_k and (3.27) and (3.36).

To give an upper bound on the right-hand side of (3.35), we notice that there exist $i_0 \in \Omega \setminus S_{k-1}$ and $j_0 \in \Omega^c$ such that

$$\|\Phi_{\Omega \setminus S_{k-1}}^\top \Phi_{S_{k-1}}^\perp \mathbf{v}\|_\infty = |\Phi_{i_0}^\top \Phi_{S_{k-1}}^\perp \mathbf{v}| \quad \text{and} \quad \|\Phi_{\Omega^c}^\top \Phi_{S_{k-1}}^\perp \mathbf{v}\|_\infty = |\Phi_{j_0}^\top \Phi_{S_{k-1}}^\perp \mathbf{v}|.$$

Therefore, we have

$$
\begin{aligned}
\|\Phi_{\Omega \setminus S_{k-1}}^\top \Phi_{S_{k-1}}^\perp \mathbf{v}\|_\infty + \|\Phi_{\Omega^c}^\top \Phi_{S_{k-1}}^\perp \mathbf{v}\|_\infty &= |\Phi_{i_0}^\top \Phi_{S_{k-1}}^\perp \mathbf{v}| + |\Phi_{j_0}^\top \Phi_{S_{k-1}}^\perp \mathbf{v}| \\
&= \|\Phi_{i_0 \cup j_0}^\top \Phi_{S_{k-1}}^\perp \mathbf{v}\|_1 \leq \sqrt{2}\|\Phi_{i_0 \cup j_0}^\top \Phi_{S_{k-1}}^\perp \mathbf{v}\|_2 \\
&\leq \sqrt{2(1 + \delta_{s+1})}\|\Phi_{S_{k-1}}^\perp \mathbf{v}\|_2 \leq \sqrt{2(1 + \delta_{s+1})}\epsilon,
\end{aligned}
\tag{3.38}
$$

where we have used the Cauchy–Schwarz inequality.

From (3.37) and (3.38), we can see that (3.35) (or equivalently (3.30)) is guaranteed by

$$(1 - \sqrt{s+1}\delta_{s+1}) \min_{i \in \Omega} |x_i| > \sqrt{2(1 + \delta_{s+1})}\epsilon,$$

i.e., by (3.27),

$$\min_{i \in \Omega} |x_i| > \frac{\sqrt{2(1 + \delta_{s+1})}\epsilon}{1 - \sqrt{s+1}\delta_{s+1}}. \tag{3.39}$$

Thus, if (3.28) holds, then Algorithm 3.2 selects a correct index in the kth iteration.

Next we need to show that Algorithm 3.2 performs exactly $|\Omega|$ iterations, which is equivalent to showing that $\|R^k\|_2 > \epsilon$ for $1 \leq k < |\Omega|$ and $\|R^{|\Omega|}\|_2 \leq \epsilon$. Since Algorithm 3.2 selects a correct index in each iteration under (3.28), by (3.32), for $1 \leq k < |\Omega|$, we have

$$
\begin{aligned}
\|R^k\|_2 &= \|\Phi_{S_k}^\perp \Phi_{\Omega \setminus S_k} \mathbf{x}_{\Omega \setminus S_k} + \Phi_{S_k}^\perp \mathbf{v}\|_2 \\
&\geq \|\Phi_{S_k}^\perp \Phi_{\Omega \setminus S_k} \mathbf{x}_{\Omega \setminus S_k}\|_2 - \|\Phi_{S_k}^\perp \mathbf{v}\|_2 \\
&\geq \|\Phi_{S_k}^\perp \Phi_{\Omega \setminus S_k} \mathbf{x}_{\Omega \setminus S_k}\|_2 - \epsilon \\
&\geq \sqrt{1 - \delta_{|\Omega|}}\|\mathbf{x}_{\Omega \setminus S_k}\|_2 - \epsilon \geq \sqrt{1 - \delta_{s+1}} \min_{i \in \Omega} |x_i| - \epsilon,
\end{aligned}
\tag{3.40}
$$

where we have used Lemma 3.6 and the monotone property of the RIC. Therefore, if

$$\min_{i \in \Omega} |x_i| > \frac{2\epsilon}{\sqrt{1 - \delta_{s+1}}}, \tag{3.41}$$

then $\|R^k\|_2 > \epsilon$ for $1 \leq k < \Omega$. By a simple calculation, we can show that

$$\frac{2\epsilon}{1 - \sqrt{s+1}\delta_{s+1}} \geq \frac{2\epsilon}{\sqrt{1 - \delta_{s+1}}}. \tag{3.42}$$

Indeed, we have

$$1 - \sqrt{s+1}\delta_{s+1} \leq 1 - \delta_{s+1} \leq \sqrt{1 - \delta_{s+1}}. \tag{3.43}$$

Thus, (3.42) follows. Therefore, by (3.41) and (3.42), if (3.28) holds, then $\|R^k\|_2 > \epsilon$ for $1 \leq k < \Omega$, i.e., Algorithm 3.2 does not terminate before the $|\Omega|$th iteration.

Similarly, by (3.32) and when $S_{|\Omega|} = |\Omega|$, we have

$$\|R^{|\Omega|}\|_2 = \|\Phi_{S_{|\Omega|}}^\perp \Phi_{\Omega \setminus S_{|\Omega|}} \mathbf{x}_{\Omega \setminus S_{|\Omega|}} + \Phi_{S_{|\Omega|}}^\perp \mathbf{v}\|_2 = \|\Phi_{S_{|\Omega|}}^\perp \mathbf{v}\|_2 \leq \epsilon.$$

So, by the stopping condition, Algorithm 3.2 terminates after performing the $|\Omega|$th iteration in total. That is, Algorithm 3.2 performs $|\Omega|$ iterations.

Next let us prove the second part, i.e., (3.29) of Theorem 3.8. Since the support of the s-sparse signal $\mathbf{x}$ can be exactly recovered by Algorithm 3.2, we get

$$\widehat{\mathbf{x}}_{|\Omega|} = (\Phi_\Omega^\top \Phi_\Omega)^{-1} \Phi_\Omega^\top \mathbf{y} = (\Phi_\Omega^\top \Phi_\Omega)^{-1} \Phi_\Omega^\top (\Phi \mathbf{x_b} + \mathbf{v}) = \mathbf{x_b} + \Phi_\Omega^\dagger \mathbf{v}$$

by a property of the pseudo-inverse. Thus, we have

$$\|\mathbf{x_b} - \widehat{\mathbf{x}}_s\|_2 = \|\Phi_\Omega^\dagger \mathbf{v}\|_2 \leq \|\mathbf{v}\|_2 \leq \epsilon.$$

This completes the proof.   □

3.4 ▪ Convergence of Algorithms 3.2 and 3.8

In this section we first discuss some basic properties of Algorithm 3.2.

Lemma 3.9. $\langle R^k, \phi_i \rangle = 0$ for $i \in S_k$.

Proof. Since $\hat{\mathbf{x}}_k$ is the optimal solution of the problem in Step 4 of Algorithm 3.2, by the first-order optimality condition, we have

$$\left\langle \mathbf{b} - \sum_{i \in S_k} x_i \phi_i, \phi_j \right\rangle = 0 \ \text{ for } \ j \in S_k,$$

which together with $R^k = \mathbf{b} - \sum_{i \in S_k} x_i \phi_i$ implies that $\langle R^k, \phi_j \rangle = 0$ for $j \in S_k$.   □

Next we show that the residual $\|R^k\|$ does not increase.

Lemma 3.10. $\|R^k\| \leq \|R^{k-1}\|$ for all $k \geq 1$.

Proof. For all $k \geq 1$, it is clear that

$$\begin{aligned}
\|R^k\|^2 &= \min_{\mathbf{x}} \left\{ \|\mathbf{b} - \Phi_{S_k} \mathbf{x}\|^2 \right\} \\
&= \min_{\mathbf{x}} \left\{ \|\mathbf{b} - \Phi_{S_{k-1} \cup \{s^k\}} \mathbf{x}\| \right\} \leq \min_{\mathbf{x}} \left\{ \|\mathbf{b} - \Phi_{S_{k-1}} \mathbf{x}\| \right\} = \|R^{k-1}\|^2,
\end{aligned}$$

as desired.   □

Recall that $R^k = \mathbf{b} - \Phi_{S_k} \hat{\mathbf{x}}_k$, and let us write $L_k(\mathbf{b}) = \Phi_{S_k} \hat{\mathbf{x}}_k = [\Phi_{S_{k-1}}, \phi_{s^k}] \hat{\mathbf{x}}_k$.

Lemma 3.11. Suppose that the sensing matrix $\Phi = [\phi_1, \phi_2, \ldots, \phi_n]$ is a dictionary. That is, $\|\phi_i\|_1 = 1$ for all $i = 1, \ldots, m$. Then $\|R^k\|^2 \leq \|R^{k-1}\|^2 - \langle R^{k-1}, \phi_{s^k} \rangle^2$ for all $k \geq 1$.

Proof. We observe that

$$\begin{aligned}
\|R^k\|^2 &= \min_{\alpha \in \mathbb{R}^k} \left\| \mathbf{b} - \sum_{i \in S_{k-1}} \alpha_i \phi_i - \alpha_k \phi_{s^k} \right\|^2 \\
&\leq \min_{\alpha_2 \in \mathbb{R}} \|\mathbf{b} - L_{k-1}(\mathbf{b}) - \alpha_2 \phi_{s^k}\|^2 = \min_{\alpha_2 \in \mathbb{R}} \|R^{k-1} - \alpha_2 \phi_{s^k}\|^2.
\end{aligned}$$

This has a closed-form solution because $\alpha_2^* = \frac{\langle R^{k-1}, \phi_{s^k} \rangle}{\langle \phi_{s^k}, \phi_{s^k} \rangle} = \langle R^{k-1}, \phi_{s^k} \rangle$. Plugging this optimum α_2^* back into the formulation, we get

$$\|R^k\|^2 \leq \|R^{k-1}\|^2 - 2\langle R^{k-1}, \phi_{s^k} \rangle \alpha_2^* + (\alpha_2^*)^2 \|\phi_{s^k}\|^2$$

$$= \|R^{k-1}\|^2 - (\langle R^{k-1}, \phi_{s^k} \rangle)^2.$$

This completes the proof. $\square$

Next we show the following result.

Lemma 3.12. *Suppose that $R^{k-1} \neq 0$ for some $k \geq 1$. Then the vectors $\phi_i, i \in S_k$, are linearly independent.*

Proof. We shall use induction. Since $R^0 = b \neq 0$, we clearly have $\phi_{s_1} \neq 0$, which is linearly independent, so the conclusion holds for $k = 1$. We next assume that it holds for $k < i$ and need to show that it also holds for $k = i$. By the induction hypothesis, $\{\phi_j, j \in S_{i-1}\}$ are linearly independent. Suppose that $R^{i-1} \neq 0$. Let S_i be the set of indices in Step 3 of Algorithm 3.2. Suppose for contradiction that $\phi_j, j \in S_i$, are not linearly independent. That is, there exist $\theta_1, \ldots, \theta_{i-1} \in \mathbb{R}$ which are not all zero such that

$$\phi_{s^i} = \sum_{j \in S_{i-1}} \theta_j \phi_j.$$

We now use Lemma 3.9 to obtain $\langle R^{i-1}, \phi_{s^i} \rangle = 0$. It follows that

$$0 = |\langle \phi_{s^i}, R^{i-1} \rangle| = \max_{1 \leq j \leq n} |\langle \phi_j, R^{i-1} \rangle|,$$

which implies that $R^{i-1} = 0$ because $\text{span}\{\phi_1, \ldots, \phi_n\} = \mathbb{R}^m$. This contradicts the fact that $R^{i-1} \neq 0$. Therefore, $\{\phi_j\}_{j \in S_i}$ is linearly independent, which completes the proof. $\square$

We finally show that the sequence obtained from Algorithm 3.2 is linearly convergent.

Theorem 3.13. *Suppose that there is a sparse solution x_b with sparsity $s = \|x_b\|_0$ satisfying $\Phi x_b = b$. Suppose that Φ has the $2s$ RIC $\delta_{2s} < 1$. Then each iteration in Algorithm 3.2 satisfies*

$$\|R^{k+1}\|^2 \leq \left(1 - \frac{1 - \delta_{2s}^\Phi}{2s}\right)^k \|b\|^2 \ \forall k \geq 0,$$

where δ_{2s}^Φ is the RIC of Φ.

Proof. Let us write $R^k = b - L_k(b) = \Phi(x_b - \hat{x}_k)$, with the entries of $\hat{x}_k$ being zero for indices in S_k^c, the complement of S_k. Then

$$\|R^k\|^2 = \langle R^k, \Phi(x_b - \hat{x}_k) \rangle = \left\langle R^k, \sum_{i=1}^{n} (x_b(i) - \hat{x}_k(i))\phi_i \right\rangle.$$

For simplicity, we write $\alpha \in \mathbb{R}^n$ with $\alpha(i) = x_b(i) - \hat{x}_k(i)$ for $i \in \{1, 2, \ldots, n\}$ with sparsity $\leq 2s$. Then

$$\|R^k\|^2 \leq \|\alpha\|_1 \cdot \max_i |\langle R^k, \phi_i \rangle| \leq \sqrt{2s} \|\alpha\|_2 \cdot |\langle R^k, \phi_{s_{k+1}} \rangle|, \tag{3.44}$$

where the first inequality follows from the Cauchy–Schwarz inequality $|\langle a, b\rangle| \le \|a\|_1 \cdot \|b\|_\infty$ and Step 2 in Algorithm 3.2, and the second inequality follows from $\|\alpha\|_1 \le \sqrt{2s}\|\alpha\|_2$ because $\alpha \in \mathbb{R}^n$ is a $2s$-sparse vector. On the other hand,

$$\|R^k\|^2 = \langle \mathbf{b} - L_k(\mathbf{b}), \mathbf{b} - L_k(\mathbf{b})\rangle = \|\Phi\alpha\|^2 \ge \left(1 - \delta_{2s}^\Phi\right)\|\alpha\|_2^2, \qquad (3.45)$$

where the last inequality follows from the RIP. Combining (3.44) and (3.45) leads to

$$|\langle R^k, \phi_{s_{k+1}}\rangle|^2 \ge \frac{1 - \delta_{2s}^\Phi}{2s}\|R^k\|^2.$$

Using this inequality and Lemma 3.11, we obtain

$$\|R^{k+1}\|^2 \le \|R^k\|^2 - (\langle R^k, \phi_{s_{k+1}}\rangle)^2 \le \|R^k\|^2 - \frac{1 - \delta_{2s}^\Phi}{2s}\|R^k\|^2 = \left(1 - \frac{1 - \delta_{2s}^\Phi}{2s}\right)\|R^k\|^2.$$

Note that since $R^0 = \mathbf{b}$, we obtain

$$\|R^{k+1}\|^2 \le \left(1 - \frac{1 - \delta_{2s}^\Phi}{2s}\right)^k \|\mathbf{b}\|^2.$$

This completes the proof.   □

In the previous section, we explained how to recover the sparse vector $\mathbf{x_b}$ with sparsity s in $s + 1$ iterations of Algorithm 3.2. The condition for this recovery is $\delta_{s+1} < 1/\sqrt{s+1}$, which is very small for large s. The results in the theorem above show that we can simply do more iterations to recover the sparse solution when $\delta_{2s} < 1$.

The above study can be carried out for Algorithm 3.8. Indeed, we have similar properties and convergence. Mainly, we have Lemmas 3.32, 3.33, and 3.34 and convergence results in Theorem 3.36 in the exercise section. We leave the proof of these properties and convergence to the reader.

The analysis in the rest of this section is motivated by the work in [2]. It mainly shows that the limit of the sequence obtained from Algorithm 3.2 is a global minimizer. Consider the s-sparse class of the span of the dictionary $\mathcal{D}$:

$$\Sigma_s := \left\{ \sum_{c_i, i \in S} c_i \phi_i : \quad S \subset \mathcal{D}, \#(S) \le s \right\}. \qquad (3.46)$$

For any $\mathbf{y} \in \operatorname{span}\{\phi_i, \quad \phi_i \in \mathcal{D}\}$, let

$$\sigma_s(\mathbf{y}) = \inf_{g \in \Sigma_s} \|\mathbf{y} - g\|_2. \qquad (3.47)$$

We first have the following result.

Lemma 3.14. *Let $\mathcal{H}$ be a Hilbert space spanned by a dictionary D whose sensing matrix $\Phi = [\phi_1, \dots, \phi_n]$ satisfies the RIP $\delta_{2s} < 1$. Then*

- *the set Σ_s is closed in $\mathbb{R}^m$;*

- *for any $\mathbf{y} \in \mathbb{R}^m$, for any $\epsilon > 0$, there exists a $g_\epsilon \in \Sigma_s$ such that*

$$\|\mathbf{y} - g_\epsilon\| \le (1 + \epsilon)\sigma_s(\mathbf{y}). \qquad (3.48)$$

Proof. Suppose that $g_k \in \Sigma_s$, $k \geq 1$, converge to g^*. We claim $g^* \in \Sigma_s$. Since $g_k = \sum_i c_{k,i} \phi_i$ with only the s coefficients $c_{k,i}$ nonzero, letting $\mathbf{x}_k = (c_{k,i}, i = 1, \ldots, n\}$, we use the RIP to get

$$\|\mathbf{x}_k - \mathbf{x}_m\|^2 \leq \frac{1}{1 - \delta_{2s}} \|g_k - g_m\|^2 \to 0$$

since $\{g_k, k = 1, \ldots\}$ is a Cauchy sequence. The coefficient sequence $\{\mathbf{x}_k, k \geq 1\}$ is a Cauchy sequence, and hence $\mathbf{x}_k \to \mathbf{x}^*$. It is easy to see that $\mathbf{x}^* \in \Sigma_s$.

For any $\mathbf{y} \in \mathbb{R}^m$, let $g_k \in \Sigma_s$ be a sequence of approximations of $\mathbf{y}$ by the least squares method. Then we can find g_ϵ to achieve (3.48). $\qquad\square$

We next show the following result.

Lemma 3.15. *Fix an $\epsilon > 0$. Let $g_\epsilon \in \Sigma_s$ satisfy (3.48). Let S_ϵ be the support set for g_ϵ. That is, $g_\epsilon = \Phi(\mathbf{x_b} + \eta)$ with S_ϵ being the support of $\mathbf{x_b} + \eta$. Let $\hat{\mathbf{x}}_k$, $k = 1, 2, \ldots$, be the sequence from Algorithm 3.2, with S_k being its support set. Then*

$$\max\{0, \|R^k\|^2 - \|\mathbf{y} - g_\epsilon\|^2\} \leq \left(1 - \frac{1 - \delta}{\#(S_\epsilon \backslash S_k)}\right) \max\{0, \|R^{k-1}\|^2 - \|\mathbf{y} - g_\epsilon\|^2\}, \quad (3.49)$$

where $\delta \leq \delta_{S_\epsilon \cup S_k} \leq \delta_{2s} < 1$.

Proof. Using Algorithm 3.2, we have

$$\|R^{k+1}\|^2 = \|\mathbf{y} - L_{k+1}(\mathbf{y})\|^2 = \|\mathbf{y} - L_k(\mathbf{y})\|^2 - \|L_k(\mathbf{y}) - L_{k+1}(\mathbf{y})\|^2 \quad (3.50)$$
$$\leq \|R^k\|^2 - |\langle R^k, \phi_{i_{k+1}}\rangle|^2. \quad (3.51)$$

Writing $\mathbf{y}_k = L_{k+1}(\mathbf{y})$ for convenience, since $R^k = \mathbf{y} - \mathbf{y}_k$, we use the Cauchy–Schwarz inequality to obtain

$$2\|\mathbf{y}_k - g_\epsilon\|\sqrt{\|R^k\|^2 - \|\mathbf{y} - g_\epsilon\|^2} \leq \|\mathbf{y}_k - g_\epsilon\|^2 + \|R^k\|^2 - \|\mathbf{y} - g_\epsilon\|^2$$
$$= \|\mathbf{y}_k - g_\epsilon\|^2 + \|R^k\|^2 - \|\mathbf{y}_k - g_\epsilon + R^k\|^2$$
$$\leq 2|\langle \mathbf{y}_k - g_\epsilon, R^k\rangle| = 2|\langle g_\epsilon, R^k\rangle|$$

because $\mathbf{y}_k$ is orthogonal to R^k. Thus, it follows that

$$\|R^k\|^2 - \|\mathbf{y} - g_\epsilon\|^2 \leq \frac{|\langle g_\epsilon, R^k\rangle|^2}{\|g_\epsilon - \mathbf{y}_k\|^2}. \quad (3.52)$$

On the other hand, if $g_\epsilon = \Phi\mathbf{x_b} + \epsilon = \Phi\mathbf{x_b} + \Phi\eta = \Phi(\mathbf{x_b} + \eta)$ for some vector η, with S_ϵ being the support of $\mathbf{x_b} + \eta$,

$$|\langle g_\epsilon, R^k\rangle| = |\langle \mathbf{y}_k - g_\epsilon, R^k\rangle| = |\langle \Phi_{S_k}\hat{\mathbf{x}}_k - g_\epsilon, R^k\rangle| = |\langle \mathbf{x_b} + \eta - \hat{\mathbf{x}}_k, \Phi^\top R^k\rangle|$$
$$\leq \|\mathbf{x_b} + \eta - \hat{\mathbf{x}}_k\|_1 \|\Phi^\top R^k\|_\infty$$
$$\leq \sqrt{\#(S_\epsilon \cup S_k)}\|\mathbf{x_b} + \eta - \hat{\mathbf{x}}_k\|_2 |\langle R^k, \phi_{i_{k+1}}\rangle|.$$

Letting $\delta = \delta_{\#(S_k \cup S_\epsilon)}$ for short, we have

$$\|g_\epsilon - \mathbf{y}_k\|^2 = \|\Phi(\mathbf{x_b} + \eta) - \Phi\hat{\mathbf{x}}_k\|^2 \geq (1 - \delta)\|\mathbf{x_b} + \eta - \hat{\mathbf{x}}_k\|^2,$$

and hence

$$\|R^k\|^2 - \|\mathbf{y} - g_\epsilon\|^2 \leq \frac{\#(S_\epsilon \cup S_k)|\langle R^k, \phi_{i_{k+1}}\rangle|^2\|\mathbf{x_b} + \eta - \hat{\mathbf{x}}_k\|^2}{(1 - \delta)\|\mathbf{x_b} + \eta - \hat{\mathbf{x}}_k\|^2}$$
$$= \frac{\#(S_\epsilon \cup S_k)}{1 - \delta}|\langle R^k, \phi_{i_{k+1}}\rangle|^2.$$

Or we have

$$|\langle R^k, \phi_{i_{k+1}} \rangle|^2 \geq \frac{1-\delta}{\#(S_\epsilon \cup S_k)} \left(\|R^k\|^2 - \|\mathbf{y} - g_\epsilon\|^2 \right).$$

Hence, using (3.51), we obtain

$$\|R^{k+1}\|^2 \leq \|R^k\|^2 - \frac{1-\delta}{\#(S_\epsilon \cup S_k)} \left(\|R^k\|^2 - \|\mathbf{y} - g_\epsilon\|^2 \right).$$

If $\|R^{k+1}\| \leq \|\mathbf{y} - g_\epsilon\|$, we are done. Otherwise, rearranging the above inequality and subtracting $\|\mathbf{y} - g_\epsilon\|^2$ from both sides, we finally get the monotone relation

$$\|R^{k+1}\|^2 - \|\mathbf{y} - g_\epsilon\|^2 \leq \left(1 - \frac{1-\delta}{\#(S_\epsilon \cup S_k)} \right) \left(\|R^k\|^2 - \|\mathbf{y} - g_\epsilon\|^2 \right) \tag{3.53}$$

for $k \geq 1$. We combine both cases to complete the proof. □

To state our main theorem on the convergence of Algorithm 3.2, we recall the definition of the sensing matrix. It was defined before in Chapter 2, but we redefine it here for convenience.

Definition 3.16. *A sensing matrix* Φ *of size* $m \times n$ *is of completely full rank if any submatrix* A *of size* $m \times m$ *of* Φ *is full rank.*

Therefore, we conclude with the following result.

Theorem 3.17 (Cohen, Dahmen, and DeVore, 2017 [2]). *Suppose that* Φ *is of completely full rank. Let* R^k *be the* k*th residual vector from Algorithm 3.2. Then either* $\|R^k\| \leq \|\mathbf{y} - g_\epsilon\|$ *or*

$$\|R^k\| \to \|\mathbf{y} - g_\epsilon\|, \quad k \to \infty. \tag{3.54}$$

Proof. If for some k we have $\|R^k\| \leq \|\mathbf{y} - g_\epsilon\|$, we are done. Otherwise, the completely full rankness of Φ implies that $1 - \delta \geq \delta_0 > 0$ for all $\delta_{\#(S_\epsilon \cup S_k)}$ and $\#(S_\epsilon \cup S_k) \leq k + s$, where $s = |S_\epsilon|$. It is easy to see that

$$1 - \frac{1-\delta}{\#(S_\epsilon \cup S_k)} \leq 1 - \frac{\delta_0}{k+s}$$

and

$$\|R^{k+1}\|^2 - \|\mathbf{y} - g_\epsilon\|^2 \leq \prod_{j=0}^{k} \left(1 - \frac{\delta_0}{j+s} \right) \left(\|R^0\|^2 - \|\mathbf{y} - g_\epsilon\|^2 \right).$$

Since $\prod_{j=0}^{k} \left(1 - \frac{\delta_0}{j+s} \right) \to 0$ as $k \to \infty$, we conclude the result in this theorem. □

In fact, we have

$$\|R^{k+1}\|^2 \leq \|\mathbf{y} - g_\epsilon\|^2 + \prod_{j=0}^{k} \left(1 - \frac{\delta_0}{j+s} \right) \left(\|R^0\|^2 - \|\mathbf{y} - g_\epsilon\|^2 \right) \leq 2\|\mathbf{y} - g_\epsilon\|^2$$

if $\|\mathbf{y} - g_\epsilon\|^2 > 0$ and k is large enough. For example, we may take k large enough such that

$$\prod_{j=0}^{k} \left(1 - \frac{\delta_0}{j+s} \right) \|R^0\|^2 \leq \|\mathbf{y} - g_\epsilon\|^2.$$

It is easy to see that $\mathbf{y}^k$ are bounded and hence have a convergent subsequence with cluster point $\mathbf{y}^*$. By Lemma 3.14, we know that $\mathbf{y}^* \in \Sigma_s$, and hence that $\mathbf{y}^*$ is a sparse solution.

3.5 ▪ Convergence Analysis of Algorithm 3.5

In this section we present a convergence analysis of Algorithm 3.5. That is, we shall give a proof of Theorem 3.20 below. We begin with a basic lemma which is a different version of Lemma 2.30.

Lemma 3.18. *Suppose that the sensing matrix Φ satisfies the RIP with RIC $\delta_s < 1$. Then for any index set T with $\#(T) \leq s$, the symmetric matrix $(\Phi_T)^\top \Phi_T$ is positive definite, the largest eigenvalue $\lambda_1 \leq 1 + \delta_s$, and the smallest eigenvalue $\lambda_s \geq 1 - \delta_s$.*

Proof. Let $\mathbf{y} \in \mathbb{R}^s$ be an eigenvector associated with the smallest eigenvalue λ_s. We claim that $\lambda_s > 0$. Indeed, since $(\Phi_T)^\top \Phi_T \mathbf{y} = \lambda_s \mathbf{y}$, we have

$$\|\Phi_T \mathbf{y}\|_2^2 = \mathbf{y}^\top (\Phi_T)^\top \Phi_T \mathbf{y} = \lambda_s \mathbf{y}^\top \mathbf{y} = \lambda_s \|\mathbf{y}\|_2^2.$$

Extend $\mathbf{y}$ to be a vector in $\mathbb{R}^n$ whose entries in the index set T are $\mathbf{y}$. For convenience, we still use $\mathbf{y}$ to denote this vector. The above equality can be rewritten as

$$\lambda_s \|\mathbf{y}_T\|_2^2 = \|\Phi_T \mathbf{y}_T\|_2^2 \geq (1 - \delta_s) \|\mathbf{y}_T\|_2^2$$

by using the RIP. Thus, $\lambda_s \geq 1 - \delta_s > 0$. The proof for the largest eigenvalue is similar.   □

Next we need another elementary lemma.

Lemma 3.19 (DeVore and Temlyakov, 1996 [4]). *Suppose we have two sequences of nonnegative numbers, $\{a_k, k \geq 1\}$ and $\{\beta_k, k \geq 1\}$, satisfying $\beta_k > 0$ and $a_1 = 1$ and*

$$a_{m+1} \leq a_m (1 - a_m \beta_m) \quad \forall m \geq 1. \tag{3.55}$$

Then

$$a_{m+1} \leq \frac{1}{1 + \sum_{k=1}^m \beta_k} \quad \forall m \geq 1. \tag{3.56}$$

Proof. Clearly, $a_{m+1} \leq a_m$ for all $m \geq 1$. If $a_{m+1} > 0$, so are a_k for all $1 \leq k \leq m$. If $a_{m+1} = 0$, so are a_k for all $k \geq m + 1$ and hence the inequality in (3.56) holds for all $k \geq m + 1$. Without loss of generality, we may assume that $a_k > 0$ for all $k \geq 1$. From inequality (3.55), we have

$$\frac{1}{a_{m+1}} \geq \frac{1}{a_m} \frac{1}{1 - \beta_m a_m} \geq \frac{1}{a_m} (1 + \beta_m a_m) \geq \frac{1}{a_m} + \beta_m.$$

It follows that

$$\frac{1}{a_{m+1}} \geq \cdots \geq \frac{1}{a_1} + \sum_{k=1}^m \beta_m \geq 1 + \sum_{k=1}^m \beta_m,$$

and hence the inequality in (3.56) holds.   □

Note that for any $\mathbf{y} \in \mathcal{D}$, we can normalize $\mathbf{y}$ by $\mathbf{y}/\|\mathbf{y}\|_{\Phi,1}$, where $\|\mathbf{y}\|_{\Phi,1} = \min\{\sum_{i=1}^n |c_i|,$ $\mathbf{y} = \Phi\mathbf{c}, \mathbf{c} = \{c_i, i = 1, \ldots, n\}\}$. Note that when $\mathcal{D}$ is a dictionary, $\mathbf{b} \in \mathcal{D}$ implies that $\|\mathbf{y}\|_2 \leq 1$ if $\|\mathbf{b}\|_{\Phi,1} \leq 1$. Let

$$A_1(\mathcal{D}) = \{\mathbf{y} \in \text{span}(\mathcal{D}), \|\mathbf{y}\|_{\Phi,1} \leq 1\}. \tag{3.57}$$

Our main result in this section is as follows.

Theorem 3.20 (Convergence of Algorithm 3.5). *Let* $\mathbf{y} \in A_1(\mathcal{D})$. *Suppose that* Φ *satisfies the RIP of order* $s > 1$ *with* $\delta_s < 1$. *With* $n_\ell \leq s$,

$$\|\widehat{R^\ell}(\mathbf{y})\|^2 \leq \frac{1}{1 + \sum_{j=1}^{\ell} t_j^2 \frac{n_j - n_{j-1}}{|S_{n_j}|}} \leq \frac{2}{2 + \hat{t}^2 n_\ell}, \tag{3.58}$$

where $\hat{t} = \min\{t_1, \ldots, t_\ell\}$.

Proof. Without loss of generality, we may assume that $S_k = \{1, 2, \ldots, n_k\}$ with index $n_k \leq m$. We first observe that the best approximation $L_{S_k}(\mathbf{y})$ of $\widehat{R^k}(\mathbf{y})$ is

$$L_{S_k}(\mathbf{y}) = \Phi_{S_k}(\Phi_{S_k}^\top \Phi_{S_k})^{-1} \left[\langle \widehat{R^{k-1}}(\mathbf{y}), \phi_i \rangle, 1 \leq i \leq n_k \right]^\top.$$

We next note that for $i \in S_k \backslash S_{k-1}$,

$$|\langle \widehat{R^{k-1}}(\mathbf{y}), \phi_i \rangle| \geq t_k M_k,$$

with

$$M_k = \max_{i \notin S_{k-1}} |\langle \widehat{R^{k-1}}(\mathbf{y}), \phi_i \rangle| = \max_{i=1,\ldots,n} |\langle \widehat{R^{k-1}}(\mathbf{y}), \phi_i \rangle|.$$

It follows that

$$|\langle \widehat{R^{k-1}}(\mathbf{y}), \mathbf{y} \rangle| = \left| \sum_{i=1}^{n} c_i \langle \widehat{R^{k-1}}(\mathbf{y}), \phi_i \rangle \right| \leq \sum_{i=1}^{n} |c_i| |\langle \widehat{R^{k-1}}(\mathbf{y}), \phi_i \rangle| \leq M_k,$$

since $\sum_{i=1}^{n} |c_i| \leq 1$ due to the assumption on $\mathbf{y}$. Hence,

$$M_k \geq |\langle \widehat{R^{k-1}}(\mathbf{y}), \mathbf{y} \rangle| = \|\widehat{R^{k-1}}(\mathbf{y})\|^2$$

by the property of the best approximation of $\widehat{R^{k-1}}(\mathbf{y})$. Thus we have

$$\|\widehat{R^k}(\mathbf{y})\|^2 = \langle \widehat{R^{k-1}}(\mathbf{y}) - L_{S_k}(\mathbf{y}), \widehat{R^{k-1}}(\mathbf{y}) - L_{S_k}(\mathbf{y}) \rangle = \|\widehat{R^{k-1}}(\mathbf{y})\|^2 - \|L_{S_k}(\mathbf{y})\|^2.$$

On the other hand,

$$\|L_{S_k}(\mathbf{y})\|_2^2 = \left[\langle \widehat{R^{k-1}}(\mathbf{y}), \phi_i \rangle, i = 1, \ldots, n_k \right] (\Phi_{S_k}^\top \Phi_{S_k})^{-1} \left[\langle \widehat{R^{k-1}}(\mathbf{y}), \phi_i \rangle, i = 1, \ldots, n_k \right]^\top$$

$$\geq \frac{1}{\lambda_{n_k}} \left\| \left[\langle \widehat{R^{k-1}}(\mathbf{y}), \phi_i \rangle, i = 1, \ldots, n_k \right]^\top \right\|^2$$

$$\geq \frac{1}{\lambda_{n_k}} (n_k - n_{k-1}) t_k^2 M_k^2 \geq \frac{1}{1 + \delta_{n_k}} t_k^2 (n_k - n_{k-1}) \|\widehat{R^{k-1}}(\mathbf{y})\|_2^4$$

by Lemma 3.18 on the estimate of the positive definite symmetric matrix $\Phi_{S_k}^\top \Phi_{S_k}$. That is,

$$\|\widehat{R^k}(\mathbf{y})\|_2^2 = \|\widehat{R^{k-1}}(\mathbf{y})\|_2^2 - \|L_{S_k}(\mathbf{y})\|_2^2$$

$$\leq \|\widehat{R^{k-1}}(\mathbf{y})\|_2^2 - \frac{1}{1 + \delta_{n_k}} t_k^2 (n_k - n_{k-1}) \|\widehat{R^{k-1}}(\mathbf{y})\|_2^4$$

$$\leq \|\widehat{R^{k-1}}(\mathbf{y})\|_2^2 \left(1 - \frac{t_k^2 (n_k - n_{k-1})}{2} \|\widehat{R^{k-1}}(\mathbf{y})\|_2^2 \right)$$

because $\delta_{n_k} \leq 1$. We now use Lemma 3.19, more precisely the inequality in (3.56), to conclude that

$$\|\widehat{R^\ell}(\mathbf{y})\|_2^2 \leq \frac{1}{1 + \sum_{k=1}^{\ell} \frac{t_k^2(n_k - n_{k-1})}{2}} \leq \frac{2}{2 + \hat{t}^2 n_\ell}.$$

This completes the proof of the convergence of Algorithm 3.5.   □

The results in Lemma 3.19 can be slightly reformulated as follows.

Lemma 3.21. *Suppose we have two sequences of nonnegative numbers, $\{a_k, k \geq 1\}$ and $\{\beta_k, k \geq 1\}$, satisfying $\beta_k \geq \beta_0$ for all $k \geq 1$ and (3.55). Then*

$$\beta_0 a_{m+1}^2 = o\left(\frac{1}{m+1}\right) \quad \forall m \geq 1. \tag{3.59}$$

Proof. Rewriting inequality (3.55) as $\beta_0 a_m^2 + a_{m+1} \leq a_m$, we have

$$\sum_{m=1}^{i} \beta_0 a_m^2 \leq a_1$$

for all i and hence $\sum_{m \geq 1} a_m^2 < \infty$. Furthermore, we know that $a_{m+1} \leq a_m$ for all m, and hence

$$\ell \beta_0 a_{2\ell}^2 \leq \sum_{m=\ell}^{2\ell} \beta_0 a_m^2 \longrightarrow 0.$$

It follows that $\beta_0 a_{2\ell}^2 = o(1/\ell) = o(1/(2\ell))$. The proof is similar for odd indices.   □

The convergence of Theorem 3.20 can also be improved.

Theorem 3.22 (Convergence of Algorithm 3.5). *Suppose that $\Phi = [\phi_1, \ldots, \phi_n]$ is a sensing matrix with $\|\phi_i\| = 1$ for $i = 1, \ldots, n$. Let $\mathbf{x_b}$ be a sparse solution satisfying $\Phi\mathbf{x_b} = \mathbf{b}$. Suppose that Algorithm 3.5 uses a fixed number $t_\ell = t_0$ for all iterations. Then*

$$\|R^\ell(\mathbf{b})\|^2 = o\left(\frac{1}{\sqrt{\ell+1}}\right) \tag{3.60}$$

for $\ell \to \infty$.

Proof. It is easy to see that since $R^k(\mathbf{b}) = \mathbf{b} - \sum_{j=1}^{k} \Phi_{S_j} L_j(\mathbf{b})$, we know that $\langle R^k(\mathbf{b}), \phi_j \rangle = 0$ for all $j \in S_k$ by the best approximation property of L_k over $H_k = \text{span}\{\phi_j, j \in S_k\}$. It follows that $\|R^\ell(\mathbf{b})\|^2 = \langle R^\ell(\mathbf{b}), \mathbf{b} - \Phi_{j \in S_\ell} L_j(\mathbf{b}) \rangle = \langle R^\ell, \mathbf{b} \rangle$. Writing $\mathbf{b} = \Phi\mathbf{x_b}$ with $\mathbf{x_b} = (x_1, \ldots, x_n)^\top$ and $\|\mathbf{x_b}\|_0 \leq s$, we have

$$\|R^\ell(\mathbf{b})\|^2 = \sum_{i=1}^{n} x_i \langle R^\ell, \phi_j \rangle \leq \|\mathbf{x_b}\|_1 \max_{j=1,\ldots,n} |\langle R^\ell, \phi_j \rangle| \leq t_\ell^{-1} \|\mathbf{x_b}\|_1 |\langle R^\ell, \phi_{n^\ell} \rangle|.$$

On the other hand, as before,

$$\begin{aligned}
\|R^k(\mathbf{b})\|^2 &= \min_{\alpha \in \mathbb{R}^k} \|\mathbf{b} - \sum_{i \in S_{k-1}} \alpha_i \phi_i - \alpha_k \phi_{s^k}\|^2 \\
&\leq \min_{\alpha_2 \in \mathbb{R}} \|\mathbf{b} - L_{k-1}(\mathbf{b}) - \alpha_2 \phi_{s^k}\|^2 \\
&= \min_{\alpha_2 \in \mathbb{R}} \|R^{k-1} - \alpha_2 \phi_{s^k}\|^2 = \|R^{k-1}\|^2 - (\langle R^{k-1}, \phi_{s^k} \rangle)^2.
\end{aligned}$$

Combining the two estimates above, we have

$$
\begin{aligned}
\|R^k(\mathbf{b})\|^2 &= \|R^{k-1}(\mathbf{b})\|^2 - (\langle R^{k-1}(\mathbf{b}), \phi_{s^k}\rangle)^2 \leq \|R^{k-1}(\mathbf{b})\|^2 - \frac{t_{k-1}^2}{\|\mathbf{x_b}\|_1^2}\|R^{k-1}(\mathbf{b})\|^4 \\
&= \|R^{k-1}(\mathbf{b})\|^2\left(1 - \frac{t_0^2}{\|\mathbf{x_b}\|_1^2}\|R^{k-1}(\mathbf{b})\|^2\right).
\end{aligned}
$$

Hence, we can use Lemma 3.21 to conclude that

$$
\|R^k(\mathbf{b})\|^4 = o\left(\frac{\|\mathbf{x_b}\|_1^2}{(k+1)t_0^2}\right)
$$

for $k \to \infty$. This completes the proof. ☐

Similarly, we can use the method in the proof of Theorem 3.22 to give another convergence analysis of Algorithms 3.2, 3.4, and 3.5. We leave these to the interested reader.

3.6 • Convergence of Algorithms 3.6 and 3.7

In this section, we first present a convergence analysis of Algorithm 3.6. For convenience, let Λ be the index set of S_k for some k.

Theorem 3.23. *Suppose that m is the cardinality of the set Λ and $0 < a_m \leq \|\Phi_\Lambda^\top \Phi_\Lambda\|_2 \leq b_m < 2$. Given $\mathbf{b} \in \mathcal{D} = span\{\phi_1, \ldots, \phi_n\}$, let $L_{\ell,\Lambda}(\mathbf{b})$ be the output of the ℓth iteration of Algorithm 3.6. Then the sequence $L_{\ell,\Lambda}(\mathbf{b})$ converges to the least squares solution $L_\Lambda(\mathbf{b})$ in the following sense:*

$$
\|L_{\ell,\Lambda}(\mathbf{b}) - L_\Lambda(\mathbf{b})\| \leq |1 - \gamma|^\ell \|L_\Lambda(\mathbf{b})\|,
$$

where γ is a nonzero eigenvalue of the symmetric positive definite matrix $\Phi_\Lambda \Phi_\Lambda^\top$ defined in Theorem 3.24 below.

Proof. Recall that $G_k(\mathbf{b}) = G_{k-1}(\mathbf{b}) - \Phi P_\Lambda \Phi^\top G_{k-1}(\mathbf{b})$ and $L_{k,\Lambda}(\mathbf{b}) = L_{k-1,\Lambda}(\mathbf{b}) + \Phi P_\Lambda \Phi^\top G_{k-1}(\mathbf{b})$ for $k = 0, \ldots, \ell$ with $G_0 = \mathbf{b}$ and $L_{0,\Lambda}(\mathbf{b}) = 0$. Let g_Λ^k be the best approximation of $G_k(\mathbf{b})$ in the space S_Λ spanned by the columns from Φ_Λ. Clearly, $g_\Lambda^0 = L_\Lambda(\mathbf{b})$. For $k \geq 1$, $g_\Lambda^k = g_\Lambda^{k-1} - \Phi P_\Lambda \Phi^\top g_\Lambda^{k-1}$, since the best approximation operator is a linear operator. Similarly, let $\mathbf{y}_\Lambda^k = L_{k,\Lambda}(\mathbf{b})$ in S_Λ. Then we have $\mathbf{y}_\Lambda^k = \mathbf{y}_\Lambda^{k-1} + \Phi P_\Lambda \Phi^\top g_\Lambda^{k-1}$. Thus,

$$
\mathbf{y}_\Lambda^k + g_\Lambda^k = \mathbf{y}_\Lambda^{k-1} + g_\Lambda^{k-1} = \cdots = \mathbf{y}_\Lambda^0 + g_\Lambda^0 = L_\Lambda(\mathbf{b}).
$$

It follows that

$$
\|\mathbf{y}^k - L_\Lambda(\mathbf{b})\|_2 = \|g_\Lambda^k\| = \|(I - \Phi P_\Lambda \Phi^\top)g_\Lambda^{k-1}\|.
$$

To estimate the norm on the right-hand side of the equality above, we claim that S_Λ is perpendicular to the null space of $\Phi P_\Lambda \Phi^\top$. Indeed, for any vector $\mathbf{a}$ in the null space of $\Phi P_\Lambda \Phi^\top = \Phi_\Lambda \Phi_\Lambda^\top$, we have $\Phi_\Lambda \Phi_\Lambda^\top \mathbf{a} = 0$, and hence $\Phi_\Lambda^\top \mathbf{a} = 0$. Thus, for any $\mathbf{s} = \Phi_\Lambda \mathbf{c} \in S_\Lambda$ and for any $\mathbf{a}$ in the null space of $\Phi_\Lambda \Phi_\Lambda^\top$, we have $\mathbf{s}^\top \mathbf{a} = \mathbf{c}^\top \Phi_\Lambda^\top \mathbf{a} = 0$. It is clear that the eigenvalues $\Phi P_\Lambda \Phi^\top$ are nonnegative and the eigenvalues of $I - \Phi P_\Lambda \Phi^\top$ are $1 - \lambda_i$, $i = 1, \ldots, n$. Since all vectors spanned by eigenvectors whose eigenvalues are associated with $\lambda_i = 0$ are in the null space of $\Phi_\Lambda \Phi_\Lambda^\top$, any vector $\mathbf{s} \in S_\Lambda$ is spanned by the eigenvectors associated with $\lambda_i \neq 0$. Thus, the norm of $I - \Phi P_\Lambda \Phi^\top$ restricted over the subspace S_Λ can be estimated by

$$
\|I - \Phi P_\Lambda \Phi^\top\| \leq \max\{|1 - \gamma_i|, \quad \gamma_i \neq 0 \text{ eigenvalues of } \Phi P_\Lambda \Phi^\top\}.
$$

Let γ_Λ be the eigenvalue such that $|1 - \gamma_\Lambda|$ achieves the maximum on the right-hand side of the above inequality. Then we have $\|(I - \Phi P_\Lambda \Phi^\top) g_\Lambda^{k-1}\| \le |1 - \gamma_\Lambda| \|g_\Lambda^{k-1}\|$.

Next, since $\|\Phi_\Lambda^\top \Phi_\Lambda\| \le b_m < 2$, we know that $0 < \lambda_\Lambda < 2$, and hence $|1 - \gamma_\Lambda| < 1$. Therefore,

$$\|\mathbf{y}^k - L_\Lambda(\mathbf{b})\|_2 = \|g_\Lambda^k\| \le |1 - \gamma_\Lambda| \|g_\Lambda^{k-1}\|$$
$$\le \cdots \le |1 - \gamma_\Lambda|^k \|g_\Lambda^0\| = |1 - \gamma_\Lambda|^k \|L_\Lambda(\mathbf{b})\|.$$

This completes the proof. $\square$

Finally, we are ready to establish the convergence of Algorithm 3.7.

Theorem 3.24. *Let $\mathbf{b} \in A_1(\mathcal{D})$ as in Theorem 3.20. Suppose that there exists an integer $m > 0$ such that a_m and b_m in Theorem 3.23 satisfy $a_m > 0$ and $b_m < 2$. Then, for $j = 1, 2, \ldots, m$,*

$$\|R^j(\mathbf{b})\|^2 \le \frac{\gamma_0 \|\mathbf{b}\|}{1 + \sum_{k=0}^{j} \frac{t_k^2 (n_k - n_{k-1}) \|\mathbf{b}\|}{\lambda_k^2}}, \tag{3.61}$$

where γ_k is a nonzero eigenvalue of $\Phi_{S_k}^\top \Phi_{S_k}$ such that

$$|1 - \gamma_k| = \max\left\{ |1 - \gamma| : \text{ all nonzero eigenvalue } \lambda \text{ of } \Phi_{S_k}^\top \Phi_{S_k} \right\}.$$

Proof. We are now ready to combine Theorems 3.20 and 3.23 to give a convergent analysis of Algorithm 3.7. First we note that

$$\|R^k(\mathbf{b})\|^2$$
$$= \langle R^{k-1}(\mathbf{b}) - L_{\ell_k, S_k}(\mathbf{b}), R^{k-1}(\mathbf{b}) - L_{\ell_k, S_k}(\mathbf{b}) \rangle$$
$$= \langle R^{k-1}(\mathbf{b}) - L_{S_k}(\mathbf{b}), R^{k-1}(\mathbf{b}) - L_{\ell_k, S_k}(\mathbf{b}) \rangle + \langle L_{S_k}(\mathbf{b}) - L_{\ell_k, S_k}(\mathbf{b}), R^{k-1}(\mathbf{b}) - L_{\ell_k, S_k}(\mathbf{b}) \rangle$$
$$= \langle R^{k-1}(\mathbf{b}) - L_{S_k}(\mathbf{b}), R^{k-1}(\mathbf{b}) - L_{S_k}(\mathbf{b}) \rangle + \langle L_{S_k}(\mathbf{b}) - L_{\ell_k, S_k}(\mathbf{b}), L_{S_k}(\mathbf{b}) - L_{\ell_k, S_k}(\mathbf{b}) \rangle$$
$$= \|R^{k-1}(\mathbf{b})\|^2 - \|L_{S_k}(\mathbf{b})\|^2 + \|L_{\ell_k, S_k}(\mathbf{b}) - L_{S_k}(\mathbf{b})\|^2$$

by using the best approximation property of $L_{S_k}(\mathbf{b})$. As in the proof of Theorem 3.20, we have

$$\|L_{S_k}(\mathbf{b})\|^2 \ge \frac{t_k^2}{\lambda_{n_k}^2} (n_k - n_{k-1}) \|R^{k-1}(\mathbf{b})\|^4 \quad \text{and} \quad \|L_{S_k}(\mathbf{b})\| \le \|R^{k-1}(\mathbf{b})\|^2,$$

and hence by Theorem 3.23 with γ_k being the γ for S_k,

$$\|R^k(\mathbf{b})\|^2 = \|R^{k-1}(\mathbf{b})\|^2 - \|L_{S_k}(\mathbf{b})\|^2 + (1 - \gamma_k)^{\ell_k} \|L_{S_k}(\mathbf{b})\|^2$$
$$\le \|R^{k-1}(\mathbf{b})\|^2 \left(1 - \frac{t_k^2}{\lambda_{n_k}^2} (n_k - n_{k-1}) \|R^{k-1}(\mathbf{b})\|^2\right) + (1 - \gamma_k)^{\ell_k} \|R^{k-1}(\mathbf{b})\|^2$$
$$= \|R^{k-1}(\mathbf{b})\|^2 \left(1 + |1 - \gamma_k|^{\ell_k} - \frac{t_k^2}{\lambda_k^2} (n_k - n_{k-1}) \|R^{k-1}(\mathbf{b})\|^2\right). \tag{3.62}$$

Iterating the inequality in (3.62), we obtain the estimate in (3.61) by using Lemma 3.25 below. This finishes the proof of Theorem 3.24. $\square$

Lemma 3.25. *Suppose we have two sequences of nonnegative numbers, $\{a_k, k \ge 1\}$ and $\{\alpha_k, k \ge 1\}$, satisfying $\alpha_k > 0$, $a_1 \le 1$, and*

$$a_{m+1} \le a_m(\beta_m - a_m \alpha_m) \quad \forall m \ge 1, \tag{3.63}$$

where $\beta_m = 1 + \theta^m$. Then there exists $\gamma_0 \ge 1$ such that

$$a_{m+1} \le \frac{\gamma_0 a_0}{1 + \sum_{k=1}^{m} \alpha_k a_0} \quad \forall m \ge 1. \tag{3.64}$$

Proof. The proof is similar to that of Lemma 3.19. We leave it to the interested reader. $\square$

3.7 ▪ Quasi-orthogonal Matching Pursuit

In this section, we study the convergence of the QOMP algorithm, which greatly enhances the performance of the classical orthogonal matching pursuit (OMP) algorithm, at some cost of computational complexity. We are able to show that under some sufficient conditions of mutual coherence of the sensing matrix, the QOMP algorithm succeeds in recovering the s-sparse signal vector $\mathbf{x}$ within s iterations where a total of $2s$ columns are selected under both noiseless and noisy settings. In addition, we will show that for the Gaussian sensing matrix, the norm of the residual of each iteration will go to zero linearly, depending on the size of the matrix, with high probability.

In this section, we shall study a version of the QOMP which is slightly different from Algorithm 3.9, given in section 3.2. We leave the proof of Algorithm 3.9 to the interested reader, or see [8].

ALGORITHM 3.10
Quasi-orthogonal Matching Pursuit (QOMP) Algorithm

1: **Input:** $\Phi_{m \times n}$, $\mathbf{b}_{n \times 1}$, sparsity $s \geq 2$, maximum iterations $k_{\max} \leq m/2$, and tolerance ϵ.
2: **Initialization:** $\mathcal{S}_0 = \emptyset$, $\mathbf{r}_0 = \mathbf{b}$, $k = 0$, $\Psi_{m \times n} = \Phi_{m \times n}$.
3: **while** $k < k_{\max}$ and $|\mathbf{r_k}| > \epsilon$:

$k = k + 1$;

$Res_{(i,j)}(\mathbf{r}_{k-1}) = \min_{u,v \in \mathbb{R}}\{\|\psi_i u + \psi_j v - \mathbf{r}_{k-1}\|_2\}$;

$(i_k, j_k) = \arg\min_{1 \leq i \leq n, 1 \leq j \leq n}\{Res_{(i,j)}(\mathbf{r}_{k-1})\}$;

$\mathcal{S}_k = \mathcal{S}_{k-1} \cup \{i_k, j_k\}$;

$\mathbf{r_k} = \mathbf{b} - \Phi_{\mathcal{S}_k}\Phi_{\mathcal{S}_k}^{\dagger}\mathbf{b}$;

$\Psi_{\{i_k, j_k\}} = \mathbf{0}$;

4: **end**
5: **Output:** $\mathcal{S} = \mathcal{S}_k$, $\mathbf{x}_S = \Phi_S^{\dagger}\mathbf{b}$, and $\mathbf{x}_{S^c} = \mathbf{0}$.

We first show that each step of the iterations, there is at least one index which is correct.

Theorem 3.26. *Suppose the mutual coherence $M(\Phi)$ satisfies $M(\Phi) \leq \frac{1}{f(m)}$ for a function f which satisfies $f(m) = o(\sqrt{m})$ as $m \to \infty$. If the sparsity s of the true signal $\mathbf{x}$ satisfies $2 \leq s \leq \frac{f(m)}{5}$, then for a large m, among the two indices selected from the column indices of Φ in the first iteration of Algorithm 3.10, at least one index is the correct one.*

Proof. Without loss of generality, let us assume each column ϕ_i is normalized, and assume the support set of $\mathbf{x}$ is $\Omega = \{1, 2, \dots, s\}$. Then $\mathbf{b} = \Phi\mathbf{x} = x_1\phi_1 + x_2\phi_2 + \cdots + x_s\phi_s = \sum_{k=1}^{s} x_k\phi_k$, where ϕ_i is the ith column of Φ. In the first iteration, for each $1 \leq i, j \leq n$, minimizing $\|\phi_i u + \phi_j v - \mathbf{b}\|_2$ is equivalent to maximizing the projection of $\mathbf{b}$ onto the hyperplane spanned by $\mathbf{a}_i$ and $\mathbf{a}_j$, which is

$$Proj(\mathbf{b}) = \begin{bmatrix} \phi_i & \phi_j \end{bmatrix} \begin{bmatrix} \phi_i^\top \phi_i & \phi_i^\top \phi_j \\ \phi_j^\top \phi_i & \phi_j^\top \phi_j \end{bmatrix}^{-1} \begin{bmatrix} \phi_i^\top \\ \phi_j^\top \end{bmatrix} \cdot \mathbf{b}$$

$$= \begin{bmatrix} \phi_i & \phi_j \end{bmatrix} \begin{bmatrix} 1 & \phi_i^\top \phi_j \\ \phi_j^\top \phi_i & 1 \end{bmatrix}^{-1} \begin{bmatrix} \phi_i^\top \\ \phi_j^\top \end{bmatrix} \cdot \left(\sum_{k=1}^{s} x_k\phi_k\right)$$

because each column ϕ_i is normalized. We further have

$$Proj(\mathbf{b}) = \frac{1}{1-|\phi_i^\top \phi_j|^2} \begin{bmatrix} \phi_i & \phi_j \end{bmatrix} \begin{bmatrix} 1 & -\phi_i^\top \phi_j \\ -\phi_j^\top \phi_i & 1 \end{bmatrix} \begin{bmatrix} \phi_i^\top \\ \phi_j^\top \end{bmatrix} \cdot \left(\sum_{k=1}^{s} x_k \phi_k \right)$$

$$= \frac{1}{1-|\phi_i^\top \phi_j|^2} (\phi_i \phi_i^\top + \phi_j \phi_j^\top - (\phi_i^\top \phi_j)\phi_j \phi_i^\top - (\phi_j^\top \phi_i)\phi_i \phi_j^\top) \cdot \left(\sum_{k=1}^{s} x_k \phi_k \right).$$

In order to show the result, we only need to show that $\|Proj(\mathbf{b})\|$ is not maximized when $i \notin S$ and $j \notin S$. We can do this by showing that $\|Proj(\mathbf{b})\|$ when both $i,j \notin S$ is strictly less than $\|Proj(\mathbf{b})\|$ when either $i \in S$ or $j \in S$.

Next suppose both $i,j \notin S$. By applying the triangle inequality together with the assumption on mutual coherence, $M(\Phi) \leq \frac{1}{f(m)}$, we get

$$\|Proj(\mathbf{b})_{i,j \notin S}\|$$
$$= \left\| \frac{1}{1-|\phi_i^\top \phi_j|^2} (\phi_i \phi_i^\top + \phi_j \phi_j^\top - (\phi_i^\top \phi_j)\phi_j \phi_i^\top - (\phi_j^\top \phi_i)\phi_i \phi_j^\top) \cdot \left(\sum_{k=1}^{s} x_k \phi_k \right) \right\|$$
$$\leq \frac{1}{1-|\phi_i^\top \phi_j|^2} \sum_{k=1}^{s} |x_k|((|\phi_i^\top \phi_k| + |\phi_i^\top \phi_j||\phi_j^\top \phi_k|)\cdot\|\phi_i\| + (|\phi_j^\top \phi_k| + |\phi_j^\top \phi_i||\phi_i^\top \phi_k|)\cdot\|\phi_j\|)$$
$$= \frac{1}{1-|\phi_i^\top \phi_j|^2} \sum_{k=1}^{s} |x_k|(|\phi_i^\top \phi_k| + |\phi_i^\top \phi_j||\phi_j^\top \phi_k| + |\phi_j^\top \phi_k| + |\phi_j^\top \phi_i||\phi_i^\top \phi_k|)$$
$$\leq \frac{1}{1-1/f^2(m)} (1/f(m) + 1/f^2(m) + 1/f(m) + 1/f^2(m)) \left(\sum_{k=1}^{s} |x_k| \right)$$
$$= \frac{f^2(m)}{f^2(m)-1} \left(\frac{2}{f(m)} + \frac{2}{f^2(m)} \right) \left(\sum_{k=1}^{s} |x_k| \right) = \frac{2f(m)+2}{f^2(m)-1} \cdot \left(\sum_{k=1}^{s} |x_k| \right).$$

Second, suppose $i \in S$ or $j \in S$. Without loss of generality, let us assume $i \in S$, $i = 1$ and $|x_1| = \max_{1 \leq i \leq s} |x_i|$ is one of the largest entries in absolute value. By applying the triangle inequality together with the assumption $M(\Phi) \leq \frac{1}{f(m)}$, we get

$$\|Proj(\mathbf{b})_{i \in S}\|$$
$$= \left\| \frac{1}{1-|\Phi_1^\top \phi_j|^2} (\phi_1 \phi_1^\top + \phi_j \phi_j^\top - (\phi_1^\top \phi_j)\phi_j \phi_1^\top - (\phi_j^\top \phi_1)\phi_1 \phi_j^\top) \cdot \left(\sum_{k=1}^{s} x_k \phi_k \right) \right\|$$
$$\geq \frac{1}{1-|\phi_1^\top \phi_j|^2} (|x_1|(\|\phi_1\| - |\phi_j^\top \phi_1|\cdot\|\phi_j\| - |\phi_1^\top \phi_j|\cdot\|\phi_j\| - |\phi_1^\top \phi_j||\phi_j^\top \phi_1|\cdot\|\phi_1\|))$$
$$- \frac{1}{1-|\phi_1^\top \phi_j|^2} (2/f(m) + 2/f^2(m)) \cdot \left(\sum_{k=2}^{s} |x_k| \right)$$
$$\geq \frac{1}{1-|\phi_1^\top \phi_j|^2} \left[(1 - 2/f(m) - 1/f^2(m))|x_1| - (2/f(m) + 2/f^2(m)) \cdot \left(\sum_{k=2}^{s} |x_k| \right) \right].$$

It follows that showing $\|Proj(\mathbf{b})_{i \in S}\| \geq \|Proj(\mathbf{b})_{i,j \notin S}\|$ is equivalent to showing

$$\frac{1}{1-|\phi_1^\top \phi_j|^2} \left(\left(1 - \frac{2}{f(m)} - \frac{1}{f^2(m)} \right) |x_1| - \left(\frac{2}{f(m)} + \frac{1}{f^2(m)} \right) \left(\sum_{k=2}^{s} |x_k| \right) \right) \quad (3.65)$$

$$\geq \frac{2f(m)+2}{f^2(m)-1} \cdot \sum_{k=1}^{s} |x_k|.$$

Since $|x_1| = \max_{1 \leq i \leq s} |x_i|$ and $1 - |\phi_1^\top \phi_j|^2 \leq 1$, it suffices to show

$$(1 - 2/f(m) - 1/f^2(m)) - (2/f(m) + 2/f^2(m)) \cdot (s-1) \geq \frac{2f(m)+2}{f^2(m)-1} \cdot s,$$

which is equivalent to

$$1 - \frac{2s}{f(m)} - \frac{2s-1}{f^2(m)} \geq \frac{2f(m)+2}{f^2(m)-1} s = \frac{2s}{f(m)-1}. \tag{3.66}$$

Because $f(m) \to \infty$ as $m \to \infty$, we can see that the left-hand side is near 1 and the right-hand side is near 0. Hence there is a threshold m_0 such that (3.66) holds as long as $m \geq m_0$. In particular, when $s \leq f(m)/5$, (3.66) is true, and hence the desired result follows. ☐

Theorem 3.27. *Under the same condition as Theorem 3.26, the exact recovery of the s-sparse signal* **x** *can be guaranteed in s iterations by using Algorithm 3.10.*

Proof. By Theorem 3.26 or from its proof, we know that the first iteration will pick at least one correct column index. Without loss of generality, suppose the first correct index that is picked in the first iteration is the first column, and the other column which is picked together with the first column is the jth column. Then in the second iteration, the residual vector gets updated to $\mathbf{r}_1 = \mathbf{b} - x_1 \phi_1 - x_j \phi_j$, where $[x_1, x_j]^\top = \Phi_{S_1}^\dagger \mathbf{r}_1$. The matrix Φ gets updated to Ψ, where Ψ is the matrix Φ but with either the first column or the first and jth column being replaced by **0** vectors because of the update step $\Psi_{\{i_k, j_k\}} = \mathbf{0}$ in Algorithm 3.10. By the same analysis, we can conclude that the second iteration will also pick at least one correct column index which is different from those being picked in the first iteration. Thus each iteration will pick at least one correct column index which is different from those picked in previous iterations, and hence the support set S is recovered within s total iterations. ☐

Let us study the rate of convergence of Algorithm 3.10. We shall give two results, one based on the standard RIP and the other based on the mutual coherence of Φ. Let $\mathbf{r}_k = \mathbf{b} - L_k$ be the residual of the kth iteration. It is easy to see that $\|\mathbf{r}_k\| \leq \|\mathbf{b}\|$ and $\|\mathbf{r}_{k+1}\| \leq \|\mathbf{r}_k\|$ for all $k \geq 1$. We leave this to the interested reader. Next it is easy to see that

$$\|\mathbf{r}_k\| = \min \|\mathbf{b} - \alpha L_{k-1} - \beta \phi_{i_k} - \gamma \phi_{j_k}\| \leq \min \|\mathbf{r}_{k-1} - \beta \phi_{i_k} - \gamma \phi_{j_k}\|.$$

Letting $\mathbf{Proj}_{i_k, j_k}(\mathbf{r}_{k-1})$ be the solution of the minimization on the right-hand side, we have

$$\|\mathbf{r}_k\|^2 \leq \|\mathbf{r}_{k-1}\|^2 - \|\mathbf{Proj}_{i_k, j_k}(\mathbf{r}_{k-1})\|^2. \tag{3.67}$$

Lemma 3.28. *Suppose that Φ is a dictionary, i.e., each column of Φ is normalized to have norm 1. Then*

$$\|\mathbf{Proj}_{i_k, j_k}(\mathbf{r}_k)\|^2 \geq \frac{1}{2}(|\langle \phi_{i_k}, \mathbf{r}_k \rangle|^2 + |\langle \phi_{j_k}, \mathbf{r}_k \rangle|^2).$$

Proof. We use Gram–Schmidt orthonormalization of ϕ_{i_k}, ϕ_{j_k} to get ϕ_{i_k}, ψ_{j_k} such that $\phi_{i,k}$ is orthogonal to ψ_{j_k}. Then

$$\mathbf{Proj}_{i_k, j_k}(\mathbf{r}_k) = \langle \mathbf{r}_k, \phi_{i_k} \rangle \phi_{i_k} + \langle \mathbf{r}_k, \psi_{j_k} \rangle \psi_{j_k},$$

and consequently

$$\|\mathbf{Proj}_{i_k, j_k}(\mathbf{r}_k)\|^2 = |\langle \mathbf{r}_k, \phi_{i_k} \rangle|^2 + |\langle \mathbf{r}_k, \psi_{j_k} \rangle|^2 \geq |\langle \mathbf{r}_k, \phi_{i_k} \rangle|^2.$$

Similarly, we have

$$\|\mathbf{Proj}_{i_k,j_k}(\mathbf{r}_k)\|^2 \geq |\langle \mathbf{r}_k, \phi_{j_k}\rangle|^2.$$

Combining the above two inequalities, we have the desired inequality.   □

Let $\phi_{k_{max}} = \{\phi_i : \max_i\{|\phi_i^\top \mathbf{r}_{k-1}|\}\}$. We have the following important property.

Lemma 3.29. *We have* $|\mathbf{Proj}_{i_k,j_k}(\mathbf{r}_{k-1})^\top \mathbf{r}_{k-1}| \geq |\phi_{i_{max}}^\top \mathbf{r}_{k-1}|.$

Proof. Otherwise, we have $|\phi_{i_{max}}^\top \mathbf{r}_{k-1}| > |\mathbf{Proj}_{i_k,j_k}(\mathbf{r}_{k-1})^\top \mathbf{r}_{k-1}|$. Then, by using the same proof as in Lemma 3.28, we would have

$$\|\mathbf{Proj}_{(i_k,i_{max})}(\mathbf{r}_{k-1})\| \geq |\phi_{i_{max}}^\top \mathbf{r}_{k-1}| > |\mathbf{Proj}_{i_k,j_k}(\mathbf{r}_{k-1})^\top \mathbf{r}_{k-1}| = \|\mathbf{Proj}_{(i_k,j_k)}(\mathbf{r}_{k-1})\|,$$

which contradicts the choice of the pair (i_k, j_k).   □

We mainly need to show that the sequence obtained from Algorithm 3.10 is linearly convergent.

Theorem 3.30. *Suppose that there is a sparse solution* $\mathbf{x_b}$ *with sparsity* $s = \|\mathbf{x_b}\|_0$ *satisfying* $\mathbf{\Phi x_b} = \mathbf{b}$. *Suppose that* $\mathbf{\Phi}$ *has* $2s$ *RIC* $\delta_{2s} \in (0,1)$. *Then each iteration in Algorithm 3.10 satisfies*

$$\|\mathbf{r}_k\|^2 \leq \left(1 - \frac{1 - \delta_{2s}^\Phi}{2s}\right)^k \|\mathbf{b}\|^2 \ \forall k \geq 0,$$

where δ_{2s}^Φ *is the RIC of* Φ.

Proof. Let us write $\mathbf{r}_k = \mathbf{b} - L_k = \mathbf{\Phi}(\mathbf{x_b} - \hat{\mathbf{x}}_k)$ since L_k can be written as $\mathbf{\Phi}\hat{\mathbf{x}}_k$ for some sparse vector $\hat{\mathbf{x}}_k$ supported on S_k and zero for indices in S_k^c, the complement of S_k. For simplicity, we write $\alpha \in \mathbb{R}^n$ with $\alpha(i) = \mathbf{x_b}(i) - \hat{\mathbf{x}}_k(i)$ for $i \in \{1, 2, \ldots, n\}$ with sparsity $\leq 2s$. Then it follows that

$$\|\mathbf{r}_k\| \leq \|\alpha\|_1 \cdot \max_i |\langle \mathbf{r}_k, \phi_i\rangle| \leq \sqrt{2s}\|\alpha\|_2 \cdot |\langle \mathbf{r}_k, \phi_{i_{max}}\rangle| \leq \sqrt{2s}\|\alpha\|_2 \cdot |\mathbf{Proj}_{(i_{k+1},j_{k+1})}(\mathbf{r}_k)^\top \mathbf{r}_k|$$
$$(3.68)$$

by using Lemma 3.29. On the other hand,

$$\|\mathbf{r}_k\|^2 = \langle \mathbf{b} - L_k(\mathbf{b}), \mathbf{b} - L_k(\mathbf{b})\rangle = \|\mathbf{\Phi}\alpha\|^2 \geq (1 - \delta_{2s}^\Phi)\|\alpha\|_2^2, \qquad (3.69)$$

where the last inequality follows from the RIP. Combining (3.68) and (3.69) leads to

$$|\mathbf{Proj}_{(i_{k+1},j_{k+1})}(\mathbf{r}_k)^\top \mathbf{r}_k|^2 \geq \frac{1 - \delta_{2s}^\Phi}{2s}\|\mathbf{r}_k\|^2.$$

Using (3.67), we obtain

$$\|\mathbf{r}_{k+1}\|^2 \leq \|\mathbf{r}_k\|^2 - (|\mathbf{Proj}_{(i_{k+1},j_{k+1})}(\mathbf{r}_k)^\top \mathbf{r}_k|)^2 \leq \|\mathbf{r}_k\|^2 - \frac{1 - \delta_{2s}^\Phi}{2s}\|\mathbf{r}_k\|^2 = \left(1 - \frac{1 - \delta_{2s}^\Phi}{2s}\right)\|\mathbf{r}_k\|^2.$$

Noting that $\mathbf{r}_0 = \mathbf{b}$, we can get

$$\|\mathbf{r}_k\|^2 \leq \left(1 - \frac{1 - \delta_{2s}^{\Phi}}{2s}\right)^k \|\mathbf{b}\|^2.$$

This completes the proof.   □

Next we give another proof of the rate of convergence. It uses the mutual coherence instead of the RIP. We recall from Lemma 2.19 that for a random matrix Φ of size $m \times n$ with $n \geq 2m$, the mutual coherence of Φ is at least $1/\sqrt{2m}$.

Theorem 3.31. *Suppose $n \geq 2m$, let $\Phi \in \mathbb{R}^{m \times n}$ be a sensing matrix with normalized columns, and let $\mathbf{x}$ be a sparse solution with sparsity $K \leq \frac{m}{2}$. Suppose that the mutual coherence $M(\Phi)$ satisfies the lower bound*

$$M(\Phi) = \max_{1 \leq i,j \leq m, i \neq j} |\phi_{\mathbf{i}}^{\top} \phi_{\mathbf{j}}| \geq \sqrt{\frac{n - m}{m(n - 1)}}. \tag{3.70}$$

Then sparse signal $\mathbf{x}$ can be exactly recovered by using Algorithm 3.10, and for all $1 \leq k \leq K$, there exists a constant $\alpha \in (0, 1)$ such that

$$\|\mathbf{r}_k\|^2 \leq \alpha \|\mathbf{r}_{k-1}\|^2. \tag{3.71}$$

Proof. Notice that for $n > m$ and for the kth iteration, the submatrix $\Phi_{\mathcal{S}_{k-1}^c}$ is of size $m \times (n - 2k + 2)$. If $k \leq K$, then $M(\Phi_{\mathcal{S}_{k-1}^c}) \geq \sqrt{\frac{n-2k-m+2}{m(n-2k-1+2)}} \geq \sqrt{\frac{n-2K-m+2}{m(n-2K-1+2)}}$. Hence, for large m, we have

$$\frac{|\phi_{k_{max}}^{\top} \mathbf{r}_{k-1}|^2}{\|\mathbf{r}_{k-1}\|^2} \geq M^2([\Phi_{\mathcal{S}_{k-1}^c}, \mathbf{r}_{k-1}]) \geq M^2(\Phi_{\mathcal{S}_{k-1}^c}) \geq \frac{n - 2K - m + 2}{m(n - 2K + 1)}. \tag{3.72}$$

By Lemma 3.29 and (3.72), we have

$$\|\mathbf{Proj}_{(i_k,j_k)}(\mathbf{r}_{k-1})\|^2 = |\mathbf{Proj}_{(i_k,j_k)}(\mathbf{r}_{k-1})^{\top} \mathbf{r}_{k-1}|^2 \geq |\phi_{k_{max}}^{\top} \mathbf{r}_{k-1}|^2 \tag{3.73}$$

$$\geq \frac{n - 2K - m + 2}{m(n - 2K + 1)} \cdot \|\mathbf{r}_{k-1}\|^2. \tag{3.74}$$

By plugging (3.73) back into (3.67), we have, for $1 \leq k \leq K$,

$$\|\mathbf{r}_k\|^2 \leq \|\mathbf{r}_{k-1}\|^2 - \|\mathbf{Proj}_{(i_k,j_k)}(\mathbf{r}_{k-1})\|^2 \leq \|\mathbf{r}_{k-1}\|^2 - \frac{n - 2K - m + 2}{m(n - 2K + 1)} \|\mathbf{r}_{k-1}\|^2$$

$$= \left(1 - \frac{(n - 2K - m + 2)}{m(n - 2K + 1)}\right) \|\mathbf{r}_{k-1}\|^2$$

for large m. With $\alpha = 1 - \frac{(n-2K-m+2)}{m(n-2K+1)}$, we will be able to finish the proof.

Indeed, since the number of iterations will always be less than or equal to $m/2$ as the algorithm chooses two columns in every iteration and does not repick the same columns already chosen from the previous iterations and since $n \geq 2m$, we have $n - 2K - m + 2 > 0$, i.e., $\alpha < 1$. Hence, this α can finish the proof.   □

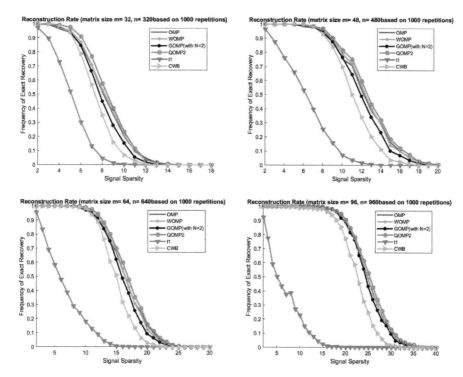

Figure 3.1. *Frequencies of recovery by OMP, GOMP, QOMP, WOMP, CWB, and ℓ_1 magic for various sizes of underdetermined linear systems based on 1000 repetitions.*

3.8 ▪ Experimental Results

In this section, we compare the performance between QOMP, i.e., Algorithm 3.9, and several standard algorithms, including the OMP Algorithm 3.2, Algorithm 3.5 (GOMP), Algorithm 3.4 with $g = 2$, the magic ℓ_1 method, and the iteratively reweighted ℓ_1 algorithm called CWB, i.e., Algorithm 6.13, to be discussed in Chapter 6. As we know from previous sections, if the signal sparsity is s, then the OMP algorithm can recover the signal within s iterations if the RIC of the sensing matrix satisfies a very restricted condition. In addition, more iterations will help recover more sparse signals when $\delta_{2s} < 1$. Thus, we shall use the OMP Algorithm 3.2 with $m/2$ iterations for rectangular linear systems of size $m \times n$ for the experiment.

We first experiment with these algorithms to solve sparse solutions for a Gaussian random sensing matrix of size $m \times 10m$ for $m = 32, 48, 64, 96$. The frequency of exact recovery of signals by each method is reported in Figure 3.1. For each method and for each fixed sparsity, a frequency of exact recovery is computed which is the ratio of the number of exact recovery (within maximum error $1e-5$) from solving the underdetermined linear system over 1000 repetitions. We only show the results for $2 \leq s < 0.4m$ because the exact recovery rate is very low for many algorithms if $s \geq 0.4m$.

From the curves in Figure 3.1, we can see that all of these greedy algorithms work quite well and the performance of Algorithm 3.9 is the best.

However, when $n = km$ for a small integer $1 \leq k < 10$, Algorithm 3.9 performs well, but not the best. We refer the reader to Figure 3.2 for the frequency of the exact recovery for sensing matrices of size 32×128 over 1000 repeated runs. From Figure 3.2, we can see that the CWB, i.e., the iteratively reweighted ℓ_1 algorithm, is much better. We will study this algorithm in Chapter 6.

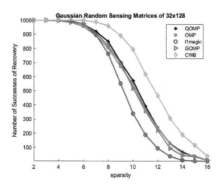

Figure 3.2. *Frequencies of exact reconstruction of signal within s iterations for underdetermined linear systems of size $m \times n$ with $m = 32$ and $n = 4m = 128$.*

Note that the computational times for all of these methods are not reported here. Clearly, the computational time for Algorithm 3.9 is the slowest. We recommend implementing Algorithm 3.9 in a GPU setting to speed up its computational time.

3.9 ▪ Exercises

Exercise 1. If $\mathcal{D}$ is a dictionary, for any $\mathbf{y} \in H = \mathrm{span}(\mathcal{D})$, define

$$\|\mathbf{y}\|_{\mathcal{D},1} = \min \left\{ \sum_{i=1}^{n} |c_i|, \mathbf{y} = \Phi \mathbf{c}, \mathbf{c} = (c_1, \dots, c_n)^\top \right\}.$$

Show that $\|\mathbf{y}\|_{\mathcal{D},1}$ is a norm. That is, show that $\| \cdot \|_{\mathcal{D},1}$ satisfies the three norm properties. Also show that if $\|\mathbf{b}\|_{\mathcal{D},1} \leq 1$, then $\|\mathbf{b}\|_2 \leq 1$.

Exercise 2. Let $\mathbf{x}$ be an s-sparse vector. Suppose that Φ has an RIP with $0 < \delta_s < 1$. Show that for any $T \subset S := \mathrm{supp}(\mathbf{x})$,

$$(1 - \delta_s)\|\mathbf{x}_{T^c}\|_2 \leq \|\Phi_{T^c \cap S}^\top (I - \Phi_T \Phi_T^\dagger)\Phi_{T^c \cap S}\mathbf{x}_{T^c \cap S}\|_2 \leq (1 + \delta_s)\|\mathbf{x}_{T^c}\|_2,$$

where $\Phi^\dagger$ is the pseudo-inverse of Φ.

Exercise 3. Let $\mathbf{x}$ be an s-sparse vector. Suppose that Φ has an RIP with $0 < \delta_s < 1$. Show that for any $T \subset S := \mathrm{supp}(\mathbf{x})$,

$$(1 - \delta_{s+1})\|\mathbf{x}|_{T^c}\|_2 \leq \|\Phi(I - \Phi_T)\Phi\mathbf{x}\|_2.$$

Exercise 4. Prove Theorem 3.5 directly.

Exercise 5. Prove Theorem 3.2.

Exercise 6. Regarding the GOMP algorithm (choosing the top g components), show that it can perfectly reconstruct an s-sparse signal if the sensing matrix Φ has an RIP with $\delta_{gs} < \frac{\sqrt{g}}{\sqrt{s}+3\sqrt{g}}$. (*Hint:* See [18] for details.)

Exercise 7. Suppose that $g_m \in \Sigma_s$ which is convergent to g_*. Show that $g_* \in \Sigma_s$.

Exercise 8. Prove Lemma 3.25.

Exercise 9. Write a MATLAB code to implement Algorithm 3.2. Use it to solve linear systems of equations for various sizes, e.g., $m < n$, $m = n$, and $m > n$. Experiment with several different types of matrices to see if Algorithm 3.2 works. In particular, choose $\mathbf{b} = \Phi_S \mathbf{x}_S$ with $|S| = s$, say, $s = 100$, and Φ_S with s columns of linear independent vectors from $\mathbb{R}^m$ with $m \gg 100$, say $m = 1000$. Add $n - s$ random columns to form Φ with $n > m$. Use your code to find $\mathbf{x}_S$ from $\mathbf{b}$.

Exercise 10. Write a MATLAB code to solve a nonsquared linear system using Algorithm 3.4. Experiment with several different types of matrices to see if Algorithm 3.4 works. In particular, choose entries of Φ from Gaussian random numbers and test the accuracy of the recovery for various sparse solutions. That is, choose a sparse solution $\mathbf{x}_b$ with sparsity $s > 0$ and let $\mathbf{b} = \Phi \mathbf{x}_b$. Use your implementation to see if you can recover $\mathbf{x}_b$ accurately from Φ and $\mathbf{b}$. Repeat your experiment 100 times to see the rate of success. Compare with Algorithm 3.2.

Exercise 11. Write a MATLAB code to implement Algorithm 3.5. Experiment with several different types of matrices to see if Algorithm 3.5 works. In particular, choose entries of Φ from Gaussian random numbers and test the accuracy of the recovery for various sparse solutions. That is, choose a sparse solution $\mathbf{x}_b$ with sparsity $s > 0$ and let $\mathbf{b} = \Phi \mathbf{x}_b$. Use your implementation to see if you can recover $\mathbf{x}_b$ accurately. Repeat your experiment 100 times to see the success rates for various sparsities $s = 1, 2, \ldots$. Compare with Algorithm 3.2 and Algorithm 3.4 to see which one is the best.

Exercise 12. Write a MATLAB code to solve a nonsquared linear system using Algorithm 3.7. Experiment with linear systems of large size to see if Algorithm 3.7 works.

Exercise 13. Let R^k be the residual vector in Algorithm 3.8. That is, $R^k = \mathbf{b} - L_k(\mathbf{b})$ and $L_k = \alpha_1^k L_{k-1} + \alpha_2^k \phi_{s^k}$. Show the following result.

Lemma 3.32. $\langle R^k, \phi_{s^k} \rangle = 0$ and $\langle R^k, L_{k-1} \rangle = 0$ for $k \geq 1$. Hence, $\langle R^k, L_k \rangle = 0$.

Exercise 14. Show the following result.

Lemma 3.33. Suppose that the sensing matrix $\Phi = [\phi_1, \phi_2, \ldots, \phi_n]$ is a dictionary. That is, $\|\phi_i\|_1 = 1$ for all $i = 1, \ldots, m$. Let R^k be the residual vector in Algorithm 3.8. Show that $\|R^k\|^2 \leq \|R^{k-1}\|^2 - (\langle R^{k-1}, \phi_{s^k} \rangle)^2$ for all $k \geq 1$.

Exercise 15. Prove the following lemma.

Lemma 3.34. Let R^k be the residual vector in Algorithm 3.8. That is, $R^k = \mathbf{b} - L_k(\mathbf{b})$. Then $\|R^k\|^2 = \|\mathbf{b}\|^2 - \|L_k(\mathbf{b})\|^2$ for all $k \geq 1$.

Exercise 16. Prove the following lemma, letting R^k be the residual vector in Algorithm 3.8. Show that the following lemma holds.

Lemma 3.35. Suppose that $R^{k-1} \neq 0$ for some $k \geq 1$. Then the vectors ϕ_{j_k} are linearly independent of L_{k-1}.

Exercise 17. Further show that the sequence obtained from Algorithm 3.8 is linearly convergent. That is, prove the following theorem.

Theorem 3.36. *Suppose that there is a sparse solution* $\mathbf{x_b}$ *with sparsity* $s = \|\mathbf{x_b}\|_0$ *satisfying* $\Phi \mathbf{x_b} = \mathbf{b}$. *Suppose that* Φ *has RIC* $\delta_{2s} \in (0,1)$. *Then each iteration in Algorithm 3.8 satisfies*

$$\|R^{k+1}\|^2 \leq \left(1 - \frac{1 - \delta_{2s}^\Phi}{2s}\right)^k \|\mathbf{b}\|^2 \quad \forall k \geq 0,$$

where δ_{2s}^Φ *is the RIC of* Φ.

Exercise 18. The sequence from Algorithm 3.8 also satisfies a similar property in Lemma 3.15. State the result and prove it.

Exercise 19. Show that the sequence from Algorithm 3.8 has a convergence result similar to the one in Theorem 3.17.

Exercise 20. Study the convergence of Algorithms 3.2 and 3.8 under the condition of the mutual coherence of the sensing matrix. (*Hint*: See the proof of Theorem 3.31.)

Exercise 21. Recall that the graph Laplacian matrix of a weighted graph $G = (V, E, \mathbf{w})$ is defined by

$$L_G = \sum_{(u,v) \in E} w_{(u,v)} (\mathbf{e}_u - \mathbf{e}_v)(\mathbf{e}_u - \mathbf{e}_v)^\top,$$

where $w_{(u,v)} \geq 0$ is the weight of edge (u, v) and $\mathbf{e}_u \in \mathbb{R}^{|V|}$ is the characteristic vector of vertex u (with a 1 at coordinate u and zeros elsewhere). Define $\mathbf{v}$ similarly. The graph sparsifier problem is to look for a graph $H = (V, \tilde{E}, \tilde{w})$ such that

$$a\mathbf{x}^\top L_G \mathbf{x} \leq \mathbf{x}^\top L_H \mathbf{x} \leq b\mathbf{x}^\top L_G \mathbf{x} \tag{3.75}$$

for all $\mathbf{x} \in \mathbb{R}^n$, where $a > 0, b > 0$, and $b/a \leq \kappa$ for a given $\kappa > 1$. Here, H is called the κ-approximation of G or the κ-sparsifier of G. Develop an algorithm similar to Algorithm 3.2 or Algorithm 3.8 to find L_H. Also show that your algorithm is convergent with a linear convergence rate.

Exercise 22. Let $V = \{\mathbf{v}_1, \dots, \mathbf{v}_m\} \subset \mathbb{R}^n$ be a collection of vectors with $m \gg n$. We consider

$$B = \sum_{i=1}^m \mathbf{v}_i \mathbf{v}_i^\top, \tag{3.76}$$

which is clearly nonnegative definite. We look for a general sparsifier $\mathbf{s} = (s_i, i = 1, \dots, m) \in \mathbb{R}^m$ with $\|\mathbf{s}\|_0 \leq n/\epsilon$ such that

$$(1 - \epsilon)^2 B \leq \sum_{i=1}^m s_i \mathbf{v}_i \mathbf{v}_i^\top \leq (1 + \epsilon)^2 B. \tag{3.77}$$

Develop an algorithm similar to Algorithm 3.2 or Algorithm 3.8 to find the sparsifier $\mathbf{s}$. Also show that your algorithm is convergent with a linear convergence rate under some conditions on $\mathbf{v}_i \in \mathbb{R}^n$ for $i = 1, \dots, m$.

Exercise 23. Let $V = \{\mathbf{v}_1, \dots, \mathbf{v}_m\} \subset \mathbb{R}^n$ be a collection of vectors with $m \gg n$. We consider B satisfying (3.76). We look for a nonnegative sparsifier $\mathbf{s} = (s_i, i = 1, \dots, m) \in \mathbb{R}_+^m$ with $\|\mathbf{s}\|_0 \leq n/\epsilon$ such that

$$(1 - \epsilon)^2 B \leq \sum_{i=1}^m s_i \mathbf{v}_i \mathbf{v}_i^\top \leq (1 + \epsilon)^2 B, s_i \geq 0, \quad i = 1, \dots, m. \tag{3.78}$$

Develop an algorithm similar to Algorithm 3.8 to find the sparsifier $\mathbf{s} \in \mathbb{R}_+^m$. Also establish convergence of your algorithm under some sufficient conditions.

Exercise 24. Prove Lemma 3.6.

Bibliography

[1] S. Chen, S. A. Billings, and W. Luo, Orthogonal least squares methods and their application to non-linear system identification, Internat. J. Control, 50 (1980), 1873–1896. (Cited on p. 57)

[2] A. Cohen, W. Dahman, and R. DeVore, Orthogonal matching pursuit under the restricted isometry property, Constr. Approx., 45 (2016), 113–127. (Cited on pp. 70, 72)

[3] G. Davis, S. Mallat, and Z. Zhang, Adaptive time-frequency decompositions, Opt. Eng., 33 (1994), 2183–2191. (Cited on p. 57)

[4] R. A. DeVore and V. N. Temlyakov, Some remarks on greedy algorithms, Adv. Comput. Math., 5 (1996), 173–187. (Cited on p. 73)

[5] D. L. Donoho, Y. Tsaig, I. Drori, and J.-L. Starck, Sparse solution of underdetermined linear equations by stagewise orthogonal matching pursuit, IEEE Trans. Inform. Theory, 58 (2012), 1094–1121. (Cited on p. 57)

[6] J. H. Friedman and W. Stuetzle, Projection pursuit regressions, J. Amer. Stat. Soc., 76 (1981), 817–823. (Cited on p. 57)

[7] S. V. Konyagin and V. N. Temlyakov, Rate of convergence of pure greedy algorithm, East J. Approx., 5 (1999), 493–499. (Cited on p. 57)

[8] M.-J. Lai and Z. Shen, A Quasi-Orthogonal Matching Pursuit Algorithm for Compressive Sensing, preprint, arXiv:2007.09534v1, 18 July, 2020. (Cited on pp. 60, 78)

[9] S. Mallat and Z. Zhang. Matching pursuit in a time-frequency dictionary, IEEE Trans. Signal Process., 41 (1993), 3397–3415. (Cited on p. 57)

[10] Q. Mo, A Sharp Restricted Isometry Constant Bound of Orthogonal Matching Pursuit, preprint, arXiv:1501.01708v1, 2015. (Cited on p. 63)

[11] Q. Mo and Y. Shen, A remark on the restricted isometry property in orthogonal matching pursuit, IEEE Trans. Inform. Theory, 58 (2012), 3654–3656. (Cited on p. 63)

[12] Y. C. Pati, R. Rezaiifar, and P. S. Krishnaprasad, Orthogonal matching pursuit: Recursive function approximation with applications to wavelet decomposition, in Proceedings of the 27th Annual Asilomar Conference on Signals, Systems and Computers, 1993, 40–44. (Cited on p. 57)

[13] A. Petukhov, Fast implementation of orthogonal greedy algorithm for tight wavelet frames, Signal Process., 86 (2006), 471–479. (Cited on pp. 59, 60)

[14] V. N. Temlyakov, A criterion for convergence of weak greedy algorithms, Adv. Comput. Math., 17 (2002), 269–280. (Cited on p. 57)

[15] V. N. Temlyakov, Nonlinear methods of approximation, Found. Comput. Math., 3 (2003), 33–107. (Cited on p. 57)

[16] V. N. Temlyakov, Greedy Approximation, Cambridge University Press, 2011. (Cited on pp. 56, 57)

[17] J. A. Tropp, Greed is good: Algorithmic results for sparse approximation, IEEE Trans. Inform. Theory, 50 (2004), 2231–2242. (Cited on pp. 57, 61)

[18] J. Wang, S. Kwon, and B. Shim, Generalized Orthogonal Matching Pursuit, preprint, arXiv: 1111.6664v2, 2014. (Cited on pp. 57, 84)

[19] J. Wen, Z. Zhou, Z. Liu, M.-J. Lai, and X. Tang, Sharp sufficient conditions for stable recovery of block sparse signals by block orthogonal matching pursuit, J. Appl. Comput. Harmon. Anal., 47 (2019), 948–974. (Cited on pp. 63, 65)

Chapter 4

Hard Thresholding and Other Approaches

In this chapter, we assume that the sparsity of the solution of an underdetermined linear system is known or that we can make a reasonable guess. This information is critical in determining the convergence of the following algorithms. As we shall see, with this information, we are able to design several convergent algorithms with a known convergence rate and an estimate of how many iterative steps are needed to achieve accuracy under the assumptions of the RIP.

4.1 ▪ Iterative Hard Thresholding Algorithm

The iterative hard thresholding algorithm was first introduced and studied for sparse recovery problems in [2, 3]. Elementary analyses show that there are good theoretical guarantees for this algorithm. It is built on simple intuitions: First, solving the rectangular system $\Phi \mathbf{x} = \mathbf{b}$ amounts to solving the normal equation $\Phi^\top \Phi \mathbf{x} = \Phi^\top \mathbf{b}$, and, second, a classical iterative method suggests defining a sequence $\mathbf{x}^{(k)}$ by the recursion $\mathbf{x}^{(k+1)} = (I - \Phi^\top \Phi)\mathbf{x}^{(k)} + \Phi^\top \mathbf{b}$. Since we are looking for a sparse solution, each step should use the hard thresholding operator $H_s : \mathbb{R}^n \mapsto \mathbb{R}^n$ defined by

$$H_s(\mathbf{x}) \in \mathbb{R}^n, \tag{4.1}$$

which is the vector keeping the s largest in magnitude components of the vector $\mathbf{x}$ and setting the other components of $\mathbf{x}$ to zero. These heuristic arguments lead to the following algorithm.

ALGORITHM 4.1
Iterative Hard Thresholding Algorithm [2]

1: From the $\mathbf{x}^{(0)}$ s-sparse initial vector, we iteratively compute

$$\mathbf{x}^{(k+1)} = H_s(\mathbf{x}^{(k)} - \Phi^\top(\Phi\mathbf{x}^{(k)} - \mathbf{b})), \tag{4.2}$$

where H_s is a nonlinear operator mapping vector $\mathbf{x} \in \mathbb{R}^n$ into a new vector $\mathbf{b}$ which has the same s largest (in modulus) entries of $\mathbf{x}$ and zero in other entries.

Note that $H_s(\mathbf{x})$ is a best s-term approximation to $\mathbf{x}$, but it is not necessarily unique. We first show the following result.

Theorem 4.1 (Blumensath and Davies, 2009 [2]). *Given a noisy observation* $\mathbf{b} = \Phi\mathbf{x} + \mathbf{e}$, *where* $\mathbf{x}$ *is an arbitrary vector, let* $\mathbf{x}_S$ *be an approximation to* $\mathbf{x}$ *with no more than s nonzero elements for which* $\|\mathbf{x} - \mathbf{x}_S\|_2$ *is minimal. If* Φ *has an RIP with* $\delta_{3s} < 1/\sqrt{32}$, *then the kth* *iteration of Algorithm 4.1 will produce an approximation* $\mathbf{x}^{(k)}$ *satisfying*

$$\|\mathbf{x} - \mathbf{x}^{(k)}\|_2 \leq \left(\frac{1}{2}\right)^k \|\mathbf{x} - \mathbf{x}_S\|_2 + 6\epsilon,$$

where $\epsilon = \|\mathbf{x} - \mathbf{x}_S\|_2 + \frac{1}{\sqrt{s}}\|\mathbf{x} - \mathbf{x}_S\|_1 + \|\mathbf{e}\|_2$. *Furthermore, after at most*

$$k^* = \left\lceil \log_2\left(\frac{\|\mathbf{x}_S\|_2}{\epsilon}\right)\right\rceil \tag{4.3}$$

iterations, $\mathbf{x}^{(k^*)}$ *approximates* $\mathbf{x}$ *with accuracy*

$$\|\mathbf{x} - \mathbf{x}^{(k*)}\|_2 \leq 7\left[\|\mathbf{x} - \mathbf{x}_S\|_2 + \frac{1}{\sqrt{s}}\|\mathbf{x} - \mathbf{x}_S\|_1 + \|\mathbf{e}\|_2\right].$$

Proof. For convenience, we only prove the exact case. We leave the proof in the noisy setting to Exercise 1. That is, assume that $\mathbf{e} = 0$ and $\mathbf{x} = \mathbf{x}_S$. First we note that the error $\mathbf{x}_S - \mathbf{x}^{(k+1)}$ is supported on the set $B_{k+1} = S \cup \Gamma_{k+1}$, where Γ_{k+1} is the support set for $\mathbf{x}^{(k+1)}$. For convenience, let $\mathbf{b}^{(k+1)} = \mathbf{x}^{(k)} - \Phi^\top(\Phi\mathbf{x}^{(k)} - \mathbf{b})$. Then we have

$$\|\mathbf{x}_S - \mathbf{x}^{(k+1)}\|_2 \leq \|\mathbf{x}_S - \mathbf{b}_{B_{k+1}}^{(k+1)}\|_2 + \|\mathbf{x}_{B_{k+1}}^{(k+1)} - \mathbf{b}_{B_{k+1}}^{(k+1)}\|_2.$$

Note that by the thresholding operation, $\mathbf{x}^{(k+1)}$ is a best s-term approximation of $\mathbf{b}_{B_{k+1}}^{(k+1)}$. This implies that

$$\|\mathbf{x}_{B_{k+1}}^{(k+1)} - \mathbf{b}_{B_{k+1}}^{(k+1)}\|_2 \leq \|\mathbf{x}_S - \mathbf{b}_{B_{k+1}}^{(k+1)}\|_2,$$

and hence $\|\mathbf{x}_S - \mathbf{x}^{(k+1)}\|_2 \leq 2\|\mathbf{x}_S - \mathbf{b}_{B_{k+1}}^{(n+1)}\|_2$. We now expand

$$\mathbf{b}_{B_{k+1}}^{(k+1)} = \mathbf{x}_{B_{k+1}}^{(k)} + \Phi_{B_{k+1}}^\top \Phi(\mathbf{x}_S - \mathbf{x}^{(k)})$$

and let $\mathbf{r}^{(k)} = \mathbf{x}_S - \mathbf{x}^{(k)}$. Thus,

$$\begin{aligned}
\|\mathbf{x}_S - \mathbf{b}_{B_{k+1}}^{(k+1)}\|_2 &= \|\mathbf{r}_{B_{k+1}}^{(k)} - \Phi_{B_{k+1}}^\top \Phi(\mathbf{r}^{(k)})\|_2 \\
&\leq \|(I - \Phi_{B_{k+1}}^\top \Phi_{B_{k+1}})(\mathbf{r}_{B_{k+1}}^{(k)})\|_2 + \|\Phi_{B_{k+1}}^\top(\Phi_{B_k\backslash B_{k+1}})(\mathbf{r}_{B_k\backslash B_{k+1}}^{(k)})\| \\
&\leq \delta_{2s}\|\mathbf{r}_{B_{k+1}}^{(k)}\|_2 + \delta_{3s}\|\mathbf{r}_{B_k\backslash B_{k+1}}^{(k)}\|_2 \\
&\leq \delta_{3s}\sqrt{2}\|\mathbf{r}_{B_{k+1}}^{(k)} + \mathbf{r}_{B_k\backslash B_{k+1}}^{(k)}\|_2 = \sqrt{2}\delta_{3s}\|\mathbf{x}_S - \mathbf{x}^{(k)}\|_2,
\end{aligned}$$

where we have used the facts that $B_k\backslash B_{k+1}$ is disjoint from B_{k+1}, $|B_k \cup B_{k+1}| \leq 3s$, and $\delta_{2s} \leq \delta_{3s}$. It follows that

$$\|\mathbf{x}_S - \mathbf{x}^{(k+1)}\|_2 \leq 2\sqrt{2}\delta_{3s}\|\mathbf{x}_S - \mathbf{x}^{(k)}\|_2 \leq \cdots \leq (\sqrt{8}\delta_{3s})^k\|\mathbf{x}_S - \mathbf{x}^{(1)}\|_2.$$

If $\delta_{3s} < 1/\sqrt{32}$, we have

$$\|\mathbf{x}_S - \mathbf{x}^{(k+1)}\|_2 \leq \left(\frac{1}{2}\right)^k \|\mathbf{x}_S - \mathbf{x}^{(1)}\|_2.$$

This completes the proof.   □

We next show an improved version of the convergence of Algorithm 4.1.

Theorem 4.2 (Foucart, 2011 [9]). *Suppose that the 3s-th order restricted isometry constant (RIC) of the matrix Φ satisfies*

$$\delta_{3s} < 1/2. \tag{4.4}$$

If $\mathbf{x} \in \mathbb{R}^n$ is an s-sparse vector, then the sequence $\{\mathbf{x}^{(k)}, k \geq 1\}$ defined by Algorithm 4.1 with $\mathbf{b} = \Phi\mathbf{x}^*$ converges to the vector $\mathbf{x}^*$. More generally, if S denotes an index set of the s largest (in modulus) entries of a vector $\mathbf{x}^* \in \mathbb{R}^n$ and if $\mathbf{b} = \Phi\mathbf{x}^* + \mathbf{e}$ for some error term $\mathbf{e} \in \mathbb{R}^n$, then

$$\|\mathbf{x}^{(k)} - \mathbf{x}_S^*\|_2 \leq \rho^k \|\mathbf{x}^{(0)} - \mathbf{x}_S^*\|_2 + \tau \|\Phi\mathbf{x}_{S^c} + \mathbf{e}\|_2, \tag{4.5}$$

where $\rho = 2\delta_{3k} < 1$ and $\tau = \frac{2\sqrt{1+\delta_{2s}}}{1-2\delta_{3s}}$.

Proof. Write $\mathbf{b}^{(k)} := \mathbf{x}^{(k)} - \Phi^\top(\Phi\mathbf{x}^{(k)} - \mathbf{b})$. By using the nonlinear operator H_s, we know $\mathbf{x}^{(k+1)}$ is the best s-term approximation to $\mathbf{b}^{(k)}$ which is better than $\mathbf{x}_S$. It follows that

$$\|\mathbf{x}^{(k+1)} - \mathbf{b}^{(k)}\|_2^2 \leq \|\mathbf{x}_S - \mathbf{b}^{(k)}\|_2^2. \tag{4.6}$$

The left-hand side is

$$\|\mathbf{x}^{(k+1)} - \mathbf{x}_S + \mathbf{x}_S - \mathbf{b}^{(k)}\|_2^2$$
$$= \|\mathbf{x}^{(k+1)} - \mathbf{x}_S\|_2^2 + 2\langle \mathbf{x}^{(k+1)} - \mathbf{x}_S, \mathbf{x}_S - \mathbf{b}^{(k)}\rangle + \|\mathbf{x}_S - \mathbf{b}^{(k)}\|_2^2.$$

Using (4.6), we have

$$\|\mathbf{x}^{(k+1)} - \mathbf{x}_S\|_2^2 \leq 2\langle \mathbf{x}^{(k+1)} - \mathbf{x}_S, \mathbf{b}^{(k)} - \mathbf{x}_S\rangle.$$

Since

$$\mathbf{b}^{(k)} - \mathbf{x}_S = \mathbf{x}^{(k)} - \mathbf{x}_S + \Phi^\top(\mathbf{b} - \Phi\mathbf{x}^{(k)})$$
$$= \mathbf{x}^{(k)} - \mathbf{x}_S + \Phi^\top\Phi(\mathbf{x}_S - \mathbf{x}^{(k)}) + \Phi^\top(\Phi\mathbf{x}_{S^c} + \mathbf{e})$$
$$= (I - \Phi^\top\Phi)(\mathbf{x}^{(k)} - \mathbf{x}_S) + \Phi^\top(\Phi\mathbf{x}_{S^c} + \mathbf{e}),$$

we have

$$\langle \mathbf{x}^{(k+1)} - \mathbf{x}_S, \mathbf{b}^{(k)} - \mathbf{x}_S\rangle \leq \delta_{3s}\|\mathbf{x}^{(k+1)} - \mathbf{x}_S\|\|\mathbf{x}^{(k)} - \mathbf{x}_S\|_2$$
$$+ \|\mathbf{x}^{(k+1)} - \mathbf{x}_S\|\|\Phi^\top(\Phi\mathbf{x}_{S^c} + \mathbf{e})\|_2.$$

Combining the above two equalities and simplifying using $\|\mathbf{x}^{(k+1)} - \mathbf{x}_S\|_2$, we have

$$\|\mathbf{x}^{(k+1)} - \mathbf{x}_S\|_2 \leq 2\delta_{3s}\|\mathbf{x}^{(k)} - \mathbf{x}_S\|_2 + \|\Phi^\top(\Phi\mathbf{x}_{S^c} + \mathbf{e})\|_2. \tag{4.7}$$

This easily implies the estimate (4.5). In particular, if $\mathbf{x}$ is an s-sparse vector, i.e., $\mathbf{x}_{S^c} = 0$, and if the measurements are accurate ($\mathbf{e} = 0$), then we have

$$\|\mathbf{x}^{(k+1)} - \mathbf{x}_S\|_2 \leq 2\delta_{3s}\|\mathbf{x}^{(k)} - \mathbf{x}_S\|_2 \leq \cdots \leq \rho^k\|\mathbf{x}^{(0)} - \mathbf{x}_S\|_2. \tag{4.8}$$

This completes the proof. □

The above result can be further improved immediately by using δ_{2s} to replace δ_{3s} in (4.4).

Theorem 4.3. *Suppose that the 2sth order RIC of the matrix Φ satisfies*

$$\delta_{2s} < \frac{1}{2}. \tag{4.9}$$

If $\mathbf{x}^* \in \mathbb{R}^n$ *is an s-sparse vector, then the sequence* $\{\mathbf{x}^{(k)}, k \geq 1\}$ *defined by Algorithm* 4.1 *with* $\mathbf{b} = \Phi\mathbf{x}^*$ *converges to the vector* $\mathbf{x}^*$. *More generally, if* S *denotes an index set of the s largest (in modulus) entries of a vector* $\mathbf{x}^* \in \mathbb{R}^n$ *and if* $\mathbf{b} = \Phi\mathbf{x}^* + \mathbf{e}$ *for some error term* $\mathbf{e} \in \mathbb{R}^n$, *then*

$$\|\mathbf{x}^{(k)} - \mathbf{x}_S^*\|_2 \leq \rho^k \|\mathbf{x}^{(0)} - \mathbf{x}_S^*\|_2 + \tau\|\Phi\mathbf{x}_{S^c} + \mathbf{e}\|_2, \tag{4.10}$$

where $\rho = 2\delta_{2k} < 1$ *and* $\tau = \frac{2}{1-2\delta_{2s}}$.

Proof. Let $\mathbf{b}^{(k)} := \mathbf{x}^{(k)} - \Phi^\top(\Phi\mathbf{x}^{(k)} - \mathbf{b})$. Similar to the proof of the previous theorem, we know that $\mathbf{x}^{(k+1)}$ is the best s-term approximation to $\mathbf{b}^{(k)}$ and

$$\|\mathbf{x}^{(k+1)} - \mathbf{b}^{(k)}\|_2^2 \leq \|\mathbf{x}_S - \mathbf{b}^{(k)}\|_2^2. \tag{4.11}$$

It is easy to see that

$$\|\mathbf{x}^{(k+1)} - \mathbf{b}^{(k)}\|_2^2 = \|\mathbf{x}^{(k+1)} - \mathbf{x}_S\|_2^2 + 2\langle\mathbf{x}^{(k+1)} - \mathbf{x}_S, \mathbf{x}_S - \mathbf{b}^{(k)}\rangle + \|\mathbf{x}_S - \mathbf{b}^{(k)}\|_2^2.$$

Using (4.11), we have

$$\|\mathbf{x}^{(k+1)} - \mathbf{x}_S\|_2^2 \leq 2\langle\mathbf{x}^{(k+1)} - \mathbf{x}_S, \mathbf{b}^{(k)} - \mathbf{x}_S\rangle.$$

Since

$$\begin{aligned}
\mathbf{b}^{(k)} - \mathbf{x}_S &= \mathbf{x}^{(k)} - \mathbf{x}_S + \Phi^\top(\mathbf{b} - \Phi\mathbf{x}^{(k)}) \\
&= \mathbf{x}^{(k)} - \mathbf{x}_S + \Phi^\top\Phi(\mathbf{x}_S - \mathbf{x}^{(k)}) + \Phi^\top(\Phi\mathbf{x}_{S^c} + \mathbf{e}) \\
&= (I - \Phi^\top\Phi)(\mathbf{x}^{(k)} - \mathbf{x}_S) + \Phi^\top(\Phi\mathbf{x}_{S^c} + \mathbf{e}),
\end{aligned}$$

we now use the Cauchy–Schwarz inequality in matrix format (see (1.69) in Chapter 1) to set

$$\begin{aligned}
&\langle\mathbf{x}^{(k+1)} - \mathbf{x}_S, \mathbf{b}^{(k)} - \mathbf{x}_S\rangle \\
&= \langle\mathbf{x}^{(k+1)} - \mathbf{x}_S, (I - \Phi^\top\Phi)(\mathbf{x}^{(k)} - \mathbf{x}_S)\rangle + \langle\mathbf{x}^{(k+1)} - \mathbf{x}_S, \Phi^\top(\Phi\mathbf{x}_{S^c} + \mathbf{e})\rangle \\
&\leq \left(\langle\mathbf{x}^{(k+1)} - \mathbf{x}_S, (I - \Phi^\top\Phi)(\mathbf{x}^{(k+1)} - \mathbf{x}_S)\rangle\right)^{1/2}\left(\langle(\mathbf{x}^{(k)} - \mathbf{x}_S), (I - \Phi^\top\Phi)(\mathbf{x}^{(k)} - \mathbf{x}_S)\rangle\right)^{1/2} \\
&\quad + \|\mathbf{x}^{(k+1)} - \mathbf{x}_S\|_2\|\Phi^\top(\Phi\mathbf{x}_{S^c} + \mathbf{e})\|_2 \\
&\leq \sqrt{\delta_{2s}}\|\mathbf{x}^{(k+1)} - \mathbf{x}_S\|_2\sqrt{\delta_{2s}}\|\mathbf{x}^{(k)} - \mathbf{x}_S\|_2 + \|\mathbf{x}^{(k+1)} - \mathbf{x}_S\|_2\|\Phi^\top(\Phi\mathbf{x}_{S^c} + \mathbf{e})\|_2.
\end{aligned}$$

Therefore, we have

$$\|\mathbf{x}^{(k+1)} - \mathbf{x}_S\|_2 \leq 2\delta_{2s}\|\mathbf{x}^{(k)} - \mathbf{x}_S\|_2 + 2\|\Phi^\top(\Phi\mathbf{x}_{S^c} + \mathbf{e})\|_2.$$

We repeat the above recursive relation to obtain

$$\|\mathbf{x}^{(k+1)} - \mathbf{x}_S\|_2 \leq (2\delta_{2s})^k\|\mathbf{x}^{(1)} - \mathbf{x}_S\|_2 + \frac{2}{1-2\delta_{2s}}\|\Phi^\top(\Phi\mathbf{x}_{S^c} + \mathbf{e})\|_2.$$

This completes the proof. □

Similarly, we may use a variation of Algorithm 4.1 to find sparse solutions. See Algorithm 4.2. We leave the convergence analysis of Algorithm 4.2 to the interested reader.

ALGORITHM 4.2

Blumensath and Davies, 2009 [2]

1: Guess a sparsity $s > 0$. Let $\nu > 0$ be a real number.
2: From the $\mathbf{x}^{(0)}$ s-sparse initial vector, we iteratively compute

$$\mathbf{x}^{(k+1)} = H_s(\mathbf{x}^{(k)} - \nu\Phi^\top(\Phi\mathbf{x}^{(k)} - \mathbf{b})) \tag{4.12}$$

for $k \geq 0$.
3: Also, we may use $\nu_k > 0$ dependent on k.

4.2 ▪ Hard Thresholding Pursuit Algorithm

In this section, we study the so-called hard thresholding pursuit algorithm proposed in [9]. The algorithm is described as follows.

ALGORITHM 4.3

Hard Thresholding Pursuit Algorithm [9]

1: Fix $\nu > 0$. Start with an s-sparse vector $\mathbf{x}_0 \in \mathbb{R}^n$, typically $\mathbf{x}_0 = 0$.
2: Iteratively find

$$S_{k+1} = \text{ the indices of the } s \text{ largest entries of } \mathbf{x}^{(k)} + \nu \Phi^\top(\mathbf{b} - \Phi\mathbf{x}^{(k)})$$

and solve

$$\mathbf{x}^{(k+1)} := \text{argmin}\{\|\mathbf{b} - \Phi\mathbf{z}\|_2; \; \text{supp}(\mathbf{z}) \subset S_{k+1}\}$$

until a stopping criterion is met.
3: Output the last iterative solution $\mathbf{x}^{(k+1)}$.

A natural criterion to stop the iteration is $S_{k+1} = S_k$. How to choose ν can be seen from the result in Theorem 4.4 below. We now study the convergence of Algorithm 4.3. First of all, we note that when k is large enough, S_k appears already in one of the previous steps because the number of distinct indices of a set with cardinality s is less than or equal to $\binom{n}{s}$. When $k > \binom{n}{s}$, S_k must be appear in one of the previous support sets. That is, $S_k = S_{k-j}$ for an integer $j < k$. With this knowledge, we can prove the following theorem.

Theorem 4.4 (Foucart, 2011 [9]). *The sequence $\{\mathbf{x}^{(k)}, k \geq 1\}$ defined by Algorithm 4.3 converges in a finite number of iterations provided that $\nu\|\Phi\|_2 < 1$.*

Proof. For convenience, let H_s be the hard thresholding operator on $\mathbb{R}^n$ to $\mathbb{R}^n$ which keeps the s largest entries in magnitude of vector $\mathbf{x} \in \mathbb{R}^n$ while setting the remaining entries of $\mathbf{x}$ to zero. Let $\mathbf{u}^{(k+1)} = H_s(\mathbf{x}^{(k)} + \Phi^\top(\mathbf{b} - \Phi\mathbf{x}^{(k)}))$. Then it is easy to see that by the property of $\mathbf{u}^{(k+1)}$,

$$\|\mathbf{u}^{(k+1)} - (\mathbf{x}^{(k)} + \nu\Phi^\top(\mathbf{b} - \Phi\mathbf{x}^{(k)}))\|_2^2 \leq \|\mathbf{x}^{(k)} - (\mathbf{x}^{(k)} + \nu\Phi^\top(\mathbf{b} - \Phi\mathbf{x}^{(k)}))\|_2^2 \quad (4.13)$$
$$= \nu^2\|\Phi^\top(\mathbf{b} - \Phi\mathbf{x}^{(k)})\|_2^2.$$

The left-hand side can be expanded to obtain

$$-2\nu\langle\mathbf{u}^{(k+1)} - \mathbf{x}^{(k)}, \Phi^*(\mathbf{b} - \Phi\mathbf{x}^{(k)})\rangle + \nu^2\|\Phi^*(\mathbf{b} - \Phi\mathbf{x}^{(k)})\|_2^2 + \|\mathbf{u}^{(k+1)} - \mathbf{x}^{(k)}\|_2^2.$$

Together with (4.13), it follows that

$$-2\langle\mathbf{u}^{(k+1)} - \mathbf{x}^{(k)}, \Phi^\top(\mathbf{b} - \Phi\mathbf{x}^{(k)})\rangle \leq -\frac{1}{\nu}\|\mathbf{u}^{(k+1)} - \mathbf{x}^{(k)}\|_2^2. \quad (4.14)$$

On the other hand, we have

$$\|\mathbf{b} - \Phi\mathbf{x}^{(k+1)}\|_2 \leq \|\mathbf{b} - \Phi\mathbf{u}^{(k+1)}\|_2$$

or

$$\|\mathbf{b} - \Phi\mathbf{x}^{(k+1)}\|_2^2 - \|\mathbf{b} - \Phi\mathbf{x}^{(k)}\|_2^2 \leq \|\mathbf{b} - \Phi\mathbf{u}^{(k+1)}\|_2^2 - \|\mathbf{b} - \Phi\mathbf{x}^{(k)}\|_2^2,$$

which can be further simplified as follows:

$$\|\mathbf{b} - \Phi\mathbf{u}^{k+1}\|_2^2 = \|\mathbf{b} - \Phi\mathbf{x}^{(k)} - \Phi(\mathbf{u}^{(k+1)} - \mathbf{x}^{(k)})\|_2^2$$
$$= \|\mathbf{b} - \Phi\mathbf{x}^{(k)}\|_2^2 - 2\langle\mathbf{b} - \Phi\mathbf{x}^{(k)}, \Phi(\mathbf{u}^{(k+1)} - \mathbf{x}^{(k)})\rangle$$
$$+ \|\Phi(\mathbf{u}^{(k+1)} - \mathbf{x}^{(k)})\|_2^2,$$

and, by (4.14),

$$\|\mathbf{b} - \Phi\mathbf{x}^{(k+1)}\|_2^2 - \|\mathbf{b} - \Phi\mathbf{x}^{(k)}\|_2^2 \le \|\Phi(\mathbf{u}^{(k+1)} - \mathbf{x}^{(k)})\|_2^2 - \frac{1}{\nu}\|\mathbf{u}^{(k+1)} - \mathbf{x}^{(k)}\|_2^2$$

$$\le \left(\|\Phi\|_2^2 - \frac{1}{\nu}\right)\|\mathbf{u}^{(k+1)} - \mathbf{x}^{(k)}\|_2^2$$

$$\le \frac{\nu\|\Phi\|_2^2 - 1}{\nu}\|\mathbf{u}^{(k+1)} - \mathbf{x}^{(k)}\|_2^2 \le 0.$$

It follows that

$$\|\mathbf{b} - \Phi\mathbf{x}^{(k+1)}\|_2 \le \|\mathbf{b} - \Phi\mathbf{x}^{(k)}\|_2 \qquad (4.15)$$

for all k. The sequence decreases and hence is convergent. Note that the periodicity of S_k implies that the sequence will be constant after a finite number of iterations. It follows that $\mathbf{u}^{(k+1)} = \mathbf{x}^{(k)}$ or $\mathbf{x}^{(k)}$ is a sparse solution.   □

In fact, when the cardinality of S_{k+1} is less than or equal to s, we can use the RIP of Φ. That is,

$$\|\Phi(\mathbf{u}^{(k+1)} - \mathbf{x}^{(k)})\|_2^2 \le (1 + \delta_{2s})\|\mathbf{u}^{(k+1)} - \mathbf{x}^{(k)}\|_2^2.$$

This inequality can be used in the previous proof. We can immediately obtain the following result.

Theorem 4.5. *The sequence $\{\mathbf{x}^{(k)}, k \ge 0\}$ defined by Algorithm 4.3 converges in a finite number of iterations if $\nu(1 + \delta_{2s}) < 1$. In particular, $\nu < 1/2$ will make the sequence converge.*

Proof. We leave the proof to Exercise 14.   □

Next we study the convergence rate. We have the following theorem.

Theorem 4.6 (Foucart, 2011 [9]). *Suppose that the sensing matrix Φ satisfies $\delta_{2s} < 1/\sqrt{3}$. Then for any s-sparse solution $\mathbf{x}^*$ for $\Phi\mathbf{x}^* = \mathbf{b}$, the sequence $\{\mathbf{x}^{(k)}, k \ge 0\}$ defined by Algorithm 4.3 converges geometrically with a rate given by*

$$\|\mathbf{x}^{(k)} - \mathbf{x}^*\|_2 \le \rho^k\|\mathbf{x}^{(0)} - \mathbf{x}^*\|_2, \text{ where } \rho = \sqrt{\frac{2\delta_{2s}^2}{1 - \delta_{2s}^2}} < 1.$$

Proof. We note that $\Phi\mathbf{x}^{(k+1)}$ is the best ℓ_2 approximation to $\mathbf{b}$ from the space $\{\Phi\mathbf{z},\ \mathrm{supp}\,(\mathbf{z}) \subset S_{k+1}\}$. Thus,

$$\langle \Phi\mathbf{x}^{(k+1)} - \mathbf{b}, \Phi\mathbf{z} \rangle = 0 \quad \forall\ \mathrm{supp}\,(\mathbf{z}) \subset S_{k+1}.$$

In other words,

$$\langle \mathbf{x}^{(k+1)} - \mathbf{x}^*, \Phi^\top\Phi\mathbf{z} \rangle = 0 \quad \forall\ \mathrm{supp}\,(\mathbf{z}) \subset S_{k+1}.$$

We further derive

$$\|(\mathbf{x}^{(k+1)} - \mathbf{x}^*)_{S_{k+1}}\|_2^2 = \langle \mathbf{x}^{(k+1)} - \mathbf{x}^*, (\mathbf{x}^{(k+1)} - \mathbf{x}^*)_{S_{k+1}} \rangle$$

$$= \langle \mathbf{x}^{(k+1)} - \mathbf{x}^*, (I - \Phi^\top\Phi)(\mathbf{x}^{(k+1)} - \mathbf{x}^*)_{S_{k+1}} \rangle$$

$$\le \delta_{2s}\|\mathbf{x}^{(k+1)} - \mathbf{x}^*\|_2\,\|(\mathbf{x}^{(k+1)} - \mathbf{x}^*)_{S_{k+1}}\|_2.$$

That is, $\|(\mathbf{x}^{(k+1)} - \mathbf{x}^*)_{S_{k+1}}\|_2 \leq \delta_{2s}\|\mathbf{x}^{(k+1)} - \mathbf{x}^*\|_2$. It follows that

$$\|\mathbf{x}^{(k+1)} - \mathbf{x}^*\|_2^2 = \|(\mathbf{x}^{(k+1)} - \mathbf{x}^*)_{S_{k+1}}\|_2^2 + \|(\mathbf{x}^{(k+1)} - \mathbf{x}^*)_{S_{k+1}^c}\|_2^2$$

$$\leq \delta_{2s}^2\|\mathbf{x}^{(k+1)} - \mathbf{x}^*\|_2^2 + \|(\mathbf{x}^{(k+1)} - \mathbf{x}^*)_{S_{k+1}^c}\|_2^2.$$

After a rearrangement, we obtain

$$\|(\mathbf{x}^{(k+1)} - \mathbf{x}^*)\|_2^2 \leq \frac{1}{1 - \delta_{2s}^2}\|(\mathbf{x}^{(k+1)} - \mathbf{x}^*)_{S_{k+1}^c}\|_2^2. \tag{4.16}$$

Next we let $S = \text{supp}(\mathbf{x}^*)$, and it is easy to see that

$$\|(\mathbf{x}^{(k)} + \Phi^\top(\mathbf{x}^* - \Phi\mathbf{x}^{(k)}))_S\|_2^2 \leq \|(\mathbf{x}^{(k)} + \Phi^\top(\mathbf{x}^* - \Phi\mathbf{x}^{(k)}))_{S_{k+1}}\|_2^2$$

by the definition of S_{k+1}. It follows that

$$\|(\mathbf{x}^{(k)} + \Phi^\top(\mathbf{x}^* - \Phi\mathbf{x}^{(k)}))_{S\setminus S_{k+1}}\|_2^2 \leq \|(\mathbf{x}^{(k)} + \Phi^\top(\mathbf{x}^* - \Phi\mathbf{x}^{(k)}))_{S_{k+1}\setminus S}\|_2^2. \tag{4.17}$$

The right-hand side can be further rewritten as

$$\|(\mathbf{x}^{(k)} + \Phi^\top(\mathbf{x}^* - \Phi\mathbf{x}^{(k)}))_{S_{k+1}\setminus S}\|_2^2 = \|((I - \Phi^\top\Phi)(\mathbf{x}^{(k)} - \mathbf{x}^*))_{S_{k+1}\setminus S}\|_2^2.$$

As for the left-hand side, we have

$$\|(\mathbf{x}^{(k)} + \Phi^\top(\mathbf{x}^* - \Phi\mathbf{x}^{(k)}))_{S\setminus S_{k+1}}\|_2$$

$$= \|(\mathbf{x}^* - \mathbf{x}^{(k+1)})_{S_{k+1}^c} + ((I - \Phi^\top\Phi)(\mathbf{x}^{(k)} - \mathbf{x}^*))_{S\setminus S_{k+1}}\|_2$$

$$\geq \|(\mathbf{x}^* - \mathbf{x}^{(k+1)})_{S_{k+1}^c}\|_2 - \|((I - \Phi^\top\Phi)(\mathbf{x}^{(k)} - \mathbf{x}^*))_{S\setminus S_{k+1}}\|_2.$$

That is, by using (4.17), we use the Cauchy–Schwarz inequality to obtain

$$\|(\mathbf{x}^* - \mathbf{x}^{(k+1)})_{S_{k+1}^c}\|_2$$

$$\leq \|(\mathbf{x}^{(k)} + \Phi^\top(\mathbf{x}^* - \Phi\mathbf{x}^{(k)}))_{S\setminus S_{k+1}}\|_2 + \|((I - \Phi^\top\Phi)(\mathbf{x}^{(k)} - \mathbf{x}^*))_{S\setminus S_{k+1}}\|_2$$

$$\leq \|(\mathbf{x}^{(k)} + \Phi^\top(\mathbf{x}^* - \Phi\mathbf{x}^{(k)}))_{S_{k+1}\setminus S}\|_2 + \|((I - \Phi^\top\Phi)(\mathbf{x}^{(k)} - \mathbf{x}^*))_{S\setminus S_{k+1}}\|_2$$

$$\leq \sqrt{2}\|((I - \Phi^\top\Phi)(\mathbf{x}^{(k)} - \mathbf{x}^*))_{S\triangle S_{k+1}}\|_2 \leq \sqrt{2}\delta_{2s}\|\mathbf{x}^{(k)} - \mathbf{x}^*\|_2.$$

Here $S\triangle S_{k+1} := (S\setminus S_{k+1})\bigcup(S_{k+1}\setminus S)$ is the symmetric difference of S and S_{k+1}. We combine (4.16) with the above estimate to obtain

$$\|\mathbf{x}^{(k+1)} - \mathbf{x}\|_2 \leq \sqrt{\frac{2\delta_{2s}^2}{1 - \delta_{2s}^2}}\|\mathbf{x}^{(k)} - \mathbf{x}^*\|_2.$$

This completes the proof.   □

The number of iterations to find the sparse solution can be estimated as follows.

Theorem 4.7. *Suppose that the matrix Φ satisfies $\delta_{2s} < 1/\sqrt{3}$. Then any s-sparse vector $\mathbf{x} \in \mathbb{R}^n$ is recovered by Algorithm 4.3 with $\mathbf{b} = \Phi\mathbf{x}$ in at most $k > K_0$ steps with*

$$K_0 := \frac{\ln(\sqrt{2/3}\|\mathbf{x}^0 - \mathbf{x}\|_2/\xi)}{\ln(1/\rho)} + 1, \tag{4.18}$$

where ξ is the absolute value of the smallest nonzero entries of $\mathbf{x}$ in modulus and $\rho = \sqrt{\frac{2\delta_{2s}^2}{1 - \delta_{2s}^2}}$.

Proof. Our goal is to show that there is a k such that $S_k = S$, where S_k and S are index sets of the supports of $\mathbf{x}^{(k)}$ and the sparse solution $\mathbf{x}$. Thus, we need to show that for all $j \in S$ and all $\ell \in S^c$ we have

$$|(\mathbf{x}^{(k-1)} + \Phi^\top\Phi(\mathbf{x} - \mathbf{x}^{(k-1)}))_j| > |(\mathbf{x}^{(k-1)} + \Phi^\top\Phi(\mathbf{x} - \mathbf{x}^{(k-1)}))_\ell|. \qquad (4.19)$$

Note that

$$|(\mathbf{x}^{(k-1)} + \Phi^\top\Phi(\mathbf{x} - \mathbf{x}^{(k-1)}))_j| = |x_j + ((I - \Phi^\top\Phi)(\mathbf{x}^{(k-1)} - \mathbf{x}))_j|$$
$$\geq \xi - |((I - \Phi^\top\Phi)(\mathbf{x}^{(k-1)} - \mathbf{x}))_j|.$$

Then it follows that

$$\xi \leq |(\mathbf{x}^{(k-1)} + \Phi^\top\Phi(\mathbf{x} - \mathbf{x}^{(k-1)}))_j| + |((I - \Phi^\top\Phi)(\mathbf{x}^{(k-1)} - \mathbf{x}))_j|. \qquad (4.20)$$

On the other hand, we have

$$|((I - \Phi^\top\Phi)(\mathbf{x}^{(k-1)} - \mathbf{x}))_j| + |((I - \Phi^\top\Phi)(\mathbf{x}^{(k-1)} - \mathbf{x}))_\ell|$$
$$\leq \sqrt{2}\|((I - \Phi^\top\Phi)(\mathbf{x}^{(k-1)} - \mathbf{x}))_{j,\ell}\|_2 \leq \sqrt{2}\delta_{2s}\|\mathbf{x}^{(k-1)} - \mathbf{x}\|_2$$
$$< \sqrt{2/3}\rho^{k-1}\|\mathbf{x}^0 - \mathbf{x}\|_2$$

and

$$|(\mathbf{x}^{(k-1)} - \Phi^\top\Phi(\mathbf{x}^{(k-1)} - \mathbf{x}))_\ell| = |((I - \Phi^\top\Phi)(\mathbf{x}^{(k-1)} - \mathbf{x}))_\ell|.$$

We see that (4.19) holds as long as

$$\xi \geq \sqrt{2/3}\rho^{k-1}\|\mathbf{x}^0 - \mathbf{x}\|_2.$$

The smallest integer k is the one given in (4.18). This completes the proof. $\qquad\square$

Finally, we study the approximate recovery of vectors from noisy measurements. Mainly, we solve

$$\min_{\mathbf{x}\in\mathbb{R}^n} \{\|\mathbf{x}\|_0, \quad \|\Phi\mathbf{x} - \mathbf{b}\|_2 \leq \epsilon\}.$$

We begin with the following lemma.

Lemma 4.8. *For any index set* $S \subset \{1, 2, \ldots, N\}$ *with* $\#(S) \leq s$, *for any* $\mathbf{e} \in \mathbb{R}^m$,

$$\|(\Phi^\top\mathbf{e})_S\|_2 \leq \sqrt{1 + \delta_s}\|\mathbf{e}\|_2. \qquad (4.21)$$

Proof. It is straightforward to obtain

$$\|(\Phi^\top\mathbf{e})_S\|_2^2 \leq \langle \Phi^\top\mathbf{e}, (\Phi^\top\mathbf{e})_S \rangle = \langle \mathbf{e}, \Phi(\Phi^\top\mathbf{e})_S \rangle$$
$$\leq \|\mathbf{e}\|_2\|\Phi(\Phi^\top\mathbf{e})_S\|_2 \leq \|\mathbf{e}\|_2\sqrt{1 + \delta_s}\|(\Phi^\top\mathbf{e})_S\|_2.$$

The inequality in (4.21) follows. $\qquad\square$

Theorem 4.9. *Suppose that* Φ *satisfies* $\delta_{2s} < 1/\sqrt{3}$. *Then for any* $\mathbf{x} \in \mathbb{R}^n$ *and any* $\mathbf{e} \in \mathbb{R}^m$, *if* S *is the index set of the* s *largest (in magnitude) entries of* $\mathbf{x}$, *the sequence* $\{\mathbf{x}^{(k)}, k \geq 0\}$ *from Algorithm 4.3 with* $\mathbf{b} = \Phi\mathbf{x} + \mathbf{e}$ *satisfies*

$$\|\mathbf{x}^{(k)} - \mathbf{x}_S\|_2 \leq \rho^k\|\mathbf{x}^0 - \mathbf{x}_S\|_2 + \tau\frac{1 - \rho^k}{1 - \rho}\|\Phi\mathbf{x}_S + \mathbf{e}\|_2 \qquad (4.22)$$

for all $k \geq 1$, where ρ is as defined before and

$$\tau = \frac{\sqrt{2(1 - \delta_{2s})} + \sqrt{1 + \delta_s}}{1 - \delta_{2s}} \leq 5.15.$$

Proof. We first notice that the least squares minimization yields

$$\langle \mathbf{x}^{(k+1)} - \mathbf{x}_S, \Phi^\top \Phi \mathbf{z} \rangle = \langle \Phi \mathbf{x}_{S^c} + \mathbf{e}, \Phi \mathbf{z} \rangle =: \langle \widetilde{\mathbf{e}}, \Phi \mathbf{z} \rangle$$

for all $\mathbf{z} \in \mathbb{R}^n$ with support in S_{k+1}, where $\widetilde{\mathbf{e}} = \Phi \mathbf{x}_{S^c} + \mathbf{e}$. We now have

$$\|(\mathbf{x}^{(k+1)} - \mathbf{x}_S)_{S_{k+1}}\|_2^2 = \langle \mathbf{x}^{(k+1)} - \mathbf{x}_S, (\mathbf{x}^{(k+1)} - \mathbf{x}_S)_{S_{k+1}} \rangle$$
$$= \langle \mathbf{x}^{(k+1)} - \mathbf{x}_S, (I - \Phi^\top \Phi)(\mathbf{x}^{(k+1)} - \mathbf{x}_S)_{S_{k+1}} \rangle + \langle \widetilde{\mathbf{e}}_{S_{k+1}}, \Phi(\mathbf{x}^{(k+1)} - \mathbf{x}_S) \rangle$$
$$\leq \delta_{2s} \|\mathbf{x}^{(k+1)} - \mathbf{x}_S\|_2 \|(\mathbf{x}^{(k+1)} - \mathbf{x}_S)_{S_{k+1}}\|_2 + \|\widetilde{\mathbf{e}}\|_2 \sqrt{1 + \delta_s} \|(\mathbf{x}^{(k+1)} - \mathbf{x}_S)_{S_{k+1}}\|_2.$$

It follows that

$$\|(\mathbf{x}^{(k+1)} - \mathbf{x}_S)_{S_{k+1}}\|_2 \leq \delta_{2s} \|\mathbf{x}^{(k+1)} - \mathbf{x}_S\|_2 + \|\widetilde{\mathbf{e}}\|_2 \sqrt{1 + \delta_s}.$$

Hence, we have

$$\|\mathbf{x}^{(k+1)} - \mathbf{x}_S\|_2^2 = \|(\mathbf{x}^{(k+1)} - \mathbf{x}_S)_{S_{k+1}^c}\|_2^2 + \|(\mathbf{x}^{(k+1)} - \mathbf{x}_S)_{S_{k+1}}\|_2^2$$
$$\leq \|(\mathbf{x}^{(k+1)} - \mathbf{x}_S)_{S_{k+1}^c}\|_2^2 + \left(\delta_{2s} \|\mathbf{x}^{(k+1)} - \mathbf{x}_S\|_2 + \|\widetilde{\mathbf{e}}\|_2 \sqrt{1 + \delta_s}\right)^2.$$

Let $t = \|\mathbf{x}^{(k+1)} - \mathbf{x}_S\|_2$. The above inequality can be rewritten as $P(t) \leq 0$ for a quadratic polynomial defined by

$$P(t) = (1 - \delta_{2s}^2)t^2 - (2\delta_{2s}\sqrt{1 + \delta_s}\|\widetilde{\mathbf{e}}\|_2)t - \left(\|(\mathbf{x}^{(k+1)} - \mathbf{x}_S)_{S_{k+1}^c}\|_2^2 + (1 + \delta_s)\|\widetilde{\mathbf{e}}\|_2^2\right).$$

Hence, $\|\mathbf{x}^{(k+1)} - \mathbf{x}_S\|_2$ is bounded by the largest root of $P(t)$. An easy calculation can find the bound for $\|\mathbf{x}^{(k+1)} - \mathbf{x}_S\|_2$ from $P(t)$. The conclusion of this theorem follows. We refer the reader to [9] for details. □

In general, we may consider a slightly more general hard thresholding pursuit algorithm (HTP_ν) with the factor ν dependent on $k \geq 1$.

ALGORITHM 4.4
General Hard Thresholding Pursuit Algorithm [9]
1: Start with an s-sparse vector $\mathbf{x}_0 \in \mathbb{R}^n$, typically $\mathbf{x}_0 = 0$, and iteratively find

$$S_{k+1} = \text{ the indices of the } s \text{ largest entries of } \mathbf{x}^{(k)} + \nu_k \Phi^\top (\mathbf{b} - \Phi \mathbf{x}^{(k)}),$$

where $\nu_k = \|\Phi^\top(\mathbf{b} - \Phi\mathbf{x}^{(k)})\|_2 / \|\Phi(\Phi^\top(\mathbf{b} - \Phi\mathbf{x}^{(k)}))\|_2$, and solve

$$\mathbf{x}^{(k+1)} := \text{argmin} \|\mathbf{b} - \Phi\mathbf{z}\|_2; \ \text{supp}(\mathbf{z}) \subset S_{k+1}$$

until a stopping criterion is met.

We leave the study of Algorithm 4.4 to the interested reader. Additional studies can be found in [4].

4.3 ▪ Compressive Sensing Matching Pursuit Algorithm

The compressive sampling matching pursuit (CoSaMP) algorithm is proposed by Needell and Tropp in [13]. They do not provide better theoretical guarantees than the simple iterative hard thresholding algorithm discussed in section 4.1, but they do offer better empirical performance. It was devised to enhance the orthogonal matching pursuit (OMP) and orthogonal greedy algorithms (OGAs) discussed in Chapter 3. The basic idea consists of choosing a good candidate for the support and then finding the vector with this support that best fits the measurements. The algorithm can be expressed in the following way: let $\mathbf{b} = \Phi \mathbf{x_b} + \mathbf{e}$ be a given vector with noise.

ALGORITHM 4.5

Compressive Sensing Matching Pursuit (CoSaMP) Algorithm [13]

1: Starting with an s-sparse $\mathbf{x}^0$, e.g., $\mathbf{x}^0 = 0$, we iteratively let

$$U_k = \text{supp}(\mathbf{x}^{(k)}) \cup \{\text{indices of } 2s \text{ largest entries of } \Phi^\top(\mathbf{b} - \Phi\mathbf{x}^{(k)})\}, \qquad (4.23)$$

$$\mathbf{u}^{(k)} = \text{argmin}\{\|\mathbf{b} - \Phi\mathbf{z}\|_2, \; \text{supp}(\mathbf{z}) \subset U_k\} \qquad (4.24)$$

and find

$$\mathbf{x}^{(k+1)} = H_s(\mathbf{u}^{(k)}) \qquad (4.25)$$

for $k \geq 0$, where H_s is the hard thresholding operator defined in Algorithm 4.1.

The ideas of Algorithm 4.5 can be explained as follows. We first identify a possible support of the sparse solution. That is, we find the indices of the $2s$ largest entries of $\Phi^\top(\mathbf{y}^{(k)})$ which could be potential entries of the sparse solution. It is possible that the support of the previous iteration, $\mathbf{x}^{(k-1)}$, contains the potential entries. We merge them to get the first step, (4.23), in Algorithm 4.5. Then we estimate by finding a least squares solution over the support U_k in (4.24). Finally, we prune the least squares solution $\mathbf{u}^{(k)}$ to get the updated $\mathbf{x}^{(k+1)}$ in (4.25).

We now examine the convergence of the above algorithm. We shall provide two kinds of analysis: one is by the creators of this algorithm (cf. [13]) and the other is by Simon Foucart in [9]. Let us first fix some notation. Let $\mathbf{r}^k = \mathbf{x_b} - \mathbf{x}^{(k)}$ be the residual vector, and let $H_{2s}(\mathbf{x})$ be a hard thresholding operator which truncates $\mathbf{x}$ to a new vector $H_{2s}(\mathbf{x})$ by keeping the $2s$ largest entries of $\mathbf{x}$ in magnitude unchanged and setting the remaining entries to zero. We start with the following lemmas.

Lemma 4.10. *Let Λ_k be the support of $\mathbf{y}^k := H_{2s}(\Phi^\top(\mathbf{b} - \Phi\mathbf{x}^k))$. Then*

$$\|\mathbf{r}^k|_{\Lambda_k^c}\|_2 \leq \frac{\delta_{2s} + \delta_{4s}}{1 - \delta_{2s}} \|\mathbf{r}^k\|_2 + \frac{2\sqrt{1 + \delta_{2s}}}{1 - \delta_{2s}} \|\mathbf{e}\|_2,$$

where Λ_k^c is the complement of Λ_k in $\{1, 2, \ldots, n\}$.

Proof. Let R_k be the set of indices of the support of $\mathbf{r}^k$. It is known that the cardinality $\#(R_k)$ is less than or equal to $2s$. Because $\#(\Lambda_k) = 2s$, we know $\|\mathbf{y}^k|_{R_k}\|_2 \leq \|\mathbf{y}^k|_{\Lambda_k}\|_2$. It follows that $\|\mathbf{y}^k|_{R_k \setminus \Lambda_k}\|_2 \leq \|\mathbf{y}^k|_{\Lambda_k \setminus R_k}\|_2$.

On one hand, we have

$$\|\mathbf{y}^k|_{\Lambda_k \setminus R_k}\|_2 = \|(\Phi|_{\Lambda_k \setminus R_k})^\top(\mathbf{b} - \Phi\mathbf{x}^k)\|_2 = \|(\Phi|_{\Lambda_k \setminus R_k})^\top(\Phi\mathbf{r}^k + \mathbf{e})\|_2$$
$$\leq \|(\Phi|_{\Lambda_k \setminus R_k})^\top \Phi\mathbf{r}^k\|_2 + \|(\Phi|_{\Lambda_k \setminus R_k})^\top \mathbf{e}\|_2$$
$$\leq \delta_{4s}\|\mathbf{r}^k\|_2 + \sqrt{1 + \delta_{2s}}\|\mathbf{e}\|_2$$

by Exercise 9 in Chapter 2. On the other hand, we have

$$
\begin{aligned}
\|\mathbf{y}^k|_{R_k\backslash\Lambda_k}\|_2 &= \|(\Phi|_{R_k\backslash\Lambda_k})^\top(\Phi\mathbf{r}^k + \mathbf{e})\|_2 \\
&\geq \|(\Phi|_{R_k\backslash\Lambda_k})^\top\Phi\mathbf{r}^k\|_2 - \|(\Phi|_{R_k\backslash\Lambda_k})^\top\mathbf{e}\|_2 \\
&\geq \|(\Phi|_{R_k\backslash\Lambda_k})^\top\Phi\mathbf{r}^k|_{R_k\backslash\Lambda_k}\| - \|(\Phi|_{R_k\backslash\Lambda_k})^\top\Phi\mathbf{r}^k|_{\Lambda^k}\| - \sqrt{1+\delta_{2s}}\|\mathbf{e}\|_2 \\
&\geq (1-\delta_{2s})\|\mathbf{r}^k|_{R_k\backslash\Lambda_k}\|_2 - \delta_{2s}\|\mathbf{r}^k\|_2 - \sqrt{1+\delta_{2s}}\|\mathbf{e}\|_2.
\end{aligned}
$$

Combining the two inequalities, we have

$$
\begin{aligned}
\|\mathbf{r}^k|_{R_k\backslash\Lambda_k}\| &\leq \frac{\delta_{2s}}{1-\delta_{2s}}\|\mathbf{r}^k\|_2 + \frac{1}{1-\delta_{2s}}\|\mathbf{y}^k|_{R_k\backslash\Lambda_k}\|_2 + \frac{\sqrt{1+\delta_{2s}}}{1-\delta_{2s}}\|\mathbf{e}\|_2 \\
&\leq \frac{\delta_{2s}}{1-\delta_{2s}}\|\mathbf{r}^k\|_2 + \frac{1}{1-\delta_{2s}}\|\mathbf{y}^k|_{\Lambda_k\backslash R_k}\| + \frac{\sqrt{1+\delta_{2s}}}{1-\delta_{2s}}\|\mathbf{e}\|_2 \\
&\leq \frac{\delta_{2s}+\delta_{4s}}{1-\delta_{2s}}\|\mathbf{r}^k\|_2 + \frac{2\sqrt{1+\delta_{2s}}}{1-\delta_{2s}}\|\mathbf{e}\|_2.
\end{aligned}
$$

This completes the proof. $\quad\square$

Next, recall that $U_k = \Lambda_k \cup \operatorname{support}(\mathbf{x}^{(k)})$ from (4.23). The cardinality $\#(U_k) \leq 3s$.

Lemma 4.11.

$$
\|\mathbf{x_b}|_{U_k^c}\|_2 \leq \|\mathbf{r}^k|_{\Lambda_k^c}\|_2. \tag{4.26}
$$

Proof. The proof is easy due to

$$
\|\mathbf{x_b}|_{U_k^c}\|_2 = \|(\mathbf{x_b}-\mathbf{x}^{(k)})|_{U_k^c}\|_2 = \|\mathbf{r}^k|_{U_k^c}\|_2 \leq \|\mathbf{r}^k|_{\Lambda_k^c}\|_2. \quad\square
$$

Furthermore, we have the following results.

Lemma 4.12.

$$
\|\mathbf{x_b}-\mathbf{u}^k\|_2 \leq \left(1+\frac{\delta_{4s}}{1-\delta_{3s}}\right)\|\mathbf{x_b}|_{U_k^c}\|_2 + \frac{1}{\sqrt{1-\delta_{3s}}}\|\mathbf{e}\|_2. \tag{4.27}
$$

Proof. We use the following sequence of inequalities. First of all, we have

$$
\begin{aligned}
\|\mathbf{x_b}-\mathbf{u}^{(k)}\|_2 &\leq \|(\mathbf{x_b}-\mathbf{u}^{(k)})_{U_k}\|_2 + \|(\mathbf{x_b}-\mathbf{u}^{(k)})_{U_k^c}\|_2 \\
&= \|(\mathbf{x_b}-\mathbf{u}^{(k)})_{U_k}\|_2 + \|(\mathbf{x_b})_{U_k^c}\|_2.
\end{aligned}
$$

Now we use an estimate of the pseudo-inverse of Φ, i.e., (2.41) in Chapter 2, to obtain

$$
\begin{aligned}
\|(\mathbf{x_b}-\mathbf{u}^{(k)})_{U_k}\|_2 &= \|\mathbf{x_b}|_{U_k} - \Phi_{U_k}^\dagger(\Phi\mathbf{x_b}+\mathbf{e})|_{U_k}\|_2 \\
&= \|\Phi_{U_k}^\dagger(\Phi_{U_k^c}\mathbf{x_b}|_{U_k^c}+\mathbf{e}|_{U_k})\|_2 \\
&\leq \|\Phi_{U_k}^\dagger(\Phi_{U_k^c}\mathbf{x_b}|_{U_k^c})\|_2 + \|\Phi_{U_k}^\dagger\mathbf{e}|_{U_k}\|_2 \\
&\leq \frac{1}{1-\delta_{3s}}\|\Phi|_{U_k^c}\mathbf{x_b}|_{U_k^c}\|_2 + \frac{1}{\sqrt{1-\delta_{3s}}}\|\mathbf{e}\|_2 \\
&\leq \frac{\delta_{4s}}{1-\delta_{3s}}\|\mathbf{x_b}|_{U_k^c}\|_2 + \frac{1}{\sqrt{1-\delta_{3s}}}\|\mathbf{e}\|_2.
\end{aligned}
$$

Now we combine the above two estimates to conclude the inequality in (4.27). $\quad\square$

Lemma 4.13.

$$\|\mathbf{x_b} - H_s(\mathbf{u}^k)\|_2 \le 2\|\mathbf{x_b} - \mathbf{u}^k\|_2. \tag{4.28}$$

Proof. It is easy to see that

$$\|\mathbf{x_b} - H_s(\mathbf{u}^k)\|_2 \le \|\mathbf{x_b} - \mathbf{u}^k\|_2 + \|\mathbf{u}^k - H_s(\mathbf{u}^k)\|_2 \le 2\|\mathbf{x_b} - \mathbf{u}^k\|_2,$$

where we have used the fact that $\|\mathbf{u}^k - H_s(\mathbf{u}^k)\|_2 \le \|\mathbf{u}^k - \mathbf{x_b}\|_2$. That is, $H_s(\mathbf{u}^k)$ is the best s-term approximation of $\mathbf{u}^k$. $\square$

We are now ready to present a proof of the convergence of Algorithm 4.5.

Theorem 4.14 (Needel and Tropp, 2009 [13]). *Suppose that the RICs* $\delta_{2s}, \delta_{3s},$ *and* δ_{4s} *satisfy*

$$\rho := 2\left(1 + \frac{\delta_{4s}}{1 - \delta_{3s}}\right)\frac{\delta_{3s} + \delta_{4s}}{1 - \delta_{2s}} < 1.$$

Then Algorithm 4.5 converges and

$$\|\mathbf{x_b} - \mathbf{x}^{(k)}\|_2 \le \rho^k\|\mathbf{x_b} - \mathbf{x}^{(0)}\|_2 + \frac{C}{1 - \rho}\|\mathbf{e}\|_2$$

for all $k \ge 1$.

Proof. Let $C_1 = \delta_{4s}/(1 - \delta_{3s})$ and $C_2 = 1/\sqrt{1 - \delta_{3s}}$ be the two positive constants in inequality (4.27). We use the inequalities in (4.28), (4.27), and (4.26) to get

$$\begin{aligned}
\|\mathbf{x_b} - \mathbf{x}^{(k+1)}\|_2 &= \|\mathbf{x_b} - H_s(\mathbf{u}^{(k)})\|_2 \le 2\|\mathbf{x_b} - \mathbf{u}^{(k)}\|_2 \\
&\le 2C_1\|\mathbf{x_b}|_{U^c_{k+1}}\|_2 + 2C_2\|\mathbf{e}\|_2 \le 2C_1\|\mathbf{r}^k|_{\Lambda^c_k}\|_2 + 2C_2\|\mathbf{e}\|_2 \\
&\le 2C_1\frac{\delta_{3s} + \delta_{4s}}{1 - \delta_{3s}}\|\mathbf{r}^k\|_2 + 2C_1\frac{1}{1 - \delta_{2s}}\|\mathbf{e}\|_2 + 2C_2\|\mathbf{e}\|_2 \\
&= 2\left(1 + \frac{\delta_{4s}}{1 - \delta_{3s}}\right)\frac{\delta_{3s} + \delta_{4s}}{1 - \delta_{2s}}\|\mathbf{x_b} - \mathbf{x}^k\|_2 + C\|\mathbf{e}\|_2,
\end{aligned}$$

where $C = 2C_1/(1 - \delta_{2s}) + 2C_2$ is a positive constant. If

$$\rho = 2\left(1 + \frac{\delta_{4s}}{1 - \delta_{3s}}\right)\frac{\delta_{3s} + \delta_{4s}}{1 - \delta_{2s}} < 1,$$

then we have

$$\|\mathbf{x_b} - \mathbf{x}^{(k+1)}\|_2 \le \rho\|\mathbf{x_b} - \mathbf{x}^k\|_2 + C\|\mathbf{e}\|_2 \le \cdots \le \rho^{k+1}\|\mathbf{x_b} - \mathbf{x}^{(0)}\|_2 + C\sum_{j=0}^{\infty}\rho^j\|\mathbf{e}\|_2.$$

This completes the proof. $\square$

In particular, when $\delta_{4s} \le 1/9$, we have $\rho < 1$ (see Exercise 2 in this chapter). The above discussion establishes Algorithm 4.5. We now further study the conditions to ensure the convergence of this algorithm. The following analysis of Algorithm 4.5 is based on [9].

Theorem 4.15. *Suppose that the RIC of the matrix Φ satisfies*

$$\delta_{4s} < \sqrt{\frac{2}{5 + \sqrt{73}}} \approx 0.38427.$$

If $\mathbf{b} = \Phi\mathbf{x_b}$ with a sparse solution $\mathbf{x_b}$ with sparsity s, then the sequence $\mathbf{x}^{(k)}, k \geq 0$, obtained from Algorithm 4.5 converges and

$$\|\mathbf{x}^{(k)} - \mathbf{x_b}\|_2 \leq \rho^k \|\mathbf{x}^{(0)} - \mathbf{x_b}\|_2 \tag{4.29}$$

for all $k \geq 0$, where $\rho = \sqrt{\frac{1+3\delta_{4s}^2}{1-\delta_{4s}^2}}(\delta_{2s} + \delta_{4s}) < 1$.

Proof. Since $\Phi\mathbf{u}^{(k)}$ is the best ℓ_2 approximation to $\mathbf{b}$ from the subspace $\{\Phi\mathbf{z}, \text{supp}(\mathbf{z}) \subset U_k\}$, we know that

$$\langle \Phi\mathbf{u}^{(k)} - \mathbf{b}, \Phi\mathbf{z} \rangle = 0, \qquad \text{supp}(\mathbf{z}) \subset U_k.$$

In other words, we have

$$\langle \mathbf{u}^{(k)} - \mathbf{x_b}, \Phi^\top \Phi\mathbf{z} \rangle = 0 \tag{4.30}$$

and

$$\begin{aligned}
\|(\mathbf{u}^{(k)} - \mathbf{x_b})_{U_k}\|_2^2 &= \langle \mathbf{u}^{(k)} - \mathbf{x_b}, (\mathbf{u}^{(k)} - \mathbf{x_b})_{U_k} \rangle \\
&= \langle \mathbf{u}^{(k)} - \mathbf{x_b}, (I - \Phi^\top \Phi)(\mathbf{u}^{(k)} - \mathbf{x_b})_{U_k} \rangle \\
&\leq \delta_{4s}\|\mathbf{u}^{(k)} - \mathbf{x_b}\|_2 \|(\mathbf{u}^{(k)} - \mathbf{x_b})_{U_k}\|_2.
\end{aligned}$$

It follows that $\|(\mathbf{u}^{(k)} - \mathbf{x_b})_{U_k}\|_2 \leq \delta_{4s}\|\mathbf{u}^{(k)} - \mathbf{x_b}\|_2$. Since

$$|\mathbf{u}^{(k)} - \mathbf{x_b}\|_2^2 = \|(\mathbf{u}^{(k)} - \mathbf{x_b})_{U_k}\|_2^2 + \|(\mathbf{u}^{(k)} - \mathbf{x_b})_{(U_k)^c}\|_2^2,$$

we have $(1 - \delta_{4s}^2)\|\mathbf{u}^{(k)} - \mathbf{x_b}\|_2^2 \leq \|(\mathbf{u}^{(k)} - \mathbf{x_b})_{(U_k)^c}\|_2^2$.

To estimate $\|\mathbf{x}^{(k+1)} - \mathbf{x_b}\|_2$, we note that

$$\begin{aligned}
\|\mathbf{x}^{(k+1)} - \mathbf{x_b}\|_2^2 &= \|\mathbf{x}^{(k+1)} - \mathbf{u}^{(k)} + \mathbf{u}^{(k)} - \mathbf{x_b}\|_2^2 \\
&= \|\mathbf{u}^{(k)} - \mathbf{x}^{(k+1)}\|_2^2 + \|\mathbf{u}^{(k)} - \mathbf{x_b}\|_2^2 - 2\langle \mathbf{x}^{(k+1)} - \mathbf{u}^{(k)}, \mathbf{u}^{(k)} - \mathbf{x_b} \rangle.
\end{aligned}$$

Equation (4.25) in Algorithm 4.5 says that $\mathbf{x}^{(k+1)}$ is a better s-term approximation to $\mathbf{u}^{(k)}$ than $(\mathbf{x_b})_{U_k}$. So we have

$$\|\mathbf{u}^{(k)} - \mathbf{x}^{(k+1)}\|_2 \leq \|\mathbf{u}^{(k)} - (\mathbf{x_b})_{U_k}\|_2 = \|(\mathbf{u}^{(k)} - \mathbf{x_b})_{U_k}\|_2.$$

That is,

$$\|\mathbf{u}^{(k)} - \mathbf{x}^{(k+1)}\|_2 \leq \delta_{4s}\|\mathbf{u}^{(k)} - \mathbf{x_b}\|_2.$$

Furthermore, in view of (4.30) and the fact that $\text{supp}(\mathbf{u}^{(k)} - \mathbf{x}^{(k+1)}) \subset U_k$, we have

$$\begin{aligned}
|\langle \mathbf{x}^{(k+1)} - \mathbf{u}^{(k)}, \mathbf{u}^{(k)} - \mathbf{x_b} \rangle| &\leq |\langle (I - \Phi^\top \Phi)(\mathbf{x}^{(k+1)} - \mathbf{u}^{(k)}), \mathbf{u}^{(k)} - \mathbf{x_b} \rangle| \\
&\leq \delta_{4s}\|\mathbf{x}^{(k+1)} - \mathbf{u}^{(k)}\|_2 \|\mathbf{u}^{(k)} - \mathbf{x_b}\|_2.
\end{aligned}$$

Combining the above estimates, we have

$$\begin{aligned}
\|\mathbf{x}^{(k+1)} - \mathbf{x_b}\|_2^2 &\leq \|\mathbf{u}^{(k)} - \mathbf{x}^{(k+1)}\|_2^2 + \|\mathbf{u}^{(k)} - \mathbf{x_b}\|_2^2 + 2\delta_{4s}\|\mathbf{x}^{(k+1)} - \mathbf{u}^{(k)}\|_2 \|\mathbf{u}^{(k)} - \mathbf{x_b}\|_2 \\
&\leq (\delta_{4s}^2 + 1)\|\mathbf{u}^{(k)} - \mathbf{x_b}\|_2^2 + 2\delta_{4s}^2\|\mathbf{u}^{(k)} - \mathbf{x_b}\|_2^2 \\
&= (1 + 3\delta_{4s}^2)\|\mathbf{u}^{(k)} - \mathbf{x_b}\|_2^2
\end{aligned}$$

and

$$\|\mathbf{x}^{(k+1)} - \mathbf{x}^*\|_2 \le \sqrt{\frac{1 + 3\delta_{4s}^2}{1 - \delta_{4s}^2}} \|(\mathbf{u}^{(k)} - \mathbf{x_b})_{(U_k)^c}\|_2. \tag{4.31}$$

It remains to estimate $\|(\mathbf{u}^{(k)} - \mathbf{x_b})_{(U_k)^c}\|_2$, which can be given in terms of $\|\mathbf{x}^{(k)} - \mathbf{x_b}\|_2$. Since $\mathbf{x}^{(k)}_{(U_k)^c} = 0 = \mathbf{u}^{(k)}_{(U_k)^c}$, we have

$$\|(\mathbf{u}^{(k)} - \mathbf{x_b})_{(U_k)^c}\|_2 = \|(\mathbf{x}^{(k)} - \mathbf{x_b})_{(U_k)^c}\|_2 \le \|(\mathbf{x}^{(k)} - \mathbf{x_b})_{(T_k)^c}\|_2 = \|(\mathbf{x}^{(k)} - \mathbf{x_b})_{(S \cup S_k) \setminus T_k}\|_2, \tag{4.32}$$

where T_k denotes the indices of the 2s largest entries in magnitude of $\Phi^\top \Phi(\mathbf{x_b} - \mathbf{x}^{(k)})$, S is the index set of the support of $\mathbf{x_b}$, and S_k is the index set of the support of $\mathbf{x}^{(k)}$.

It is straightforward to obtain

$$\|(\Phi^\top \Phi(\mathbf{x_b} - \mathbf{x}^{(k)}))_{S \cup S_k \setminus T_k}\|_2$$
$$\ge \|(\mathbf{x_b} - \mathbf{x}^{(k)})_{S \cup S_k \setminus T_k}\|_2 - \|((\Phi^\top \Phi - I)(\mathbf{x_b} - \mathbf{x}^{(k)}))_{S \cup S_k \setminus T_k}\|_2$$
$$\ge \|(\mathbf{x_b} - \mathbf{x}^{(k)})_{S \cup S_k \setminus T_k}\|_2 - \delta_{2s} \|\mathbf{x_b} - \mathbf{x}^{(k)}\|_2.$$

That is, we obtain

$$\|(\mathbf{x}^* - \mathbf{x}^{(k)})_{S \cup S_k \setminus T_k}\|_2 \le \delta_{2s} \|\mathbf{x_b} - \mathbf{x}^{(k)}\|_2 + \|(\Phi^\top \Phi(\mathbf{x_b} - \mathbf{x}^{(k)}))_{S \cup S_k \setminus T_k}\|_2. \tag{4.33}$$

Furthermore, since $(\Phi^\top \Phi(\mathbf{x_b} - \mathbf{x}^{(k)}))_{T_k}$ is the best 2s-term approximation to $\Phi^\top \Phi(\mathbf{x_b} - \mathbf{x}^{(k)})$, we have

$$\|(\Phi^\top \Phi(\mathbf{x_b} - \mathbf{x}^{(k)}))_{S \cup S_k \setminus T_k}\|_2 \le \|(\Phi^\top \Phi(\mathbf{x_b} - \mathbf{x}^{(k)}))_{T_k \setminus (S \cup S_k)}\|_2$$
$$= \|((\Phi^\top \Phi - I)(\mathbf{x_b} - \mathbf{x}^{(k)}))_{T_k \setminus (S \cup S_k)}\|_2$$
$$\le \delta_{4s} \|\mathbf{x_b} - \mathbf{x}^{(k)}\|_2.$$

Together with (4.33), we have

$$\|(\mathbf{x_b} - \mathbf{x}^{(k)})_{S \cup S_k \setminus T_k}\|_2 \le (\delta_{2s} + \delta_{4s}) \|\mathbf{x_b} - \mathbf{x}^{(k)}\|_2.$$

Therefore, using (4.31) and (4.32), we finally obtain

$$\|\mathbf{x}^{(k+1)} - \mathbf{x_b}\|_2 \le \sqrt{\frac{1 + 3\delta_{4s}^2}{1 - \delta_{4s}^2}} (\delta_{2s} + \delta_{4s}) \|\mathbf{x_b} - \mathbf{x}^{(k)}\|_2 = \rho \|\mathbf{x_b} - \mathbf{x}^{(k)}\|_2,$$

with $\rho = \sqrt{\frac{1 + 3\delta_{4s}^2}{1 - \delta_{4s}^2}} (\delta_{2s} + \delta_{4s}) < 1$ since

$$\frac{1 + 3\delta_{4s}^2}{1 - \delta_{4s}^2} 4\delta_{4s}^2 < 1$$

when $\delta_{4s} < \sqrt{2/(5 + \sqrt{73})}$, whose square is the positive root of the quadratic equation

$$12t^2 + 5t - 1 = 0. \qquad \square$$

We leave the discussion of the noisy recovery to the interested reader, or see [8] for details.

4.4 ▪ Nonconvex Minimizations

In this section, we discuss several nonconvex regularizers, interpolated between the ℓ_0 norm and the ℓ_1 norm to better approximate the ℓ_0 norm minimization. They include the ℓ_q norm ($0 < q < 1$) [10], the smoothly clipped absolute deviation (SCAD) [7], the log-sum penalty (LSP) [5], the minimax concave penalty (MCP) [18], and the capped-ℓ_1 penalty [17]. However, it is challenging to numerically solve the corresponding nonconvex optimization problems. We shall provide two general schemes to use nonconvex minimization for the sparse solution of underdetermined linear systems. The first scheme is called convex-concave programming (CCCP) (cf. [16]). The second one is a general iterative shrinkage and thresholding (GIST) algorithm (cf. [11]).

4.4.1 ▪ The CCCP Algorithm

The CCCP algorithm is also called difference of convex (DC) programming. Let us use the ℓ_q minimization for the sparse solution to explain the CCCP algorithm. We rewrite the ℓ_q minimization as the following new minimization:

$$\min\{\|\mathbf{x}\|_q^q : \quad \Phi\mathbf{x} = \mathbf{b}\} = \min_{\substack{\mathbf{x}\in\mathbb{R}^n \\ \Phi\mathbf{x}=\mathbf{b}}} \|\mathbf{x}\|_1 - (\|\mathbf{x}\|_1 - \|\mathbf{x}\|_q^q), \tag{4.34}$$

where both $\|\mathbf{x}\|_1$ and $(\|\mathbf{x}\|_1 - \|\mathbf{x}\|_q^q)$ are convex functions for $q \in (0, 1)$. In general, we consider

$$\min_{\substack{\mathbf{x}\in\mathbb{R}^n \\ \Phi\mathbf{x}=\mathbf{b}}} f_1(\mathbf{x}) - f_2(\mathbf{x}), \tag{4.35}$$

where $f_1(\mathbf{x})$ and $f_2(\mathbf{x})$ are two convex functions.

The CCCP algorithm (cf. [16]) solves problem (4.35) by generating a sequence $\{\mathbf{x}^{(k)}\}$ as

$$\mathbf{x}^{(k+1)} = \arg\min_{\substack{\mathbf{x}\in\mathbb{R}^n \\ \Phi\mathbf{x}=\mathbf{b}}} f_1(\mathbf{x}) - f_2(\mathbf{x}^{(k)}) - \langle \mathbf{s}_2(\mathbf{x}^{(k)}), \mathbf{x} - \mathbf{x}^{(k)}\rangle, \tag{4.36}$$

where $\mathbf{s}_2(\mathbf{x}^{(k)})$ denotes the gradient or a subgradient (see Chapter 5) of $f_2(\mathbf{x})$ at $\mathbf{x}^{(k)}$. Obviously, the objective function in problem (4.36) is convex, and hence $\mathbf{x}^{(k+1)}$ exists and can be found numerically, e.g., by the gradient descent method. The following is the main result.

Theorem 4.16. *For $\mathbf{x}^{(k)}$ from (4.36) we have*

$$f_1(\mathbf{x}^{(k+1)}) - f_2(\mathbf{x}^{(k+1)}) \leq f_1(\mathbf{x}^{(k)}) - f_2(\mathbf{x}^{(k)}) \quad \forall k \geq 1 \tag{4.37}$$

and the subgradients

$$0 \in \partial f_1(\mathbf{x}^{(k)}) - \partial f_2(\mathbf{x}^{(k)}) \quad \forall k \geq 1. \tag{4.38}$$

Proof. By the convexity of f_2 and the minimizer $\mathbf{x}^{(k+1)}$, we have

$$\begin{aligned} f_1(\mathbf{x}^{(k+1)}) - f_2(\mathbf{x}^{(k+1)}) &\leq f_1(\mathbf{x}^{(k+1)}) - f_2(\mathbf{x}^{(k)}) - \langle \mathbf{s}_2(\mathbf{x}^{(k)}), \mathbf{x}^{(k+1)} - \mathbf{x}^{(k)}\rangle \\ &= \min_{\substack{\mathbf{x}\in\mathbb{R}^n \\ \Phi\mathbf{x}=\mathbf{b}}} f_1(\mathbf{x}) - \langle \mathbf{s}_2(\mathbf{x}^{(k)}), \mathbf{x} - \mathbf{x}^{(k)}\rangle - f_2(\mathbf{x}^{(k)}) \\ &\leq f_1(\mathbf{x}^{(k)}) - f_2(\mathbf{x}^{(k)}) \end{aligned}$$

for all $k \geq 1$.

Next, since $\mathbf{x}^{(k+1)}$ is a minimizer of (4.36) which is convex, we have (4.38). This completes the proof. □

For convenience, we may assume that $f(\mathbf{x}) = f_1(\mathbf{x}) - f_2(\mathbf{x})$ is bounded from below. This is indeed the case for (4.34) since $f(\mathbf{x}) = \|\mathbf{x}\|_q^q \geq 0$. Then the monotonicity of $f(\mathbf{x}^{(k)}), k \geq 1$, implies the convergence of the values $f(\mathbf{x}^{(k)})$. Suppose that $\mathbf{x}^{(k)}, k \geq 1$, are bounded. Then we have a convergent subsequence which converges to a cluster point $\mathbf{x}^*$. By using a result in [15], we know $\partial f_1(\mathbf{x}^*) - \partial f_2(\mathbf{x}^*) = 0$ if f_2 is continuously differentiable, and hence $\mathbf{x}^*$ is a critical point of (4.35) and satisfies

$$f_1(\mathbf{x}^*) - f_2(\mathbf{x}^*) \leq f_1(\mathbf{x}^{(k)}) - f_2(\mathbf{x}^{(k)})$$

for all $k \geq 1$. Returning to the ℓ_q minimization discussed before, i.e., (4.34), the above inequality implies $\|\mathbf{x}^*\|_q^q \leq \|\mathbf{x}^{(k)}\|_q^q$ for all $k \geq 1$. As pointed out in Chapter 1, there are only finitely many local minimizers for the ℓ_q minimization. We leave the discussion of when an minimizer of the ℓ_q minimization subject to $\Phi\mathbf{x} = \mathbf{b}$ is a sparse solution to Chapter 7. That is, when Φ satisfies some RIP condition, the minimizer of (4.34) will be a sparse solution.

Certainly, we may use this CCCP algorithm for the minimization

$$\min \|\mathbf{x}\|_1 - \frac{\lambda}{2}\|\mathbf{x}\|_2^2, \quad \Phi\mathbf{x} = \mathbf{b}, \tag{4.39}$$

and many other possible combinations of two convex functions. In general, if f_2 is strongly convex, e.g., f_2 in (4.39), we have

$$f_2(\mathbf{x}^{(k+1)}) \geq f_2(\mathbf{x}^{(k)}) + \nabla f_2(\mathbf{x}^{(k)})^\top(\mathbf{x}^{(k+1)} - \mathbf{x}^{(k)}) + \frac{\nu}{2}\|\mathbf{x}^{(k+1)} - \mathbf{x}^{(k)}\|^2$$

for strong convexity $\nu > 0$. Then we use the idea as in the proof above to get

$$f(\mathbf{x}^{(k+1)}) \leq f(\mathbf{x}^{(k)}) - \frac{\nu}{2}\|\mathbf{x}^{(k+1)} - \mathbf{x}^{(k)}\|^2. \tag{4.40}$$

Summing the above inequalities for $k \geq 1$ yields

$$\frac{\nu}{2}\sum_{k=1}^m \|\mathbf{x}^{(k+1)} - \mathbf{x}^{(k)}\|^2 \leq f(\mathbf{x}^{(1)}) - f(\mathbf{x}^{(m+1)}) \leq f(\mathbf{x}^{(1)}) \leq f_1(\mathbf{x}^{(1)}). \tag{4.41}$$

This shows that $\mathbf{x}^{(k)}$ are very close to each other. If f_1 has some nice properties, more study can be carried out. See Chapter 11 for more discussion of the DC algorithm for phase retrieval.

4.4.2 ▪ The GIST Algorithm

In this subsection, we study a general iterative shrinkage and thresholding (GIST) algorithm to solve the nonconvex optimization problem for a large class of nonconvex penalties. The discussion in this subsection is largely based on the study in [11]. The GIST algorithm iteratively solves a proximal operator problem, which in turn has a closed-form solution for many commonly used penalties. At each outer iteration of the algorithm, we use a nonmonotone line search, and we use the Barzilai–Borwein rule [1] to find an appropriate step size. These two steps greatly accelerate the computational speed. We also present a convergence analysis of the GIST algorithm.

We mainly solve the general problem

$$\min_{\mathbf{x}\in\mathbb{R}^n} \{f(\mathbf{x}) = g(\mathbf{x}) + r(\mathbf{x})\}, \tag{4.42}$$

Table 4.1. *List of regularizers.*

Name	$r_i(x_i), r(\mathbf{x}) = \sum_{i=1}^{n} r_i(x_i)$
ℓ_1 norm	$\lambda\|x_i\|$
ℓ_q quasi-norm	$\lambda\|x_i\|^q, \quad q = 1/2$
LSP	$\lambda \log(1 + \|w_i\|/\theta), \theta > 0$
SCAD	$\lambda \int_0^{\|x_i\|} \min\left(1, \frac{[\theta\lambda - x]_+}{(\theta-1)\lambda}\right) dx, \quad \theta > 2$
MCP	$\lambda \int_0^{\|x_i\|} \left[1 - \frac{x}{\theta\lambda}\right]_+ dx, \quad \theta > 0$
Capped ℓ_1	$\lambda \min(\|x_i\|, \theta), \quad \theta > 0$

where g and r satisfy one of the following assumptions:

A1 $g(\mathbf{x})$ is continuously differentiable with Lipschitz continuous gradient, i.e., there exists a positive constant $\beta(g)$ such that

$$\|\nabla g(\mathbf{v}) - \nabla g(\mathbf{u})\| \le \beta(g)\|\mathbf{v} - \mathbf{u}\| \quad \forall \mathbf{v}, \mathbf{u} \in \mathbb{R}^n.$$

A2 $r(\mathbf{x})$ is a continuous function which is possibly *nonsmooth* and *nonconvex* but can be rewritten as the difference of two convex functions, $r_1(\mathbf{x})$ and $r_2(\mathbf{x})$, i.e.,

$$r(\mathbf{x}) = r_1(\mathbf{x}) - r_2(\mathbf{x}).$$

A3 $f(\mathbf{x})$ is bounded from below.

We say that $\mathbf{x}^\star$ is a critical point of problem (4.42) if

$$\mathbf{0} \in \nabla g(\mathbf{x}^\star) + \partial r_1(\mathbf{x}^\star) - \partial r_2(\mathbf{x}^\star), \tag{4.43}$$

where $\partial r_1(\mathbf{x}^\star)$ and $\partial r_2(\mathbf{x}^\star)$ are the subdifferentials of the functions $r_1(\mathbf{x})$ and $r_2(\mathbf{x})$ at $\mathbf{x} = \mathbf{x}^\star$.

The standard least squares and logistic loss functions are two common functionals for g which satisfy assumption A1:

$$g(\mathbf{x}) = \frac{1}{2}\|\Phi\mathbf{x} - \mathbf{b}\|^2 \quad \text{or} \quad g(\mathbf{x}) = \frac{1}{N}\sum_{i=1}^{n} \log\left(1 + \exp(-y_i \phi_i^\top \mathbf{x})\right),$$

where $\Phi = [\phi_1, \dots, \phi_n] \in \mathbb{R}^{m \times n}$ is a sensing matrix and $\mathbf{y} = [y_1, \dots, y_n]^\top \in \mathbb{R}^n$ is a target vector. Some of the regularizers (penalties) which satisfy assumption A2 are presented in Table 4.1.

They can be nonconvex (except the ℓ_1 norm). However, each $r(x)$ can be rewritten as $r(x) = r_1(x) - r_2(x)$ for two convex functions r_1 and r_2. For example, for the capped ℓ_1 regularizer, we have

$$r(x) = \lambda\|x\| - \lambda(\|x\| - \min(\|x_i\|, \theta)).$$

It is easy to rewrite the regularizers in Table 4.1 as the difference of two convex functions like the capped ℓ_1 regularizer. The details are left to Exercise 11.

The functions $g(\mathbf{x})$ and $r(\mathbf{x})$ mentioned above are nonnegative. Hence, f is bounded from below and satisfies assumption A3. We are now ready to present the main algorithm in this section.

ALGORITHM 4.6
General Iterative Shrinkage and Thresholding (GIST) Algorithm [11]

1: Choose parameters $\eta > 1$ and $t_{\min}, t_{\max}$ with $0 < t_{\min} < t_{\max}$.
2: Initialize iteration counter $k \leftarrow 0$ and a bounded starting point $\mathbf{x}^{(0)}$.
3: **repeat**
4:     $t^{(k)} \in [t_{\min}, t_{\max}]$.
5:     **repeat**
6:         $\mathbf{x}^{(k+1)} = \arg\min_{\mathbf{x}} \{g(\mathbf{x}^{(k)}) + \langle \nabla g(\mathbf{x}^{(k)}), \mathbf{x} - \mathbf{x}^{(k)} \rangle + \frac{t^{(k)}}{2}\|\mathbf{x} - \mathbf{x}^{(k)}\|^2 + r(\mathbf{x})\}$.
7:         $t^{(k)} \leftarrow \eta t^{(k)}$
8:     **until** some line search criterion is satisfied, e.g., (4.47).
9:     $k \leftarrow k + 1$.
10: **until** some stopping criterion is satisfied.

Our GIST algorithm, i.e., Algorithm 4.6, solves problem (4.42) by generating a sequence $\{\mathbf{x}^{(k)}\}$ via

$$\mathbf{x}^{(k+1)} = \arg\min_{\mathbf{x}} g(\mathbf{x}^{(k)}) + \langle \nabla g(\mathbf{x}^{(k)}), \mathbf{x} - \mathbf{x}^{(k)} \rangle + \frac{t^{(k)}}{2}\|\mathbf{x} - \mathbf{x}^{(k)}\|^2 + r(\mathbf{x}), \quad (4.44)$$

which is equivalent to the proximal operator problem

$$\mathbf{x}^{(k+1)} = \arg\min_{\mathbf{x}} \frac{1}{2}\|\mathbf{x} - \mathbf{u}^{(k)}\|^2 + \frac{1}{t^{(k)}} r(\mathbf{x}), \quad (4.45)$$

where $\mathbf{u}^{(k)} = \mathbf{x}^{(k)} - \nabla g(\mathbf{x}^{(k)})/t^{(k)}$. Thus, the GIST algorithm first performs a gradient descent along the direction $-\nabla g(\mathbf{x}^{(k)})$ with step size $1/t^{(k)}$ and then solves a proximal operator problem.

For all the regularizers listed in Table 4.1, problem (4.44) has a closed-form solution (details are left to Exercise 12) and hence has an efficient performance. For example, for the ℓ_1 and capped ℓ_1 regularizers, we have closed-form solutions as follows:

$$\ell_1 : x_i^{(k+1)} = \text{sign}(u_i^{(k)}) \max\left(0, |u_i^{(k)}| - \lambda/t^{(k)}\right),$$

$$\text{capped } \ell_1 : x_i^{(k+1)} = \begin{cases} y_1 & \text{if } h_i(y_1) \leq h_i(y_2), \\ y_2 & \text{otherwise,} \end{cases}$$

where $y_1 = \text{sign}(u_i^{(k)}) \max(|u_i^{(k)}|, \theta)$, $y_2 = \text{sign}(u_i^{(k)}) \min(\theta, [|u_i^{(k)}| - \lambda/t^{(k)}]_+)$, and $h_i(x) = 0.5(x - u_i^{(k)})^2 + \lambda/t^{(k)} \min(|x|, \theta)$.

Let us explain the two remaining issues in Algorithm 4.6: how to initialize $t^{(k)}$ (in line 4) and how to select a line search criterion (in line 8) at each outer iteration, which will be explained in the following two subsections.

The Step Size and Line Search Criterion

Intuitively, a good step size initialization strategy at each outer iteration can greatly reduce the line search cost (lines 5–8) and hence is essential for the fast convergence of Algorithm 4.6. We shall initialize the step size by adopting the Barzilai–Borwein rule [1], which uses a diagonal matrix $t^{(k)}I$ to approximate the Hessian matrix $\nabla^2 g(\mathbf{w})$ at $\mathbf{w} = \mathbf{x}^{(k)}$ although it may not exist at all the points. Denote

$$\mathbf{w}^{(k)} = \mathbf{x}^{(k)} - \mathbf{x}^{(k-1)}, \quad \mathbf{z}^{(k)} = \nabla g(\mathbf{x}^{(k)}) - \nabla g(\mathbf{x}^{(k-1)}).$$

Then $t^{(k)}$ is initialized at the outer iteration k as

$$t^{(k)} = \arg\min_t \|t\mathbf{w}^{(k)} - \mathbf{z}^{(k)}\|^2 = \frac{\langle \mathbf{w}^{(k)}, \mathbf{z}^{(k)} \rangle}{\langle \mathbf{z}^{(k)}, \mathbf{z}^{(k)} \rangle}. \tag{4.46}$$

This is the step size we will use in the computation.

Next let us discuss a line search criterion. A natural line search criterion is to reduce the objective function value monotonically. More specifically, we accept the step size $1/t^{(k)}$ at the outer iteration k if the following monotone line search criterion is satisfied:

$$f(\mathbf{x}^{(k+1)}) \le f(\mathbf{x}^{(k)}) - \frac{\sigma}{2} t^{(k)} \|\mathbf{x}^{(k+1)} - \mathbf{x}^{(k)}\|^2, \tag{4.47}$$

where σ is a constant in the interval $(0,1)$. Certainly, there are other line search criteria, e.g., a nonmonotone line search criterion. For example, we can accept the step size $1/t^{(k)}$ if $\mathbf{x}^{(k+1)}$ makes the objective function value smaller than the average over the previous m iterations, i.e.,

$$f(\mathbf{x}^{(k+1)}) \le \frac{1}{m} \sum_{i=\max(0,k-m+1),\ldots,k} f(\mathbf{x}^{(i)}) - \frac{\sigma}{2} t^{(k)} \|\mathbf{x}^{(k+1)} - \mathbf{x}^{(k)}\|^2, \tag{4.48}$$

where $\sigma \in (0,1)$.

Convergence Analysis

We are now ready to present a convergence analysis under the monotone search criterion. We shall leave the study of the nonmonotone line search criterion to Exercise 13. We first present a result which ensures that the monotone line search criterion in (4.47) is satisfied.

Lemma 4.17. *Suppose that assumptions A1–A3 hold and the constant* $\sigma \in (0,1)$ *is given. Then for any integer* $k \ge 1$, *the monotone line search criterion in* (4.47) *is satisfied whenever* $t^{(k)} \ge \beta(g)/(1-\sigma)$.

Proof. Since $\mathbf{x}^{(k+1)}$ is a minimizer of problem (4.44), we have

$$\langle \nabla g(\mathbf{x}^{(k)}), \mathbf{x}^{(k+1)} - \mathbf{x}^{(k)} \rangle + \frac{t^{(k)}}{2} \|\mathbf{x}^{(k+1)} - \mathbf{x}^{(k)}\|^2 + r(\mathbf{x}^{(k+1)}) \le r(\mathbf{x}^{(k)}). \tag{4.49}$$

It follows from assumption A1 that

$$g(\mathbf{x}^{(k+1)}) \le g(\mathbf{x}^{(k)}) + \langle \nabla g(\mathbf{x}^{(k)}), \mathbf{x}^{(k+1)} - \mathbf{x}^{(k)} \rangle + \frac{\beta(g)}{2} \|\mathbf{x}^{(k+1)} - \mathbf{x}^{(k)}\|^2. \tag{4.50}$$

Combining (4.49) and (4.50), we have

$$g(\mathbf{x}^{(k+1)}) + r(\mathbf{x}^{(k+1)}) \le g(\mathbf{x}^{(k)}) + r(\mathbf{x}^{(k)}) - \frac{t^{(k)} - \beta(g)}{2} \|\mathbf{x}^{(k+1)} - \mathbf{x}^{(k)}\|^2.$$

In other words, we have

$$f(\mathbf{x}^{(k+1)}) \le f(\mathbf{x}^{(k)}) - \frac{t^{(k)} - \beta(g)}{2} \|\mathbf{x}^{(k+1)} - \mathbf{x}^{(k)}\|^2.$$

Therefore, the line search criterion in (4.47) is satisfied whenever $(t^{(k)} - \beta(g))/2 \ge \sigma t^{(k)}/2$, i.e., $t^{(k)} \ge \beta(g)/(1-\sigma)$. This completes the proof of the lemma. □

Next, we summarize the boundedness of $t^{(k)}$ in the following lemma.

Lemma 4.18. *For any* $k \geq 0$, $t^{(k)}$ *is bounded under the monotone line search criterion in* (4.47).

Proof. It is trivial to show that $t^{(k)}$ is bounded from below, since $t^{(k)} \geq t_{\min}$ ($t_{\min}$ is defined in Algorithm 4.6). Next we prove that $t^{(k)}$ is bounded from above by contradiction. Assume that $t^{(k)}$ is unbounded from above. Without loss of generality, we assume that $t^{(k)}$ increases monotonically to $+\infty$ and $t^{(k)} \geq \eta \beta(g)/(1 - \sigma)$ for some $k > 0$. Thus, the value $t = t^{(k)}/\eta \geq \beta(g)/(1 - \sigma)$ must have been tried at iteration k and does not satisfy the line search criterion in (4.47). But Lemma 4.17 states that $t = t^{(k)}/\eta \geq \beta(l)/(1 - \sigma)$ is guaranteed to satisfy the line search criterion in (4.47). This leads to a contradiction. Thus, $t^{(k)}$ is bounded from above.   □

Based on Lemmas 4.17 and 4.18, we have the following convergence results.

Theorem 4.19 (Gong, Zhang, Lu, Huang, and Ye, 2013 [11]). *Suppose that assumptions* A1–A3 *hold and suppose that the monotone line search criterion in* (4.47) *is satisfied. Then all cluster points of the sequence* $\{\mathbf{x}^{(k)}\}$ *generated by Algorithm 4.6 are critical points of problem* (4.42).

Proof. Based on Lemma 4.17, the monotone line search criterion in (4.47) is satisfied and hence

$$f(\mathbf{x}^{(k+1)}) \leq f(\mathbf{x}^{(k)}) \quad \forall k \geq 0,$$

which implies that the sequence $\{f(\mathbf{x}^{(k)})\}_{k=0,1,\dots}$ is monotonically decreasing. Let $\mathbf{x}^\star$ be a cluster point of the sequence $\{\mathbf{x}^{(k)}\}$, i.e., there exists a subsequence $k_1, \dots, k_n, \dots$ such that

$$\lim_{n \to \infty} \mathbf{x}^{(k_n)} = \mathbf{x}^\star.$$

Since f is bounded from below, together with the fact that $\{f(\mathbf{x}^{(k_n)})\}$ is monotonically decreasing, $\lim_{n \to \infty} f(\mathbf{x}^{(k_n)})$ exists. We observe that if f is continuous, then

$$\lim_{k \to \infty} f(\mathbf{x}^{(k)}) = \lim_{n \to \infty} f(\mathbf{x}^{(k_n)}) = f(\mathbf{x}^\star).$$

Taking limits on both sides of (4.47) with k, we have

$$\lim_{k \to \infty} \|\mathbf{x}^{(k+1)} - \mathbf{x}^{(k)}\| = 0. \tag{4.51}$$

Considering that the minimizer $\mathbf{x}^{(k+1)}$ is also a critical point of problem (4.44) and $r(\mathbf{x}) = r_1(\mathbf{x}) - r_2(\mathbf{x})$, we have

$$\mathbf{0} \in \nabla l(\mathbf{x}^{(k)}) + t^{(k)}(\mathbf{x}^{(k+1)} - \mathbf{x}^{(k)}) + \partial r_1(\mathbf{x}^{(k+1)}) - \partial r_2(\mathbf{x}^{(k+1)}).$$

Taking limits on both sides of the above equation with $k_n \to \infty$, by using the semicontinuity of $\partial r_1(\cdot)$ and $\partial r_2(\cdot)$, the boundedness of $t^{(k)}$ (based on Lemma 4.18), and (4.51), we obtain

$$\mathbf{0} \in \nabla g(\mathbf{x}^\star) + \partial r_1(\mathbf{x}^\star) - \partial r_2(\mathbf{x}^\star).$$

Therefore, $\mathbf{x}^\star$ is a critical point of problem (4.42). This completes the proof.   □

Based on (4.51), we know that $\lim_{k \to \infty} \|\mathbf{x}^{(k+1)} - \mathbf{x}^{(k)}\|^2 = 0$ is a necessary optimality condition of Algorithm 4.6. Thus, $\|\mathbf{x}^{(k+1)} - \mathbf{x}^{(k)}\|^2$ is a quantity that measures the convergence of the sequence $\{\mathbf{x}^{(k)}\}$ to a critical point. We present the convergence rate in terms of $\|\mathbf{x}^{(k+1)} - \mathbf{x}^{(k)}\|^2$ in the following theorem.

Theorem 4.20. *Let $\{\mathbf{x}^{(k)}\}$ be the sequence generated by Algorithm 4.6 with the monotone line search criterion in (4.47) satisfied. Then for every $n \geq 1$ we have*

$$\min_{0 \leq k \leq n} \|\mathbf{x}^{(k+1)} - \mathbf{x}^{(k)}\|^2 \leq \frac{2(f(\mathbf{x}^{(0)}) - f(\mathbf{x}^\star))}{n \sigma t_{\min}},$$

where $\mathbf{x}^\star$ is a cluster point of the sequence $\{\mathbf{x}^{(k)}\}$.

Proof. Based on (4.47) with $t^{(k)} \geq t_{\min}$, we have

$$\frac{\sigma t_{\min}}{2} \|\mathbf{x}^{(k+1)} - \mathbf{x}^{(k)}\|^2 \leq f(\mathbf{x}^{(k)}) - f(\mathbf{x}^{(k+1)}).$$

Summing the above inequality over $k = 0, \ldots, n$, we obtain

$$\frac{\sigma t_{\min}}{2} \sum_{k=0}^{n} \|\mathbf{x}^{(k+1)} - \mathbf{x}^{(k)}\|^2 \leq f(\mathbf{x}^{(0)}) - f(\mathbf{x}^{(n+1)}),$$

which implies that

$$\min_{0 \leq k \leq n} \|\mathbf{x}^{(k+1)} - \mathbf{x}^{(k)}\|^2 \leq \frac{2(f(\mathbf{x}^{(0)}) - f(\mathbf{x}^{(n+1)}))}{n \sigma t_{\min}} \leq \frac{2(f(\mathbf{x}^{(0)}) - f(\mathbf{x}^\star))}{n \sigma t_{\min}}.$$

This completes the proof of the theorem. □

4.5 ▪ Alternating Projection Methods

Let $\mathcal{D}_s(\mathbb{R}^n)$ denote the collection of all s-sparse vectors in $\mathbb{R}^n$, i.e.,

$$\mathcal{D}_s(\mathbb{R}^n) := \{\mathbf{x} \in \mathbb{R}^n \mid \quad \|\mathbf{x}\|_0 = s\},$$

and let $\mathcal{P}_{\mathcal{D}_s}$ and $\mathcal{P}_{\mathcal{A}}$ denote the projection onto the set $\mathcal{D}_s(\mathbb{R}^n)$ and the affine space $\mathcal{A} := \{\mathbf{x} : \Phi\mathbf{x} = \mathbf{b}\}$, respectively. It is easy to see that $\mathcal{A} = \text{Null}(\Phi) + \mathbf{x}_0$, where $\mathbf{x}_0 \in \mathbb{R}^n$ satisfying $\Phi\mathbf{x}_0 = \mathbf{b}$ is a particular solution. Note that the projection $\mathcal{P}_{\mathcal{D}_s}(\mathbf{x}_k)$ can be computed easily by setting the smallest $n - s$ components of the vector $\mathbf{x}_k$ to zero. The computation of the projection $\mathcal{P}_{\mathcal{A}}$ will be explained later.

An alternating projection algorithm for the minimization (1.4) in Chapter 1 can be stated as follows.

ALGORITHM 4.7

Alternating Projection for ℓ_0 Minimization [12]

1: Input the sparsity s of the solution $\mathbf{x}_b$ and a tolerance $\epsilon > 0$.
2: Initialize $\mathbf{x}_1$ to a random vector in the affine space $\mathcal{A}$.
3: For $k \geq 1$, do
4: **Step 1:** $\mathbf{y}_k = \mathcal{P}_{\mathcal{D}_s}(\mathbf{x}_k)$
5: **Step 2:** $\mathbf{x}_{k+1} = \mathcal{P}_{\mathcal{A}}(\mathbf{y}_k)$
6: until the smallest $n - s$ components of $\mathbf{x}_{k+1}$ have magnitude less than ϵ.

We will now discuss the convergence of Algorithm 4.7. The discussion is divided into three subsections: preliminary on projections, local convergence, and global convergence.

4.5.1 ▪ Preliminary on Projections

In this subsection we will explain a theory of alternate projections on affine linear spaces and their convergence. We begin with the *angle* between two linear spaces.

Definition 4.21. *The angle between two affine linear spaces $\mathbf{a} + K$ and $\mathbf{b} + L$, where K and L are vector spaces, is defined to be the angle $\gamma := \gamma(K, L)$ between 0 and $\frac{\pi}{2}$ whose cosine is given by*

$$\cos(\gamma) := \sup \left\{ \frac{\langle k, l \rangle}{\|k\|\|l\|} \; : \; k \in K \cap (K \cap L)^{\perp}, l \in L \cap (K \cap L)^{\perp} \right\}.$$

We first have the following lemma.

Lemma 4.22. *For any two distinct subspaces K and L of a finite-dimensional subspace of a Euclidean space,*

$$\cos(\gamma(K, L)) < 1.$$

Proof. Using the Cauchy–Schwarz inequality, it is clear that $\cos(\gamma(K, L)) \leq 1$. Now, assume, on the contrary, that $\cos(\gamma(K, L)) = 1$. This then implies that

$$\sup \left\{ \frac{\langle k, l \rangle}{\|k\|\|l\|} \; : \; k \in K \cap (K \cap L)^{\perp}, l \in L \cap (K \cap L)^{\perp} \right\} = 1. \tag{4.52}$$

Since $K \cap (K \cap L)^{\perp}$ and $L \cap (K \cap L)^{\perp}$ are subspaces of a finite-dimensional Euclidean space, they are closed. Hence, the supremum in equation (4.52) is attained, which implies that there exist $k \in K \cap (K \cap L)^{\perp}$ and $l \in L \cap (K \cap L)^{\perp}$ such that $\frac{\langle k, l \rangle}{\|k\|\|l\|} = 1$. It follows that $k, l \neq 0$ and $\langle k, l \rangle^2 = \|k\|^2 \|l\|^2$, which is the equality in a Cauchy–Schwarz inequality. Therefore, k and l are linearly dependent. Hence, we deduce that $k, l \in (K \cap L) \cap (K \cap L)^{\perp} = \{0\}$. So, $k = l = 0$, contradicting the fact that $k, l \neq 0$. $\quad\square$

Let us consider an alternating projection between affine linear spaces A and B, which is a classic algorithm that has appeared many times in the literature, e.g., [14] and [6].

ALGORITHM 4.8
Alternating Projection Algorithm [14]

1: Input a vector $\mathbf{x}_1 \in A$ and a tolerance $\epsilon > 0$.
2: For $k \geq 1$,
3: **Step 1:** $\mathbf{y}_k = \mathcal{P}_B(\mathbf{x}_k)$
4: **Step 2:** $\mathbf{x}_{k+1} = \mathcal{P}_A(\mathbf{y}_k)$
5: until $\|\mathbf{x}_{k+1} - \mathbf{x}_k\|$ is within the tolerance ϵ.

We now state and prove the convergence of Algorithm 4.8.

Theorem 4.23. *Let A and B be two distinct closed affine subspaces of a finite-dimensional normed linear space, e.g., $\mathbb{R}^n$, say $A = \mathbf{a} + K$, $B = \mathbf{b} + L$ for vectors $\mathbf{a}, \mathbf{b} \in \mathbb{R}^n$ and closed subspaces K, L. Then the alternating projection Algorithm 4.8 converges linearly with rate $\cos(\gamma(K, L))$, independent of the starting point. More specifically, if $\mathbf{x}$ is a starting point, then the algorithm converges to $\mathcal{P}_{K \cap L}(\mathbf{x})$ with a linear rate.*

To prove this result, we need two lemmas.

Lemma 4.24. *Let K and L be two subspaces in the statement of Theorem 4.23. Letting $M = K \cap L$, $\mathcal{P}_K(\mathbf{x}_{M^\perp}) \in K \cap M^\perp$, where $\mathbf{x} = \mathbf{x}_M + \mathbf{x}_{M^\perp} \in \mathbb{R}^n$.*

Proof. Let $m \in M = K \cap L$ be an arbitrary element. Then

$$\langle \mathcal{P}_K(\mathbf{x}_{M^\perp}), m \rangle = \langle \mathbf{x}_{M^\perp} - \mathcal{P}_{K^\perp}(\mathbf{x}_{M^\perp}), m \rangle$$
$$= \langle \mathbf{x}_{M^\perp}, m \rangle - \langle \mathcal{P}_{K^\perp}(\mathbf{x}_{M^\perp}), m \rangle$$
$$= 0 - 0 = 0.$$

The first summation in the next-to-last step is zero because $\mathbf{x}_{M^\perp} \in M^\perp$ and $m \in M$. The second summation is zero because $\mathcal{P}_{K^\perp}(\mathbf{x}_{M^\perp}) \in K^\perp$ and $m \in K$. □

Lemma 4.25. *Let K and L be two subspaces in the statement of Theorem 4.23 and $M = K \cap L$. Then we have $\mathcal{P}_L \mathcal{P}_K(\mathbf{x}_{M^\perp}) \in L \cap M^\perp$.*

Proof. It is clear that $\mathcal{P}_L \mathcal{P}_K(\mathbf{x}_{M^\perp}) \in L$. To prove $\mathcal{P}_L \mathcal{P}_K(\mathbf{x}_{M^\perp}) \in M^\perp$, we proceed as follows. As above, let $m \in M = K \cap L$ be an arbitrary element of $M = K \cap L$. Then

$$\langle \mathcal{P}_L \mathcal{P}_K(\mathbf{x}_{M^\perp}), m \rangle = \langle \mathcal{P}_K(\mathbf{x}_{M^\perp}) - \mathcal{P}_{L^\perp} \mathcal{P}_K(\mathbf{x}_{M^\perp}), m \rangle$$
$$= \langle \mathcal{P}_K(\mathbf{x}_{M^\perp}), m \rangle - \langle \mathcal{P}_{L^\perp} \mathcal{P}_K(\mathbf{x}_{M^\perp}), m \rangle$$
$$= \langle \mathcal{P}_K(\mathbf{x}_{M^\perp}), m \rangle - 0$$
$$= \langle \mathbf{x}_{M^\perp} - \mathcal{P}_{K^\perp}(\mathbf{x}_{M^\perp}), m \rangle$$
$$= \langle \mathbf{x}_{M^\perp}, m \rangle - \langle \mathcal{P}_{K^\perp}(\mathbf{x}_{M^\perp}), m \rangle = 0 - 0 = 0.$$

The term in the third step is zero because $\mathcal{P}_{L^\perp} \mathcal{P}_K(\mathbf{x}_{M^\perp}) \in L^\perp$ and $m \in L$. □

Proof of Theorem 4.23. We will prove the case when $A = K$ and $B = L$ are vector spaces. The general case follows from this case because the normed distances involved in the proof remain invariant under a translation. So we may assume $\mathbf{a} = 0 = \mathbf{b}$. We will denote by $\mathcal{P}_K$ and $\mathcal{P}_L$ the projections onto spaces K and L, respectively.

As above, let $M := K \cap L$. We begin by decomposing the vector $\mathbf{x}$ as

$$\mathbf{x} = \mathbf{x}_M + \mathbf{x}_{M^\perp},$$

where $\mathbf{x}_M \in M$ and $\mathbf{x}_{M^\perp} \in M^\perp$. So

$$\mathcal{P}_L \mathcal{P}_K(\mathbf{x}) = \mathcal{P}_L \mathcal{P}_K(\mathbf{x}_M) + \mathcal{P}_L \mathcal{P}_K(\mathbf{x}_{M^\perp}) = \mathbf{x}_M + \mathcal{P}_L \mathcal{P}_K(\mathbf{x}_{M^\perp}).$$

Hence, by repeated application of the operator $\mathcal{P}_L \mathcal{P}_K$, we obtain, for $n = 1, 2, \ldots$,

$$\left(\mathcal{P}_L \mathcal{P}_K \right)^n (\mathbf{x}) = \mathbf{x}_M + \left(\mathcal{P}_L \mathcal{P}_K \right)^n (\mathbf{x}_{M^\perp}). \tag{4.53}$$

We proceed by noting that

$$\| \mathcal{P}_L \mathcal{P}_K(\mathbf{x}_{M^\perp}) \|^2 = \langle \mathcal{P}_L \mathcal{P}_K(\mathbf{x}_{M^\perp}), \mathcal{P}_L \mathcal{P}_K(\mathbf{x}_{M^\perp}) \rangle$$
$$= \langle \mathcal{P}_K(\mathbf{x}_{M^\perp}) - \mathcal{P}_{L^\perp} \mathcal{P}_K(\mathbf{x}_{M^\perp}), \mathcal{P}_L \mathcal{P}_K(\mathbf{x}_{M^\perp}) \rangle$$
$$= \langle \mathcal{P}_K(\mathbf{x}_{M^\perp}), \mathcal{P}_L \mathcal{P}_K(\mathbf{x}_{M^\perp}) \rangle$$
$$\leq \cos(\gamma(K, L)) \| \mathcal{P}_K(\mathbf{x}_{M^\perp}) \| \| \mathcal{P}_L \mathcal{P}_K(\mathbf{x}_{M^\perp}) \|$$
$$\leq \cos(\gamma(K, L)) \| \mathbf{x}_{M^\perp} \| \| \mathcal{P}_L \mathcal{P}_K(\mathbf{x}_{M^\perp}) \|$$

by using Lemmas 4.24 and 4.25. After canceling $\|\mathcal{P}_L \mathcal{P}_K(\mathbf{x}_{M^\perp})\|$ from both sides of the above inequality, we obtain

$$\|\mathcal{P}_L \mathcal{P}_K(\mathbf{x}_{M^\perp})\| \le \cos(\gamma(K,L))\|\mathbf{x}_{M^\perp}\|.$$

By simple induction, it follows that, for $n = 1, 2, \ldots$,

$$\|\left(\mathcal{P}_L \mathcal{P}_K\right)^n (\mathbf{x}_{M^\perp})\| \le \cos(\gamma(K,L))^n \|\mathbf{x}_{M^\perp}\|.$$

Using the fact that $\cos(\gamma(K,L)) < 1$ from Lemma 4.22, it follows that

$$\|\left(\mathcal{P}_L \mathcal{P}_K\right)^n (\mathbf{x}_{M^\perp})\| \to 0.$$

Hence, from (4.53), we get $\left(\mathcal{P}_L \mathcal{P}_K\right)^n (\mathbf{x}) \to \mathbf{x}_M = \mathcal{P}_{K \cap L}(\mathbf{x})$. This completes the proof. □

Next we discuss how to compute the projection to an affine linear space. For simplicity we consider A as a hyperplane first. That is,

$$A = \{\mathbf{x} \in \mathbb{R}^n : \langle \mathbf{a}, \mathbf{x} \rangle = c\}$$

for a fixed nonzero vector $\mathbf{a} \in H$ and a real number c. Then we have the following result.

Theorem 4.26. *For the set A of the hyperplane defined above, the projection $\mathcal{P}_A$ is given by*

$$\mathcal{P}_A(\mathbf{y}) = \mathbf{y} - \|\mathbf{a}\|^{-2}(\langle \mathbf{a}, \mathbf{y} \rangle - c)\mathbf{a} \tag{4.54}$$

for any $\mathbf{y} \in H$.

Proof. It is clear that $\mathcal{P}_A(\mathbf{y}) \in A$ since we have

$$\langle \mathbf{a}, \mathcal{P}_A(\mathbf{y}) \rangle = \langle \mathbf{a}, \mathbf{y} \rangle - \|\mathbf{a}\|^{-2}(\langle \mathbf{a}, \mathbf{y} \rangle - c)\langle \mathbf{a}, \mathbf{a} \rangle = \langle \mathbf{a}, \mathbf{y} \rangle - (\langle \mathbf{a}, \mathbf{y} \rangle - c) = c.$$

The distance from $\mathbf{y}$ to A is

$$\operatorname{dist}(\mathbf{y}, A) = \min_{\mathbf{x} \in A} \|\mathbf{x} - \mathbf{y}\| \le \|\mathcal{P}_A(\mathbf{y}) - \mathbf{y}\| = \|\mathbf{a}\|^{-2}|\langle \mathbf{a}, \mathbf{y} \rangle - c|\|\mathbf{a}\| = \|\mathbf{a}\|^{-1}|\langle \mathbf{a}, \mathbf{y} \rangle - c|. \tag{4.55}$$

On the other hand, for any $\mathbf{x} \in A$, we have $\langle \mathbf{a}, \mathbf{x} \rangle = c$ and

$$\|\mathbf{a}\|^{-1}|\langle \mathbf{a}, \mathbf{y} \rangle - c| = \|\mathbf{a}\|^{-1}|\langle \mathbf{a}, \mathbf{y} - \mathbf{x} \rangle| \le \|\mathbf{x} - \mathbf{y}\|.$$

By taking the minimal value of $\mathbf{x} \in A$, we have

$$\|\mathbf{a}\|^{-1}|\langle \mathbf{a}, \mathbf{y} \rangle - c| \le \operatorname{dist}(\mathbf{y}, A).$$

This shows that $\mathcal{P}_A(\mathbf{y})$ achieves the minimal value and is the projection of $\mathbf{y}$ onto A. □

Let us consider a slightly more general setting. Let us write $\Phi = [\psi_1^\top; \psi_2^\top; \cdots; \psi_m^\top]$ with vectors $\psi_i, i = 1, \ldots, m$, and $\mathbf{b} = [b_1; \cdots; b_m]$. That is, $\psi_i^\top$ is the ith row vector of Φ. We can define hyperplanes $H_i = \{\mathbf{x} : \langle \psi_i, \mathbf{x} \rangle = b_i\}$ for $i = 1, \ldots, m$. Then the solution space $\mathcal{A} = \{\mathbf{x} \in \mathbb{R}^n : \Phi \mathbf{x} = \mathbf{b}\}$ is the intersection of the projections $\mathcal{P}_{H_i}$. That is,

$$\mathcal{A} = \{\mathbf{x} \in \mathbb{R}^n : \Phi \mathbf{x} = \mathbf{b}\} = \bigcap_{i=1}^{m} H_i. \tag{4.56}$$

It follows that

$$\mathcal{P}_A(\mathbf{x}) = \bigcap_{i=1}^{m} \mathcal{P}_{H_i}(\mathbf{x}) \tag{4.57}$$

for all $\mathbf{x} \in \mathbb{R}^n$. We can extend the alternating projection between two affine linear spaces A and B to the setting of multiple affine linear spaces $H_1, \ldots, H_m$. We have the following result.

Theorem 4.27 (Halperin Theorem [6]). *Let $V_1, \ldots, V_m$ be closed subspaces in a Hilbert space $\mathbb{R}^n$. Suppose that $V^* = \cap_{i=1}^{m} V_i \neq \emptyset$. Then*

$$\lim_{n \to \infty} (\mathcal{P}_{V_1} \cdots \mathcal{P}_{V_m})^n(\mathbf{x}) = \mathcal{P}_{V^*}(\mathbf{x})$$

for each $\mathbf{x} \in X$.

Proof. We leave the proof to the interested reader, or see [6] for details. □

In fact, we know the convergence rate.

Theorem 4.28. *Let $V_1, \ldots, V_m$ be closed subspaces in a Hilbert space $\mathbb{R}^n$. Suppose that $V^* = \cap_{i=1}^{m} V_i \neq \emptyset$. Let c_i be the angle between V_i and $\cap_{j=i+1}^{m} V_j$ for $i = 1, \ldots, m$. Then for each $\mathbf{x} \in X$,*

$$\|(\mathcal{P}_{V_1} \cdots \mathcal{P}_{V_m})^n(\mathbf{x}) - \mathcal{P}_{V^*}(\mathbf{x})\| \leq c^n \|\mathbf{x}\|,$$

where

$$c = \left[1 - \prod_{i=1}^{m-1} (1 - c_i^2)\right]^{1/2}.$$

Proof. We leave the proof to the interested reader, or see [6] for details. □

Certainly, we can extend the results in the two theorems above from the subspaces $V_i, i = 1, \ldots, m$, to the affine linear space $H_i, i = 1, \ldots, m$, if $\cap_{i=1}^{m} H_i \neq \emptyset$. Indeed, there exists a vector $\mathbf{x}_0 \in \cap_{i=1}^{m} H_i$ such that $H_i = V_i + \mathbf{x}_0$ for a linear space V_i. The above results on the convergence and convergence rate also hold for affine linear spaces. We leave the proofs to Exercises 18 and 19.

Another way to find the projection to $\mathcal{A} = \{\mathbf{x} \in \mathbb{R}^n, \Phi\mathbf{x} = \mathbf{b}\}$ is the following.

Theorem 4.29 (Lai and Varghese, 2017 [12]). *Suppose that Φ is of full row rank. Then the projection $\mathcal{P}_A$ is given by*

$$\mathcal{P}_A(\mathbf{x}) = \Phi^\top (\Phi\Phi^\top)^{-1} \mathbf{b} - (\Phi^\top (\Phi\Phi^\top)^{-1} \Phi - I_{n \times n})\mathbf{x} \quad \forall \mathbf{x} \in \mathbb{R}^n. \tag{4.58}$$

Proof. It is clear that we have $\Phi(\mathcal{P}_A(\mathbf{x})) = \mathbf{b}$. That is, $\mathcal{P}_A(\mathbf{x}) \in \mathcal{A}$ for any $\mathbf{x} \in \mathbb{R}^n$. Next we consider the distance

$$\text{dist}(\mathbf{x}, \mathcal{A}) = \min_{\mathbf{y} \in \mathcal{A}} \|\mathbf{x} - \mathbf{y}\| \leq \|\mathbf{x} - (\mathcal{P}_A(\mathbf{x}))\| = \|\Phi^\top (\Phi\Phi^\top)^{-1}(\mathbf{b} - \Phi\mathbf{x})\|$$

$$= \|\Phi^\top (\Phi\Phi^\top)^{-1}\Phi(\mathbf{y} - \mathbf{x})\| \leq \|\Phi^\top (\Phi\Phi^\top)^{-1}\Phi\| \, \|\mathbf{y} - \mathbf{x}\|$$

for $\mathbf{y} \in \mathcal{A}$ since $\mathbf{b} = \Phi\mathbf{y}$. Using the SVD of Φ because Φ is of full row rank, we can show that $\|\Phi^\top (\Phi\Phi^\top)^{-1}\Phi\| \leq 1$. That is, we have

$$\text{dist}(\mathbf{x}, \mathcal{A}) \leq \|\mathbf{y} - \mathbf{x}\|$$

for any $\mathbf{y} \in \mathcal{A}$. It follows that $\text{dist}(\mathbf{x}, \mathcal{A}) = \|\mathbf{x} - \mathcal{P}_A(\mathbf{x})\|$. □

Both methods of finding the projection to the solution set $\mathcal{A} = \{\mathbf{x}, \Phi\mathbf{x} = \mathbf{b}\}$ in (4.57) and (4.58) have their own advantages. When m is small versus n, (4.58) is better. We can find the projection by using the formula directly. When m is large, the computation of the inverse of $(\Phi\Phi^\top)$ is not easy. We can use (4.57) based on the Halperin Theorem 4.27. When Φ is not of full row rank, we have to use (4.57) or the pseudo-inverse of $\Phi\Phi^\top$. Also, when $m > n$, the formula in (4.58) does not work. We leave the problem of finding a formula for the case $m > n$ to the interested reader.

4.5.2 ▪ Local Convergence

First of all, we will prove the local convergence of Algorithm 4.7. That is, if an initial guess is sufficiently close to a sparse solution, then Algorithm 4.7 will converge. Let us begin with some elementary results.

Lemma 4.30. *The sparse set $\mathcal{D}_s(\mathbb{R}^n)$ is the union of the collections of cardinal s-dimensional subspaces in $\mathbb{R}^n$:*

$$\mathcal{L}_s(\mathbb{R}^n) = \bigsqcup_{\mathcal{I}}\{\mathbf{x} \in \mathbb{R}^n \mid x_j = 0 \quad \forall j \in \mathcal{I}^c\},$$

where the index set $\mathcal{I}$ ranges over all the subsets of $\{1, 2, \dots, n\}$ with cardinality s, where $\bigsqcup_{\mathcal{I}}$ stands for the disjoint union over $\mathcal{I}$. That is, $\mathcal{D}_s(\mathbb{R}^n)$ consists of a disjoint union of affine spaces.

Proof. It is easy to see that the statement is correct. $\quad\square$

Lemma 4.31. *The set of vectors in $\mathbb{R}^n$ for which $\mathcal{P}_{\mathcal{D}_s}(x)$ is single valued is given by the open set*

$$V_s = \{\mathbf{x} \in \mathbb{R}^n \mid |x_{i_1}| \geq |x_{i_2}| \geq \cdots \geq |x_{i_n}|, |x_{i_{s+1}}| \neq |x_{i_s}|\}$$

consisting of vectors with the property that if we arrange the components in decreasing order of magnitude, then the sth and $(s+1)$th entries are distinct.

Proof. We start by noting that the projection $\mathcal{P}_{\mathcal{D}_s}(\mathbf{x})$ is obtained by setting the smallest $n - s$ components in magnitude of the vector $\mathbf{x}$ to zero. Hence, the projection is single valued if the $n - s$ smallest components of $\mathbf{x}$ are in unique positions (indices). Hence we must have that the $(n-s)$th and $(n-s+1)$th components of $\mathbf{x}$ are distinct. Now we will show that the set V_s is an open set. Let $\mathbf{x} = (x_1, x_2, \dots, x_n) \in \mathbb{R}^n$ with $|x_{i_1}| \geq |x_{i_2}| \geq \cdots \geq |x_{i_n}|, |x_{i_{s+1}}| \neq |x_{i_s}|$. Let

$$\epsilon := \frac{||x_{i_{s+1}}| - |x_{i_s}||}{4} > 0.$$

Consider an open ball $B_\epsilon(\mathbf{x})$ centered at $\mathbf{x}$ of radius ϵ. We have, for all $\mathbf{y} \in B_\epsilon(\mathbf{x})$ and $j \in \{1, 2, \dots, n\}$,

$$||y_j| - |x_j|| \leq |y_j - x_j| \leq \|\mathbf{y} - \mathbf{x}\| < \epsilon.$$

Therefore, we have

$$|y_{i_{j+1}}| \leq |x_{i_{j+1}}| + ||y_{i_{j+1}}| - |x_{i_{j+1}}|| < |x_{i_{s+1}}| + \epsilon < \frac{|x_{i_{s+1}}| + |x_{i_s}|}{2}$$

for $j \geq s$. Similarly,

$$|y_{i_j}| \geq |x_{i_j}| - ||x_{i_j}| - |y_{i_j}|| > |x_{i_s}| - \epsilon > \frac{|x_{i_{s+1}}| + |x_{i_s}|}{2}$$

for $j \leq s$. Hence, we deduce that, for all $\mathbf{y} \in B_\epsilon(\mathbf{x})$ and $j \in \{1, 2, \dots, s\}$, $|y_{i_j}| \geq |y_{i_{j+1}}|$, which implies that $\mathbf{y} \in V_s$ and, therefore, $B_\epsilon(\mathbf{x}) \subset V_s$. $\quad\square$

Next let us recall the convergence result of the alternating projection algorithm, i.e., Theorem 4.23 from subsection 4.5.1. With the above preparation, we are ready to prove the convergence of Algorithm 4.7.

Theorem 4.32. *Let* $\mathbf{x}^\star$ *be a sparse solution to* $\Phi \mathbf{x} = \mathbf{b}$ *with sparsity* s. *If* $\mathbf{x}^\star$ *is an isolated point of* $\mathcal{D}_s(\mathbb{R}^n) \cap \mathcal{A}$, *then Algorithm 4.7 will locally converge to* $\mathbf{x}^\star$ *linearly.*

Proof. Let $\mathcal{I} = \text{Supp}(\mathbf{x}^\star)$ be the support of $\mathbf{x}^\star$ and $s = \|\mathbf{x}^\star\|_0$. Consider an open set V_s of vectors with the property that their $n - s$ smallest components are in unique positions (indices). In fact, V_s can be concretely described as

$$V_s = \left\{ \mathbf{x} \in \mathbb{R}^n \mid \quad |x_{i_1}| \geq |x_{i_2}| \geq \cdots \geq |x_{i_n}|, |x_{i_s}| \neq |x_{i_{s+1}}| \right\}.$$

Clearly $\mathbf{x}^\star \in V_s$. Let $B(\mathbf{x}^\star, r)$ be an open ball centered at $\mathbf{x}^\star$ and of radius r completely contained inside V_s. Since $B(\mathbf{x}^\star, r) \subseteq V_s$, for any $\mathbf{x} \in B(\mathbf{x}^\star, r)$, the projection $\mathcal{P}_{\mathcal{D}_s}(\mathbf{x})$ is uniquely defined.

Since affine spaces in a finite-dimensional Euclidean space are closed, we can shrink the ball $B(\mathbf{x}, r)$, if necessary, such that the restriction $\mathcal{D}_s(\mathbb{R}^n)|_{B(\mathbf{x}^\star, r)}$ of the set of s-sparse vectors to the open set $B(\mathbf{x}^\star, r)$ is an affine space. Then, under the assumption of the hypothesis in this theorem, the convergence result follows from Theorem 4.23.   □

A sufficient condition that guarantees that $\mathbf{x}^\star$ is an isolated point is that the tangent spaces of $\mathcal{D}_s(\mathbb{R}^N)$ and $\mathcal{A}$ intersect trivially.

Lemma 4.33. *Assume that the sensing matrix* Φ *has the following property:* $\text{spark}(\Phi) > s$, *i.e.,*

$$\mathcal{D}_s(\mathbb{R}^n) \cap \text{Null}(\Phi) = \{0\}, \tag{4.59}$$

where $\text{Null}(\Phi)$ *is the null space of* Φ. *Furthermore, assume that* $\mathbf{x}^\star \in \mathcal{D}_s(\mathbb{R}^n) \cap \mathcal{A}$. *Then* $\mathbf{x}^\star$ *is an isolated point of* $\mathcal{D}_s(\mathbb{R}^n) \cap \mathcal{A}$.

Proof. Let us suppose, on the contrary, that $\mathbf{x}^\star$ is not an isolated point of the set $\mathcal{D}_s(\mathbb{R}^n) \cap \mathcal{A}$. Then, since $\mathcal{A}$ and $\mathcal{D}_s(\mathbb{R}^n)$ are locally affine spaces, there exists a linear space L of dimension greater than or equal to 1 such that $L + \mathbf{x}^\star \subseteq \mathcal{D}_s(\mathbb{R}^n) \cap \mathcal{A}$. Since each of the intersecting spaces is an affine space locally, L must also lie in the intersection of their tangent spaces. Hence,

$$L \subseteq T_{\mathcal{D}_s(\mathbb{R}^n)}(\mathbf{x}^\star) \cap \text{Null}(\Phi),$$

where $T_{\mathcal{D}_s(\mathbb{R}^n)}(\mathbf{x}^\star)$ stands for the tangent space to $\mathcal{D}_s(\mathbb{R}^n)$ at the point $\mathbf{x}^\star$. Now, since $\mathcal{D}_s(\mathbb{R}^n)$ is a union of linear spaces by Lemma 4.30, we may assume that $\mathbf{x}^\star \in L_0 \subseteq \mathcal{D}_s(\mathbb{R}^n)$ lies in a linear space L_0 contained in $\mathcal{D}_s(\mathbb{R}^n)$. Therefore, we have

$$L \subseteq T_{L_0}(\mathbf{x}^\star) \cap \text{Null}(\Phi) = L_0 \cap \text{Null}(\Phi) \subseteq \mathcal{D}_s(\mathbb{R}^n) \cap \text{Null}(\Phi) = \{0\},$$

which leads to a contradiction because L is of dimension greater than or equal to 1. Note that, in order to derive the equality in the last equation, we have used the fact that the tangent space of a linear space is the linear space itself.   □

The discussion above leads to the following conclusion.

Theorem 4.34. *Under the assumption* (4.59) *in Lemma 4.33, Algorithm 4.7 will locally converge linearly.*

Proof. We simply combine Lemma 4.33 and Theorem 4.32 to obtain this result.   □

4.5.3 ▪ Global Convergence

Next we will investigate conditions under which Algorithm 4.7 will converge globally. That is, Algorithm 4.7 will converge starting from any initial guess. We shall assume, without loss of generality, that Φ has full row rank. We say that Φ satisfies the η-condition if Φ satisfies the following property: There exists a positive number $\eta_s = \eta_s(\Phi) < 1$ such that

$$\max_{M_{s \times s} \subset \Phi^\top (\Phi \Phi^\top)^{-1} \Phi - I_{n \times n}} \sigma_1(M_{s \times s}) \leq \eta_s, \tag{4.60}$$

where $\sigma_1(M)$ is the largest singular value of matrix M.

We will abuse the notation a bit and let

$$\eta_s(\Phi) := \max_{M_{s \times s} \subset \Phi^\top (\Phi \Phi^\top)^{-1} \Phi - I_{n \times n}} \sigma_1(M_{s \times s}).$$

Note that the η_s-condition is different from the classic RIC δ_s. We can rewrite η_s in the following way. For any index sets S_1, S_2 with $\#(S_1) \leq s$ and $\#(S_2) \leq s$ and for any vectors $\mathbf{x}, \mathbf{y} \in \mathbb{R}^n$, we see that

$$\begin{aligned}
(\Phi_{S_1} \mathbf{x}_{S_1})^\top (\Phi \Phi^\top)^{-1} (\Phi_{S_2} \mathbf{y}_{S_2}) - \mathbf{x}_{S_1}^\top \mathbf{y}_{S_2} &= \langle (\Phi_{S_1}^\top (\Phi \Phi^\top)^{-1} \Phi_{S_2} - I_{S_1,S_2}) \mathbf{x}_{S_1}, \mathbf{y}_{S_2} \rangle \\
&\leq \mathbf{x}_{S_1}^\top (M_{s \times s} - I_{S_1,S_2}) \mathbf{y}_{S_2} \\
&\leq \|\mathbf{x}_{S_1}\|_2 \sigma_1(M_{s \times s}) \|\mathbf{y}_{S_2}\|_2,
\end{aligned}$$

where I_{S_1,S_2} is the restriction of the identity matrix of size $n \times n$ over the row index set S_1 and column index set S_2. That is,

$$\max_{\mathbf{x}_{S_1} \neq 0, \mathbf{y}_{S_2} \neq 0} \frac{\langle (\Phi_{S_1}^\top (\Phi \Phi^\top)^{-1} \Phi_{S_2} - I_{S_1,S_2}) \mathbf{x}_{S_1}, \mathbf{y}_{S_2} \rangle}{\|\mathbf{x}_{S_1}\|_2 \|\mathbf{y}_{S_2}\|} \leq \sigma_1(M_{s \times s}) \leq \eta_s(\Phi). \tag{4.61}$$

Therefore, we obtain the following characterization for $\eta_s(\Phi)$.

Theorem 4.35. *We have*

$$\eta_s(\Phi) = \max_{\substack{S_1, S_2 \subset \{1, \ldots, n\} \\ \#(S_1) \leq s, \#(S_2) \leq s}} \|\Phi_{S_1}^\top (\Phi \Phi^\top)^{-1} \Phi_{S_2} - (I)_{S_1,S_2}\|_{2 \to 2}. \tag{4.62}$$

Let $\mathbf{x}, \mathbf{y} \in \mathbb{R}^n$ be two vectors with sparsity $\|\mathbf{x}\|_0 \leq s$ and $\|\mathbf{y}\|_0 \leq s$. If $\sup(\mathbf{x}) \cap \sup(\mathbf{y}) = \emptyset$, then

$$\begin{aligned}
|\langle (\Phi \Phi^\top)^{-1} \Phi_{S_2} \mathbf{y}, \Phi_{S_1} \mathbf{x} \rangle| &= |\langle (\Phi \Phi^\top)^{-1} \Phi_{S_2} \mathbf{y}, \Phi_{S_1} \mathbf{x} \rangle - \langle I_{S_1,S_2} \mathbf{y}, \mathbf{x} \rangle| \\
&= |\langle (\Phi_{S_1}^\top (\Phi \Phi^\top)^{-1} \Phi_{S_2} - I_{S_1,S_2}) \mathbf{y}, \mathbf{x} \rangle| \\
&\leq \|\Phi_{S_1}^\top (\Phi \Phi^\top)^{-1} \Phi_{S_2} - I_{S_1,S_2}\|_{2 \to 2} \|\mathbf{x}\|_2 \|\mathbf{y}\|_2 \\
&\leq \eta_s \|\mathbf{x}\|_2 \|\mathbf{y}\|_2.
\end{aligned}$$

Let us recall Definition 2.33, the (s,t)-restricted orthogonality condition $\theta_{s,t} = \theta_{s,t}(M)$ of a matrix $M \in \mathbb{R}^{m \times n}$, which is the smallest $\theta \geq 0$ such that

$$|\langle M\mathbf{u}, M\mathbf{v} \rangle| \leq \theta \|\mathbf{u}\|_2 \|\mathbf{v}\|_2. \tag{4.63}$$

That is,

$$\theta_{s,t} = \max\{\|M_T^\top M_S\|_{2 \to 2}, S \cap T = \emptyset, \#(S) \leq s, \#(T) \leq s\}. \tag{4.64}$$

We are now ready to establish the following result on the η-condition.

Theorem 4.36. *Assume that Φ is of full row rank. Define by $M = (\Phi\Phi^\top)^{-1/2}\Phi$ a normalized sensing matrix. Then*

$$\eta_s(\Phi) = \theta_{s,s}(M). \tag{4.65}$$

Proof. Indeed, we have

$$
\begin{aligned}
\|M_T^\top M_S\|_{2\to 2} &= \max\left\{ \frac{|\langle M_T\mathbf{u}_T, M_S\mathbf{v}_S\rangle|}{\|\mathbf{u}_T\|_2\|\mathbf{v}_S\|_2},\ s\text{-sparse vectors } \mathbf{u}_T, \mathbf{v}_S \right\} \\
&= \max\left\{ \frac{|\langle \Phi_S^\top(\Phi\Phi^\top)^{-1}\Phi_T\mathbf{u}_T, \mathbf{v}_S\rangle|}{\|\mathbf{u}_T\|_2\|\mathbf{v}_S\|_2},\ s\text{-sparse vectors } \mathbf{u}_T, \mathbf{v}_S \right\} \\
&\leq \|\Phi_S^\top(\Phi\Phi^\top)^{-1}\Phi_T - (\mathbf{I})_{S,T}\|_{2\to 2} \leq \eta_s.
\end{aligned}
$$

It thus follows that $\theta_{s,s} \leq \eta_s$. Similarly, we have the other direction. This completes the proof of (4.65). $\quad\square$

Since many properties of $\theta_{s,s}$ are known, the relation in (4.65) helps us understand η_s better. For example, $\theta_{s,s} \leq \delta_{2s}$ and $\delta_{2s} \leq \delta_s + \theta_{s,s}$.

Finally, we are now ready to establish the following convergence result.

Theorem 4.37. *Suppose that Φ is of full row rank and satisfies the η-condition with $\eta_{2s}(\Phi) < 1/2$. Then Algorithm 4.7 will globally converge linearly.*

Proof. Let $\mathbf{x}^* \in \mathbb{R}^n$ be a solution of (1.4) in Chapter 1 with $\mathbf{x}^* \in \mathcal{D}_s(\mathbb{R}^n) \cap \mathcal{A}$. Starting with an initial guess $\mathbf{y}_0$, we find

$$\mathbf{x}_{k+1} = \mathcal{P}_\mathcal{A}(\mathbf{y}_k) = \Phi^\top(\Phi\Phi^\top)^{-1}\mathbf{b} - (\Phi^\top(\Phi\Phi^\top)^{-1}\Phi - I_{n\times n})\mathbf{y}_k.$$

Also, we have

$$\|\mathbf{y}_{k+1} - \mathbf{x}_{k+1}\|^2 = \|\mathbf{y}_{k+1} - \mathbf{x}^*\|^2 + \|\mathbf{x}^* - \mathbf{x}_{k+1}\|^2 + 2\langle \mathbf{y}_{k+1} - \mathbf{x}^*, \mathbf{x}^* - \mathbf{x}_{k+1}\rangle.$$

For convenience, we write $B = \mathcal{D}_s(\mathbb{R}^n)$. Then

$$\|\mathbf{y}_{k+1} - \mathbf{x}_{k+1}\|^2 = \|\mathcal{P}_B(\mathbf{x}_{k+1}) - \mathbf{x}_{k+1}\|^2 \leq \|\mathbf{x}^* - \mathbf{x}_{k+1}\|^2.$$

Combining the right-hand sides of the above two equations leads to

$$
\begin{aligned}
\|\mathbf{y}_{k+1} - \mathbf{x}^*\|^2 &\leq 2\langle \mathbf{y}_{k+1} - \mathbf{x}^*, \mathbf{x}_{k+1} - \mathbf{x}^*\rangle \\
&= 2\langle \mathbf{y}_{k+1} - \mathbf{x}^*, \Phi^\top(\Phi\Phi^\top)^{-1}\mathbf{b} - (\Phi^\top(\Phi\Phi^\top)^{-1}\Phi - I_{n\times n})\mathbf{y}_k - \mathbf{x}^*\rangle \\
&= 2\langle \mathbf{y}_{k+1} - \mathbf{x}^*, (\Phi^\top(\Phi\Phi^\top)^{-1}\Phi - I_{n\times n})(\mathbf{x}^* - \mathbf{y}_k)\rangle \\
&\leq 2(\langle \mathbf{y}_{k+1} - \mathbf{x}^*, (\Phi^\top(\Phi\Phi^\top)^{-1}\Phi - I_{n\times n})(\mathbf{y}_{k+1} - \mathbf{x}^*)\rangle)^{1/2} \\
&\quad \times(\langle \mathbf{y}_k - \mathbf{x}^*, (\Phi^\top(\Phi\Phi^\top)^{-1}\Phi - I_{n\times n})(\mathbf{y}_k - \mathbf{x}^*)\rangle)^{1/2} \\
&\leq 2\|\mathbf{y}_{k+1} - \mathbf{x}^*\|\eta_{2s}(\Phi)\|\mathbf{y}_k - \mathbf{x}^*\|,
\end{aligned}
$$

where we have used the Cauchy–Schwarz inequality in matrix format, i.e., (1.69). In other words, we have

$$\|\mathbf{y}_{k+1} - \mathbf{x}^*\| \leq 2\eta_{2s}(\Phi)\|\mathbf{y}_k - \mathbf{x}^*\|$$

for all $k \geq 1$. Since, by hypothesis, $\gamma := 2\eta_{2s}(\Phi) < 1$, we have

$$\|\mathbf{y}_{k+1} - \mathbf{x}^*\| \leq \gamma\|\mathbf{y}_k - \mathbf{x}^*\|, \quad \text{and hence} \quad \|\mathbf{y}_{k+1} - \mathbf{x}^*\| \leq \gamma^k\|\mathbf{y}_1 - \mathbf{x}^*\|$$

for all $k \geq 1$. This completes the proof. $\quad\square$

4.6 ▪ Exercises

Exercise 1. Prove Theorem 4.1. That is, give a proof of the result in Theorem 4.1 in the noisy setting.

Exercise 2. Show that if $\delta_{4s} \leq 1/9$, then

$$\rho := 2\left(1 + \frac{\delta_{4s}}{1 - \delta_{3s}}\right)\frac{\delta_{3s} + \delta_{4s}}{1 - \delta_{2s}} < 1.$$

Exercise 3. Let $\mathbf{x}$ be an s-sparse vector. Suppose that Φ has a RIP with $0 < \delta_s < 1$. Then for any $T \subset \mathrm{supp}(\mathbf{x})$, show that $\|\Phi_T^\top \Phi_{T^c}\mathbf{x}\|_2 \leq \delta_s\|\mathbf{x}_{T^c}\|_2$ and $\|(\Phi_T^\top \Phi_{T^c})^{-1}\mathbf{x}\|_2 \leq \frac{1}{1-\delta_s}\|\mathbf{x}_{T^c}\|_2$.

Exercise 4. Prove Theorem 4.5.

Exercise 5. Write a MATLAB code to solve an underdetermined linear system by using Algorithm 4.2. Experiment with several different types of matrices to see if Algorithm 4.2 works.

Exercise 6. Study the convergence of Algorithm 4.2.

Exercise 7. Write a MATLAB code to solve an underdetermined linear system using Algorithm 4.3. Experiment with several different types of matrices to see if Algorithm 4.3 works.

Exercise 8. Study the convergence of Algorithm 4.4.

Exercise 9. Write a MATLAB code to solve an underdetermined linear system using Algorithm 4.4. Experiment with several different types of matrices to see if Algorithm 4.4 works.

Exercise 10. Write a MATLAB code to solve an underdetermined linear system using Algorithm 4.5. Experiment with several different types of matrices to see if Algorithm 4.5 works.

Exercise 11. Show that all functions in Table 4.1 can be written as a difference of two convex functions.

Exercise 12. Show that for all the regularizers listed in Table 4.1, problem (4.44) has a closed-form solution.

Exercise 13. Study the convergence of Algorithm 4.6 using the nonmonotone line search strategy.

Exercise 14. Write a MATLAB code to project any vector $\mathbf{x}$ to the solution set $\mathcal{A} = \{\mathbf{y}, \Phi\mathbf{y} = \mathbf{b}\}$ of an underdetermined linear system using (4.57) and Theorem 4.27.

Exercise 15. Write a MATLAB code to project any vector $\mathbf{x}$ to the solution set $\mathcal{A} = \{\mathbf{y}, \Phi\mathbf{y} = \mathbf{b}\}$ of an underdetermined linear system using (4.58).

Exercise 16. Implement the alternating projection Algorithm 4.8 for the sparse solution of an underdetermined linear system.

Exercise 17. Let Φ be a matrix of size $m \times n$ and $m < n$. Suppose that Φ is of full row rank. Show that $\|\Phi^\top(\Phi\Phi^\top)^{-1}\Phi\| \leq 1$.

Exercise 18. Prove Theorem 4.27.

Exercise 19. Prove Theorem 4.28.

Bibliography

[1] J. Barzilai and J. M. Borwein, Two-point step size gradient methods, IMA J. Numer. Anal., 8 (1988), 141–148. (Cited on pp. 104, 106)

[2] T. Blumensath and M. E. Davies, Iterative hard thresholding for compressed sensing, Appl. Comput. Harmon. Anal., 27 (2009), 265–274. (Cited on pp. 89, 92)

[3] T. Blumensath and M. E. Davies, Normalized iterative hard thresholding: Guaranteed stability and performance, IEEE J. Sel. Top. Signal Processing, 4 (2010), 298–309. (Cited on p. 89)

[4] J.-L. Bouchot, S. Foucart, and P. Hitczenko, Hard thresholding pursuit algorithms: Number of iterations. Appl. Comput. Harmon. Anal., 41 (2016), 412–435. (Cited on p. 97)

[5] E. J. Candés, M. B. Wakin, and S. P. Boyd, Enhancing sparsity by reweighted ℓ_1 minimization, J. Fourier Anal. Appl., 14 (2008), 877–905. (Cited on p. 103)

[6] F. Deutsch, Best Approximation in Inner Product Spaces, Canadian Mathematical Society, 2001. (Cited on pp. 110, 113)

[7] J. Fan and R. Li, Variable selection via nonconcave penalized likelihood and its oracle properties, J. Amer. Stat. Assoc., 96 (2001), 1348–1360. (Cited on p. 103)

[8] S. Foucart, A note on guaranteed sparse recovery via ℓ_1-minimization, Appl. Comput. Harmon. Anal., 29 (2010), 97–103. (Cited on p. 102)

[9] S. Foucart, Hard thresholding pursuit: An algorithm for compressive sensing, SIAM J. Numer. Anal., 49 (2011), 2543–2563. (Cited on pp. 91, 93, 94, 97, 98, 100)

[10] S. Foucart and M.-J. Lai, Sparsest solutions of underdetermined linear systems via ℓ_q-minimization for $0 \leq q \leq 1$, Appl. Comput. Harmon. Anal., 26 (2009), 395–407. (Cited on p. 103)

[11] P. Gong, C. Zhang, Z. Lu, J. Huang, and J. Ye, A general iterative shrinkage and thresholding algorithm for non-convex regularized optimization problems, in Proceedings of the 30th International Conference on Machine Learning, 28, 2013, 37–45. (Cited on pp. 103, 104, 106, 108)

[12] M.-J. Lai and A. Varghese, On Convergence of the Alternating Projection Method for Matrix Completion and Sparse Recovery Problems, preprint, arXiv:1711.02151. (Cited on pp. 109, 113)

[13] D. Needell and J. A. Tropp, CoSaMP: Iterative signal recovery from incomplete and inaccurate samples, Appl. Comput. Harmon. Anal., 26 (2009), 301–332. (Cited on pp. 98, 100)

[14] J. Von Neumann, Functional Operators, Volume 2: The Geometry of Orthogonal Spaces, Princeton University Press, 1950. (Cited on p. 110)

[15] R. T. Rockafellar, Convex Analysis, Princeton University Press, 1970. (Cited on p. 104)

[16] A. L. Yuille and A. Rangarajan, The concave-convex procedure, Neural Comput., 15(4) (2003), 915–936. (Cited on p. 103)

[17] C. H. Zhang, Nearly unbiased variable selection under minimax concave penalty, Ann. Stat., 38(2) (2010), 894–942. (Cited on p. 103)

[18] T. Zhang, Analysis of multi-stage convex relaxation for sparse regularization, J. Mach. Learn. Res., 11 (2010), 1081–1107. (Cited on p. 103)

Chapter 5

Unconstrained Minimization for Sparse Recovery

We begin by discussing how to solve the unconstrained minimization problem

$$\min\{f(\mathbf{x}) : \quad \mathbf{x} \in \mathbb{R}^n\}, \tag{5.1}$$

where $f(\mathbf{x})$ is a convex function, e.g., $f(\mathbf{x}) = \lambda \|\mathbf{x}\|_1 + \frac{1}{2}\|\Phi\mathbf{x} - \mathbf{b}\|^2$. More precisely, we solve

$$\min\left\{\|\mathbf{x}\|_1 + \frac{1}{2\lambda}\|\Phi\mathbf{x} - \mathbf{b}\|_2^2\right\} \tag{5.2}$$

as an approximation of our main problem (1.4) in Chapter 1. This is the well-known lasso problem in statistics (cf. [13]), dating back to 1996. Note that $\lambda\|\mathbf{x}\|_1$ is not differentiable, so the standard gradient descent method cannot be directly applied. We shall use the so-called proximal gradient descent method instead. In addition, there is a rich theory for various algorithms, their convergence analysis, and their accelerated techniques. To this end, we start with a review of convex analysis, introducing a few methods: gradient descent methods, accelerated gradient descent methods, and proximal gradient descent methods, as well as their accelerated versions. Finally, we explain how to use these algorithms to compute sparse solutions of underdetermined systems of linear equations.

5.1 ▪ Convex Sets

We shall use the standard Euclidean space $\mathbb{R}^n$ of dimension $n \geq 1$ with standard norm $\|\mathbf{x}\| = \sqrt{\sum_{i=1}^n x_i^2}$ for any point $\mathbf{x} \in \mathbb{R}^n$. Then $V = \mathbb{R}^n$ is a normed vector space.

Let $K \subset \mathbb{R}^n$ be a subset. K is called a convex set if for any $\mathbf{x}^1, \mathbf{x}^2 \in K$ and $0 \leq \lambda \leq 1$, $\lambda\mathbf{x}^1 + (1-\lambda)\mathbf{x}^2 \in K$. There are many examples of convex sets. We shall let $V \subset \mathbb{R}^n$ be a finite-dimensional space.

- Any ball $B_r = \{\mathbf{x} \in V, \|\mathbf{x}\|_V \leq r\}$ with radius $r > 0$ in a normed vector space is a convex set.

- Any hyperplane $H = \{\mathbf{x} \in V, \mathbf{a} \cdot \mathbf{x} = b\}$ for a nonzero vector $\mathbf{a} \in V$ is convex, where b is a scalar.

- Every affine set or linear manifold $A = \{\mathbf{x} : \mathbf{x} \in \mathbb{R}^d, \Phi\mathbf{x} = \mathbf{b}\}$ is convex, and so is every half space $K = \{\mathbf{x} : \mathbf{x} \in \mathbb{R}^d, \mathbf{b} \cdot \mathbf{x} < 0\}$, where Φ is a matrix of size $m \times d$, $\mathbf{b}$ is a vector in $\mathbb{R}^d$, and $\mathbf{0}$ is a vector in $\mathbb{R}^m$.

- Any subspace of a vector space is convex.

- A cone C with vertex at $\mathbf{x}_0$ is the set of the points in $\mathbb{R}^d$ in the form $\mathbf{x}_0 + \alpha(\mathbf{x} - \mathbf{x}_0)$ for any $\mathbf{x} \in C, \alpha > 0$.

- Any ellipsoid, i.e., a set $E = \{\mathbf{x} \in \mathbb{R}^n, \mathbf{x}^\top C \mathbf{x} \leq a\}$ with C being a symmetric positive definite matrix of size $n \times n$, is convex.

- Any polyhedron, i.e., a set $P = \{\mathbf{x} : \mathbf{x} \in \mathbb{R}^n, \Phi \mathbf{x} \leq \mathbf{b}\}$, where Φ is a matrix of size $m \times n$ and $\mathbf{b}$ is a vector in $\mathbb{R}^m$, is convex. That is, P is the intersection of a finite number of closed half spaces.

Next, let $L_2(\Omega)$ be the space of all functions which are square integrable over a bounded domain $\Omega \subset \mathbb{R}^n$, i.e., a function $f \in L_2(\Omega)$ if and only if $\int_\Omega |f(\mathbf{x})|^2 dx < \infty$. We shall use the norm $\|f\|_2 = \sqrt{\int_\Omega |f(\mathbf{x})|^2 dx}$ and the inner product $\langle f, g \rangle = \int_\Omega f(\mathbf{x})g(\mathbf{x})dx$. We shall also use $\langle \mathbf{x}, \mathbf{y} \rangle = \mathbf{x}^\top \mathbf{y}$ for the standard inner product on $\mathbb{R}^n$. There exist many normed linear spaces V with a given norm $\| \cdot \|_V$ and several Hilbert spaces, which are closed normed linear spaces with a given inner product such as $L_2(\Omega)$. Then we can show the following:

- Any ball $B_r = \{f \in L_2(\Omega), \|f\|_2 \leq r\}$ with radius $r > 0$ is a convex set.

- Any closed half space, i.e., a set $E_g = \{f \in L_2(\Omega), \langle f, g \rangle \leq a\}$ with a a real number, is convex.

The following is a list of standard results on convex sets. The proof is left to the interested reader.

Theorem 5.1. *The following is a list of basic properties:*

(1) *If K_1, K_2 are convex sets in a vector space V, then $\alpha K_1 = \{\mathbf{z} : \mathbf{z} = \alpha \mathbf{x}, \mathbf{x} \in K\}$ is convex and $K_1 + K_2 = \{\mathbf{x} + \mathbf{y} : \mathbf{x} \in K_1, \mathbf{y} \in K_2\}$ is convex.*

(2) *If $K_1, \ldots, K_m$ are convex sets in a vector space V, then $\cap_{i=1}^m K_i$ is convex.*

(3) *If K_1, K_2 are convex, $K_1 \cup K_2$ might not be convex.*

5.2 ▪ Convex Functions and Functionals

Next we discuss convex functions and functionals.

Definition 5.2. *Let $K \subset \mathbb{R}^n$ be a convex set. Let $f : K \mapsto \mathbb{R}$ be a function defined on K. f is said to be convex on K if*

$$f(\alpha \mathbf{v} + (1 - \alpha)\mathbf{u}) \leq \alpha f(\mathbf{v}) + (1 - \alpha)f(\mathbf{u}) \quad \forall \mathbf{v}, \mathbf{u} \in K, \alpha \in [0, 1].$$

Similarly, let $K \subset L_2(\Omega)$ be a convex set of functions and $F : K \mapsto \mathbb{R}$ be a functional on K. If F satisfies the property

$$F(\alpha f + (1 - \alpha)g) \leq \alpha F(f) + (1 - \alpha)F(g) \quad \forall f, g \in K, \alpha \in [0, 1],$$

then F is a convex functional.

Example 5.3. The following are some examples of convex functions:

- $f(x) = ax + b$ for $x \in \mathbb{R}$ or $f(\mathbf{x}) = \|A\mathbf{x} + \mathbf{b}\|$ for $\mathbf{x} \in \mathbf{R}^d$;

- $f(x) = x^2$ for $x \in \mathbb{R}$ or $f(\mathbf{x}) = \|\mathbf{x}\|_2^2$ for $\mathbf{x} \in \mathbf{R}^d$;

- $f(x) = \sqrt{1 + x^2}$ for $x \in \mathbb{R}$ or $f(\mathbf{x}) = \|\mathbf{x}\|_2$ for $\mathbf{x} \in \mathbf{R}^d$;

- $f(x) = |x|$ for $x \in \mathbb{R}$ or $f(\mathbf{x}) = \|\mathbf{x}\|_p$ for $\mathbf{x} \in \mathbf{R}^d$ for $p \geq 1$.

Also, $f(\mathbf{x}) = \|\mathbf{x}\|_p^p$ for $\mathbf{x} \in \mathbf{R}^d$ is a convex function when $p \geq 1$.

Example 5.4. The following are examples of convex functionals:

- $F(f) = \int_\Omega (af(\mathbf{x}) + b)d\mathbf{x}$ for any fixed real numbers a, b, where $f \in L_2(\Omega)$;

- $F(f) = \int_\Omega |f(\mathbf{x})|d\mathbf{x}$ for $f \in K \subset L_2(\Omega)$, where K is a convex set;

- $F(f) = \int_\Omega \sqrt{1 + f(\mathbf{x})^2}d\mathbf{x}$ for $f \in K \subset L_2(\Omega)$, where K is a convex set;

- $F(f) = \int_\Omega \sqrt{1 + |\nabla f(\mathbf{x})|^2}d\mathbf{x}$ for all $f \in H^1(\Omega) \subset L_2(\Omega)$, where ∇f stands for the gradient vector of f and $H^1(\Omega) = \{f \in L_2(\Omega), \|\nabla f\|_2 < \infty\}$ is a Sobolev space.

Here are a few more properties which are written for convex set K in a normed vector space V, e.g., $V = \mathbb{R}^n$.

Theorem 5.5. *We have the following additional properties of convex functions:*

- *Suppose that f is convex on $K \subset V$. Then for any $x_1, \ldots, x_m \in K$ and $\alpha_1 \geq 0, \ldots, \alpha_m \geq 0$ with $\sum_{i=1}^m \alpha_i = 1$,*

$$f\left(\sum_{i=1}^m \alpha_i x_i\right) \leq \sum_{i=1}^m \alpha_i f(x_i).$$

- *Let $f_i, i = 1, \ldots, m$, be convex functions on K. Then*

$$g(\mathbf{x}) = \sup_{1 \leq i \leq m} f_i(\mathbf{x}), \quad \mathbf{x} \in K,$$

is a convex function.

- *Let f and g be convex functions. Then $f(g(\mathbf{x}))$ is a convex on K.*

- *Let f be a convex function on V, and let P be an operator from V to V (i.e., P is a $d \times d$ matrix if $V = \mathbb{R}^d$). Then $g(\mathbf{x}) = f(P\mathbf{x} + \mathbf{b})$ is convex for any $\mathbf{b} \in V$.*

- *Let*

$$L(f, \alpha) = \{\mathbf{x} \in K, f(\mathbf{x}) \leq \alpha\}$$

be the level set for a fixed real number α. Then $L(f, \alpha)$ is a convex set.

One easy property of a convex function f is the following.

Lemma 5.6. *Suppose that f is convex over a domain $\Omega \subset \mathbb{R}^n$. Suppose that f has a gradient over Ω. Then*

$$f(\mathbf{y}) - f(\mathbf{x}) \geq \langle \nabla f(\mathbf{x}), \mathbf{y} - \mathbf{x} \rangle \quad \forall \mathbf{x}, \mathbf{y} \in \Omega. \tag{5.3}$$

This property will be used again and again later.

To check whether a function f is convex, the following is a sufficient condition.

Theorem 5.7. *Suppose that f is twice differentiable on $\Omega \subset \mathbb{R}^n$. If its Hessian*

$$H(f)(\mathbf{x}) := [\nabla^2 f(\mathbf{x})] = \left[\frac{\partial^2}{\partial x_i \partial x_j} f(\mathbf{x}) \right]_{i,j=1,\ldots,n} \tag{5.4}$$

is nonnegative definite over Ω, then f is convex.

Proof. Mainly, we use Taylor expansion at $\mathbf{z} = \alpha\mathbf{x} + (1-\alpha)\mathbf{y}$ to get

$$f(\mathbf{x}) = f(\mathbf{z}) + \nabla f(\mathbf{z})(\mathbf{x} - \mathbf{z}) + \frac{1}{2}(\mathbf{x} - \mathbf{z})^\top H(f)(\xi)(\mathbf{x} - \mathbf{z}).$$

We proceed in a similar way for $f(\mathbf{y})$. It follows that

$$\alpha f(\mathbf{x}) + (1-\alpha)f(\mathbf{y})$$
$$= f(\mathbf{z}) + \frac{\alpha}{2}(\mathbf{x} - \mathbf{z})^\top H(f)(\xi)(\mathbf{x} - \mathbf{z}) + \frac{1-\alpha}{2}(\mathbf{y} - \mathbf{z})^\top H(f)(\eta)(\mathbf{y} - \mathbf{z})$$
$$= f(\alpha\mathbf{x} + (1-\alpha)\mathbf{y}) + \frac{\alpha(1-\alpha)}{2}(\alpha(\mathbf{x}-\mathbf{y})^\top H(f)(\eta)(\mathbf{x}-\mathbf{y}) + (1-\alpha)(\mathbf{x}-\mathbf{y})^\top H(f)(\xi)(\mathbf{x}-\mathbf{y}))$$
$$\geq f(\alpha\mathbf{x} + (1-\alpha)\mathbf{y}),$$

since the last two terms are nonnegative because $H(f)$ is nonnegative. □

Next we introduce the concept of strongly convex functions.

Definition 5.8. *A function f defined on a convex domain $\Omega \subset \mathbb{R}^n$ is said to be μ-strongly convex with respect to a norm $\|\cdot\|$ on $\mathbb{R}^m$ if for any $\alpha \in [0,1]$ and x, y in Ω,*

$$f(\alpha\mathbf{x} + (1-\alpha)\mathbf{y}) + \mu\frac{\alpha(1-\alpha)}{2}\|\mathbf{x} - \mathbf{y}\|^2 \leq \alpha f(\mathbf{x}) + (1-\alpha)f(\mathbf{y}).$$

When f is a differentiable function defined on $\Omega \subset \mathbb{R}^n$, it is easy to verify that if f satisfies the condition

$$f(\mathbf{y}) \geq f(\mathbf{x}) + \langle \nabla f(\mathbf{x}), \mathbf{x} - \mathbf{y} \rangle + \frac{\mu}{2}\|\mathbf{x} - \mathbf{y}\|_2^2 \quad \forall \mathbf{x}, \mathbf{y} \in \Omega, \tag{5.5}$$

then f is a μ-strongly convex function, where ∇f is the gradient vector of f. We leave the verification to the interested reader.

Also, when a convex function f defined on $\Omega \subset \mathbb{R}^n$ satisfies

$$\langle \mathbf{y}^\top \nabla^2 f(\mathbf{x}), \mathbf{y} \rangle \geq \mu\|\mathbf{y}\|_2^2 \quad \forall \mathbf{y} \in \Omega, \tag{5.6}$$

f is μ-strongly convex at $\mathbf{x}$, where $\nabla^2 f$ stands for the Hessian of f at $\mathbf{x} \in \Omega$.

There are many other properties on the strong convexity of f. Let us state the following three properties and leave the proofs to the interested reader. See [12] for more on convex analysis.

Theorem 5.9. *Suppose that f is continuously differentiable on Ω. Then f is strongly convex if and only if*

$$\langle \nabla f(\mathbf{x}) - \nabla f(\mathbf{y}), \mathbf{x} - \mathbf{y} \rangle \geq \nu\|\mathbf{x} - \mathbf{y}\|^2 \quad \forall \mathbf{x}, \mathbf{y} \in \Omega. \tag{5.7}$$

Theorem 5.10. *Suppose that f is continuously differentiable on Ω. Suppose that f is strongly convex. Then*

$$f(\mathbf{y}) \le f(\mathbf{x}) + \langle \nabla f(\mathbf{x}), \mathbf{y} - \mathbf{x} \rangle + \frac{1}{2\nu} \|\nabla f(\mathbf{x}) - \nabla f(\mathbf{y})\|^2 \tag{5.8}$$

and

$$\langle \nabla f(\mathbf{x}) - \nabla f(\mathbf{y}), \mathbf{x} - \mathbf{y} \rangle \le \frac{1}{\nu} \|\mathbf{x} - \mathbf{y}\|^2 \quad \forall \mathbf{x}, \mathbf{y} \in \Omega. \tag{5.9}$$

Theorem 5.11. *Suppose that a functional F is Lipschitz differentiable with constant L (see Definition 5.17 below). Then*

$$\frac{1}{L} \|\nabla F(\mathbf{x}) - \nabla F(\mathbf{y})\|^2 \le \langle \nabla F(\mathbf{x}) - \nabla F(\mathbf{y}), \mathbf{x} - \mathbf{y} \rangle \tag{5.10}$$

for all $\mathbf{x}, \mathbf{y} \in \mathbb{R}^n$.

5.3 ▪ Subdifferentiation

Let V be a normed vector space. We define the dual V^* of V to be the space of all continuous linear functionals on V. We can easily show that the dual of $\mathbb{R}^n$ is itself. Also, the dual of $L_2(\Omega)$ is equivalent to itself by the Riesz representation theorem. For a linear functional $u^* \in V^*$, we shall use $\langle u^*, v \rangle$ to denote the action of u^* on $v \in V$.

For a differentiable function f on a normed vector space V whose dual space is V^*, $\nabla f(\mathbf{u})$ is a vector in V^*. When f is convex, we know that

$$f(\mathbf{v}) - f(\mathbf{u}) \ge \langle \nabla f(\mathbf{u}), \mathbf{v} - \mathbf{u} \rangle \quad \forall \mathbf{v} \in V$$

For a convex functional f which is not differentiable, we need the concepts of subgradient and subdifferential.

Definition 5.12. *Suppose that $f(\mathbf{u})$ is convex on V. For a fixed vector $\mathbf{u} \in V$, any vector $\mathbf{u}^*$ in V^* such that*

$$f(\mathbf{v}) - f(\mathbf{u}) \ge \langle \mathbf{u}^*, \mathbf{v} - \mathbf{u} \rangle \quad \forall \mathbf{v} \in V \tag{5.11}$$

is called a subgradient of f at $\mathbf{u}$. The collection of all subgradients of f at $\mathbf{u}$ is called the subdifferentials of f at $\mathbf{u}$ and is denoted by $\partial f(\mathbf{u})$.

We now present subgradients of two convex functions.

Example 5.13. Consider $f(u) = |u|$ for $u \in \mathbb{R}$. Then

$$\partial f(u) = \begin{cases} +1 & \text{if } u > 0, \\ [-1, 1] & \text{if } u = 0, \\ -1 & \text{if } u < 0. \end{cases}$$

Similarly, consider $f(\mathbf{u}) = \|\mathbf{u}\|_1$ for $\mathbf{u} \in \mathbb{R}^n$. Then

$$\partial f(\mathbf{u}) = [\partial|u_1|, \partial|u_2|, \dots, \partial|u_n|]$$

if $\mathbf{u} = [u_1, u_2, \dots, u_n]^\top \in \mathbb{R}^n$.

Example 5.14. Consider $f(u) = |u| + \frac{\lambda}{2}|u - a|^2$. Then

$$\partial f(u) = \begin{cases} 1 + \lambda(u - a) & \text{if } u > 0, \\ [-1, 1] + \lambda(-a) & \text{if } u = 0, \\ -1 + \lambda(u - a) & \text{if } u < 0. \end{cases} \tag{5.12}$$

Subgradients are different from the usual derivatives, Fréchet derivatives, and Gâteaux derivatives. We can show that

$$\partial(\lambda f(u)) = \lambda \partial f(u) \quad \forall \lambda > 0. \tag{5.13}$$

However, this equality may not hold for $\lambda < 0$. In addition, we have only

$$\partial f_1(u) + \partial f_2(u) \subset \partial(f_1(u) + f_2(u)) \tag{5.14}$$

for any general convex functionals f_1 and f_2. That is, we do not have equality unless f_1 and f_2 satisfy additional properties.

The following important theorem on the existence of minimizers will be useful.

Theorem 5.15. *Suppose that f is convex and $f(\mathbf{x}) > -\infty$ for all $\mathbf{x} \in C$, where C is a convex set. Then $\mathbf{x}$ is a minimizer of f over C, i.e., $\mathbf{x} = \arg\min_{\mathbf{y} \in C} f(\mathbf{y})$ if and only if $0 \in \partial f(\mathbf{x})$.*

Proof. It is easy to see that if 0 is a subgradient of f at $\mathbf{x}$, we have $f(\mathbf{y}) \geq f(\mathbf{x}) + \langle 0, \mathbf{y} - \mathbf{x} \rangle = f(\mathbf{x})$ for all $\mathbf{y} \in C$. Thus, $\mathbf{x}$ is a minimizer of f. On the other hand, by definition of the subgradient, we have $f(\mathbf{y}) - f(\mathbf{x}) \geq \langle \partial f(\mathbf{x}), \mathbf{y} - \mathbf{x} \rangle$ for all $\mathbf{y}$. Since $\mathbf{x}$ is a minimizer of f, we have $f(\mathbf{y}) \geq f(\mathbf{x})$ for all $\mathbf{y}$, i.e., $0 \in \partial f(\mathbf{x})$. □

The following monotone inequality is useful.

Lemma 5.16. *Suppose that $f(\mathbf{x})$ is convex over Ω. For any $\mathbf{x}, \mathbf{y} \in \Omega$, we have*

$$\langle s - t, \mathbf{x} - \mathbf{y} \rangle \geq 0 \quad \forall s \in \partial f(\mathbf{x}), t \in \partial f(\mathbf{y}). \tag{5.15}$$

Here $\partial f(\mathbf{x})$ stands for the subdifferentiation of f at $\mathbf{x}$.

Proof. By definition, for any $s \in \partial f(\mathbf{x})$ and $t \in \partial f(\mathbf{y})$, we have

$$\langle s, \mathbf{y} - \mathbf{x} \rangle \leq f(\mathbf{y}) - f(\mathbf{x}),$$
$$\langle t, \mathbf{x} - \mathbf{y} \rangle \leq f(\mathbf{x}) - f(\mathbf{y}).$$

Adding these two inequalities together yields (5.15). □

5.4 ▪ Lipschitz Differentiability

Next we discuss Lipschitz differentiability.

Definition 5.17. *A function f is said to be Lipschitz differentiable on $\Omega \subset \mathbb{R}^n$ if and only if f is differentiable and*

$$\|\nabla f(\mathbf{x}) - \nabla f(\mathbf{y})\| \leq L\|\mathbf{x} - \mathbf{y}\| \quad \forall \mathbf{x}, \mathbf{y} \in \Omega, \tag{5.16}$$

where $\|\cdot\|$ is a norm on $\mathbb{R}^n$.

It is easy to see that when f is twice differentiable, f is Lipschitz differentiable if $\langle \nabla^2 f(x)y, y \rangle \leq L\|y\|^2$. One important fact about L-Lipschitz differentiability is contained in the following lemma.

Lemma 5.18. *If f is L-Lipschitz differentiable over its domain Ω, then*

$$f(y) \leq f(x) + \langle \nabla f(x), y - x \rangle + \frac{L}{2}\|y - x\|^2 \quad \forall x, y \in \Omega.$$

Proof. It is easy to see that

$$\begin{aligned}
f(y) - f(x) - \langle \nabla f(x), y - x \rangle &= \int_0^1 \langle \nabla f(x + t(y - x)) - \nabla f(x), y - x \rangle dt \\
&\leq \int_0^1 \|\nabla f(x + t(y - x)) - \nabla f(x)\| \|y - x\| dt \\
&\leq L \int_0^1 \|t(y - x)\| \|y - x\| dt = \frac{L}{2}\|y - x\|^2.
\end{aligned}$$

This completes the proof. □

Furthermore, we have a few basic results.

Lemma 5.19. *Suppose that f is convex and Lipschitz differentiable with Lipschitz constant L. Then*

$$f(x) + \langle \nabla f(x), y - x \rangle + \frac{1}{2L}\|\nabla f(x) - \nabla f(y)\|^2 \leq f(y)$$

for all $x, y \in \mathbb{R}^n$.

Proof. Fix $x \in \mathbb{R}^n$ and consider a new function

$$g(y) = f(y) - \langle \nabla f(x), y \rangle.$$

Then g is a convex function and achieves its minimum at x since $\nabla g(x) = 0$. Thus, we use Lemma 5.18:

$$\begin{aligned}
g(x) \leq g(y - \nabla g(y)/L) &\leq g(y) + \langle \nabla g(y), -\nabla g(y)/L \rangle + \frac{L}{2}\|\nabla g(y)/L\|^2 \\
&= f(y) - \langle \nabla f(x), y \rangle - \frac{1}{2L}\|\nabla g(y)\|^2.
\end{aligned}$$

Since $\nabla g(x) = 0$ and $\nabla g(y) = \nabla g(y) - \nabla g(x) = \nabla f(y) - \nabla f(x)$, we have

$$f(x) - \langle f(x), x \rangle \leq f(y) - \langle \nabla f(x), y \rangle - \frac{1}{2L}\|\nabla f(y) - \nabla f(x)\|^2,$$

which is the desired inequality after a simplification. □

The most fundamental fact in optimization is the following statement.

Theorem 5.20 (First-Order Optimality Condition). *Consider the unconstrained minimization problem (5.1). Suppose that $f(\mathbf{x})$ is a differentiable function. Let $\mathbf{x}^*$ be a local minimizer. Then*

$$\nabla f(\mathbf{x}^*) = 0. \tag{5.17}$$

Proof. Since $\mathbf{x}^*$ is a local minimum of f, there exists an $\epsilon > 0$ such that for all $\mathbf{y} \in \Omega$ with $\|\mathbf{y} - \mathbf{x}^*\| \leq \epsilon$, we have $f(\mathbf{y}) \geq f(\mathbf{x}^*)$. Since f is differentiable, we have

$$f(\mathbf{y}) = f(\mathbf{x}^*) + \langle \nabla f(\mathbf{x}^*), \mathbf{y} - \mathbf{x}^* \rangle + o(\|\mathbf{y} - \mathbf{x}\|) \geq f(\mathbf{x}^*).$$

It follows that $\langle \nabla f(\mathbf{x}^*), \mathbf{y} - \mathbf{x}^* \rangle \geq 0$ since $\epsilon > 0$ can be sufficiently small. Write $\mathbf{u} = \mathbf{y} - \mathbf{x}^*$ to be the point in the ball $B(\epsilon)$ of radius ϵ. $-\mathbf{u}$ is also in the same ball $B(\epsilon)$. Therefore, we have

$$\langle \nabla f(\mathbf{x}^*), \mathbf{u} \rangle = 0 \quad \forall \mathbf{u} \in B(\epsilon).$$

That is, $\nabla f(\mathbf{x}^*) = 0$. $\square$

Next we present the following result.

Theorem 5.21 (Second-Order Optimality Condition). *Let $\mathbf{x}^*$ be a local minimum of twice differentiable function $f(\mathbf{x}), \mathbf{x} \in \Omega$. Then*

$$\nabla f(\mathbf{x}^*) = 0 \quad and \quad \nabla^2 f(\mathbf{x}^*) \geq 0, \tag{5.18}$$

where the Hessian $\nabla^2 f(\mathbf{x}^*)$ of f at $\mathbf{x}^*$ is nonnegative definite.

Proof. We use the proof above to obtain

$$f(\mathbf{y}) = f(\mathbf{x}^*) + \langle \nabla f(\mathbf{x}^*), \mathbf{y} - \mathbf{x}^* \rangle + o(\|\mathbf{y} - \mathbf{x}^*\|) = f(\mathbf{x}^*) + o(\|\mathbf{y} - \mathbf{x}^*\|) \geq f(\mathbf{x}^*).$$

Note that the little o term is
$$\langle \mathbf{u}, \nabla^2 f(\mathbf{x}^*)\mathbf{u} \rangle \geq 0$$

for all $\mathbf{u} \in B(\epsilon)$. Hence, the Hessian $\nabla^2 f(\mathbf{x}^*)$ of f at $\mathbf{x}^*$ is nonnegative definite. $\square$

5.5 ▪ Standard Gradient Descent Methods

To numerically solve the convex minimization (5.1) with differentiable minimizing functional $f(\mathbf{x})$, one usually uses the so-called gradient descent method. With a step size $h > 0$ or variable step size h_k (as indicated below), one uses the following standard gradient descent method.

ALGORITHM 5.1
Gradient Descent Algorithm

1: Start with an initial guess $\mathbf{x}^{(1)}$.
2: For $k = 1, 2, \ldots$, compute

$$\mathbf{x}^{(k+1)} = \mathbf{x}^{(k)} - h_k \nabla f(\mathbf{x}^{(k)}) \tag{5.19}$$

until the maximal number of iterations is achieved or $\|\mathbf{x}^{(k+1)} - \mathbf{x}^{(k)}\|$ is less than a given tolerance.

There are several choices for step h_k. For example,

(1) $h_k = h > 0$ for all $k \geq 1$. For example, $h = 1/L$, where L is the Lipschitz constant of the gradient vector of f. See Algorithm 5.2 (below). For another example, $h = \nu/L^2$ in Theorem 5.22 (below), where $\nu > 0$ is the ν-strong convexity.

(2) $h_k = [\nabla^2 f(\mathbf{x}^{(k)})]^{-1}$, which is the well-known Newton method, where f has to be twice differentiable and the inverse of the Hessian of f at $\mathbf{x}^{(k)}$ can be efficiently computed.

(3) Letting $\mathbf{y}_k = \nabla f(\mathbf{x}^{(k)}) - \nabla f(\mathbf{x}^{(k-1)})$ and $\mathbf{s}_k = \mathbf{x}^{(k)} - \mathbf{x}^{(k-1)}$,

$$h_k = \|\mathbf{s}_k\|^2 / \langle \mathbf{s}_k, \mathbf{y}_k \rangle, \tag{5.20}$$

where $k > 1$. This is the so-called Barzilai–Borwein technique (cf. [3]).

(4) $h_k = 1/(k+1)$ for all $k \geq 1$, where $h > 0$ is a constant. For example, $h_k = a_k$, as in the proof of Theorem 5.27 (below).

(5) $h_k = \arg\min_{h>0} f(\mathbf{x}^{(k)} - h\nabla f(\mathbf{x}^{(k)}))$, which is called the standard steepest descent method.

(6) Use a backtracking step to find h_k such that

$$\alpha \langle \nabla f(\mathbf{x}^k), \mathbf{x}^k - \mathbf{x}^{k+1} \rangle \leq f(\mathbf{x}^k) - f(\mathbf{x}^{k+1}) \tag{5.21}$$

for $\alpha \in (0,1)$ and

$$\beta \langle \nabla f(\mathbf{x}^k), \mathbf{x}^k - \mathbf{x}^{k+1} \rangle \geq f(\mathbf{x}^k) - f(\mathbf{x}^{k+1}) \tag{5.22}$$

for $\beta \in (0,1)$ with $\alpha < \beta$. These are related to the so-called Wolfe condition (cf. [11]).

We refer the reader to [4] for more on the theoretical study of optimization and to [11] for numerical aspects of optimization. We now discuss the convergence of Algorithm 5.1. Let us begin with the following theorem.

Theorem 5.22. *Suppose that f is L-Lipschitz differentiable and is ν-strongly convex. Letting $\mathbf{x}^*$ be the minimizer of* (5.1)*, we have*

$$\|\mathbf{x}^{(k+1)} - \mathbf{x}^*\| \leq \gamma^k \|\mathbf{x}^{(1)} - \mathbf{x}^*\| \quad \forall k \geq 1, \tag{5.23}$$

where $\gamma \in (0,1)$ is a constant independent of k if the step size $h_k \leq h < 1$ such that $1 - 2h\nu + h^2 L^2 < 1$.

Proof. Using (5.19), we have

$$\|\mathbf{x}^{(k+1)} - \mathbf{x}^*\|^2 = \|\mathbf{x}^{(k)} - \mathbf{x}^*\|^2 - 2\langle h_k \nabla f(\mathbf{x}^{(k)}), \mathbf{x}^{(k)} - \mathbf{x}^* \rangle + h_k^2 \|\nabla f(\mathbf{x}^{(k)})\|^2. \tag{5.24}$$

Since $\nabla f(\mathbf{x}^*) = 0$, we use the Lipschitz condition of ∇f and strong convexity to get

$$\|\mathbf{x}^{(k+1)} - \mathbf{x}^*\|^2 \leq \|\mathbf{x}^{(k)} - \mathbf{x}^*\|^2 - 2h_k\nu\|\mathbf{x}^{(k)} - \mathbf{x}^*\|^2 + L_k^2 h_k^2 \|\mathbf{x}^{(k)} - \mathbf{x}^*\|^2 \tag{5.25}$$
$$\leq \|\mathbf{x}^{(k)} - \mathbf{x}^*\|^2 (1 - 2h_k\nu + h_k^2 L^2) \leq \cdots \leq \gamma^{2k} \|\mathbf{x}^{(1)} - \mathbf{x}^*\|^2,$$

with $\gamma = (1 - 2h\nu + h^2 L^2)^{1/2}$. $\square$

Note that for any given tolerance ϵ, we need $k = O(\log(\epsilon)/\log(\gamma))$. In particular, $\log(\gamma) \approx -\nu^2/L^2$. Indeed, one usually chooses $h = \nu/L^2$ to have the minimal value of $1 - 2h\nu + h^2 L^2$, and hence $\gamma = (1 - \nu^2/L^2)^{1/2}$ and $\log(\gamma) = \frac{1}{2}\log(1 - \nu^2/L^2) \approx -\frac{\nu^2}{2L^2}$. Thus, we need

$$k \approx \frac{L^2}{2\nu^2} \log(1/\epsilon) \tag{5.26}$$

iterative steps. To speed up the process, we can use the following Nesterov technique (cf. [8], [9]). Let us state the accelerated gradient algorithm first and then establish the convergence with a new approximate constant which explains the reduction of the number of iterations.

ALGORITHM 5.2

Nesterov Accelerated Gradient Descent Algorithm [8]

1: Given an initial guess $\mathbf{x}^{(1)}$, let $\mathbf{y}^{(1)} = \mathbf{x}^{(1)}$.
2: For $k \geq 1$, we update

$$\mathbf{y}^{(k+1)} := \mathbf{x}^{(k)} - h_k \nabla f(\mathbf{x}^{(k)}), \tag{5.27}$$
$$\mathbf{x}^{(k+1)} = \mathbf{y}^{(k+1)} - q(\mathbf{y}^{(k+1)} - \mathbf{y}^{(k)}) \tag{5.28}$$

until a maximum number of iterations is achieved, where $h_k = 1/L$ and

$$q = (\sqrt{L/\nu} - 1)/(\sqrt{L/\nu} + 1).$$

The following is the convergence result.

Theorem 5.23. *Suppose that $f(\mathbf{x})$ is ν-strongly convex and L-Lipschitz differentiable. Then Algorithm 5.2 converges and we have*

$$f(\mathbf{y}_k) - f(\mathbf{x}^*) \leq \frac{L + \nu}{2} \|\mathbf{x}^{(1)} - \mathbf{x}^*\|^2 \left(1 - \frac{1}{Q}\right)^k \tag{5.29}$$

for all $k \geq 1$, where $Q = \sqrt{L/\nu}$.

We remark that with the second step in (5.27), we don't need as many iterations, i.e.,

$$k \approx \sqrt{\frac{L}{\nu}} \log\left(\frac{(L + \nu)C}{\epsilon}\right), \tag{5.30}$$

which is significantly better than (5.26), where C is a positive constant dependent on the initial value $\mathbf{x}^{(1)}$.

Proof. Define a sequence of quadratic polynomials:

$$F_1(\mathbf{x}) = f(\mathbf{x}^{(1)}) + \frac{\nu}{2}\|\mathbf{x} - \mathbf{x}^{(1)}\|^2,$$

$$F_{k+1}(\mathbf{x}) = \left(1 - \frac{1}{Q}\right) F_k(\mathbf{x}) + \frac{1}{Q}\left(f(\mathbf{x}^{(k)}) + (\nabla f(\mathbf{x}^{(k)}))^\top(\mathbf{x} - \mathbf{x}^{(k)}) + \frac{\nu}{2}\|\mathbf{x} - \mathbf{x}^{(k)}\|^2\right)$$

for $k \geq 1$. Since $f(\mathbf{x})$ is ν-strongly convex, we have

$$f(\mathbf{x}^{(k)}) + (\nabla f(\mathbf{x}^{(k)}))^\top(\mathbf{x} - \mathbf{x}^{(k)}) + \frac{\nu}{2}\|\mathbf{x} - \mathbf{x}^{(k)}\|^2 \leq f(\mathbf{x})$$

for all $\mathbf{x} \in \mathbb{R}^n$. It follows that

$$F_{k+1}(\mathbf{x}) - f(\mathbf{x}) \leq \left(1 - \frac{1}{Q}\right)(F_k(\mathbf{x}) - f(\mathbf{x})) \leq \cdots \leq \left(1 - \frac{1}{Q}\right)^k (F_1(\mathbf{x}) - f(\mathbf{x})). \tag{5.31}$$

Next we claim that

$$f(\mathbf{y}^{(k)}) \leq \min_{\mathbf{y} \in \mathbb{R}^n} F_k(\mathbf{y}) \tag{5.32}$$

for all $k \geq 1$. If we have the claim, then letting $\mathbf{x}^*$ be the minimizer, we have

$$f(\mathbf{y}^{(k+1)}) - f(\mathbf{x}^*) \leq F_{k+1}(\mathbf{x}^*) - f(\mathbf{x}^*) \leq \left(1 - \frac{1}{Q}\right)^k (F_1(\mathbf{x}^*) - f(\mathbf{x}^*))$$

$$= \left(1 - \frac{1}{Q}\right)^k \frac{\nu + L}{2}\|\mathbf{x}^* - \mathbf{x}^{(1)}\|^2,$$

which finishes the proof.

Therefore, all we need to do is to prove the claim. Let us simply use induction to establish the claim. For $k = 1$, we have $\mathbf{y}^{(1)} = \mathbf{x}^{(1)}$, and hence (5.32) holds for $k = 1$. Assume the claim holds for $k \geq 1$. Now let us study the case $k+1$. By the Lipschitz differentiability and the convexity of f,

$$
\begin{aligned}
f(\mathbf{y}^{(k+1)}) &\leq f(\mathbf{x}^{(k)}) + (\nabla f(\mathbf{x}^{(k)}))^\top (\mathbf{y}^{(k+1)} - \mathbf{x}^{(k)}) + \frac{L}{2} \|\mathbf{y}^{(k+1)} - \mathbf{x}^{(k)}\|^2 \\
&= f(\mathbf{x}^{(k)}) - \frac{1}{2L} \|\nabla f(\mathbf{x}^{(k)})\|^2 \\
&= \left(1 - \frac{1}{Q}\right) f(\mathbf{y}^{(k)}) + \frac{1}{Q} f(\mathbf{x}^{(k)}) + \left(1 - \frac{1}{Q}\right) (f(\mathbf{x}^{(k)}) \\
&\quad - f(\mathbf{y}^{(k)})) - \frac{1}{2L} \|\nabla f(\mathbf{x}^{(k)})\|^2 \\
&\leq \left(1 - \frac{1}{Q}\right) \min_{\mathbf{y}} F_k(\mathbf{y}) + \frac{1}{Q} f(\mathbf{x}^{(k)}) \\
&\quad + \left(1 - \frac{1}{Q}\right) (\nabla f(\mathbf{x}^{(k)}))^\top (\mathbf{x}^{(k)} - \mathbf{y}^{(k)}) - \frac{1}{2L} \|\nabla f(\mathbf{x}^{(k)})\|^2, \quad (5.33)
\end{aligned}
$$

where we have used the induction hypothesis. We need to show that the right-hand side of the above is less than or equal to $\min_{\mathbf{y}} F_{k+1}(\mathbf{y}) =: F_{k+1}^*$. Since F_k is a quadratic polynomial, we can always write it as

$$
F_k(\mathbf{x}) = F_k^* + \frac{\nu}{2} \|\mathbf{x} - \mathbf{v}^{(k)}\|^2,
$$

where $\mathbf{v}^{(k)}$ is the location of the minimum of F_k or the root of $\nabla F_k(\mathbf{x}) = 0$. Let us compute the root $\mathbf{v}^{(k+1)}$ of $\nabla F_{k+1} = 0$. From the definition of F_{k+1}, we have

$$
\nabla F_{k+1}(\mathbf{x}) = \nu \left(1 - \frac{1}{Q}\right)(\mathbf{x} - \mathbf{v}^{(k)}) + \frac{1}{Q}\nabla f(\mathbf{x}^{(k)}) + \frac{\nu}{Q}(\mathbf{x} - \mathbf{x}^{(k)}) = 0.
$$

Then the root $\mathbf{v}^{(k+1)}$ can be found. In fact,

$$
\mathbf{v}^{(k+1)} = \left(1 - \frac{1}{Q}\right)\mathbf{v}^{(k)} + \frac{1}{Q}\mathbf{x}^{(k)} - \frac{1}{\nu Q}\nabla f(\mathbf{x}^{(k)}). \tag{5.34}
$$

So at $\mathbf{x} = \mathbf{x}^{(k)}$, we have

$$
\begin{aligned}
F_{k+1}(\mathbf{x}^{(k)}) &= \left(1 - \frac{1}{Q}\right) F_k(\mathbf{x}^{(k)}) + \frac{1}{Q} f(\mathbf{x}^{(k)}) \\
&= \left(1 - \frac{1}{Q}\right)\left(F_k^* + \frac{\nu}{2}\|\mathbf{x}^{(k)} - \mathbf{v}^{(k)}\|^2\right) + \frac{1}{Q} f(\mathbf{x}^{(k)})
\end{aligned}
$$

by the definition of F_{k+1} and the quadratic form of F_k. Similarly, the quadratic form of F_{k+1} gives

$$
\begin{aligned}
F_{k+1}(\mathbf{x}^{(k)}) &= F_{k+1}^* + \frac{\nu}{2}\|\mathbf{x}^{(k)} - \mathbf{v}^{(k+1)}\|^2 \\
&= F_{k+1}^* + \frac{\nu}{2}\left\|\left(1 - \frac{1}{Q}\right)(\mathbf{v}^{(k)} - \mathbf{x}^{(k)}) - \frac{1}{\nu Q}\nabla f(\mathbf{x}^{(k)})\right\|^2.
\end{aligned}
$$

We shall prove another claim—$(\mathbf{v}^{(k)} - \mathbf{x}^{(k)}) = Q(\mathbf{x}^{(k)} - \mathbf{y}^{(k)})$—in just a minute. Now let us use the second claim to continue the proof of the first claim. With the above two equalities, we have

$$
\begin{aligned}
F_{k+1}^* &= \left(1 - \frac{1}{Q}\right)\left(F_k^* + \frac{\nu}{2}\|\mathbf{x}^{(k)} - \mathbf{v}^{(k)}\|^2\right) + \frac{1}{Q}f(\mathbf{x}^{(k)}) - \frac{\nu}{2}\|\mathbf{x}^{(k)} - \mathbf{v}^{(k+1)}\|^2 \\
&= \left(1 - \frac{1}{Q}\right)\left(F_k^* + \frac{\nu}{2}\|\mathbf{x}^{(k)} - \mathbf{v}^{(k)}\|^2\right) + \frac{1}{Q}f(\mathbf{x}^{(k)}) \\
&\quad - \frac{\nu}{2}\left\|\left(1 - \frac{1}{Q}\right)(\mathbf{v}^{(k)} - \mathbf{x}^{(k)}) - \frac{1}{\nu Q}\nabla f(\mathbf{x}^{(k)})\right\|^2 \\
&= \left(1 - \frac{1}{Q}\right)\left(F_k^* + \frac{\nu}{2}\|\mathbf{x}^{(k)} - \mathbf{v}^{(k)}\|^2\right) + \frac{1}{Q}f(\mathbf{x}^{(k)}) - \frac{\nu}{2}\left(1 - \frac{1}{Q}\right)^2\|\mathbf{v}^{(k)} - \mathbf{x}^{(k)}\|^2 \\
&\quad + \nu\left(1 - \frac{1}{Q}\right)(\mathbf{v}^{(k)} - \mathbf{x}^{(k)})^{\top}\frac{1}{\nu Q}\nabla f(\mathbf{x}^{(k)}) - \frac{1}{2\nu Q^2}\|\nabla f(\mathbf{x}^{(k)})\|^2 \\
&\geq \left(1 - \frac{1}{Q}\right)F_k^* + \frac{1}{Q}f(\mathbf{x}^{(k)}) + \left(1 - \frac{1}{Q}\right)(\mathbf{x}^{(k)} - \mathbf{y}^{(k)})^{\top}\nabla f(\mathbf{x}^{(k)}) - \frac{1}{2L}\|\nabla f(\mathbf{x}^{(k)})\|^2,
\end{aligned}
$$

which is the right-hand side of (5.33), where we have used the second claim in the last step.

Now we take our time to prove the second claim. We shall use induction to do so. Clearly, when $k = 1$, we have the second claim. Let us consider $k \geq 1$. Using the formula (5.34) for $\mathbf{v}_{k+1}$, we have

$$
\begin{aligned}
\mathbf{v}^{(k+1)} - \mathbf{x}^{(k+1)} &= \left(1 - \frac{1}{Q}\right)\mathbf{v}^{(k)} + \frac{1}{Q}\mathbf{x}^{(k)} - \frac{1}{\nu Q}\nabla f(\mathbf{x}^{(k)}) - \mathbf{x}^{(k+1)} \\
&= \left(1 - \frac{1}{Q}\right)(\mathbf{x}^{(k)} + Q(\mathbf{x}^{(k)} - \mathbf{y}^{(k)})) + \frac{1}{Q}\mathbf{x}^{(k)} \\
&\quad + \frac{1}{\nu Q}L(\mathbf{y}^{(k+1)} - \mathbf{x}^{(k)}) - \mathbf{x}^{(k+1)} \\
&= Q\mathbf{x}^{(k)} + (1 - Q)\mathbf{y}^{(k)} + Q(\mathbf{y}^{(k+1)} - \mathbf{x}^{(k)}) - \mathbf{x}^{(k+1)} \\
&= Q\left(\mathbf{y}^{(k+1)} + \frac{1-Q}{Q}\mathbf{y}^{(k)} - \frac{1}{Q}\mathbf{x}^{(k+1)}\right) \\
&= Q(\mathbf{x}^{(k+1)} - \mathbf{y}^{(k+1)}),
\end{aligned}
$$

since $\mathbf{x}^{(k+1)} = \frac{Q}{1+Q}(2\mathbf{y}^{(k+1)} + \frac{1-Q}{Q}\mathbf{y}^{(k)})$, which is another form for the second formula in (6.27). □

Example 5.24. Consider the least squares problem $\Phi\mathbf{x} \approx \mathbf{b}$ with full rank matrix Φ. We formulate the problem as a minimization problem:

$$
\min_{\mathbf{x}\in\mathbb{R}^n} \frac{1}{2}\|\Phi\mathbf{x} - \mathbf{b}\|^2. \tag{5.35}
$$

We use Algorithms 5.1 and 5.2 to solve the minimization (5.35). Comparing the number of iterations of these two algorithms for the solution of the same least squares problem shows that Algorithm 5.2 is indeed faster.

When f is not strongly convex, e.g., $f(\mathbf{x}) = \frac{1}{2}\|\Phi\mathbf{x} - \mathbf{b}\|^2$ with Φ not full rank, we do not have such a nice result in Theorem 5.22 and in the accelerated version, i.e., Algorithm 5.2. Instead we have the following result.

Theorem 5.25. *Suppose that f is L-Lipschitz differentiable. Let $\mathbf{x}^*$ be a minimizer of (5.1). Choose $h = 1/L$. Then the gradient method generates a convergent sequence $\mathbf{x}^{(k)}, k \geq 1$, such that*

$$f(\mathbf{x}^{(k)}) - f(\mathbf{x}^*) \leq \frac{2L\|\mathbf{x}^{(0)} - \mathbf{x}^*\|_2^2}{k+1}. \tag{5.36}$$

Furthermore, we have

$$f(\mathbf{x}^{(k)}) - f(\mathbf{x}^*) = o\left(\frac{1}{k}\right) \quad or \quad k(f(\mathbf{x}^{(k)}) - f(\mathbf{x}^*)) \to 0 \quad as \ k \to \infty. \tag{5.37}$$

Proof. Using (5.19), we have (5.24). Using the L-Lipschitz differentiability and $\nabla f(\mathbf{x}^*) = 0$, we use Theorem 5.11 to obtain

$$\|\mathbf{x}^{(k+1)} - \mathbf{x}^*\|^2 \leq \|\mathbf{x}^{(k)} - \mathbf{x}^*\|^2 - h\left(\frac{2}{L} - h\right)\|\nabla f(\mathbf{x}^{(k)}) - \nabla f(\mathbf{x}^*)\|^2 \tag{5.38}$$

$$= \|\mathbf{x}^{(k)} - \mathbf{x}^*\|^2 - \frac{1}{L^2}\|\nabla f(\mathbf{x}^{(k)})\|^2, \tag{5.39}$$

where we have used Theorem 5.11, i.e., the inequality in (5.10). It follows that

$$\|\mathbf{x}^{(k)} - \mathbf{x}^*\| \leq C < \infty \quad and \quad \sum_{k=1}^{\infty} \|\nabla f(\mathbf{x}^{(k)})\|^2 < \infty. \tag{5.40}$$

The boundedness of $\mathbf{x}^{(k)}$ implies that there exists a convergent subsequence $\mathbf{x}^{(k_j)}$ and $\widehat{\mathbf{x}}$ such that $\mathbf{x}^{(k_j)} \to \widehat{\mathbf{x}}$. In addition, we have $\nabla f(\widehat{\mathbf{x}}) = 0$ by the second inequality in (5.40).

Now using Lemma 5.18, we have

$$f(\mathbf{x}^{(k+1)}) \leq f(\mathbf{x}^{(k)}) + \langle \nabla f(\mathbf{x}^{(k)}), \mathbf{x}^{(k+1)} - \mathbf{x}^{(k)} \rangle + \frac{L}{2}\|\mathbf{x}^{(k+1)} - \mathbf{x}^{(k)}\|^2$$

$$= f(\mathbf{x}^{(k)}) - h\langle \nabla f(\mathbf{x}^{(k)}), \nabla f(\mathbf{x}^{(k)}) \rangle + \frac{L}{2}h^2\|\nabla f(\mathbf{x}^{(k)})\|^2$$

$$= f(\mathbf{x}^{(k)}) - \frac{1}{2L}\|\nabla f(\mathbf{x}^{(k)})\|^2,$$

where we have used the gradient descent method with $h = 1/L$. That is, we have the monotone property

$$f(\mathbf{x}^{(k+1)}) - f(\mathbf{x}^*) \leq f(\mathbf{x}^{(k)}) - f(\mathbf{x}^*) - \frac{1}{2L}\|\nabla f(\mathbf{x}^{(k)})\|^2, \tag{5.41}$$

which will be used later. It is easy to see that

$$f(\mathbf{x}^{(k)}) - f(\mathbf{x}^*) \leq \langle \nabla f(\mathbf{x}^{(k)}), \mathbf{x}^{(k)} - \mathbf{x}^* \rangle \leq \|\nabla f(\mathbf{x}^{(k)})\| \, \|\mathbf{x}^{(k)} - \mathbf{x}^*\| \leq \|\nabla f(\mathbf{x}^{(k)})\| \, \|\mathbf{x}^{(0)} - \mathbf{x}^*\|$$

by using (5.38). Letting $\delta_k = f(\mathbf{x}^{(k)}) - f(\mathbf{x}^*)$, we have

$$\delta_{k+1} \leq \delta_k - \frac{1}{2L}\|\nabla f(\mathbf{x}^{(k)})\|^2 \leq \delta_k - \frac{1}{2L}\frac{\delta_k^2}{r_0^2},$$

where $r_0 = \|\mathbf{x}^{(0)} - \mathbf{x}^*\|$. In other words, we have

$$\frac{1}{\delta_{k+1}} \geq \frac{1}{\delta_k} + \frac{1}{2Lr_0^2}\frac{\delta_k}{\delta_{k+1}} \geq \frac{1}{\delta_k} + \frac{1}{2Lr_0^2}$$

by using (5.38). It thus follows that

$$\frac{1}{\delta_{k+1}} \geq \frac{1}{\delta_0} + \frac{k+1}{2Lr_0^2} \quad \text{or} \quad \delta_{k+1} \leq \frac{2Lr_0^2}{k+1}.$$

This establishes the big-O estimate in (5.36). We further show the small-o estimate as follows. We again use Lemma 5.18 to obtain

$$
\begin{aligned}
f(\mathbf{x}^{(k+1)}) &\leq f(\mathbf{x}^{(k)}) + \langle \nabla f(\mathbf{x}^{(k)}), \mathbf{x}^{(k+1)} - \mathbf{x}^{(k)} \rangle + \frac{L}{2}\|\mathbf{x}^{(k+1)} - \mathbf{x}^{(k)}\|^2 \\
&= f(\mathbf{x}^{(k)}) + \langle \nabla f(\mathbf{x}^{(k)}), \mathbf{x}^* - \mathbf{x}^{(k)} \rangle + \langle \nabla f(\mathbf{x}^{(k)}), \mathbf{x}^{(k+1)} - \mathbf{x}^* \rangle \\
&\quad + \frac{L}{2}\|\mathbf{x}^{(k+1)} - \mathbf{x}^{(k)}\|^2 \\
&\leq f(\mathbf{x}^*) + \frac{L}{2}\|\mathbf{x}^* - \mathbf{x}^{(k)}\|^2 - \frac{L}{2}\|\mathbf{x}^{(k+1)} - \mathbf{x}^*\|^2,
\end{aligned}
$$

where we have used the gradient descent iteration (5.19) with $h_k = 1/L$. That is, we have

$$\eta_{k+1} \leq \frac{L}{2}\|\mathbf{x}^* - \mathbf{x}^{(k)}\|^2 - \frac{L}{2}\|\mathbf{x}^{(k+1)} - \mathbf{x}^*\|^2. \tag{5.42}$$

It follows that

$$\sum_{k=1}^{\infty} \eta_k \leq \frac{L}{2}\|\mathbf{x}^* - \mathbf{x}^{(0)}\|^2 < \infty. \tag{5.43}$$

Since the error terms η_k are monotonically decreasing by (5.41), we have

$$N\eta_{2N} \leq \sum_{k=N}^{2N} \eta_k \to 0 \quad \text{as } N \to \infty.$$

That is, $\eta_{2N} = o(\frac{1}{2N})$, and similarly for η_{2N+1}. Thus, we have the estimate in (5.37). This completes the proof.　　□

The proof of (5.37) can be found in [5].

Example 5.26. Consider the least squares problem $\Phi\mathbf{x} \approx \mathbf{b}$ whose observation matrix Φ is not full rank. We again formulate the problem as a minimization problem:

$$\min_{\mathbf{x} \in \mathbb{R}^n} \frac{1}{2}\|\Phi\mathbf{x} - \mathbf{b}\|^2. \tag{5.44}$$

We use Algorithms 5.1 and 5.3 given in the next section to solve the minimization (5.44). Comparing the number of iterations of these two algorithms for the solution of the same least squares problem shows that Algorithm 5.3 is indeed faster.

5.6 ▪ Accelerated Gradient Descent Methods

In fact, we can improve the convergence rate from $o(\frac{1}{k+1})$ to $\frac{C}{(k+1)^2}$. We shall present an accelerated gradient descent algorithm.

ALGORITHM 5.3

Nesterov's Accelerated Gradient Descent Method [7]

1: Suppose that L is the Lipschitz constant of ∇f.
2: Let $\mathbf{x}^{(1)}, \mathbf{y}^{(1)}$ be two initial guesses and $a_1 = 1$.
3: Compute the following minimization and iterative steps:

$$\mathbf{x}^{(k+1)} = \mathbf{y}^{(k)} - \frac{1}{L}\nabla f(\mathbf{y}^{(k)}), \tag{5.45}$$

4: Let

$$a_{k+1} = \left(1 + \sqrt{4a_k^2 + 1}\right)/2 \tag{5.46}$$

and define

$$\mathbf{y}^{(k+1)} = \mathbf{x}^{(k+1)} + \frac{a_k - 1}{a_{k+1}}(\mathbf{x}^{(k+1)} - \mathbf{x}^{(k)}) \tag{5.47}$$

for all $k \geq 1$ until a given number of iterations is achieved.

Then the convergence rate of these $\mathbf{x}^k$, $k \geq 1$, from Algorithm 5.3 is accelerated, as indicated in the following.

Theorem 5.27 (Nesterov, 1983 [7]). *Suppose that $\mathbf{x}^{(k)}$, $k \geq 1$, is a sequence from Algorithm 5.3. Let $\eta_k = f(\mathbf{x}^{(k)}) - f(\mathbf{x}^*)$. Then*

$$\eta_k \leq \frac{C}{(k+1)^2}$$

for all $k \geq 1$, where $C > 0$ is a constant.

To prove this theorem, we need the following gradient mapping lemma.

Lemma 5.28 (Gradient Mapping Property). *For all $\mathbf{x} \in \mathbb{R}^n$, we have*

$$f(\mathbf{x}) \geq f(\mathbf{x}^{(k+1)}) + \frac{L}{2}\|\mathbf{x}^{(k+1)} - \mathbf{y}^{(k)}\|^2 + L\langle \mathbf{y}^{(k)} - \mathbf{x}, \mathbf{x}^{(k+1)} - \mathbf{y}^{(k)}\rangle. \tag{5.48}$$

Proof. Using (5.19), we have

$$\langle \mathbf{x} - \mathbf{x}^{k+1}, L(\mathbf{x}^{k+1} - \mathbf{y}^k) + \nabla f(\mathbf{y}^k)\rangle = 0 \quad \forall \mathbf{x} \in \mathbb{R}^n. \tag{5.49}$$

Letting $\phi(\mathbf{x}) = f(\mathbf{y}^k) + \langle \nabla f(\mathbf{y}^{(k)}), \mathbf{x} - \mathbf{y}^{(k)}\rangle + \frac{L}{2}\|\mathbf{x} - \mathbf{y}^{(k)}\|^2$ be a functional, we use the convexity of f and the equality in (5.49) to get

$$\begin{aligned}
f(\mathbf{x}) &\geq f(\mathbf{y}^{(k)}) + \langle \nabla f(\mathbf{y}^{(k)}), \mathbf{x} - \mathbf{y}^{(k)}\rangle \\
&= f(\mathbf{y}^{(k)}) + \langle \nabla f(\mathbf{y}^{(k)}), \mathbf{x}^{k+1} - \mathbf{y}^{(k)}\rangle + \langle \nabla f(\mathbf{y}^{(k)}), \mathbf{x} - \mathbf{x}^{(k+1)}\rangle \\
&= \phi(\mathbf{x}^{(k+1)}) - \frac{L}{2}\|\mathbf{x}^{(k+1)} - \mathbf{y}^{(k)}\|^2 + \langle \nabla f(\mathbf{y}^{(k)}), \mathbf{x} - \mathbf{x}^{k+1}\rangle \\
&= \phi(\mathbf{x}^{(k+1)}) - \frac{L}{2}\|\mathbf{x}^{(k+1)} - \mathbf{y}^{(k)}\|^2 + \langle L(\mathbf{y}^k - \mathbf{x}^{k+1}), \mathbf{x} - \mathbf{x}^{(k+1)}\rangle \\
&= \phi(\mathbf{x}^{(k+1)}) + \frac{L}{2}\|\mathbf{x}^{(k+1)} - \mathbf{y}^{(k)}\|^2 + L\langle \mathbf{y}^k - \mathbf{x}^{k+1}, \mathbf{x} - \mathbf{y}^{(k)}\rangle.
\end{aligned}$$

Note that $\phi(\mathbf{x}^{(k+1)}) \geq f(\mathbf{x}^{(k+1)})$ by the L-Lipschitz differentiation property of f, i.e., Lemma 5.18. We have established the desired inequality. □

Proof of Theorem 5.27. The choice of a_k in Algorithm 5.3 seems unintuitive in the above discussion. The following argument can explain where a_k arises naturally. Recall the gradient mapping property, which can be rewritten as

$$f(\mathbf{x}) \geq f(\mathbf{x}^{k+1}) + \frac{L}{2}\|\mathbf{x}^{(k+1)} - \mathbf{x}\|^2 - \frac{L}{2}\|\mathbf{y}^{(k)} - \mathbf{x}\|^2 \tag{5.50}$$

for any $\mathbf{x} \in \mathbb{R}^n$. We consider $\mathbf{x} = (1 - \beta_k)\mathbf{x}^{(k)} + \beta_k\mathbf{x}^*$ in (5.50) for some $\beta_k \in (0, 1)$. Then (5.50) becomes

$$
\begin{aligned}
f(\mathbf{x}^{(k+1)}) &\leq f(\mathbf{x}) + \frac{L}{2}\|\mathbf{y}^{(k)} - \mathbf{x}\|^2 - \frac{L}{2}\|\mathbf{x}^{(k+1)} - \mathbf{x}\|^2 \\
&\leq (1 - \beta_k)f(\mathbf{x}^{(k)}) + \beta_k f(\mathbf{x}^*) \\
&\quad + \frac{L}{2}\|(1 - \beta_k)\mathbf{x}^{(k)} + \beta_k\mathbf{x}^* - \mathbf{y}^{(k)}\|^2 - \frac{L}{2}\|\mathbf{x}^{(k+1)} - (1 - \beta_k)\mathbf{x}^{(k)} - \beta_k\mathbf{x}^*\|^2 \\
&\leq (1 - \beta_k)f(\mathbf{x}^{(k)}) + \beta_k f(\mathbf{x}^*) + \frac{L\beta_k^2}{2}\|\mathbf{x}^* + (\beta_k^{-1} - 1)\mathbf{x}^{(k)} - \beta_k^{-1}\mathbf{y}^{(k)}\|^2 \\
&\quad - \frac{L\beta_k^2}{2}\|\mathbf{x}^* + (\beta_k^{-1} - 1)\mathbf{x}^{(k)} - \beta_k^{-1}\mathbf{x}^{(k+1)}\|^2.
\end{aligned}
\tag{5.51}
$$

Choose $\mathbf{y}^{(k)}$ in the second term on the right-hand side above such that

$$\mathbf{x}^* + (\beta_k^{-1} - 1)\mathbf{x}^{(k)} - \beta_k^{-1}\mathbf{y}^{(k)} = \mathbf{x}^* + (\beta_{k-1}^{-1} - 1)\mathbf{x}^{(k-1)} - \beta_{k-1}^{-1}\mathbf{x}^{(k)}. \tag{5.52}$$

Letting $\mathbf{z}^k = \beta_k^{-1}\mathbf{x}^{(k)} - (\beta_k^{-1} - 1)\mathbf{x}^{(k-1)}$, the inequality in (5.51) yields

$$\eta_{k+1} \leq (1 - \beta_k)\eta_k + \frac{L\beta_k^2}{2}\|\mathbf{x}^* - \mathbf{z}^k\|^2 - \frac{L\beta_k^2}{2}\|\mathbf{x}^* - \mathbf{z}^{k+1}\|^2.$$

Dividing by β_k^2 and choosing β_k such that $\frac{1}{\beta_k^2} = \frac{1-\beta_{k+1}}{\beta_{k+1}^2}$, we have

$$\frac{1 - \beta_{k+1}}{\beta_{k+1}^2}\eta_{k+1} \leq \frac{1 - \beta_k}{\beta_k^2}\eta_k + \frac{L}{2}\|\mathbf{x}^* - \mathbf{z}^k\|^2 - \frac{L}{2}\|\mathbf{x}^* - \mathbf{z}^{k+1}\|^2.$$

Summing over k for $k = 1, \ldots, K$, we have

$$\frac{1 - \beta_{K+1}}{\beta_{K+1}^2}\eta_{K+1} \leq \frac{1 - \beta_1}{\beta_1^2}\eta_1 + \frac{L}{2}\|\mathbf{x}^* - \mathbf{z}^1\|^2.$$

With $\beta_1 = 1$, $\eta_K \leq \frac{\beta_K^2}{1 - \beta_K}C$. It is easy to see that $\frac{1}{\beta_k^2} = \frac{1 - \beta_{k+1}}{\beta_{k+1}^2}$ implies that $a_k = \frac{1}{\beta_k}$. This proves Theorem 5.27 since $\alpha_k \geq (k + 1)/2$ or $\beta_k \leq 2/(k + 1)$. Equation (5.52) leads to the updated rule in (5.47) of Algorithm 5.3. $\quad\square$

The computation above shows that we have many other choices of a_k. Any choice of β_k such that $\frac{1}{\beta_k^2} \geq \frac{1-\beta_{k+1}}{\beta_{k+1}^2}$ will lead to the same conclusion that $\eta_K \leq C\beta_K^2/(1 - \beta_K)$. For example, the parameters in (5.46) can be replaced by

$$\alpha_k = (k + 1)/2, \quad k \geq 2, \tag{5.53}$$

and the update rule in (5.47) may be simplified as

$$
\mathbf{y}^{(k+1)} = \mathbf{x}^{(k+1)} + \frac{k-1}{k+2}(\mathbf{x}^{(k+1)} - \mathbf{x}^{(k)}). \tag{5.54}
$$

Then the modified algorithm is also convergent with convergence rate $O(1/k^2)$. We leave the investigation to the interested reader.

Furthermore, following a recent development (cf. [1]), we can simply replace $\frac{a_k-1}{a_{k+1}}$ by $\frac{k-1}{k+\alpha}$ for $\alpha > 2$ in (5.47) to get an algorithm with an even faster rate. Let us present this new algorithm (cf. [1]) in Algorithm 5.4.

ALGORITHM 5.4

Attouch–Peypouquet Accelerated Gradient Algorithm [1]

1: Fix $\beta \le 1/L$, the Lipschitz constant of ∇f.
2: Let $\mathbf{x}^{(1)}, \mathbf{y}^{(1)}$ be two initial guesses.
3: Compute the following minimization and iterative steps:

$$
\mathbf{x}^{(k+1)} = \arg\min\left\{ \theta(\mathbf{x}) + \langle \nabla f(\mathbf{y}^{(k)}), \mathbf{x} - \mathbf{y}^{(k)} \rangle + \frac{1}{2\beta}\|\mathbf{x} - \mathbf{y}^{(k)}\|^2 \right\}, \tag{5.55}
$$

and define

$$
\mathbf{y}^{(k+1)} = \mathbf{x}^{(k+1)} + \frac{k}{k+1+\alpha}(\mathbf{x}^{(k+1)} - \mathbf{x}^{(k)}) \tag{5.56}
$$

for all $k \ge 1$ until a given number of iterations is achieved, where $\alpha > 2$.

The rate of convergence of Algorithm 5.4 can be established.

Theorem 5.29 (Attouch and Peypouquet [1]). *Suppose that* $\mathbf{x}^{(k)}, k \ge 1$, *is a sequence from Algorithm 5.4. Let* $\eta_k = f(\mathbf{x}^{(k)}) - f(\mathbf{x}^*)$. *Then*

$$
\eta_k = o\left(\frac{1}{k^2}\right), \quad i.e., \quad k^2\eta_k \to 0, \tag{5.57}
$$

for $k \to \infty$.

Proof of Theorem 5.29. As before, the gradient mapping property in Lemma 5.28 can be rewritten as

$$
f(\mathbf{x}) \ge f(\mathbf{x}^{(k+1)}) + \frac{L}{2}\|\mathbf{x}^{(k+1)} - \mathbf{x}\|^2 - \frac{L}{2}\|\mathbf{y}^{(k)} - \mathbf{x}\|^2 \tag{5.58}
$$

for any $\mathbf{x} \in C \subset \mathbb{R}^n$. We first choose $\mathbf{x} = (1 - \beta_k)\mathbf{x}^{(k)} + \beta_k\mathbf{x}^*$ in (5.58) for some $\beta_k \in (0, 1)$. Then (5.58) becomes

$$
\begin{aligned}
f(\mathbf{x}^{(k+1)}) &\le f(\mathbf{x}) + \frac{L}{2}\|\mathbf{y}^{(k)} - \mathbf{x}\|^2 - \frac{L}{2}\|\mathbf{x}^{(k+1)} - \mathbf{x}\|^2 \\
&\le (1 - \beta_k)F(\mathbf{x}^{(k)}) + \beta_k f(\mathbf{x}^*) \\
&\quad + \frac{L}{2}\|(1 - \beta_k)\mathbf{x}^{(k)} + \beta_k\mathbf{x}^* - \mathbf{y}^{(k)}\|^2 - \frac{L}{2}\|\mathbf{x}^{(k+1)} - (1 - \beta_k)\mathbf{x}^{(k)} - \beta_k\mathbf{x}^*\|^2 \\
&\le (1 - \beta_k)f(\mathbf{x}^{(k)}) + \beta_k f(\mathbf{x}^*) + \frac{L\beta_k^2}{2}\|\mathbf{x}^* + (\beta_k^{-1} - 1)\mathbf{x}^{(k)} - \beta_k^{-1}\mathbf{y}^{(k)}\|^2 \\
&\quad - \frac{L\beta_k^2}{2}\|\mathbf{x}^* + (\beta_k^{-1} - 1)\mathbf{x}^{(k)} - \beta_k^{-1}\mathbf{x}^{(k+1)}\|^2. \tag{5.59}
\end{aligned}
$$

Next we choose $\mathbf{y}^{(k)} = \mathbf{x}^{(k)} + (1 - \beta_k)(\mathbf{x}^{(k)} - \mathbf{x}^{(k-1)})$ so that the third term on the right-hand side above can be rewritten as

$$\frac{L}{2}\|\beta_k\mathbf{x}^* + (1 - \beta_k)\mathbf{x}^{(k)} - (\mathbf{x}^{(k)} + (1 - \beta_k)(\mathbf{x}^{(k)} - \mathbf{x}^{(k-1)}))\|^2$$

$$= \frac{L}{2}\|\beta_k\mathbf{x}^* + (1 - \beta_k)\mathbf{x}^{(k)} - (\mathbf{x}^{(k)} + (1 - \beta_k)(\mathbf{x}^{(k)} - \mathbf{x}^{(k-1)}))\|^2$$

$$\leq \frac{L}{2}\|\beta_k\mathbf{x}^* + (1 - \beta_k)\mathbf{x}^{(k)} - \mathbf{x}^{(k)} - (1 - \beta_k)(\mathbf{x}^{(k)} - \mathbf{x}^{(k-1)})\|^2$$

$$= \frac{L\beta_k^2}{2}\|\beta_k\mathbf{x}^* - \beta_k^{-1}\mathbf{x}^{(k)} + (\beta_k^{-1} - 1)\mathbf{x}^{(k-1)}\|^2.$$

Letting $\mathbf{z}^k = \beta_k^{-1}\mathbf{x}^{(k)} - (\beta_k^{-1} - 1)\mathbf{x}^{(k-1)}$, the inequality in (5.59) yields

$$\eta_{k+1} \leq (1 - \beta_k)\eta_k + \frac{L\beta_k^2}{2}\|\mathbf{x}^* - \mathbf{z}^k\|^2 - \frac{L\beta_k^2}{2}\|\mathbf{x}^* - \mathbf{z}^{k+1}\|^2.$$

Choosing $\beta_k = \alpha/(k + \alpha)$ and letting $\xi_k = \frac{L}{2}\|\mathbf{x}^* - \mathbf{z}^k\|^2$, the above inequality can be rewritten as

$$(k + \alpha)^2\eta_{k+1} \leq k(k + \alpha)\eta_k + \alpha^2(\xi_k - \xi_{k+1}), \tag{5.60}$$

which can be rewritten as

$$(k + \alpha)^2\eta_{k+1} \leq (k - 1 + \alpha)^2\eta_k - [(\alpha - 2)k + (\alpha - 1)^2]\eta_k + \alpha^2(\xi_k - \xi_{k+1}).$$

Hence, we have

$$(k + \alpha)^2\eta_{k+1} + (\alpha - 2)k\eta_k + (\alpha - 1)^2\eta_k \leq (k - 1 + \alpha)^2\eta_k + \alpha^2(\xi_k - \xi_{k+1}). \tag{5.61}$$

Summing the above inequalities for $k = 1, \ldots, K$, we have

$$(K + \alpha)^2\eta_{K+1} + (\alpha - 2)\sum_{k=1}^{K} k\eta_k \leq \alpha^2\eta_1 + \alpha^2\xi_1 =: A_0.$$

When $\alpha \geq 2$, we have $\eta_{K+1} \leq A_0/(K + \alpha)^2$. Furthermore, when $\alpha > 2$, we have

$$\sum_{k=1}^{K} k\eta_k \leq A_0/(\alpha - 2) \quad \forall K > 1. \tag{5.62}$$

That is,

$$\sum_{k=1}^{K} \frac{1}{k}(k^2\eta_k) \leq A_0/(\alpha - 2) \quad \forall K > 1. \tag{5.63}$$

If $k^2\eta_k$ converges to c, then c must be zero since $\sum_{k=1}^{K} \frac{1}{k}$ diverges.

In order to show $\eta_{K+1} = o(1/(K + \alpha)^2)$, we need to use (5.58) again. We choose $\mathbf{x} = \mathbf{x}^{(k)}$ and $\mathbf{y}^{(k)} = \mathbf{x}^{(k)} + (1 - \beta_k)(\mathbf{x}^{(k)} - \mathbf{x}^{(k-1)})$ in (5.58) for the same $\beta_k = \alpha/(k + \alpha) \in (0, 1)$. It follows that

$$\eta_{k+1} \leq \eta_k + \frac{L}{2}(1 - \beta_k)^2\|\mathbf{x}^{(k)} - \mathbf{x}^{(k-1)}\|^2 - \frac{L}{2}\|\mathbf{x}^{(k+1)} - \mathbf{x}^{(k)}\|^2. \tag{5.64}$$

Letting $d_k = L\|\mathbf{x}^{(k)} - \mathbf{x}^{(k-1)}\|^2/2$ and noting that $k + \alpha - 1 \geq k + 1$, we have $1 - \beta_k = k/(k + \alpha)$, and the above inequality can be simplified to

$$(k + \alpha)^2 d_{k+1} - k^2 d_k \leq (k + \alpha)^2(\eta_k - \eta_{k+1}).$$

The left-hand side of the above inequality can be simplified to $(k+1)^2 d_{k+1} - k^2 d_k + [2(k+1)(\alpha - 1) + (\alpha - 1)^2] d_{k+1}$, while the right-hand side is $(k+\alpha)^2 \eta_k - (k+1+\alpha)^2 \eta_{k+1} + (2k+2\alpha+1)\eta_{k+1}$. Combining these two equations yields

$$(k+1)^2 d_{k+1} + (k+1+\alpha)^2 \eta_{k+1} + 2(k+1)(\alpha-1)d_{k+1} \le k^2 d_k + (k+\alpha)^2 \eta_k + 2(k+1+\alpha)\eta_{k+1}. \tag{5.65}$$

By (5.61), we know that the summation $\sum_{k=1}^{\infty} 2(k+\alpha+1)\eta_{k+1} \le A_1 < \infty$ for a positive constant A_1. It follows that

$$\sum_{k=1}^{\infty} 2(k+1)d_{k+1} \le d_1 + (1+\alpha)^2 \eta_1 + A_1. \tag{5.66}$$

Inequality (5.65) can be simplified as follows:

$$(k+1)^2 d_{k+1} + (k+1+\alpha)^2 \eta_{k+1} \le k^2 d_k + (k+\alpha)^2 \eta_k + 2(k+1+\alpha)\eta_{k+1}. \tag{5.67}$$

The last term above goes to zero, and the sequence $\{k^2 d_k + (k+\alpha)^2 \eta_k, k \ge 1\}$ is nearly monotone and is bounded below. Thus, the sequence is convergent.

Finally, we see that

$$\sum_{k=1}^{K} \frac{1}{k}[k^2 d_k + (k+\alpha)^2 \eta_k] \le \sum_{k=1}^{K} k d_k + \left(k + \alpha + \frac{\alpha}{k}\right)\eta_k < \infty$$

by using (5.66) and (5.62). Since the sequence $k^2 d_k + (k+\alpha)^2 \eta_k$ has a limit, we conclude that the limit has to be zero. That is, $k^2 d_k + (k+\alpha)^2 \eta_k \to 0$ since $\sum_{k=1}^{K} \frac{1}{k}$ diverges. $\qquad\square$

5.7 ▪ Proximal Gradient Descent Methods

In this section, we consider $f(\mathbf{x}) = g(\mathbf{x}) + \theta(\mathbf{x})$, where $g(\mathbf{x})$ is Lipschitz differentiable and $\theta(\mathbf{x})$ is just a convex function. For example, $f(\mathbf{x}) = \frac{1}{2}\|\Phi\mathbf{x} - \mathbf{b}\|^2 + \lambda\|\mathbf{x}\|_1$, where $\theta(\mathbf{x}) = \lambda\|\mathbf{x}\|_1$. The minimization (5.1) is

$$\min g(\mathbf{x}) + \theta(\mathbf{x}) \quad \forall \mathbf{x} \in \mathbb{R}^n. \tag{5.68}$$

In this case, the standard gradient methods in previous sections do not work. Let us review a common strategy to derive another algorithm to approximate the minimizer $\mathbf{x}^*$. To this end, we suppose that we have $\mathbf{x}^k$ and look for $\mathbf{x}^{k+1}$ by solving a new minimization problem to approximate the minimizer:

$$\min_{\mathbf{x} \in \mathbb{R}^n} \left\{ \theta(\mathbf{x}) + \langle \nabla g(\mathbf{x}^k), \mathbf{x} - \mathbf{x}^k \rangle + \frac{1}{2\beta}\|\mathbf{x} - \mathbf{x}^k\|^2 \right\}, \tag{5.69}$$

where $\beta \le 1/L$ and L is the Lipschitz constant of the differentiation of f. This is a quadratic polynomial approximation of the convex function $f(\mathbf{x})$ at the kth step. Since $\theta(\mathbf{x})$ is convex, the above problem has a unique solution because of its strong convexity. Let that solution be $\mathbf{x}^{k+1}$, which satisfies

$$\frac{1}{\beta}(\mathbf{x}^{k+1} - \mathbf{x}^k) + \partial\theta(\mathbf{x}^{k+1}) + \nabla g(\mathbf{x}^k) = 0, \tag{5.70}$$

where $\partial\theta(\mathbf{x}^{k+1})$ stands for the subderivative of θ at $\mathbf{x}^{k+1}$. Therefore, we have the proximal gradient descent algorithm.

ALGORITHM 5.5

Proximal Gradient Descent Algorithm

1: Fix $\beta \leq 1/L$. Suppose that we are given $\mathbf{x}^{(1)}$.

2: For $k \geq 1$, we solve the minimization in (5.69) and let the minimizer be $\mathbf{x}^{k+1}$. That is,

$$\mathbf{x}^{k+1} := \arg\min_{\mathbf{x}} \left\{ \theta(\mathbf{x}) + \langle \nabla g(\mathbf{x}^k), \mathbf{x} - \mathbf{x}^k \rangle + \frac{1}{2\beta} \|\mathbf{x} - \mathbf{x}^k\|^2 \right\}. \tag{5.71}$$

3: We repeat the above computation until a maximum number of iterations is achieved.

Let us briefly discuss how to solve (5.71). Define a proximal operator by

$$\text{Prox}_\theta(\mathbf{x}) := \arg\min_{\mathbf{y}} \left\{ \frac{1}{2} \|\mathbf{x} - \mathbf{y}\|^2 + \lambda\theta(\mathbf{y}) \right\}. \tag{5.72}$$

Then it is easy to verify that $\mathbf{x}^{k+1}$ in (5.71) is given by

$$\mathbf{x}^{k+1} = \text{Prox}_\theta(\mathbf{x}^k - \beta\nabla g(\mathbf{x}^k)). \tag{5.73}$$

We leave the verification to the reader. The following three cases enable us to find the solution $\mathbf{x}^{(k+1)}$.

(1) When $\theta \equiv 0$, the proximal operator is the identity.

(2) When $\theta(\mathbf{x}) = \lambda\|\mathbf{x}\|_1$, the solution is given by using the shrinkage operator. See the next section for details.

(3) When $\theta(\mathbf{x}) = \lambda\mathbf{x}_+$, the positive part of the function, the solution is $\text{Prox}_\theta(x) = x - \lambda$ if $x \geq \lambda$, $\text{Prox}_\theta(x) = x$ if $x \leq 0$, and $\text{Prox}_\theta(x) = 0$ if $0 < x < \lambda \leq 1$.

See Exercises 26 and 27 for some properties of the proximal operators and more examples.

Now we assume that the problem in (5.71) can be found. We shall show the convergence of this algorithm.

Theorem 5.30 (Convergence). *Suppose that ∇f is L-Lipschitz continuous, i.e., f is L-Lipschitz differentiable. Let $\mathbf{x}^{(k)}, k \geq 1$, be the iterative solutions of Algorithm 5.5. Let $\eta_k = f(\mathbf{x}^{(k)}) - f(\mathbf{x}^*)$, where $\mathbf{x}^*$ is a minimizer of the minimization problem (5.1). Then*

$$\eta_k = o(1/k), \quad k \to \infty. \tag{5.74}$$

We first need the following lemma in order to prove the result in Theorem 5.30.

Lemma 5.31 (Monotone Property). *Fix $\beta > 0$ satisfying $\beta \leq 1/L$. Suppose that f is L-Lipschitz continuously differentiable. Let $\mathbf{x}^{(k)}, k \geq 1$, be the iterative solutions of Algorithm 5.5. Then*

$$f(\mathbf{x}^{(k+1)}) \leq f(\mathbf{x}^{(k)}) \quad \forall k \geq 1.$$

Proof. We use the Lipschitz differentiability, i.e., Lemma 5.18, to get

$$g(\mathbf{x}^{k+1}) \leq g(\mathbf{x}^k) + \langle \nabla g(\mathbf{x}^k), \mathbf{x}^{k+1} - \mathbf{x}^k \rangle + \frac{L}{2} \|\mathbf{x}^{k+1} - \mathbf{x}^k\|^2$$

$$\leq g(\mathbf{x}^k) + \langle \nabla g(\mathbf{x}^k), \mathbf{x}^{k+1} - \mathbf{x}^k \rangle + \frac{1}{2\beta} \|\mathbf{x}^{k+1} - \mathbf{x}^k\|^2.$$

It follows that

$$
\begin{aligned}
f(\mathbf{x}^{(k+1)}) &= \theta(\mathbf{x}^{k+1}) + g(\mathbf{x}^{k+1}) \\
&\leq g(\mathbf{x}^{(k)}) + \theta(\mathbf{x}^{k+1}) + \langle \nabla g(\mathbf{x}^k), \mathbf{x}^{(k+1)} - \mathbf{x}^{(k)} \rangle + \frac{1}{2\beta} \| \mathbf{x}^{(k+1)} - \mathbf{x}^{(k)} \|^2 \\
&\leq g(\mathbf{x}^{(k)}) + \langle \nabla g(\mathbf{x}^{(k)}), \mathbf{x}^{(k)} - \mathbf{x}^{(k)} \rangle + \theta(\mathbf{x}^{(k)}) + \frac{1}{2\beta} \| \mathbf{x}^{(k)} - \mathbf{x}^{(k)} \|^2 = f(\mathbf{x}^{(k)})
\end{aligned}
$$

by the minimization property of $\mathbf{x}^{(k+1)}$. This is valid for all $k \geq 1$. Thus we complete the proof. □

Next we are ready to prove Theorem 5.30.

Proof of Theorem 5.30. We first prove

$$
\eta_k \leq \frac{L}{2} \frac{\| \mathbf{x}^* - \mathbf{x}^{(1)} \|^2}{k+1} \quad \forall k \geq 1.
$$

Since g is Lipschitz differentiable, we use Lemma 5.18 to get

$$
\begin{aligned}
f(\mathbf{x}^{k+1}) &= g(\mathbf{x}^{(k+1)}) + \theta(\mathbf{x}^{k+1}) \\
&\leq g(\mathbf{x}^{(k)}) + \langle \nabla g(\mathbf{x}^k), \mathbf{x}^{(k+1)} - \mathbf{x}^{(k)} \rangle + \frac{L}{2} \| \mathbf{x}^{(k+1)} - \mathbf{x}^{(k)} \|^2 + \theta(\mathbf{x}^{k+1}). \quad (5.75)
\end{aligned}
$$

By the optimization condition (5.70), we have

$$
\left\langle \mathbf{x}^* - \mathbf{x}^{(k+1)}, \frac{1}{\beta}(\mathbf{x}^{(k+1)} - \mathbf{x}^{(k)}) + \partial\theta(\mathbf{x}^{k+1}) + \nabla g(\mathbf{x}^k) \right\rangle = 0.
$$

Adding the above to both sides of the inequality in (5.75), we use the convexity of θ, i.e., $\theta(\mathbf{x}^*) - \theta(\mathbf{x}^{k+1}) \geq \langle \partial\theta(\mathbf{x}^{k+1}), \mathbf{x}^* - \mathbf{x}^{k+1} \rangle$, to obtain

$$
\begin{aligned}
f(\mathbf{x}^{k+1}) &\\
&\leq g(\mathbf{x}^{(k)}) + \langle \nabla g(\mathbf{x}^{(k)}), \mathbf{x}^* - \mathbf{x}^{(k)} \rangle + \langle \mathbf{x}^* - \mathbf{x}^{k+1}, (\mathbf{x}^{(k+1)} - \mathbf{x}^{(k)})/\beta \rangle \\
&\quad + \frac{L}{2} \| \mathbf{x}^{k+1} - \mathbf{x}^{(k)} \|^2 + \theta(\mathbf{x}^*) \\
&\leq f(\mathbf{x}^*) + \frac{1}{2\beta} \| \mathbf{x}^* - \mathbf{x}^{(k)} \|^2 - \frac{1}{2\beta} \| \mathbf{x}^{(k+1)} - \mathbf{x}^* \|^2
\end{aligned}
$$

by using the condition $\beta \leq 1/L$, where we have used the convexity of f and an elementary equality: $(b-c)^2 - (a-c)^2 = (b-a)^2 + 2(b-a)(a-c)$. It follows that

$$
\eta_{k+1} \leq \frac{1}{2\beta} \| \mathbf{x}^* - \mathbf{x}^{(k)} \|^2 - \frac{1}{2\beta} \| \mathbf{x}^{(k+1)} - \mathbf{x}^* \|^2
$$

for all $k \geq 1$. Adding the above inequalities together from $k = 1, 2, \ldots, K$, we use Lemma 5.31 to obtain

$$
\begin{aligned}
(K+1)\eta_{K+1} &\leq \sum_{k=1}^{K} \eta_{k+1} \leq \sum_{k=1}^{K} \frac{1}{2\beta} \| \mathbf{x}^* - \mathbf{x}^{(k)} \|^2 - \frac{1}{2\beta} \| \mathbf{x}^{(k+1)} - \mathbf{x}^* \|^2 \\
&\leq \frac{1}{2\beta} \| \mathbf{x}^* - \mathbf{x}^{(1)} \|^2.
\end{aligned}
$$

That is, we have

$$\eta_{K+1} \leq \frac{1}{2\beta(K+1)}\|\mathbf{x}^* - \mathbf{x}^{(1)}\|^2 \quad \forall K \to \infty.$$

Furthermore, since $\sum_{j=1}^{2K} \eta_{j+1} \leq L\|\mathbf{x}^* - \mathbf{x}^{(1)}\|^2/2 < \infty$, we have $(K+1)\eta_{2K+1} \leq \sum_{k=K}^{2K} \eta_{k+1} \to 0$ as $K \to \infty$. That is,

$$\eta_{2K+1} = o\left(\frac{1}{2K+1}\right) \quad \forall K \to \infty.$$

The process is similar for η_{2K} and hence we have (5.74).   □

ALGORITHM 5.6

Accelerated Proximal Gradient Algorithm

1: Fix $\beta \leq 1/L$, the Lipschitz constant of ∇f.
2: Let $\mathbf{x}^{(1)}, \mathbf{y}^{(1)}$ be two initial guesses and $a_1 = 1$.
3: Compute the following minimization and iterative steps:

$$\mathbf{x}^{(k+1)} = \arg\min \left\{ \theta(\mathbf{x}) + \langle \nabla g(\mathbf{y}^{(k)}), \mathbf{x} - \mathbf{y}^{(k)} \rangle + \frac{1}{2\beta}\|\mathbf{x} - \mathbf{y}^{(k)}\|^2 \right\}. \qquad (5.76)$$

4: Let

$$a_{k+1} = \left(1 + \sqrt{4a_k^2 + 1}\right)/2$$

and define

$$\mathbf{y}^{(k+1)} = \mathbf{x}^{(k+1)} + \frac{a_k - 1}{a_{k+1}}(\mathbf{x}^{(k+1)} - \mathbf{x}^{(k)})$$

for all $k \geq 1$ until a given number of iterations is achieved.

Similar to the previous section, we can also speed up the computation of the proximal gradient method. See Algorithm 5.6. Then the convergence of $\mathbf{y}^{(k)}, k \geq 1$, from Algorithm 5.6 with a rate $O(1/k^2)$ can be established. We leave the details to the interested reader. Furthermore, if we use $k/(k+1+\alpha)$ in place of $\frac{a_k-1}{a_{k+1}}$ above, the convergence is even faster. Again we can establish the convergence rate $o(1/k^2)$.

When the number of components of variable $\mathbf{x}$ is large, i.e., $\mathbf{x} \in \mathbb{R}^n$ with $n \gg 1$, the computation of the gradient vector is expensive. We should use the coordinate descent method (cf., e.g., [10]) or the stochastic gradient descent method. Details are beyond the scope of this book.

Finally, we point out that when f is strongly convex, the convergence of Algorithm 5.5 is linear.

Theorem 5.32. *Suppose that f is L-Lipschitz continuous and μ-strongly convex. Let $\mathbf{x}_*$ be the minimizer of (5.68). Let $\mathbf{x}^{(k)}, k \geq 1$, be the iterative solutions of Algorithm 5.5. Then*

$$\|\mathbf{x}^{(k)} - \mathbf{x}^*\| \leq C\gamma^k$$

for a positive constant C dependent on the initial value and $\gamma \in (0,1)$ if $\beta > 0$ is sufficiently small.

Proof. First of all, let us recall a property of the proximal operator. For any $\mathbf{x}, \mathbf{y} \in \mathbb{R}^n$,

$$\|\text{Prox}_\theta(\mathbf{x}) - \text{Prox}_\theta(\mathbf{y})\| \leq \|\mathbf{x} - \mathbf{y}\|. \qquad (5.77)$$

We leave the proof to Exercise 27. Using (5.73), we have

$$
\begin{aligned}
&\|\mathbf{x}^{(k+1)} - \mathbf{x}^*\|^2 \\
&= \|\mathrm{Prox}_\theta(\mathbf{x}^{(k)} - \beta\nabla f(\mathbf{x}^{(k)})) - \mathrm{Prox}_\theta(\mathbf{x}^* - \beta\nabla f(\mathbf{x}^*))\|^2 \\
&\leq \|\mathbf{x}^{(k)} - \mathbf{x}^* - \beta(\nabla f(\mathbf{x}^{(k)}) - \nabla f(\mathbf{x}^*))\|^2 \\
&= \|\mathbf{x}^{(k)} - \mathbf{x}^*\|^2 - 2\beta\langle\nabla f(\mathbf{x}^{(k)}) - \nabla f(\mathbf{x}^*), \mathbf{x}^{(k)} - \mathbf{x}^*\rangle + \beta^2\|\nabla f(\mathbf{x}^{(k)}) - \nabla f(\mathbf{x}^*)\|^2.
\end{aligned}
$$

We use the Lipschitz condition of ∇f and the strong convexity of f to get

$$
\|\mathbf{x}^{(k+1)} - \mathbf{x}^*\|^2 \leq \|\mathbf{x}^{(k)} - \mathbf{x}^*\|^2(1 - 2\beta\nu + \beta^2 L^2) \leq \cdots \leq \gamma^{2k}\|\mathbf{x}^{(1)} - \mathbf{x}^*\|^2, \tag{5.78}
$$

with $\gamma = (1 - 2\beta\nu + \beta^2 L^2)^{1/2} < 1$ for an appropriate $\beta > 0$. □

5.8 ▪ Sparse Solutions

Let us return to our main problem in this book, i.e., solving (1.4). Among the many approaches mentioned in Chapter 1, one approach is to approximate the minimization in the form

$$
\min_{\mathbf{x}\in\mathbb{R}^n} \|\mathbf{x}\|_1 + \frac{1}{2\lambda}\|\Phi\mathbf{x} - \mathbf{b}\|^2. \tag{5.79}
$$

Then the above problem is an unconstrained minimization (5.1) with $f(\mathbf{x}) = \lambda\|\mathbf{x}\|_1 + \frac{1}{2}\|\Phi\mathbf{x} - \mathbf{b}\|^2$. We now apply Algorithm 5.5 to solve (5.79). The main problem remaining is how to solve (5.71) with $\theta(\mathbf{x}) = \lambda\|\mathbf{x}\|_1$ and $g(\mathbf{x}) = \frac{1}{2}\|\Phi\mathbf{x} - \mathbf{b}\|^2$. It turns out that there is a simple way to do this. Indeed, because $g(\mathbf{x}) = \frac{1}{2}\|\Phi\mathbf{x} - \mathbf{b}\|^2$, $\nabla g(\mathbf{x}) = \Phi^\top(\Phi\mathbf{x} - \mathbf{b})$. We should use $L = \|\Phi^\top\Phi\|$ for the Lipschitz differentiability of g and $\beta = 1/L$. The minimizing functional of (5.71) can be rewritten as

$$
\mathbf{x}^{(k+1)} := \min_{\mathbf{x}} \lambda\|\mathbf{x}\|_1 + \frac{L}{2}\left\|\mathbf{x} - \mathbf{x}^{(k)} + \frac{1}{L}\Phi^\top(\Phi\mathbf{x}^{(k)} - \mathbf{b})\right\|^2 - \frac{1}{2L}\|\Phi^\top\Phi(\mathbf{x}^{(k)}) - \mathbf{b})\|^2. \tag{5.80}
$$

Recall from Example 5.14 that when $f(u) = |u| + \frac{1}{2a}|u - a|^2$, the subdifferential of f is

$$
\partial f(u) = \begin{cases} 1 + \frac{1}{\alpha}(u - a) & \text{if } u > 0, \\ [-1, 1] - \frac{a}{\alpha} & \text{if } u = 0, \\ -1 + \frac{1}{\alpha}(u - a) & \text{if } u < 0. \end{cases} \tag{5.81}
$$

So the critical point such that $\partial f(u) = 0$ is $u = a - \alpha$ if $u > 0$ and $u = a + \alpha$ if $u < 0$. In other words, if $a > \alpha$, $u = a - \alpha$; if $a < -\alpha$, $u = a + \alpha$; and if $a = \alpha$ or $a = -\alpha$, $u = 0$. For convenience, let us write these as the shrinkage operator

$$
S_\alpha(a) = \begin{cases} a - \alpha & \text{if } a > \alpha, \\ 0 & \text{if } -\alpha \leq a \leq \alpha, \\ a + \alpha & \text{if } a < -\alpha. \end{cases} \tag{5.82}
$$

Letting $\mathbf{b}^{(k)} = \Phi^\top(\Phi\mathbf{x}^{(k)} - \mathbf{b})/L$, $\mathbf{a}^{(k)} = \mathbf{x}^{(k)} - \mathbf{b}^{(k)}$, and $\alpha = 2\lambda/L$, we obtain

$$
\mathbf{x}^{(k+1)} = S_{2\lambda/L}(\mathbf{a}^{(k)}) = (S_{2\lambda/L}(a_1^{(k)}), \ldots, S_{2\lambda/L}(a_n^{(k)}))^\top, \tag{5.83}
$$

which is called the shrinkage and thresholding step. Therefore, in this setting, the proximal gradient descent method and accelerated proximal gradient descent method can be explicitly given.

For convenience, let us simply write down a new version of the accelerated proximal gradient descent method. The following is the well-known fast iterative shrinkage and thresholding algorithm, or FISTA for short.

ALGORITHM 5.7
Fast Iterative Shrinkage and Thresholding Algorithm (FISTA) [2]
1: Fix $\beta = 1/L$ with $L = \|\Phi^\top \Phi\|$.
2: Let $\mathbf{x}^{(1)}, \mathbf{y}^{(1)}$ be two initial guesses and $a_1 = 1$.
3: Compute $\mathbf{b}^{(k)} = \Phi^\top (\Phi \mathbf{y}^{(k)} - \mathbf{b})/L$, $\mathbf{a}^{(k)} = \mathbf{y}^{(k)} - \mathbf{b}^{(k)}$.
4: Compute the shrinkage and thresholding step, i.e., (5.83).
5: Let $a_{k+1} = (1 + \sqrt{4a_k^2 + 1})/2$ and compute

$$\mathbf{y}^{(k+1)} = \mathbf{x}^{(k+1)} + \frac{a_k - 1}{a_{k+1}}(\mathbf{x}^{(k+1)} - \mathbf{x}^{(k)}) \tag{5.84}$$

for all $k \geq 1$ until a given number of iterations is achieved.

Then the discussion in this and previous sections shows that $\mathbf{y}^{(k)} \to \mathbf{x}^*$ in $O(1/k^2)$, where $\mathbf{x}^*$ is a minimizer of the minimizing functional $f(\mathbf{x}) = \lambda\|\mathbf{x}\|_1 + \frac{1}{2}\|\Phi\mathbf{x} - \mathbf{b}\|^2$. Furthermore, if we use (5.56) instead of (5.84), the convergence will be $o(1/k^2)$.

The algorithms provide a good method of approximating sparse solutions. One way to improve the accuracy of the iterations from these algorithms is to use least squares. That is, after performing many iterations, we can use the locations of nonzero entries of the iterative approximation of the sparse solution to formulate a least squares problem of much smaller size and solve the least squares to get an excellent approximation of the sparse solution. Our numerical experiments show that when the sparsity of a sparse solution is small, say $s \leq m/10$, these algorithms work well in the sense that there is a more than 50% chance of finding the sparse solution. The performances using (5.84) and (5.56) are very similar.

The minimization problem (5.68) is also called the lasso problem in statistics (cf. [13]). Because the measurement vector $\mathbf{b}$ may contain some noise, in practice, one mainly solves (5.68).

5.9 ▪ Exercises

Exercise 1. A function is subadditive if $f(x + y) \leq f(x) + f(y)$. Show that the following functions satisfy this subadditivity: $(|x|^2 + \epsilon)^{q/2}, 0 < q \leq 1$, and $\log(1 + |x|^2)$.

Exercise 2. Suppose that $q \in (0, 1)$. Show that $f(x) = (|x|^2 + \epsilon)^{q/2}$ is convex when $|x| \leq \sqrt{\epsilon/(1 - q)}$.

Exercise 3. Show that $f(x) = \sqrt{\epsilon + x^2}$ is a convex function.

Exercise 4. Show that $f(x) = \log(1 + \exp(-ax))$ is strictly convex for any fixed $a > 0$. Furthermore, show that it is strongly convex.

Exercise 5. Prove the inequality in (5.3).

Exercise 6. Prove the following inequality: If $u^2 \leq Au + B$, where A and B are positive, then $u \leq A + B/A$.

Exercise 7. Prove Theorem 5.1.

Exercise 8. Prove Theorem 5.5.

Exercise 9. Prove that when $d = 1$, a convex function f has the following useful properties:

(1) $f(x_2) \leq \frac{x_3 - x_2}{x_3 - x_1} f(x_1) + \frac{x_2 - x_1}{x_3 - x_1} f(x_3)$ for all $a < x_1 < x_2 < x_3 < b$.

(2) $\frac{f(x_2) - f(x_1)}{x_2 - x_1} \leq \frac{f(x_3) - f(x_1)}{x_3 - x_1}$ for all $a < x_1 < x_2 < x_3 < b$.

(3) $\frac{f(x_3) - f(x_1)}{x_3 - x_1} \leq \frac{f(x_3) - f(x_2)}{x_3 - x_2}$ for all $a < x_1 < x_2 < x_3 < b$.

(4) $f(x) \leq \frac{1}{2h} \int_{x-h}^{x+h} f(t)dt$ for all x, h with $a \leq x - h < x + h \leq b$.

(5) The determinant
$$
\begin{vmatrix}
1 & x_1 & f(x_1) \\
1 & x_2 & f(x_2) \\
1 & x_3 & f(x_3)
\end{vmatrix} \geq 0
$$
for all $a < x_1 < x_2 < x_3 < b$.

(6) If f is differentiable, then f' is increasing.

(7) If f is twice differentiable on (a, b), then f is convex if and only if $f''(x) > 0$ for all $x \in (a, b)$.

(8) f is a continuous function on (a, b).

(9) f' exists on (a, b) except for a countable set of points.

(10) f'' exists on (a, b) except on a set of points of Lebesgue measure equal to zero.

Exercise 10. Show that $f(x) = \sqrt{\epsilon + x^2}$ is convex for all $x \in \mathbb{R}$ and $\epsilon \geq 0$. Compute its subgradient at any x. Next use the definition of subgradient to show that
$$
\left(\frac{x}{\sqrt{x^2 + \epsilon}} - \frac{y}{\sqrt{y^2 + \epsilon}} \right) (x - y) \geq 0 \quad \forall x, y \in \mathbb{R}.
$$

Furthermore, show that there exists a positive constant α dependent on $\epsilon > 0$ and the bound M on x and y such that
$$
\left(\frac{x}{\sqrt{x^2 + \epsilon}} - \frac{y}{\sqrt{y^2 + \epsilon}} \right) (x - y) \geq \alpha(x - y)^2 \quad \forall x, y \in \mathbb{R}, |x| \leq M, |y| \leq M. \tag{5.85}
$$

Exercise 11. Prove Theorems 5.9 and 5.10.

Exercise 12. With the subgradient, we can define the Bregman distance associated with J at v as
$$
B(u, v) := J(u) - J(v) - \langle \partial J(v), u - v \rangle. \tag{5.86}
$$

Show that $B(u, v)$ satisfies the following properties: (1) $B(u, v) \geq 0$ for all u, (2) $B(u + w, v) \leq B(u, v) + B(w, v)$, and (3) $B(w, v)$ is between $B(u, v)$ and $B(v, v)$ if w is between u and v.

Exercise 13. Show (5.13) and (5.14).

Exercise 14. Consider the minimization problem

$$\min_{x \in \mathbb{R}} |1 - x|^p + \frac{1}{2}|x|^q \tag{5.87}$$

for $1 \le p$ and $q \in (0, 1]$. It is easy to verify that it has a unique solution in $[0, 1]$. Show that the optimal value is $1/2$ for any $q \in (0, 1)$ and for any fixed $p \ge 1$.

Exercise 15. Show that the dual of the primal problem

$$\min\{\|\mathbf{x}\|_1, \quad A\mathbf{x} = \mathbf{b}\}$$

is

$$\min\{-\mathbf{b}^\top \mathbf{y}, \quad \|A^\top \mathbf{y}\|_\infty \le 1\}.$$

Exercise 16. Show that the dual of the primal problem

$$\min_{\mathbf{x} \in \mathbb{R}^N} \left\{ \|\mathbf{x}\|_1 + \frac{1}{2\alpha}\|\mathbf{x}\|_2^2 : \quad \Phi \mathbf{x} = \mathbf{b} \right\}$$

is

$$\max_{\mathbf{y}, \mathbf{z}} \left\{ -\mathbf{b}^\top \mathbf{y} + \frac{\alpha}{2}\|\Phi^\top \mathbf{y} - \mathbf{z}\|_2^2, \quad \mathbf{z} \in [-1, 1]^N \right\}.$$

Exercise 17. Show that the dual of the primal problem

$$\inf_{\mathbf{x} \in \mathbb{R}^n} \frac{1}{2}\|A\mathbf{x} - \mathbf{b}\|^2 + \lambda\|\mathbf{x}\|_1 \tag{5.88}$$

is

$$\inf_\theta \left\{ \frac{\lambda^2}{2}\left\|\theta - \frac{\mathbf{b}}{\lambda}\right\|^2 : \quad |\mathbf{a}_i^\top \theta| \le 1, i = 1, \dots, p \right\}, \tag{5.89}$$

where $A = [\mathbf{a}_1, \dots, \mathbf{a}_n]$.

Exercise 18. Show that the dual of the primal problem

$$\min_{\mathbf{x}, c} \left\{ \frac{1}{m}\sum_{i=1}^m \log(1 + \exp(-\langle \mathbf{x}, \mathbf{y}_i \rangle - b_i c)) + \lambda\|\mathbf{x}\|_1 \right\} \tag{5.90}$$

is

$$\min_\theta \left\{ \frac{1}{m}\sum_{i=1}^m f(\theta_i) : \quad \|Y^\top \theta\|_\infty \le m\lambda, \langle \theta, \mathbf{b} \rangle = 0, \theta \in \mathcal{C} \right\}, \tag{5.91}$$

where $\mathcal{C} = \{\theta \in \mathbb{R}^m, \theta_i \in (0, 1), i = 1, \dots, m\}$ and $f(t) = t\log(t) + (1 - t)\log(1 - t)$ for $t \in (0, 1)$.

Exercise 19. Suppose that a functional F is Lipschitz differentiable with constant L. Show that

$$\langle \nabla F(\mathbf{x}) - \nabla F(\mathbf{y}), \mathbf{x} - \mathbf{y} \rangle \le L\|\mathbf{x} - \mathbf{y}\|^2$$

for all $\mathbf{x}, \mathbf{y} \in \mathbb{R}^n$.

Exercise 20. Suppose that a functional F is Lipschitz differentiable with constant L. Show that

$$\frac{1}{L}\|\nabla F(\mathbf{x}) - \nabla F(\mathbf{y})\|^2 \leq \langle \nabla F(\mathbf{x}) - \nabla F(\mathbf{y}), \mathbf{x} - \mathbf{y} \rangle \tag{5.92}$$

for all $\mathbf{x}, \mathbf{y} \in \mathbb{R}^n$. (*Hint*: See [8].)

Exercise 21. Suppose that a functional F is continuously differentiable and ν-strongly convex. Show that

$$\langle \nabla F(\mathbf{x}) - \nabla F(\mathbf{y}), \mathbf{x} - \mathbf{y} \rangle \geq \nu \|\mathbf{x} - \mathbf{y}\|^2$$

for all $\mathbf{x}, \mathbf{y} \in \mathbb{R}^n$.

Exercise 22. Suppose that a functional F is Lipschitz differentiable with constant L and ν-strongly convex. Show that

$$\langle \nabla F(\mathbf{x}) - \nabla F(\mathbf{y}), \mathbf{x} - \mathbf{y} \rangle \geq \frac{\nu L}{\nu + L}\|\mathbf{x} - \mathbf{y}\|^2 + \frac{1}{\nu + L}\|\nabla F(\mathbf{x}) - \nabla F(\mathbf{y})\|^2$$

for all $\mathbf{x}, \mathbf{y} \in \mathbb{R}^n$.

Exercise 23. Show that the minimization problem

$$\min \frac{1}{2}\mathbf{x}^\top A \mathbf{x} - \mathbf{x}^\top \mathbf{b} \tag{5.93}$$

is equivalent to the solution of the linear system $A\mathbf{x} = \mathbf{b}$, where A is a symmetric matrix.

Exercise 24. Apply the gradient descent method to the minimization problem (5.93) to solve a linear system $A\mathbf{x} = \mathbf{b}$ for a matrix of large scale with various step approaches. Generate a plot of the accuracy of the numerical solution against the exact solution versus the number of iterations for each choice of step size approach. (*Hint*: Start with a reasonable size first and make sure your code works before testing a linear system of large size.)

Exercise 25. Implement the Nesterov acceleration algorithm to solve a linear system $A\mathbf{x} = \mathbf{b}$ of large size. You can use MATLAB to generate a random matrix A of size $10{,}000 \times 10{,}000$ and $\mathbf{b}$ of size $10{,}000 \times 1$ or even larger, where A is symmetric. Compare the performance with the gradient descent method in the previous problem by plotting the accuracy of numerical solution against the exact solution versus number of iterations.

Exercise 26. Show some basic properties of the proximal operator. In particular, show that the proximal operator is a nonexpansive operator: $\|\text{Prox}_\theta(\mathbf{x})\| \leq \|\mathbf{x}\|$ for any $\mathbf{x} \in \mathbb{R}^n$, where θ is a convex function.

Exercise 27. Furthermore, show that for any $\mathbf{x}, \mathbf{y} \in \mathbb{R}^n$,

$$\|\text{Prox}_\theta(\mathbf{x}) - \text{Prox}_\theta(\mathbf{y})\| \leq \|\mathbf{x} - \mathbf{y}\|. \tag{5.94}$$

(*Hint*: See [6].)

Exercise 28. Prove the convergence of Algorithm 5.6 and establish the $O(1/k^2)$ convergence rate.

Exercise 29. Find a least squares solution

$$\min_{\mathbf{x}} \frac{1}{2}\|A\mathbf{x} - \mathbf{b}\|^2 \tag{5.95}$$

for a tall matrix A of size $m \times n$ with $m \gg n$ using the gradient descent method and the Nesterov acceleration method. Plot both accuracies of numerical solution against the exact solution versus number of iterations.

Exercise 30. Find a least squares solution

$$\min_{\mathbf{x}} \|A\mathbf{x} - \mathbf{b}\|^2 + \|\mathbf{x}\|^2 \tag{5.96}$$

for a tall matrix A of size $m \times n$ with $m \gg n$ using the gradient descent method and the Nesterov acceleration method. Plot both accuracies of numerical solution against the exact solution versus number of iterations.

Exercise 31. Find a nonnegative least squares solution

$$\min_{\mathbf{x} \geq 0} \frac{1}{2} \|A\mathbf{x} - \mathbf{b}\|^2 \tag{5.97}$$

for a tall matrix A of size $m \times n$ with $m \gg n$ using the projected gradient descent method and its acceleration method. Plot both accuracies of numerical solution against the exact solution versus number of iterations.

Bibliography

[1] H. Attouch and J. Peypouquet, The rate of convergence of Nesterov's accelerated forward-backward method is actually faster than $1/k^2$, SIAM J. Optim., 26 (2016), 1824–1834. (Cited on p. 137)

[2] A. Beck and M. Teboulle, A fast iterative shrinkage-thresholding algorithm for linear inverse problems, SIAM J. Imaging Sci., 2 (2009), 183–202. (Cited on p. 144)

[3] J. Barzilai and J. M Borwein, Two-point step size gradient methods, IMA J. Numer. Anal., 8 (1988), 141–148. (Cited on p. 129)

[4] P. Ciarlet, Introduction to Numerical Linear Algebra and Optimization, Cambridge University Press, 1989. (Cited on p. 129)

[5] W. Deng, M.-J. Lai, Z. Peng, and W. Yin, Parallel multi-block ADMM with $o(1/k)$ convergence, J. Sci. Comput., 71 (2017), 712–736. (Cited on p. 134)

[6] J. J. Moreau, Proximité et dualité dans un espace Hilbertien, Bull. Soc. Math. France, 93 (1965), 273–299. (Cited on p. 147)

[7] Y. E. Nesterov, A method for solving the convex programming problem with convergence rate $O(1/k^2)$, Dokl. Akad. Nauk SSSR, 269 (1983), 543–547 (in Russian). (Cited on p. 135)

[8] Y. Nesterov, Introductory Lectures on Convex Optimization, Kluwer Academic, 2004. (Cited on pp. 129, 130, 147)

[9] Y. Nesterov, Gradient methods for minimizing composite objective function, Math. Program., 140 (2013), 125–161. (Cited on p. 129)

[10] Y. Nesterov, Efficiency of coordinate descent methods on huge-scale optimization problems, SIAM J. Optim., 22 (2012), 341–362. (Cited on p. 142)

[11] J. Nocedal and S. Wright, Numerical Optimization, Springer, 2006. (Cited on p. 129)

[12] R. T. Rochafellar and R. Wets, Variational analysis, Grundlehren Math. Wis. 317, Springer-Verlag, Berlin, 1998. (Cited on p. 124)

[13] R. Tibshirani, Regression shrinkage and selection via the lasso. J. R. Statist. Soc. Ser. B, 58 (1996), 267–288. (Cited on pp. 121, 144)

Chapter 6

Constrained Minimization Approaches

In this chapter, we present several constrained minimization approaches for the sparse solution of underdetermined linear systems. We start with the convex ℓ_1 minimization

$$\min_{\mathbf{x} \in \mathbb{R}^n} \{\|\mathbf{x}\|_1 : \Phi\mathbf{x} = \mathbf{b}\}. \tag{6.1}$$

Note that the ℓ_1 minimization in (6.1) is equivalent to the linear programming problem

$$\min \sum_{i=1}^{n} x_i \quad \text{subject to } \Phi\mathbf{x} \geq \mathbf{b}, \tag{6.2}$$

which has been well studied in the literature and in practice. It can be solved using the simplex method and the interior point method. Although the simplex method has several variations, such as the gradient boosting algorithm (cf. [13]), the piecewise linearization method (cf. [24]), the quadratic programming approach (cf. [25]), and least angle regression [11], we first present a basic simplex method without any speed-up techniques. Because it is a classical method, we simply present a basic explanation of this method and refer the reader to any standard numerical analysis textbook for details (e.g., [6]). Next we present an interior point method for (6.1) which is based on the discussion in [17].

The minimization (6.1) is a special case of the general constrained minimization

$$\min_{\mathbf{x} \in \mathbb{R}^n} \{f(\mathbf{x}) : \Phi\mathbf{x} = \mathbf{b}\}, \tag{6.3}$$

where $f(\mathbf{x})$ is a convex function, e.g., $f(\mathbf{x}) = \lambda\|\mathbf{x}\|_1 + g(\mathbf{x})$, with differentiable convex function g. In particular, $f(\mathbf{x}) = \|\mathbf{x}\|_1 + \frac{1}{2\alpha}\|\mathbf{x}\|_2^2$ for a large parameter α is a good approximation of the original minimization (6.1). As in the previous chapter, there is also a rich theory on constrained convex minimization: algorithms and their convergence analysis. We can use the projected gradient descent method for a differentiable function f and the projected proximal gradient method as well as their accelerated versions for (6.3). The well-known Uzawa algorithm is a different approach for (6.3). We shall apply Uzawa's algorithm for the sparse solution of underdetermined linear systems. Another approach is to use duality. The dual version of the minimization with $f(\mathbf{x}) = \|\mathbf{x}\|_1 + \frac{1}{2\alpha}\|\mathbf{x}\|_2^2$ is differentiable, and hence a gradient descent method can be used directly. This leads to an iterative algorithm. We shall show that the convergence of the new minimization is linear.

Next we shall present an iteratively reweighted ℓ_1 method and an ℓ_1 greedy method. Their numerical performances are excellent and set a standard for the sparse solution of underdetermined linear systems. We shall compare all methods to see their computational efficiency and effectiveness.

When the size of an underdetermined linear system is large, we may use a parallel computational approach. We refer the reader to [10] for such a study. There is another approach, called screening rules, which identifies inactive columns from Φ to reduce the size of the linear system. See section 6.7. Many algorithms have been proposed in the literature. We will not be able to present them all in this book, although some of them will be included in section 6.11.

6.1 ▪ The Simplex Method

Let us rewrite the ℓ_1 minimization problem as a standard linear programming problem. The ℓ_1 minimization (1.20), i.e., (6.1), can be rewritten as

$$\min \sum_{i=1}^{n} c_i^1 + c_i^2 \tag{6.4}$$

subject to

$$\sum_{j=1}^{n} \phi_{ij}(c_j^1 - c_j^2) = b_i, \quad i = 1, \ldots, m,$$

$$c_i^1, c_i^2 \geq 0, \quad i = 1, \ldots, n.$$

Minimization problem (6.4) is a standard linear programming problem, as in (6.2).

On the other hand, it is easy to see that any solution of the above linear programming problem must have $c_i^1 c_i^2 = 0$ for all $i = 1, \ldots, n$ since we only need one of c_i^1, c_i^2 to be nonzero to achieve the minimal value and, hence, $c_i^1 + c_i^2 = |c_i^1 - c_i^2| = |c_i|$, and the minimization in (6.1) is the minimization of (1.20). This shows that the ℓ_1 minimization in (1.20) is equivalent to the standard linear programming problem (6.4).

Next we note that there are three canonical forms of linear programming problems. The first is

$$\min_{c_1, \ldots, c_n} \sum_{i=1}^{n} a_i c_i \tag{6.5}$$

subject to

$$\sum_{j=1}^{n} b_{ij} c_j \leq d_i, \quad i = 1, \ldots, m,$$

where $a_1, \ldots, a_n$ are given coefficients and so are b_{ij} and $d_i, i = 1, \ldots, m,$ are given values. This is the standard form. It can be rewritten in two equivalent forms as follows:

$$\min_{\hat{c}_1, \ldots, \hat{c}_n} \sum_{i=1}^{\hat{n}} \hat{a}_i \hat{c}_i \tag{6.6}$$

subject to

$$\sum_{j=1}^{\hat{n}} \hat{b}_{ij} \hat{c}_j \leq \hat{d}_i, \quad i = 1, \ldots, \hat{m}, \ \hat{c}_j \geq 0, \ j = 1, \ldots, \hat{n},$$

where $\hat{m} < \hat{n}$, and

$$\min_{\tilde{c}_1, \ldots, \tilde{c}_n} \sum_{i=1}^{\tilde{n}} \tilde{a}_i \tilde{c}_i \tag{6.7}$$

subject to

$$\sum_{j=1}^{\tilde{n}} \tilde{b}_{ij}\tilde{c}_j = \hat{d}_i, \quad i = 1,\ldots,\tilde{m}, \ \tilde{c}_j \geq 0, \ j = 1,\ldots,\tilde{n},$$

where $\tilde{m} < \tilde{n}$. The equivalence of these three forms is left to Exercise 3.

The most important point about the solutions of a linear programming problem is that no interior point can be a solution of (6.5) if $J(\mathbf{c}) = \sum_{j=1}^{n} a_j c_j \neq 0$. Indeed, if $\mathbf{c}_0$ is a solution and is an interior point of the feasible domain $U = \{\mathbf{c}, \sum_{j=1}^{n} b_{ij}c_j \leq d_i, i = 1,\ldots,m\}$, i.e., there exists a radius $r > 0$ such that $\mathbf{c}_0 + B(0, r) \subset U$, where $B(0, r) = \{\mathbf{c}, \|\mathbf{c}\|_2 \leq r\}$ is the ball of radius r at the origin, then $\mathbf{c}_1 = \mathbf{c}_0 - \frac{r}{2\|\mathbf{a}\|_2}\mathbf{a} \in U$ and

$$J(\mathbf{c}_1) = J(\mathbf{c}_0) - \frac{2}{r} < J(\mathbf{c}_0)$$

has a smaller value, which contradicts the minimal value $J(\mathbf{c}_0)$.

Thus, we need to understand the vertices of the feasible domain associated with the linear programming problem. The following is a standard discussion on the simplex method. (See similar ideas in [6].) Let

$$\mathcal{F} := \left\{ \mathbf{c} = (c_1,\ldots,c_n)^\top \in \mathbb{R}^n_+ : \sum_{j=1}^{n} c_j \mathbf{b}_j = \mathbf{d} \right\}, \tag{6.8}$$

which is a polyhedron, where $\mathbf{b} = [b_{1,j},\ldots,b_{m,j}]^\top, j = 1,\ldots,n$, and $\mathbf{d} = [d_1,\ldots,d_m]^\top$. For convenience, for any point $u = (u_1,\ldots,u_n)^\top \in \mathbb{R}^n$, let $I_+(u) = \{j, u_j > 0\}$.

Theorem 6.1. *A point u in the feasible domain $\mathcal{F}$ other than the origin is a vertex of $\mathcal{F}$ if and only if the vectors $\mathbf{b}_j, j \in I_+(u)$, are linearly independent.*

Proof. Let $u \in \mathcal{F}$ be a point. Suppose that $\mathbf{b}_i, i \in I_+(u)$, are linearly dependent. Then there exist $w = (w_1,\ldots,w_n)^\top$ with $w_j = 0$ if $j \notin I_+(u)$ such that

$$\sum_{i=1}^{n} w_j \mathbf{b}_j = 0.$$

Since $u \in \mathcal{F}$, we have

$$\mathbf{d} = \sum_{i \in I_+(u)} u_i \mathbf{b}_i = \sum_{i \in I_+(u)} u_i \mathbf{b}_i - \theta \sum_{i=1}^{n} w_j \mathbf{b}_j = \sum_{i \in I_+(u)} (u_i - \theta w_i)\mathbf{b}_i$$

for any θ. It is easy to see that there is a $\delta > 0$ such that for all $\theta \in [-\delta, \delta]$ we have $u_i \pm \delta w_i \geq 0$ for $i \in I_+(u)$. Thus, u is not a vertex of $\mathcal{F}$.

On the other hand, assume that $\mathbf{b}_i, i \in I_+(u)$, are linearly independent. Then u has to be a vertex of $\mathcal{F}$. Otherwise, there are $v, w \in \mathcal{F}$ with $v \neq w$ such that $u = \alpha v + (1 - \alpha)w$ for $\alpha \in (0, 1)$. Then $I_+(v) \subset I_+(u)$ and $I_+(w) \subset I_+(u)$ since $v \geq 0$ and $w \geq 0$. Let $z = w - v \neq 0$. Note that $z_i = 0$ for all $i \notin I_+(u)$. Then

$$\sum_{i \in I_+(u)} z_i \mathbf{b}_i = \sum_{i \in I_+(u)} w_i \mathbf{b}_i - \sum_{i \in I_+(u)} v_i \mathbf{b}_i = \mathbf{d} - \mathbf{d} = 0.$$

This implies that $\mathbf{b}_i, i \in I_+(u)$, are linearly dependent, which contradicts the assumption. $\quad\square$

It is easy to see that the number of vertices of the polyhedron $\mathcal{F}$ is finite. Also, a linear programming problem may not have a solution if the feasible domain is unbounded. For example, consider $J(\mathbf{c}) = c_1$ over $\mathbf{U} = \mathbb{R}^n$. Thus, we need to check the coerciveness of $J(\mathbf{c})$.

Theorem 6.2. *Let $J(\mathbf{c})$ be a linear functional on $\mathbb{R}^n$, and let $U \subset \mathbb{R}^n$ be a nonempty set. If* $\inf_{\mathbf{c} \in U} J(\mathbf{c}) > -\infty$, *then the linear programming problem*

$$\inf\{J(\mathbf{c}), \quad \mathbf{c} \in U\}$$

has at least one solution.

Proof. The proof is left as to Exercise 4. $\quad\square$

The key observation is the following.

Theorem 6.3. *If a linear programming problem of the form (6.7) has a solution, then at least one vertex of the feasible domain is also a solution.*

Proof. Let $I_+(u) = \{j : u_j > 0\}$ for a solution $u = (u_1, \ldots, u_n)^\top$. If $I_+(u) = \emptyset$, then $u = 0 \in \mathbb{R}_+^n$ is a vertex of the feasible domain $\mathcal{F}$. When $I_+(u) \neq \emptyset$, there are two cases: (1) If $\mathbf{b}_i, i \in I_+(u)$, are linearly independent, then we use Theorem 6.1 to see that u is a vertex of $\mathcal{F}$. (2) If $\mathbf{b}_i, i \in I_+(u)$, are linearly dependent, then there exist $w = (w_1, \ldots, w_n)^\top$ with $w_j = 0$ if $j \notin I_+(u)$ such that

$$\sum_{i=1}^n w_j \mathbf{b}_j = 0.$$

Without loss of generality, we may assume that $\max_j w_j > 0$. Now consider $u - \theta w$ with θ small enough so that $u - \theta w \in \mathcal{F}$. Indeed, since $u \in \mathcal{F}$, we have

$$\mathbf{d} = \sum_{i \in I_+(u)} u_i \mathbf{b}_i = \sum_{i \in I_+(u)} u_i \mathbf{b}_i - \theta \sum_{i=1}^n w_j \mathbf{b}_j = \sum_{i \in I_+(u)} (u_i - \theta w_i) \mathbf{b}_i$$

for any θ. It is easy to see that there is a $\delta > 0$ such that for all $\theta \in [-\delta, \delta]$ we have $u_i - \theta w_i \geq 0$ for $i \in I_+(u)$. For all such θ, we let $u' = (u_1', \ldots, u_n')^\top$ with

$$u_i' = u_i - \theta w_i, \quad i \in I_+(u), \quad \text{and} \quad u_0' = 0, i \notin I_+(u).$$

Since $J(u)$ is the minimal,

$$J(u') = J(u - \theta w) = J(u) - \theta J(w) \geq J(u).$$

This forces $J(w) = 0$. Let us fix a $\theta \in [-\delta, \delta]$ such that at least one of $u_i - \theta w_i, i \in I_+(u)$, is zero and the rest are positive. Then $I_+(u') \subset I_+(u)$ and the cardinality of $I_+(u')$ is strictly less than $I_+(u)$. Since $J(u') = J(u)$, we can repeat the above argument whether $\mathbf{b}_i, i \in I_+(u')$, are linearly independent or not. Because $I_+(u)$ is finite, we are eventually led to a vertex of $\mathcal{F}$. $\quad\square$

We are now in a position to describe the simplex method. The general idea of the simplex method is as follows: Starting with a vertex u_0 of the feasible domain and evaluating the linear functional J at u_0, construct a sequence $u_k, k = 1, 2, 3, \ldots$, from the vertices of the feasible domain $\mathcal{F}$ such that we have

$$J(u_0) > J(u_1) > \cdots > J(u_k) > J(u_{k+1}), \quad k \geq 0. \tag{6.9}$$

Then this procedure leads to a vertex of $\mathcal{F}$ which is a solution in a finite number of iterations because the total number of vertices is finite. Let us carefully construct such a sequence satisfying (6.9).

Suppose we have u_k. By Theorem 6.1, $\mathbf{b}_j, j \in I_+(u_k)$, are linearly independent. Without loss of generality, we may assume that they do not form a basis for $\mathbb{R}^m$. Then we have to add more vectors from $\mathbf{b}_i, i = 1, \ldots, n$, to form a basis. Let us denote the new index set by $I(u_k)$. For each $j \notin I(u_k)$, we find γ_{ij} such that

$$\mathbf{b}_j = \sum_{i \in I(u_k)} \gamma_{ij} \mathbf{b}_i.$$

Writing $u_k = [u_{k,1}, \ldots, u_{k,n}]^\top$, we note that $\sum_{i \in I_+(u_k)} u_{k,i} \mathbf{b}_i = \mathbf{d}$. Thus, for any $\theta_j \in \mathbb{R}$,

$$\sum_{i \in I(u_k)} (u_{k,i} - \theta_j \gamma_{ij}) \mathbf{b}_i + \theta_j \mathbf{b}_j = \mathbf{d}.$$

Now we choose θ_j carefully such that $\tilde{u} = (\tilde{u}_1, \tilde{u}_2, \ldots, \tilde{u}_n)^\top$ with

$$\tilde{u}_i = \begin{cases} u_{k,i} - \theta_j \gamma_{ij} & \text{if } i \in I_+(u_k), \\ -\theta_j \gamma_{ij} & \text{if } i \in I(u_k) \backslash I_+(u_k), \\ \theta_j & \text{if } i = j, \\ 0 & \text{if } i \notin I(u_k) \end{cases} \tag{6.10}$$

is in $\mathbb{R}^n_+$, and hence $\tilde{u} \in \mathcal{F}$. Since the objective functional $J(u) = \sum_{i=1}^n a_i u_i$, we have

$$J(\tilde{u}_j) = J(u_k) + \theta_j \left(a_j - \sum_{i \in I(u_k)} a_i \gamma_{ij} \right).$$

If we can choose an index $j \notin I(u_k)$ and $\theta_j > 0$ such that $\tilde{u}_j \in \mathbb{R}^n_+$ and $J(\tilde{u}_j) < J(u_k)$, then we advance u_k to $u_{k+1} := \tilde{u}_j$. The existence of such a u_{k+1} is dependent on the following condition:

$$a_j - \sum_{i \in I(u_k)} a_i \gamma_{ij} < 0, \text{ and there exists } \theta_j > 0 \text{ such that } u_{k,i} - \theta_j \gamma_{ij} \geq 0 \ \forall i \in I(u_k)$$

for some $j \notin I(u_k)$. Otherwise, we have $a_j - \sum_{i \in I(u_k)} a_i \gamma_{ij} \geq 0$ for all $j \notin I(u_k)$. Then we look for the sign of $a_j - \sum_{i \in I(u_k)} a_i \gamma_{ij}$ with $j \in I(u_k)$.

Therefore, we can divide the situation into two cases. The first case is

$$a_j - \sum_{i \in I(u_k)} a_i \gamma_{ij} \geq 0 \quad \forall j \in I(u_k). \tag{6.11}$$

This is an easy case. In this case, we stop the iteration and use u_k for the solution of the linear programming problem since we have the following result.

Theorem 6.4. *Suppose that we have* (6.11). *Then* u_k *is a minimizer.*

Proof. For all $j \in I(u_k)$, let $\gamma_{ij} = \delta_{ij}$ so that we have $\mathbf{b}_j = \sum_{i \in I(u_k)} \gamma_{ij} \mathbf{b}_i$ for $j \in I(u_k)$. Thus, we have

$$a_j - \sum_{i \in I(u_k)} a_i \gamma_{ij} \geq 0 \quad \forall 1 \leq j \leq n. \tag{6.12}$$

For any $\mathbf{v} \in \mathcal{F}$, we have

$$\mathbf{d} = \sum_{j=1}^{n} v_j \mathbf{b}_j = \sum_{j=1}^{n} v_j \sum_{i \in I(u_k)} \gamma_{ij} \mathbf{b}_i = \sum_{i \in I(u_k)} \sum_{j=1}^{n} v_j \gamma_{ij} \mathbf{b}_i.$$

Since $\mathbf{d} = \sum_{i \in I_+(u_k)} u_{k,i} \mathbf{b}_i$ and $\mathbf{b}_i, i \in I(u_k)$, are linearly independent and form a basis for $\mathbb{R}^m$, we have $\sum_{j=1}^{n} v_j \gamma_{ij} = u_{k,i}$ for $i \in I(u_k)$. Now

$$J(\mathbf{v}) - J(u_k) = \sum_{j=1}^{n} a_j v_j - \sum_{i \in I(u_k)} a_i u_{k,i} = \sum_{j=1}^{n} v_j \left(a_j - \sum_{i \in I(u_k)} a_i \gamma_{ij} \right) \geq 0$$

since $v_j \geq 0, j = 1, \ldots, n$. Thus, $J(\mathbf{v}) \geq J(u_k)$ for all $\mathbf{v} \in \mathcal{F}$, or u_k is a global minimizer. $\quad\square$

The second case is that $a_j - \sum_{i \in I(u_k)} a_i \gamma_{ij} < 0$ for at least one $j \notin I(u_k)$. This case can be further divided into three subcases:

(1) For an index $j \notin I(u_k)$ with $a_j - \sum_{i \in I(u_k)} a_i \gamma_{ij} < 0$, we have $\gamma_{ij} \leq 0$ for all $i \in I(u_k)$. In this subcase, any $\theta_j > 0$ will make $\tilde{u} \in \mathcal{F}$ by (6.10), and hence the minimum of J is $-\infty$ since θ_j can be chosen as large as possible.

(2) For each $j \notin I(u_k)$ with $a_j - \sum_{i \in I(u_k)} a_i \gamma_{ij} < 0$, there is a $\gamma_{ij} > 0$ for $i \in I(u_k) \backslash I_+(u_k)$. Then no $\theta_j > 0$ can be chosen such that $\tilde{u}_j$ is in the feasible domain. This is called the cycling phenomenon. It is better to start with another initial vertex u_0 of the feasible domain $\mathcal{F}$ with $\mathbf{b}_i, i \in I_+(u_0)$, forming a basis for $\mathbb{R}^m$.

(3) For an index $j \notin I(u_k)$ with $a_j - \sum_{i \in I(u_k)} a_i \gamma_{ij} < 0$, we have all $\gamma_{ij} \leq 0$ for $i \in I(u_k) \backslash I_+(u_k)$ and at least one $\gamma_{ij} > 0$ for $i \in I_+(u_k)$. Let

$$\theta_j^+ = \min \left\{ \frac{u_{k,i}}{\gamma_{ij}}, \gamma_{ij} > 0, i \in I_+(u_k) \right\} > 0.$$

We choose $\theta_j = \theta_i^+$. Then $\tilde{u}_j \in \mathbb{R}_+^n$ by (6.10). In this subcase, we are sure that $J(\tilde{u}_j) < J(u_k)$. Furthermore, we can prove that $\tilde{u}_j$ is a vertex of $\mathcal{F}$. See Lemma 6.5 below. Hence, we let $u_{k+1} = \tilde{u}_j$.

Note that when $I(u_k) = I_+(u_k)$, we have either subcase (1) or subcase (3). Thus, the cycling subcase (2) will not happen when $\mathbf{b}_i, i \in I_+(u_k)$, span the whole space $\mathbb{R}^m$.

Lemma 6.5. *Let $\tilde{u}$ be the vector defined in (6.10) above. Then $\mathbf{b}_i, i \in I_+(\tilde{u})$, are linearly independent.*

Proof. It is clear that the cardinality of $I_+(\tilde{u})$ is less than or equal to the cardinality of $I(u_k)$. Indeed, let i_0 be the index such that

$$\frac{u_{k,i_0}}{\gamma_{i_0,j}} = \theta_j^+.$$

Then $u_{k,i_0} - \theta_j^+ \gamma_{i_0,j} = 0$, and hence $I_+(\tilde{u})$ does not include i_0. However, $I_+(\tilde{u})$ includes j. Because $\mathbf{b}_j$ is a linear combination of $\mathbf{b}_i, i \in I(u_k)$, we know $\mathbf{b}_i, i \in I_+(\tilde{u}_j)$, are linearly independent. By Theorem 6.1, $\tilde{u}_j$ is a vertex of the feasible domain $\mathcal{F}$. $\quad\square$

The discussion above provides the key ingredients for the simplex method. However, the number of vertices increases exponentially as the number of constraints and variables m, n increases moderately. Thus, a linear programming problem needs nonpolynomial time to solve in general. The cycling phenomenon above cannot be avoided. Nevertheless, the discussion above provides a framework for linear programming algorithms based on which there are many speeding-up techniques available in the literature and in practice. They have been tested successfully in practice millions of times although the explanation in this section does not provide an efficient algorithm.

MATLAB code to solve (1.20) or (6.1) using the simplex method is called magicl1 and can be found at Professor Emmanuel Candés's website. A comparison of the performance for recovery of sparse solutions with other methods will be given at the end of this chapter.

6.2 ▪ An Interior Point Method

We first present an algorithm based on the interior point method, i.e., the Karmarkar algorithm for linear programming. It is a variation of the interior point method that we use to find the sparse solution. It starts with an interior point which is feasible. In fact, each iteration is a feasible solution. Then we use the gradient method to search around to find a better solution and then repeat the iteration until a stopping criterion is met. This can be compared with the standard simplex method, which computes the vertices of the feasible set. Although this may not be an efficient algorithm, it is very easy to implement and its performance of sparse recovery is excellent. So we present it here.

To motivate this method, we consider, by using the Lagrange multiplier method,

$$L(\mathbf{x}, \lambda) = \|\mathbf{x}\|_1 + \lambda^\top (\Phi \mathbf{x} - \mathbf{b}).$$

The gradient of $L(\mathbf{x}, \lambda)$ is $\mathrm{diag}(\mathrm{sign}(x_1), \dots, \mathrm{sign}(x_n)) + \Phi^\top \lambda$. Thus, we have two systems of linear equations to solve. That is, we let Sign be a matrix operator such that $\mathrm{Sign}(\mathbf{x}) = \mathrm{diag}(\mathrm{sign}(x_1), \dots, \mathrm{sign}(x_n))$ and Z be a zero of matrix of size $m \times m$,

$$\begin{bmatrix} \mathrm{Sign} & \Phi^\top \\ \Phi & Z \end{bmatrix} \begin{bmatrix} \mathbf{x} \\ \lambda \end{bmatrix} = \begin{bmatrix} \mathbf{z} \\ \mathbf{b} \end{bmatrix},$$

where $\mathbf{z}$ is a zero vector of size $n \times 1$. Discretizing the matrix operator Sign is an art. Following the ideas of [17], we present our algorithm as follows.

ALGORITHM 6.1
An Interior Point Algorithm

1: Step 1. Fix a tolerance $\epsilon > 0$. Start with a vector $\mathbf{x}^0$ satisfying $\Phi \mathbf{x}^0 = \mathbf{b}$ and $\alpha_0 = \|\mathbf{x}^0\|_\infty$. Let $\mathbf{w}^0 := (w_1, \dots, w_n) = \frac{2}{3\alpha_0} \mathbf{x}^0$.

2: Step 2. While $k = 1, 2, \dots$, let D_k be the diagonal matrix whose ith entry is $1 - |w_i|$, where $\mathbf{w}^{k-1} = (w_1, w_2, \dots, w_n)$. Let

$$\mathbf{A} = \begin{bmatrix} D_k^2 & \Phi^\top \\ \Phi & Z \end{bmatrix} \quad \text{and} \quad B = \begin{bmatrix} \mathbf{z} \\ \mathbf{b} \end{bmatrix}, \tag{6.13}$$

where Z is a zero matrix of size $m \times m$ and $\mathbf{z}$ is a zero vector of size $n \times 1$.

3: 2.1 Compute $\mathbf{y}^k = \mathbf{A} \backslash B$ using the pseudo-inverse and let $\mathbf{x}^k$ be the vector consisting of the first n components of $\mathbf{y}^k$.

4: 2.2 Compute $\mathbf{p}^k = D_k^2 \mathbf{x}^k$. Writing $\mathbf{p}^k = (p_1, \dots, p_n)$, let

$$\alpha_k = \max_i \left\{ \max \left(\frac{p_i}{1 - w_i}, -\frac{p_i}{1 + w_i} \right) \right\}.$$

5: 2.3 Compute $\mathbf{w}^k = \mathbf{w}^{k-1} + \frac{2}{3\alpha_k} \mathbf{p}^k$.

6: Step 3. If $\|\mathbf{x}^k\|_1 - (\mathbf{x}^k)^\top \mathbf{w}^k \leq \epsilon$, then stop, and $\mathbf{x}^k$ is an approximation of the ℓ_1 norm minimizer.

7: Step 4. Otherwise, go back to Step 2.

We now explain why Algorithm 6.1 works. Note that in Step 2.1 we find $\mathbf{x}^k$ satisfying $\Phi\mathbf{x}^k = \mathbf{b}$. Indeed, writing $\mathbf{y}^k = [\mathbf{x}^k; \lambda^k]^\top$, we have $\mathbf{A}\mathbf{y}^k = B$, and hence $\Phi\mathbf{x}^k = \mathbf{b}$. Thus, we have the following result.

Lemma 6.6. $\mathbf{x}^k$ *is feasible for all k.*

Next we have the following lemma.

Lemma 6.7. $\mathbf{w}^k$ *satisfies* $-1 \leq (\mathbf{w}^k)_i \leq 1$ *for all i. Hence,* $(\mathbf{w}^k)^\top \mathbf{x}^k \leq \|\mathbf{x}^k\|_1$ *for all k.*

Proof. We use induction. Clearly, $\mathbf{w}^0$ satisfies this condition. Assume that $\mathbf{w}^{k-1}$ satisfies it as well. Writing $\mathbf{g} = (g_1, \ldots, g_m)^\top = D_k\mathbf{x}_k$, we write $\mathbf{p}^k = (p_1, \ldots, p_n)$ with $p_i = \min(1 + w_i, 1 - w_i)g_i$, where $(w_1, w_2, \ldots, w_n) = \mathbf{w}^{k-1}$. Then

$$
\begin{aligned}
\alpha_k &= \max_i \left[\max \left(\frac{\min(1 + w_i, 1 - w_i)g_i}{1 - w_i}, -\frac{\min(1 + w_i, 1 - w_i)g_i}{1 + w_i} \right) \right] \\
&= \max_i \left[|g_i| \max \left(\frac{\min(1 + w_i, 1 - w_i)}{1 - w_i}, \frac{\min(1 + w_i, 1 - w_i)}{1 + w_i} \right) \right] \\
&= \max_i |g_i| \leq \|\mathbf{g}\|_2 = \|D_k\mathbf{x}^k\|_2.
\end{aligned}
$$

Thus, the ith component of $\frac{2\mathbf{p}^k}{3\alpha_k}$ is

$$
\frac{2}{3} \frac{(1 - |w_i|)g_i}{\|\mathbf{g}\|_\infty},
$$

and the ith component of $\mathbf{w}^k$ is $w_i + \frac{2}{3}\frac{(1-|w_i|)g_i}{\|\mathbf{g}\|_\infty}$, which is shown below to satisfy

$$
w_i + \frac{2}{3}\frac{(1 - |w_i|)g_i}{\|\mathbf{g}\|_\infty} \leq w_i + \frac{2}{3}(1 - |w_i|) \leq \frac{2}{3} + \frac{1}{3}|w_i| \leq 1
$$

and

$$
w_i + \frac{2}{3}\frac{(1 - |w_i|)g_i}{\|\mathbf{g}\|_\infty} \geq w_i - \frac{2}{3}(1 - |w_i|) \geq -\frac{2}{3} - \frac{1}{3}|w_i| \geq -1.
$$

Thus, we have $-1 \leq (\mathbf{w}^k)_i \leq 1$ for $i = 1, 2, \ldots, n$.   □

As in the proof above, we actually have that α_k is bounded above.

Lemma 6.8. *For all k,*

$$
\alpha_k \leq \|D_k\mathbf{x}^k\|_2 \leq \|\mathbf{x}^k\|_2.
$$

The boundedness of $\mathbf{w}^k$ in Lemma 6.7 implies that there exists a subsequence which converges. For simplicity, let us assume that the whole sequence $\mathbf{w}^k$ converges to $\mathbf{w}^*$. (It is not the case that the whole sequence converges. This is the point that one is not able to find sparse solutions for all underdetermined linear systems by this algorithm.) Then D_k converges and so does $\mathbf{x}^k$ by Steps 2.1 and 2.2. Let $\mathbf{x}^*$ be the limit of the sequence $\{\mathbf{x}^k, k \geq 1\}$. Note that if $-1 < (\mathbf{w}^*)_i < 1$, then $(\mathbf{x}^*)_i = 0$ by Step 2.3. If $(\mathbf{w}^*)_i = 1$, we claim that $(\mathbf{x}^*)_i \geq 0$. Otherwise, $(\mathbf{x}^*)_i < 0$, there exists $k > 0$ such that $(\mathbf{x}^k)_i < 0$, and we have

$$(\mathbf{w}^k)_i - (\mathbf{w}^{k-1})_i = \frac{2(1 - |(\mathbf{w}^{k-1})_i|)^2}{3\alpha_k}(\mathbf{x}^k)_i < 0$$

for all k sufficiently large. This contradicts the fact that $(\mathbf{w}^k)_i \to 1$. Similarly, if $(\mathbf{w}^*)_i = -1$, then $(\mathbf{x}^*)_i \leq 0$. Thus, it follows that

$$(\mathbf{w}^*)^\top \mathbf{x}^* = \|\mathbf{x}^*\|_1.$$

By Lemma 6.7, we have $(\mathbf{w}^k)^\top \mathbf{x}^k \leq \|\mathbf{x}^k\|_1$. We expect that $(\mathbf{w}^k)^\top \mathbf{x}^k \to (\mathbf{w}^*)^\top \mathbf{x}^* = \|\mathbf{x}^*\|_1$. This is the rationale for the stopping criterion in Algorithm 6.1 above.

Finally, let us present our MATLAB implementation of Algorithm 6.1 and some numerical experimental results.

```
function x=L1min(C,b,tol,x0)
%      x=L1min(C,b,tol,x0)
% This function min ||x||_1, subject to Cx=b.
% tol=, e.g., 1e-4 and x0 is an initial guess  satisfying Cx0=b.
% It is written by Dr. Ming-Jun Lai on June 1, 2008
% based on an interior point method for linear programming described in
% the paper:  M. J. Lai and P. Wenston,  $L^1$ Spline Methods for
% Scattered Data Interpolation and Approximation, Advances in Computational
% Mathematics, vol. 21 (2004) pp. 293--315.
[k,m]=size(C);
x=x0;
alpha=norm(b,inf);
w=(2/(3*alpha))*x;
it_count=0;
max_it=25;
%The above number may be adjusted for the accuracy and the problem size.
Z=zeros(m,1);
cvg=0;
while ~cvg & it_count <=max_it
D=spdiags(1-abs(w),0,m,m);
xnew=[[(D)'*(D),C'];[C,sparse(k,k)];sparse(1,m+k)]\[Z;b;0];
xnew=xnew(1:m);
x=xnew;
p=D^2*x;
alpha=max(max(p'./(1-w'),-p'./(1+w')));
w=w+(2/(3*alpha))*p;
err=norm(x,1)-w'*x;
cvg=err<tol;
it_count=it_count+1;
end;
```

Let us present a numerical experimental result. Consider a matrix Φ of size 64×128 with uniform random variables as its entries. Let $\mathbf{x}_\mathbf{b}$ be a vector of sparsity s with nonzero entries which are uniform random values. For $\mathbf{b} = \Phi\mathbf{x}_\mathbf{b}$, we use the above MATLAB code to solve $\mathbf{x}^*$ and measure the maximum norm. If the maximum norm $\|\mathbf{x}_\mathbf{b} - \mathbf{x}^*\|_\infty < 10^{-3}$, we view it as a

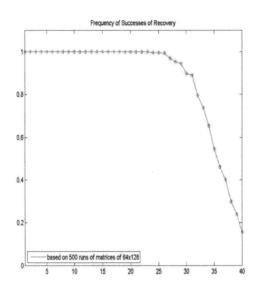

Figure 6.1. *Numerical results based on Algorithm* 6.1.

successful recovery. Otherwise, we say that the algorithm does not find the solution. We perform 500 independent runs of this recovery for sparsity $s = 1, \ldots, 40$. The percentage of recovery (or frequency of successes of recovery) is shown in Figure 6.1.

A detailed MATLAB program is given below.

```
%This is a test code written by Dr. Ming-Jun Lai on March 5, 2009.
recovery=zeros(40,1);
N=500; %Total number of independent runs.
for j=1:40
for i=1:N
A=rand(64,128);
x=zeros(128,1);
Ind=randperm(128);
x(Ind(1:j))=rand(j,1);
b=A*x;
x0=pinv(A)*b;
xnew=L1min(A,b,1e-5,x0);
e=norm(xnew-x,inf);
if e<=1e-3
recovery(j)=recovery(j)+1;
end
end
[j, recovery(j)]
end
plot(recoveryt/N)
axis([1 40 0 1.1])
hold on
plot(recovery/N,'r*')
title('Frequency of Successes of Recovery')
legend('based on 500 runs of matrices of 64x128', 'Location','SouthWest')
```

Similarly, we tested matrices with entries consisting of Gaussian random variables. That is, replace A=rand(64,128) by A=randn(64,128)/8 in the above. We obtained similar numerical results. The interested reader is encouraged to carry out the experiment and see the performance.

6.3 ▪ Projected Gradient Descent Methods

We are interested in solving the minimization problem

$$\min\{\theta(\mathbf{x}) + g(\mathbf{x}), \mathbf{x} \in C\}, \tag{6.14}$$

where C is a convex set in $\mathbb{R}^n$, $g(\mathbf{x})$ is convex, and so is the regularizer $\theta(\mathbf{x})$, e.g., $g(\mathbf{x}) \equiv 0$, $\theta = \|\mathbf{x}\|_1$, and $C = \{\mathbf{x} \in \mathbb{R}^n, \Phi\mathbf{x} = \mathbf{b}\}$ for a given matrix Φ of size $m \times n$ and a given vector $\mathbf{b} \in \mathbb{R}^m$, which is our problem (1.20) or (6.1). For another example, let $C = \mathbb{R}^n_+ \cap \{\mathbf{x} \in \mathbb{R}^n, \Phi\mathbf{x} = \mathbf{b}\}$, the set of nonnegative solutions for $\Phi\mathbf{x} = \mathbf{b}$, where $\mathbb{R}^n_+$ is the positive octant of the Euclidean space $\mathbb{R}^n$.

Let us start by recalling some useful notation and definitions. Since g is Lipschitz differentiable, let $\nabla g(\mathbf{x})$ be the gradient of $g(\mathbf{x})$. It is known that problem (6.14) has a solution $\mathbf{x}^* \in C$. Then there exists a subdifferentiable $\partial\theta(\mathbf{x}^*)$ such that

$$\langle \mathbf{x} - \mathbf{x}^*, \partial\theta(\mathbf{x}^*) + \nabla g(\mathbf{x}^*)\rangle \geq 0 \quad \forall \mathbf{x} \in C. \tag{6.15}$$

Let us review a common strategy for deriving an algorithm to approximate $\mathbf{x}^*$. To this end, we suppose that we have $\mathbf{x}^k \in C$ and look for $\mathbf{x}^{k+1}$. Recall that $\mathcal{P}_C(\mathbf{x})$ is the projection of $\mathbf{x} \in \mathbb{R}^n$ onto C defined by

$$\mathcal{P}_C(\mathbf{x}) := \arg\min\{\|\mathbf{x} - \mathbf{z}\|_2 : \quad \mathbf{z} \in C\}. \tag{6.16}$$

We consider a new minimization problem to approximate the minimizer in (6.14) at the kth step:

$$\mathbf{x}^{(k+1)} := \arg\min_{\mathbf{x}\in C}\left\{\theta(\mathbf{x}) + \langle\nabla g(\mathbf{x}^{(k)}), \mathbf{x} - \mathbf{x}^k\rangle + \frac{1}{2\beta}\|\mathbf{x} - \mathbf{x}^k\|^2\right\}, \tag{6.17}$$

where $\beta \geq 1/L > 0$. Since $\theta(\mathbf{x}) + g(\mathbf{x})$ is convex and the constraint set C is convex and nonempty, problem (6.17) has a unique solution because of its strong convexity, say $\mathbf{x}^{k+1}$, which satisfies

$$\left\langle \mathbf{x} - \mathbf{x}^{k+1}, \frac{1}{\beta}(\mathbf{x}^{k+1} - \mathbf{x}^k) + \partial\theta(\mathbf{x}^{k+1}) + \nabla g(\mathbf{x}^k)\right\rangle \geq 0 \quad \forall \mathbf{x} \in C, \tag{6.18}$$

where $\partial\theta(\mathbf{x}^{k+1})$ stands for the subderivative of θ at $\mathbf{x}^{k+1}$. In fact, $\mathbf{x}^{(k+1)}$ can be found by using the projection method. That is, let $\mathbf{z}^k$ be the global minimizer of

$$\min_{\mathbf{x}\in\mathbb{R}^n}\left\{\theta(\mathbf{x}) + \langle\nabla g(\mathbf{x}^k), \mathbf{x} - \mathbf{x}^k\rangle + \frac{1}{2\beta}\|\mathbf{x} - \mathbf{x}^k\|^2\right\}. \tag{6.19}$$

Then we know

$$\mathbf{x}^{(k+1)} := \mathcal{P}_C(\mathbf{z}^k). \tag{6.20}$$

This approximating minimization is called the projected proximal gradient method and has been used for many years. See, e.g., [22].

Let us explain how to solve (6.17) when $\theta(\mathbf{x}) = \lambda\|\mathbf{x}\|_1$ and $g(\mathbf{x}) \equiv 0$ for our purposes. Letting $\psi_i = \Phi(i, :)$ be the ith row of Φ and $\mathbf{b} = [b_1, \ldots, b_m]^\top$, we define

$$H_i = \{\mathbf{x} \in \mathbb{R}^m : \quad \langle\mathbf{x}, \phi_i\rangle = b_i\} \tag{6.21}$$

for $i = 1, \ldots, m$. The feasible set $C = \{\mathbf{x} : \Phi\mathbf{x} = \mathbf{b}\} = \bigcap_{i=1}^{m} H_i$. Note that each H_i is a hyperplane. Recall some basic properties of hyperplanes and their projections.

Definition 6.9 (Hyperplanes). *A hyperplane in $\mathbb{R}^n$ is any set of the form*

$$H = \{\mathbf{x} \in \mathbb{R}^n : \quad \langle \mathbf{x}, \mathbf{h} \rangle = c\} \tag{6.22}$$

for a vector $\mathbf{h} \neq 0$ and $c \in \mathbb{R}$.

Note that a hyperplane H is never empty. Indeed, since $\mathbf{h} \neq 0$, there exists a vector $\mathbf{y}$ such that $\langle \mathbf{y}, \mathbf{h} \rangle = d \neq 0$. It follows that $\mathbf{x} = c\mathbf{y}/d \in \mathbb{R}^n$.

Distance to Hyperplane Let H be a hyperplane as given in (6.22). Then the distance of any point $\mathbf{p} \in X$ to the hyperplane H is

$$\text{dist}(\mathbf{p}, H) = \frac{1}{\|\mathbf{h}\|} |\langle \mathbf{p}, \mathbf{h} \rangle - c|. \tag{6.23}$$

Let us find the projection operator $\mathcal{P}_H$ as follows. For any $\mathbf{y} \in H$, let

$$\mathcal{P}_H(\mathbf{y}) = \mathbf{y} - \mathbf{h}\frac{\langle \mathbf{y}, \mathbf{h} \rangle - c}{\|\mathbf{h}\|^2}, \tag{6.24}$$

which is the projection of $\mathbf{y}$ onto H. Indeed, we can easily check that $\|\mathbf{y} - \mathcal{P}_H(\mathbf{y})\| = \text{dist}(\mathbf{y}, H)$.

To project any vector $\mathbf{x}$ to the intersection of finitely many hyperplanes, we can use the alternating projection method described in Algorithm 6.2 (cf. [9]).

ALGORITHM 6.2
Method of Alternating Projection

  1: Let $\mathbf{x}_1 = \mathbf{x}$.
  2: For $k = 1, 2, \ldots,$
  3:        For $i = 1, \ldots, m$, $\mathbf{x}_{k+i/m} = \mathcal{P}_{H_i}(\mathbf{x}_{k+(i-1)/m})$.
  4: until $\mathbf{x}_{k+1}$ is very close to $\mathbf{x}_k$.
  5: Output $\mathbf{x}_{k+1}$, which is an approximation of the projection of $\mathbf{x}$ on C.

The algorithm above will converge, and its convergence rate is linear and depends on the angles between hyperplanes. See section 4.5 and [9] for details.

On the other hand, if $\Phi\Phi^\top$ is of full rank, the projection $\mathcal{P}_C$ has an explicit formula, where $C = \{\mathbf{x} \in \mathbb{R}^n, \Phi\mathbf{x} = \mathbf{b}\}$. Indeed, we have the following result.

Theorem 6.10. *For any $\mathbf{p} \in \mathbb{R}^n$, the projection $\mathcal{P}_C(\mathbf{p})$ is given by*

$$\mathcal{P}_C(\mathbf{p}) = \mathbf{p} - \Phi^\top(\Phi\Phi^\top)^{-1}(\Phi\mathbf{p} - \mathbf{b}) \tag{6.25}$$

if Φ is of full row rank. If Φ is not of full row rank, we write $\Phi = U\Sigma V^\top$ and obtain

$$\mathcal{P}_C(\mathbf{p}) = \mathbf{p} - \Phi^\top U(\Sigma_+^{-2})U^\top(\Phi\mathbf{p} - \mathbf{b}). \tag{6.26}$$

Proof. It is easy to see that $\mathcal{P}_C(\mathbf{p}) \in C$ by a straightforward calculation of $\Phi(\mathcal{P}_C(\mathbf{p})) = \mathbf{b}$. Next let $\text{Null}(A) = \{\mathbf{x} \in \mathbb{R}^n : \Phi\mathbf{x} = 0\}$ be the null space of Φ. For any nonzero $\mathbf{x} \in \text{Null}(\Phi)$ and $\alpha \in \mathbb{R}$, we know that $\mathbf{p}^* + \alpha\mathbf{x} \in C$. Now we claim that

$$\|\mathbf{p} - \mathbf{p}^*\| = \min_{\alpha} \|\mathbf{p} - (\mathbf{p}^* + \alpha\mathbf{x})\|,$$

where $\mathbf{p}^*$ is the vector on the right-hand side of (6.25). Indeed, we can easily verify that

$$\langle \mathbf{p} - \mathbf{p}^*, \mathbf{x} \rangle = 0,$$

since $\mathbf{x} \in \text{Null}(A)$. Therefore, we have the claim. □

In the rest of this section we may assume that the minimizer $\mathbf{x}^{k+1}$ satisfying (6.18) can be found. Then we can present the projected proximal gradient method in Algorithm 6.3.

ALGORITHM 6.3
Projected Proximal Gradient Algorithm

1: Fix $\beta \leq 1/L$, the Lipschitz constant of ∇g. Suppose that we are given $\mathbf{x}^{(1)} \in C$.
2: For $k \geq 1$, we solve the minimization in (6.17) and let the minimizer be $\mathbf{x}^{k+1}$. That is,

$$\mathbf{x}^{k+1} := \arg\min_{\mathbf{x} \in C} \left\{ \theta(\mathbf{x}) + \langle \nabla f(\mathbf{x}^k), \mathbf{x} - \mathbf{x}^k \rangle + \frac{1}{2\beta} \|\mathbf{x} - \mathbf{x}^k\|^2 \right\}. \tag{6.27}$$

3: We repeat the above computation until a maximum number of iterations is achieved.

We shall show the convergence of Algorithm 6.3.

Theorem 6.11 (Convergence). *Let $f(\mathbf{x}) = g(\mathbf{x}) + \theta(\mathbf{x})$ be the minimizing functional in (6.14). Suppose that ∇f is L-Lipschitz continuous over C. Let $\mathbf{x}^{(k)}, k \geq 1$, be the iterative solutions of Algorithm 6.3. Let $\eta_k = F(\mathbf{x}^{(k)}) - F(\mathbf{x}^*)$, where $\mathbf{x}^*$ is a minimizer of the minimization problem (6.14). Then*

$$\eta_k = o(1/k), \quad k \to \infty. \tag{6.28}$$

That is, $k\eta_k \to 0$ as $k \to \infty$.

The proof is similar to that of Theorem 5.30. We leave the proof to Exercise 13.

We next extend Nesterov's accelerated gradient method for unconstrained minimization (cf. [20], [23], [22], [21]) to our constrained minimization problem (6.14) to speed up the computation.

ALGORITHM 6.4
Accelerated Proximal Gradient Algorithm

1: Fix $\beta \leq 1/L$, the Lipschitz constant of g.
2: Let $\mathbf{x}^{(1)}, \mathbf{y}^{(1)}$ be two initial guesses in the convex set C and $a_1 = 1$.
3: Compute the following minimization and iterative steps:

$$\mathbf{x}^{(k+1)} = \arg\min \left\{ \theta(\mathbf{x}) + \langle \nabla g(\mathbf{y}^{(k)}), \mathbf{x} - \mathbf{y}^{(k)} \rangle + \frac{1}{2\beta} \|\mathbf{x} - \mathbf{y}^{(k)}\|^2, \text{ subject to } \mathbf{x} \in C \right\}. \tag{6.29}$$

4: Let

$$a_{k+1} = \frac{1}{2} \left(1 + \sqrt{4a_k^2 + 1} \right) \tag{6.30}$$

and define

$$\mathbf{y}^{(k+1)} = \mathcal{P}_C \left(\mathbf{x}^{(k+1)} + \frac{a_{k+1} - 1}{a_{k+1}} (\mathbf{x}^{(k+1)} - \mathbf{x}^{(k)}) \right) \tag{6.31}$$

for all $k \geq 1$ until a given number of iterations is achieved.

Then the convergence rate of these $\mathbf{x}^k, k \geq 1$, is accelerated as indicated in the following result.

Theorem 6.12. *Suppose that* $\mathbf{x}^{(k)}, k \geq 1$, *is a sequence from Algorithm* 6.4. *Let* $\eta_k = f(\mathbf{x}^{(k)}) - f(\mathbf{x}^*)$. *Then*

$$\eta_k \leq \frac{A}{k^2} \tag{6.32}$$

for all $k \geq 1$, *where* $A > 0$ *is a constant.*

In fact, the parameters in (6.30) can be replaced by

$$\alpha_k = \frac{k}{2}, \quad k \geq 2, \tag{6.33}$$

and the update rule in (6.31) may be simplified as

$$\mathbf{y}^{(k+1)} = \mathcal{P}_C \left(\mathbf{x}^{(k+1)} + \frac{k-1}{k+1}(\mathbf{x}^{(k+1)} - \mathbf{x}^{(k)}) \right). \tag{6.34}$$

Then the new algorithm is also convergent with convergence rate $O(1/k^2)$. We leave the proof to Exercise 14.

Furthermore, following a recent development (cf. [1]), we should simply replace $\frac{a_{k+1}-1}{a_{k+1}}$ by $\frac{k-1}{k+\alpha}$ for $\alpha > 2$ in (6.31) to have an algorithm with an even faster rate. See Algorithm 6.5.

ALGORITHM 6.5

Attouch–Peypouquet Accelerated Proximal Gradient Algorithm

1: Fix $\beta \leq 1/L$, the Lipschitz constant of $g = \nabla f$.
2: Let $\mathbf{x}^{(1)}, \mathbf{y}^{(1)}$ be two initial guesses in the convex set C.
3: Compute the following minimization and iterative steps:

$$\mathbf{x}^{(k+1)} = \arg\min \left\{ \theta(\mathbf{x}) + \langle \nabla g(\mathbf{y}^{(k)}), \mathbf{x} - \mathbf{y}^{(k)} \rangle + \frac{1}{2\beta}\|\mathbf{x} - \mathbf{y}^{(k)}\|^2, \text{ subject to } \mathbf{x} \in C \right\} \tag{6.35}$$

and define

$$\mathbf{y}^{(k+1)} = \mathcal{P}_C \left(\mathbf{x}^{(k+1)} + \frac{k+1}{k+1+\alpha}(\mathbf{x}^{(k+1)} - \mathbf{x}^{(k)}) \right) \tag{6.36}$$

for all $k \geq 1$ until a given number of iterations is achieved, where $\alpha > 2$.

We are able to prove the following theorem.

Theorem 6.13. *Suppose that* $\mathbf{x}^{(k)}, k \geq 1$, *is a sequence from Algorithm* 6.5. *Let* $\eta_k = F(\mathbf{x}^{(k)}) - F(\mathbf{x}^*)$. *Then*

$$\eta_k = o\left(\frac{1}{k^2}\right), \quad i.e., \quad k^2\eta_k \to 0, \tag{6.37}$$

for $k \to \infty$.

The proof of the results (6.32) and (6.37) is based on extending the ideas in [2] and [1] from the unconstrained minimization setting to the constrained minimization setting. The counterparts in the previous chapter will be useful to establish the above results. In particular, we need to use inequality (5.49). We leave the details to the interested reader.

Let us return to our main problem in this book and solve the following ℓ_1 minimization instead:

$$\min\{\|\mathbf{x}\|_1 \quad \text{subject to } \Phi\mathbf{x} = \mathbf{b}\}. \tag{6.38}$$

One approach is to rewrite the minimization in the form

$$\min\left\{\|\mathbf{x}\|_1 + \frac{1}{2\lambda}\|\Phi\mathbf{x} - \mathbf{b}\|^2 \quad \text{subject to } \Phi\mathbf{x} = \mathbf{b}\right\}. \tag{6.39}$$

Then the above problem is the constrained minimization (6.14) with $F(\mathbf{x}) = \lambda\|\mathbf{x}\|_1 + \frac{1}{2}\|\Phi\mathbf{x} - \mathbf{b}\|^2$ and convex set $C = \{\mathbf{x} \in \mathbb{R}^n : \Phi\mathbf{x} = \mathbf{b}\}$. We can solve it using our Algorithm 6.4 or Algorithm 6.5. Note that in this case, we have $\mathbf{y}_{k+1} = \mathbf{x}^{(k+1)} + \frac{\alpha_k-1}{\alpha_k}(\mathbf{x}^{(k)} - \mathbf{x}^{(k+1)})$ in C without using any projection. We leave the implementation to the interested reader.

6.4 ▪ Nonnegative Sparse Solution

ALGORITHM 6.6

Half-Shrinkage Algorithm

1: Fix $\beta \le 1/L$, the Lipschitz constant of g. Suppose that we are given $\mathbf{x}^{(1)} \in C$.
2: For $k \ge 1$, we solve the minimization in (6.17) and let the minimizer be $\mathbf{x}^{k+1}$. That is,

$$\mathbf{x}^{k+1} = H_\lambda\left(\mathbf{x}^k - \frac{1}{\beta}A^\top(\Phi\mathbf{x}^k - \mathbf{b})\right). \tag{6.40}$$

3: We repeat the above computation until a maximum number of iterations is achieved.

Next we consider the problem

$$\min\left\{\lambda\|\mathbf{x}\|_1 + \frac{1}{2}\|\Phi\mathbf{x} - \mathbf{b}\|_2^2 : \quad \mathbf{x} \in \mathbb{R}_+^n\right\}. \tag{6.41}$$

See the motivation and computation in [12] and [15]. This minimization fits into our model (6.14) since $g(\mathbf{x}) = \frac{1}{2}\|\Phi\mathbf{x} - \mathbf{b}\|^2$ is clearly Lipschitz differentiable with $L = \|\Phi^\top\Phi\|$ and $C = \mathbb{R}_+^n$ is a convex set. Now we see that the solution of (6.17) can be solved rather easily using the half-shrinkage operator $H_\lambda(\mathbf{x})$:

$$H_\lambda(x) = \max\{x, \lambda\} \text{ for } x \in \mathbb{R} \text{ and } H_\lambda(\mathbf{x}) = (H_\lambda(x_1), \dots, H_\lambda(x_n))^\top. \tag{6.42}$$

Thus, the computation of (6.17) is

$$\mathbf{x}^{k+1} = \arg\min_{\mathbf{x}\in\mathbb{R}_+^n} \lambda\|\mathbf{x}\|_1 + \frac{1}{2}\left\|\mathbf{x} - \left(\mathbf{x}^k - \frac{1}{\beta}A^\top(\Phi\mathbf{x}^k - \mathbf{b})\right)\right\|^2 = H_\lambda\left(\mathbf{x}^k - \frac{1}{\beta}A^\top(\Phi\mathbf{x}^k - \mathbf{b})\right).$$

The computation for (6.29) is similar. In terms of the half-shrinkage operator, Algorithms 6.3 and 6.4 can be reformulated as Algorithm 6.7.

ALGORITHM 6.7

Accelerated Half-Shrinkage Algorithm

1: Fix $\beta \le 1/L$, the Lipschitz constant of g.
2: Let $\mathbf{x}^{(1)}, \mathbf{y}^{(1)}$ be two initial guesses in the convex set C and $a_1 = 1$.
3: Compute the following minimization and iterative steps:

$$\mathbf{x}^{(k+1)} = H_\lambda\left(\mathbf{y}^k - \frac{1}{\beta}A^\top(A\mathbf{y}^k - \mathbf{b})\right), \tag{6.43}$$

let

$$a_{k+1} = \left(1 + \sqrt{4a_k^2 + 1}\right)/2,$$

and define

$$\mathbf{y}^{(k+1)} = \left(\mathbf{x}^{(k+1)} + \frac{a_k - 1}{a_{k+1}}(\mathbf{x}^{(k+1)} - \mathbf{x}^{(k)})\right)_+$$

for all $k \geq 1$ until a given number of iterations is achieved.

Then, as before, we can establish the convergence for Algorithms 6.6 and 6.7. The convergences will be $o(1/k)$ and $O(1/k^2)$, respectively.

Our method can be compared with the modified hard thresholding algorithm for (6.41). We leave the implementation and numerical study to the interested reader.

6.5 ▪ A Scheme Based on Uzawa's Method

In this section, we derive a numerical scheme based on the ideas of the proof of the convergence of the Uzawa iterative algorithm to find the minimizers of a convex minimization subject to some inequality constraints. We mainly consider the minimization

$$\min_{\mathbf{x} \in \mathbb{R}^n} \left\{ \|\mathbf{x}\|_1 + \frac{1}{2\alpha}\|\mathbf{x}\|_2^2, \quad \Phi\mathbf{x} = \mathbf{b} \right\}, \tag{6.44}$$

where $\alpha > 0$ is a parameter. Usually, we choose α large enough so that the above minimization approximates the minimization (1.20). However, we can show that as long as $\alpha \geq 10\|\mathbf{x_b}\|_\infty$, the minimizer will be the sparse solution when Φ satisfies some RIP conditions.

It is easy to understand that minimization (6.44) has a unique minimizer since the minimizing functional is strictly convex when $\alpha < \infty$. Let us first provide a numerical scheme to compute the minimizer. We let

$$L(\mathbf{x}, \lambda) = \|\mathbf{x}\|_1 + \frac{1}{2\alpha}\|\mathbf{x}\|_2^2 + \lambda^\top(\mathbf{b} - \Phi\mathbf{x})$$

for a parameter vector λ and use the well-known Uzawa algorithm (cf. [6]) to obtain Algorithm 6.8.

ALGORITHM 6.8
The Uzawa Algorithm [6]

1: Fix a tolerance $\epsilon > 0$. Initially set $\lambda^0 = 0$ and choose $\rho \in (0, 2/(\alpha\|\Phi\Phi^\top\|))$.
2: We compute

$$\mathbf{x}^{(k+1)} = \arg\min_{\mathbf{x} \in \mathbb{R}^n} L(\mathbf{x}, \lambda_k) \tag{6.45}$$

and

$$\lambda_{k+1} = \lambda_k + \rho(\mathbf{b} - \Phi\mathbf{x}^{(k)}) \tag{6.46}$$

for $k = 0, 1, 2, \ldots$ until $|\lambda_{k+1} - \lambda_k| \leq \epsilon$.

To find the minimizer $\mathbf{x}^{(k+1)}$ in (6.45), we use the subdifferential to find

$$0 \in \partial L(\mathbf{x}, \lambda) = \sum_{i=1}^n \partial \left(|x_i| + \frac{1}{2\alpha}x_i^2 - \lambda^\top \phi_i x_i \right),$$

where $\Phi = [\phi_1, \phi_2, \ldots, \phi_n]$. It follows that

$$\mathbf{x}^{(k+1)} = \alpha \, \text{Shrinkage}(\Phi^\top \lambda_k, 1),$$

where $\text{Shrinkage}(\mathbf{y}, 1) = [\text{Shrinkage}(y_1, 1), \ldots, \text{Shrinkage}(y_n, 1)]^\top$ with

$$\text{Shrinkage}(y, 1) = \begin{cases} y - 1 & \text{if } y > 1, \\ 0 & \text{if } -1 \le y \le 1, \\ y + 1 & \text{if } y < -1. \end{cases}$$

This leads to the following new version of the Uzawa algorithm.

ALGORITHM 6.9
Jia, Zhang, and Zhang's Version of the Uzawa Algorithm, 2009 [14]
  1: Initially set $\lambda_0 = 0$ and choose $\rho \in (0, 1/(2\alpha\|\Phi\Phi^\top\|))$.
  2: We compute
$$\mathbf{x}^{(k+1)} = \alpha \, \text{Shrinkage}(\Phi^\top \lambda_k, 1) \tag{6.47}$$

and
$$\lambda_{k+1} = \lambda_k + \rho(\mathbf{b} - \Phi\mathbf{x}^{(k)}) \tag{6.48}$$

for $k = 0, 1, 2, \ldots$ until $|\lambda_{k+1} - \lambda_k| \le \epsilon$.

The advantage of Algorithm 6.9 is that there is no matrix inversion involved. Only matrix-vector multiplications are needed. Thus, Algorithm 6.9 is very efficient and, in particular, good for problems of large scale. We now establish the convergence of Algorithm 6.9.

Theorem 6.14. *Let* $\mathbf{x}^*$ *be the minimizer of* (6.44), *and let* $\mathbf{x}^{(k)}$ *be iterative solutions from Algorithm 6.9. If* $0 < \rho < 2/(\alpha\|\Phi\Phi^\top\|)$, *then* $\lim_{k \to \infty} \mathbf{x}^{(k)} = \mathbf{x}^*$.

Proof. Let $g^k = \partial\|\mathbf{x}^{(k)}\|_1$ and $g^* = \partial\|\mathbf{x}^*\|_1$. Then, by the definition of subgradient, we have

$$\|\mathbf{x}^*\|_1 - \|\mathbf{x}^{(k)}\|_1 \ge \langle g^k, \mathbf{x}^* - \mathbf{x}^{(k)}\rangle \quad \text{and} \quad \|\mathbf{x}^{(k)}\|_1 - \|\mathbf{x}^*\|_1 \ge \langle g^*, \mathbf{x}^{(k)} - \mathbf{x}^*\rangle.$$

The summation of the above two inequalities yields

$$\langle g^k - g^*, \mathbf{x}^{(k)} - \mathbf{x}^*\rangle \ge 0. \tag{6.49}$$

Next, by (6.45), $\mathbf{x}^{(k)}$ is the minimizer of $L(\mathbf{x}, \lambda_k)$. Also, let $\mathbf{x}^*$ be the minimizer of $L(\mathbf{x}, \lambda^*)$, $\lambda^* > 0$. Thus,

$$g^k + \mathbf{x}^{(k)}/\alpha - \Phi^\top \lambda_k = 0 \quad \text{and} \quad g^* + \mathbf{x}^*/\alpha - \Phi^\top \lambda^* = 0.$$

Their difference is $(g^k - g^*) + (\mathbf{x}^{(k)} - \mathbf{x}^*)/\alpha - \Phi^\top(\lambda^k - \lambda^*) = 0$. For convenience, we have $e_k = \mathbf{x}^{(k)} - \mathbf{x}^*$ and $d_k = \lambda_k - \lambda^*$. Then

$$\langle \Phi e_k, d_k\rangle = \langle e_k, \Phi^\top d_k\rangle = \langle e_k, (g^k - g^*) + e_k/\alpha\rangle.$$

By (6.49), we have $\langle e_k, g^k - g^*\rangle \ge 0$ and

$$\langle \Phi e_k, d_k\rangle \ge \|e_k\|_2^2/\alpha. \tag{6.50}$$

On the other hand, we have

$$\lambda_{k+1} = \lambda_k + \rho(\mathbf{b} - \Phi\mathbf{x}^{(k)}) \quad \text{or} \quad d_{k+1} = d_k - \rho\Phi e_k.$$

Taking the inner product of both sides of the second equation above with $d_{k+1} + d_k$, we obtain

$$\|d_{k+1}\|_2^2 - \|d_k\|_2^2 + \rho\langle \Phi e_k, d_{k+1} + d_k\rangle = 0$$

or

$$\|d_k\|_2^2 - \|d_{k+1}\|_2^2 = \rho\langle\Phi e_k, d_{k+1} - d_k\rangle + 2\rho\langle\Phi e_k, d_k\rangle$$
$$\geq -\rho^2\|\Phi e_k\|_2^2 + 2\rho\|e_k\|_2^2/\alpha \geq (-\rho\|\Phi\Phi^\top\| + 2/\alpha)\|e_k\|_2^2\rho \geq 0. \quad (6.51)$$

That is, if ρ satisfies the assumption in Algorithm 6.9, then $\|d_k\|$ is decreasing, and hence convergent. It follows that $\|e_k\|_2 \to 0$. That is, $\mathbf{x}^{(k)} \to \mathbf{x}^*$. This completes the proof. $\quad\square$

Let us explain the stopping criterion in Algorithm 6.8. As we saw above,

$$\|e_k\|_2^2 \leq \frac{1}{\rho(-\rho\|\Phi\Phi^\top\| + 2/\alpha)}(\|d_k\| - \|d_{k+1}\|)(\|d_k\| + \|d_{k+1}\|).$$

Since $\|d_k\|$ is decreasing, we have $(\|d_k\| + \|d_{k+1}\|) \leq 2\|d_0\|$ and

$$\|d_k\| - \|d_{k+1}\| \leq \|d_k - d_{k+1}\| = \|\lambda_k - \lambda_{k+1}\|.$$

It follows that for a positive constant C dependent on $\alpha, \rho, \|\Phi\Phi^\top\|$, and λ^*, we have

$$\|e_k\|_2 \leq C\|\lambda_k - \lambda_{k+1}\|^{1/2}.$$

This gives the stopping criterion in Algorithm 6.9. If we use (6.48),

$$\|e_k\|_2^2 \leq \frac{1}{(-\rho\|\Phi\Phi^\top\| + 2/\alpha)}\|\mathbf{b} - \Phi\mathbf{x}^{(k)}\|.$$

That is, if $\|\Phi\mathbf{x}^{(k)} - \mathbf{b}\|$ is within a tolerance, $\mathbf{x}^{(k)}$ is close to the minimizer of (6.44).

Next we need to show when the minimizer is the sparse solution (cf. [19]).

Theorem 6.15 (Lai and Yin, 2013 [19]). *Let* $\mathbf{x_b} \in \mathbb{R}^n$ *be s-sparse and* $\mathbf{b} := \Phi\mathbf{x_b}$. *Suppose that* Φ *satisfies the RIP with* $\delta_{2s} \leq 0.4404$. *Then* $\mathbf{x_b}$ *is the unique minimizer of* (6.44) *as long as* $\alpha \geq 10\|\mathbf{x_b}\|_\infty$.

Proof. Let $S := \mathrm{supp}(\mathbf{x_b})$ and S^c be the complement of S in $\{1,\ldots,n\}$. For any *nonzero* $h \in \mathrm{Null}(\Phi)$,

$$\|\mathbf{x_b} + h\|_1 + \frac{1}{2\alpha}\|\mathbf{x_b} + h\|^2$$
$$= \|(\mathbf{x_b})_S + h_S\|_1 + \frac{1}{2\alpha}\|(\mathbf{x_b})_S + h_S\|^2 + \|h_{S^c}\|_1 + \frac{1}{2\alpha}\|h_{S^c}\|^2$$
$$\geq \|(\mathbf{x_b})_S\|_1 - \|h_S\|_1 + \frac{1}{2\alpha}\|(\mathbf{x_b})_S\|^2 - \frac{1}{\alpha}\langle(\mathbf{b_b})_S, h_S\rangle + \frac{1}{2\alpha}\|h_S\|^2 + \|h_{S^c}\|_1 + \frac{1}{2\alpha}\|h_{S^c}\|^2$$
$$\geq \left[\|(\mathbf{x_b})_S\|_1 + \frac{1}{2\alpha}\|(\mathbf{x_b})_S\|^2\right] + \left[\|h_{S^c}\|_1 - \|h_S\|_1 - \frac{\|(\mathbf{x_b})_S\|_\infty}{\alpha}\|h_S\|_1 + \frac{1}{2\alpha}\|h\|^2\right]$$
$$= \left[\|\mathbf{x_b}\|_1 + \frac{1}{2\alpha}\|\mathbf{x_b}\|^2\right] + \left[\|h_{S^c}\|_1 - \left(1 + \frac{\|(\mathbf{x_b})_S\|_\infty}{\alpha}\right)\|h_S\|_1 + \frac{1}{2\alpha}\|h\|^2\right],$$

where the first inequality follows from the triangle inequality, and the second one follows from $\|h_S\|^2 + \|h_{S^c}\|^2 = \|h\|^2$ and $\langle(\mathbf{x_b})_S, h_S\rangle \leq \|(\mathbf{x_b})_S\|_\infty\|h_S\|_1$.

Hence, $\mathbf{x_b}$ is the unique minimizer of (6.44) provided that the second block of the above equation is strictly positive, which holds if

$$\left(1 + \frac{\|(\mathbf{x_b})_S\|_\infty}{\alpha}\right)\|h_S\|_1 \leq \|h_{S^c}\|_1. \quad (6.52)$$

By (7.14) in the next chapter, any $h \in \mathrm{Null}(\Phi)$ satisfies

$$\|h_S\|_1 \leq \theta_{2s}\|h_{S^c}\|_1,$$

where

$$\theta_{2s} := \sqrt{\frac{4(1 + 5\delta_{2s} - 4\delta_{2s}^2)}{(1 - \delta_{2s})(32 - 25\delta_{2s})}}. \tag{6.53}$$

Hence, (6.52) holds if

$$\left(1 + \frac{\|(\mathbf{x_b})_S\|_\infty}{\alpha}\right)^{-1} \geq \theta_{2s}$$

or, in light of $\theta_{2s} < 1$,

$$\alpha \geq \left(\theta_{2s}^{-1} - 1\right)^{-1} \|\mathbf{x_b}\|_\infty. \tag{6.54}$$

For $\delta_{2s} = 0.4404$, we obtain α around 9.9849, which proves the desired result.   □

Remark 6.5.1. *From inequality (6.54), we can associate each δ_{2s} with a bound on α. For example, if $\delta_{2s} \leq 0.4715$, then exact recovery is guaranteed for $\alpha \geq 25\|\mathbf{x_b}\|_\infty$. If the better condition $\delta_{2k} \leq 0.1273$ holds, then we only need $\alpha \geq \|\mathbf{x}^0\|_\infty$.*

Next we study the noisy recovery case.

Theorem 6.16. *Let $\mathbf{x_b} \in \mathbb{R}^n$ be an arbitrary vector, and let $\sigma_s(\mathbf{x_b})$ be its best s-term approximation. Let $\mathbf{b} := \Phi\mathbf{x_b} + \epsilon$, where ϵ is an arbitrary error vector. If Φ satisfies the RIP with $\delta_{2s} \leq 0.3814$, then the solution $\mathbf{x}^*$ of Algorithm 6.9 with any $\alpha \geq 10\|\mathbf{x}^0\|_\infty$ satisfies the error bounds*

$$\|\mathbf{x}^* - \mathbf{x_b}\|_1 \leq C_1\sqrt{s}\|\epsilon\| + C_2\|\mathbf{x_b} - \sigma_s(\mathbf{x_b})\|_1, \tag{6.55}$$

$$\|\mathbf{x}^* - \mathbf{x_b}\|_2 \leq \bar{C}_1\|\epsilon\| + \bar{C}_2\|\mathbf{x_b} - \sigma_s(\mathbf{b_b})\|_1/\sqrt{s}, \tag{6.56}$$

where C_1, C_2, $\bar{C}_1$, and $\bar{C}_2$ depend only on δ_{2k} and are defined in (6.61), (6.62), (6.64), and (6.65), respectively.

Proof. Clearly, $\|\mathbf{x_b} - \sigma_k(\mathbf{x_b})\| = \|(\mathbf{x_b})_{S^c}\|$. Letting $h := \mathbf{x}^* - \mathbf{x_b}$, we have

$$\|\mathbf{x_b} + h\|_1 + \frac{1}{2\alpha}\|\mathbf{x_b} + h\|^2 \leq \|\mathbf{x_b}\|_1 + \frac{1}{2\alpha}\|\mathbf{x_b}\|^2 \tag{6.57}$$

and

$$\|\mathbf{x_b} + h\|_1 + \frac{1}{2\alpha}\|\mathbf{x_b} + h\|^2$$

$$= \|(\mathbf{x_b})_S + h_S\|_1 + \frac{1}{2\alpha}\|(\mathbf{x_b})_S + h_S\|^2 + \|(\mathbf{x_b})_{S^c} + h_{S^c}\|_1 + \frac{1}{2\alpha}\|(\mathbf{x_b})_{S^c} + h_{S^c}\|^2$$

$$\geq \|(\mathbf{x_b})_S\|_1 - \|h_S\|_1 + \frac{1}{2\alpha}\|(\mathbf{x_b})_S\|^2 - \frac{1}{\alpha}\langle(\mathbf{x_b})_S, h_S\rangle + \frac{1}{2\alpha}\|h_S\|^2$$

$$\quad + \|h_{S^c}\|_1 - \|(\mathbf{x_b})_{S^c}\|_1 + \frac{1}{2\alpha}\|(\mathbf{x_b})_{S^c}\|^2 - \frac{1}{\alpha}\langle(\mathbf{x_b})_{S^c}, h_{S^c}\rangle + \frac{1}{2\alpha}\|h_{S^c}\|^2$$

$$= \left(\|\mathbf{x_b}\|_1 + \frac{1}{2\alpha}\|\mathbf{x_b}\|^2\right) - 2\|(\mathbf{x_b})_{S^c}\|_1 - \left(\|h_S\|_1 + \frac{1}{\alpha}\langle(\mathbf{x_b})_S, h_S\rangle\right)$$

$$\quad + \left(\|h_{S^c}\|_1 - \frac{1}{\alpha}\langle(\mathbf{x_b})_{S^c}, h_{S^c}\rangle\right) + \frac{1}{2\alpha}\|h\|^2$$

$$\geq \left(\|\mathbf{x_b}\|_1 + \frac{1}{2\alpha}\|\mathbf{x_b}\|^2\right) - 2\|(\mathbf{x_b})_{S^c}\|_1 - \left(1 + \frac{\|(\mathbf{x_b})_S\|_\infty}{\alpha}\right)\|h_S\|_1$$

$$\quad + \left(1 - \frac{\|(\mathbf{x_b})_{S^c}\|_\infty}{\alpha}\right)\|h_{S^c}\|_1 + \frac{1}{2\alpha}\|h\|^2,$$

where the first inequality follows from the triangle inequality and the second is due to $\langle a, b \rangle \leq \|a\|_\infty \|b\|_1$. Combining this with (6.57), we get

$$\left(1 - \frac{\|(\mathbf{x_b})_{S^c}\|_\infty}{\alpha}\right) \|h_{S^c}\|_1 + \frac{1}{2\alpha}\|h\|^2 \leq \left(1 + \frac{\|(\mathbf{x_b})_S\|_\infty}{\alpha}\right) \|h_S\|_1 + 2\|(\mathbf{x_b})_{S^c}\|_1$$

and, after dropping the nonnegative term $\frac{1}{2\alpha}\|h\|^2$,

$$\|h_{S^c}\|_1 \leq C_3\|h_S\|_1 + C_4\|(\mathbf{x_b})_{S^c}\|_1, \tag{6.58}$$

where

$$C_3 := \frac{\alpha + \|(\mathbf{x_b})_S\|_\infty}{\alpha - \|(\mathbf{x_b})_{S^c}\|_\infty} \quad \text{and} \quad C_4 := \frac{2\alpha}{\alpha - \|(\mathbf{x_b})_{S^c}\|_\infty}.$$

Since $\|\Phi h\| = \|\Phi \mathbf{x}^* - \Phi \mathbf{x_b}\| = \|\Phi \mathbf{x}^* - \mathbf{b} + \epsilon\| \leq \|\Phi \mathbf{x}^* - \mathbf{b}\| + \|\epsilon\| \leq 2\|\epsilon\|$ and $\delta_{2s} < 2/3$, we use the proof of Theorem 7.11 to obtain

$$\|h_S\|_1 \leq \frac{2\sqrt{s}}{\sqrt{1 - \delta_{2s}}}\|\epsilon\| + \theta_{2s}\|h_{S^c}\|_1, \tag{6.59}$$

where θ_{2s} is defined in (6.53) as a function of δ_{2s}. Indeed, because we know that $\Phi(h_S + h_{T_1}) = \Phi h - \Phi(\sum_{j\geq 2} h_{T_j})$, $\|\Phi(h_S + h_{T_1})\|_2 \leq 2\|\epsilon\| + \|\Phi(\sum_{j\geq 2} h_{T_j})\|$. We use the proof of Lemmas 7.15 and 7.14 in Chapter 7 to obtain

$$\frac{\sqrt{1 - \delta_{2s}}}{\sqrt{s}}(\tau(h)^2 + t^2)^{1/2}\|h_{S^c}\|_1 \leq 2\|\epsilon\| + \left(\frac{(1-t)t + \delta_{2s}(1 - 3t/4)^2}{s}\right)^{1/2}\|h_{S^c}\|_1,$$

where $\|h_S\|_1 = \tau(h)\|h_{S^c}\|_1$ and $\|h_{T_1}\|_1 = t\|h_{(S+T_1)^c}\|_1$. It follows that

$$\|h_S\|_1 \leq (\tau(h)^2 + t^2)^{1/2}\|h_{S^c}\|_1 \leq \frac{\sqrt{s}}{\sqrt{1 - \delta_{2s}}}2\|\epsilon\| + \theta_{2s}\|h_{S^c}\|_1,$$

which is (6.59). For the choice of $\delta_{2s} = 0.3814$, it is easy to verify that $C_3\theta_{2s} < 1$ for all nonzero $\mathbf{x_b}$ and for α satisfying the assumption in this theorem. Hence, combining (6.58) and (6.59) yields the bound of $\|h_{S^c}\|_1$:

$$\|h_S\|_1 \leq (1 - C_3\theta_{2s})^{-1}\left(\frac{2\sqrt{s}}{\sqrt{1 - \delta_{2s}}}\|\epsilon\| + C_4\|(\mathbf{x_b})_{S^c}\|_1\right). \tag{6.60}$$

Applying (6.59) and (6.60) gives us (6.55) or

$$\|\mathbf{x}^* - \mathbf{x_b}\|_1 = \|h\|_1 = \|h_S\|_1 + \|h_{S^c}\|_1 \leq C_1\sqrt{s}\|\epsilon\| + C_2\|\mathbf{x_b} - \sigma_s(\mathbf{x_b})\|_1,$$

where

$$C_1 = \frac{2(1 + C_3)}{\sqrt{1 - \delta_{2s}}(1 - C_3\theta_{2s})}, \tag{6.61}$$

$$C_2 = \frac{C_4(1 + C_3)}{1 - C_3\theta_{2s}} + C_4. \tag{6.62}$$

To prove (6.56), by the proof of Theorem 7.11 again we obtain the inequality

$$\|h_S\|_2 \leq \frac{2}{\sqrt{1 - \delta_{2s}}}\|\epsilon\| + \sqrt{\frac{4(8 - \delta_{2s})(9 - \delta_{2s})}{(1 - \delta_{2s})}}\frac{1}{(8 - 9\delta_{2s})}\frac{\|h_{S^c}\|_1}{\sqrt{s}}. \tag{6.63}$$

Indeed, we use Lemmas 7.15 and 7.14 from Chapter 7 to get

$$\|h\|^2 = \|h_{S+T_1}\|^2 + \sum_{i \geq 2} \|h_{T_i}\|^2 \leq \frac{4\|\epsilon\|^2}{1 - \delta_{2s}} + \frac{1}{1 - \delta_{2s}} \frac{1}{s} (2(1-t)t + \delta_{2s}(1 - 3t/4)^2) \|h_{S^c}\|_1^2.$$

The term on the far right achieves its maximum at $t = (2 - 3\delta_{2s})/(4 - 9\delta_{2s}) < 1$. After a simplification, we get (6.63). That is,

$$\|\mathbf{x}^* - \mathbf{x_b}\|_2 = \|h\|_2 \leq \bar{C}_1 \|\epsilon\| + \bar{C}_2 \|\mathbf{x_b} - \sigma_s(\mathbf{x_b})\|_1 / \sqrt{s},$$

where

$$\bar{C}_1 := \frac{2}{\sqrt{1 - \delta_{2s}}} \left(\frac{4C_3}{1 - C_3\theta_{2s}} \sqrt{\frac{2 - \delta_{2s}}{(1 - \delta_{2s})(32 - 25\delta_{2s})}} + 1 \right), \tag{6.64}$$

$$\bar{C}_2 := \frac{2C_4}{1 - C_3\theta_{2s}} \sqrt{\frac{2(2 - \delta_{2s})}{(1 - \delta_{2s})(32 - 25\delta_{2s})}}. \tag{6.65}$$

This completes the proof.   □

6.6 ▪ A Linearly Convergent Algorithm

In this section, we continue the study of the minimization problem in the previous section. This time, we consider the dual version of the minimization problem (6.44). In fact, the dual minimization problem is unconstrained with smooth minimizing functional. We then apply the standard gradient descent algorithm to the new minimization problem and show that the algorithm is globally linearly convergent. It is easy to see that the minimizer $\mathbf{x}^*$ of (6.44) is unique due to the strictly convex minimizing functional.

Next let us recall the dual problem of the minimization (6.44), which can be obtained as follows. Recall $L(\mathbf{x}, \mathbf{y})$ defined in the previous section:

$$\max_{\mathbf{y}} \min_{\mathbf{x}} L(\mathbf{x}, \mathbf{y}) = \min_{\mathbf{x}} \max_{\mathbf{y}} \|\mathbf{x}\|_1 + \frac{1}{2\alpha} \|\mathbf{x}\|_2^2 - \mathbf{y}^\top (\Phi\mathbf{x} - \mathbf{b})$$

$$= \min_{\mathbf{x}} \max_{\mathbf{y},\mathbf{z}} \left\{ \mathbf{x}^\top \mathbf{z} + \frac{1}{2\alpha} \|\mathbf{x}\|_2^2 - \mathbf{y}^\top \Phi\mathbf{x} + \mathbf{y}^\top \mathbf{b} : \|\mathbf{z}\|_\infty \leq 1 \right\}$$

$$= \max_{\mathbf{y},\mathbf{z}} \min_{\mathbf{x}} \left\{ \mathbf{x}^\top \mathbf{z} + \frac{1}{2\alpha} \|\mathbf{x}\|_2^2 - \mathbf{y}^\top \Phi\mathbf{x} + \mathbf{b}^\top \mathbf{y} : \|\mathbf{z}\|_\infty \leq 1 \right\}$$

$$= \max_{\mathbf{y}, \|\mathbf{z}\|_\infty \leq 1} \mathbf{y}^\top \mathbf{b} - \frac{\alpha}{2} \|\Phi^\top \mathbf{y} - \mathbf{z}\|^2, \quad \text{since } \mathbf{x}^* = \alpha(\Phi^\top \mathbf{y} - \mathbf{z})$$

$$= \min_{\mathbf{y},\mathbf{z}} \left\{ -\mathbf{b}^\top \mathbf{y} + \frac{\alpha}{2} \|\Phi^\top \mathbf{y} - \mathbf{z}\|_2^2 : \|\mathbf{z}\|_\infty \leq 1 \right\}.$$

Eliminating $\mathbf{z}$ from the last equation gives the dual problem

$$\min_{\mathbf{y}} \left\{ -\mathbf{b}^\top \mathbf{y} + \frac{\alpha}{2} \|\Phi^\top \mathbf{y} - \text{Proj}_{[-1,1]^n} (\Phi^\top \mathbf{y})\|_2^2 \right\}. \tag{6.66}$$

Since the well-known shrinkage (or soft-thresholding) operator satisfies $\text{Shrinkage}(\mathbf{z}) = \mathbf{z} - \text{Proj}_{[-1,1]^n}(\mathbf{z})$, the second term in (6.66) equals $(\alpha/2) \|\text{Shrinkage}(\Phi^\top \mathbf{y})\|_2^2$.

This leads to its advantage of being unconstrained and differentiable (despite the use of projection). Furthermore, for any solution $\mathbf{y}^*$ to (6.66), we can recover the unique minimizer by

$$\mathbf{x}^* = \alpha \, \text{Shrinkage}(\Phi^\top \mathbf{y}^*). \tag{6.67}$$

The computation of $\mathbf{y}^*$ based on the minimization (6.66) can be carried out easily using the following simple gradient descent method on the new minimization problem:

$$\min_{\mathbf{y}} -\mathbf{b}^\top \mathbf{y} + \frac{\alpha}{2}\|\text{Shrinkage}(\Phi^\top \mathbf{y})\|_2^2. \tag{6.68}$$

Indeed, letting

$$f(\mathbf{y}) = -\mathbf{b}^\top \mathbf{y} + \frac{\alpha}{2}\|\text{Shrinkage}(\Phi^\top \mathbf{y})\|_2^2, \tag{6.69}$$

it is easy to see that $\nabla f(\mathbf{y}) = -\mathbf{b} + \alpha \Phi \text{Shrinkage}(\Phi^\top \mathbf{y})$. The standard gradient descent method for (6.68) yields Algorithm 6.10.

ALGORITHM 6.10

Dual Gradient Descent Algorithm [19]

1: We simply apply the gradient descent method to the minimization (6.66). That is, starting with a good initial guess $\mathbf{y}^{(0)}$, we compute

$$\mathbf{y}^{(k)} = \mathbf{y}^{(k-1)} + h(\mathbf{b} - \alpha \Phi \text{Shrinkage}(\Phi^\top \mathbf{y}^{(k-1)})) \tag{6.70}$$

for $k \geq 1$ until the maximum iteration number is achieved, where $h > 0$ is the step size.

We shall show that Algorithm 6.10 is convergent linearly in the sense of Theorems 6.17 and 6.18 (below). For point $\mathbf{z} \in \mathbb{R}^m$ and a point set $\mathcal{Z} \subset \mathbb{R}^m$, we define

$$\text{dist}(z, \mathcal{Z}) := \min_{\mathbf{z}'}\{\|\mathbf{z} - \mathbf{z}'\|_2 : \mathbf{z}' \in \mathcal{Z}\}.$$

Then the set of minimizers of (6.66), i.e., by (6.67), is given by

$$\mathcal{Y}^* = \{\mathbf{y}' \in \mathbb{R}^m : \alpha \text{Shrinkage}(\Phi^\top \mathbf{y}') = \mathbf{x}^*\}, \tag{6.71}$$

which is clearly a convex set. Next we define

$$\nu := \left(\min_{i \in \text{support}(\mathbf{x}^*)} \frac{\alpha|x_i^*|}{|x_i^*| + 2\alpha}\right)\left(\min_{S:|S|=m} \lambda_{\min}(\Phi_S \Phi_S^\top)\right), \tag{6.72}$$

where Φ_S is the submatrix of Φ with column indices in S. When Φ is of completely full rank, we have that Φ_S is of full rank for any $S \subset \{1, \ldots, N\}$ with $|S| = m$, and hence $\nu > 0$.

The main objective in this section is to establish the following convergent result.

Theorem 6.17 (Lai and Yin, 2013 [19]). *Suppose that the sensing matrix Φ is of completely full rank. Algorithm 6.10 with step size $0 < h \leq 2\nu/(\alpha^2\|\Phi\|^4)$ generates a globally linearly converging sequence $\{\mathbf{y}^{(k)}, k \geq 1\}$ in the following sense:*

$$\text{dist}(\mathbf{y}^{(k)}, \mathcal{Y}^*) \leq \left(1 - 2h\nu + h^2\alpha^2\|\Phi\|^4\right)^{k/2} \text{dist}(\mathbf{y}^{(0)}, \mathcal{Y}^*) \tag{6.73}$$

and

$$f(\mathbf{y}^{(k)}) - f^* \leq L\left(1 - 2h\nu + h^2\alpha^2\|\Phi\|^4\right)^k \text{dist}(\mathbf{y}^{(0)}, \mathcal{Y}^*)^2, \tag{6.74}$$

where $\mathcal{Y}^*$ is given in (6.71)) the restricted strong convexity constant ν is given in (6.72), and $L > 0$ is a positive constant. Furthermore, $\{\mathbf{x}^{(k)}\}$ is a globally linear converging sequence since

$$\|\mathbf{x}^{(k)} - \mathbf{x}^*\|_2 \leq \alpha\|\Phi\|\text{dist}(\mathbf{y}^{(k)}, \mathcal{Y}^*). \tag{6.75}$$

The proof in [19] is very complicated. We shall provide another strategy to prove the linear convergence. Suppose that we look for a solution within a tolerance ϵ. Let

$$\mathcal{Y}_\epsilon^* = \mathcal{Y}^* + B(0, \epsilon), \tag{6.76}$$

where $B(0, \epsilon)$ is the standard open ball of radius ϵ at 0.

Theorem 6.18. *Fix a tolerance $\epsilon > 0$. There exists a positive constant $\mu > 0$ which is dependent on ϵ such that Algorithm 6.10 with step size $0 < h \leq 2\mu/(\alpha^2\|\Phi\|^4)$ generates a globally linearly converging sequence $\{\mathbf{y}^{(k)}, k \geq 1\}$ in the following senses: either $\mathbf{y}^{(k+1)} \in \mathcal{Y}_\epsilon^*$ or*

$$\text{dist}(\mathbf{y}^{(k+1)}, \mathcal{Y}^*) \leq \left(1 - 2h\mu + h^2\alpha^2\|\Phi\|^4\right)^{1/2}\text{dist}(\mathbf{y}^{(k)}, \mathcal{Y}^*) \tag{6.77}$$

and

$$f(\mathbf{y}^{(k+1)}) - f^* \leq L\left(1 - 2h\mu + h^2\alpha^2\|\Phi\|^4\right)^{1/2}\text{dist}(\mathbf{y}^{(k)}, \mathcal{Y}^*), \tag{6.78}$$

where $\mathcal{Y}^*$ is given in (6.71).

The proof of the above linear convergences needs a few preparatory results, which are given in the next subsection.

6.6.1 ▪ Preparatory Results

In this subsection, we prove a few results that will be used in the main convergence analysis. A typical tool for obtaining global convergence at a linear rate (namely, global geometric convergence) is the strong convexity of the objective function. The concept of strong convexity was defined in Chapter 4. Let us recall it here. A function g is strongly convex with a constant c if it satisfies

$$\langle \mathbf{y} - \mathbf{y}', \nabla f(\mathbf{y}) - \nabla f(\mathbf{y}')\rangle \geq c\|\mathbf{y} - \mathbf{y}'\|^2 \quad \forall \mathbf{y}, \mathbf{y}' \in \text{domain of } f. \tag{6.79}$$

Intuitively, a strongly convex function is supported below by a quadratic function in the form $q(\mathbf{z}) = \frac{c}{2}\|\mathbf{z} - \mathbf{y}\|^2$ with $c > 0$ at every point $\mathbf{y}$ in its domain. The strong convexity, however, does not hold for our $f(\mathbf{y})$ since $\nabla f(\mathbf{y}^*) = \mathbf{0} \ \forall \mathbf{y}^* \in \mathcal{Y}^*$ and $\mathcal{Y}^*$ is not a singleton. Fortunately, we have the following result.

Lemma 6.19. *Let* Shrinkage *be a shrinkage operator mapping any real number to a real number such that* Shrinkage$(s) = \text{sign}(s)\max\{|s| - 1, 0\}$. *Then the inequality*

$$(s - s^*) \cdot (\text{Shrinkage}(s) - \text{Shrinkage}(s^*)) \geq \frac{|\text{Shrinkage}(s^*)|}{|\text{Shrinkage}(s^*)| + 2} \cdot (s - s^*)^2 \geq 0 \tag{6.80}$$

holds for all $s, s^* \in \mathbb{R}$. The first equality holds when $s = -\text{sign}(s^*)$.

Proof. The first inequality in (6.80) can be proved by elementary case-by-case analysis. The second one is trivial. We leave the details to Exercise 24. □

Next we point out that the set $\mathcal{Y}^*$ of minimizers can be rewritten in terms of the indices of the unique minimizer $\mathbf{x}^*$ of (6.44). That is, let $\mathbf{x}_+^*, \mathbf{x}_-^*$, and $\mathbf{x}_0^*$ be the positive, negative, and zero components of $\mathbf{x}^*$, respectively. Let $\mathcal{S}_+, \mathcal{S}_-, \mathcal{S}_0$ be the index sets corresponding to the indices of $\mathbf{x}_+^*, \mathbf{x}_-^*, \mathbf{x}_0^*$, respectively. Finally, let $\Phi_+ := \Phi_{\mathcal{S}_+}, \Phi_- := \Phi_{\mathcal{S}_-}, \Phi_0 := \Phi_{\mathcal{S}_0}$ be the submatrices of Φ with indices in $\mathcal{S}_+, \mathcal{S}_-, \mathcal{S}_0$. Then

$$\mathcal{Y}^* = \{\mathbf{y}' : \Phi_+^\top\mathbf{y}' - \mathbf{1} = \mathbf{x}_+^*/\alpha, \Phi_-^\top\mathbf{y}' + \mathbf{1} = \mathbf{x}_-^*/\alpha, -\mathbf{1} \leq \Phi_0^\top\mathbf{x}_0^* \leq \mathbf{1}\}, \tag{6.81}$$

where $\mathbf{1}$ is the vector with components equal to 1.

We now use Lemma 6.19 to establish the following.

Lemma 6.20. *Suppose that Φ is of completely full rank. Recall that $\nabla f(\mathbf{y}) = -\mathbf{b} + \alpha\Phi\text{Shrinkage}(\Phi^{\top}\mathbf{y})$. Then if $\mathbf{y} \notin \mathcal{Y}_{\epsilon}^{*}$,*

$$\langle \mathbf{y} - \mathbf{y}', \nabla f(\mathbf{y}) - \nabla f(\mathbf{y}') \rangle \geq \mu\|\Phi(\mathbf{y} - \mathbf{y}')\|^{2} \quad \forall \mathbf{y} \notin \mathcal{Y}_{\epsilon}^{*}, \tag{6.82}$$

where $\mathbf{y}' \in \mathcal{Y}^{*}$ is the projection of $\mathbf{y}$ in $\mathcal{Y}^{*}$ and $\mu > 0$ is a positive constant which is dependent on the first factor defined in (6.72) and $\epsilon > 0$.

Proof. Since $\nabla f(\mathbf{y}) = -\mathbf{b} + \alpha\Phi\text{Shrinkage}(\Phi^{\top}(\mathbf{y}))$, we know that $\nabla f(\mathbf{y}') = 0$. Let us write $\Phi = [\Phi_{+}, \Phi_{-}, \Phi_{0}] = [\phi_{1}, \ldots, \phi_{n}]$. Then Lemma 6.19 implies

$$
\begin{aligned}
&\langle \mathbf{y} - \mathbf{y}', \nabla f(\mathbf{y}) - \nabla f(\mathbf{y}') \rangle \\
&= \langle \Phi^{\top}\mathbf{y} - \Phi^{\top}\mathbf{y}', \alpha(\text{Shrinkage}(\Phi^{\top}\mathbf{y}) - \text{Shrinkage}(\Phi^{\top}\mathbf{y}')) \rangle \\
&= \alpha\sum_{i=1}^{n}(\phi_{i}^{\top}\mathbf{y} - \phi_{i}^{\top}\mathbf{y}')(\text{Shrinkage}(\phi_{i}^{\top}\mathbf{y}) - \text{Shrinkage}(\phi_{i}^{\top}\mathbf{y}')) \\
&\geq \alpha\sum_{i\in\mathcal{S}_{+}\cup\mathcal{S}_{-}\cup\mathcal{S}_{0}}(\phi_{i}^{\top}\mathbf{y} - \phi_{i}^{\top}\mathbf{y}')(\text{Shrinkage}(\phi_{i}^{\top}\mathbf{y}) - \text{Shrinkage}(\phi_{i}^{\top}\mathbf{y}')) \\
&\geq \alpha\sum_{i\in\mathcal{S}_{+}\cup\mathcal{S}_{-}}\frac{\text{Shrinkage}(\mathbf{x}_{i}^{*})/\alpha}{\text{Shrinkage}(\mathbf{x}_{i}^{*})/\alpha + 2}(\phi_{i}^{\top}\mathbf{y} - \phi_{i}^{\top}\mathbf{y}')^{2} \\
&\quad + \sum_{i\in\mathcal{S}_{0}}\alpha(\phi_{i}^{\top}(\mathbf{y} - \mathbf{y}'))(\text{Shrinkage}(\phi_{i}^{\top}\mathbf{y}) - \text{Shrinkage}(\phi_{i}^{\top}\mathbf{y}')) \\
&\geq \nu_{1}\|[\Phi_{+}\,\Phi_{-}]^{\top}(\mathbf{y} - \mathbf{y}')\|^{2} + \alpha\sum_{i\in\mathcal{S}_{0}}(\phi_{i}^{\top}(\mathbf{y} - \mathbf{y}'))(\text{Shrinkage}(\phi_{i}^{\top}\mathbf{y}) - \text{Shrinkage}(\phi_{i}^{\top}\mathbf{y}'))
\end{aligned}
$$
$$\tag{6.83}$$

for a positive constant ν_{1}. Next we decompose $\mathcal{S}_{0}$ into the following subsets:

$$
\begin{aligned}
\mathcal{S}_{01} &:= \{i \in \mathcal{S}_{0} : \phi_{i}^{\top}\mathbf{y} > 1\}, \\
\mathcal{S}_{02} &:= \{i \in \mathcal{S}_{0} : 0 < \phi_{i}^{\top}\mathbf{y} \leq 1\}, \\
\mathcal{S}_{03} &:= \{i \in \mathcal{S}_{0} : \phi_{i}^{\top}\mathbf{y} < -1\}, \\
\mathcal{S}_{04} &:= \{i \in \mathcal{S}_{0} : -1 \leq \phi_{i}^{\top}\mathbf{y} < 0\}, \\
\mathcal{S}_{05} &:= \{i \in \mathcal{S}_{0} : \phi_{i}^{\top}\mathbf{y} = 0\}.
\end{aligned}
$$

For $i \in \mathcal{S}_{01}$, $\phi_{i}^{\top}\mathbf{y}' = 1$, and for $i \in \mathcal{S}_{03}$, $\phi_{i}^{\top}\mathbf{y}' = -1$, because of the projection property of $\mathbf{y}'$. That is, for $i \in \mathcal{S}_{01}$,

$$(\phi_{i}^{\top}(\mathbf{y} - \mathbf{y}'))(\text{Shrinkage}(\phi_{i}^{\top}\mathbf{y}) - \text{Shrinkage}(\phi_{i}^{\top}\mathbf{y}')) = (\phi_{i}^{\top}(\mathbf{y}) - 1)^{2} = (\phi_{i}^{\top}(\mathbf{y} - \mathbf{y}'))^{2}.$$

For $i \in \mathcal{S}_{03}$,

$$(\phi_{i}^{\top}(\mathbf{y} - \mathbf{y}'))(\text{Shrinkage}(\phi_{i}^{\top}\mathbf{y}) - \text{Shrinkage}(\phi_{i}^{\top}\mathbf{y}')) = (\phi_{i}^{\top}(\mathbf{y}) + 1)^{2} = (\phi_{i}^{\top}(\mathbf{y} - \mathbf{y}'))^{2}.$$

Together, we can write the terms with indices in $\mathcal{S}_{01} \cup \mathcal{S}_{03}$ as well as $\mathcal{S}_{05}$ by

$$
\begin{aligned}
&\sum_{i\in\mathcal{S}_{01}\cup\mathcal{S}_{03}\cup\mathcal{S}_{05}}(\phi_{i}^{\top}(\mathbf{y} - \mathbf{y}'))(\text{Shrinkage}(\phi_{i}^{\top}\mathbf{y}) - \text{Shrinkage}(\phi_{i}^{\top}\mathbf{y}')) \\
&= \|\Phi_{01}^{\top}(\mathbf{y} - \mathbf{y}')\|^{2} + \|\Phi_{03}^{\top}(\mathbf{y} - \mathbf{y}')\|^{2} + \|\Phi_{05}^{\top}(\mathbf{y} - \mathbf{y}')\|^{2}.
\end{aligned}
\tag{6.84}
$$

Let us summarize the discussion above by letting

$$A = [\Phi_+ \; \Phi_- \; \Phi_{01} \; \Phi_{03} \; \Phi_{05}]$$

and writing the terms in (6.83) and (6.84) together as follows:

$$\langle \mathbf{y} - \mathbf{y}', \nabla f(\mathbf{y}) - \nabla f(\mathbf{y}') \rangle \geq \min\{\nu_1, \alpha\} \|\Phi^\top(\mathbf{y} - \mathbf{y}')\|^2. \tag{6.85}$$

Finally, we claim that the ratio below is bounded from below:

$$\min_{\mathbf{y} \in (\mathcal{Y}_\epsilon^*)^c} \frac{\|[\Phi^\top(\mathbf{y} - \mathbf{y}')\|^2}{\|\Phi(\mathbf{y} - \mathbf{y}')\|^2} \geq \mu > 0, \tag{6.86}$$

where $(\mathcal{Y}_\epsilon^*)^c$ is the complement of $\mathcal{Y}_\epsilon^*$, A is dependent on $\mathbf{y}$ for the indices in $\mathcal{S}_0$, and $\mathbf{y}'$ is the projection of $\mathbf{y}$ in $\mathcal{Y}^*$. If $\mu = 0$, there exists a sequence $\{\mathbf{y}_k \in (\mathcal{Y}_\epsilon^*)^c, k \geq 1\}$ and a limit point $\mathbf{y}_0$ such that $\mathbf{y}_k \to \mathbf{y}_0$ and $\mu_k \to 0$. Since $(\mathcal{Y}_\epsilon^*)^c$ is a closed set, $\mathbf{y}_0 \in (\mathcal{Y}_\epsilon^*)^c$. We have $\|\mathbf{y}_0 - \mathbf{y}_0'\| \geq \epsilon$. Also, it is easy to see that $\|\Phi^\top(\mathbf{y}_0 - \mathbf{y}_0')\| \neq \infty$. Thus, by (6.86),

$$\|A^\top(\mathbf{y}_0 - \mathbf{y}_0')\|^2 = 0.$$

We have $\Phi_+^\top(\mathbf{y}_0 - \mathbf{y}_0') = 0$, so $\Phi_+(\mathbf{y}_0) = 1 + \mathbf{x}_+^*/\alpha$ because of $\mathbf{y}_0' \in \mathcal{Y}^*$. Similarly, $\Phi_-^\top(\mathbf{y}_0) = -1 + \mathbf{x}_-^*/\alpha$. In addition, $\Phi_{0j}^\top(\mathbf{y}_0 - \mathbf{y}_0') = 0$ for $j = 1, 3, 5$. That is, $\Phi_{0j}^\top(\mathbf{y}_0) = \Phi_{0j}^\top(\mathbf{y}_0')$. So $-1 \leq \Phi_{0j}^\top(\mathbf{y}_0) \leq 1$ for $j = 1, 3, 5$. Because $0 \leq \Phi_{02}^\top(\mathbf{y}_0) \leq 1$ and $-1 \leq \Phi_{04}^\top(\mathbf{y}_0) \leq 0$, it follows that $\mathbf{y}_0 \in \mathcal{Y}^*$. This is a contradiction since $\mathbf{y}_0 \in (\mathcal{Y}_\epsilon^*)^c$. $\quad\square$

Finally, we need to show that the shrinkage operator is a nonexpansive operator. For convenience, we let

$$\text{Shrinkage}(x, \epsilon) = \begin{cases} x - \epsilon & \text{if } x > \epsilon, \\ 0 & \text{if } |x| < \epsilon, \\ x + \epsilon & \text{if } x < -\epsilon. \end{cases} \tag{6.87}$$

Lemma 6.21. *The shrinkage operator* $\text{Shrinkage}(x, \epsilon)$ *is nonexpansive, i.e.,*

$$|\text{Shrinkage}(x, \epsilon) - \text{Shrinkage}(y, \epsilon)| \leq |x - y| \quad \forall x, y \in \mathbb{R}, \tag{6.88}$$

where $\epsilon > 0$.

Proof. We leave the proof to the interested reader, or see (2.16) in [7]. $\quad\square$

6.6.2 ▪ Proof of Linear Convergence

In this subsection, we show that the sequence obtained from Algorithm 6.10 gives a globally linearly convergent sequence $\{\mathbf{y}_k\}$.

Proof of Theorem 6.18. We let $\mathbf{y}_k$ be the kth iteration from Algorithm 6.10, and let $\mathbf{y}_k' \in \mathcal{Y}^*$ be the projection of $\mathbf{y}_k$. Since $\nabla f(\mathbf{y}_k') = 0$, we have $\mathbf{b} = \alpha\Phi\text{Shrinkage}(\Phi^\top\mathbf{y}_k')$ and use the projection property to get

$$\|\mathbf{y}_k - \mathbf{y}_k'\|^2 \leq \|\mathbf{y}_k - \mathbf{y}_{k-1}'\|^2$$
$$= \|\mathbf{y}_{k-1} - \mathbf{y}_{k-1}' - h\alpha\Phi(\text{Shrinkage}(\Phi^\top\mathbf{y}_{k-1}) - \text{Shrinkage}(\Phi^\top\mathbf{y}_{k-1}'))\|^2.$$

It follows that

$$
\begin{aligned}
\|\mathbf{y}_k - \mathbf{y}_k'\|^2 &\leq \|\mathbf{y}_{k-1} - \mathbf{y}_{k-1}'\|^2 \\
&\quad -2h\alpha\langle \mathbf{y}_{k-1} - \mathbf{y}_{k-1}', \Phi(\text{Shrinkage}(\Phi^\top \mathbf{y}_{k-1}) - \text{Shrinkage}(\Phi^\top \mathbf{y}_{k-1}'))\rangle \\
&\quad +\alpha^2 h^2 \|\Phi(\text{Shrinkage}(\Phi^\top \mathbf{y}_{k-1}) - \text{Shrinkage}(\Phi^\top \mathbf{y}_{k-1}'))\|^2 \\
&\leq \|\mathbf{y}_{k-1} - \mathbf{y}_{k-1}'\|^2 - 2h\alpha\mu\|\Phi^\top(\mathbf{y}_{k-1} - \mathbf{y}_{k-1}')\|^2 \\
&\quad +\alpha^2 h^2 \|\Phi\|^2 \|\text{Shrinkage}(\Phi^\top \mathbf{y}_{k-1}) - \text{Shrinkage}(\Phi^\top \mathbf{y}_{k-1}')\|^2 \\
&\leq \|\mathbf{y}_{k-1} - \mathbf{y}_{k-1}'\|^2 - 2h\alpha\mu\|\Phi\|\|\mathbf{y}_{k-1} - \mathbf{y}_{k-1}'\|^2 \\
&\quad +\alpha^2 h^2 \|\Phi\|^2 \|\Phi^\top(\mathbf{y}_{k-1} - \mathbf{y}_{k-1}')\|^2 \\
&\leq \|\mathbf{y}_{k-1} - \mathbf{y}_{k-1}'\|^2 (1 - 2h\alpha\mu + \alpha^2 h^2 \|\Phi\|^4),
\end{aligned}
$$

where we have used Lemmas 6.20 and 6.21. By choosing $h > 0$ small enough,

$$
\theta = 1 - 2h\alpha\mu + \alpha^2 h^2 \|\Phi\|^4 < 1,
$$

we have the linear convergence (6.77) if $\mathbf{y}_{k-1} \notin \mathcal{Y}_\epsilon^*$.

Next, for the minimal value f^*, we have $f^* = f(\mathbf{y}_k')$ for any $\mathbf{y}_k$, the kth iterative solution from Algorithm 6.10, where $\mathbf{y}' \in \mathcal{Y}^*$ is the projection of $\mathbf{y}_k$ in $\mathcal{Y}^*$. Hence, we have

$$
\begin{aligned}
f(\mathbf{y}_k) - f^* &= -\mathbf{b}^\top(\mathbf{y}_k - \mathbf{y}_k') \\
&\quad -\frac{\alpha}{2}(\text{Shrinkage}(\Phi^\top \mathbf{y}_k) - \text{Shrinkage}(\Phi^\top \mathbf{y}_k'))^\top(\text{Shrinkage}(\Phi^\top \mathbf{y}_k) + \text{Shrinkage}(\Phi^\top \mathbf{y}_k')) \\
&\leq \|\mathbf{b}\|\|(\mathbf{y}_k - \mathbf{y}_k')\| + \frac{\alpha}{2}\|\Phi^\top(\mathbf{y}_k - \mathbf{y}_k')\| \, \|\text{Shrinkage}(\Phi^\top \mathbf{y}_k) + \text{Shrinkage}(\Phi^\top \mathbf{y}_k')\| \\
&\leq (\mathbf{b} + \alpha M\|\Phi\|)\|\mathbf{y}_k - \mathbf{y}_k'\| =: L\|\mathbf{y}_k - \mathbf{y}_k'\|
\end{aligned}
$$

for a positive constant L since $\|\text{Shrinkage}(\Phi^\top \mathbf{y}_k)\| \leq M$ for all $k \geq 1$, where we have used Lemma 6.21. This completes the proof. □

6.7 ▪ Screening Rules for Sparse Solutions

In this section, we explain some screening rules to remove the inactive columns from Φ, which enables us to reduce the size of an underdetermined linear system and hence help find a sparse solution. Several simple rules will be provided to identify the columns from Φ which will be inactive, i.e., where the corresponding component of a sparse solution $\mathbf{x}_b$ is zero. Remove as many inactive columns as possible so that the linear system is reduced to a smaller size and so that any good sparse algorithm discussed in the previous sections can find the sparse solution more effectively. The following presentation is based on the ideas in [26] with some simplification.

Recall that Φ is a sensing matrix of size $m \times n$. We are interested in finding the sparse solution $\mathbf{x}$:

$$
\inf_{\mathbf{x} \in \mathbb{R}^n} \|\mathbf{x}\|_0, \quad \Phi\mathbf{x} = \mathbf{b}. \tag{6.89}
$$

In general, since the measurement $\mathbf{b}$ may contains some noise, it is more reasonable to find

$$
\inf_{\mathbf{x} \in \mathbb{R}^n} \frac{1}{2}\|\Phi\mathbf{x} - \mathbf{b}\|^2 + \lambda\|\mathbf{x}\|_0 \tag{6.90}
$$

for some $\lambda > 0$. It is standard to use $\|\mathbf{x}\|_1$ to approximate $\|\mathbf{x}\|_0$, which leads to the convex minimization problem

$$
\inf_{\mathbf{x} \in \mathbb{R}^n} \frac{1}{2}\|\Phi\mathbf{x} - \mathbf{b}\|^2 + \lambda\|\mathbf{x}\|_1. \tag{6.91}
$$

Following the approach in [3], we consider the new minimization problem

$$\inf_{\substack{\mathbf{x} \in \mathbb{R}^n \\ \mathbf{z} \in \mathbb{R}^m}} \frac{1}{2}\|\mathbf{z}\|^2 + \lambda\|\mathbf{x}\|_1 \quad \text{subject to } \mathbf{z} = \Phi\mathbf{x} - \mathbf{b}, \tag{6.92}$$

which is clearly equivalent to (6.91). Now we let η be a dual variable and let

$$L(\mathbf{z}, \mathbf{x}, \eta) = \frac{1}{2}\|\mathbf{z}\|^2 + \lambda\|\mathbf{x}\|_1 + \eta^\top(\mathbf{b} - \Phi\mathbf{x} - \mathbf{z})$$

be the Lagrange function. Now let us define the dual function by

$$f^*(\eta) = \inf_{\mathbf{x}, \mathbf{z}} L(\mathbf{z}, \mathbf{x}, \eta).$$

We can rewrite the above infinitum as

$$f^*(\eta) = \inf_{\mathbf{x}} -\eta^\top \Phi\mathbf{x} + \lambda\|\mathbf{x}\|_1 + \inf_{\mathbf{z}} \frac{1}{2}\|\mathbf{z}\|^2 - \eta^\top\mathbf{z} + \eta^\top\mathbf{b}.$$

That is, we need to solve the two minimization problems

$$\inf_{\mathbf{x}} -\eta^\top \Phi\mathbf{x} + \lambda\|\mathbf{x}\|_1 \quad \text{and} \quad \inf_{\mathbf{z}} \frac{1}{2}\|\mathbf{z}\|^2 - \eta^\top\mathbf{z}.$$

The second subproblem is easy to solve because

$$\inf_{\mathbf{z}} \frac{1}{2}\|\mathbf{z}\|^2 - \eta^\top\mathbf{z} = \inf_{\mathbf{z}} \frac{1}{2}\|\mathbf{z} - \eta\|^2 - \frac{1}{2}\|\eta\|^2 = -\frac{1}{2}\|\eta\|^2.$$

The function associated with the first subproblem is nonsmooth. We have to compute its subgradient to get

$$-\Phi^\top\eta + \lambda\mathbf{d},$$

where $\mathbf{d}$ is a subgradient of $\|\mathbf{x}\|_1$ which satisfies

$$\mathbf{d}^\top\mathbf{x}^* = \|\mathbf{x}^*\|_1 \quad \text{and} \quad \|\mathbf{d}\|_\infty \le 1 \tag{6.93}$$

for any minimizer $\mathbf{x}^*$. The necessary condition for the first subproblem is

$$0 \in \{-\Phi^\top\eta + \lambda\mathbf{d}\}. \tag{6.94}$$

That is, $\mathbf{d} = \Phi^\top\eta/\lambda$. So $\|\Phi^\top\eta\|_\infty \le \lambda$. It now follows that

$$\inf_{\mathbf{x}} -\eta^\top\Phi\mathbf{x} + \lambda\|\mathbf{x}\|_1 = -\eta^\top\Phi\mathbf{x}^* + \lambda\mathbf{d}^\top\mathbf{x}^*$$
$$= -\eta^\top\Phi\mathbf{x}^* + \lambda(\Phi^\top\eta/\lambda)^\top\mathbf{x}^* = 0.$$

Therefore, we have

$$f^*(\eta) = \eta^\top\mathbf{b} - \frac{1}{2}\|\eta\|^2 = -\frac{1}{2}\|\eta - \mathbf{b}\|^2 + \frac{1}{2}\|\mathbf{b}\|^2, \tag{6.95}$$

with $\|\Phi^\top\eta\|_\infty \le \lambda$, which is bounded above. We now take the maximum of the dual function f^* over the feasible set of η to get

$$\max_{\substack{\eta \\ \|\Phi^\top\eta\|_\infty \le \lambda}} f^*(\eta) = \inf_{\eta} \frac{1}{2}\|\eta - \mathbf{b}\|^2 - \frac{1}{2}\|\mathbf{b}\|^2 \quad \text{subject to } \|\Phi^\top\eta\|_\infty \le \lambda. \tag{6.96}$$

It is standard to show that the minimization problem (6.92) is equivalent to (6.96). The problem in (6.96) is called the dual formulation of the primal problem (6.92), which is equivalent to (6.91).

We now use the dual formulation to screen the active and inactive components of primal variable $\mathbf{x}$ and find the sparse solution by removing the inactive columns from Φ and then solving an equivalent problem of smaller size. This is the major point of this approach.

For convenience, letting $\theta = \eta/\lambda$, we can rewrite the minimization problem (6.96) in the form

$$\inf_{\theta} \frac{\lambda^2}{2} \left\| \theta - \frac{\mathbf{b}}{\lambda} \right\|^2 - \frac{1}{2}\|\mathbf{b}\|^2 \quad \text{subject to } \|\Phi^\top \theta\|_\infty \leq 1. \tag{6.97}$$

Letting θ^* be the minimizer of (6.97), we see that θ^* is the projection of $\mathcal{P}_C(\frac{\mathbf{b}}{\lambda})$ over a convex set $C = \{\theta \in (-1,1)^n, -1 \leq \phi_i^\top \theta \leq 1\}$, where $[\phi_1, \ldots, \phi_N] = \Phi$. Equations (6.93) and (6.94) together imply that $(\mathbf{x}^*)^\top \Phi^\top \theta^* = \|\mathbf{x}^*\|_1$. It follows that

$$\phi_i^\top \theta^* \in \begin{cases} \text{sign}((\mathbf{x}^*)_i) & \text{if } \mathbf{x}_i^* \neq 0, \\ [-1,1] & \text{if } \mathbf{x}_i^* = 0 \end{cases} \tag{6.98}$$

for all $i = 1, \ldots, N$. Hence, we obtain the following result.

Proposition 6.22. *Let $\theta^*(\lambda)$ be the unique minimizer of (6.97) for a fixed parameter $\lambda > 0$, and let $\mathbf{x}^*(\lambda)$ be a minimizer of (6.92). If $|\phi_i^\top \theta^*(\lambda)| < 1$, then $\mathbf{x}_i^*(\lambda) = 0$. That is, ϕ_i is an inactive component in the matrix Φ.*

When λ is sufficiently large, we can expect that the minimizer of (6.92) is the trivial solution, i.e., $\mathbf{x}^* \equiv 0$. Indeed, we have the following result.

Theorem 6.23. *Let*

$$\lambda_{max} = \max_{i=1,\ldots,N} \{|\phi_i^\top \mathbf{b}|\}.$$

If $\lambda > \lambda_{max}$, then the solution $\mathbf{x}^*(\lambda) \equiv 0$ is trivial.

Proof. Consider $\lambda > \lambda_{max}$ first. It is easy to see that $\theta^* = \mathbf{b}/\lambda$ is the solution of (6.97) in this case. Thus, $|\phi_i^\top \theta^*| < 1$ for all i. Thus, $\mathbf{x}_i^*(\lambda) = 0$ by (6.98) for all $i = 1, \ldots, N$. Hence, $\mathbf{x}^*(\lambda)$ is trivial. It is easy to see that the minimizer $\mathbf{x}^*(\lambda)$ is continuous with respect to λ. It follows that $\mathbf{x}^*(\lambda_{max}) = 0$. $\square$

Next we need to use the projection operator $\mathcal{P}_C$ for any convex set $C \subset \mathbb{R}^n$. That is, for any $\mathbf{x} \in \mathbb{R}^n$,

$$\mathcal{P}_C(\mathbf{x}) = \arg\min_{\mathbf{y}\in C} \frac{1}{2}\|\mathbf{x} - \mathbf{y}\|^2. \tag{6.99}$$

We note that the projection operator $\mathcal{P}_C$ is firmly nonexpansive. That is, $\mathcal{P}_C$ satisfies the inequality

$$\|\mathcal{P}_C(\mathbf{x}) - \mathcal{P}_C(\mathbf{y})\|^2 + \|\mathbf{x} - \mathbf{y} - (\mathcal{P}_C(\mathbf{x}) - \mathcal{P}_C(\mathbf{y}))\|^2 \leq \|\mathbf{x} - \mathbf{y}\|^2 \tag{6.100}$$

for all $\mathbf{x}, \mathbf{y} \in \mathbb{R}^m$. We leave the proof to the interested reader. This firm nonexpansive property of $\mathcal{P}_C$ implies the following.

Proposition 6.24. *For any $\lambda_1 > \lambda_2 \geq 0$, we have*

$$\|\theta^*(\lambda_1) - \theta^*(\lambda_2)\|^2 + \left\| \frac{\mathbf{b}}{\lambda_2} - \frac{\mathbf{b}}{\lambda_1} - (\theta^*(\lambda_2) - \theta^*(\lambda_1)) \right\|^2 \leq \left(\frac{1}{\lambda_2} - \frac{1}{\lambda_1} \right)^2 \|\mathbf{b}\|^2.$$

If $\theta^*(\lambda_1)$ is known, e.g., $\theta^*(\lambda_{max}) = \mathbf{b}/\lambda_{max}$, we can use it to estimate $\theta(\lambda_2)$ using Proposition 6.24 for $\lambda_2 < \lambda_1$. Indeed, we have one of the main results in this section.

Theorem 6.25. *Suppose that $\theta^*(\lambda_1)$ is known. Let $\lambda_2 < \lambda_1$ be a positive tuning parameter. Then $\mathbf{x}^*(\lambda_2)_i = 0$ if*

$$|\phi_i^\top(\theta^*(\lambda_1))| < 1 - \rho\|\phi_i\|, \tag{6.101}$$

or if

$$|\phi_i^\top(\theta^*(\lambda_1) - \mathbf{b}(\lambda_2^{-1} - \lambda_1^{-1}))| < 1 - \rho\|\phi_i\|, \tag{6.102}$$

where $\rho = \|\mathbf{b}\|\left|\frac{1}{\lambda_2} - \frac{1}{\lambda_1}\right|$ and $\mathbf{x}^*(\lambda_2)$ is a minimizer of (6.97) for tuning parameter λ_2.

Proof. Let us consider condition (6.101) first. Letting $\mathbf{v} = \theta(\lambda_2) - \theta(\lambda_1)$, we have

$$|\phi_i^\top\theta(\lambda_2)| = |\phi_i^\top(\theta(\lambda_1) + \mathbf{v})| \le |\phi_i^\top(\theta(\lambda_1))| + \rho\|\phi_i\|$$

by Proposition 6.24. It follows that $|\phi_i^\top\theta(\lambda_2)| < 1$. By (6.98), we know that $\mathbf{x}^*(\lambda_2)_i = 0$. Similarly, we can establish (6.102). This completes the proof. $\square$

Next we can improve conditions (6.101) and (6.102) further by using the well-known equality

$$\left\|\frac{1}{2}(\mathbf{a}+\mathbf{b}) - \mathbf{c}\right\|^2 = \frac{1}{2}\|\mathbf{a}-\mathbf{c}\|^2 + \frac{1}{2}\|\mathbf{b}-\mathbf{c}\|^2 - \frac{1}{4}\|\mathbf{a}-\mathbf{b}\|^2. \tag{6.103}$$

By choosing $\mathbf{c} = \theta^*(\lambda_1)$, $\mathbf{a} = \theta^*(\lambda_2)$, and $\mathbf{b} = \theta^*(\lambda_2) - \mathbf{b}(\lambda_2^{-1} - \lambda_1^{-1})$, we use the firm nonexpansive property above to obtain

$$\left\|\theta^*(\lambda_2) - \frac{1}{2}\mathbf{b}(\lambda_2^{-1} - \lambda_1^{-1}) - \theta^*(\lambda_1)\right\|^2$$

$$= \frac{1}{2}\|\theta^*(\lambda_2) - \theta^*(\lambda_1)\|^2 + \frac{1}{2}\|\theta^*(\lambda_2) - \mathbf{b}(\lambda_2^{-1} - \lambda_1^{-1}) - \theta^*(\lambda_1)\|^2 - \frac{1}{4}\|\mathbf{b}(\lambda_2^{-1} - \lambda_1^{-1})\|^2$$

$$\le \frac{1}{2}\|\mathbf{b}(\lambda_2^{-1} - \lambda_1^{-1})\|^2 - \frac{1}{4}\|\mathbf{b}(\lambda_2^{-1} - \lambda_1^{-1})\|^2 = \frac{1}{4}\|\mathbf{b}(\lambda_2^{-1} - \lambda_1^{-1})\|^2. \tag{6.104}$$

With the above preparation, we can establish the following result.

Theorem 6.26. *Suppose that $\theta^*(\lambda_1)$ is known. Let $\lambda_2 < \lambda_1$ be a positive tuning parameter. Then $\mathbf{x}^*(\lambda_2)_i = 0$ if*

$$|\phi_i^\top(\theta^*(\lambda_1) + \mathbf{b}(\lambda_2^{-1} - \lambda_1^{-1})/2)| < 1 - \frac{\rho}{2}\|\phi_i\|, \tag{6.105}$$

where $\rho = \|\mathbf{b}\|\left|\frac{1}{\lambda_2} - \frac{1}{\lambda_1}\right|$ and $\mathbf{x}^*(\lambda_2)$ is a minimizer of (6.97) for tuning parameter λ_2.

Proof. Indeed, we have

$$|\phi_i^\top\theta^*(\lambda_2)| \le |\phi_i^\top(\theta^*(\lambda_1) + \mathbf{b}(\lambda_2^{-1} - \lambda_1^{-1})/2)| + |\phi_i^\top(\theta^*(\lambda_2) - \theta^*(\lambda_1) - \mathbf{b}(\lambda_2^{-1} - \lambda_1^{-1})/2)|$$

$$< 1 - \frac{\rho}{2}\|\phi_i\| + \|\phi_i\|\sqrt{\frac{1}{4}\|\mathbf{b}(\lambda_2^{-1} - \lambda_1^{-1})\|^2} = 1$$

by (6.104). Then, using (6.98), we conclude that $\mathbf{x}_i^*(\lambda_2) = 0$. $\square$

The results in Theorem 6.25 lead to the following sequential screening algorithm.

ALGORITHM 6.11

Wang and Ye's Screening Rules Algorithm [26]

1: Let $\lambda_{max} = \lambda_1 > \lambda_2 > \cdots > \lambda_m$. Suppose that $\mathbf{x}^*(\lambda_k)$ is known for $k \geq 1$.
2: Then we can conclude that $\mathbf{x}^*(\lambda_{k+1})_i = 0$ if any one of the following three conditions happens:

$$(1) \quad |\phi_i^\top (\mathbf{b} - \Phi \mathbf{x}^*(\lambda_k))/\lambda_k| < 1 - \left(\frac{1}{\lambda_{k+1}} - \frac{1}{\lambda_k}\right) \|\mathbf{b}\|\|\phi_i\|.$$

$$(2) \quad |\phi_i^\top (\mathbf{b} - \Phi(\mathbf{x}^*(\lambda_k)/\lambda_k + \mathbf{b}(\lambda_{k+1}^{-1} - \lambda_k^{-1})/2))| < 1 - \frac{1}{2}\left(\frac{1}{\lambda_{k+1}} - \frac{1}{\lambda_k}\right)\|\mathbf{b}\|\|\phi_i\|.$$

$$(3) \quad |\phi_i^\top (\mathbf{b} - \Phi \mathbf{x}^*(\lambda_k))/\lambda_k - \mathbf{b}(\lambda_{k+1}^{-1} - \lambda_k^{-1})| < 1 - \left(\frac{1}{\lambda_{k+1}} - \frac{1}{\lambda_k}\right)\|\mathbf{b}\|\|\phi_i\|.$$

3: Use a good lasso solver to find the remaining $\mathbf{x}^*(\lambda_{k+1})_i$ for those i for which none of the above three inequalities holds.

Next we shall improve the performance of Algorithm 6.11. We assume that for λ_0, we know $\theta^*(\lambda_0)$ and $\mathbf{x}^*(\lambda_0)$. Also, we assume that $\theta^*(\lambda_0) \neq \mathbf{b}/\lambda_0$. In this case, we let $\theta(t) = \theta^*(\lambda_0) + t(\frac{\mathbf{b}}{\lambda_0} - \theta^*(\lambda_0))$. Then it is known that $\mathcal{P}_C(\theta(t)) = \theta^*(\lambda_0)$ for all $t \geq 0$. It follows that for any $\lambda_1 \in (0, \lambda_0)$,

$$\|\theta^*(\lambda_1) - \theta^*(\lambda_0)\| = \left\|\mathcal{P}_C\left(\frac{\mathbf{b}}{\lambda_1}\right) - \mathcal{P}_C(\theta(t))\right\|,$$

$$\|\theta^*(\lambda_1) - \theta^*(\lambda_0)\|^2 + \left\|\frac{\mathbf{b}}{\lambda_1} - \theta^*(\lambda_1) - (\theta(t) - \theta^*(\lambda_0))\right\|^2 \leq \left\|\frac{\mathbf{b}}{\lambda_1} - \theta(t)\right\|^2$$

for all $t \geq 0$. We now minimize the right-hand side of the above inequality over $t \geq 0$, i.e.,

$$\min_{t \geq 0} \left\|\frac{\mathbf{b}}{\lambda_1} - \theta(t)\right\| = \min_{t \geq 0} \left\|\frac{\mathbf{b}}{\lambda_1} - \theta^*(\lambda_0) - t\left(\frac{\mathbf{b}}{\lambda_0} - \theta^*(\lambda_0)\right)\right\|.$$

For convenience, let $\mathbf{v}_0 = \frac{\mathbf{b}}{\lambda_0} - \theta^*(\lambda_0)$ and $\mathbf{v}_1 = \frac{\mathbf{b}}{\lambda_1} - \theta^*(\lambda_0)$. The minimization yields

$$\min_{t \geq 0} \left\|\frac{\mathbf{b}}{\lambda_1} - \theta(t)\right\|^2 = \left\|\mathbf{v}_1 - \frac{\langle \mathbf{v}_0, \mathbf{v}_1 \rangle \mathbf{v}_0}{\|\mathbf{v}_0\|^2}\right\|^2 =: \|\mathbf{v}_2\|^2, \tag{6.106}$$

where $\mathbf{v}_2 = \mathbf{v}_1 - \langle \mathbf{v}_0, \mathbf{v}_1 \rangle \mathbf{v}_0 / \|\mathbf{v}_0\|^2$ for simplicity. For this particular t, we have

$$\|\theta^*(\lambda_1) - \theta^*(\lambda_0)\|^2 + \left\|\frac{\mathbf{b}}{\lambda_1} - \theta^*(\lambda_1) - (\theta(t) - \theta^*(\lambda_0))\right\|^2$$

$$= \|\theta^*(\lambda_1) - \theta^*(\lambda_0)\|^2 + \|\mathbf{v}_2 + \theta^*(\lambda_0) - \theta^*(\lambda_1)\|^2 \leq \|\mathbf{v}_2\|^2.$$

Now we use equality (6.103) to obtain

$$\left\|\theta^*(\lambda_1) - \left(\theta^*(\lambda_0) + \frac{1}{2}\mathbf{v}_2\right)\right\|^2 \leq \frac{1}{2}\|\mathbf{v}_2\|^2 - \frac{1}{4}\|\mathbf{v}_2\|^2 = \frac{1}{4}\|\mathbf{v}_2\|^2. \tag{6.107}$$

The discussion above leads to the following result.

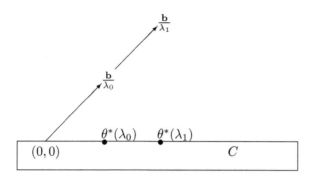

Figure 6.2. *The geometry of $\theta^*(\lambda_0)$ and $\theta^*(\lambda_1)$.*

Theorem 6.27. *Assume that the dual optimal solution $\theta^*(\lambda_0)$ at $\lambda_0 \in (0, \lambda_{max}]$ is known. Then for each $\lambda \in (0, \lambda_0)$, suppose one of the following three conditions holds:*

$$\left| \phi_i^\top \left(\theta^*(\lambda_0) + \frac{1}{2}\mathbf{v}_2 \right) \right| < 1 - \frac{1}{2}\|\mathbf{v}_2\|\|\phi_i\|,$$

$$|\phi_i^\top (\theta^*(\lambda_0))| < 1 - \|\mathbf{v}_2\|\|\phi_i\|,$$

or

$$|\phi_i^\top (\theta^*(\lambda_0) + \mathbf{v}_2)| < 1 - \|\mathbf{v}_2\|\|\phi_i\|.$$

Then we have $\mathbf{x}^*(\lambda)_i = 0$.

Proof. Let us consider the first condition. Indeed, we have

$$|\phi_i^\top \theta^*(\lambda_1)| \le \left| \phi_i^\top (\theta^*(\lambda_0)) - \left(\theta^*(\lambda_0) + \frac{1}{2}\mathbf{v}_2 \right) \right| + \left| \phi_i^\top \left(\theta^*(\lambda_0) + \frac{1}{2}\mathbf{v}_2 \right) \right|$$

$$< 1 - \frac{1}{2}\|\mathbf{v}_2\|\|\phi_i\| + \|\phi_i\|\sqrt{\frac{1}{4}\|\mathbf{v}_2\|^2} = 1$$

by (6.107). Then, using (6.98), we conclude that $\mathbf{x}_i^*(\lambda_2) = 0$. Similarly, we can establish the other two conditions. $\square$

The geometry of these arguments is illustrated in Figure 6.2. From Figure 6.2, we can see that $\|\mathbf{v}_2\| \le \|\mathbf{b}\|(\lambda_2^{-1} - \lambda_1^{-1})$. That is why the result in Theorem 6.27 is an improvement on the results in Theorems 6.25 and 6.26.

The results in Theorems 6.25 and 6.27 and the discussion of Figure 6.2 lead to the following improved sequential screening algorithm.

ALGORITHM 6.12

Improved Screening Rules Algorithm [26]

1: Let $\lambda_{max} = \lambda_1 > \lambda_2 > \cdots > \lambda_m$. Suppose that $\mathbf{x}^*(\lambda_k)$ is known for $k \ge 1$.
2: Then we can conclude that $\mathbf{x}^*(\lambda_{k+1})_i = 0$ if any one of the following three conditions happens:

(1) $|\phi_i^\top (\mathbf{b} - \Phi\mathbf{x}^*(\lambda_k))/\lambda_k| < 1 - \|\mathbf{v}_2\|\|\phi_i\|;$

(2) $|\phi_i^\top (\mathbf{b} - \Phi(\mathbf{x}^*(\lambda_k))/\lambda_k + \mathbf{v}_2(\lambda_k, \lambda_{k+1}))| < 1 - \|\mathbf{v}_2(\lambda_k, \lambda_{k+1})\|\|\phi_i\|;$

(3) $\left| \phi_i^\top \left(\mathbf{b} - \Phi(\mathbf{x}^*(\lambda_k))/\lambda_k + \frac{1}{2}\mathbf{v}_2(\lambda_k, \lambda_{k+1}) \right) \right| < 1 - \frac{1}{2}\|\mathbf{v}_2(\lambda_k, \lambda_{k+1})\|\|\phi_i\|,$

where $\mathbf{v}_2(\lambda_k, \lambda_{k+1}) = \mathbf{v}_1 - \langle \mathbf{v}_0, \mathbf{v}_1 \rangle \mathbf{v}_0 / \|\mathbf{v}_0\|^2$, $\mathbf{v}_0 = \mathbf{b}/\lambda_k - \theta^*(\lambda_k)$, and $\mathbf{v}_1 = \mathbf{b}/\lambda_{k+1} - \theta^*(\lambda_k)$.

3: Use a good lasso solver to find the remaining $\mathbf{x}^*(\lambda_{k+1})_i$ for those i for which none of the above three inequalities holds.

Remark 6.7.1. *One of the three results in Theorem 6.26 is the same as the one in Theorem 14 in [26]. One of the two remaining results is the same as the one in Theorem 19 in [26]. The remaining result in Theorem 6.26 is new. It can be obtained from one of the arguments in [26]. Again we provide a slightly different approach.*

6.8 ▪ Iteratively Reweighted ℓ_1 Algorithm

An improved ℓ_1 minimization approach for (1.20) was developed in [5]. It is called the iteratively reweighted ℓ_1 algorithm, and it performs much better than the standard ℓ_1 method as well as other iteratively reweighted algorithms, e.g., in [18] and in [8]. Let us begin with Algorithm 6.13.

ALGORITHM 6.13
Iteratively Reweighted ℓ_1 Algorithm [5]
We fix $\epsilon > 0$, e.g., $\epsilon = 0.1$, and start with a weight vector $W^{(1)} \geq 0$, say, $W^{(1)} = (1, 1, \ldots, 1)^\top$.

1: **while**
2: **do** For $k = 1, 2, \ldots, 25$, we solve the reweighted ℓ_1 minimization problem
3:

$$\mathbf{x}^{(k)} := \arg \min_{\mathbf{x} \in \mathbb{R}^n} \{ (W^{(k)})^\top |\mathbf{x}| : \quad \Phi \mathbf{x} = \mathbf{b} \}, \tag{6.108}$$

where $|\mathbf{x}| = (|x_1|, |x_2|, \ldots, |x_n|)^\top$ is a vector of absolute values of $\mathbf{x} = (x_1, \ldots, x_n)^\top$ and
4: $(W^{(k)})^\top |\mathbf{x}| = \sum_{i=1}^n W_i^{(k)} |x_i|$, with $W_i^{(k)}$ being the ith component of $W^{(k)}$.
5: And we update

$$W^{(k+1)} = (W_1^{(k+1)}, \ldots, W_n^{(k+1)})^\top, \tag{6.109}$$

with $W_i^{(k+1)} = 1/(|\mathbf{x}_i^{(k)}| + \epsilon)$ for $i = 1, \ldots, n$.
6: Terminate the iteration when k reaches the maximum number of iteration, or $\|\mathbf{x}^{(k)} - \mathbf{x}^{(k+1)}\|_1$ reaches a given tolerance.
7: **end while**

We now analyze the above iteratively reweighted ℓ_1 algorithm. To establish the convergence of Algorithm 6.13, we introduce the function

$$L(\mathbf{x}, W) = \sum_{i=1}^n W_i |x_i| + \epsilon W_i - \log W_i. \tag{6.110}$$

It is easy to see that the minimization (6.108) is equivalent to

$$\mathbf{x}^{(k)} := \arg \min \{ L(\mathbf{x}, W^{(k)}), \quad \Phi \mathbf{x} = \mathbf{b} \}.$$

On the other hand, step (6.109) for updating $W^{(k+1)}$ is equivalent to

$$W^{(k+1)} := \arg \min_W L(\mathbf{x}^{(k)}, W). \tag{6.111}$$

Note that $L(\mathbf{x}^{(k)}, W)$ is a convex function of W. Based on the above discussion, we have the following crucial observation:

$$L(\mathbf{x}^{(k+1)}, W^{(k+1)}) \leq L(\mathbf{x}^{(k)}, W^{(k+1)})$$
$$= \min_W L(\mathbf{x}^{(k)}, W) \leq L(\mathbf{x}^{(k)}, W^{(k)})$$

It follows that $L(\mathbf{x}^{(k+1)}, W^{(k+1)}) \leq \cdots \leq L(\mathbf{x}^{(1)}, W^{(1)}) = \|\mathbf{x}^{(1)}\|_1$. In addition, we also have

$$L(\mathbf{x}^{(k)}, W^{(k+1)}) \leq L(\mathbf{x}^{(k)}, W^{(k)}) \tag{6.112}$$
$$= \min_{\mathbf{x}} L(\mathbf{x}, W^{(k)}) \leq L(\mathbf{x}^{(k-1)}, W^{(k)}).$$

We first shows that the sequence $\mathbf{x}^{(k)}, k \geq 1$, is bounded so that it has a convergent subsequence.

Lemma 6.28. *For any fixed $\epsilon > 0$, the sequence $\mathbf{x}^{(k)}, k \geq 1$, from Algorithm 6.13 is bounded, and hence has a convergent subsequence.*

Proof. If $\mathbf{x}^{(k)}, k \geq 1$, are unbounded, we may assume that $x_1^{(k)} \to \infty$. Clearly, $\|W^{(k)}\|_1 \leq n/\epsilon$ for all $k \geq 1$. Let W^* be the limit of any convergent subsequence of $W^{(k)}$. For simplicity, we say $W^{(k)} \to W^*$. In this case, $W_1^* = 0$. Clearly, each component of W^* is nonnegative. By (6.112), we have

$$L(\mathbf{x}^{(k)}, W^{(k+1)}) \leq \cdots \leq L(\mathbf{x}^{(1)}, W^{(2)}) < \infty. \tag{6.113}$$

However, $L(\mathbf{x}^{(k)}, W^{(k+1)}) = \sum_{i=1}^{n} |x_i^{(k)}| W_i^{(k+1)} + \epsilon W_i^{(k+1)} - \log W_i^{(k+1)}$ is unbounded. This is because $-\log W_1^{(k+1)} = \log(|x_1^{(k)}| + \epsilon) \to \infty$. This contradicts the boundedness of $\|W^{(k)}\|_1$. □

The following is another property of the sequence $\{\mathbf{x}^{(k)}, k \geq 1\}$.

Lemma 6.29. *Let $\mathbf{x}^*$ be the limit of a convergent subsequence from $\mathbf{x}^{(k)}$. If $\mathbf{y}^*$ is the limit of another convergent subsequence from $\{\mathbf{x}^{(k)}, k \geq 1\}$, then*

$$\prod_{i=1}^{n}(|\mathbf{x}_i^*| + \epsilon) = \prod_{i=1}^{n}(|\mathbf{y}_i^*| + \epsilon). \tag{6.114}$$

Proof. By (6.112), we have

$$L(\mathbf{x}^*, W^*) = L(\mathbf{y}^*, U^*),$$

where W^* is the weight vector associated with $\mathbf{x}^*$ and U^* is the weight vector associated with $\mathbf{y}^*$. That is,

$$W^* = \left(\frac{1}{|\mathbf{x}_1^*| + \epsilon}, \frac{1}{|\mathbf{x}_2^*| + \epsilon}, \cdots, \frac{1}{|\mathbf{x}_n^*| + \epsilon}\right)^{\top},$$

and similar for U^*. By using the definition of $L(\mathbf{x}, W)$, we have

$$\sum_{i=1}^{n} W_i^* |\mathbf{x}_i^*| + \epsilon W_i^* - \log W_i^* = \sum_{i=1}^{n} U_i^* |\mathbf{y}_i^*| + \epsilon U_i^* - \log U_i^*.$$

It is easy to see that $\sum_{i=1}^{n} W_i^* |\mathbf{x}_i^*| + \epsilon W_i^* = n$, and so is $\sum_{i=1}^{n} U_i^* |\mathbf{y}_i^*| + \epsilon U_i^*$. It follows that

$$\log \prod_{i=1}^{n} W_i^* = \log \prod_{i=1}^{n} U_i^*.$$

Hence, equality (6.114) follows. □

Similarly, from $L(\mathbf{x}^{(k)}, W^{(k+1)}) \leq L(\mathbf{x}^{(k-1)}, W^{(k)})$, we can conclude the following.

Lemma 6.30. *For $\mathbf{x}^{(k)}, k \geq 1$, we have*

$$\prod_{i=1}^{n}(|\mathbf{x}_i^{(k+1)}| + \epsilon) \leq \prod_{i=1}^{n}(|\mathbf{x}_i^{(k)}| + \epsilon) \tag{6.115}$$

for all $k \geq 1$.

We now give some sufficient conditions for $\mathbf{x}^*$ to be a sparse solution. First of all, if the dynamic range of the nonzero entries of $\widehat{\mathbf{x}}$ is not very big, say $|\widehat{\mathbf{x}}_i|$ with $\widehat{\mathbf{x}}_i \neq 0$ are around 1, then $\mathbf{x}^*$ will be a sparse solution if the whole sequence $\mathbf{x}^{(k)}, k \geq 1$, converges.

Theorem 6.31. *Suppose that $\|\widehat{\mathbf{x}}\|_1 \leq \|\widehat{\mathbf{x}}\|_0 + \delta$ for some $\delta < 1$. Choose $\epsilon > 0$ such that $n\epsilon < 1 - \delta$. Suppose that $\mathbf{x}^{(k)}, k \geq 1$, converges and let $\mathbf{x}^*$ be the limit. If $|\mathbf{x}_i^*| \geq 1 - \epsilon$ for all $\mathbf{x}_i^* \neq 0$, then $\mathbf{x}^*$ is a sparse solution of $\Phi\mathbf{x} = \mathbf{b}$. Furthermore, if the restricted isometry constant (RIC) δ_{2s} of Φ satisfies $\delta_{2s} < 1$, then $\mathbf{x}^* = \widehat{\mathbf{x}}$.*

Proof. Since $\widehat{\mathbf{x}}$ is a sparse solution, we have $\|\widehat{\mathbf{x}}\|_0 \leq \|\mathbf{x}^*\|_0$. Since the $\mathbf{x}^{(k)}$'s are convergent to $\mathbf{x}^*$, $\mathbf{x}^{(k)} \to \mathbf{x}^*$ and $\mathbf{x}^{(k-1)} \to \mathbf{x}^*$ when $k \to \infty$. We have

$$\|\mathbf{x}^*\|_0 = \sum_{\mathbf{x}_i^* \neq 0} \frac{|\mathbf{x}_i^*| + \epsilon}{|\mathbf{x}_i^*| + \epsilon} \leq \sum_{\mathbf{x}_i^* \neq 0} \frac{|\mathbf{x}_i^*|}{|\mathbf{x}_i^*| + \epsilon} + \epsilon n$$

$$\leq \lim_{k \to \infty} \sum_{i=1}^{n} \frac{|\mathbf{x}_i^{(k)}|}{|\mathbf{x}_i^{(k-1)}| + \epsilon} + \epsilon n = \lim_{k \to \infty} \min_{\mathbf{x}} \left\{ \sum_{i=1}^{n} W_i^{(k)} |\mathbf{x}_i| \right\} + \epsilon n$$

$$\leq \lim_{k \to \infty} \sum_{i=1}^{n} W_i^{(k)} |\widehat{\mathbf{x}}_i| + \epsilon n = \sum_{\widehat{\mathbf{x}}_i \neq 0} \frac{|\widehat{\mathbf{x}}_i|}{|\mathbf{x}_i^*| + \epsilon} + \epsilon n$$

$$< \sum_{\widehat{\mathbf{x}}_i \neq 0} |\widehat{\mathbf{x}}_i| + 1 - \delta \leq \|\widehat{\mathbf{x}}\|_0 + \delta + 1 - \delta = \|\widehat{\mathbf{x}}\|_0 + 1.$$

That is, $\|\mathbf{x}^*\|_0 < \|\widehat{\mathbf{x}}\|_0 + 1$. We conclude that $\|\mathbf{x}^*\|_0 = \|\widehat{\mathbf{x}}\|_0$. When $\delta_{2s} < 1$, we have $\mathbf{x}^* = \widehat{\mathbf{x}}$. $\quad\square$

Next, for convenience, let $f(\mathbf{x}) := \sum_{i=1}^{n} \frac{|\mathbf{x}_i|}{|\mathbf{x}_i| + \epsilon}$ and $g(\mathbf{x}) := \sum_{i=1}^{n} \frac{1}{|\mathbf{x}_i| + \epsilon}$. Suppose that the sequence $\{\mathbf{x}^{(k)}, k \geq 1\}$ from Algorithm 6.13 converges. We can see that when k is very large, (6.108) minimizes $f(\mathbf{x})$. If $f(\mathbf{x}^*) \leq f(\mathbf{x}_\mathbf{b})$, then we can show that $\mathbf{x}^*$ is a small distance away from $\mathbf{x}_\mathbf{b}$.

Theorem 6.32. *Let $\mathbf{x}_\mathbf{b}$ be a sparse solution to $A\mathbf{x}_\mathbf{b} = \mathbf{b}$ and $\mathbf{x}^*$ be the limit of any convergent subsequence from $\{\mathbf{x}^{(k)}, k \geq 1\}$ obtained from Algorithm 6.13. If $f(\mathbf{x}^*) \leq f(\mathbf{x}_\mathbf{b})$, then*

$$\left| \|\mathbf{x}_\mathbf{b}\|_0 - \|\mathbf{x}^*\|_0 \right| \leq \epsilon g(\mathbf{x}^*). \tag{6.116}$$

Proof. Since $\mathbf{x}_\mathbf{b}$ is a sparse solution of $\Phi\mathbf{x}_\mathbf{b} = \mathbf{b}$, we have $\|\mathbf{x}_\mathbf{b}\|_0 \leq \|\mathbf{x}^*\|_0$. It is easy to see that

$$f(\mathbf{x}^*) \leq f(\mathbf{x}_\mathbf{b}) \leq \|\mathbf{x}_\mathbf{b}\|_0 \leq \|\mathbf{x}^*\|_0 \leq f(\mathbf{x}^*) + \epsilon g(\mathbf{x}^*).$$

Then (6.116) follows. $\quad\square$

Furthermore, we note that $L(\mathbf{x}, W)$ is a function of variables $\mathbf{x}$ and W. For each k, $\mathbf{x}^{(k)}$ is the minimizer of $L(\mathbf{x}, W^{(k)})$ and $W^{(k+1)}$ is the minimizer of $L(\mathbf{x}^{(k)}, W)$. By (6.112), we know that if $\mathbf{x}^*$ is the limit of a convergent subsequence of $\mathbf{x}^{(k)}, k \geq 1$, writing W^* as the weight vector associated with $\mathbf{x}^*$, we have

$$L(\mathbf{x}^*, W^*) \leq L(\mathbf{x}^{(k)}, W^{(k)}) \qquad \forall k \geq 1.$$

If $(\mathbf{x}^*, W^*)$ is a global minimizer pair of $L(\mathbf{x}, W)$, then $L(\mathbf{x}^*, W^*) \leq L(\mathbf{x}_\mathbf{b}, W_\mathbf{b})$, where $W_\mathbf{b}$ is the weight vector associated with the sparse solution $\mathbf{x}_\mathbf{b}$. In this case, we can show that $\mathbf{x}^*$ will be equal to $\mathbf{x}_\mathbf{b}$ under an additional condition.

Denote by T_0 the index set of the support of the sparse solution $\mathbf{x_b}$. Consider $h = \mathbf{x}^* - \mathbf{x_b}$. Let T_1 be the index set of the s largest entries (in magnitude) of $h|_{T_0^c}$. For convenience, let $T_{01} = T_0 \cup T_1$. We are now ready to prove the following theorem.

Theorem 6.33. *Suppose that the sensing matrix Φ satisfies the RIP with RIC $\delta_{2s} < 1$. Suppose that $(\mathbf{x}^*, W^*)$ is a global minimizer pair of the function $L(\mathbf{x}, W)$. Let $h = \mathbf{x}^* - \mathbf{x_b}$. Define*

$$\beta = \frac{\sqrt{1 - \delta_{2s}}}{\sqrt{2s}(L - 1)},$$

where L is the smallest integer bigger than or equal to n/s. Letting

$$|x_{\min}| = \min_{\hat{x}_i \neq 0} |\mathbf{x}_{\mathbf{b},i}|,$$

if $|x_{\min}| \geq \epsilon(1/\beta - 1) + \|h_{T_{01}}\|_2$, then $\mathbf{x}^*$ is the sparse solution.

We need the following lemma to prove the above result.

Lemma 6.34. *For any nonzero $h \in \mathrm{Null}(\Phi)$, the s largest (in magnitude) entries of $h|_{T_0^c}$ satisfy the inequality*

$$|h_i| \geq \beta \|h_{T_{01}}\|_2 \quad \forall i \in T_1, \tag{6.117}$$

Proof. It is easy to see that $\|\Phi h_{T_{01}^c}\|_2 = \|\Phi h_{T_{01}}\|_2$. Let $T_i, i \geq 2$, be the index sets of $h_{T_{01}^c}$ with $\#(T_i) = s$, where the entries in T_i are decreasing in magnitude in T_i and decreasing in all $T_i, i = 2, 3, \ldots, L$. Such a decomposition was used many times in Chapter 7.

We use the well-known RIP of Φ to obtain

$$0 = \|\Phi h_{T_{01}^c}\|_2 - \|\Phi h_{T_{01}}\|_2 \leq \sum_{i=2}^{L} \|\Phi h_{T_i}\|_2 - \sqrt{1 - \delta_{2s}} \|h_{T_{01}}\|_2$$

$$\leq \sqrt{1 + \delta_s} \sum_{i=2}^{L} \|h_{T_i}\|_2 - \sqrt{1 - \delta_{2s}} \|h_{T_{01}}\|_2$$

$$\leq \sqrt{2}(L - 1)\|h_{T_2}\|_2 - \sqrt{1 - \delta_{2s}} \|h_{T_{01}}\|_2$$

$$\leq \sqrt{2s}(L - 1)|h_{T_2^1}| - \sqrt{1 - \delta_{2s}} \|h_{T_{01}}\|_2,$$

where T_2^1 is the first index in T_2. It follows that

$$|h_i| \geq |h_{T_2^1}| \geq \frac{\sqrt{1 - \delta_{2s}}}{\sqrt{2s}(L - 1)} \|h_{T_{01}}\|_2$$

for all $i \in T_1$. $\square$

Corollary 6.35. *Suppose that Φ possesses a RIP with RIC $\delta_{2s} < 1$. Let $\mathbf{x}^*$ be a solution to $\Phi \mathbf{x} = \mathbf{b}$. If $\|\mathbf{x}_{T_0^c}\|_0 \leq s$, then $\mathbf{x}^*$ is a sparse solution.*

Proof. Let $h = \mathbf{x}^* - \mathbf{x_b}$, with $\mathbf{x_b}$ being a sparse solution with sparsity s. As in the proof above, we have $\sqrt{1 - \delta_{2s}}\|h_{T_{01}}\|_2 \leq \sqrt{2s}(L - 1)\|h_{T_2}\|_2$. Since $\|\mathbf{x}_{T_0^c}\|_0 \leq s$, we have $h_{T_2} = 0$. It follows that $\|h_{T_{01}}\|_2 = 0$ or $h = 0$. That is, $\mathbf{x}^* = \mathbf{x_b}$. $\square$

Proof of Theorem **6.33.** Since $\mathbf{x}^*$ satisfies $\mathbf{b} = \Phi\mathbf{x}^*$, then $h = \mathbf{x}^* - \mathbf{x}_\mathbf{b}$ is in the null space of Φ. Since $L(\mathbf{x}^*, W^*) \leq L(\mathbf{x}_\mathbf{b}, W_\mathbf{b})$, by using the same argument as in Lemma 6.29, we have

$$\prod_{i=1}^{n}(|\mathbf{x}_i^*| + \epsilon) = \prod_{i=1}^{n}(|\widehat{\mathbf{x}}_i + h_i| + \epsilon) \leq \prod_{i=1}^{n}(|\widehat{\mathbf{x}}_i| + \epsilon),$$

where $\widehat{\mathbf{x}} := \mathbf{x}_\mathbf{b}$. That is,

$$\prod_{i \in T_0}\left(\frac{|\widehat{\mathbf{x}}_i + h_i| + \epsilon}{|\widehat{\mathbf{x}}_i| + \epsilon}\right) \leq \prod_{i \in T_0^c}\left(\frac{\epsilon}{|h_i| + \epsilon}\right). \tag{6.118}$$

Using Lemma 6.34, we now show that $h = 0$. Indeed, because each term on the right-hand side of (6.118) is less than or equal to 1, we have

$$\prod_{i \in T_0^c}\left(\frac{\epsilon}{|h_i| + \epsilon}\right) \leq \left(\frac{\epsilon}{\beta\|h_{T_{01}}\|_2 + \epsilon}\right)^s \tag{6.119}$$

by Lemma 6.34. But, on the other hand, we claim that

$$\prod_{i \in T_0}\left(\frac{|\widehat{\mathbf{x}}_i + h_i| + \epsilon}{|\widehat{\mathbf{x}}_i| + \epsilon}\right) \geq \prod_{i \in T_0}\left(\frac{|\widehat{\mathbf{x}}_i| - |h_i| + \epsilon}{|\widehat{\mathbf{x}}_i| + \epsilon}\right) > \left(\frac{\epsilon}{\beta\|h_{T_{01}}\|_2 + \epsilon}\right)^s. \tag{6.120}$$

Indeed, for each $i \in T_0$, if $h_i = 0$, we have

$$\frac{|\widehat{\mathbf{x}}_i| - |h_i| + \epsilon}{|\widehat{\mathbf{x}}_i| + \epsilon} = 1 > \frac{\epsilon}{\beta\|h_{T_{01}}\|_2 + \epsilon}.$$

If $h_i \neq 0$, we can show that

$$\frac{|\widehat{\mathbf{x}}_i| - |h_i| + \epsilon}{|\widehat{\mathbf{x}}_i| + \epsilon} > \frac{\epsilon}{\beta|h_i| + \epsilon} \geq \frac{\epsilon}{\beta\|h_{T_{01}}\|_2 + \epsilon}.$$

A simple elementary calculation shows that the first inequality in the above can be rewritten as

$$1 - \frac{\epsilon}{\beta|h_i| + \epsilon} > \frac{|h_i|}{\epsilon + |\widehat{\mathbf{x}}_i|} \quad \text{or} \quad |\widehat{\mathbf{x}}_i| > |h_i| + \epsilon\left(\frac{1}{\beta} - 1\right).$$

The second assumption ensures that the second inequality in the above equation holds. This establishes the claim. Using (6.119) and (6.120) together with (6.118), we get a contradiction. Thus, $h \equiv 0$ or $\mathbf{x}^* \equiv \widehat{\mathbf{x}}$. □

A MATLAB implementation of Algorithm 6.13 can be requested from Professor E. Candés. Its numerical performance will be shown at the end of this chapter.

6.9 ▪ The ℓ_1 Greedy Algorithm

In this section we present another ℓ_1 minimization algorithm which improves the performance of Algorithm 6.13. It is called the ℓ_1 greedy algorithm and was introduced in [16]. Let us begin with Algorithm 6.14 (below). To explain the computation, let us introduce two functions: For $\mathbf{x} = (x_1, x_2, \ldots, x_n)^\top$, $I_M(\mathbf{x})$ and $J_M(\mathbf{x})$ are two vectors defined by

$$I_M(\mathbf{x}) = (I_M(x_1), I_M(x_2), \ldots, I_M(x_n))^\top, J_M(\mathbf{x}) = (J_M(x_1), J_M(x_2), \ldots, J_M(x_n))^\top \tag{6.121}$$

with

$$I_M(x_i) = \begin{cases} 1 & \text{if } |x_i| < M, \\ 0 & \text{otherwise,} \end{cases} \quad \text{and} \quad J_M(x_i) = \begin{cases} 1 & \text{if } |x_i| \geq M, \\ 0 & \text{otherwise} \end{cases} \tag{6.122}$$

for $i = 1, \ldots, n$.

ALGORITHM 6.14

The ℓ_1 Greedy Algorithm [16]

We start with a weight vector $W^{(0)}$ with nonnegative entries, e.g., $W^{(0)} = (1, 1, \ldots, 1)^\top$.

1: **while** For $k = 1, \ldots, 25$ we compute **do**

    (1) Compute $A_w^{(k-1)} = \Phi \cdot /W^{(k-1)}$, where $\cdot$ is a MATLAB command for columnwise division.

    (2) Solve $\mathbf{x}_w = \min\{\|\mathbf{x}\|_1, \quad A_w^{(k-1)}\mathbf{x} = \mathbf{b}\}$ using an ℓ_1 minimization method.

    (3) $\mathbf{x}^{(k)} = \mathbf{x}_w \cdot /W^{(k-1)}$,

    where $\|\mathbf{x}\|_1 = \sum_{i=1}^n |x_i|$ is the standard ℓ_1 norm.

2:     If $k = 1$, let $M_k = \|\mathbf{x}^{(k)}\|_\infty$. Otherwise, $M_k = 0.87 M_{k-1}$.

3:     Let $W^{(k)} = I_{M_k}(\mathbf{x}^{(k)}) + 0.1 J_{M_k}(\mathbf{x}^{(k)})$.

4: **end while**

End the for loop unless the difference of the consecutive solutions is within a given tolerance.

We now give a convergence analysis of Algorithm 6.14. It is easy to see that steps (1) to (3) above are equivalent to solving

$$\mathbf{x}^{(k)} := \arg \min\{(W^{(k-1)})^\top |\mathbf{x}|, \quad \Phi\mathbf{x} = \mathbf{b}\}, \tag{6.123}$$

where $|\mathbf{x}| = (|x_1|, |x_2|, \ldots, |x_n|)^\top$ denotes the absolute value of $\mathbf{x} = (x_1, x_2, \ldots, x_n)^\top$. Note that $W^{(k-1)}$ divides the indices of $\mathbf{x}$ into two groups: one is the less important portion of the indices collected in $J_{M_{k-1}}$, which is scaled by 0.1, and the other is the more important portion of the indices denoted by $I_{M_{k-1}}$. Heuristically, in each step the larger components of the iterative solution $\mathbf{x}^{(k)}$ are found and moved into the less important group, while the smaller components of $\mathbf{x}^{(k)}$ need to be computed more accurately and hence are moved to the important group.

To study the convergence of the iterative solutions $\mathbf{x}^{(k)}$, we first show that $\|\mathbf{x}^{(k)}\|_1$ for all $k \geq 1$ are bounded. To this end, we define three functions:

$$L_M(\mathbf{x}) = \sum_{i=1}^n g_M(x_i) + 0.1 f_M(x_i), \tag{6.124}$$

where $g_M(x) = \min\{|x|, M\}$ and $f_M(x) = \max\{|x|, M\}$ for any $x \in (-\infty, \infty)$. Note that for each $x \in (0, \infty)$,

$$L_M(x) = g_M(x) + 0.1 f_M(x)$$

is concave. This can be seen in Figure 6.3.

It is easy to see that $L_M(x) \leq L_N(x)$ if $M < N$. A crucial observation is that the subgradients of L_M, g_M, and f_M are connected in the following way:

$$\partial L_M(x) = \partial g_M(x) + 0.1 \partial f_M(x) = I_M(x) + 0.1 J_M(x) \tag{6.125}$$

for each $x \in (-\infty, \infty)$. Also, $L_M(\mathbf{x}) = (\partial L_M(\mathbf{x}))^\top |\mathbf{x}|$. Steps (1) to (3) of Algorithm 6.14,

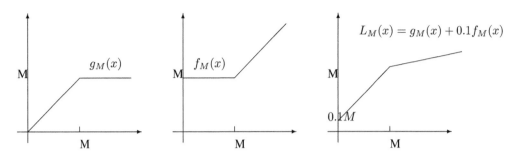

Figure 6.3. *Functions g_M, f_M, and L_M.*

i.e., (6.123), are

$$\mathbf{x}^{(k)} := \min_{\mathbf{x}}\{\partial L_{M_{k-1}}(\mathbf{x}^{k-1})^\top |\mathbf{x}|, \quad \Phi\mathbf{x} = \mathbf{b}\}, \tag{6.126}$$

where $|\mathbf{x}| = (|x_1|, |x_2|, \ldots, |x_n|)^\top$ for any $\mathbf{x} = (x_1, x_2, \ldots, x_n)^\top$.

We now claim that

$$L_{M_k}(\mathbf{x}^{(k+1)}) \le L_{M_{k-1}}(\mathbf{x}^{(k)}) \tag{6.127}$$

for all $k \ge 1$. Indeed, due to the concavity of L_M and (6.126), we have

$$
\begin{aligned}
L_{M_k}(\mathbf{x}^{(k+1)}) &\le L_{M_k}(\mathbf{x}^{(k)}) + \partial L_{M_k}(\mathbf{x}^{(k)})^\top (|\mathbf{x}^{(k+1)}| - |\mathbf{x}^{(k)}|) \\
&= L_{M_k}(\mathbf{x}^{(k)}) + \min_{\mathbf{x}} \partial L_{M_k}(\mathbf{x}^{(k)})^\top (|\mathbf{x}| - |\mathbf{x}^{(k)}|) \\
&\le L_{M_k}(\mathbf{x}^{(k)}) \le L_{M_{k-1}}(\mathbf{x}^{(k)})
\end{aligned}
$$

since $M_k \le M_{k-1}$. We therefore have the following result.

Lemma 6.36. *Suppose that $\|\mathbf{x}^{(2)}\|_1$ is bounded. Then there exists a convergent subsequence from $\mathbf{x}^{(k)}, k \ge 1$, and a limit $\mathbf{x}^*$ such that $\mathbf{x}^{(k_j)} \to \mathbf{x}^*$ as $j \to \infty$.*

Proof. By using (6.127), we have

$$
\begin{aligned}
0.1\|\mathbf{x}^{(k)}\|_1 &\le 0.1(g_{M_{k-1}}(\mathbf{x}^{(k)}) + f_{M_{k-1}}(\mathbf{x}^{(k)})) \le L_{M_{k-1}}(\mathbf{x}^{(k)}) \\
&\le \cdots \le L_{M_1}(\mathbf{x}^{(2)}) \le \|\mathbf{x}^{(2)}\|_1
\end{aligned}
$$

for each $k \ge 1$. It follows that $\mathbf{x}^{(k)}, k \ge 1$, are bounded, and hence there exists a convergent subsequence of $\mathbf{x}^{(k)}, k \ge 1$, and a limit $\mathbf{x}^*$ such that $\mathbf{x}^{(k_j)} \to \mathbf{x}^*$ for $j \to \infty$.   □

Lemma 6.37. *Let $\widehat{\mathbf{x}}$ be the sparsest vector which satisfies $\Phi\mathbf{x} = \mathbf{b}$. Then the limit $\mathbf{x}^*$ of any subsequence of $\mathbf{x}^{(k)}$ satisfies*

$$\|\mathbf{x}^*\|_1 \le \|\widehat{\mathbf{x}}\|_1. \tag{6.128}$$

Furthermore, if $\mathbf{x}^*$ and $\mathbf{y}^*$ are the two limits of the subsequences of $\mathbf{x}^{(k)}$, then $\|\mathbf{x}^*\|_1 = \|\mathbf{y}^*\|_1$.

Proof. Let $\alpha = \min_{\widehat{x}_i \neq 0} |\widehat{x}_i| > 0$. For k large enough, we have $M_k < \alpha$, and hence $L_{M_k}(\mathbf{x}^{(k+1)}) \le L_{M_k}(\widehat{\mathbf{x}}) = 0.1\|\widehat{\mathbf{x}}\|_1$. It follows that $0.1\|\mathbf{x}^*\|_1 \le 0.1\|\widehat{\mathbf{x}}\|_1$ since $L_{M_{k_j}}(\mathbf{x}^*) \to 0.1\|\mathbf{x}^*\|_1$. Thus, we have (6.128).

Similarly, we have $0.1\|\mathbf{x}^*\|_1 \le L_{M_{k_j}}(\mathbf{y}^*)$ for $j \to \infty$. That is, $0.1\|\mathbf{x}^*\|_1 \le 0.1\|\mathbf{y}^*\|_1$. This statement can be reversed. This completes the proof.   □

Therefore, we have obtained the following result.

Theorem 6.38. *Suppose that the sparse solution* $\widehat{\mathbf{x}}$ *is solved by the standard* ℓ_1 *minimization*

$$\widehat{\mathbf{x}} := \min_{\mathbf{x} \in \mathbb{R}^n} \{\|\mathbf{x}\|_1 : \quad \Phi\mathbf{x} = \mathbf{b}\}. \tag{6.129}$$

For example, the RIC δ_{2s} *of* Φ *satisfies* $\delta_{2s} < 1$ *or* δ_s *of* Φ *satisfies* $\delta_s < 1/3$ *(see Chapter 7 or [4]). Then Algorithm* 6.14 *converges and the limit* $\mathbf{x}^*$ *is equal to* $\widehat{\mathbf{x}}$.

Proof. By Lemma 6.37 above, the limit $\mathbf{x}^*$ of any subsequence from $\mathbf{x}^{(k)}$ obtained from Algorithm 6.14 satisfies $\|\mathbf{x}^*\|_1 \leq \|\widehat{\mathbf{x}}\|_1$ and $\Phi\mathbf{x}^* = \mathbf{b}$. It follows that $\mathbf{x}^* = \widehat{\mathbf{x}}$. Thus, Algorithm 6.14 converges.   □

A MATLAB implementation of Algorithm 6.14 based on Algorithm 6.1 can be found below. Its numerical performance will be demonstrated at the end of this chapter.

```
function x0=lai2012(A,y)
%   x0=lai2012(A,y)
%This is a MATLAB program to find the sparsest
%solution of rectangular matrix A of size m x n with
% m <n such that A x=y based on Kozlov and Petukhov's strategy together
%  with Lai's L1min.m  and starting with zero initial guess.
%It was implemented by Ming-Jun Lai in Aug. 2012.
NIt=15; %This number is dependent on the size of sensing matrix A.
[m,n]=size(A); AW=A;
W=ones(n,1);
j0=1;
iv=zeros(n,1); x0=iv;
for i=1:NIt
x0=L1min(AW,y,1e-9,iv);
x0=x0./W;
if i==1
[Mx j0]=max(abs(x0));
now;
Mx=Mx*0.87;
 W=(abs(x0)>Mx)/10+(abs(x0)<=Mx);
for j=1:n
    AW(:,j)=A(:,j)/W(j);
end
end
```

6.10 ▪ Numerical Results

In this section, we present a numerical test to find sparse solutions by the various methods discussed so far. For simplicity, we use Gaussian random matrices of size 64×128 and 64×256 with sparsity from 1 to 45. We test Algorithm 6.14 based on magicl1 (called KP for short in Figures 6.4 and 6.5), Algorithm 6.14 based on Algorithm 6.1 (called Lai for short), Algorithm 6.13 (called CWB for short), Algorithm 6.4 (called FISTA for short), and the simplex L_1 method discussed in this chapter.

Since all methods can find sparse solutions when the sparsity is less than 10, we only show the performance for sparsities from 10 to 45. Most methods fail to find a sparse solution when

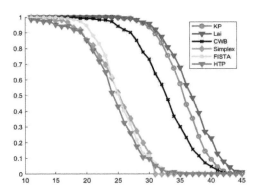

Figure 6.4. *Frequency of successes of various algorithms for a sensing matrix of size* 64×128.

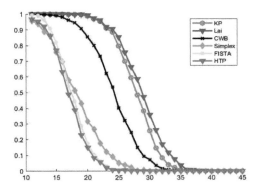

Figure 6.5. *Frequency of successes of various algorithms for a sensing matrix of size* 64×256.

the sparsity is larger than 45. If a method finds a solution within the maximum norm 10^{-4}, the method is said to succeed for this run. Otherwise, the method is not successful in searching for sparse solutions. We repeated this test for 1000 runs to compute the frequency of success, which is the percentage of successes over the 1000 runs.

Successes for the KP, Lai, CWB, SimplexL1, FISTA, and HTP methods are shown in Figures 6.4 and 6.5. We can see that the Lai method performs the best for most cases of sparsity, although it is the slowest one due to the repeated solution of linear systems.

From Figures 6.4 and 6.5, we can see that when n is larger, the performance of all methods decreases. So when $m \ll n$, the methods discussed so far are challenged. More study and research are needed.

6.11 ▪ Exercises

Exercise 1. In Algorithm 6.1, show that one can replace Step 2.2 by

$$\mathbf{w}^k = \mathbf{w}^{k-1} + \frac{\beta}{\alpha_k}\mathbf{p}^k$$

for any fixed $\beta \in (0,1)$ and the resulting algorithm still works.

Exercise 2. Adapt Algorithm 6.1 to solve the least absolute deviation (LAD) regression problem

$$\min\{\|\Psi\mathbf{x} - \mathbf{b}\|_1, \quad H\mathbf{x} = 0\}$$

for a matrix Ψ of size $n \times m$ with $n > m$ and a vector $\mathbf{b}$ of size $n \times 1$, where H is a matrix of size $k \times n$.

Exercise 3. Show that all three forms of linear programming problem (6.5), (6.6), (6.7) are equivalent.

Exercise 4. Prove Theorem 6.2.

Exercise 5. Implement Algorithm 6.13 in MATLAB and experiment with your code numerically for sparse solutions of underdetermined linear systems.

Exercise 6. Implement Algorithm 6.3 in MATLAB and experiment with your code numerically for sparse solutions of underdetermined linear systems.

Exercise 7. Implement Algorithm 6.4 in MATLAB and experiment with your code numerically for sparse solutions of underdetermined linear systems.

Exercise 8. Implement Algorithm 6.2 in MATLAB and experiment with your code for computing the projection of a vector $\mathbf{x}$ to the convex set $\{\mathbf{x} \in \mathbb{R}^n, \Phi\mathbf{x} = \mathbf{b}\}$.

Exercise 9. Show that if in Algorithm 6.4, the parameters in (5.46) are replaced by

$$\alpha_k = \beta k, \quad k \text{ large enough}, \tag{6.130}$$

where $\beta \in (0, 1/2]$, and the update rule in (5.47) is replaced by

$$\mathbf{y}^{(k+1)} = P_C\left(\mathbf{x}^{(k+1)} + \frac{\beta k - 1 + \beta}{\beta k + \beta}(\mathbf{x}^{(k+1)} - \mathbf{x}^{(k)})\right), \tag{6.131}$$

then the new algorithm is also convergent with convergence rate $O(1/k^2)$.

Exercise 10. Implement Algorithm 6.6 in MATLAB and perform some numerical experiments.

Exercise 11. Implement Algorithm 6.7 in MATLAB and perform some numerical experiments. Compare the performance with Algorithm 6.6 to confirm that this one is faster.

Exercise 12. Implement Algorithm 6.9 in MATLAB and experiment with your code numerically for sparse solutions of underdetermined linear systems.

Exercise 13. Prove Theorem 6.11.

Exercise 14. Prove Theorem 6.12.

Exercise 15. Prove Theorem 6.13.

Exercise 16. Consider the constrained problem

$$\min \left\{ \begin{array}{c} \frac{1}{2\lambda}F(\mathbf{x}) + \sum_{i=1}^n t_i \\ \text{such that } t_i^2 = |\mathbf{x}_i|, i = 1, \ldots, n, \\ t_i \geq 0, i = 1, \ldots, n \end{array} \right\}, \tag{6.132}$$

where $F(\mathbf{x}) = \|\Phi\mathbf{x} - \mathbf{b}\|_2^2$. Show that this is equivalent to

$$\min \sum_{i=1}^n |\mathbf{x}_i|^{1/2} + \frac{1}{2\lambda}\|\Phi\mathbf{x} - \mathbf{b}\|_2^2. \tag{6.133}$$

Exercise 17. Relax the minimization problem in (6.132) to

$$
\min \left\{
\begin{array}{c}
\frac{1}{2\lambda} F(\mathbf{x}) + \sum_{i=1}^{n} t_i \\
\text{such that } t_i^2 \mathbf{x}_i \geq 0, i = 1, \ldots, n, \\
t_i^2 + \mathbf{x}_i \geq 0, i = 1, \ldots, n, \\
t_i \geq 0, i = 1, \ldots, n
\end{array}
\right\}.
\tag{6.134}
$$

Show that the two minimization problems (6.132) and (6.134) are equivalent. More precisely, show the following.

Lemma 6.39. *If* $\mathbf{x}^*$ *is a minimizer of* (6.133), *it is a solution of* (6.134). *Also, if* $\mathbf{x}^*$ *is a solution of* (6.134), *it is a solution of* (6.133).

Exercise 18. Use the well-known Uzawa algorithm (cf. [6]) for the minimization (1.28) with $q = 1$. That is, consider

$$
\min_{\mathbf{x} \in \mathbb{R}^n} \{ \|\mathbf{x}\|_{1,\epsilon}, \quad \Phi \mathbf{x} = \mathbf{b} \},
\tag{6.135}
$$

where we recall that $\|\mathbf{x}\|_{1,\epsilon^2} = \sum_{i=1}^{n} \sqrt{|x_i|^2 + \epsilon^2}$. Note that we rewrite the minimization problem (6.135) as

$$
\min_{\mathbf{x} \in \mathbb{R}^n} \{ \|\mathbf{x}\|_{1,\epsilon}, \quad \Phi \mathbf{x} \leq \mathbf{b}, -\Phi \mathbf{x} \leq -\mathbf{b} \}.
\tag{6.136}
$$

Implement Uzawa's algorithm for (6.136) to find sparse solutions of underdetermined linear systems for various $\epsilon > 0$.

Exercise 19. An iteratively reweighted least squares algorithm was introduced in [8] to find sparse solutions of underdetermined linear systems. Let

$$
\mathcal{J}(\mathbf{x}, \omega, \epsilon) = \frac{1}{2} \left(\sum_{i=1}^{n} (\mathbf{x}_i)^2 \omega_i + \sum_{j=1}^{n} (\epsilon^2 \omega_j + 1/\omega_j) \right).
$$

Recall that $r(\mathbf{z})$ is the nonincreasing rearrangement of the absolute values of the entries of $\mathbf{z}$.

ALGORITHM 6.15
Iteratively Reweighted Least Squares Algorithm [8]
1: Initializing by $\omega^{(0)} = (1, 1, \ldots, 1)^\top \in \mathbb{R}^n$, $\epsilon_0 = 1$, we iteratively compute

$$
\mathbf{x}^{(k+1)} := \arg\min\{ \mathcal{J}(\mathbf{z}, \omega^{(k)}, \epsilon_k), \quad \Phi \mathbf{z} = \mathbf{b} \}
\tag{6.137}
$$

and

$$
\epsilon_{k+1} = \min \left\{ \epsilon_k, \frac{r(\mathbf{x}^{(k+1)})_{s+1}}{N} \right\},
\tag{6.138}
$$

where s is a fixed integer and

$$
\omega^{(k+1)} = \arg\min_{\omega > 0} \mathcal{J}(\mathbf{x}^{(k+1)}, \omega, \epsilon_{k+1})
\tag{6.139}
$$

for $k \geq 0$.

Study the convergence of the algorithm. More precisely, show that

$$
\mathcal{J}(\mathbf{x}^{(k)}, \omega^{(k)}, \epsilon_k) = \left(\sum_{i=1}^{n} \sqrt{(\mathbf{x}_i^{(k)})^2 + \epsilon_k^2} \right)
\tag{6.140}
$$

for all $k \geq 1$. Next, show that

$$\mathcal{J}(\mathbf{x}^{(k+1)}, \omega^{k+1}, \epsilon_{k+1}) \leq \mathcal{J}(\mathbf{x}^{(k+1)}, \omega^{(k)}, \epsilon_{k+1}) \leq \mathcal{J}(\mathbf{x}^{(k+1)}, \omega^{(k)}, \epsilon_k) \leq \mathcal{J}(\mathbf{x}^{(k)}, \omega^{(k)}, \epsilon_k)$$
(6.141)

and $\epsilon_{k+1} \leq \epsilon_k$ for all $k \geq 0$. Finally, show that there exists a constant $C = \mathcal{J}(\mathbf{x}^{(1)}, \omega^{(0)}, \omega_0)$ such that

$$\|\mathbf{x}^{(k)}\|_1 \leq \mathcal{J}(\mathbf{x}^{(k)}, \omega^{(k)}, \epsilon_k) \leq C \quad \text{and} \quad \omega_j^{(k)} \geq 1/C \quad \forall j = 1, \ldots, N.$$
(6.142)

Exercise 20. Furthermore, prove the following lemma.

Lemma 6.40. *For any* $\mathbf{b} \in \mathbb{R}^n$, *let* $\mathbf{x}^{(k)}, k \geq 1$, *from Algorithm 6.15. Then*

$$\sum_{k \geq 1} \|\mathbf{x}^{(k+1)} - \mathbf{x}^{(k)}\|_2^2 \leq 2C^2,$$

where C *is the constant in Exercise 19.*

Exercise 21. Show that if $\mathbf{x}^{(k)}, \mathbf{x}^{(k+1)}, \omega^k$ are from Algorithm 6.15, then

$$\langle \mathbf{x}^{(k)} - \mathbf{x}^{(k+1)}, \mathbf{x}^{(k+1)} \odot \omega^k \rangle = 0.$$

Exercise 22. Show that the sequence $\mathbf{x}^{(k)}$ from Algorithm 6.15 converges. That is, prove the following theorem.

Theorem 6.41 (Daubechies, DeVore, Fornasier, and Gurtunk, 2010 [8]). *Suppose that* Φ *satisfies the NSP of order* s *with* $\gamma < 1$. *Let* $\epsilon_* = \lim_{k \to \infty} \epsilon_k$. *Then the sequence* $\mathbf{x}^{(k)}, k \geq 0$, *from Algorithm 6.15 converges. Let* $\mathbf{x}^{\epsilon_*}$ *be the limit. If* $\epsilon_* = 0$, $\mathbf{x}^{\epsilon_*}$ *is a sparse solution. If* $\epsilon_* > 0$, *we have*

$$\|\mathbf{x_b} - \mathbf{x}^{\epsilon_*}\|_1 \leq \frac{1 + \gamma_s}{1 - \gamma_s} \sigma_s(\mathbf{x}^{\epsilon_*})_1$$

and, for $t < s$ *such that* $\mu = \frac{1+\gamma}{1-\gamma} \frac{1}{s+1-t} < 1$,

$$\|\mathbf{x_b} - \mathbf{x}^{\epsilon_*}\|_1 \leq \frac{\mu}{1 - \mu} \sigma_t(\mathbf{x_b})_1,$$

which implies that $\mathbf{x}^{\epsilon_*}$ *is the sparse solution of* $\|\mathbf{x_b}\|_0 \leq t$.

Exercise 23. Implement Algorithm 6.15 in MATLAB and experiment with your code numerically for sparse solutions of underdetermined linear systems.

Exercise 24. Prove Lemma 6.19.

Exercise 25. Prove Lemma 6.21.

Exercise 26. Prove the nonexpansive property (6.88) of the proximal operator. Furthermore, show that the following inequality is true:

$$(x - y)(\text{Prox}_\phi(x) - \text{Prox}_\phi(y)) \geq (\text{Prox}_\phi(x) - \text{Prox}_\phi(y))^2.$$
(6.143)

Exercise 27. Extend the results in the above exercise to the multidimensional setting.

Exercise 28. The purpose of this exercise is to understand restricted strong convexity (RSC) better. Let g be a convex function on $W \subset \mathbb{R}^n$. Let W^* be the collection of all minimizers of g over W. We define RSC as follows.

Definition 6.42. *A Lipschitz differentiable function g is RSC with respect to W^* over a convex set W if there exists a positive constant ν_g such that*

$$\langle \nabla g(\mathbf{x}) - \nabla g(\mathbf{x}^*), \mathbf{x} - \mathbf{x}^* \rangle \geq \nu_g \|\mathbf{x} - \mathbf{x}^*\|^2, \quad \mathbf{x} \in W. \tag{6.144}$$

Show that if g is strongly convex, g is RSC with respect to any convex set in W. Next find an example of an RSC function g which is not strongly convex.

Exercise 29. Prove the following lemma.

Lemma 6.43. *Let $f = g + r$, where g is convex and Lipschitz differentiable and r is convex only over W. Let W^* be the collection of the minimizers of f over W. Suppose that g is RSC with respect to W^*. Then there exists a positive constant θ_f such that*

$$f(\mathbf{x}) - f(\mathbf{x}^*) \geq \theta_f \|\mathbf{x} - \mathbf{x}^*\|^2 \quad \forall \mathbf{x} \in W. \tag{6.145}$$

We say that f is semistrongly convex (SSC) if f satisfies (6.145).

Exercise 30. For any $\mathbf{x} \in \mathbb{R}^n$, let

$$f(\mathbf{x}) = \max\{\|\mathbf{x}\|^2 - 1, 0\}. \tag{6.146}$$

Show that $f(\mathbf{x})$ is a convex function, but not strongly convex. Furthermore, show that f satisfies the SSC property (6.145).

Exercise 31. Furthermore, show that the minimal functional $f(\mathbf{x}) = \frac{1}{2}\|\Phi\mathbf{x} - \mathbf{b}\|^2 + \lambda\|\mathbf{x}\|_1$ associated with the lasso problem satisfies the SSC condition.

Exercise 32. Prove the nonexpansive property in (6.100).

Exercise 33. Prove Proposition 6.24.

Exercise 34. Implement Algorithm 6.11 in MATLAB and experiment with the percentage of columns that can be removed for various λ and $\mathbf{b}$.

Exercise 35. Show that $L_M(\mathbf{x}) = (\partial L_M(\mathbf{x}))^\top |\mathbf{x}|$.

Bibliography

[1] H. Attouch and J. Peypouquet, The rate of convergence of Nesterov's accelerated forward-backward method is actually faster than $1/k^2$, SIAM J. Optim., 26 (2016), 1824–1834. (Cited on p. 162)

[2] A. Beck and M. Teboulle, A fast iterative shrinkage-thresholding algorithm for linear inverse problems, SIAM J. Imaging Sci., 2 (2009), 183–202. (Cited on p. 162)

[3] S. Boyd and L. Vandenberghe, Convex Optimization, Cambridge University Press, 2004. (Cited on p. 175)

[4] T. T. Cai and A. Zhang, Sharp RIP bound for sparse signal and low-rank matrix recovery, Appl. Comput. Harmon. Anal., 35 (2013), 74–93. (Cited on p. 187)

[5] E. J. Candès, M. B. Wakin, and S. Boyd, Enhancing sparsity by reweighted ℓ_1 minimization, J. Fourier Anal. Appl., 14 (2008), 877–905. (Cited on p. 180)

[6] P. Ciarlet, Introduction to Numerical Linear Algebra and Optimization, Cambridge University Press, 1989. (Cited on pp. 149, 151, 164, 190)

[7] P. L. Combettes and V. R. Wajs, Signal recovery by proximal forward-backward splitting, Multiscale Model. Simul., 4 (2005), 1168–1200. (Cited on p. 173)

[8] I. Daubechies, R. DeVore, M. Fornasier, and C. S. Gunturk, Iteratively re-weighted least squares minimization for sparse recovery, Commun. Pure Appl. Math., 63 (2010), 1–38. (Cited on pp. 180, 190, 191)

[9] F. Deutsch, Best Approximation in Inner Product Spaces, Canadian Mathematical Society, 2001. (Cited on p. 160)

[10] W. Deng, M.-J. Lai, Z. Peng, and W. Yin, Parallel multi-block ADMM with $o(1/k)$ convergence, J. Sci. Comput., 71 (2017), 712–736. (Cited on p. 150)

[11] B. Efron, T. Hastie, I. Johnstoe, and R. Tibshirani, Least angle regression, Ann. Statist., 32 (2004), 407–499. (Cited on p. 149)

[12] S. Foucart and D. Koslicki, Sparse recovery by means of nonnegative least squares, IEEE Signal Process. Lett., 21/4 (2014), 498–502. (Cited on p. 163)

[13] J. Friedman, T. Hastie, and R. Tibshirani, Additive logistic regression, a statistical view of boosting, Ann. Statist., 28 (2002), 337–407. (Cited on p. 149)

[14] R. Q. Jia, H. Q. Zhang, and W. Zhang, Convergence analysis of the Bregman method for the variational model of image denoising, Appl. Comput. Harmon. Anal., 27 (2009), 367–379. (Cited on p. 165)

[15] D. Koslicki, S. Foucart, and G. Rosen, Quikr: A method for rapid reconstruction of bacterial communities via compressive sensing, Bioinform., 29 (2013), 2096–2102. (Cited on p. 163)

[16] I. Kozlov and A. Petukhov, Sparse solution of underdetermined linear systems, in Handbook of Geomathematics, W. Freeden, M. Z. Nashed, and T. Sonar (Eds.), Springer, 2010, pp. 1243–1259. (Cited on pp. 184, 185)

[17] M.-J. Lai and P. Wenston, L^1 spline methods for scattered data interpolation and approximation, Adv. Comput. Math., 21 (2004), 293–315. (Cited on pp. 149, 155)

[18] M.-J. Lai, Y. Y. Xu, and W. T. Yin, Improved iteratively reweighted least squares for unconstrained smoothed ℓ_q minimization, SIAM J. Numer. Anal., 51 (2013), 927–957. (Cited on p. 180)

[19] M.-J. Lai and W. T. Yin, Augmented ℓ_1 and nuclear-norm models with a globally linearly convergent algorithm, SIAM J. Imaging Sci., 6 (2013), 1059–1091. (Cited on pp. 166, 170, 171)

[20] Y. E. Nesterov, A method for solving the convex programming problem with convergence rate $O(1/k^2)$, Dokl. Akad. Nauk SSSR, 269 (1983), 543–547. (Cited on p. 161)

[21] Y. Nesterov, Efficiency of coordinate descent methods on huge-scale optimization problems, SIAM J. Optim., 22 (2012), 341–362. (Cited on p. 161)

[22] Y. Nesterov, Gradient methods for minimizing composite objective function, Math. Program. Ser. B, 140 (2013), 125–161. (Cited on pp. 159, 161)

[23] Y. Nesterov, Introductory Lectures on Convex Optimization, Kluwer Academic, 2004. (Cited on p. 161)

[24] S. Ross and J. Zhu, Piecewise linear regularization solution paths, Ann. Statist., 35 (2007), 1012–1030. (Cited on p. 149)

[25] R. Tibshirani, Regression shrinkage and selection via the lasso, J. Roy. Statist. Soc. Ser. B, 58 (1996), 267–288. (Cited on p. 149)

[26] J. Wang and J. Ye, Lasso screening rules via dual polytope projection, J. Mach. Learn. Res., 16 (2015), 1063–1101. (Cited on pp. 174, 178, 179, 180)

Chapter 7

Recoverability of Sparse Solutions

In this chapter, we present some sufficient conditions in terms of RIP which ensure that the minimizers of (1.20) or (1.27) are the sparse solution of underdetermined linear system (1.4). We start with several easy cases and gradually present more complicated cases. We shall present the optimal bounds of δ_{ts} with $t > 0$ and sparsity $s > 0$ for the ℓ_1 minimization and show that these bounds cannot be improved. Each proof is based on completely different ideas and techniques. Furthermore, we extend some results based on δ_{2s} to the setting of the ℓ_q minimization (1.27).

We shall also discuss the noisy recovery setting. All of the recovery conditions for δ_{ts} are sufficient for recovering the sparse solution in the noisy setting. We present some results in a few settings and leave most cases to the interested reader and/or exercises. Finally, we present some miscellaneous conditions for the minimizer of (1.20) to be a sparse solution.

7.1 ▪ Conditions Based on δ_{3s} for ℓ_1 Minimizers

We start with conditions for δ_{3s} to ensure that ℓ_1 minimization finds the s-sparse solution exactly. That is, we consider noiseless recovery.

Theorem 7.1 (An Easy Case). *Suppose that for an integer $s \geq 1$, Φ satisfies the RIP with*

$$\delta_{3s} < \frac{1}{2}.$$

Let $\mathbf{x_b} \in \mathbb{R}^n$ be a vector with $\|\mathbf{x_b}\|_0 \leq s$. Then for $\mathbf{b} = \Phi\mathbf{x_b}$, the solution $\mathbf{x}^*$ of minimization (1.20) is unique and equal to $\mathbf{x_b}$.

Proof. Let $h = \mathbf{x_b} - \mathbf{x}^* \in \mathcal{N}(\Phi)$, the null space of matrix Φ. Suppose that $h \neq 0$. We are going to use the null space property (NSP), i.e., Theorem 2.5. Let $S_0 \subset \{1, \ldots, n\}$ be an index set of size s. Consider sets $S_1, S_2, \ldots, S_j, \ldots$ of s indices ordered by decreasing magnitude of entries of $h_{S_0^c}$, where S_0^c is the complement of S_0 in $\{1, 2, \ldots, n\}$. It is easy to see that $\Phi h_{S_0 \cup S_1} = \Phi h - \sum_{j \geq 2} \Phi h_{S_j} = -\sum_{j \geq 2} \Phi h_{S_j}$. Then by the RIP (cf. (2.23) in Chapter 2),

$$\|h_{S_0 \cup S_1}\|_2^2 \leq \frac{1}{1 - \delta_{2s}} \|\Phi(h_{S_0 \cup S_1})\|_2^2 = \frac{1}{1 - \delta_{2s}} \sum_{j \geq 2} \langle \Phi(h_{S_0 \cup S_1}), -\Phi(h_{S_j}) \rangle$$

$$\leq \frac{\delta_{3s}}{1 - \delta_{2s}} \|h_{S_0 \cup S_1}\|_2 \sum_{j \geq 2} \|h_{S_j}\|_2,$$

where we have used Lemma 2.31. Dividing by $\|h_{S_0 \cup S_1}\|_2$ and using the inequality $\|h_{S_j}\|_2 \leq \frac{1}{\sqrt{s}} \|h_{S_{j-1}}\|_1$ for $j \geq 2$, we obtain

$$\|h_{S_0 \cup S_1}\|_2 \leq \frac{\delta_{3s}}{1 - \delta_{2s}} \frac{1}{\sqrt{s}} \|h_{S_0^c}\|_1.$$

That is, $\|h_{S_0}\|_1 \leq \sqrt{s}\|h_{S_0}\|_2 \leq \sqrt{s}\|h_{S_0 \cup S_1}\|_2 \leq \frac{\delta_{3s}}{1-\delta_{3s}}\|h_{S_0^c}\|_1$ by the above inequality with $\delta_{2s} \leq \delta_{3s} < 1$. Whenever $\delta_{3s} < 1/2$, we have $\frac{\delta_{3s}}{1-\delta_{3s}} < 1$ and hence

$$\|h_{S_0}\|_1 < \|h_{S_0^c}\|_1.$$

Because S_0 is an arbitrary subset of $\{1, \ldots, n\}$, we use the NSP, i.e., Theorem 2.5, and the ℓ_1 minimization gives the unique sparse solution. Thus, $\mathbf{x}^* = \mathbf{x_b}$. $\square$

Next we have the following theorem.

Theorem 7.2 (Candès and Tao, 2005 [7]). *Let* $\mathbf{x} \in \mathbb{R}^n$ *have sparsity* $\|\mathbf{x_b}\|_0 = s$. *Suppose* Φ *is a matrix satisfying*

$$\delta_{3s} + 3\delta_{4s} < 2. \tag{7.1}$$

Then $\mathbf{x_b}$ *is the unique minimizer of* (1.20).

Proof. Let $\mathbf{x}^*$ be a minimizer of (1.20) and $h = \mathbf{x}^* - \mathbf{x_b}$ be a vector in the null space of Φ. Next, let T_0 be the support of $\mathbf{x_b}$. It is easy to see that

$$\left| \|\mathbf{x_b}\|_1 - \|h_{T_0}\|_1 \right| \leq \|\mathbf{x_b} + h_{T_0}\|_1.$$

Denote by T_0^c the complement of T_0 in $\{1, 2, \ldots, n\}$. We have

$$\|\mathbf{x_b}\|_1 - \|h_{T_0}\|_1 + \|h_{T_0^c}\|_1 \leq \|\mathbf{x_b} + h_{T_0} + h_{T_0^c}\|_1 = \|\mathbf{x}^*\|_1 \leq \|\mathbf{x_b}\|_1.$$

It follows that $\|h_{T_0^c}\|_1 \leq \|h_{T_0}\|_1$.

Let $a \geq 1$ be an integer. Next, we divide T_0^c into $T_1, T_2, \ldots,$ with T_1 being the set of the indices of the as largest entries in magnitude of $h_{T_0^c}$, T_2 being the set of the indices of the next as largest entries, and so on. Note that the last index set T_J for some J may have fewer than as entries. We now decompose Φh as follows:

$$0 = \|\Phi h\|_2 = \|\Phi(h_{T_0 \cup T_1}) + \sum_{j \geq 2} \Phi h_{T_j}\|_2 \geq \|\Phi(h_{T_0 \cup T_1})\|_2 - \left\| \sum_{j \geq 2} \Phi h_{T_j} \right\|_2$$

$$\geq \|\Phi(h_{T_0 \cup T_1})\|_2 - \sum_{j \geq 2} \|\Phi h_{T_j}\|_2 \geq \sqrt{1 - \delta_{(1+a)s}} \|h_{T_0 \cup T_1}\|_2 - \sqrt{1 + \delta_{as}} \sum_{j \geq 2} \|h_{T_j}\|_2.$$

Note that $\|h_{T_{j+1}}\|_2 \leq \sqrt{as} \|h_{T_{j+1}}\|_\infty \leq \sqrt{as} \frac{\|h_{T_j}\|_1}{as} = \frac{\|h_{T_j}\|_1}{\sqrt{as}}$. It follows that

$$0 \geq \sqrt{1 - \delta_{(1+a)s}} \|h_{T_0 \cup T_1}\|_2 - \sqrt{1 + \delta_{as}} \sum_{j \geq 1} \frac{\|h_{T_j}\|_1}{\sqrt{as}}$$

$$\geq \sqrt{1 - \delta_{(1+a)s}} \|h_{T_0}\|_2 - \sqrt{1 + \delta_{as}} \|h_{T_0^c}\|_1 \frac{1}{\sqrt{as}}$$

$$\geq \sqrt{1 - \delta_{(1+a)s}} \|h_{T_0}\|_2 - \sqrt{1 + \delta_{as}} \|h_{T_0}\|_1 \frac{1}{\sqrt{as}};$$

where we have used the inequality $\|h_{T_0^c}\|_1 \le \|h_{T_0}\|_1$ as proved above. By using the Cauchy–Schwarz inequality, we have $\|h_{T_0}\|_1 \le \sqrt{s}\|h_{T_0}\|_2$, and hence

$$0 \ge \left(\sqrt{1 - \delta_{(1+a)s}} - \sqrt{1 + \delta_{as}}\, \frac{\sqrt{s}}{\sqrt{as}} \right) \|h_{T_0}\|_2. \tag{7.2}$$

By the condition in (7.1), with $a = 3$, we have

$$1 - \delta_{4s} > (1 + \delta_{3s})/3.$$

The coefficient of $\|h_{T_0}\|_2$ in (7.2) is strictly positive. It follows that $\|h_{T_0}\|_2 = 0$. Using $\|h_{T_0^c}\|_1 \le \|h_{T_0}\|_1$, we conclude that $h \equiv 0$ or $\mathbf{x}^* = \mathbf{x_b}$. □

The above proof can be extended to ℓ_q minimization. We have the following result, whose proof is left to the interested reader; see also [8].

Theorem 7.3 (Chartrand and Steneva, 2008 [8]). *Let* $\mathbf{x} \in \mathbb{R}^n$ *have sparsity* $\|\mathbf{x_b}\|_0 = s$. *Suppose* $0 < q \le 1$ *and* $b > 1$. *Let* $a = b^{q/(2-q)}$. *Suppose* Φ *is a matrix satisfying*

$$\delta_{as} + b\delta_{(a+1)s} < b - 1. \tag{7.3}$$

Then the ℓ_q *minimization* (1.27) *has a unique minimizer which is exactly* $\mathbf{x_b}$.

We remark that since b is very large and q is very small, we have $a \approx 1$, and the inequality in (7.3) shows that

$$\delta_{2s} < \frac{b-1}{b} - \frac{\delta_s}{b} \approx 1.$$

Of course, the above is just a heuristic argument. We shall make it more precise in section 7.5.

7.2 ▪ Conditions Based on δ_{2s} for ℓ_1 Minimizers

We now develop conditions for δ_{2s} to ensure that the ℓ_1 minimization finds the s-sparse solution exactly. We begin with an easy condition.

Theorem 7.4. *If the restricted isometry constant (RIC)* δ_{2s} *of a matrix* Φ *satisfies* $\delta_{2s} < 1/3$, *then the NSP of order* s *of* Φ *is satisfied and every* s-sparse *vector* $\mathbf{x_b} \in \mathbb{R}^n$ *is recovered by* ℓ_1 *minimization.*

Proof. As before, let $h \in \mathcal{N}(\Phi)$, the null space of matrix Φ. Suppose that $h \ne 0$. Let S_0 be the index set of the s largest nonzero entries in absolute value of h. Consider sets $S_1, \dots, S_j, \dots$ of s indices ordered by decreasing magnitude of entries of $h_{S_0^c}$, where S_0^c is the complement of S in $\{1, 2, \dots, n\}$. It is easy to see that $\Phi h_{S_0} = \Phi h - \sum_{j\ge 1} \Phi h_{S_j} = -\sum_{j\ge 1} \Phi h_{S_j}$. Then

$$\|h_{S_0}\|_2^2 \le \frac{1}{1-\delta_s} \|\Phi(h_{S_0})\|_2^2 \le \frac{1}{1-\delta_{2s}} \sum_{j\ge 1} \langle \Phi(h_{S_0}), -\Phi(h_{S_j}) \rangle$$

$$\le \frac{\delta_{2s}}{1-\delta_{2s}} \|h_{S_0}\|_2 \sum_{j\ge 1} \|h_{S_j}\|_2,$$

where we have used Lemma 2.31. Dividing by $\|h_{S_0}\|_2$ and using the inequality $\|h_{S_j}\|_2 \le \frac{1}{\sqrt{s}}\|h_{S_{j-1}}\|_1$ for $j \ge 1$, we obtain

$$\|h_{S_0}\|_1 \le \sqrt{s}\|h_{S_0}\|_2 \le \frac{\sqrt{s}\delta_{2s}}{1-\delta_{2s}} \frac{1}{\sqrt{s}} (\|h_{S_0}\|_1 + \|h_{S_0^c}\|_1).$$

That is, $(1 - \frac{\delta_{2s}}{1-\delta_{2s}})\|h_{S_0}\|_1 \le \frac{\delta_{2s}}{1-\delta_{2s}}\|h_{S_0^c}\|_1$. That is,

$$\|h_{S_0}\|_1 \le \frac{\delta_{2s}}{1 - 2\delta_{2s}}\|h_{S_0^c}\|_1 < \|h_{S_0^c}\|_1,$$

since $\delta_{2s} < 1/3$. Now for any set S_0 of size s, let S_1 be the set of indices in S_0^c of the s largest nonzero entries in absolute value of h. Similarly, we let $S_2, \dots, S_j, \dots$. Then we have

$$\|h_{S_0 \cup S_1}\|_2^2 \le \frac{1}{1 - \delta_{2s}} \sum_{j \ge 2} \langle \Phi(h_{S_0}) + \Phi(h_{S_1}), -\Phi(h_{S_j}) \rangle$$

$$\le \frac{\delta_{2s}}{1 - \delta_{2s}}(\|h_{S_0}\|_2 + \|h_{S_1}\|_2) \sum_{j \ge 2} \|h_{S_j}\|_2 \le \frac{2\delta_{2s}}{1 - \delta_{2s}}\|h_{S_1}\|_2 \sum_{j \ge 2} \|h_{S_j}\|_2.$$

That is, $\|h_{S_1}\|_2 \le \frac{2\delta_{2s}}{1-\delta_{2s}} \sum_{j \ge 2} \|h_{S_j}\|_2$. It follows that

$$\|h_{S_0}\|_1 \le \|h_{S_1}\|_1 \le \sqrt{s}\|h_{S_1}\|_2 \le \frac{2\sqrt{s}\delta_{2s}}{1 - \delta_{2s}}\frac{1}{\sqrt{s}} \sum_{j \ge 1} \|h_{S_j}\|_1 = \frac{2\delta_{2s}}{1 - \delta_{2s}}\|h_{S_0^c}\| < \|h_{S_0^c}\|_1.$$

Thus, h satisfies the NSP. Therefore, the ℓ_1 minimization gives the unique sparse solution. □

Next we present the following well-known result.

Theorem 7.5 (Candès, 2008 [6]). *Suppose that for $s \ge 1$, Φ satisfies*

$$\delta_{2s} < \sqrt{2} - 1 \approx 0.4142.$$

Let $\mathbf{x_b} \in \mathbb{R}^n$ be a vector with $\|\mathbf{x_b}\|_0 \le s$. Then for $\mathbf{b} = \Phi\mathbf{x_b}$, the solution $\mathbf{x}^*$ of minimization (1.20) *is unique and equal to $\mathbf{x_b}$.*

Proof. Let $h = \mathbf{x}^* - \mathbf{x_b}$. Then h is in the null space of Φ. That is, $\Phi h = 0$. Rewrite h as a sum of vectors $h_{T_0}, h_{T_1}, h_{T_2}, \dots$, each of sparsity at most s. Here, T_0 corresponds to the locations of the s entries of $\mathbf{x_b}$, T_1 to the locations of the s largest entries of $h_{T_0^c}$, T_2 to the locations of the next s largest entries of $h_{T_0^c}$, and so on, where T_0^c stands for the complement index set of T_0 in $\{1, 2, \dots, n\}$. As before, it is easy to see that

$$\|h_{T_j}\|_2 \le s^{1/2}\|h_{T_j}\|_\infty \le s^{-1/2}\|h_{T_{j-1}}\|_1.$$

It follows that

$$\sum_{j \ge 2} \|h_{T_j}\|_2 \le s^{-1/2}(\|h_{T_1}\|_2 + \|h_{T_2}\|_1 + \cdots) = s^{-1/2}\|h_{T_0^c}\|_1. \tag{7.4}$$

Recall from (2.4) that we have

$$\|h_{T_0^c}\|_1 \le \|h_{T_0}\|_1 + 2\|(\mathbf{x_b})_{T_0^c}\|_1 = \|h_{T_0}\|_1. \tag{7.5}$$

Together with (7.4), we use the Cauchy–Schwarz inequality for bounding $\|h_{T0}\|_1$ by $s^{1/2}\|h_{T0}\|_2$ and get

$$\sum_{j \ge 2} \|h_{T_j}\|_2 \le s^{-1/2}\|h_{T_0}\|_1 \le \|h_{T_0}\|_2. \tag{7.6}$$

Since $\|h_{T_0}\|_2 \leq \|h_{T_0 \cup T_1}\|_2$ and $\|h_{(T_0 \cup T_1)^c}\|_2 \leq \sum_{j \geq 2} \|h_{T_j}\|_2$, the above estimate can be rewritten as

$$\sum_{j \geq 2} \|h_{T_j}\|_2 \leq \|h_{T_0 \cup T_1}\|_2. \tag{7.7}$$

Next, as seen before, $\Phi(h_{T_0 \cup T_1}) = \Phi h - \Phi(\sum_{j \geq 2} h_{T_j}) = -\Phi(h_{(T_0 \cup T_1)^c})$. Thus, we apply Lemma 2.31 and (7.7) to get

$$\begin{aligned}
\|\Phi(h_{T_0 \cup T_1})\|_2^2 &= -\langle \Phi(h_{T_0 \cup T_1}), \Phi(h_{(T_0 \cup T_1)^c}) \rangle \\
&= -\sum_{j \geq 2} \langle \Phi(h_{T_0}), \Phi(h_{T_j}) \rangle - \sum_{j \geq 2} \langle \Phi(h_{T_1}), \Phi(h_{T_j}) \rangle \\
&\leq \delta_{2s}(\|h_{T_0}\|_2 + \|h_{T_1}\|_2) \sum_{j \geq 2} \|h_{T_j}\|_2 \\
&\leq \delta_{2s} \sqrt{2} \|h_{T_0 \cup T_1}\|_2 \|h_{T_0 \cup T_1}\|_2 = \delta_{2s} \sqrt{2} \|h_{T_0 \cup T_1}\|_2^2.
\end{aligned}$$

We now use the RIP to conclude that

$$(1 - \delta_{2s}) \|h_{T_0 \cup T_1}\|_2^2 \leq \|\Phi(h_{T_0 \cup T_1})\|_2^2 \leq \delta_{2s} \sqrt{2} \|h_{T_0 \cup T_1}\|_2^2.$$

That is,

$$\|h_{T_0 \cup T_1}\|_2 \leq \frac{\delta_{2s} \sqrt{2}}{1 - \delta_{2s}} \|h_{T_0 \cup T_1}\|_2^2. \tag{7.8}$$

That is, if $\delta_{2s} < 1/(1 + \sqrt{2}) = \sqrt{2} - 1$, the above is a contradiction unless $h_{T_0 \cup T_1} \equiv 0$. This is the desired result of Theorem 7.5. $\square$

The proof above can easily be improved. Instead of $\|h_{T_0}\|_2 + \|h_{T_1}\|_2 \leq \sqrt{2}\|h_{T_0} + h_{T_1}\|_2$, we use the following argument. By using (7.6),

$$\begin{aligned}
\|\Phi(h_{T_0 \cup T_1})\|_2^2 &= -\langle \Phi(h_{T_0 \cup T_1}), \Phi(h_{(T_0 \cup T_1)^c}) \rangle \\
&= -\sum_{j \geq 2} \langle \Phi(h_{T_0 \cup T_1}), \Phi h_{T_j} \rangle \\
&= -\sum_{j \geq 2} \langle \Phi(h_{T_0}), \Phi h_{T_j} \rangle - \sum_{j \geq 2} \langle \Phi(h_{T_1}), \Phi h_{T_j} \rangle \\
&\leq \delta_{2s}(\|h_{T_0}\|_2 + \|h_{T_1}\|_2) \sum_{j \geq 2} \|h_{T_j}\|_2 \\
&\leq \delta_{2s}(\|h_{T_0}\|_2 + \|h_{T_1}\|_2)\|h_{T_0}\|_2 \\
&\leq \delta_{2s}(\|h_{T_0}\|_2^2 + \|h_{T_0}\|_2\|h_{T_1}\|_2).
\end{aligned}$$

Using the RIP, we have

$$(1 - \delta_{2s})(\|h_{T_0}\|_2^2 + \|h_{T_1}\|_2^2) = (1 - \delta_{2s})\|h_{T_0 \cup T_1}\|_2^2 \leq \|\Phi(h_{T_0 \cup T_1})\|_2^2.$$

It follows that

$$(1 - 2\delta_{2s})\|h_{T_0}\|_2^2 + (1 - \delta_{2s})\|h_{T_1}\|_2^2 \leq \delta_{2s}\|h_{T_0}\|_2\|h_{T_1}\|_2. \tag{7.9}$$

Using the standard inequality $xy \leq \alpha x^2 + \frac{1}{4\alpha} y^2$ for any positive α, the right-hand side of inequality (7.9) can be rewritten as

$$(1 - 2\delta_{2s})\|h_{T_0}\|_2^2 + (1 - \delta_{2s})\|h_{T_1}\|_2^2 \leq \alpha \delta_{2s}\|h_{T_0}\|_2^2 + \frac{1}{4\alpha} \delta_{2s}\|h_{T_1}\|_2^2, \tag{7.10}$$

and hence

$$(1 - (2 + \alpha)\delta_{2s})\|h_{T_0}\|_2^2 + \left(1 - \left(1 + \frac{1}{4\alpha}\right)\delta_{2s}\right)\|h_{T_1}\|_2^2 \le 0. \qquad (7.11)$$

By choosing α such that $(2 + \alpha) = 1 + \frac{1}{4\alpha}$, we have $\alpha = (\sqrt{2} - 1)/2$. Therefore, with $2 + \alpha = (\sqrt{2} + 3)/2$, we have

$$\left(1 - \frac{\sqrt{2} + 3}{2}\delta_{2s}\right)\|h_{T_0 \cup T_1}\|_2^2 \le 0.$$

If $1 > \frac{\sqrt{2}+3}{2}\delta_{2s}$ or $\delta_{2s} < 2/(3 + \sqrt{2}) = 0.4531$, we conclude that $\|h_{T_0 \cup T_1}\|_2 = 0$, and hence $h \equiv 0$ and $\mathbf{x}^*$ is the exact sparse solution $\mathbf{x}$. This completes the proof of the following theorem.

Theorem 7.6 (Foucart and Lai, 2009 [10]). *Suppose that $s \ge 1$ such that*

$$\delta_{2s} < \frac{2}{3 + \sqrt{2}} \approx 0.4531,$$

and let $\mathbf{x_b} \in \mathbb{R}^n$ be a vector with $\|\mathbf{x_b}\|_0 \le s$. Then for $\mathbf{b} = \Phi\mathbf{x_b}$, the solution of (1.20) is unique and equal to $\mathbf{x_b}$.

These results can be further improved. To describe the improved results, we need the following shift inequality.

Lemma 7.7 (Shift Inequality). *Let q, r be positive integers satisfying $q \le 3r$. Then any nonincreasing sequence of real numbers*

$$a_1 \ge a_2 \ge \cdots \ge a_r \ge b_1 \ge \cdots \ge b_q \ge c_1 \ge \cdots \ge c_r \ge 0$$

satisfies

$$\sqrt{\sum_{i=1}^q b_i^2 + \sum_{i=1}^r c_i^2} \le \frac{\sum_{i=1}^r a_i + \sum_{i=1}^q b_i}{\sqrt{r + q}}.$$

A particular case is any nonincreasing sequence of real numbers $a_1 \ge \cdots \ge a_q \ge 0$ that satisfies

$$\sqrt{\sum_{i=1}^q b_i^2 + rb_q^2} \le \frac{rb_1 + \sum_{i=1}^q b_i}{\sqrt{r + q}},$$

where $3r \ge q$.

Proof. Let us prove a special case as follows. Let us write $b_i = b_{i+1} + d_i$ for $i = 1, \ldots, q - 1$. Then

$$(q + r)\left(\sum_{i=1}^q b_i^2 + rb_q^2\right) = (r + q)\left(rb_q^2 + \sum_{i=1}^q \left(b_q + \sum_{j=i}^{q-1} d_j\right)^2\right)$$

$$= (r + q)\left((r + q)b_q^2 + 2b_q\sum_{i=1}^q \sum_{j=i}^{q-1} d_j + \sum_{i=1}^q \left(\sum_{j=i}^{q-1} d_j\right)^2\right)$$

$$= (r + q)^2 b_q^2 + 2b_q(r + q)\sum_{i=1}^{q-1} id_i + (r + q)\sum_{i=1}^q \left(\sum_{j=i}^{q-1} d_j\right)^2.$$

On the other hand, we have

$$\left(rb_1 + \sum_{i=1}^{q} b_i\right)^2 = \left((r+q)b_q + \sum_{i=1}^{q-1}(r+i)d_i\right)^2$$

$$= (q+r)^2 b_q^2 + 2(q+r)b_q \sum_{i=1}^{q-1}(r+i)d_i + \left(\sum_{i=1}^{q-1}(r+i)d_i\right)^2.$$

Since $d_i \geq 0$, we have $2(q+r)b_q \sum_{i=1}^{q-1}(r+i)d_i \geq 2b_q(r+q)\sum_{i=1}^{q-1} id_i$. Also, we claim that $(\sum_{i=1}^{q-1}(r+i)d_i)^2 \geq (r+q)\sum_{i=1}^{q}(\sum_{j=i}^{q-1} d_j)^2$. Indeed, let us look at the coefficient of term $d_i d_j$ on both sides for $1 \leq i < j \leq q-1$. On the left, we have $2(r+i)(r+j)$, and on the right, $2(q+r)i$. When $q \leq 3r$, we have

$$(r+i)(r+j) \geq (r+i)^2 = (r-i)^2 + 4ri \geq 4ri \geq (q+r)i,$$

which establishes the claim. Then the inequality for the special case is established.

We leave the proof for the general case to the interested reader. □

With the above inequality we can show the following result.

Theorem 7.8 (Cai, Wang, and Xu, 2010 [1]). *Suppose that $s \geq 1$ such that*

$$\delta_{s+a} + \sqrt{\frac{s}{b}}\theta_{s+a,b} < 1$$

for positive numbers a and b with $b \leq 4a$. For any $\mathbf{x_b} \in \mathbb{R}^n$ with $\|\mathbf{x_b}\|_0 \leq s$, the solution of (1.20) *is unique for $\mathbf{b} = \Phi\mathbf{x_b}$ and is equal to $\mathbf{x_b}$. Furthermore,*

$$\|\mathbf{x} - \mathbf{x}^*\|_2 \leq C_0 b^{-1/2}\|\mathbf{x}_{T_0^c}\|_1$$

for a constant $C_0 > 0$ that depends on $\delta_{s+a}, \theta_{s+a,b}$.

Proof. Let $h = \mathbf{x} - \mathbf{x}^*$ be a vector in the null space of Φ. We fix integers $a > 0$ and $b > 0$ with $b \leq 4a$ and divide h into a sum of vectors $h_{T_0}, h_{T_1}, h_{T_2}, \ldots$, where T_0 corresponds to the locations of the s largest entries of $\mathbf{x}$, T_1 to the locations of the a largest entries of $h_{T_0^c}$, T_2 to the locations of the next b largest entries of $h_{(T_0 \cup T_1)^c}$, T_3 to the locations of the next b largest entries of $h_{(T_0 \cup T_1 \cup T_2)^c}$, and so on, where T_0^c stands for the complement index set of T_0 in $\{1, 2, \ldots, n\}$, and similar for $(T_0 \cup T_1)^c$. We further divide T_j into two parts: $T_{j,1}$ and $T_{j,2}$, with $T_{j,1}$ being the index set of the $b-a$ largest entries in T_j and $T_{j,2}$ being the complement of $T_{j,1}$ in T_j for $j \geq 2$. Applying the shift inequality (Lemma 7.7), we have

$$\|h_{T_2}\|_2 \leq \frac{\|h_{T_1}\|_1 + \|h_{T_{2,1}}\|_1}{\sqrt{b}} \text{ and } \|h_{T_j}\|_2 \leq \frac{\|h_{T_{j-1,2}}\|_1 + \|h_{T_{j,1}}\|_1}{\sqrt{b}} \quad \forall j \geq 3.$$

Adding these inequalities yields

$$\sum_{j \geq 2} \|h_j\|_2 \leq \frac{1}{\sqrt{b}}\left(\|h_{T_1}\|_1 + \sum_{j \geq 2}\|h_{T_j}\|_1\right) = \frac{\|h_{T_0^c}\|_1}{\sqrt{b}}$$

$$\leq \frac{\|h_{T_0}\|_1 + 2\|\mathbf{x}_{T_0^c}\|_1}{\sqrt{t}} \leq \sqrt{\frac{s}{b}}\|h_{T_0}\|_2 + \frac{2\|\mathbf{x}_{T_0^c}\|_1}{\sqrt{b}},$$

where we have used the property (2.4). As before, we use $\Phi h = 0$ and the inequality above to obtain

$$
(1 - \delta_{s+a})\|h_{T_0 \cup T_1}\|_2^2 \leq -\sum_{j \geq 2} \langle \Phi(h_{T_0 \cup T_1}), \Phi(h_{T_j}) \rangle
$$

$$
\leq \theta_{s+a,b}\|h_{T_0 \cup T_1}\|_2 \|h_{(T_0 \cup T_1)^c}\|_2
$$

$$
\leq \theta_{s+a,b}\sqrt{\frac{s}{b}}\|h_{T_0 \cup T_1}\|_2^2 + \theta_{s+a,b}\|h_{T_0 \cup T_1}\|_2 \frac{2\|\mathbf{x}_{T_0^c}\|_1}{\sqrt{b}}
$$

or

$$
(1 - (\delta_{s+a} + \sqrt{s/b}\theta_{s+a,b}))\|h_{T_0 \cup T_1}\|_2 \leq \theta_{s+a,b}\frac{2\|\mathbf{x}_{T_0^c}\|_1}{\sqrt{b}}.
$$

That is, if $\delta_{s+a} + \sqrt{\frac{s}{b}}\theta_{s+a,b} < 1$, we have

$$
\|h_{T_0 \cup T_1}\|_2 \leq \frac{\theta_{s+a,b}}{1 - (\delta_{s+a} + \sqrt{\frac{s}{b}}\theta_{s+a,b})} \frac{2\|\mathbf{x}_{T_0^c}\|_1}{\sqrt{b}}. \tag{7.12}
$$

Finally, we have

$$
\|h\|_2 \leq \|h_{T_0 \cup T_1}\|_2 + \|h_{(T_0 \cup T_1)^c}\|_2 \leq \left(1 + \sqrt{\frac{s}{b}}\right)\|h_{T_0 \cup T_1}\|_2 + \frac{2\|\mathbf{x}_{T_0^c}\|_1}{\sqrt{b}}.
$$

Using (7.12), we have

$$
\|h\|_2 \leq \left(\frac{\theta_{s+a,b}(1 + \sqrt{\frac{s}{b}})}{1 - (\delta_{s+a} + \sqrt{\frac{s}{b}}\theta_{s+a,b})} + 1\right)\frac{2\|\mathbf{x}_{T_0^c}\|_1}{\sqrt{b}}. \qquad \square
$$

Here are a few examples of how to use Theorem 7.8.

Corollary 7.9. *Suppose s is an integer divisible by 4, i.e., $s = 4r$ for some positive integer r. Suppose that $\delta_{2s} < \frac{1}{1+\sqrt{1.25}} \approx 0.4751$. Then the solution of (1.20) is the sparse solution.*

Proof. Using $a = s/4$ and $b = s$, we have $\theta_{s+a,b} = \theta_{1.25s,s} \leq \sqrt{1.25}\delta_{2s}$ by Lemmas 2.35 and 2.31. Because $\delta_{1.25s} \leq \delta_{2s}$, we have

$$
\delta_{1.25s} + \sqrt{\frac{s}{b}}\theta_{1.25k,b} \leq \delta_{2s}(1 + \sqrt{1.25}) < 1.
$$

Hence we can apply Theorem 7.8 to conclude the result. $\qquad \square$

Corollary 7.10. *Suppose s is an even integer and suppose that $\delta_{3s} < 2(2 - \sqrt{3})$. Then the solution of (1.20) is the sparse solution.*

Proof. Using $a = s/2$ and $b = 2s$, we have the condition $\delta_{s+a} + \sqrt{s/b}\theta_{s+a,b} = \delta_{1.5s} + \frac{1}{\sqrt{2}}\theta_{1.5s,2s}$. We use the monotone property $\delta_{1.5s} \leq \delta_{3s}$ and Lemma 2.35 to get

$$
\delta_{s+a} + \sqrt{s/b}\theta_{s+a,b} < \delta_{3s} + \sqrt{1.5/2}\theta_{s,2s} = \delta_{3s}\left(1 + \frac{\sqrt{3}}{2}\right) < 1
$$

by the assumption. Thus, Theorem 7.8 implies that the minimizer of (1.20) is the sparse solution. $\qquad \square$

Next we improve the sufficient conditions on δ_{2s} as follows.

Theorem 7.11 (Mo and Li, 2011 [12]). *Suppose $\delta_{2s} < 0.4931$. Then the solution $\mathbf{x}^*$ of the minimization problem (1.20) is the sparse solution $\mathbf{x_b}$ satisfying $\Phi\mathbf{x_b} = \mathbf{b}$. Furthermore, $\mathbf{x}^*$ satisfies*

$$\|\mathbf{x_b} - \mathbf{x}^*\|_2 \leq C_0 \||\mathbf{x}^*|_{T_0^c}\|_1, \tag{7.13}$$

where C_0 is a positive constant depending on δ_{2s}, T_0 is the index set of the support of $\mathbf{x_b}$, and T_0^c is the complement of T_0 in $\{1, 2, \ldots, n\}$.

To prove this result, we need some more notation and a series of lemmas. Let T_0 be the supporting index of a sparse solution of $\Phi\mathbf{x} = \mathbf{b}$ with $\#(T_0) = s$. For example, T_0 could be the support of $\mathbf{x_b}$. Recall that $\mathbf{x}^*$ is the minimizer of (1.20). Let $h = \mathbf{x_b} - \mathbf{x}^*$ and $\tau(h)$ be a positive number satisfying

$$\|h_{T_0}\|_1 = \tau(h)\|h_{T_0^c}\|_1. \tag{7.14}$$

The purpose of the study is to show how to make $\tau(h) < 1$ for all nonzero vectors h in the null space of Φ so that the property (2.4) can be applied to get a contradiction: $\|h_{T_0^c}\|_1 < \|h_{T_0^c}\|_1$.

For any nonzero vector h in the null space of Φ, we rewrite h as a sum of vectors h_{T_0}, h_{T_1}, h_{T_2}, $\ldots$, each of sparsity at most s. Here, T_0 corresponds to the locations of the s largest entries of $\mathbf{x_b}$, T_1 to the locations of the s largest entries of $h_{T_0^c}$, T_2 to the locations of the next s largest entries of $h_{T_0^c}$, and so on, where T_0^c stands for the complement index set of T_0 in $\{1, 2, \ldots, n\}$. Without loss of generality, we may assume that $h = (h_{T_0}, h_{T_1}, h_{T_2}, \ldots)^\top$, with the cardinality of T_i being equal to s for all $i = 0, 1, 2, \ldots$. Let $t \in [0, 1]$ be another positive number such that $\|h_{T_1}\|_1 = t \sum_{i\geq 1} \|h_{T_i}\|_1$. First of all, we have the following lemma.

Lemma 7.12.

$$\sum_{i\geq 2} \|h_{T_i}\|_2^2 \leq \frac{1}{s}(1-t)t \left(\sum_{i\geq 1} \|h_{T_i}\|_1\right)^2. \tag{7.15}$$

Proof. It is easy to see that

$$\sum_{i\geq 2} \|h_{T_i}\|_2^2 \leq |h_{2s+1}| \sum_{i\geq 2} \|h_{T_i}\|_1 \leq \left(\frac{\|h_{T_1}\|_1}{s}\right) \sum_{i\geq 2} \|h_{T_i}\|_1$$

$$\leq \left(\frac{\|h_{T_1}\|_1}{s}\right)\frac{1-t}{t}\|h_{T_1}\|_1 \leq \frac{1}{s}(1-t)t \left(\sum_{i\geq 1} \|h_{T_i}\|_1\right)^2.$$

The result in (7.15) follows. $\quad\square$

Next we have the following result.

Lemma 7.13.

$$\sum_{i\geq 2} \|h_{T_i}\|_2 \leq \frac{1 - 3t/4}{s^{1/2}} \left(\sum_{i\geq 1} \|h_{T_i}\|_1\right). \tag{7.16}$$

Proof. By Lemma 7.24 in section 7.6, we have

$$s^{1/2}\|h_{T_i}\|_2 \leq \|h_{T_i}\|_1 + \frac{s}{4}(|h_{is+1}| - |h_{is+s}|)$$

for $i \geq 2$. It follows that

$$s^{1/2} \sum_{i \geq 2} \|h_{T_i}\|_2 \leq \sum_{i \geq 2} \|h_{T_i}\|_1 + \frac{1}{4}\|h_{T_1}\|_1 \leq \left(1 - \frac{3t}{4}\right) \sum_{i \geq 1} \|h_{T_i}\|_1.$$

This completes the proof.   □

On the other hand, we have the following lemma.

Lemma 7.14.

$$\|\Phi(h_{T_0} + h_{T_1})\|_2^2 \geq \frac{1 - \delta_{2s}}{s}(\tau(h)^2 + t^2)\left(\sum_{i \geq 1} \|h_{T_i}\|_1\right)^2. \tag{7.17}$$

Proof. By the definition of δ_{2s} and using (7.14), we have

$$\|\Phi(h_{T_0} + h_{T_1})\|_2^2 \geq (1 - \delta_{2s})\|h_{T_0} + h_{T_1}\|_2^2 = (1 - \delta_{2s})(\|h_{T_0}\|_2^2 + \|h_{T_1}\|_2^2)$$

$$\geq \frac{1 - \delta_{2s}}{s}(\tau(h)^2 + t^2)\left(\sum_{i \geq 1} \|h_{T_i}\|_1\right)^2.$$

Therefore, we have the desired inequality.   □

As before, we know that $\Phi(h_{T_0} + h_{T_1}) = \Phi h - \Phi(\sum_{j \geq 2} h_{T_j}) = -\Phi(\sum_{i \geq 2} h_{T_i})$. We have the following estimate.

Lemma 7.15.

$$\|\Phi(h_{T_0} + h_{T_1})\|_2^2 = \left\|\Phi\left(\sum_{j \geq 2} h_{T_j}\right)\right\|_2^2 \leq \left(\frac{(1 - t)t + \delta_{2s}(1 - 3t/4)^2}{s}\right)\left(\sum_{i \geq 1} \|h_{T_i}\|_1\right)^2. \tag{7.18}$$

Proof. A straightforward calculation shows that

$$\left\|\Phi\left(\sum_{j \geq 2} h_{T_j}\right)\right\|_2^2 = \sum_{i,j \geq 2} \langle \Phi(h_{T_i}), \Phi(h_{T_j})\rangle$$

$$= \sum_{j \geq 2} \langle \Phi(h_{T_j}), \Phi(h_{T_j})\rangle + 2 \sum_{2 \leq i < j} \langle \Phi(h_{T_i}), \Phi(h_{T_j})\rangle$$

$$\leq (1 + \delta_s) \sum_{i \geq 2} \|h_{T_i}\|_2^2 + 2\delta_{2s} \sum_{i > j \geq 2} \|h_{T_i}\|_2 \|h_{T_j}\|_2$$

$$\leq \sum_{i \geq 2} \|h_{T_i}\|_2^2 + \delta_{2s} \left(\sum_{i \geq 2} \|h_{T_i}\|_2\right)^2$$

$$\leq \left(\frac{(1 - t)t}{s} + \frac{\delta_{2s}(1 - 3t/4)^2}{s}\right)\left(\sum_{i \geq 1} \|h_{T_i}\|_1\right)^2.   □$$

By using (7.17) and (7.18), we have

$$(1 - \delta_{2s})(\tau(h)^2 + t^2) \leq (1 - t)t + \delta_{2s}(1 - 3t/4)^2$$

or

$$\tau(h)^2 \leq (\delta_{2s} + (1 - 3\delta_{2s}/2)t - (2 - 25\delta_{2s}/16)t^2)/(1 - \delta_{2s}). \qquad (7.19)$$

It is easy to find the maximum of the right-hand side as a function of $t \in [0, 1]$ when $\delta_{2s} < 2/3$. The maximal value at $t = (1 - 3\delta_{2s}/2)/(4 - 25\delta_{2s}/8)$ is

$$\tau(h) \leq \sqrt{\frac{4(1 + 5\delta_{2s} - 4\delta_{2s}^2)}{(1 - \delta_{2s})(32 - 25\delta_{2s})}} < 1,$$

when $\delta_{2s} \approx 0.4931$. These establish the first part of Theorem 7.11. The second part, i.e., inequality (7.13), is left as Exercise 4. These complete the proof of Theorem 7.11.

The result is further improved in [3] for $\delta_{2s} < 1/2$ as in the following.

Theorem 7.16 (Cai and Zhang, 2013 [3]). *Suppose that for an integer $s \geq 1$, Φ satisfies the RIP with*

$$\delta_{2s} < \frac{1}{2}.$$

Let $\mathbf{x_b} \in \mathbb{R}^n$ be a vector with $\|\mathbf{x_b}\|_0 \leq s$. Then for $\mathbf{b} = \Phi\mathbf{x_b}$, the solution $\mathbf{x}^*$ of minimization (1.20) is unique and equal to $\mathbf{x_b}$.

The proof is left to Exercise 13. We refer the reader to [9], [2], and [4] for more related results. Furthermore, the sufficient condition is improved to be $\delta_{2s} < 1/\sqrt{2}$, which is the best possible. It will be explained and proved in the next section.

7.3 ▪ Conditions Based on δ_{ts} with $t \geq 4/3$

Theorem 7.17 (Cai and Zhang, 2014 [5]). *Suppose that for an integer $s \geq 1$, Φ satisfies the RIP with*

$$\delta_{ts} < \sqrt{(t - 1)/t},$$

where $t \geq 4/3$. Let $\mathbf{x_b} \in \mathbb{R}^n$ be a vector with $\|\mathbf{x_b}\|_0 \leq s$. Then for $\mathbf{b} = \Phi\mathbf{x_b}$, the solution $\mathbf{x}^*$ of minimization (1.20) is unique and equal to $\mathbf{x_b}$.

Proving this result requires an interesting result on polytopes and an identity. To describe the polytope result, we need to define a special polytope. For a positive number a and a positive integer s, define the polytope $T(a, s)$ by

$$T(a, s) = \{\mathbf{v} \in \mathbb{R}^n : \|\mathbf{v}\|_\infty \leq a, \|\mathbf{v}\|_1 \leq sa\}.$$

Next, define the set of sparse vectors $U(a, s, \mathbf{v}) \subset \mathbb{R}^n$ for each $\mathbf{v} \in \mathbb{R}^n$ by

$$U(a, s, \mathbf{v}) = \{\mathbf{u} \in \mathbb{R}^n : \text{supp}(\mathbf{u}) \subset \text{supp}(\mathbf{v}), \|\mathbf{u}\|_0 \leq s, \|\mathbf{u}\|_1 = \|\mathbf{v}\|_1, \|\mathbf{u}\|_\infty \leq a\}.$$

Lemma 7.18 (Sparse Representation of a Polytope). $\mathbf{v} \in T(a, s)$ *if and only if $\mathbf{v}$ is in the convex hull of $U(a, s, \mathbf{v})$. That is, any $\mathbf{v} \in T(a, s)$ can be expressed as*

$$\mathbf{v} = \sum_{i=1}^{n} \lambda_i \mathbf{u}_i, \quad 0 \leq \lambda_i \leq 1, \quad \sum_{i=1}^{n} \lambda_i = 1, \quad and \quad \mathbf{u}_i \in U(a, s, \mathbf{v}).$$

Proof. An induction proof can be used to established this result. That is, assume that $\mathbf{v}$ is an ℓ-sparse vector, i.e., $\|\mathbf{v}\|_0 = \ell \geq 1$. For $\ell \leq s$, the result is true. For $\ell > s$, we use induction on ℓ to establish the statement of this lemma. The details are left to the interested reader   □

We are ready to prove Theorem 7.17.

Proof of Theorem 7.17. By the NSP, i.e., Theorem 2.5, we need to show, for any nonzero $\mathbf{v} \in \mathcal{N}(\Phi)$, that
$$\|\mathbf{v}_S\|_1 < \|\mathbf{v}_{S^c}\|_1 \quad \forall S \subset \{1, \ldots, N\} \text{ and } |S| \leq s$$
under the assumption of δ_{ts}. Suppose the opposite is true, that there exists a nonzero $\mathbf{v} \in \mathcal{N}(\Phi)$ such that the above inequality does not hold. Letting $a = \|\mathbf{v}_S\|_1/s$, we divide $h = \mathbf{v}_{S^c}$ into two parts, $h = h^{(1)} + h^{(2)}$, with
$$h^{(1)} = \{h_i, |h_i| > a/(t-1)\} \quad \text{and} \quad h^{(2)} = \{h_i, |h_i| \leq a/(t-1)\}. \qquad (7.20)$$

It is easy to see that $\|h^{(1)}\|_1 \leq \|h\|_1 \leq \|\mathbf{v}_S\|_1 = sa$ since the vector $\mathbf{v}$ does not satisfy the NSP. Let us say that $\|h^{(1)}\|_0 = m$. Since $\mathbf{v} \in U(a, s, \mathbf{v})$, it must satisfy
$$\sum_i \lambda_i \|\mathbf{u}_i\|_2^2 \leq \sum_i \lambda_i \|\mathbf{u}_i\|_1 \|\mathbf{u}_i\|_\infty \leq sa^2.$$

Then $sa \geq \|h^{(1)}\|_1 \geq ma/(t-1)$. That is, $m \leq (t-1)s$. In addition, we have
$$\|h^{(2)}\|_1 = \|h\|_1 - \|h^{(1)}\|_1 \leq \|\mathbf{v}_S\|_1 - \frac{ma}{t-1} \leq sa - \frac{ma}{t-1} = (s(t-1) - m)\frac{a}{t-1}.$$

Since $\|h^{(2)}\|_\infty \leq a/(t-1)$, we apply Lemma 7.18 to $h^{(2)}$ with sparsity $s(t-1) - m$. So $h^{(2)}$ can be written as a convex combination of some sparse vectors. That is,
$$h^{(2)} = \sum_i \lambda_i \mathbf{u}_i,$$

with $\lambda_i \geq 0$, $\sum_i \lambda_i = 1$, $\mathbf{u}_i \subset h^{(2)}$, $\|\mathbf{u}_i\|_1 = \|h^{(2)}\|_1$, and $\|\mathbf{u}_i\|_\infty \leq a/(t-1)$, where $\|\mathbf{u}_i\|_0 \leq s(t-1) - m$ for all i.
It follows that
$$\|h^{(2)}\|_2 \leq \sqrt{\|h^{(2)}\|_0} \|h^{(2)}\|_\infty \leq a/(t-1)\sqrt{s(t-1) - m} \leq a\sqrt{s/(t-1)}. \qquad (7.21)$$

Next, for convenience, we let $\nu = \sqrt{t(t-1)} - (t-1)$ and choose
$$\beta_i = \mathbf{v}_S + h^{(1)} + \nu \mathbf{u}_i, \quad i = 1, \ldots, n.$$

Then we have
$$\sum_{j=1}^n \lambda_j \beta_j - \frac{1}{2}\beta_i = \mathbf{v}_S + h^{(1)} + \nu h^{(2)} - \frac{1}{2}\beta_i = \left(\frac{1}{2} - \nu\right)(\mathbf{v}_S + h^{(1)}) - \frac{1}{2}\nu \mathbf{u}_i + \nu \mathbf{v}.$$

Since $\Phi \mathbf{v} = 0$, we have
$$\Phi\left(\sum_{j=1}^n \lambda_j \beta_j - \frac{1}{2}\beta_i\right) = \Phi\left(\left(\frac{1}{2} - \nu\right)(\mathbf{v}_S + h^{(1)}) - \frac{1}{2}\nu \mathbf{u}_i\right),$$

and

$$\|\beta_i\|^2 = \|\mathbf{v}_S + h^{(1)}\|^2 + \nu^2\|\mathbf{u}_i\|^2, \tag{7.22}$$

due to the nonoverlapping supports. Since the sparsity of $(\frac{1}{2} - \nu)(\mathbf{v}_S + h^{(1)}) - \frac{1}{2}\nu\mathbf{u}_i)$ is st, we use the RIP to obtain

$$\left\| \Phi \left(\sum_{j=1}^{n} \lambda_j \beta_i - \frac{1}{2}\beta_i \right) \right\|^2 \leq (1 + \delta_{st}) \left(\left(\frac{1}{2} - \nu\right)^2 \|\mathbf{v}_S + h^{(1)}\|^2 + \frac{1}{4}\nu^2\|\mathbf{u}_i\|^2 \right). \tag{7.23}$$

Furthermore, we can easily construct the equality

$$\sum_{i=1}^{n} \lambda_i \left\| \Phi \left(\sum_{j=1}^{n} \lambda_j \beta_j - \frac{1}{2}\beta_i \right) \right\|^2 = \frac{1}{4} \sum_{i=1}^{n} \lambda_i \|\Phi\beta_i\|^2. \tag{7.24}$$

Then (7.23) and (7.24) imply

$$(1 + \delta_{st}) \quad \left(\left(\frac{1}{2} - \nu\right)^2 \|\mathbf{v}_S + h^{(1)}\|^2 + \sum_{i=1}^{n} \frac{1}{4}\nu^2\lambda_i\|\mathbf{u}_i\|^2 \right)$$

$$\geq \sum_{i=1}^{n} \lambda_i \left\| \Phi \left(\sum_{j=1}^{n} \lambda_j \beta_j - \frac{1}{2}\beta_i \right) \right\|^2 = \sum_{i=1}^{n} \frac{1}{4}\lambda_i \|\Phi\beta_i\|^2$$

$$\geq \sum_{i=1}^{n} \frac{1}{4}\lambda_i(1 - \delta_{st})\|\beta_i\|^2$$

$$= \sum_{i=1}^{n} \frac{1}{4}\lambda_i(1 - \delta_{st})(\|\mathbf{v}_S + h^{(1)}\|^2 + \nu^2\|\mathbf{u}_i\|^2)$$

$$= \frac{1 - \delta_{ts}}{4} \left(\|\mathbf{v}_S + h^{(1)}\|^2 + \nu^2 \sum_{i=1}^{n} \lambda_i\|\mathbf{u}_i\|^2 \right),$$

where we have used the lower-bound estimate for $\|\Phi\beta_i\|^2$. In other words, we have

$$0 \leq \|\mathbf{v}_S + h^{(1)}\|^2 \left((1 + \delta_{st}) \left(\frac{1}{2} - \nu\right)^2 - \frac{1 - \delta_{st}}{4} \right) + \sum_{i=1}^{n} \lambda_i\|\mathbf{u}_i\|^2\nu^2 \left(\frac{1 + \delta_{st}}{4} - \frac{1 - \delta_{st}}{4} \right).$$

Note that by (7.21), we have

$$\|\mathbf{u}_i\| \leq \|h^{(2)}\| \leq a\sqrt{s/(t-1)} \leq \|\mathbf{v}_S\|_1/\sqrt{s(t-1)} \leq \|\mathbf{v}_S\|_2/\sqrt{t-1} \leq \|\mathbf{v}_S + h^{(1)}\|_2/\sqrt{t-1}.$$

The right-hand side of the above inequality can be estimated as

$$0 \leq \|\mathbf{v}_S + h^{(1)}\|^2 \left((1 + \delta_{st}) \left(\frac{1}{2} - \nu\right)^2 - \frac{1 - \delta_{st}}{4} + \frac{\delta_{st}\nu^2}{2(t-1)} \right)$$

$$= \|\mathbf{v}_S + h^{(1)}\|^2 \left(\delta_{ts} \left(\frac{1}{2} - \nu + 1 + \frac{1}{2(t-1)}\nu^2\right) + \nu^2 - \nu \right)$$

$$= \|\mathbf{v}_S + h^{(1)}\|^2 (\delta_{ts}(t(2t-1) - 2t\sqrt{t(t-1)}) - ((2t-1)\sqrt{t(t-1)} - 2t(t-1))) < 0$$

if $\delta_{ts} < \sqrt{(t-1)/t}$. This contradiction shows that the ℓ_1 minimization problem satisfies the NSP.

The case that ts is not an integer can be proved similarly by considering $t' = \lceil ts \rceil / s$ with $t' > t$. Then

$$\delta_{t's} = \delta_{ts} \le \sqrt{(t-1)/t} \le \sqrt{(t'-1)/t'},$$

since the function $f(t) = \sqrt{(t-1)/t}$ is increasing. We use the same argument for $\delta_{t's}$ to finish the proof. $\square$

We simply choose $t = 2$ in Theorem 7.17 to conclude that $\delta_{2s} < 1/\sqrt{2}$ will imply that the minimizer $\mathbf{x}^*$ of the ℓ_1 minimization is the sparse solution.

7.4 • Conditions Based on δ_{ts} with $t < 4/3$

Theorem 7.19 (Zhang and Li, 2018 [14]). *Suppose that for an integer s with $ts \ge 2$, Φ satisfies the RIP with*

$$\delta_{ts} < \frac{t}{4-t},$$

where $0 < t < \frac{4}{3}$. Let $\mathbf{x_b} \in \mathbb{R}^n$ be a vector with $\|\mathbf{x_b}\|_0 \le s$. Then for $\mathbf{b} = \Phi\mathbf{x_b}$, the solution $\mathbf{x}^*$ of minimization (1.20) is unique and equal to $\mathbf{x_b}$.

One interesting case is $t = 1$. According to the result above, when $\delta_s < 1/3$, the minimization of (1.20) is unique and is equal to $\mathbf{x_b}$. The conclusion for $t = 1$ can be established easily as follows.

Proof of the Special Case $t = 1$ of Theorem 7.19. To motivate the proof, we begin with the easy case that $t = 1$ and s is an even integer. Suppose that the NSP is not satisfied, i.e., we have a nonzero vector $h \in \mathcal{N}(\Phi)$ with $T \subset \{1, \ldots, n\}$ and $|T| \le s$ such that $\|h_{T^c}\|_1 \le \|h_T\|_1$. Furthermore, we can choose T such that not only does $\|h_{T^c}\|_1 \le \|h_T\|_1$, but also $\|h_{T^c}\|_\infty \le \|h_T\|_1/s$. In this case, we shall apply Lemma 7.18 with $a = 2\|h_T\|_1/s$. Decompose $h_{T^c} = \sum \lambda_i \mathbf{u}_i$ with each u_i being $s/2$-sparse. Notice that $\sum_i \lambda_i \|\mathbf{u}_i\|_2^2 \le 2\|h_T\|_2^2$.

Divide T into two sets, T_1 and T_2, with $|T_1| = |T_2| = s/2$. Then we have the equality

$$\sum \lambda_i \|\Phi(2h_{T_1} - \mathbf{u}_i)\|_2^2 + \sum \lambda_i \|\Phi(2h_{T_2} - \mathbf{u}_i)\|_2^2$$
$$= \sum \lambda_i \|\Phi(2h_{T_1} + \mathbf{u}_i)\|_2^2 + \sum \lambda_i \|\Phi(2h_{T_2} + \mathbf{u}_i)\|_2^2 + 8\|\Phi h_T\|_2^2.$$

Then the left-hand side is upper bounded as

$$\text{LHS} \le (1 + \delta_s) \left(4\|h_{T_1}\|_2^2 + 4\|h_{T_2}\|_2^2 + 2\sum \lambda_i \|\mathbf{u}_i\|_2^2 \right),$$
$$= (1 + \delta_s) \left(4\|h_T\|_2^2 + 2\sum \lambda_i \|\mathbf{u}_i\|_2^2 \right)$$

and the right-hand side is lower bounded as

$$\text{RHS} \ge (1 - \delta_s) \left(2\sum \lambda_i \|\mathbf{u}_i\|_2^2 + 12\|h_T\|_2^2 \right).$$

Combining the above two inequalities yields

$$0 \le -8\|h_T\|_2^2 + \left(16\|h_T\|_2^2 + 4\sum \lambda_i \|u_i\|_2^2 \right) \delta_s$$
$$\le -8\|h_T\|_2^2 + 24\|h_T\|_2^2 \delta_s = 8\|h_T\|^2 (3\delta_s - 1).$$

So if $\delta_s < \frac{1}{3}$, we get a contradiction. $\square$

To prove the general case of Theorem 7.19, we need a few elementary identities, which will be presented in Lemmas 7.20 and 7.21 below.

Lemma 7.20. *Fix vectors* $\{\mathbf{v}_i : i \in T\} \subset \mathbb{R}^n$, *where* T *is an index set and* $|T| = s$. *Choose all subsets* $T_i \subset T$ *with* $|T_i| = l$, $i \in I$, *with* $|I| = \binom{s}{l}$. *Then we have*

$$\sum_{i \in I} \sum_{p \in T_i} \mathbf{v}_p = \binom{s-1}{l-1} \sum_{p \in T} \mathbf{v}_p \ (l \geq 1) \tag{7.25}$$

and

$$\sum_{i \in I} \sum_{p \neq q \in T_i} \langle \mathbf{v}_p, \mathbf{v}_q \rangle = \binom{s-2}{l-2} \sum_{p \neq q} \langle \mathbf{v}_p, \mathbf{v}_q \rangle \ (l \geq 2). \tag{7.26}$$

Proof. Notice that there are exactly $\binom{s-1}{l-1}$ sets T_i that contain a fixed $p \in T$, and that there are exactly $\binom{s-2}{l-2}$ sets T_i that contain a fixed $p \neq q \in T$. □

With the two identities (7.25) and (7.26), we present the following lemma.

Lemma 7.21. *Fix a matrix* $\Phi \in \mathbb{R}^{m \times s}$ *and a vector* $x \in \mathbb{R}^s$ ($s \geq 2$), *and set* $T = \{1, 2, \ldots, s\}$. *Choose all subsets* $T_i \subset T$ *with* $|T_i| = a$, $i \in I$, *with* $|I| = \binom{s}{a}$, *and all subsets* $S_j \subset T$ *with* $|S_j| = b$, $j \in J$, *with* $|J| = \binom{s}{b}$. *Then we have*

$$\frac{s-b}{a|I|} \sum_{i \in I} \|\Phi \mathbf{x}_{T_i}\|_2^2 - \frac{s-a}{b|J|} \sum_{j \in J} \|\Phi \mathbf{x}_{S_j}\|_2^2 = (a-b)\frac{\|\Phi \mathbf{x}\|_2^2}{s}, \tag{7.27}$$

and when $s \geq a + b$,

$$\frac{1}{|I|\binom{s-a}{b}} \sum_{T_i \cap S_j = \emptyset} \left(\|\Phi(\mathbf{x}_{T_i} + \mathbf{x}_{S_j})\|_2^2 - \frac{s-a-b}{abs} \|\Phi(b\mathbf{x}_{T_i} - a\mathbf{x}_{S_j})\|_2^2 \right) = \left(\frac{a+b}{s}\right)^2 \|\Phi \mathbf{x}\|_2^2. \tag{7.28}$$

Proof. The proof is quite straightforward, and we only calculate the sum of square terms on the left-hand sides of (7.27) and (7.28). Denote $\mathbf{x}_i$ to be the projection of $\mathbf{x}$ onto the ith canonical basis vector and then $\mathbf{x} = \sum_{i=1}^s \mathbf{x}_i$. For identity (7.27), we have

$$\sum_{i \in I} \|\Phi \mathbf{x}_{T_i}\|_2^2 = \sum_{i \in I} \left(\sum_{p \in T_i} \|\Phi \mathbf{x}_p\|_2^2 + \sum_{p \neq q \in T_i} \langle \Phi \mathbf{x}_p, \Phi \mathbf{x}_q \rangle \right)$$

$$= \binom{s-1}{a-1} \sum_{i=1}^s \|\Phi \mathbf{x}_i\|_2^2 + \binom{s-2}{a-2} \sum_{i \neq j} \langle \Phi \mathbf{x}_i, \Phi \mathbf{x}_j \rangle \tag{7.29}$$

and

$$\sum_{j \in J} \|\Phi \mathbf{x}_{S_j}\|_2^2 = \sum_{j \in J} \left(\sum_{p \in S_j} \|\Phi \mathbf{x}_p\|_2^2 + \sum_{p \neq q \in S_j} \langle \Phi \mathbf{x}_p, \Phi \mathbf{x}_q \rangle \right)$$

$$= \binom{s-1}{b-1} \sum_{i=1}^s \|\Phi \mathbf{x}_i\|_2^2 + \binom{s-2}{b-2} \sum_{i \neq j} \langle \Phi \mathbf{x}_i, \Phi \mathbf{x}_j \rangle, \tag{7.30}$$

where the second line of the above equalities uses the identities (7.25) and (7.26). Then the left-hand side of (7.27) is equal to

$$\frac{s-b}{a\binom{s}{a}}\left[\binom{s-1}{a-1}\sum_{i=1}^{s}\|\Phi\mathbf{x}_i\|_2^2 + \binom{s-2}{a-2}\sum_{i\neq j}\langle\Phi\mathbf{x}_i,\ \Phi\mathbf{x}_j\rangle\right]$$

$$-\frac{s-a}{b\binom{s}{b}}\left[\binom{s-1}{b-1}\sum_{i=1}^{s}\|\Phi\mathbf{x}_i\|_2^2 + \binom{s-2}{b-2}\sum_{i\neq j}\langle\Phi\mathbf{x}_i,\ \Phi\mathbf{x}_j\rangle\right]$$

$$=\frac{a-b}{s}\left(\sum_{i=1}^{s}\|\Phi\mathbf{x}_i\|_2^2 + \sum_{i\neq j}\langle\Phi\mathbf{x}_i,\ \Phi\mathbf{x}_j\rangle\right)=\frac{a-b}{s}\|\Phi\mathbf{x}\|_2^2.$$

For identity (7.28), we have

$$\sum_{T_i\cap S_j=\emptyset}\|\Phi\left(\mathbf{x}_{T_i}+\mathbf{x}_{S_j}\right)\|_2^2 = \sum_{T_i\cap S_j=\emptyset}\left(\|\Phi\mathbf{x}_{T_i}\|_2^2+\|\Phi\mathbf{x}_{S_j}\|_2^2+2\langle\Phi\mathbf{x}_{T_i},\ \Phi\mathbf{x}_{S_j}\rangle\right)$$

$$=\binom{s-a}{b}\sum_{i\in I}\|\Phi\mathbf{x}_{T_i}\|_2^2 + \binom{s-b}{a}\sum_{j\in J}\|\Phi\mathbf{x}_{S_j}\|_2^2$$

$$+2\binom{s-2}{a+b-2}\binom{a+b-2}{a-1}\sum_{p\neq q}\langle\Phi\mathbf{x}_p,\ \Phi\mathbf{x}_q\rangle$$

$$=\frac{a+b}{s}\binom{s}{a+b}\binom{a+b}{a}\left[\sum_{i=1}^{s}\|\Phi\mathbf{x}_i\|_2^2 \ + \frac{a+b-1}{s-1}\sum_{i\neq j}\langle\Phi\mathbf{x}_i,\ \Phi\mathbf{x}_j\rangle\right],$$

where the square terms of the second equality use the fact that there are exactly $\binom{s-a}{b}$ sets S_j for a fixed T_i and there are exactly $\binom{s-b}{a}$ sets T_i for a fixed S_j. The inner product terms follow from the fact that there are exactly $\binom{s-2}{a+b-2}$ sets $T_i\cup S_j$ with $T_i\cap S_j=\emptyset$ for fixed $p\neq q$ such that $p,q\in T_i\cup S_j$, and there are $\binom{a+b-2}{a-1}$ sets (T_i,S_j) for such fixed $T_i\cup S_j$. The last line of the above equality uses the identities (7.29) and (7.30). Similarly, we have

$$\sum_{T_i\cap S_j=\emptyset}\|\Phi\left(b\mathbf{x}_{T_i}-a\mathbf{x}_{S_j}\right)\|_2^2$$

$$=\frac{a+b}{s}ab\binom{s}{a+b}\binom{a+b}{a}\left[\sum_{i=1}^{s}\|\Phi\mathbf{x}_i\|_2^2 - \frac{1}{s-1}\sum_{i\neq j}\langle\Phi\mathbf{x}_i,\ \Phi\mathbf{x}_j\rangle\right].$$

Combining the above equalities, we have identity (7.28).   □

These two lemmas are sufficient to establish Theorem 7.19.

Proof of Theorem 7.19. Suppose ts is an integer. We assume that the NSP (see Theorem 2.5 with $q=1$) is not satisfied, i.e., there exists $h\in\mathcal{N}(\Phi)\backslash\{0\}$ such that $\|h_{T^c}\|_1\leq\|h_T\|_1$, where h_T consists of the s largest absolute value elements of h. Define $\alpha=\|h_T\|_1/s$.

Choose positive integers a and b with $a+b=ts$ and $b\leq a\leq s$. Denote all the possible index sets $T_i,\ S_j\subset\{1,2,\ldots,s\}$ with $|T_i|=a$, $|S_j|=b$, respectively, where $i\in I\ (|I|=\binom{s}{a})$ and $j\in J\ (|J|=\binom{s}{b})$. Since

$$\|h_{T^c}\|_1\leq\|h_T\|_1=s\alpha=b\cdot\frac{s}{b}\alpha,\quad \|h_{T^c}\|_\infty\leq\alpha\leq\frac{s}{b}\alpha,$$

we can apply Lemma 7.18 and decompose h_{T^c} as the convex sum of b-sparse vectors, i.e., $h_{T^c} = \sum_l \lambda_l \mathbf{u}_l$, with

$$\sum_l \lambda_l \|\mathbf{u}_l\|_2^2 \leq b\left(\frac{s}{b}\alpha\right)^2 = \frac{s^2}{b}\alpha^2. \tag{7.31}$$

Similarly, h_{T^c} can also be decomposed as $h_{T^c} = \sum_l \lambda'_l \mathbf{v}_l$ and $h_{T^c} = \sum_l \lambda''_l \mathbf{w}_l$, with $\mathbf{v}_l$ being a-sparse and $\mathbf{w}_l$ being $(t-1)s$-sparse (in the case of $t > 1$) such that

$$\sum_l \lambda'_l \|\mathbf{v}_l\|_2^2 \leq \frac{s^2}{a}\alpha^2 \tag{7.32}$$

and

$$\sum_l \lambda''_l \|\mathbf{w}_l\|_2^2 \leq \frac{s\alpha^2}{t-1}. \tag{7.33}$$

For notational convenience, we set $Y_{a,b}$ to be

$$Y_{a,b} := \frac{s-b}{a\binom{s}{a}} \sum_{i \in I,l} \lambda_l \left[a^2 \left\| \Phi\left(h_{T_i} + \frac{b}{s}\mathbf{u}_l\right) \right\|_2^2 - b^2 \left\| \Phi\left(h_{T_i} - \frac{a}{s}\mathbf{u}_l\right) \right\|_2^2 \right]$$
$$+ \frac{s-a}{b\binom{s}{b}} \sum_{j \in J,l} \lambda'_l \left[b^2 \left\| \Phi\left(h_{S_j} + \frac{a}{s}\mathbf{v}_l\right) \right\|_2^2 - a^2 \left\| \Phi\left(h_{S_j} - \frac{b}{s}\mathbf{v}_l\right) \right\|_2^2 \right]. \tag{7.34}$$

We claim that when $4/3 > t \geq 1$,

$$(4 - 3t)Y_{a,b} - 2t^3 \left[ab - (t-1)s^2 \right] \langle \Phi h_T, \ \Phi h_T \rangle$$
$$= C_{t,a,b} \sum_l \lambda''_l \left[\|\Phi\left(h_T + (t-1)\mathbf{w}_l\right)\|_2^2 - (t-1)^2 \|\Phi(h_T - \mathbf{w}_l)\|_2^2 \right], \tag{7.35}$$

and when $0 < t < 1$,

$$tY_{a,b} - 2t^2(2-t)ab\langle \Phi h_T, \ \Phi h \rangle$$
$$= \frac{C_{t,a,b}}{\binom{s}{a}\binom{s-a}{b}} \sum_{T_i \cap S_j = \emptyset} \left(\|\Phi\left(h_{T_i} + h_{S_j}\right)\|_2^2 - \frac{1-t}{ab}\|\Phi\left(bh_{T_i} - ah_{S_j}\right)\|_2^2 \right) \tag{7.36}$$

both hold for $C_{t,a,b} = (a-b)^2 - 2(2-t)ab$. The proof of the above two equalities is left as Exercise 9. We can set $a = b = \frac{ts}{2}$ when ts is even and $a = b+1 = \frac{ts+1}{2}$ when ts is odd. Both cases indicate that $C_{t,a,b} < 0$. These two identities are elaborately constructed so that we can apply the RIP of the measurement matrix Φ.

Since h_{T_i}, $\mathbf{v}_l$ are a-sparse and h_{S_j}, $\mathbf{u}_l$ are b-sparse, and $a + b = ts$, we can apply the RIP of order ts to $Y_{a,b}$. That is,

$$Y_{a,b} \geq \frac{s-b}{a\binom{s}{a}} \sum_{i \in I,l} \lambda_l \left[a^2(1-\delta_{ts}) \left\| h_{T_i} + \frac{b}{s}\mathbf{u}_l \right\|_2^2 - b^2(1+\delta_{ts}) \left\| h_{T_i} - \frac{a}{s}\mathbf{u}_l \right\|_2^2 \right]$$
$$+ \frac{s-a}{b\binom{s}{b}} \sum_{j \in J,l} \lambda'_l \left[b^2(1-\delta_{ts}) \left\| h_{S_j} + \frac{a}{s}\mathbf{v}_l \right\|_2^2 - a^2(1+\delta_{ts}) \left\| h_{S_j} - \frac{b}{s}\mathbf{v}_l \right\|_2^2 \right]$$

$$= (a^2 - b^2) \left(\frac{s-b}{a\binom{s}{a}} \sum_{i \in I} \|h_{T_i}\|_2^2 - \frac{s-a}{b\binom{s}{b}} \sum_{j \in J} \|h_{S_j}\|_2^2 \right)$$

$$- (a^2 + b^2) \delta_{ts} \left(\frac{s-b}{a\binom{s}{a}} \sum_{i \in I} \|h_{T_i}\|_2^2 + \frac{s-a}{b\binom{s}{b}} \sum_{j \in J} \|h_{S_j}\|_2^2 \right)$$

$$- 2\delta_{ts} \frac{ab^2(s-b)}{s^2} \sum_l \lambda_l \|\mathbf{u}_l\|_2^2 - 2\delta_{ts} \frac{a^2 b(s-a)}{s^2} \sum_l \lambda_l' \|\mathbf{v}_l\|_2^2,$$

where we have used the fact that the support of $\mathbf{u}_l(\mathbf{v}_l)$ is disjoint with $h_{T_i}(h_{S_j})$. Notice that $\sum_{i \in I} \|h_{T_i}\|_2^2 = \binom{s-1}{a-1} \|h_T\|_2^2$ and $\sum_{j \in J} \|h_{S_j}\|_2^2 = \binom{s-1}{b-1} \|h_T\|_2^2$ (an application of identity (7.25)); then we have

$$Y_{a,b} \geq \frac{a^2 - b^2}{s} (a-b) \|h_T\|_2^2 - \frac{a^2 + b^2}{s} \delta_{ts} (2-t) s \|h_T\|_2^2$$

$$- 2\delta_{ts} \frac{ab^2(s-b)}{s^2} \cdot \frac{s^2}{b} \alpha^2 - 2\delta_{ts} \frac{a^2 b(s-a)}{s^2} \cdot \frac{s^2}{a} \alpha^2$$

$$= \left[t(a-b)^2 - (2-t)(a^2+b^2)\delta_{ts} \right] \|h_T\|_2^2 - 2(2-t)abs\delta_{ts}\alpha^2, \qquad (7.37)$$

where inequalities (7.31) and (7.32) are applied in the second line above.

Next, since h_T is s-sparse and $\mathbf{w}_l$ is $(t-1)s$-sparse, under the condition $C_{t,a,b} < 0$, we have

$$C_{t,a,b} \sum_l \lambda_l'' [\| \Phi (h_T + (t-1)\mathbf{w}_l) \|_2^2 - (t-1)^2 \|\Phi(h_T - \mathbf{w}_l)\|_2^2]$$

$$\leq C_{t,a,b} \sum_l \lambda_l'' [(1 - \delta_{ts})\|h_T + (t-1)\mathbf{w}_l\|_2^2 - (t-1)^2(1+\delta_{ts})\|h_T - \mathbf{w}_l\|_2^2]$$

$$\leq C_{t,a,b} \left[\left((1-\delta_{ts}) - (t-1)^2(1+\delta_{ts}) \right) \|h_T\|_2^2 - 2(t-1)^2 \delta_{ts} \frac{s\alpha^2}{t-1} \right], \qquad (7.38)$$

where the last inequality follows from (7.33).

Note that the terms of the inner product of the two equalities (7.35) and (7.36) vanish since $\Phi h = 0$ in this case. When $1 \leq t < 4/3$, the right-hand side of identity (7.35) will be less than or equal to 0 under the condition $C_{t,a,b}$, with $C_{t,a,b} = (a-b)^2 - 2(2-t)ab$. Inequalities (7.37) and (7.38) give

$$0 \geq (4 - 3t)Y_{a,b} - C_{t,a,b} \left[\left((1-\delta_{ts}) - (t-1)^2(1+\delta_{ts}) \right) \|h_T\|_2^2 - 2(t-1)^2 \delta_{ts} \frac{s\alpha^2}{t-1} \right]$$

$$= (4-3t) \left[t(a-b)^2 - (2-t)(a^2+b^2)\delta_{ts} \right] \|h_T\|_2^2$$

$$- \left((a-b)^2 - 2(2-t)ab \right) \left((1-\delta_{ts}) - (t-1)^2(1+\delta_{ts}) \right) \|h_T\|_2^2$$

$$- 2(2-t)(4-3t)abs\delta_{ts}\alpha^2 + \left((a-b)^2 - 2(2-t)ab \right) 2(t-1)\delta_{ts}s\alpha^2.$$

Surprisingly, we find that they have a common factor $2t^2(ab - (t-1)s^2)$. Then the above can be simplified to

$$0 \geq 2t^2 \left(ab - (t-1)s^2 \right) \left[(t - (3-t)\delta_{ts}) \|h_T\|_2^2 - s\delta_{ts}\alpha^2 \right]. \qquad (7.39)$$

Since $ab \geq \frac{(ts)^2 - 1}{4} = \frac{(2-t)^2}{4} s^2 - \frac{1}{4} + (t-1)s^2 > (t-1)s^2$ and $\alpha^2 = \left(\frac{\|h_T\|_1}{s} \right)^2 \leq \frac{\|h_T\|_2^2}{s}$, $\delta_{ts} < \frac{t}{4-t}$ leads to $h_T = 0$.

When $0 < t < 1$, we apply an RIP of order ts to the right-hand side of (7.36), and then we have

$$\frac{C_{t,a,b}}{\binom{s}{b}\binom{s-a}{b}} \sum_{T_i \cap S_j = \emptyset} \|\Phi\left(h_{T_i} + h_{S_j}\right)\|_2^2 - \frac{1-t}{ab}\|\Phi\left(bh_{T_i} - ah_{S_j}\right)\|_2^2 \tag{7.40}$$

$$\leq \frac{C_{t,a,b}}{\binom{s}{a}\binom{s-a}{b}} \sum_{T_i \cap S_j = \emptyset} (1 - \delta_{ts})\|h_{T_i} + h_{S_j}\|_2^2 - (1 + \delta_{ts})\frac{1-t}{ab}\|bh_{T_i} - ah_{S_j}\|_2^2$$

$$= \frac{C_{t,a,b}}{\binom{s}{b}\binom{s-a}{b}} \left[(1 - \delta_{ts})\left(\binom{s-a}{b}\binom{s-1}{a-1} + \binom{s-b}{a}\binom{s-1}{b-1}\right)\|h_T\|_2^2\right.$$

$$\left. -(1 + \delta_{ts})\frac{1-t}{ab}\left(b^2\binom{s-a}{b}\binom{s-1}{a-1} + a^2\binom{s-b}{a}\binom{s-1}{b-1}\right)\|h_T\|_2^2\right]$$

$$= C_{t,a,b}t\left[t - (2 - t)\delta_{ts}\right]\|h_T\|_2^2, \tag{7.41}$$

where in the first equality we have used identity (7.25) and $C_{t,a,b} < 0$. Thus, inequalities (7.37) and (7.41) and identity (7.36) together yield

$$0 \geq t\{[t(a-b)^2 - (2-t)(a^2+b^2)\delta_{ts}]\|h_T\|_2^2 - 2(2-t)abs\delta_{ts}\alpha^2\}$$
$$-C_{t,a,b}t\left[t - (2-t)\delta_{ts}\right]\|h_T\|_2^2$$
$$= 2t(2-t)ab\left[(t - (3-t)\delta_{ts})\|h_T\|_2^2 - s\delta_{ts}\alpha^2\right],$$

which means that $\delta_{ts} < \frac{t}{4-t}$ can lead to $h_T = 0$.

When ts is not an integer, notice that

$$\delta_{\lceil ts \rceil} = \delta_{ts} < \frac{t}{4-t} < \begin{cases} \frac{\lceil ts \rceil/s}{4 - \lceil ts \rceil/s}, & \frac{\lceil ts \rceil}{s} < \frac{4}{3}; \\ \sqrt{\frac{\lceil ts \rceil/s - 1}{\lceil ts \rceil/s}}, & \frac{\lceil ts \rceil}{s} \geq \frac{4}{3}, \end{cases}$$

where the first line is the same as the former case and the second line is the same as the case in [5]. Thus, we have finished the proof of Theorem 7.19. □

This estimate on $\delta_{ts} < t/(4-t)$ is the best possible according to [3], where a counterexample can be found.

7.5 ▪ Conditions on δ_{2s} for ℓ_q Minimizers

All the studies in the previous sections can be extended to the approach of ℓ_q minimization (1.27).

Theorem 7.22 (Zhang and Li, 2019 [15]). *For any $0 < q \leq 1$, consider the sparse recovery model (1.27). If*

$$\delta_{2s} < \frac{\eta}{2 - q - \eta} := \delta(q), \tag{7.42}$$

where $\eta \in (1 - q, 1 - \frac{q}{2})$ is the only positive solution of the equation

$$\frac{q}{2}\eta^{\frac{2}{q}} + \eta - 1 + \frac{q}{2} = 0, \tag{7.43}$$

then $\mathbf{x}$ can be exactly recovered by (1.27). When η is irrational and $q < 1$, "$<$" in (7.42) can be replaced by "$\leq$".

Before proving the theorem, we need a useful lemma, which is an extension of Lemma 7.18.

Lemma 7.23. *For* $\mathbf{x} \in \mathbb{R}^n$ *which satisfies* $|\text{supp}(\mathbf{x})| = n$, $\|\mathbf{x}\|_q^q \leq s\alpha^q$, *and* $\|\mathbf{x}\|_\infty \leq \alpha$ *with* $s \leq n$ *being a positive integer,* α *being a positive constant, and* $0 < q \leq 1$, $\mathbf{x}$ *can be represented as a convex combination of s-sparse vectors, i.e.,*

$$\mathbf{x} = \sum_i \lambda_i \mathbf{u}_i,$$

where $\lambda_i > 0$, $\sum_i \lambda_i = 1$, *and* $\|\mathbf{u}_i\|_0 \leq s$. *Furthermore,*

$$\sum_i \lambda_i \|\mathbf{u}_i\|_2^2 \leq \min\left\{\frac{n}{s}\|\mathbf{x}\|_2^2, \; \alpha^q \|\mathbf{x}\|_{2-q}^{2-q}\right\}. \tag{7.44}$$

Proof. For simplicity, we write $\mathbf{x} = \sum_{i=1}^n x_i \mathbf{e}_i$, where $\mathbf{e}_i$ denotes the unit vector whose ith coordinate is 1 or -1, and $x_1 \geq x_2 \geq \cdots \geq x_n > 0$. If $n = s$, we can set $\mathbf{u}_1 = \mathbf{x}$ and $\lambda_1 = 1$, and then $\|\mathbf{u}_1\|_2^2 = \|\mathbf{x}\|_2^2 \leq \alpha^q \|\mathbf{x}\|_{2-q}^{2-q}$. Now we want to show that the conclusions also hold for $n > s$. Set $x_0 = \alpha$, and fix $a = (a_1, a_2, \ldots, a_n) \in \mathbb{R}_+^n$, where $a_i = x_i^{q-1}$, $i = 0, 1, \ldots, n$. Then we have $\sum_{i=1}^n a_i x_i \leq s\alpha^q$ and $\alpha^q = a_0 x_0 \geq a_1 x_1 \geq \cdots \geq a_n x_n$. We first decompose the vector $\mathbf{x}$ as the convex combination of $(n-1)$-sparse vectors.

Step 1: Decompose the n-sparse vector as the convex combination of $(n-1)$-sparse vectors. Denote the set as

$$S = \left\{j \in \{1, \ldots, n-1\} : \sum_{i=j}^n a_i x_i \leq (n-j)a_{j-1}x_{j-1}\right\}. \tag{7.45}$$

It is easy to see that $1 \in S$, so S is not empty. Then we take $j = \max S$, which implies

$$\sum_{i=j}^n a_i x_i \leq (n-j)a_{j-1}x_{j-1}, \qquad \sum_{i=j+1}^n a_i x_i > (n-j-1)a_j x_j. \tag{7.46}$$

It follows that

$$(n-j)a_j x_j < \sum_{i=j}^n a_i x_i \leq (n-j)a_{j-1}x_{j-1}.$$

Set

$$\mathbf{y}_w = \sum_{i=1}^{j-1} x_i \mathbf{e}_i + \frac{\sum_{i=j}^n a_i x_i}{n-j} \sum_{i=j, i \neq w}^n a_i^{-1} \mathbf{e}_i,$$

$$\xi_w = 1 - \frac{n-j}{\sum_{i=j}^n a_i x_i} \cdot a_w x_w > 0,$$

where $w = j, j+1, \ldots, n$. Then by a simple calculation, we can see that $\sum_{w=j}^n \xi_w = 1$ and

$$\mathbf{x} = \sum_{w=j}^n \xi_w \mathbf{y}_w, \tag{7.47}$$

where $\mathbf{y}_w$ is $(n-1)$-sparse for all $w = j, j+1, \ldots, n$. Furthermore,

$$\frac{\sum_{i=j}^n a_i x_i}{n-j} \sum_{w=j}^n \xi_w \sum_{t=j, t \neq w}^n a_t^{-2} = \frac{\sum_{i=j}^n a_i x_i}{n-j}\left(\sum_{t=j}^n a_t^{-2} - \sum_{w=j}^n \xi_w a_w^{-2}\right) = \sum_{w=j}^n \frac{x_w}{a_w}. \tag{7.48}$$

The representation of $\mathbf{y}_w = \sum_{i=1}^n y_{w,i}\mathbf{e}_i$ implies that

$$\sum_{i=1}^n a_i \mathbf{y}_{w,i} = \sum_{i=1}^n a_i x_i \leq s\alpha^q, \quad \alpha = y_{w,0} \geq y_{w,1} \geq y_{w,2} \geq \cdots \geq y_{w,n},$$

and

$$\alpha^q = a_0 y_{w,0} \geq a_1 y_{w,1} \geq a_2 y_{w,2} \geq \cdots \geq a_n y_{w,n},$$

where we have omitted the items $y_{w,w}$ in the inequalities and set $y_{w,0} = \alpha$.

Next replace $\mathbf{x}$ with $\mathbf{y}_w$, change n to $n-1$, and continue the process above, and so on. Finally, we have the l-sparse vectors $(l \geq s+1)$, which have the form

$$\mathbf{v} = \sum_{i=1}^{j'-1} x_i \mathbf{e}_i + \frac{\sum_{i=j'}^n a_i x_i}{l+1-j'} \sum_{i \in Q} a_i^{-1} \mathbf{e}_i,$$

where $j' \in \{1, \ldots, l\}$ and $Q \subset \{j', j'+1, \ldots, n\}$ $(|Q| = l+1-j')$ are determined by the process. Now we need to decompose $\mathbf{v}$ as the convex combination of $(l-1)$-sparse vectors.

Step 2. Decompose the l-sparse vector as the convex combination of $(l-1)$-sparse vectors. Define the set

$$S_0 = \left\{ 1 \leq j \leq l-1 : \sum_{i=j}^n a_i x_i \leq (l-j) a_{j-1} x_{j-1} \right\}.$$

Then S_0 is not empty since $1 \in S_0$. Denote $j_0 = \max S_0 \leq j'$, which yields that

$$\sum_{i=j_0}^n a_i x_i \leq (l-j_0) a_{j_0-1} x_{j_0-1} \quad \text{and} \quad \sum_{i=j_0+1}^n a_i x_i > (l-1-j_0) a_{j_0} x_{j_0}, \quad (7.49)$$

which means that

$$(l-j_0) a_{j_0} x_{j_0} < \sum_{i=j_0}^n a_i x_i \leq (l-j_0) a_{j_0-1} x_{j_0-1}.$$

Set $P = Q \cup \{j_0, \ldots, j'-1\}$ and

$$u_t = \sum_{i=1}^{j_0-1} x_i \mathbf{e}_i + \frac{\sum_{i=j_0}^n a_i x_i}{l-j_0} \sum_{i \in P, i \neq t} a_i^{-1} \mathbf{e}_i,$$

$$\lambda_t = \begin{cases} 1 - \dfrac{l-j_0}{\sum_{i=j_0}^n a_i x_i} \cdot a_t x_t, & t \in \{j_0, \ldots, j'-1\}, \\ 1 - \dfrac{\sum_{i=j'}^n a_i x_i}{l+1-j'} \cdot \dfrac{l-j_0}{\sum_{i=j_0}^n a_i x_i}, & t \in Q, \end{cases}$$

where $t \in P$. Then, as we expect, $\lambda_t > 0$,

$$\sum_{t \in P} \lambda_t = \sum_{t \in Q} \lambda_t + \sum_{t=j_0}^{j'-1} \lambda_t = l+1-j_0 - \frac{l-j_0}{\sum_{i=j_0}^n a_i x_i} \sum_{t=j_0}^n a_t x_t = 1$$

and

$$\sum_{t \in P} \lambda_t u_t = \sum_{i=1}^{j_0-1} x_i \mathbf{e}_i + \sum_{t=j_0}^{j'-1} x_i \mathbf{e}_i + \frac{\sum_{i=j'}^n a_i x_i}{l+1-j'} \sum_{t \in Q} a_t^{-1} \mathbf{e}_t$$

$$= \sum_{i=1}^{j'-1} x_i \mathbf{e}_i + \frac{\sum_{i=j'}^n a_i x_i}{l+1-j'} \sum_{t \in Q} a_t^{-1} \mathbf{e}_t = \mathbf{v}.$$

Now we have a similar equality to (7.48):

$$\frac{\sum_{i=j_0}^n a_i x_i}{l - j_0} \sum_{t \in P} \lambda_t \sum_{s \in P, s \neq t} a_s^{-2} = \frac{\sum_{i=j_0}^n a_i x_i}{l - j_0} \left(\sum_{t \in P} a_t^{-2} - \sum_{t \in P} \lambda_t a_t^{-2} \right)$$

$$= \sum_{t=j_0}^{j'-1} \frac{x_t}{a_t} + \frac{\sum_{i=j'}^n a_i x_i}{l + 1 - j'} \sum_{t \in Q} a_t^{-2}. \qquad (7.50)$$

Since the above equality is always true when $l = n - 1, n - 2, \ldots, s + 1$, we redefine j', j_0, and the set P, Q in the case of $l = s + 1$.

We estimate $\sum_{t \in P} \lambda_t \|\mathbf{u}_t\|_2^2$ as follows:

$$\sum_{t \in P} \lambda_t \|\mathbf{u}_t\|_2^2 = \sum_{i=1}^{j_0-1} x_i^2 + \left(\frac{\sum_{i=j_0}^n a_i x_i}{s + 1 - j_0} \right)^2 \sum_{i \in P} a_i^{-2} - \left(\frac{\sum_{i=j_0}^n a_i x_i}{s + 1 - j_0} \right)^2 \sum_{t \in P} \lambda_t a_t^{-2}$$

$$= \sum_{i=1}^{j_0-1} x_i^2 + \frac{\sum_{i=j_0}^n a_i x_i}{s + 1 - j_0} \sum_{i=j_0}^{j'-1} \frac{x_i}{a_i} + \frac{\sum_{i=j_0}^n a_i x_i}{s + 1 - j_0} \cdot \frac{\sum_{i=j'}^n a_i x_i}{s + 2 - j'} \sum_{t \in Q} a_t^{-2}.$$

So with some abuse of notation, we can write x as the combination of s-sparse vectors $x = \sum_i \lambda_i \mathbf{u}_i$, where $\sum_i \lambda_i = 1$ and $\lambda_i > 0$.

Step 3. Calculate $\sum_i \lambda_i \|\mathbf{u}_i\|_2^2$.

Furthermore, by using equality (7.48) combined with (7.50) for $l = s + 2, \ldots, n - 1$, we can obtain

$$\sum_i \lambda_i \|\mathbf{u}_i\|_2^2 = \sum_{i=1}^{j_0-1} x_i^2 + \frac{\sum_{i=j_0}^n a_i x_i}{s + 1 - j_0} \sum_{i=j_0}^{j'-1} \frac{x_i}{a_i} + \frac{\sum_{i=j_0}^n a_i x_i}{s + 1 - j_0} \cdot \sum_{t=j'}^n \frac{x_t}{a_t}$$

$$= \sum_{i=1}^{j_0-1} x_i^2 + \frac{\sum_{i=j_0}^n x_i^q}{s + 1 - j_0} \sum_{i=j_0}^n x_i^{2-q}. \qquad (7.51)$$

The above equality is precisely what we want, and we estimate it with two types of bounds. With the help of the basic norm inequality, we have

$$\sum_i \lambda_i \|\mathbf{u}_i\|_2^2 \leq \sum_{i=1}^{j_0-1} x_i^2 + \frac{n - j_0 + 1}{s + 1 - j_0} \sum_{i=j_0}^n x_i^2 = \sum_{i=1}^{j_0-1} x_i^2 + \left(\frac{n}{s} + \frac{(n-s)(j_0-1)}{s(s+1-j_0)} \right) \sum_{i=j_0}^n x_i^2$$

$$\leq \sum_{i=1}^{j_0-1} x_i^2 + \frac{n}{s} \sum_{i=j_0}^n x_i^2 + \frac{(n-s)(j_0-1)}{s} x_{j_0-1}^2 \leq \frac{n}{s} \sum_{i=1}^n x_i^2,$$

where we have used $\sum_{i=j_0}^n x_i^2 \leq x_{j_0-1}^{2-q} \sum_{i=j_0}^n x_i^q \leq x_{j_0-1}^{2-q} \cdot (s+1-j_0) x_{j_0-1}^q = (s+1-j_0) x_{j_0-1}^2$ (from (7.49)) in the second inequality and $(j_0 - 1) x_{j_0-1}^2 \leq \sum_{i=1}^{j_0-1} x_i^2$ in the third inequality. Instead of this upper bound, we can also have

$$\sum_i \lambda_i \|\mathbf{u}_i\|_2^2 \leq \sum_{i=1}^{j_0-1} x_i^2 + x_{j_0-1}^q \sum_{i=j_0}^n x_i^{2-q} \leq \alpha^q \sum_{i=1}^n x_i^{2-q},$$

where we used inequality (7.49) in the first line and $\alpha \geq x_i$ for all i. Thus we complete the proof of Lemma 7.23.   ☐

Now we prove Theorem 7.22 using the key tool, Lemma 2.2.

Proof of Theorem 7.22. First, we consider equation (7.43) and write $f(\eta) = \frac{q}{2}\eta^{\frac{2}{q}} + \eta - 1 + \frac{q}{2}$, where $\eta \in (0,1)$. Since $f'(\eta) = \eta^{\frac{2}{q}-1} + 1 > 0$, the function $f(\eta)$ is monotone increasing. At the point $\eta = 1 - q$, we have $f(1-q) = \frac{q}{2}(1-q)^{\frac{2}{q}} - \frac{q}{2} < 0$, and at the point $\eta = 1 - \frac{q}{2}$, $f(\eta) = \frac{q}{2}\left(1 - \frac{q}{2}\right)^{\frac{2}{q}} > 0$. So the unique solution of the equation is between $1 - q$ and $1 - \frac{q}{2}$, and with some abuse of notation, we regard it as η. We should keep in mind that η is a function of q, and here we just simplify the notation. Notice that $\eta \in (1-q, 1-\frac{q}{2})$ can be rational or irrational (in the case of $q = 1$, we have that $\eta = \sqrt{2} - 1$ is irrational).

Assume that the NSP (2.5) is not satisfied, i.e., there exists a vector h in the null space of Φ such that $\|h_{T^c}\|_q^q \leq \|h_T\|_q^q$, where h_T denotes the restriction of h on the index set T with the $|T|$ largest absolute value elements of h and zero everywhere else. Here we assume $|\text{supp}(h_{T^c})| = n$. Denote $\alpha^q = \|h_T\|_q^q / k$; then $\|h_{T^c}\|_q^q \leq k\alpha^q$ and $\|h_{T^c}\|_\infty \leq \alpha$. We can apply Lemma 7.23 to h_{T^c}, which means that there exist positive coefficients λ_i and k-sparse vectors $\mathbf{u}_i$, where $\sum_i \lambda_i = 1$, such that $h_{T^c} = \sum_i \lambda_i \mathbf{u}_i$. The definition of α also yields

$$\|h_T\|_2^2 \geq k^{1-\frac{2}{q}}\|h_T\|_q^2 = k^{1-\frac{2}{q}}(k\alpha^q)^{\frac{2}{q}} = k\alpha^2. \tag{7.52}$$

For $\mu \in \mathbb{R}$, let $\gamma_j = h_T + \mu u_j$, and then we have $\gamma_i - \gamma_j = \mu(\mathbf{u}_i - u_j)$ and

$$\sum_j \lambda_j \gamma_j - \frac{q}{2}\gamma_i = \mu h + \left(1 - \mu - \frac{q}{2}\right)h_T - \frac{q}{2}\mu \mathbf{u}_i.$$

It is easy to establish the equality

$$\sum_i \lambda_i \left\|\Phi\left(\sum_j \lambda_j \gamma_j - \frac{q}{2}\gamma_i\right)\right\|_2^2 + \frac{1-q}{2}\sum_{i,j}\lambda_i\lambda_j\|\Phi(\gamma_i - \gamma_j)\|_2^2$$

$$= \left\|\Phi\left(\sum_j \lambda_j \gamma_j\right)\right\|_2^2 - q\left\|\Phi\left(\sum_j \lambda_j \gamma_j\right)\right\|_2^2 + \frac{q^2}{4}\sum_i \lambda_i\|\Phi\gamma_i\|_2^2$$

$$+ \frac{1-q}{2}\left(2\sum_i \lambda_i\|\Phi\gamma_i\|_2^2 - 2\left\|\Phi\left(\sum_i \lambda_i \gamma_i\right)\right\|_2^2\right)$$

$$= \left(1 - \frac{q}{2}\right)^2 \sum_i \lambda_i\|\Phi\gamma_i\|_2^2, \tag{7.53}$$

where $\sum_i \lambda_i = 1$.

Since h_T and $\mathbf{u}_i$ are both k-sparse, we can apply the RIP of order $2k$ associated with Φ. The left-hand side of equality (7.53) can be upper bounded as

$$\text{LHS} \leq (1 + \delta_{2k})\left[\left(1 - \mu - \frac{q}{2}\right)^2\|h_T\|_2^2 + \frac{q^2}{4}\mu^2\sum_i \lambda_i\|\mathbf{u}_i\|_2^2\right]$$

$$+ (1 + \delta_{2k})\left[\frac{1-q}{2}\mu^2\sum_i \lambda_i\lambda_j\|\mathbf{u}_i - u_j\|_2^2\right]$$

$$= (1 + \delta_{2k})\left[\left(1 - \mu - \frac{q}{2}\right)^2\|h_T\|_2^2 + \frac{q^2}{4}\mu^2\sum_i \lambda_i\|\mathbf{u}_i\|_2^2\right]$$

$$+ (1 + \delta_{2k})\left[(1-q)\mu^2\left(\sum_i \lambda_i\|\mathbf{u}_i\|_2^2 - \left\|\sum_i \lambda_i \mathbf{u}_i\right\|_2^2\right)\right].$$

The right-hand side of equality (7.53) can be lower bounded as

$$\text{RHS} \geq \left(1 - \frac{q}{2}\right)^2 (1 - \delta_{2k}) \sum_i \lambda_i \left(\|h_T\|_2^2 + \mu^2 \|\mathbf{u}_i\|_2^2\right).$$

Combined with the above inequalities, we have

$$\delta_{2k} \geq \frac{\left[(2 - q)\mu - \mu^2\right] \|h_T\|_2^2 + (1 - q)\mu^2 \|h_{T^c}\|_2^2}{\left[(1 - \mu - \frac{q}{2})^2 + (1 - \frac{q}{2})^2\right] \|h_T\|_2^2 + 2(1 - \frac{q}{2})^2 \mu^2 \sum_i \lambda_i \|\mathbf{u}_i\|_2^2 - (1 - q)\mu^2 \|h_{T^c}\|_2^2}.$$

In the following, we will analyze the right-hand side of the above inequality and provide a tight lower bound. Since the parameter μ can be chosen arbitrarily, we try to find the maximum of the function of μ. By elementary calculation, the maximum can be achieved at the point

$$\mu = \frac{(1 - q)\|h_{T^c}\|_2^2 - \|h_T\|_2^2 + \sqrt{[(1 - q)\|h_{T^c}\|_2^2 - \|h_T\|_2^2]^2 + (2 - q)^2 \|h_T\|_2^2 \sum_i \lambda_i \|\mathbf{u}_i\|_2^2}}{(2 - q) \sum_i \lambda_i \|\mathbf{u}_i\|_2^2}.$$

Then we have

$$\delta_{2k} \geq \left(\sqrt{\left[(1 - q)\frac{\|h_{T^c}\|_2^2}{\|h_T\|_2} - 1\right]^2 + (2 - q)^2 \frac{\sum_i \lambda_i \|\mathbf{u}_i\|_2^2}{\|h_T\|_2^2}} - (1 - q)\frac{\|h_{T^c}\|_2^2}{\|h_T\|_2^2}\right)^{-1}. \quad (7.54)$$

From Lemma 7.23 and Hölder's inequality, we can obtain

$$\sum_i \lambda_i \|\mathbf{u}_i\|_2^2 \leq \alpha^q \|h_{T^c}\|_{2-q}^{2-q} \leq \alpha^q \left(\|h_{T^c}\|_2^2\right)^{\frac{2-2q}{2-q}} \left(\|h_{T^c}\|_q^q\right)^{\frac{q}{2-q}}$$

$$\leq \left(\|h_{T^c}\|_2^2\right)^{\frac{2-2q}{2-q}} \left(k\alpha^2\right)^{\frac{q}{2-q}} \leq \left(\|h_{T^c}\|_2^2\right)^{\frac{2-2q}{2-q}} \left(\|h_T\|_2^2\right)^{\frac{q}{2-q}}, \quad (7.55)$$

where in the second line we have used the definition of α and (7.52). Much attention must be paid here if we ask when the equalities hold. The equalities can all be achieved only when the absolute value of each nonzero element in h_{T^c} is the same (say β), and the absolute value of each element in h_T equals α and $\|h_{T^c}\|_q^q = \|h_T\|_q^q$ (i.e., $n\beta^q = k\alpha^q$) in the case of $0 < q < 1$.

Next we assign the bound (7.55) to the estimate (7.54). Now we need to upper bound

$$g\left(\frac{\|h_{T^c}\|_2^2}{\|h_T\|_2^2}\right) = \sqrt{\left[(1 - q)\frac{\|h_{T^c}\|_2^2}{\|h_T\|_2} - 1\right]^2 + (2 - q)^2 \left(\frac{\|h_{T^c}\|_2^2}{\|h_T\|_2^2}\right)^{\frac{2-2q}{2-q}}} - (1 - q)\frac{\|h_{T^c}\|_2^2}{\|h_T\|_2^2}.$$

By calculating the derivative of the function $g(x)$ in the region $x \in [0, 1]$, we find that only if $x = \eta^{\frac{2-q}{q}}$ can $g(x)$ achieve its supremum:

$$g(\eta^{\frac{2-q}{q}}) = \sqrt{[(1 - q)\eta^{\frac{2-q}{q}} - 1]^2 + (2 - q)^2 \eta^{\frac{2-2q}{2-q}}} - (1 - q)\eta^{\frac{2-q}{q}} = \frac{2 - q - \eta}{\eta}.$$

Then we clearly have

$$\delta_{2k} \geq g^{-1}(\eta^{\frac{2-q}{q}}) = \frac{\eta}{2 - q - \eta}.$$

In addition, when $q < 1$, the equality holds only when

$$\eta^{\frac{2-q}{q}} = \frac{\|h_{T^c}\|_2^2}{\|h_T\|_2^2} = \frac{n\beta^2}{k\alpha^2} = \frac{n}{k\alpha^2}\left(\frac{k\alpha^q}{n}\right)^{\frac{2}{q}} = \left(\frac{k}{n}\right)^{\frac{2}{q} - 1},$$

which means η is rational.

Now we come to our conclusion. If $\delta_{2k} < \frac{\eta}{2-q-\eta}$ with $0 < q < 1$, where $\eta \in (1-q, 1-\frac{q}{2})$ satisfies (7.43), then the NSP holds true, which provides a sufficient condition for sparse signal recovery by using model (1.27) with zero noise. When η is irrational and $0 < q < 1$, we have $\delta_{2k} \leq \frac{\eta}{2-q-\eta}$ as the sufficient condition to recover sparse signals by using model (1.27) with zero noise. $\quad \square$

7.6 ▪ Conditions Based on δ_s

In this section, we present a result based on the restriction of δ_s instead of δ_{2s}. We start with the following inequality of interest.

Lemma 7.24. *For any vector* $\mathbf{x} = (x_1, \ldots, x_n)^T \in \mathbb{R}^n$,

$$\|\mathbf{x}\|_2 - \frac{\|\mathbf{x}\|_1}{\sqrt{n}} \leq \frac{\sqrt{n}}{4}(M - m),$$

where $M = \max_i |x_i|$ *and* $m = \min_i |x_i|$.

Proof. When $m = 0$, the inequality is easy to prove. Indeed,

$$\|\mathbf{x}\|_2 \leq \sqrt{\|\mathbf{x}\|_1 \|\mathbf{x}\|_\infty} \leq \frac{\alpha}{2}\|\mathbf{x}\|_1 + \frac{1}{2\alpha}\|\mathbf{x}\|_\infty$$

for $\alpha > 0$. Choose $\alpha = 2/(\sqrt{n})$ to finish the proof.

When $m \neq 0$, we have to do more. It is obvious that the result holds when the absolute values of all coordinates are equal. Without loss of generality, we may assume that $x_1 \geq x_2 \geq \cdots \geq x_n \geq 0$ and not all x_i are equal. Let

$$f(\mathbf{x}) = \|\mathbf{x}\|_2 - \frac{\|\mathbf{x}\|_1}{\sqrt{n}}.$$

Let us fix x_1 and find an upper bound for $f(\mathbf{x})$. Since

$$\frac{\partial f}{\partial x_i} = \frac{x_i}{\|\mathbf{x}\|_2} - \frac{1}{\sqrt{n}},$$

which is an increasing function, $f(\mathbf{x})$ is convex as a function of x_i with $i = 2, \ldots, n-1$. The maximum is achieved at $x_i = x_{i-1}$ or $x_i = x_{i+1}$. It follows that when f achieves its maximum, $\mathbf{x}$ must be of the form $x_1 = x_2 = \cdots = x_k$ and $x_{k+1} = \cdots = x_n$ for some $1 \leq k < n$. Thus,

$$f(\mathbf{x}) = \sqrt{k(x_1^2 - x_n^2) + nx_n^2} - \frac{k(x_1 - x_n) + nx_n}{\sqrt{n}}.$$

To find out which k maximizes $f(\mathbf{x})$, we treat the right-hand side of the equation above as a function $g(k)$. Note that $g(n) = 0$ and $g(0) = 0$. The maximum of g must happen inside k between 1 and $N - 1$. The derivative of g is

$$g'(k) = \frac{x_1^2 - x_n^2}{2\sqrt{k(x_1^2 - x_n^2) + nx_n^2}} - \frac{(x_1 - x_n)}{\sqrt{n}} = 0.$$

The critical point is

$$k(x_1^2 - x_n^2) = n\left(\frac{x_1 + x_n}{2}\right)^2 - nx_n^2 \quad \text{or} \quad k = n\frac{\left(\frac{x_1+x_n}{2}\right)^2 - x_n^2}{x_1^2 - x_n^2}.$$

Thus,

$$g(k) \leq g\left(n\frac{\left(\frac{x_1+x_n}{2}\right)^2 - x_n^2}{x_1^2 - x_n^2}\right) = \sqrt{n}(x_1 - x_n)\left(\frac{1}{2} - \frac{x_1 + 3x_n}{4(x_1 + x_n)}\right).$$

Since $(x_1 + 3x_n)/(x_1 + x_n) \geq 1$, we have $g(k) \leq \sqrt{n}(x_1 - x_n)/4$. That is, we have the inequality in Lemma 7.24.

We can also see that the above inequality becomes an equality if and only if $x_{k+1} = \cdots = x_n = 0$ and $k = n/4$. $\quad\square$

Theorem 7.25 (Cai, Wang, and Xu, 2010 [1]). *Suppose that $\delta_s < 1/(1 + \sqrt{5}) \approx 0.307$. Then the solution of* (1.20) *is the sparse solution.*

Proof. Let $h \in N(\Phi)$ be a nonzero vector in the null space of Φ. Without loss of generality, we may assume that $h = (h_1, h_2, \ldots, h_n)$ with $|h_1| \geq |h_2| \geq \cdots \geq |h_n|$. Assume that the index set S is associated with the nonzero entries of sparse solution $\mathbf{x}$. Note that we have

$$\|h_S\|_1 \geq \|h_{S^c}\|_1, \tag{7.56}$$

with S^c being the complement index set of S in $\{1, 2, \ldots, n\}$. Let us say that the cardinality $\#(S) = s$. Letting $T = \{1, 2, \ldots, s\}$, since the cardinalities of both sets, $\#(S \cap T^c)$ and $\#(S^c \cap T)$, are the same, we have

$$\|h_{S^c \cap T}\|_1 \geq \|h_{S \cap T^c}\|_1. \tag{7.57}$$

We now claim that

$$\|h_T\|_1 \geq \|h_{T^c}\|_1. \tag{7.58}$$

Indeed, we have

$$\begin{aligned}
\|h_T\|_1 &= \|h_{S \cap T}\|_1 + \|h_{S^c \cap T}\|_1 \geq \|h_S\|_1 - \|h_{S \cap T^c}\|_1 + \|h_{S^c \cap T}\|_1 \\
&\geq \|h_{S^c}\|_1 - \|h_{S \cap T^c}\|_1 + \|h_{S^c \cap T}\|_1 \\
&= 2\|h_{S^c \cap T}\|_1 + \|h_{S^c \cap T^c}\|_1 + \|h_{S \cap T^c}\|_1 - 2\|h_{S \cap T^c}\|_1 \\
&\geq \|h_{T^c}\|_1 + 2(\|h_{S^c \cap T}\|_1 - \|h_{S \cap T^c}\|_1) \geq \|h_{T^c}\|_1,
\end{aligned}$$

where we have used (7.56) and (7.57). This shows that (7.58) is valid.

We now partition h into the following subsets: Starting with $T = (1, \ldots, s)$, let $T_1 = (s+1, \ldots, s+k_1), T_2 = (s + k_1 + 1, \ldots, s + 2k_1), \ldots$. Using Lemma 7.24, we have

$$\begin{aligned}
\sum_{i \geq 1} \|h_{T_i}\|_2 &\leq \frac{1}{\sqrt{k_1}} \sum_{i \geq 1} \|h_{T_i}\|_1 + \frac{\sqrt{k_1}}{4}|h_{s+1}| \\
&= \frac{1}{\sqrt{k_1}}\|h_{T^c}\|_1 + \frac{\sqrt{k_1}}{4}|h_{s+1}| \leq \frac{1}{\sqrt{k_1}}\|h_T\|_1 + \frac{\sqrt{k_1}}{4}|h_{s+1}| \\
&\leq \frac{\sqrt{s}}{\sqrt{k_1}}\|h_T\|_2 + \frac{\sqrt{k_1}}{4\sqrt{s}}\|h_T\|_2 = K\|h_T\|_2,
\end{aligned}$$

where $K = \frac{\sqrt{s}}{\sqrt{k_1}} + \frac{\sqrt{k_1}}{4\sqrt{s}}$.

Finally, since h is in the null space of Φ, we use the RIP to get

$$\begin{aligned}
0 = \langle \Phi h, \Phi h_T \rangle &= \langle \Phi h_T, \Phi h_T \rangle - \sum_{i \geq 1}\langle \Phi h_{T_i}, \Phi h_T \rangle \\
&\geq (1 - \delta_s)\|h_T\|_2^2 - \theta_{s,k_1} \sum_{i \geq 1} \|h_{T_i}\|_2 \|h_T\|_2 \\
&\geq (1 - \delta_s - K\theta_{s,k_1})\|h_T\|_2^2.
\end{aligned}$$

When $K\theta_{s,k_1} + \delta_s < 1$, we must have $\|h_T\|_2 = 0$, and hence, by (7.57), $h = 0$. That is, the sparse solution will be unique. To this end, let us choose $k_1 = 4s/9$. Then we have

$$K = \frac{\sqrt{s}}{\sqrt{4s/9}} + \frac{\sqrt{4s/9}}{4\sqrt{s}} = \frac{3}{2} + \frac{2}{12} = \frac{5}{3}.$$

Since $\theta_{s,4s/9} \leq \sqrt{\frac{9}{5}}\theta_{5s/9,4s/9} \leq \frac{3}{\sqrt{5}}\delta_s$ by Lemmas 2.35 and 2.34, we have

$$\delta_s + K\theta_{s,4s/9} \leq \delta_s + \frac{5}{3}\frac{3}{\sqrt{5}}\delta_s = (1+\sqrt{5})\delta_s.$$

If $\delta_s < 1/(1+\sqrt{5}) \approx 0.307$, we can conclude that $h \equiv 0$, and hence the ℓ_1 minimization finds the sparse solution. □

Next we have the following theorem.

Theorem 7.26 (Cai and Zhang, 2013 [3]). *Suppose that for an integer $s \geq 1$, Φ satisfies the RIP with $\delta_s < 1/3$. Let $\mathbf{x_b} \in \mathbb{R}^n$ be a vector with $\|\mathbf{x_b}\|_0 \leq s$. Then for $\mathbf{b} = \Phi\mathbf{x_b}$, the solution $\mathbf{x}^*$ of minimization (1.20) is unique and equal to $\mathbf{x_b}$.*

This result was established in section 7.4. In this section we shall give another proof based on the following so-called division lemma.

Lemma 7.27 (Division Lemma). *Let r and m be positive integers with $m \geq 2r$. Suppose that $a_1 \geq a_2 \geq a_3 \geq \cdots \geq a_m \geq 0$ is a sequence of nonincreasing real numbers satisfying*

$$\sum_{i=1}^{r} a_i \geq \sum_{j=r+1}^{m} a_j. \tag{7.59}$$

Then there exist nonnegative real numbers $s_{ij}, 1 \leq i \leq r, 2r+1 \leq j \leq m$, such that

$$\sum_{i=1}^{r} s_{ij} = a_j, \quad 2r+1 \leq j \leq m, \tag{7.60}$$

and

$$\frac{1}{r}\sum_{j=1}^{r} a_j \geq a_{r+i} + \sum_{j=2r+1}^{m} s_{ij} \quad \forall i = 1, \ldots, r. \tag{7.61}$$

Proof. We can prove the result by choosing each s_{ij} as follows:

$$s_{ij} = \frac{a_j}{a_{2r+1} + \cdots + a_m}\left(\frac{1}{r}\sum_{k=1}^{r} a_k - a_{r+i}\right) \text{ for } i = 1, 2, \ldots, r-1 \text{ and } j = 2r+1, \ldots, m,$$

and then

$$s_{rj} = a_j - \sum_{i=1}^{r-1} s_{ij}, \text{ for } j = 2r+1, \ldots, m.$$

Now we can easily check that conditions (7.60) and (7.61) are satisfied. □

Next we need another lemma. For convenience, we call a vector with 1 or -1 in only one entry and zeros elsewhere an indicator vector.

Lemma 7.28. *Suppose that $\{b_i, i = 1,\ldots,h\}$ and $\{c_j, j = 1,\ldots,h\}$ are two real sequences for integer $h > 0$. Suppose g, h, ℓ are nonnegative integers such that $g + h \le s$, and $\{d_i, i = 1,\ldots,g\}$, $\{e_j, j = 1,\ldots,\ell\}$, and $\{t_{ij}\}_{1\le i\le g, 1\le j\le \ell}$ are sequences of nonnegative real numbers satisfying*

$$\min_{1\le i\le g} d_i \ge \max_{1\le j\le \ell} e_j \quad and \quad \sum_{i=1}^{g} t_{ij} = e_j, \quad j = 1,\ldots,\ell.$$

Let $\{u_{11},\ldots,u_{1h}; u_{31},\ldots,u_{3g}; u_{41},\ldots,u_{4\ell}\}$ be a set of indicator vectors with distinct support in $\mathbb{R}^n$, and let $\{u_{21},\ldots,u_{2h}; u_{31},\ldots,u_{3g}; u_{41},\ldots,u_{4\ell}\}$ also be a set of indicator vectors with distinct support. Define

$$\beta_1 = \sum_{i=1}^{h} b_i u_{1i} + \sum_{j=1}^{g} d_j u_{3j} + \sum_{k=1}^{\ell} e_k u_{4k}, \quad \beta_2 = \sum_{i=1}^{h} c_i u_{2i} + \sum_{j=1}^{g} d_j u_{3j} + \sum_{k=1}^{\ell} e_k u_{4k}. \quad (7.62)$$

Then we have

$$\|\Phi\beta_1\|^2 - \|\Phi\beta_2\|^2 \ge (1 - \delta_s)\left(\sum_{i=1}^{h} |b_i|^2 + \sum_{i=1}^{g}\left(d_i + \sum_{j=1}^{\ell} t_{ij}\right)^2\right)$$

$$-(1 + \delta_s)\left(\sum_{i=1}^{h} |c_i|^2 + \sum_{i=1}^{g}\left(d_i + \sum_{j=1}^{\ell} t_{ij}\right)^2\right). \quad (7.63)$$

Proof. We can prove the lemma by using induction on ℓ. For $\ell = 0$, the u_{ij}'s are distinct supports, and using the RIP we conclude (7.63) easily.

For $\ell > 0$, we first have the identity

$$\sum_{i=1}^{h} b_i^2 + \sum_{i=1}^{g}\left(d_i + \sum_{j=1}^{l} t_{ij}\right)^2 - \left(\sum_{i=1}^{h} b_i^2 + \sum_{i=1}^{g} d_i^2 + \sum_{i=1}^{l} e_i^2\right) \quad (7.64)$$

$$= \sum_{i=1}^{h} c_i^2 + \sum_{i=1}^{g}\left(d_i + \sum_{j=1}^{l} t_{ij}\right)^2 - \left(\sum_{i=1}^{h} c_i^2 + \sum_{i=1}^{g} d_i^2 + \sum_{i=1}^{l} e_i^2\right). \quad (7.65)$$

Next let us show that both sides of the above equation are nonnegative. Let us expand the left-hand side of the above equation by using $\sum_{i=1}^{g} t_{ij} = e_j$, letting $d = \min_{1\le i\le g}\{d_i\}$ and $e = \max_{1\le j\le l}\{e_j\}$. We have

$$\sum_{i=1}^{h} b_i^2 + \sum_{i=1}^{g}\left(d_i + \sum_{j=1}^{l} t_{ij}\right)^2 - \left(\sum_{i=1}^{h} b_i^2 + \sum_{i=1}^{g} d_i^2 + \sum_{i=1}^{l} e_i^2\right)$$

$$= 2\sum_{i=1}^{g}\sum_{j=1}^{l} d_i t_{ij} + \sum_{i=1}^{g}\left(\sum_{j=1}^{l} t_{ij}\right)^2 - \sum_{i=1}^{l} e_i^2$$

$$= 2\sum_{i=1}^{g}\sum_{j=1}^{l} d_i t_{ij} + \sum_{i=1}^{g}\left(\sum_{j=1}^{l} t_{ij}\right)^2 - \sum_{j=1}^{l}\left(\sum_{i=1}^{g} t_{ij}\right)^2$$

$$= 2\sum_{j=1}^{l}\sum_{i=1}^{g}d_i t_{ij} - \sum_{j=1}^{l}\left(\sum_{i=1}^{g}t_{ij}\right)^2 + \sum_{i=1}^{g}\left(\sum_{j=1}^{l}t_{ij}\right)^2$$

$$\geq 2d\sum_{j=1}^{l}\sum_{i=1}^{g}t_{ij} - e\sum_{j=1}^{l}\sum_{i=1}^{g}t_{ij} + \sum_{i=1}^{g}\left(\sum_{j=1}^{l}t_{ij}\right)^2$$

$$= (2d-e)\sum_{j=1}^{l}\sum_{i=1}^{g}t_{ij} + \sum_{i=1}^{g}\left(\sum_{j=1}^{l}t_{ij}\right)^2 \geq 0,$$

since $2d-e \geq 0$. That is, letting A, B, C, D be the four terms in (7.64) such that $A-B=C-D$, the above inequality shows that $A > B$ and $C > D$.

We are now ready to prove (7.63). Since the u_{ij}'s are distinct supports, by the RIP property, we have

$$\|\Phi\beta_1\|^2 - \|\Phi\beta_2\|^2$$

$$\geq (1-\delta_s)\left(\sum_{i=1}^{h}b_i^2 + \sum_{i=1}^{g}d_i^2 + \sum_{i=1}^{l}e_i^2\right) - (1+\delta_s)\left(\sum_{i=1}^{h}c_i^2 + \sum_{i=1}^{g}d_i^2 + \sum_{i=1}^{l}e_i^2\right)$$

$$\geq (1-\delta_s)(A+B-A) - (1+\delta)(C+D-C)$$

$$= (1-\delta_s)A - (1+\delta_s)C + (A-B)\delta_s + (C-D)\delta_s$$

$$\geq (1-\delta_s)A - (1+\delta_s)C.$$

This completes the proof. □

We are now ready to prove Theorem 7.26.

Proof of Theorem 7.26. Suppose there exists a vector $\mathbf{v} \in \mathcal{N}(\Phi)$ such that

$$\|\mathbf{v}_S\|_1 \geq \|\mathbf{v}_{S^c}\|_1$$

for a support set S with $|S| \leq s$. Let us write

$$\mathbf{v} = \sum_{i=1}^{n} a_i \mathbf{u}_i$$

with distinct indicator vectors $\mathbf{u}_i$ and $a_1 \geq a_2 \geq \cdots \geq a_n \geq 0$. Without loss of generality, we may assume that $n = N$.

By Lemma 7.27, we can find $t_{ij}, 1 \leq i \leq s, 2s+1 \leq j \leq n$, satisfying the properties in the lemma. We first consider s to be an even integer. Then let

$$\beta_{11} = \sum_{i=1}^{s/2}a_i\mathbf{u}_i, \quad \beta_{12} = \sum_{i=1}^{s/2+1}a_i\mathbf{u}_i, \quad \beta_{21} = \sum_{i=s+1}^{3s/2}a_i\mathbf{u}_i, \quad \beta_{22} = \sum_{i=3s/2+1}^{2s}a_i\mathbf{u}_i$$

and

$$\beta_{31} = \sum_{j=2s+1}^{n}\sum_{i=1}^{s/2}t_{ij}\mathbf{u}_j, \quad \beta_{32} = \sum_{j=2s+1}^{n}\sum_{i=s/2+1}^{s}t_{ij}\mathbf{u}_j.$$

It is easy to see that $\mathbf{v} = \sum_{i=1}^{3} \sum_{j=1}^{2} \beta_{ij}$, and then $\Phi\mathbf{v} = \sum_{i=1}^{3} \sum_{j=1}^{2} \Phi\beta_{ij} = 0$, which will be used in the following estimate.

Next we use the standard parallelogram identity to obtain

$$
\|\Phi(-\beta_{11} + \beta_{22} + \beta_{32})\|^2 + \|\Phi(-\beta_{12} + \beta_{21} + \beta_{31}\|^2
$$
$$
= \frac{1}{2}\|\Phi(-\beta_{11} + \beta_{22} + \beta_{32} - \beta_{12} + \beta_{21} + \beta_{31})\|^2
$$
$$
+ \frac{1}{2}\|\Phi(-\beta_{11} + \beta_{22} + \beta_{32} + \beta_{12} - \beta_{21} - \beta_{31})\|^2
$$
$$
= \frac{1}{2}\left(\|\Phi(2\beta_{11} + 2\beta_{12})\|^2 + \frac{1}{2}\|\Phi(2\beta_{11} + 2\beta_{21} + 2\beta_{31})\|^2 + \frac{1}{2}\|\Phi(2\beta_{22} + 2\beta_{32} + 2\beta_{12})\|^2 \right)
$$
$$
= 2\|\Phi(\beta_{11} + \beta_{12})\|^2 + \|\Phi(\beta_{11} + \beta_{21} + \beta_{31})\|^2 + \|\Phi(\beta_{22} + \beta_{32} + \beta_{12})\|^2.
$$

We now move the terms on the left-hand side to the right-hand side, use the RIP on the first term, and apply Lemma 7.28 to the remaining terms to get

$$
0 = 2\|\Phi(\beta_{11} + \beta_{12})\|^2 + \|\Phi(\beta_{11} + \beta_{21} + \beta_{31})\|^2 - \|\Phi(-\beta_{12} + \beta_{21} + \beta_{31})\|^2
$$
$$
+ \|\Phi(\beta_{22} + \beta_{32} + \beta_{12})\|^2 - \|\Phi(-\beta_{11} + \beta_{22} + \beta_{32})\|^2
$$
$$
\geq 2(1 - \delta_s)\|\beta_{11} + \beta_{12}\|^2
$$
$$
+ (1 - \delta_s)\left(\sum_{i=1}^{s/2} a_i^2 + \sum_{i=s+1}^{3s/2} \left(a_i + \sum_{j=2s+1}^{n} t_{ij} \right)^2 \right)
$$
$$
- (1 + \delta_s)\left(\sum_{i=s/2+1}^{s} a_i^2 + \sum_{i=s+1}^{3s/2} \left(a_i + \sum_{j=2s+1}^{n} t_{ij} \right)^2 \right)
$$
$$
+ (1 - \delta_s)\left(\sum_{i=s/2+1}^{s} a_i^2 + \sum_{i=3s/2+1}^{2s} \left(a_i + \sum_{j=2s+1}^{n} t_{ij} \right)^2 \right)
$$
$$
- (1 + \delta_s)\left(\sum_{i=1}^{s/2} a_i^2 + \sum_{i=3s/2+1}^{2s} \left(a_i + \sum_{j=2s+1}^{n} t_{ij} \right)^2 \right)
$$
$$
= 2(1 - \delta_s)\sum_{i=1}^{s} a_i^2 - 2\delta_s \sum_{i=1}^{s} a_i^2 - 2\delta_s \left(\sum_{i=s+1}^{2s} \left(a_i + \sum_{j=2s+1}^{n} t_{ij} \right)^2 \right)
$$
$$
\geq 2(1 - 2\delta_s)\sum_{i=1}^{s} a_i^2 - 2\delta_s \left(\frac{1}{s} \sum_{i=1}^{s} a_i \right)^2 \geq 2(1 - 3\delta_s)\sum_{i=1}^{s} a_i^2,
$$

where we have used Lemma 7.27 and the Cauchy–Schwarz inequality. It thus follows that $\sum_{i=1}^{s} a_i^2 = 0$ when $\delta_s < 1/3$. Hence, $\mathbf{v} \equiv 0$, which is a contradiction.

Next we can consider the case when s is an odd integer. The proof is similar and is left to Exercise 16.   □

This estimate on $\delta_s < 1/3$ is the best possible according to [3].

7.7 · Conditions for Noisy Recovery

All conditions on δ_{2s} for noiseless recovery have a counterpart for noisy recovery. We shall present a few conditions in the noisy setting in this section and leave the remaining conditions to Exercises 6 and 11. We start with the following result.

Theorem 7.29. *Suppose that for an integer $s \geq 1$, Φ satisfies the RIP with*

$$\delta_{3s} < \frac{1}{2}.$$

Let $\mathbf{x} \in \mathbb{R}^n$ be a vector with $\|\mathbf{x}\|_0 \leq s$. Then for $\mathbf{b} = \Phi\mathbf{x}$, the solution $\mathbf{x}^*$ of minimization (1.23) satisfies the estimate

$$\|(\mathbf{x} - \mathbf{x}^*)_S\|_2 \leq \frac{1}{1-\gamma} \frac{\sqrt{2}\delta}{1-\delta_{2s}},$$

where $\gamma = \delta_{3s}/(1 - \delta_{3s}) < 1$.

Proof. Let S_0 be the index set of nonzero entries of $\mathbf{x}$. Let $h = \mathbf{x} - \mathbf{x}^*$. Let $S_1, S_2, \ldots, S_j, \ldots$ be the index sets, as in the proof of Theorem 7.1. It is easy to see $\Phi h_{S_0 \cup S_1} = \Phi h - \sum_{j \geq 2} \Phi h_{S_j}$ with $\|\Phi h\|_2 \leq \delta$. Then

$$\|h_{S_0 \cup S_1}\|_2^2 \leq \frac{1}{1-\delta_{2s}} \|\Phi(h_{S_0 \cup S_1})\|_2^2$$

$$= \frac{1}{1-\delta_{2s}} \left(\sum_{j \geq 2} \langle \Phi(h_{S_0 \cup S_1}), -\Phi(h_{S_j}) \rangle + \langle \Phi(h_{S_0 \cup S_1}), \Phi h \rangle \right)$$

$$\leq \frac{\delta_{3s}}{1-\delta_{2s}} \|h_{S_0 \cup S_1}\|_2 \sum_{j \geq 2} \|h_{S_j}\|_2 + \frac{\sqrt{1+\delta_{2s}}}{1-\delta_{2s}} \|h_{S_0 \cup S_1}\|_2 \delta,$$

where we have used Lemma 2.31 since $(S_0 \cup S_1) \cap S_j = \emptyset$ for all $j \geq 2$. Dividing by $\|h_{S_0 \cup S_1}\|_2$ and using the inequality $\|h_{S_j}\|_2 \leq \frac{1}{\sqrt{s}} \|h_{S_{j-1}}\|_1$, we obtain

$$\|h_{S_0}\|_2 \leq \frac{\delta_{3s}}{1-\delta_{3s}} \frac{1}{\sqrt{s}} \|h_{S_0^c}\|_1 \frac{\sqrt{2}\delta}{1-\delta_{2s}}$$

$$\leq \gamma \frac{1}{\sqrt{s}} \|h_{S_0}\|_1 + \frac{\sqrt{2}\delta}{1-\delta_{2s}} \leq \gamma \|h_{S_0}\|_2 + \frac{\sqrt{2}\delta}{1-\delta_{2s}},$$

where we have used the NSP, i.e., Theorem 2.5. By the Cauchy–Schwarz inequality and since $\gamma = \frac{\delta_{3s}}{1-\delta_{3s}} < 1$, we have

$$(1-\gamma)\|h_S\|_2 \leq \frac{\sqrt{2}\delta}{1-\delta_{2s}}.$$

We have thus completed the proof. $\square$

Next let us look at an improved version.

Theorem 7.30 (Candés, 2008 [6]). *Suppose that for $s \geq 1$, Φ satisfies*

$$\delta_{2s} < \sqrt{2} - 1 \approx 0.4142.$$

Let $\mathbf{x_b} \in \mathbb{R}^n$ be a vector satisfying $\mathbf{b} = \Phi\mathbf{x_b}$. Let $\mathbf{x}^*$ be a minimizer of minimization (1.23). Then

$$\|\mathbf{x_b} - \mathbf{x}^*\|_2 \leq C(s^{-1/2}\|(\mathbf{x_b})_{T_0^c}\|_1 + \delta),$$

where T_0 is the index set of the s largest entries in absolute value of $\mathbf{x_b}$ and $C > 0$ is a positive constant dependent on $\delta_{2s} < \sqrt{2} - 1$.

Proof. Let $\mathbf{x}^*$ be a solution of the minimization problem (1.23). Let $h = \mathbf{x}^* - \mathbf{x_b}$. Note that $\|\Phi h\|_2 \leq \delta$. Rewrite h as a sum of vectors $h_{T_0}, h_{T_1}, h_{T_2}, \ldots$, each of sparsity at most s, as in the proof of Theorem 7.5. Recall that we have

$$\|h_{T_j}\|_2 \leq s^{1/2}\|h_{T_j}\|_\infty \leq s^{-1/2}\|h_{T_{j-1}}\|_1$$

and

$$\sum_{j\geq 2}\|h_{T_j}\|_2 \leq s^{-1/2}(\|h_{T_1}\|_2 + \|h_{T_2}\|_1 + \cdots) = s^{-1/2}\|h_{T_0^c}\|_1.$$

We use the Cauchy–Schwarz inequality to bound $\|h_{T_0}\|_1$ by $s^{1/2}\|h_{T_0}\|_2$ and get

$$\sum_{j\geq 2}\|h_{T_j}\|_2 \leq s^{-1/2}(\|h_{T_0}\|_1 + 2\|(\mathbf{x_b})_{T_0^c}\|_1) \leq \|h_{T_0}\|_2 + 2s^{-1/2}\|(\mathbf{x_b})_{T_0^c}\|_1$$

by the NSP, i.e., Theorem 2.5. Since $\|h_{T_0}\|_2 \leq \|h_{T_0 \cup T_1}\|_2$ and $\|h_{(T_0 \cup T_1)^c}\|_2 \leq \sum_{j\geq 2}\|h_{T_j}\|_2$, the above estimate can be rewritten as

$$\|h_{(T_0 \cup T_1)^c}\|_2 \leq \|h_{T_0 \cup T_1}\|_2 + 2s^{-1/2}\|\mathbf{x}_{T_0^c}\|_1.$$

Next, since $\Phi(h_{T_0 \cup T_1}) = \Phi h - \Phi(\sum_{j\geq 2} h_{T_j})$, we apply Lemma 2.31 to get

$$\begin{aligned}
\|\Phi(h_{T_0 \cup T_1})\|_2^2 &= \langle \Phi(h_{T_0 \cup T_1}), \Phi h \rangle - \langle \Phi(h_{T_0 \cup T_1}), \Phi(h_{(T_0 \cup T_1)^c}) \rangle \\
&\leq \|\Phi(h_{T_0 \cup T_1})\|\|\Phi h\| + \delta_{2s}(\|h_{T_0}\|_2 + \|h_{T_1}\|_2)\|h_{(T_0 \cup T_1)^c}\|_2 \\
&\leq \sqrt{1 + \delta_{2s}}\|h_{T_0 \cup T_1}\|_2\delta + \delta_{2s}\sqrt{2}\|h_{T_0 \cup T_1}\|_2(\|h_{T_0 \cup T_1}\|_2 + 2s^{-1/2}\|\mathbf{x}_{T_0^c}\|_1) \\
&\leq \delta_{2s}\sqrt{2}\|h_{T_0 \cup T_1}\|_2^2 + 2\delta_{2s}\|h_{T_0 \cup T_1}\|_2 s^{-1/2}\|\mathbf{x}_{T_0^c}\|_1 + \sqrt{1 + \delta_{2s}}\|h_{T_0 \cup T_1}\|_2\delta.
\end{aligned}$$

We now use the RIP to conclude that

$$\begin{aligned}
(1 - \delta_{2s})\|h_{T_0 \cup T_1}\|_2^2 &\leq \|\Phi(h_{T_0 \cup T_1})\|_2^2 \\
&\leq \delta_{2s}\sqrt{2}\|h_{T_0 \cup T_1}\|_2^2 + 2\delta_{2s}\|h_{T_0 \cup T_1}\|_2 s^{-1/2}\|(\mathbf{x_b})_{T_0^c}\|_1 \\
&\quad + \sqrt{1 + \delta_{2s}}\|h_{T_0 \cup T_1}\|_2\delta.
\end{aligned}$$

That is,

$$(1 - (1 + \sqrt{2})\delta_{2s})\|h_{T_0 \cup T_1}\|_2 \leq 2\delta_{2s}s^{-1/2}\|(\mathbf{x_b})_{T_0^c}\|_1 + \sqrt{1 + \delta_{2s}}\delta. \qquad (7.66)$$

Summing up, we have

$$\begin{aligned}
\|h\|_2 &\leq \|h_{T_0 \cup T_1}\|_2 + \|h_{(T_0 \cup T_1)^c}\|_2 \leq 2\|h_{T_0 \cup T_1}\|_2 + 2s^{-1/2}\|\mathbf{x}_{T_0^c}\|_1 + \sqrt{1 + \delta_{2s}}\delta \\
&\leq \frac{\delta_{2s}}{1 - (1 + \sqrt{2})\delta_{2s}}\left(4s^{-1/2}\|(\mathbf{x_b})_{T_0^c}\|_1 + 2s^{-1/2}\|(\mathbf{x_b})_{T_0^c}\|_1 + \sqrt{2}\delta\right).
\end{aligned}$$

That is, if $\delta_{2s} < 1/(1 + \sqrt{2}) = \sqrt{2} - 1$, the above is the desired result of Theorem 7.30. $\qquad \square$

Next we present the study contained in [10].

Theorem 7.31 (Foucart and Lai, 2009 [10]). *Given $0 < q \le 1$, if $\delta_{2s} < 2(3 - \sqrt{2})/7$, then a solution $\mathbf{x}^*$ of (1.23) approximates the original vector $\mathbf{x_b}$ with errors*

$$\|\mathbf{x} - \mathbf{x}^*\|_q \le C_1 \sigma_s(\mathbf{x})_q + D_1 s^{1/q-1/2} \cdot \delta, \tag{7.67}$$

$$\|\mathbf{x} - \mathbf{x}^*\|_2 \le C_2 \frac{\sigma_s(\mathbf{x})_q}{t^{1/q-1/2}} + D_2 \cdot \theta. \tag{7.68}$$

The constants C_1, C_2, D_1, and D_2 depend only on q and γ_{2t}.

Proof. The proof is divided into three steps.

Step 1. Consequence of the assumption on δ_{2s}.

We consider an arbitrary index set S with $|S| \le s$. Let $\mathbf{v}$ be a given vector in $\mathbb{R}^n$. We partition the complement of S in $\{1, \ldots, N\}$ as $S^c = T_1 \cup T_2 \cup \cdots$, where

$$T_1 := \{\text{indices of the } s \text{ largest absolute value components of } \mathbf{v} \text{ in } S^c\},$$

$$T_2 := \{\text{indices of the next } s \text{ largest absolute value components of } \mathbf{v} \text{ in } S^c\},$$

$$\cdots.$$

We first observe that

$$\|\mathbf{v}_S\|_2^2 + \|\mathbf{v}_{T_1}\|_2^2 = \|\mathbf{v}_S + \mathbf{v}_{T_1}\|_2^2 \le \frac{1}{1 - \delta_{2s}} \|\Phi(\mathbf{v}_S + \mathbf{v}_{T_1})\|_2^2$$

$$= \frac{1}{1 - \delta_{2s}} \langle \Phi(\mathbf{v} - \mathbf{v}_{T_2} - \mathbf{v}_{T_3} - \cdots), \Phi(\mathbf{v}_S + \mathbf{v}_{T_1}) \rangle \tag{7.69}$$

$$= \frac{1}{1 - \delta_{2s}} \langle \Phi\mathbf{v}, \Phi(\mathbf{v}_S + \mathbf{v}_{T_1}) \rangle + \frac{1}{1 - \delta_{2s}} \sum_{k \ge 2} \left[\langle \Phi(-\mathbf{v}_{T_k}), \Phi\mathbf{v}_S \rangle + \langle \Phi(-\mathbf{v}_{T_k}), \Phi\mathbf{v}_{T_1} \rangle \right].$$

Let us renormalize the vectors $-\mathbf{v}_{T_k}$ and $\mathbf{v}_S$ so that their ℓ_2 norms equal one by setting $\mathbf{u}_k := -\mathbf{v}_{T_k}/\|\mathbf{v}_{T_k}\|_2$ and $\mathbf{u}_0 := \mathbf{v}_S/\|\mathbf{v}_S\|_2$. We then obtain

$$\frac{\langle \Phi(-\mathbf{v}_{T_k}), \Phi\mathbf{v}_S \rangle}{\|\mathbf{v}_{T_k}\|_2 \|\mathbf{v}_S\|_2} = \langle \Phi\mathbf{u}_k, \Phi\mathbf{u}_0 \rangle = \frac{1}{4} \left[\|\Phi(\mathbf{u}_k + \mathbf{u}_0)\|_2^2 - \|\Phi(\mathbf{u}_k - \mathbf{u}_0)\|_2^2 \right]$$

$$\le \frac{1}{4} \left[(1 + \delta_{2s}) \|\mathbf{u}_k + \mathbf{u}_0\|_2^2 - (1 - \delta_{2s}) \|\mathbf{u}_k - \mathbf{u}_0\|_2^2 \right] = \delta_{2s}.$$

With a similar argument with T_1 in place of S, we can derive

$$\langle \Phi(-\mathbf{v}_{T_k}), \Phi\mathbf{v}_S \rangle + \langle \Phi(-\mathbf{v}_{T_k}), \Phi\mathbf{v}_{T_1} \rangle \le \delta_{2s} \|\mathbf{v}_{T_k}\|_2 \left[\|\mathbf{v}_S\|_2 + \|\mathbf{v}_{T_1}\|_2 \right]. \tag{7.70}$$

In addition, we have

$$\langle \Phi\mathbf{v}, \Phi(\mathbf{v}_S + \mathbf{v}_{T_1}) \rangle \le \|\Phi\mathbf{v}\|_2 \cdot \|\Phi(\mathbf{v}_S + \mathbf{v}_{T_1})\|_2 \le \|\Phi\mathbf{v}\|_2 \sqrt{1 + \delta_{2s}} \left[\|\mathbf{v}_S\|_2 + \|\mathbf{v}_{T_1}\|_2 \right]. \tag{7.71}$$

Substituting inequalities (7.70) and (7.71) into (7.69), we have

$$\|\mathbf{v}_S\|_2^2 + \|\mathbf{v}_{T_1}\|_2^2 \le \left(\frac{\sqrt{1 + \delta_{2s}}}{1 - \delta_{2s}} \|\Phi\mathbf{v}\|_2 + \frac{\delta_{2s}}{1 - \delta_{2s}} \sum_{k \ge 2} \|\mathbf{v}_{T_k}\|_2 \right) \left[\|\mathbf{v}_S\|_2 + \|\mathbf{v}_{T_1}\|_2 \right].$$

With $c := \frac{\sqrt{1+\delta_{2s}}}{1-\delta_{2s}} \|\Phi\mathbf{v}\|_2$, $d := \frac{\delta_{2s}}{1-\delta_{2s}}$, and $\Sigma = \sum_{k \ge 2} \|\mathbf{v}_{T_k}\|_2$, this reads

$$\left[\|\mathbf{v}_S\|_2 - \frac{c + d\Sigma}{2} \right]^2 + \left[\|\mathbf{v}_{T_1}\|_2 - \frac{c + d\Sigma}{2} \right]^2 \le \frac{(c + d\Sigma)^2}{2}.$$

The above inequality easily implies that

$$\|\mathbf{v}_S\|_2 \leq \frac{c + d\Sigma}{2} + \frac{c + d\Sigma}{\sqrt{2}} = \frac{1 + \sqrt{2}}{2} \cdot (c + d\Sigma). \tag{7.72}$$

By the Cauchy–Schwarz inequality, we get

$$\|\mathbf{v}_S\|_q \leq s^{1/q - 1/2} \|\mathbf{v}_S\|_2 \leq \frac{1 + \sqrt{2}}{2} \cdot (c + d\Sigma) \cdot s^{1/q - 1/2}. \tag{7.73}$$

It now remains to bound Σ. Given an integer $k \geq 2$, let us consider $i \in T_k$ and $j \in T_{k-1}$. From the inequality $|v_i| \leq |v_j|$ raised to the power q, we derive that $|v_i|^q \leq s^{-1} \|\mathbf{v}_{T_{k-1}}\|_q^q$ by averaging over j. In turn, this yields the inequality $\|\mathbf{v}_{T_k}\|_2^2 \leq s^{1-2/q} \|\mathbf{v}_{T_{k-1}}\|_q^2$ by raising to the power $2/q$ and summing over i. It follows that

$$\Sigma = \sum_{k \geq 2} \|\mathbf{v}_{T_k}\|_2 \leq s^{1/2 - 1/q} \sum_{k \geq 1} \|\mathbf{v}_{T_k}\|_q \leq s^{1/2 - 1/q} \left[\sum_{k \geq 1} \|\mathbf{v}_{T_k}\|_q^q \right]^{1/q} = s^{1/2 - 1/q} \|\mathbf{v}_{\overline{S}}\|_q.$$

Combining the above inequality with (7.73), we obtain the partial conclusion

$$\|\mathbf{v}_S\|_q \leq \frac{(1 + \sqrt{2})\sqrt{1 + \delta_{2s}}}{2(1 - \delta_{2s})} \|\Phi\mathbf{v}\|_2 \cdot s^{1/q - 1/2} + \frac{(1 + \sqrt{2})\delta_{2s}}{2(1 - \delta_{2s})} \|\mathbf{v}_{\overline{S}}\|_q, \qquad \mathbf{v} \in \mathbb{R}^n, \ |S| \leq s, \tag{7.74}$$

where the constants λ and μ are given by

$$\lambda := (1 + \sqrt{2})\gamma_{2t} \quad \text{and} \quad \mu := \frac{(1 + \sqrt{2})\delta_{2s}}{2(1 - \delta_{2s})}. \tag{7.75}$$

Note that the assumption on δ_{2t} translates into the inequality $\mu < 1$.

Step 2. Consequence of the ℓ_q minimization.

Now let S be specified as the set of indices of the s largest absolute value components of $\mathbf{x}$, and let $\mathbf{v}$ be specified as $\mathbf{v} := \mathbf{x} - \mathbf{x}^*$. Because $\mathbf{x}^*$ is a minimizer of $(P_{q,\theta})$, we have

$$\|\mathbf{x}\|_q^q \geq \|\mathbf{x}^*\|_q^q, \qquad \text{i.e.,} \qquad \|\mathbf{x}_S\|_q^q + \|\mathbf{x}_{S^c}\|_q^q \geq \|\mathbf{x}_S^*\|_q^q + \|\mathbf{x}_{S^c}^*\|_q^q.$$

By the triangle inequality, we obtain

$$\|\mathbf{x}_S\|_q^q + \|\mathbf{x}_{\overline{S}}\|_q^q \geq \|\mathbf{x}_S\|_q^q - \|\mathbf{v}_S\|_q^q + \|\mathbf{v}_{\overline{S}}\|_q^q - \|\mathbf{x}_{\overline{S}}\|_q^q.$$

Rearranging the latter yields the inequality

$$\|\mathbf{v}_{S^c}\|_q^q \leq 2 \|\mathbf{x}_{S^c}\|_q^q + \|\mathbf{v}_S\|_q^q = 2\sigma_s(\mathbf{x})_q^q + \|\mathbf{v}_S\|_q^q. \tag{7.76}$$

Step 3. Error estimates.

We now take into account the bound

$$\|\Phi\mathbf{v}\|_2 = \|\Phi\mathbf{x} - \Phi\mathbf{x}^*\|_2 \leq \|\Phi\mathbf{x} - \mathbf{y}\|_2 + \|\Phi\mathbf{x}^* - \mathbf{y}\|_2 \leq 2\sqrt{1 + \delta_{2s}}\delta.$$

For the ℓ_q error, we combine the estimates (7.74) and (7.76) to get

$$\|\mathbf{v}_{S^c}\|_q^q \leq 2\sigma_s(\mathbf{x})_q^q + \lambda^q s^{1-q/2} \cdot \delta^q + \mu^q \|\mathbf{v}_{S^c}\|_q^q.$$

As a consequence of $\mu < 1$, we now obtain

$$\|\mathbf{v}_{S^c}\|_q^q \leq \frac{2}{1-\mu^q}\sigma_s(\mathbf{x})_q^q + \frac{\lambda^q}{1-\mu^q}\cdot s^{1-q/2}\cdot\theta^q.$$

Using the estimate (7.74) once more, we can derive that

$$\|\mathbf{v}\|_q = \left[\|\mathbf{v}_S\|_q^q + \|\mathbf{v}_{S^c}\|_q^q\right]^{1/q} \leq \left[(1+\mu^q)\cdot\|\mathbf{v}_{S^c}\|_q^q + \lambda^q\cdot s^{1-q/2}\cdot\delta^q\right]^{1/q}$$

$$\leq \left[\frac{2\,(1+\mu^q)}{1-\mu^q}\cdot\sigma_s(\mathbf{x})_q^q + \frac{2\,\lambda^q}{1-\mu^q}\cdot s^{1-q/2}\cdot\delta^q\right]^{1/q}$$

$$\leq 2^{1/q-1}\left[\frac{2^{1/q}\,(1+\mu^q)^{1/q}}{(1-\mu^q)^{1/q}}\cdot\sigma_s(\mathbf{x})_q + \frac{2^{1/q}\,\lambda}{(1-\mu^q)^{1/q}}\cdot s^{1/q-1/2}\cdot\theta\right],$$

where we have made use of the inequality $[a^q + b^q]^{1/q} \leq 2^{1/q-1}[a+b]$ for $a, b \geq 0$. The estimate (7.67) follows with

$$C_1 := \frac{2^{2/q-1}\,(1+\mu^q)^{1/q}}{(1-\mu^q)^{1/q}} \qquad \text{and} \qquad D_1 := \frac{2^{2/q-1}\,\lambda}{(1-\mu^q)^{1/q}}.$$

As for the ℓ_2 error, we remark that the bound (7.72) also holds for $\|\mathbf{v}_{T_1}\|_2$ in place of $\|\mathbf{v}_S\|_2$ to obtain

$$\|\mathbf{v}\|_2 = \left[\sum_{k\geq 0}\|\mathbf{v}_{T_k}\|_2^2\right]^{1/2} \leq \sum_{k\geq 0}\|\mathbf{v}_{T_k}\|_2 \leq (1+\sqrt{2})\cdot(c+d\Sigma) + \Sigma \leq \nu\cdot\Sigma + 2\lambda\cdot\delta,$$

where $\nu := (\lambda + 1 - \sqrt{2})/2$. Then, in view of the bound

$$\Sigma \leq t^{1/2-1/q}\|\mathbf{v}_{\overline{S}}\|_q \leq t^{1/2-1/q}\cdot 2^{1/q-1}\cdot\left[\frac{2^{1/q}}{(1-\mu^q)^{1/q}}\cdot\sigma_s(\mathbf{x})_q + \frac{\lambda}{(1-\mu^q)^{1/q}}\cdot s^{1/q-1/2}\cdot\theta\right],$$

we may finally conclude that

$$\|\mathbf{v}\|_2 \leq \frac{2^{2/q-1}\,\nu}{(1-\mu^q)^{1/q}}\cdot\frac{\sigma_s(\mathbf{x})_q}{s^{1/q-1/2}} + \left[\frac{2^{1/q-1}\,\lambda\,\nu}{(1-\mu^q)^{1/q}} + 2\lambda\right]\delta.$$

This leads to the estimate (7.68) with

$$C_2 := \frac{2^{2/q-2}\,(\lambda+1-\sqrt{2})}{(1-\mu^q)^{1/q}} \qquad \text{and} \qquad D_2 := \frac{2^{1/q-2}\,\lambda\,(\lambda+1-\sqrt{2})}{(1-\mu^q)^{1/q}} + 2\lambda.$$

The reader is invited to verify that the constants C_1, D_1, C_2, and D_2 depend only on q and β_{2t}. However, they grow exponentially fast when q tends to zero. □

We leave the noisy recovery based on other sufficient conditions to Exercise 11.

7.8 ▪ Miscellaneous Conditions

In this section, we present other types of sufficient conditions to ensure that the minimizers of (1.20) are sparse solutions. For any $\mathbf{x} \in \mathbb{R}^n$, let

$$\text{sign}(\mathbf{x}) = (\text{sign}(x_1),\ldots,\text{sign}(x_N))^\top \in \mathbb{R}^n,$$

where $\text{sign}(x_j) = x_j/|x_j|$ if $x_j \neq 0$ and $= 0$ if $x_j = 0$ for $j = 1, \ldots, N$. We begin with the following lemma.

Lemma 7.32 (Fuchs, 2014 [11]). *Let Φ be a sensing matrix and $\mathbf{b}$ be a given observed vector in $\mathbb{R}^m$. Suppose that there exists a vector $\mathbf{x} \in \mathbb{R}^h$ satisfying $\Phi\mathbf{x} = \mathbf{b}$ and another vector $h \in \mathbb{R}^m$ such that letting S be the set of indices where the entries of $\mathbf{x}$ are nonzero,*

$$\Phi_S^\top h = \text{sign}((\mathbf{x})_S) \quad \text{and} \quad |(\Phi^\top h)_j| < 1 \quad \forall j \in S^c. \tag{7.77}$$

Assume that Φ_S is injective. Then $\mathbf{x}$ is the unique solution to the ℓ_1 minimization problem (1.20).

Proof. Using an h which satisfies the properties in (7.77), we have

$$\|\mathbf{x}\|_1 = \langle \Phi_S^\top h, \mathbf{x} \rangle = \langle h, \Phi\mathbf{x} \rangle = \langle h, \mathbf{b} \rangle.$$

For any other $\mathbf{y} \in \mathbb{R}^n$ with $\Phi\mathbf{y} = \mathbf{b}$, we have

$$\|\mathbf{x}\|_1 = \langle h, \Phi\mathbf{y} \rangle = \langle \Phi^\top h, \mathbf{y} \rangle = \langle \Phi^\top h, \mathbf{y}_S \rangle + \langle \Phi^\top h, \mathbf{y}_{S^c} \rangle$$
$$\leq \|(\Phi^\top h)_S\|_\infty \|\mathbf{y}_S\|_1 + \|(\Phi^\top h)_{S^c}\|_\infty \|\mathbf{y}_{S^c}\|_1 < \|\mathbf{y}_S\|_1 + \|\mathbf{y}_{S^c}\|_1 = \|\mathbf{y}\|_1.$$

That is, $\mathbf{x}$ is the minimizer of (1.20). Note that the strict inequality above follows from $\|\mathbf{y}_{S^c}\|_1 > 0$. Otherwise, the vector $\mathbf{y}$ would be supported on S and the injectivity of Φ_S would imply that $\mathbf{y} = \mathbf{x}$. Therefore, $\mathbf{x}$ is the unique minimizer.   □

In fact, we can have an explicit formula for h in Lemma 7.32, as explained in the following.

Theorem 7.33 (Fuchs, 2004 [11]). *Let $\Phi = [\phi_1, \ldots, \phi_N]$ be a sensing matrix and $\mathbf{b}$ be a given observed vector in $\mathbb{R}^m$. Suppose that $\mathbf{x_b}$ satisfies $\Phi\mathbf{x_b} = \mathbf{b}$ and Φ_S is injective, where S is the set of indices where $\mathbf{x_b}$ are nonzero. If*

$$|\langle \Phi_S^\dagger \phi_j, \text{sign}(\mathbf{x_b}) \rangle| < 1 \quad \forall j \in S^c,$$

then the vector $\mathbf{x_b}$ is the unique solution to (1.20), where $\Phi_S^\dagger$ is the pseudo-inverse of Φ_S.

Proof. Let $h = (\Phi_S^\dagger)^\top \text{sign}(\mathbf{x_b})$. Then h satisfies

$$\Phi_S^\top h = \Phi_S^\top \Phi_S (\Phi_S^\top \Phi_S)^{-1} \text{sign}(\mathbf{x_b}) = \text{sign}(\mathbf{x_b})$$

and

$$|(\Phi^\top h)_j| = |\phi_j^\top (\Phi_S^\dagger)^\top \text{sign}(\mathbf{x_b})| = |\langle \Phi_S^\dagger \phi_j, \text{sign}(\mathbf{x_b}) \rangle| < 1.$$

Then Lemma 7.32 can be applied to conclude that $\mathbf{x_b}$ is the unique minimizer of (1.20).   □

Finally, in this section we present another type of analysis. It is based on the q-RIP defined in Definition 2.38.

Theorem 7.34 (Shen and Li 2012 [13]). *Let s and t be two positive integers and $0 < q \leq 1$. Suppose that Φ satisfies the q-RIP with $\delta_{s+t,q} < 1$. If*

$$\gamma = \frac{1 + \delta_{t,q}}{1 - \delta_{s+t,q}} \left(\frac{s}{t}\right)^{1-q/2} < 1,$$

then letting $\mathbf{x}^*$ *be the minimizer of* (1.27),

$$\|\mathbf{x_b} - \mathbf{x}^*\|_q^q \le 2\frac{1+\gamma}{1-\gamma}\sigma_s(\mathbf{x_b})_q^q,$$

where $\sigma_s(\mathbf{x}_q)$ *is the best s-term approximation in the quasi-norm* ℓ_q.

Proof. Let $h = \mathbf{x_b} - \mathbf{x}^*$ and S_0 be the set of indices of the s largest entries of $\mathbf{x_b}$ in magnitude. Let S_1 be the set of indices of the t largest entries of h in magnitude in S_0^c, S_2 be the set of the next t largest entries of h in magnitude, and so on. It is easy to see that

$$\sum_{j\ge 2}\|h_{S_j}\|_2^q \le t^{q/2-1}\sum_{j\ge 1}\|h_{S_j}\|_q^q = t^{q/2-1}\|h_{S_0^c}\|_q^q. \tag{7.78}$$

Note that $\Phi h = 0$. It follows that

$$0 = \|\Phi h\|_q^q = \left\|\Phi(h_{S_0} + h_{S_1}) + \sum_{j\ge 2}\Phi h_{S_j}\right\|_q^q \ge \|\Phi(h_{S_0} + h_{S_1})\|_q^q - \left\|\sum_{j\ge 2}\Phi h_{S_j}\right\|_q^q.$$

Together with the q-RIP, i.e., (2.28) and (7.78), we obtain

$$(1 - \delta_{s+t,q})\|h_{S_0} + h_{S_1}\|_q^q \le \|\Phi(h_{S_0} + h_{S_1})\|_q^q \le \left\|\sum_{j\ge 2}\Phi h_{S_j}\right\|_q^q \le (1 + \delta_{t,q})t^{q/2-1}\|h_{S_0^c}\|_q^q. \tag{7.79}$$

Hence, we can conclude that

$$\|h_{S_0}\|_q^q \le s^{1-q/2}\|h_{S_0}\|_2^q \le \frac{1+\delta_{t,q}}{1-\delta_{s+t,q}}\left(\frac{s}{t}\right)^{1-q/2}\|h_{S_0^c}\|_q^q =: \gamma\|h_{S_0^c}\|_q^q. \tag{7.80}$$

Next we recall that $\|\mathbf{x}^*\|_q^q \le \|\mathbf{x_b}\|_q^q$. By the arguments in the proof of Theorem 2.4 and, more precisely, by (7.76), we have

$$\|h_{S_0^c}\|_q^q \le 2\|\mathbf{x_b}|_{S_0^c}\|_q^q + \|h_{S_0}\|_q^q \le 2\sigma_s(\mathbf{x_b})_q^q + \gamma\|h_{S_0^c}\|_q^q$$

or $(1 - \gamma)\|h_{S_0^c}\|_q^q \le 2\sigma_s(\mathbf{x_b})_q^q$. Consequently, we get

$$\|\mathbf{x_b} - \mathbf{x}^*\|_q^q = \|h_{S_0}\|_q^q + \|h_{S_0^c}\|_q^q \le (1+\gamma)\|h_{S_0^c}\|_q^q \le 2\frac{1+\gamma}{1-\gamma}\sigma_s(\mathbf{x_b})_q^q.$$

This completes the proof. $\square$

7.9 ▪ Exercises

Exercise 1. Prove Theorem 7.3.

Exercise 2. Prove the following inequality: Let $a_1 \ge a_2 \ge \cdots \ge a_{5r} \ge 0$. Then

$$\sum_{j=r+1}^{5r} a_j^2 \le \frac{1}{4r}\left(\sum_{j=1}^{4r} a_j\right)^2.$$

Exercise 3. Finish the proof of the shift inequality. That is, prove the general case of Lemma 7.7.

Exercise 4. Prove inequality (7.13).

Exercise 5. Extend the results in Theorem 7.11 to the ℓ_p minimization for $p \in (0, 1]$.

Exercise 6. Study noisy recovery under the same condition as in Theorem 7.11.

Exercise 7. Prove Lemma 7.18.

Exercise 8. Prove the equality in (7.24).

Exercise 9. Prove the identities in (7.36) and (7.35).

Exercise 10. Prove the identity in (7.47). *Hint*: You may use *Mathematica* or another symbolic computational software.

Exercise 11. Study noisy recovery under the same condition as in Theorem 7.25.

Exercise 12. Let s and t be two positive integers and $0 < q \leq 1$. Suppose that Φ satisfies the q-RIP with $\delta_{s+t,q} < 1$. Suppose

$$\gamma = \frac{1 + \delta_{t,q}}{1 - \delta_{s+t,q}} \left(\frac{s}{t}\right)^{1-q/2} < 1.$$

Show that letting $\mathbf{x}^*$ be the minimizer of (1.27),

$$\|\mathbf{x_b} - \mathbf{x}^*\|_2^q \leq C_{2,q} \frac{1}{s^{1-q/2}} \sigma_s(\mathbf{x_b})_q^q.$$

Exercise 13. Prove Theorem 7.16 using the ideas in the proof of Theorem 7.26.

Exercise 14. Prove Lemma 7.27 for $\ell > 0$.

Exercise 15. Prove Lemma 7.28.

Exercise 16. Finish the proof of Theorem 7.26, i.e., provide a proof for the case when s is an odd integer.

Bibliography

[1] T. Cai, L. Wang, and G. Xu, Shifting inequality and recovery of sparse signals, IEEE Trans. Signal Process., 58 (2010), 1300–1308. (Cited on pp. 201, 220)

[2] T. Cai, L. Wang, and G. Xu, New bounds for restricted isometry constants, IEEE Trans. Inform. Theory, 56 (2010), 4388–4394. (Cited on p. 205)

[3] T. T. Cai and A. Zhang, Sharp RIP bound for sparse signal and low-rank matrix recovery. Appl. Comput. Harmon. Anal., 35 (2013), 74–93. (Cited on pp. 205, 213, 221, 224)

[4] T. T. Cai and A. Zhang, Compressed sensing and affine rank minimization under restricted isometry. IEEE Trans. Signal Process., 61 (2013), 3279–3290. (Cited on p. 205)

[5] T. T. Cai and A. Zhang, Sparse representation of a polytope and recovery of sparse signals and low-rank matrices. IEEE Trans. Inform. Theory, 60 (2014), 122–132. (Cited on pp. 205, 213)

[6] E. J. Candès, The restricted isometry property and its implications for compressed sensing, C. R. Math. Acad. Sci. Sér. I, 346 (2008), 589–592. (Cited on pp. 198, 225)

[7] E. Candès and T. Tao, Decoding by linear programming, IEEE Trans. Inform. Theory, 51 (2005), 4203–4215. (Cited on p. 196)

[8] R. Chartrand and V. Staneva, Restricted isometry properties and nonconvex compressive sensing, Inverse Problems, 24 (2008), 035020. (Cited on p. 197)

[9] S. Foucart, Sparse recovery algorithms: Sufficient conditions in terms of restricted isometry constants, in Proceedings of the 13th Approximation Theory Conference, Springer-Verlag, 2011, 65–77. (Cited on p. 205)

[10] S. Foucart and M.-J. Lai, Sparsest solutions of under-determined linear systems via ℓ_q minimization for $0 < q \leq 1$, Appl. Comput. Harmon. Anal., 26 (2009), 395–407. (Cited on pp. 200, 226, 227)

[11] J. J. Fuchs, On sparse representations in arbitrary redundant bases, IEEE Trans. Inform. Theory, 50 (2004), 1341–1344. (Cited on p. 230)

[12] Q. Mo and S. Li, New bounds on the restricted isometry constant δ_{2k}, Appl. Comput. Harmon. Anal., 31 (2011), 460–468. (Cited on p. 203)

[13] Y. Shen and S. Li, Restricted p-isometry property and its application for nonconvex compressive sensing, Adv. Comput. Math., 37 (2012), 441–452. (Cited on p. 230)

[14] R. Zhang and S. Li, A proof of conjecture on restricted isometry property constants δ_{tk} ($0 < t < 4/3$), IEEE Trans. Inform. Theory, 64 (2018), 1699–1705. (Cited on p. 208)

[15] R. Zhang and S. Li, Optimal RIP bounds for sparse signals recovery via ℓ_p minimization, Appl. Comput. Harmon. Anal., 47 (2019), 566–584. (Cited on p. 213)

Chapter 8

RIP Matrices

In this chapter we discuss what kinds of sensing matrices Φ satisfy the RIP. There are a couple constructions of sensing matrices based on random entries and structured matrices, such as partial Fourier matrices with randomized rows, which will satisfy the RIP condition. That is, we use probability theory to show that the matrices so constructed satisfy the RIP with overwhelming probability. More precisely, we shall establish the probability estimate

$$P\left(\left|\|\Phi\mathbf{x}\|_2^2 - \|\mathbf{x}\|_2^2\right| \leq \delta\|\mathbf{x}\|_2^2 \quad \forall\|\mathbf{x}\|_0 \leq s\right) \geq 1 - Ce^{-m\delta^2/c} \tag{8.1}$$

for some positive constants c, C and $\delta \in (0, 1)$. The first such results can be found in [5]. The study is based on the theory of the concentration property of probability measures (cf., e.g., [3], [15], [9]). We will provide an elementary approach to such results as (8.1) for various probability distributions in this chapter.

We shall begin with Gaussian random matrices, then move on to sub-Gaussian random matrices, and end with pre-Gaussian random variables including uniform and Bernoulli distributions. The discussion of structured matrices such as partial Fourier matrices (randomized rows from a Fourier matrix) will be left to the last section.

8.1 ▪ Basics for Probability

Let Ω be a sample space, e.g., $\Omega = \mathbb{R}^n$ for some $n \geq 1$. Let $\mathcal{F}$ be a collection of events satisfying (1) $\Omega \in \mathcal{F}$; (2) if $A \in \mathcal{F}$, the complement A^c of A is also in $\mathcal{F}$; and (3) if $A_i, i = 1, 2, \ldots,$ are in $\mathcal{F}$, then $\cup_{i=1}^{\infty} A_i \in \mathcal{F}$. That is, $\mathcal{F}$ is a σ algebra. Let P be a function defined on $\mathcal{F}$. Suppose P satisfies the following conditions: (1) $P(A) \geq 0$ for all $A \in \mathcal{F}$; (2) $P(\Omega) = 1$; and (3) if $A_i, i = 1, 2, \ldots,$ are in $\mathcal{F}$ and are disjoint, i.e., $A_i \cap A_j = \emptyset$ for all $i \neq j, i, j = 1, 2, \ldots,$ then

$$P(\cup_{i=1}^{\infty} A_i) = \sum_{i=1}^{\infty} P(A_i).$$

Then P is called a probability function.

Any measurable function ξ on a probability space $(\Omega, \mathcal{F}, P)$ is called a random variable. That is, ξ is a real variable function defined on Ω such that for any Borel set B,

$$\{\omega : \xi(\omega) \in B\} \in \mathcal{F}.$$

Then $P(\omega : \xi(\omega) \in B)$ is called the probability distribution of ξ. In particular, letting $B = (a, b)$, $P(a < \xi(\omega) < b) = P(\xi(\omega) < b) - P(\xi(\omega) < a)$. A probability distribution function $F(x) = P(\xi(\omega) < x)$ satisfies the following theorem.

Theorem 8.1. *Any probability distribution function $F(x)$ satisfies the following conditions:*

(1) $F(x)$ *is monotone, i.e., if $a < b$, $F(a) \leq F(b)$;*

(2) $\lim_{x \to -\infty} F(x) = 0$ *and* $\lim_{x \to \infty} F(x) = 1$; *and*

(3) $F(x)$ *is right continuous, i.e.,* $\lim_{y \to x_+} F(y) = F(x)$.

When F is an absolutely continuous function, there exists an integrable function p such that

$$F(x) = \int_{-\infty}^{x} p(t)dt \quad \text{for almost all } x \in (-\infty, \infty).$$

In this case, p is called the density function associated with F. Let us give some examples of density functions.

Example 8.2 (Uniform Density). Let

$$p(x) = \begin{cases} \frac{1}{b-a} & \text{if } x \in (a,b), \\ 0 & \text{otherwise} \end{cases}$$

be the uniform density function. Any random variable with this density function is called a uniform random variable.

Example 8.3 (Normal Density). Let $p(t) = e^{-\frac{(t-a)^2}{2\sigma^2}}$, where $\sigma > 0$ and both a and σ are constants. This is called the normal density function. In particular, when $a = 0$ and $\sigma = 1$, it is called the standard normal distribution. A random variable ξ is called a Gaussian random variable if its probability distribution is subject to

$$P(\xi < x) = \frac{1}{\sqrt{2\pi}\sigma} \int_{-\infty}^{x} e^{-\frac{(t-a)^2}{2\sigma^2}} dt$$

for all $x \in (-\infty, \infty)$. This distribution is called the normal distribution and is denoted by $N(a, \sigma^2)$.

Example 8.4 (Exponential Density). Let

$$p(x) = \begin{cases} \lambda e^{-\lambda x} & \text{if } x \geq 0, \\ 0 & \text{otherwise.} \end{cases}$$

This is called the exponential density function, where $\lambda > 0$. The distribution function is

$$F(x) = \begin{cases} 1 - e^{-\lambda x} & \text{if } x \geq 0, \\ 0 & \text{otherwise.} \end{cases}$$

A random variable ξ is called an exponential random variable if its probability distribution is subject to $P(a < \xi < b) = F(b) - F(a)$ for any real numbers $a < b$.

Example 8.5 (Γ-distribution). Let

$$p(x) = \begin{cases} \frac{\lambda^r}{\Gamma(r)} x^{r-1} e^{-\lambda x} & \text{if } x \geq 0, \\ 0 & \text{otherwise} \end{cases}$$

for $r \geq 1$ and $\lambda > 0$. Any random variable with this density function is called a Γ random variable.

If a random variable ξ with density function $p(x)$ satisfies

$$\int_{-\infty}^{\infty} xp(x)dx < \infty,$$

we can define its expectation and write $\mathbb{E}(\xi) = \int_{-\infty}^{\infty} xp(x)dx$, which is also called the mean of the random variable. Next we define $\mathbb{D}(\xi) = \mathbb{E}(\xi - \mathbb{E}(\xi))^2$, which is called the variance of ξ. It is easy to see that

$$\mathbb{D}(\xi) = \int_{-\infty}^{\infty} (x - \mathbb{E}(\xi))^2 p(x)dx = \int_{-\infty}^{\infty} (x^2 - 2x\mathbb{E}(\xi) + (\mathbb{E}(\xi))^2)p(x)dx$$
$$= \mathbb{E}(\xi^2) - (\mathbb{E}(\xi))^2.$$

When a random variable ξ has a finite variance, we can normalize ξ by letting

$$\xi^* = \frac{\xi - \mathbb{E}(\xi)}{\sqrt{\mathbb{D}(\xi)}},$$

which satisfies $\mathbb{E}(\xi^*) = 0$ and $\mathbb{D}(\xi^*) = 1$.

In the above, we use $\mathbb{E}(\xi^2)$, which is called the second moment of ξ. In general, we should consider $\mathbb{E}(\xi^m)$ for all $m \geq 1$ for the mth moment as well as the exponential moment

$$\mathbb{E}(\exp(\lambda\xi)) = \sum_{k=0}^{\infty} \frac{\lambda^k}{k!} \mathbb{E}(\xi^k).$$

Definition 8.6 (Sub-Gaussian Random Variables). *A random variable ξ is called sub-Gaussian if there exists a nonnegative number $a \in [0, \infty)$ such that*

$$\mathbb{E}(\exp(\lambda\xi)) \leq \exp(a^2\lambda^2/2)$$

for all $\lambda \in \mathbb{R}$.

It is easy to verify that any Gaussian random variable ξ with mean zero and variance $\sigma^2 = \mathbb{D}(\xi)$ satisfies

$$\mathbb{E}(\exp(\lambda\xi)) = \exp\left(\frac{\sigma^2\lambda^2}{2}\right)$$

and hence is a sub-Gaussian random variable. Also, each bounded random variable ξ with mean $\mathbb{E}(\xi) = 0$ is sub-Gaussian. See a justification in the proof of Corollary 8.11. We can show the following result.

Lemma 8.7. *Suppose that a random variable ξ is sub-Gaussian. Then for any real number $p > 0$,*
$$\mathbb{E}(|\xi|^p) < \infty.$$
Moreover, $\mathbb{E}(\xi) = 0$ and $\mathbb{E}(\xi^2) \leq \tau^2(\xi)$, where

$$\tau(\xi) = \inf\{a \geq 0, \mathbb{E}(\exp(\lambda\xi)) \leq \exp(a^2\lambda^2/2), \lambda \in \mathbb{R}\}.$$

We leave the proof to Exercise 13, or see [4]. Another property of sub-Gaussian random variables is as follows.

Lemma 8.8. *Suppose that a random variable ξ is sub-Gaussian. Then for any $x > 0$,*

$$P(|\xi| \geq x) \leq 2 \exp\left(-\frac{x^2}{\tau^2(\xi)}\right), \tag{8.2}$$

where $\tau(\xi)$ is defined in the previous lemma.

Proof. Let $p(t)$ be the density function associated with the sub-Gaussian random variable ξ. Then

$$P(\xi > x) = \int_{t \geq x} p(t)dt \leq \int_{t \geq x} \frac{t}{x}p(t)dt \leq \frac{1}{x}\int_{-\infty}^{\infty} tp(t)dt = \frac{1}{x}\mathbb{E}(\xi).$$

Thus, it follows that for $\lambda > 0$,

$$P(\xi > x) = P\left(\exp(\lambda\xi) \geq \exp(\lambda x)\right) \leq \exp(-\lambda x)\mathbb{E}\left(\exp(\lambda\xi)\right) \leq \exp(-\lambda x)\exp(\lambda^2 a^2/2)$$

for some nonnegative number a by the definition of ξ. Note that

$$\exp(-\lambda x)\exp(\lambda^2 a^2/2) = \exp\left(\frac{(\lambda a - x/a)^2}{2}\right)\exp\left(-\frac{x^2}{2a^2}\right).$$

If we choose $\lambda = x/a^2$, we have

$$P(\xi > x) \leq \exp(-x^2/(2a^2)).$$

Similarly, we can show that $P(\xi < -x) \leq \exp(-x^2/(2a^2))$. By choosing the smallest a, we conclude the inequality in (8.2). □

The inequality in (8.2) is often used to define sub-Gaussian random variables.

Next we consider multiple random variables or random vectors. Let $\xi_1, \ldots, \xi_n$ be random variables defined on the same probability space $(\Omega, \mathcal{F}, P)$. Their joint probability distribution is

$$P(\xi_1(\omega) < x_1, \xi_2(\omega) < x_2, \ldots, \xi_n(\omega) < x_n).$$

For multiple random variables, one important phenomenon is the independence of these random variables. We shall say that $\xi_1, \ldots, \xi_n$ are independent if for any $x_1, x_2, \ldots, x_n$,

$$P(\xi_1(\omega) < x_1, \xi_2(\omega) < x_2, \ldots, \xi_n(\omega) < x_n)$$
$$= P(\xi_1(\omega) < x_1)P(\xi_2(\omega) < x_2) \cdots P(\xi_n(\omega) < x_n).$$

Naturally, we may consider a matrix Φ with random variables as its entries. For example, let $\Phi = [\phi_{ij}]_{1 \leq i \leq m, 1 \leq j \leq n}$, with entry ϕ_{ij} being the independently distributed Gaussian random variable $N(0, \sigma^2)$. That is, ϕ_{ij} is a random variable in $N(0, \sigma^2)$ and $\phi_{ij}, i = 1, \ldots, m, j = 1, \ldots, n$, are independent.

Theorem 8.9. *Suppose that $\xi_1, \ldots, \xi_n$ are independent sub-Gaussian random variables. Then, letting $S_n = \sum_{i=1}^{n} \xi_i$, S_n is sub-Gaussian and*

$$\tau^2(S_n) \leq \sum_{i=1}^{n} \tau^2(\xi_i).$$

Proof. For any $\lambda \in \mathbb{R}$,

$$\mathbb{E}(\exp(\lambda S_n)) = \prod_{i=1}^{n} \mathbb{E}(\exp(\lambda \xi_i)) \leq \prod_{i=1}^{n} \exp(\tau^2(\xi_i)\lambda^2/2) = \exp\left(\sum_{i=1}^{n} \tau^2(\xi_i)\lambda^2/2\right).$$

That is, S_n is a sub-Gaussian random variable. □

Since S_n is sub-Gaussian, we use Lemma 8.8 to obtain the following result.

Theorem 8.10. *Suppose that $\xi_1, \ldots, \xi_n$ are independent sub-Gaussian random variables. Then, letting $S_n = \sum_{i=1}^n \xi_i$, we have for any $t > 0$,*

$$P(|S_n| \geq t) \leq 2 \exp\left(-\frac{t^2}{\sum_{i=1}^n \tau^2(\xi_i)}\right).$$

The following corollary is well known.

Corollary 8.11 (Hoeffding's Inequality). *Let $\xi_1, \ldots, \xi_n$ be independent zero mean random variables which are bounded almost surely. That is, $|\xi_i| \leq a_i, i = 1, \ldots, n$, almost surely. Then for any $t > 0$,*

$$P(|S_n| \geq t) \leq 2 \exp\left(-\frac{t^2}{\sum_{i=1}^n a_i^2}\right).$$

Proof. We first claim that each bounded random variable ξ_i with zero mean is sub-Gaussian. Set $f(\lambda) = \ln \mathbb{E}(\exp(\lambda \xi_i))$. Then $f(0) = 0$, $\frac{df}{d\lambda}\big|_{\lambda=0} = \frac{\mathbb{E}(\xi_i(\exp(\lambda \xi_i)))}{\mathbb{E}(\exp(\lambda \xi_i))}\big|_{\lambda=0} = \mathbb{E}\xi_i = 0$, and

$$\frac{d^2 f}{d\lambda^2} = \frac{\mathbb{E}(\xi_i^2 \exp(\lambda \xi_i))\mathbb{E}\exp(\lambda \xi_i) - (\mathbb{E}\xi_i(\exp(\lambda \xi_i)))^2}{(\mathbb{E}\exp(\lambda \xi_i))^2} = \frac{\mathbb{E}(\xi_i^2 \exp(\lambda \xi_i))}{\mathbb{E}\exp(\lambda \xi_i)} \leq a_i^2.$$

We have $f''(0) = a_i^2$. The result is similar for all higher-order derivatives of f at 0. So we have

$$f(\lambda) = \sum_{k \geq 0} \frac{f^{(k)}(0)}{k!} \lambda^k \leq \frac{a_i^2 \lambda^2}{2},$$

which means $\mathbb{E}(\exp(\lambda \xi_i)) \leq \exp(\frac{a_i^2 \lambda^2}{2})$. Thus, $\tau(\xi_i) \leq a_i$. We now use the inequality in Theorem 8.10 to conclude the inequality in this corollary. $\quad\square$

Finally, we introduce pre-Gaussian random variables.

Definition 8.12. *A random variable ξ is called pre-Gaussian if there exist a positive number $\Lambda \in (0, \infty]$ and a nonnegative number $a \in [0, \infty)$ such that the inequality*

$$\mathbb{E}(\exp(\lambda \xi)) \leq \exp(a^2 \lambda^2 / 2)$$

holds for all $\lambda \in (-\Lambda, \Lambda)$. Let $\Lambda(\xi)$ be the largest number and $\tau(\xi)$ be the smallest number a satisfying the inequality above.

Clearly, all Gaussian random variables with zero mean and sub-Gaussian random variables are pre-Gaussian. Let us consider a nontrivial pre-Gaussian random variable. Let η be a Gaussian random variable $N(0, \sigma^2)$ for $\sigma > 0$ and $\xi = \eta^2 - \mathbb{E}(\eta)^2 = \eta^2 - \sigma^2$. Then ξ is a pre-Gaussian random variable. Indeed, it is clear that

$$\mathbb{E}(\exp(\lambda \xi)) = \mathbb{E}(\exp(\lambda(\eta^2 - \sigma^2))) = \exp(-\lambda \sigma^2)\mathbb{E}(\exp(\lambda \eta^2)) = \frac{\exp(-\lambda \sigma^2)}{\sqrt{1 - 2\lambda \sigma^2}},$$

which is finite if $|\lambda| < 1/(2\sigma^2)$. By using the following result, ξ is pre-Gaussian.

Theorem 8.13. *Let ξ be a random variable with zero mean $\mathbb{E}(\xi) = 0$. The following statements are equivalent:*

(1) *ξ is pre-Gaussian.*

(2) *There exists a constant $H > 0$ such that $\mathbb{E}(\exp(\lambda\xi)) < \infty$ for all $|\lambda| < H$.*

(3) *There exists a number $a > 0$ such that $\mathbb{E}(\exp(a|\xi|)) < \infty$.*

(4) *One can find numbers $c > 0$ and $b > 0$ such that $P(|\xi| \geq x) \leq b\exp(-cx)$ for all $x > 0$.*

Proof. We mainly show that (1) and (2) are equivalent. Clearly, (1) implies (2). We now show that (2) implies (1). Since ξ has zero mean $\mathbb{E}(\xi) = 0$, we have

$$\mathbb{E}(\exp(\lambda\xi)) = 1 + \frac{\lambda^2}{2}\mathbb{E}(\xi^2) + o(\lambda^2)$$

as $\lambda \to 0$. Also, it is clear that

$$\exp\left(\frac{a^2\lambda^2}{2}\right) = 1 + \frac{\lambda^2}{2}a^2 + o(\lambda^2)$$

as $\lambda \to 0$. The boundedness of $\mathbb{E}(\exp(\lambda\xi))$ by (2) implies that there exists a such that $\mathbb{E}(\xi^2) < a^2$. It then follows from a comparison of the above two inequalities that

$$\mathbb{E}(\exp(\lambda\xi)) \leq \exp\left(\frac{a^2\lambda^2}{2}\right)$$

for $\lambda \in (-\Lambda, \Lambda)$ for some $\Lambda > 0$. We leave the other parts to the interested reader. $\quad\square$

Lemma 8.14. *Suppose that ξ is a pre-Gaussian random variable with parameters $\Lambda = \Lambda(\xi)$ and $\tau = \tau(\xi)$. For any $x > 0$, let*

$$Q(x) = \begin{cases} \exp(-\frac{x^2}{2\tau^2}) & \text{if } 0 < x \leq \Lambda\tau^2, \\ \exp(-\frac{\Lambda x}{2}) & \text{if } \Lambda\tau^2 \leq x. \end{cases}$$

Then

$$P(\xi > x) \leq Q(x), \quad P(\xi < -x) \leq Q(x), \quad \text{and} \quad P(|\xi| \geq x) \leq 2Q(x).$$

Proof. Without loss of generality, we may assume $\Lambda < \infty$ and $\tau > 0$. For any $\lambda < \Lambda$, we use the Chebyshev–Markov inequality to get

$$P(\xi \geq x) = P(\exp(\lambda\xi) \geq \exp(\lambda x)) \leq \exp(-\lambda x)\mathbb{E}(\exp(\lambda\xi))$$
$$\leq \exp(-\lambda x)\exp\left(\frac{\lambda^2\tau^2}{2}\right)$$
$$= \exp\left(\frac{\lambda^2\tau^2}{2} - \lambda x\right) =: D(\lambda, x)$$

for any $x > 0$. Let

$$L(x) = \inf_{0 < \lambda < \Lambda} D(\lambda, x).$$

For any fixed $x > 0$, the minimum in λ_{min} of $D(\lambda, x)$ is achieved at $\lambda_{min} = x/\tau^2$. If $0 < x < \Lambda\tau^2$, $\lambda_{min} < \Lambda$, we have $L(x) = \exp(-x^2/(2\tau^2)) = Q(x)$. If $\Lambda\tau^2 \leq x$, $\Lambda \leq \lambda_{min}$. Since

$D(\lambda, x)$ is monotone, $L(x) = D(\Lambda, x)$. Since $D(\Lambda, x) \leq \exp(-\Lambda x/2) = Q(x)$, this proves the first inequality. The second inequality can be proved in the same way. The third and fourth inequalities follow by combining the first two inequalities.   □

Finally, we are now ready to prove the following key inequality.

Theorem 8.15 (Bernstein Inequality). *Suppose that $\xi_i, 1 \leq i \leq m$, are independent random variables with $\mathbb{E}(\xi_i) = 0$ and $\mathbb{E}(\xi_i^2) \leq \nu_i^2 < \infty, 1 \leq i \leq m$. Let $S_m = \sum_{i=1}^m \xi_i$. Moreover, suppose there exists a constant $H > 0$ such that*

$$|\mathbb{E}(\xi_i^k)| \leq \frac{k!}{2}\nu_i^2 H^{k-2} \tag{8.3}$$

for all integer $k > 1$ and all $i = 1, \ldots, m$. Then the following probability inequality holds for all $t > 0$:

$$P(|S_m| > t) \leq 2\exp\left\{-\frac{t^2}{2(tH + \sum_{i=1}^m \nu_i^2)}\right\}.$$

Proof. By the assumption (8.3), we have

$$\mathbb{E}(\exp(\lambda\xi_i)) = 1 + \sum_{k\geq 1}\frac{\lambda^k}{k!}\mathbb{E}(\xi^k) \leq 1 + \sum_{k\geq 2}\frac{|\lambda|^k}{2}H^{k-2}\nu_i^2$$

$$\leq 1 + \frac{\nu_i^2|\lambda|^2}{2}\frac{1}{1-|\lambda|H} \leq \exp\left(\frac{\nu_i^2\lambda^2}{2(1-|\lambda|H)}\right).$$

It follows that for any $\lambda \in (0, 1/H)$,

$$\mathbb{E}(\exp(\lambda S_n)) = \prod_{i=1}^m \mathbb{E}(\exp(\lambda\xi_i)) \leq \exp\left(\frac{\lambda^2}{2(1-\lambda H)}\sum_{i=1}^m \nu_i^2\right).$$

For $t > 0$, we use the Chebyshev–Markov inequality to get

$$P(S_n > t) \leq \exp(-\lambda t)\mathbb{E}(\exp(\lambda S_n)) \leq \exp\left(\frac{\lambda^2}{2(1-\lambda H)}\sum_{i=1}^m \nu_i^2 - \lambda t\right).$$

By choosing $\lambda = \frac{t}{Ht + \sum_{i=1}^m \nu_i} < 1/H$, the exponent on the right-hand side of the inequality above gives

$$\frac{\lambda^2}{2(1-\lambda H)}\sum_{i=1}^m \nu_i^2 - \lambda t = -\frac{t^2}{2(Ht + \sum_{i=1}^m \nu_i^2)}.$$

This shows that

$$P(S_n > t) \leq \exp\left(-\frac{t^2}{2(Ht + \sum_{i=1}^m \nu_i^2)}\right).$$

Similarly, we can show that

$$P(S_n < -t) \leq \exp\left(-\frac{t^2}{2(Ht + \sum_{i=1}^m \nu_i^2)}\right).$$

Combining these two inequalities finishes the proof.   □

8.2 ▪ Gaussian Random Matrices

The following theorem shows that a Gaussian random matrix satisfies the RIP with overwhelming probability as m gets large enough.

Theorem 8.16. *Suppose that* $\Phi = [\phi_{ij}]_{1 \leq i \leq m, 1 \leq j \leq n}$ *is a matrix with entries* ϕ_{ij} *being i.i.d. Gaussian random variables with mean zero and variance* $1/\sqrt{m}$*. Then the probability inequality*

$$P\left(\left|\|\Phi\mathbf{x}\|_2^2 - \|\mathbf{x}\|_2^2\right| \leq \delta\|\mathbf{x}\|_2^2 \quad \forall \|\mathbf{x}\|_0 \leq s\right) \geq 1 - \binom{n}{s}\left(1 + \frac{24}{\delta}\right)^s e^{-m\delta^2/c} \qquad (8.4)$$

holds for any fixed number $\delta \in (0, 1)$*, where* $c > 4$ *is a constant.*

Once $\delta \in (0, 1)$ is fixed, we choose $m > s$ large enough such that $\binom{n}{s}(1 + 24/\delta)^s e^{-m\delta^2/c} < 1$. In this case we have

$$P\left(\left|\|\Phi\mathbf{x}\|_2^2 - \|\mathbf{x}\|_2^2\right| \leq \delta\|\mathbf{x}\|_2^2 \quad \forall \|\mathbf{x}\|_0 \leq s\right) > 0.$$

That is, a matrix with the RIP can be found with positive probability. If m is sufficiently large, the probability is overwhelming. Indeed, since $\binom{n}{s} = n(n-1)\cdots(n-s+1)/s! \leq n(n-1)\cdots(n-s)/e^s \leq (n/e)^s$ for $s > e$,

$$\binom{n}{s}(1 + 24/\delta)^s \exp(-m\delta^2/c) \leq e^{-m\delta^2/c + s\ln(n/e) + s\ln(1 + 24/\delta)}.$$

If $\delta = 1/2$, say, as long as $m > 4sc\ln(51n/e)$, we will have a good probability to have a matrix satisfying the RIP.

Theorem 8.16 can be proved in two steps. First we need the following probability concentration inequality.

Theorem 8.17 (Baranuik, Daveport, DeVore, and Wakin, 2008 [1]). *Suppose that we are given a sensing matrix* $\Phi = [\phi_{ij}]_{1 \leq i \leq m, 1 \leq j \leq n}$ *with entries* ϕ_{ij} *being i.i.d. Gaussian random variables with mean zero and variance* $1/\sqrt{m}$*. Then for any* $\delta > 0$*, the probability*

$$P(\left|\|\Phi\mathbf{x}\|_2^2 - \|\mathbf{x}\|_2^2\right| < \delta\|\mathbf{x}\|_2^2) \geq 1 - 2\exp\left(-\frac{\delta^2 m}{c}\right) \qquad (8.5)$$

for each $\mathbf{x} \in \mathbb{R}^n$*, where* c *is a positive constant independent of* δ *and* $\|\mathbf{x}\|_2$*.*

Proof. For a rectangular matrix $\Phi = [\phi_{ij}]_{1 \leq i \leq m, 1 \leq j \leq n}$ with ϕ_{ij} being i.i.d. Gaussian random variables with mean zero and variance $1/\sqrt{m}$, we consider a random variable $X = (X_1, \ldots, X_m)^\top$ with $X_i = (\sum_{j=1}^n \phi_{ij}x_j)^2, i = 1, \ldots, m$. Since the expectation $\mathbb{E}(\phi_{ij}) = 0$, we have $\mathbb{E}(X_i) = \sigma^2\|\mathbf{x}\|_2^2$ for all i. Let $\xi_i = X_i - \mathbb{E}(X_i)$ be a new random variable and let

$$S_m = \sum_{i=1}^m \xi_i$$

be the sum of these new independent random variables. It is easy to see that

$$S_m = \|\Phi\mathbf{x}\|_2^2 - m\sigma^2\|\mathbf{x}\|_2^2 = \|\Phi\mathbf{x}\|_2^2 - \|\mathbf{x}\|_2^2.$$

We need to study (8.3) for $\xi_i = X_i - \mathbb{E}(X_i)$ for $k \geq 3$. For $k = 2$, (8.3) is satisfied trivially. For convenience, let $\mu = \mathbb{E}(X_i) = \sigma^2 \|\mathbf{x}\|_2^2$. For $k \geq 3$, we have

$$\mathbb{E}(|\xi_i|^k) = \mathbb{E}((X_i - \mu)^k) = \sum_{j=0}^{k} \binom{k}{j} \mathbb{E}(X_i^j)(-1)^{k-j} \mu^{k-j}.$$

Let us spend some effort to compute $\mathbb{E}(X_i^j)$. We have

$$\mathbb{E}(X_i^j) = \mathbb{E}\left(\sum_{j=1}^{n} \phi_{ij} x_j \right)^{2j} = \sum_{j_1 + \cdots + j_n = 2j} \frac{(2j)!}{j_1! \cdots j_n!} \mathbb{E}(\phi_{i,1}^{j_1} \phi_{i,2}^{j_2} \cdots \phi_{i,n}^{j_n}) x_1^{j_1} \cdots x_n^{j_n}.$$

Note that $\mathbb{E}(\phi_{ij}^\ell) = 0$ for all odd integers ℓ, and straightforward calculation using integration by parts shows that

$$\mathbb{E}(\phi_{ij}^\ell) = \frac{\ell!}{2^{\ell/2}(\ell/2)!} \sigma^\ell \tag{8.6}$$

for even integers ℓ. Since ϕ_{ij} are i.i.d. random variables, we have

$$\mathbb{E}(X_i^j) = \sum_{2j_1 + \cdots + 2j_n = 2j} \frac{(2j)!}{(2j_1)! \cdots (2j_n)!} \mathbb{E}(\phi_{i,1}^{2j_1} \phi_{i,2}^{2j_2} \cdots \phi_{i,n}^{2j_n})(x_1)^{2j_1} \cdots (x_n)^{2j_n}$$

$$= \frac{(2j)!}{j!} \sum_{j_1 + \cdots + j_n = j} \frac{j!}{(2j_1)! \cdots (2j_n)!} \frac{(2j_1)! \cdots (2j_n)!}{2^j j_1! \cdots j_n!} \sigma^{2j}(x_1)^{2j_1} \cdots (x_n)^{2j_n}$$

$$= \frac{(2j)! \sigma^{2j}}{2^j j!} \left(\sum_{j=1}^{n} x_j^2 \right)^j = \frac{(2j)!}{2^j j!} \sigma^{2j} \|\mathbf{x}\|_2^{2j} = \frac{(2j)!}{2^j j!} \mu^j.$$

In particular, $\mathbb{E}(X_i^2) = 3\mu^2$, and thus $\mathbb{E}(|\xi_i|^2) = 2\mu^2 = 2\sigma^4 \|\mathbf{x}\|^4$. For $k \geq 3$,

$$\mathbb{E}(|\xi_i|^k) \leq \sum_{j=0}^{k} \binom{k}{j} \frac{(2j)!}{2^j j!} \mu^j \mu^{k-j}.$$

By using Stirling's formula, we have $\frac{(2j)!}{2^j j!} \leq 2^j j!/2 \leq 2^j k!/2$ for $j \leq k$, and hence

$$\mathbb{E}(|\xi_i|^k) \leq \sum_{j=0}^{k} \binom{k}{j} \frac{2^j k!}{2} \mu^k = \frac{k!}{2} 3^k \mu^k$$

$$= \frac{k!}{2} \mathbb{E}(\xi_i^2) \frac{9}{2} 3^{k-2} (\sigma^2 \|\mathbf{x}\|_2^2)^{k-2} \leq \frac{k!}{2} \mathbb{E}(\xi_i^2) H^{k-2},$$

where $H = 13.5\sigma^2 \|\mathbf{x}\|_2^2$. With this H, (8.3) is satisfied for $k \geq 3$. By Theorem 8.15, we have

$$P(|\|\Phi\mathbf{x}\|_2^2 - m\sigma^2\|\mathbf{x}\|_2^2| > t) \leq 2\exp\left\{ -\frac{t^2}{2(t13.5\mu + 2m\mu^2)} \right\}.$$

Choosing $t = \delta\|\mathbf{x}\|_2^2$, we have $t13.5\mu + 2m\mu^2 = \sigma^2\|\mathbf{x}\|_2^4(13.5\delta + 2m\sigma^2)$, and the above probability yields

$$P(|\|\Phi\mathbf{x}\|_2^2 - \|\mathbf{x}\|_2^2| > \delta\|\mathbf{x}\|_2^2) \leq 2\exp\left\{ -\frac{\delta^2 m}{2(13.5\delta + 2)} \right\}. \tag{8.7}$$

In other words, the desired result of Theorem 8.17 is proved.   □

For the next step to prove the main result in Theorem 8.16, we need to have the following.

Lemma 8.18. *Fix an integer $s > 0$. With $\mathcal{S} := \{\mathbf{x} \in \mathbb{R}^s : \|\mathbf{x}\|_2 = 1\}$ denoting the unit sphere of $\mathbb{R}^s$ for any $\delta > 0$, there exists a finite set $\mathcal{U} \subseteq \mathcal{S}$ with*

$$\min_{\mathbf{u} \in \mathcal{U}} \|\mathbf{x} - \mathbf{u}\|_2 \leq \delta \quad \forall \mathbf{x} \in \mathcal{S} \quad \text{and} \quad \mathrm{card}(\mathcal{U}) \leq \left(1 + \frac{2}{\delta}\right)^s.$$

Proof. Let $(\mathbf{u}_1, \ldots, \mathbf{u}_k)$ be a set of k points on the sphere $\mathcal{S}$ such that $\|\mathbf{u}_i - \mathbf{u}_j\|_2 > \delta$ for all $i \neq j$. We choose k a large integer. Thus, it is clear that

$$\min_{1 \leq i \leq k} \|\mathbf{x} - \mathbf{u}_i\|_2 \leq \delta \quad \forall \mathbf{x} \in \mathcal{S}.$$

Let $\mathcal{B} := \{\mathbf{x} \in \mathbb{R}^s : \|\mathbf{x}\|_2 \leq 1\}$. It is easy to see that the $(\delta/2)$ balls centered at $\mathbf{u}_i$,

$$\mathbf{u}_i + \frac{\delta}{2}\mathcal{B}, \quad 1 \leq i \leq k,$$

are disjoint. Indeed, if $\mathbf{x}$ were in the $(\delta/2)$ ball centered at $\mathbf{u}_i$ as well as in the $(\delta/2)$ ball centered at $\mathbf{u}_j$, we would have

$$\|\mathbf{u}_i - \mathbf{u}_j\|_2 \leq \|\mathbf{u}_i - \mathbf{x}\|_2 + \|\mathbf{u}_j - \mathbf{x}\|_2 \leq \frac{\delta}{2} + \frac{\delta}{2} = \delta,$$

which is a contradiction. In addition, it is easy to see that

$$\mathbf{u}_i + \frac{\delta}{2}\mathcal{B} \subseteq \left(1 + \frac{\delta}{2}\right)\mathcal{B}, \quad 1 \leq i \leq k.$$

By comparison of volumes, we get

$$k\mathrm{Vol}\left(\frac{\delta}{2}\mathcal{B}\right) = \sum_{i=1}^{k} \mathrm{Vol}\left(\mathbf{u}_i + \frac{\delta}{2}\mathcal{B}\right) \leq \mathrm{Vol}\left(\left(1 + \frac{\delta}{2}\right)\mathcal{B}\right).$$

Then, by homogeneity of the volumes, we have

$$k\left(\frac{\delta}{2}\right)^s \mathrm{Vol}\,(\mathcal{B}) \leq \left(1 + \frac{\delta}{2}\right)^s \mathrm{Vol}\,(\mathcal{B}),$$

which implies that $k \leq \left(1 + \frac{2}{\delta}\right)^s$. This completes the proof. $\square$

We are now ready to prove Theorem 8.16.

Proof of Theorem 8.16. We start by considering a fixed index set $T \subseteq \{1, \ldots, N\}$ of cardinality s and denote by $\mathbb{R}^T$ the space of vectors in $\mathbb{R}^n$ supported on T. In view of Lemma 8.18, we can find a subset $\mathcal{U}_T$ of the unit sphere $\mathcal{S}_T$ of $\mathbb{R}^T$ such that

$$\min_{\mathbf{u} \in \mathcal{U}_T} \|\mathbf{x} - \mathbf{u}\|_2 \leq \frac{\delta}{4} \quad \forall \mathbf{x} \in \mathcal{S}_T \quad \text{and} \quad \mathrm{card}(\mathcal{U}_T) \leq \left(1 + \frac{8}{\delta}\right)^s.$$

Using Theorem 8.17, we obtain

$$P\left(\left|\|\Phi\mathbf{u}\|_2^2 - \|\mathbf{u}\|_2^2\right| > \frac{\delta}{4}\|\mathbf{u}\|_2^2\right) \leq \left(1 + \frac{8}{\delta}\right)^s 2\exp\left\{-\frac{\delta^2 m}{2c}\right\} \tag{8.8}$$

for all $\mathbf{u} \in \mathcal{U}_T$. This means that with high probability we have

$$\left| \|\Phi\mathbf{u}\|_2^2 - \|\mathbf{u}\|_2^2 \right| < \frac{\delta}{4} \|\mathbf{u}\|_2^2 \quad \forall \mathbf{u} \in \mathcal{U}_T,$$

which trivially gives us

$$\left(1 - \frac{\delta}{4}\right) \|\mathbf{u}\|_2 \le \|\Phi\mathbf{u}\|_2 \le \left(1 + \frac{\delta}{4}\right) \|\mathbf{u}\|_2 \quad \forall \mathbf{u} \in \mathcal{U}_T.$$

Let $\tilde{\delta}$ be the smallest positive constant such that

$$\|\Phi\mathbf{x}\|_2 \le (1 + \tilde{\delta})\|\mathbf{x}\|_2 \qquad \forall \mathbf{x} \in \mathcal{S}_T. \tag{8.9}$$

We claim that $\tilde{\delta} \le \delta$. Indeed, given $\mathbf{x} \in \mathcal{S}_T$, picking $\mathbf{u} \in \mathcal{U}_T$ with $\|\mathbf{x} - \mathbf{u}\|_2 \le \delta/4$, we derive

$$\|\Phi\mathbf{x}\|_2 \le \|\Phi\mathbf{u}\|_2 + \|\Phi(\mathbf{x} - \mathbf{u})\|_2 \le \left(1 + \frac{\delta}{4}\right) + (1 + \tilde{\delta})\|\mathbf{x} - \mathbf{u}\|_2$$

$$\le \left(1 + \frac{\delta}{2}\right) + (1 + \tilde{\delta})\frac{\delta}{4}.$$

The minimality of $\tilde{\delta}$ implies that

$$1 + \tilde{\delta} \le 1 + \frac{\delta}{2} + (1 + \tilde{\delta})\frac{\delta}{4} \le 1 + \frac{3\delta}{4} + \frac{\tilde{\delta}}{4}, \qquad \text{i.e.,} \quad \tilde{\delta} \le \delta.$$

Substituting into (8.9), we obtain the upper estimate

$$\|\Phi\mathbf{x}\|_2 \le (1 + \delta)\|\mathbf{x}\|_2 \qquad \forall \mathbf{x} \in \mathbb{R}^T. \tag{8.10}$$

On the other hand, given $\mathbf{x} \in \mathcal{S}_T$, letting $\mathbf{u} \in \mathcal{U}_T$ with $\|\mathbf{x} - \mathbf{u}\|_2 \le \delta/4$, we have

$$\|\Phi\mathbf{x}\|_2 \ge \|\Phi\mathbf{u}\|_2 - \|\Phi(\mathbf{x} - \mathbf{u})\|_2 \ge \left(1 - \frac{\delta}{2}\right) - (1 + \delta)\|\mathbf{x} - \mathbf{u}\|_2$$

$$\ge \left(1 - \frac{\delta}{2} - (1 + \delta)\frac{\delta}{4}\right) \ge 1 - \delta.$$

Thus we obtain the lower estimate

$$\|\Phi\mathbf{x}\|_2 \ge (1 - \delta)\|\mathbf{x}\|_2 \qquad \forall \mathbf{x} \in \mathbb{R}^T. \tag{8.11}$$

It follows that

$$(1 - 3\delta)\|\mathbf{x}\|_2^2 \le (1 - \delta)^2\|\mathbf{x}\|_2^2 \le \|\Phi\mathbf{x}\|_2^2 \le (1 + \delta)^2\|\mathbf{x}\|_2^2 \le (1 + 3\delta)\|\mathbf{x}\|_2^2. \tag{8.12}$$

Since the above estimate holds as soon as (8.8) is fulfilled, we have

$$P\left(\left|\|\Phi\mathbf{x}\|_2^2 - \|\mathbf{x}\|_2^2\right| > 3\delta\|\mathbf{x}\|_2^2 \ \forall \mathbf{x} \in \mathbb{R}^T\right) \le 2\left(1 + \frac{8}{\delta}\right)^s \exp\left(-\frac{\delta^2 m}{c}\right).$$

We now observe that the set of all s-sparse vectors is the union of $\binom{n}{s} \le (en/s)^s$ subspaces $\mathbb{R}^T$ and deduce that

$$P\left(\left|\|\Phi\mathbf{x}\|_2^2 - \|\mathbf{x}\|_2^2\right| > 3\delta\|\mathbf{x}\|_2^2 \ \forall \mathbf{x} \in \mathbb{R}^n \text{ with } \|\mathbf{x}\|_0 \le s\right)$$

$$\le 2\binom{n}{s}\exp\left(-\frac{\delta^2 m}{c} + \frac{8s}{\delta}\right) \le 2\exp\left(-\frac{\delta^2 m}{c} + \frac{8s}{\delta} + s\ln\left(\frac{en}{s}\right)\right).$$

If we impose that

$$\frac{4s}{\delta} + s \ln\left(\frac{en}{s}\right) < \frac{1}{2}\frac{\delta^2 m}{c} \quad \text{or} \quad m > \frac{8c}{\delta^3}s + 2cs\ln(en/s),$$

we have the result of Theorem 8.16 with 3δ replaced by δ. □

In general, many other random matrices satisfy the probability concentration estimate in Theorem 8.17. Typically, matrices with sub-Gaussian random variables possess the RIP. These will be discussed in the next section. However, matrices with i.i.d. pre-Gaussian random variables do not have the standard RIP. Instead, they satisfy a modified RIP. We shall discuss this in section 8.6.

8.3 ▪ Sub-Gaussian Random Matrices (I)

In this section, we need to show that a sub-Gaussian random matrix satisfies the RIP with overwhelming probability for m large enough.

Theorem 8.19. *Suppose that $\Phi = [\phi_{ij}]_{1 \le i \le m, 1 \le j \le n}$ is a matrix with entries ϕ_{ij} being i.i.d. sub-Gaussian random variables with mean zero and variance $1/\sqrt{m}$. Then the probability*

$$P\left(\left|\|\Phi\mathbf{x}\|_2^2 - \|\mathbf{x}\|_2^2\right| \le \delta\|\mathbf{x}\|_2^2, \quad \|\mathbf{x}\|_0 \le s\right) \ge 1 - \binom{n}{s}\left(1 + \frac{24}{\delta}\right)^s e^{-m\delta^2/c} \qquad (8.13)$$

holds for any fixed number $\delta \in (0,1)$, where $c > 4$ is a constant and $\delta \in (0,1)$ is any fixed number.

The proof is almost the same as the proof in the previous section. It is divided into two steps. In particular, the second step is exactly the same as in the previous section. All we need to establish is the first step, the probability concentration inequality for sub-Gaussian random matrices. That is, we have the following theorem.

Theorem 8.20. *Suppose that $\Phi = [\phi_{ij}]_{1 \le i \le m, 1 \le j \le n}$ is a matrix with entries ϕ_{ij} being i.i.d. sub-Gaussian random variables with mean zero and variance $1/\sqrt{m}$. Then for any $\delta > 0$, the probability*

$$P(\left|\|\Phi\mathbf{x}\|_2^2 - \|\mathbf{x}\|_2^2\right| < \delta\|\mathbf{x}\|_2^2) \ge 1 - 2\exp\left(-\frac{\delta^2 m}{c}\right) \qquad (8.14)$$

holds for any $\mathbf{x} \in \mathbb{R}^n$, where c is a positive constant independent of δ and $\|\mathbf{x}\|_2$.

Proof. As in the proof of Theorem 8.17, we let $X_i = \left(\sum_{j=1}^n \phi_{ij}x_j\right)^2$ and $\xi_i = X_i - \mathbb{E}(X_i) = X_i - \nu$ with $\nu = \sigma^2\|\mathbf{x}\|_2^2$. For simplicity, we first consider the case that ξ_i has a symmetric distribution function. Then we shall follow the proof of Theorem 8.17 by considering $\mathbb{E}(|\xi_i|^k)$. One major step is to show the critical inequality

$$\mathbb{E}(|\xi_i|^\ell) \le \frac{\ell!}{2^{\ell/2}(\ell/2)!}b^\ell \qquad (8.15)$$

for a positive constant b. This inequality will play a role similar to the one in (8.6). First we note that for any $s > 0$,

$$\max_{x \ge 0} x^s \exp(-x) = (s/e)^s.$$

Thus, $x^s \le (s/e)^s \exp(x)$. For $x = \lambda|\xi|$ with $\lambda > 0$, we have

$$\mathbb{E}(|\xi|^s) \le (s/(\lambda e))^s \mathbb{E}(\exp(\lambda|\xi|)) \le (s/(\lambda e))^s (\mathbb{E}(\exp(\lambda\xi)) + \mathbb{E}(\exp(-\lambda\xi))).$$

By using the definition of the sub-Gaussian random variable,

$$\mathbb{E}(|\xi|^s) \le 2(s/(\lambda e))^s \exp(\lambda^2\tau^2/2)$$

for all $\lambda > 0$. Choosing $\lambda = \sqrt{s}/\tau$, we have

$$\mathbb{E}(|\xi|^s) \le 2(s/e)^{s/2}\tau^s. \tag{8.16}$$

Next, using Stirling's formula, $n! = \sqrt{2\pi n}(n/e)^n e^{\theta_n}$ with $|\theta_n| < 1/(12n)$, we can show that

$$\frac{2^k k!}{(2k)!} = \frac{1}{\sqrt{2}}\left(\frac{e}{2k}\right)^k \exp(\theta_k - \theta_{2k}) \le e^{9/16}\left(\frac{e}{2k}\right)^k.$$

Together with (8.16), we have

$$\frac{2^k k!}{(2k)!}\mathbb{E}(|\xi|^{2k}) \le 2e^{9/16}\tau^{2k} \le b^{2k},$$

with $b = 2e^{9/16}\tau$ for $\ell = 2k$ in (8.15).

Using the estimate above and following the proof of Theorem 8.17,

$$\mathbb{E}(X_i^j) \le \frac{(2j)!}{2^j j!}\nu^j,$$

with $\nu = b^2\|\mathbf{x}\|^2$. In particular, $\mathbb{E}(X_i^2) \le 3\nu^2$ and $\mathbb{E}(\|\xi_i\|^2) \le 3\nu^2$.

For $k \ge 3$, we recall that $\mu = \mathbb{E}(X_i) \le \sigma^2\|\mathbf{x}\|^2$. Letting $\alpha = \max\{\sigma^2, b^2\}$,

$$\mathbb{E}(|\xi_i|^k) \le \sum_{j=0}^{k}\binom{k}{j}\mathbb{E}(X_i^j)\mu^{k-j} \le \sum_{j=0}^{k}\binom{k}{j}\frac{2^j k!}{2}\alpha^k\|\mathbf{x}\|^{2k}$$

$$= \frac{k!}{2}3^k\alpha^k\|\mathbf{x}\|^{2k} \le \frac{k!}{2}(3\alpha^4\|\mathbf{x}\|^4)H^{k-2}$$

for $H = 9\alpha^2\|\mathbf{x}\|^2$. Thus, we can use the Bernstein inequality in Theorem 8.15 to conclude the proof of Theorem 8.20. $\quad\square$

***Proof of Theorem* 8.19.** We use exactly the arguments in the proof of Theorem 8.16 to finish the proof of Theorem 8.19. $\quad\square$

From the proof of Theorem 8.20, we can see that the inequality in (8.15) is crucial. For some pre-Gaussian random variables, we do not have such an inequality. Thus, we cannot use this argument to establish the RIP for random matrices with i.i.d. pre-Gaussian random variables.

8.4 ▪ RIP Matrices in the ℓ_q Setting

A generalization of the RIP in the ℓ_q quasi-norm with $0 < q \le 1$ was studied in [6] for Gaussian random matrices. We shall present these results in this and the next section. First of all, we present a basic result.

Theorem 8.21 (Chartrand and Staneva, 2008 [6]). *Suppose that* $\Phi = [\phi_{ij}]_{1 \leq i \leq m, 1 \leq j \leq n}$ *is a matrix with entries* ϕ_{ij} *being i.i.d. Gaussian random variables with mean zero and variance* $1/\sqrt{m}$*. Let* $0 < q \leq 1$ *and* $\nu_q := 2^{q/2}\sigma^q\Gamma((q+1)/2)/\sqrt{\pi}$*. Then for any* $\delta > 0$*, we have the probability*

$$P\left(\left|\|\Phi\mathbf{x}\|_q^q - m\nu_q\|\mathbf{x}\|_2^q\right| < \delta m\nu_q\|\mathbf{x}\|_2^q\right) \geq 1 - 2\exp\left(-\frac{\delta^2 m}{2(C_q\sqrt{q} + 2)}\right), \tag{8.17}$$

where C_q *is a positive constant independent of* δ *and* $\|\mathbf{x}\|_2$ *for any* $\mathbf{x} \in \mathbb{R}^n$*. In particular, there exist two positive constants* C_1 *and* C_2 *such that when*

$$m \geq C_1 s + C_2 s \ln(n/s), \tag{8.18}$$

the matrix Φ *satisfies the q-RIP*

$$(1 - \delta_s)m\nu_q\|\mathbf{x}\|_2^q \leq \|\Phi\mathbf{x}\|_q^q \leq (1 + \delta_s)m\nu_q\|\mathbf{x}\|_2^q \qquad \forall \|\mathbf{x}\|_0 \leq s$$

with overwhelming probability.

Note that in (8.18), when $q \to 0_+$, m is proportional to the sparsity s as the second term in (8.18) goes to zero. This gives an advantage of using the ℓ_q minimization for compressed sensing problems. That is, fewer measurements are needed for a sensing matrix Φ to satisfy the q-RIP. This result will be presented in the next section.

In this section, we mainly prove Theorem 8.21. We shall use Bernstein's inequality discussed in the previous section to do so. Let us begin with some preparation. Let ξ be a random variable with Gaussian density function $p(x) = \frac{1}{\sqrt{2\pi}}\exp(-x^2/(2\sigma^2))$. Let $\mathbf{x} = (x_1, \ldots, x_n)^\top$ be a vector in $\mathbf{R}^n$. For a random variable ξ with density function P, we set

$$\|\mathbf{x}\|_q := \left(\underbrace{\int_{-\infty}^{\infty} \cdots \int_{-\infty}^{\infty}}_{n} \left|\sum_{j=1}^{n} t_j x_j\right|^q p(t_1)\cdots p(t_n) dt_1 \cdots dt_n\right)^{1/q}$$

for $0 < q < \infty$. We can show that $\|\mathbf{x}\|_q$ defines a norm on $\mathbb{R}^n$ for any $q \geq 1$ and $\|\mathbf{x}\|_q^q$ is a quasi-norm when $0 < q < 1$. See Exercise 14. For a Gaussian density function, we can use a change of variables to get

$$\|\mathbf{x}\|_q^q = \|\mathbf{x}\|_2^q \underbrace{\int_{-\infty}^{\infty} \cdots \int_{-\infty}^{\infty}}_{n} |u_1|^q p(u_1)\cdots p(u_n) du_1 du_2 \cdots du_n$$

$$= \frac{2\|\mathbf{x}\|_2^q}{\sqrt{2\pi\sigma^2}} \int_0^{\infty} u^q \exp\left(-\frac{u^2}{2\sigma^2}\right) du$$

$$= \frac{2^{q/2}\sigma^q}{\sqrt{\pi}} \Gamma\left(\frac{q+1}{2}\right) \|\mathbf{x}\|_2^q = \nu_q\|\mathbf{x}\|_2^q.$$

Next we introduce the function $M_p(\cdot, \mathbf{x})$ associated with the density function P with direction $\mathbf{x} \neq 0$ as follows. For $\mathbf{x} = (x_1, \ldots, x_n)^\top$, let $M_p(\cdot; \mathbf{x})$ be a function satisfying

$$\int_{-\infty}^{\infty} M_p(u; \mathbf{x}) f(u) du = \underbrace{\int_{-\infty}^{\infty} \cdots \int_{-\infty}^{\infty}}_{n} f\left(\sum_{j=1}^{n} t_j x_j\right) p(t_1)\cdots p(t_n) dt_1 \cdots dt_n \tag{8.19}$$

for all compactly supported continuous functions f defined on $\mathbf{R}$. The following basic properties of $M_p(\cdot; \mathbf{x})$ can be proved easily. They are left to Exercise 4.

(1) $M_p(\cdot; \mathbf{x}) \geq 0$;

(2) $M_p(-u; \mathbf{x}) = M_p(u; \mathbf{x})$;

(3) $\int_{-\infty}^{\infty} M_p(u; \mathbf{x}) du = 1$; and

(4) $\lambda M_p(\lambda u; \lambda \mathbf{x}) = M_p(u; \mathbf{x})$.

For a Gaussian density function P, we have an explicit formula for $M_p(\cdot, \mathbf{x})$. Indeed, using a change of variables, we have

$$\underbrace{\int_{-\infty}^{\infty} \cdots \int_{-\infty}^{\infty}}_{n} f\left(\sum_{j=1}^{n} t_j x_j\right) P(t_1) \cdots P(t_n) dt_1 \cdots dt_n$$

$$= \underbrace{\int_{-\infty}^{\infty} \cdots \int_{-\infty}^{\infty}}_{n} f(\|\mathbf{x}\|_2 u_1) P(u_1) \cdots P(u_n) du_1 \cdots du_n$$

$$= \int_{-\infty}^{\infty} f(\|\mathbf{x}\|_2 u_1) \frac{1}{\sqrt{2\pi\sigma^2}} e^{-\frac{u_1^2}{2\sigma^2}} du_1$$

$$= \int_{-\infty}^{\infty} f(u) \frac{1}{\sqrt{2\pi\sigma^2 \|\mathbf{x}\|_2^2}} e^{-\frac{u^2}{2\sigma^2 \|\mathbf{x}\|_2^2}} du.$$

That is, we have

$$M_p(t, \mathbf{x}) = \frac{1}{\sqrt{2\pi\sigma^2 \|\mathbf{x}\|_2^2}} e^{-\frac{u^2}{2\sigma^2 \|\mathbf{x}\|_2^2}} \leq \frac{1}{\sqrt{2\pi\sigma^2 \|\mathbf{x}\|_2^2}} = \frac{(\Gamma((q+1)/2)/\sqrt{\pi})^{1/q}}{\sqrt{\pi}(\nu_q \|\mathbf{x}\|_2^q)^{1/q}}. \quad (8.20)$$

Let $X_i := |\sum_{i=1}^{n} \phi_{ij} x_j|$ with the random variables ϕ_{ij} being independent and subject to the same distribution with density function P. Then

$$\mathbb{E}(X_i^q) = \int_{-\infty}^{\infty} \cdots \int_{-\infty}^{\infty} \left|\sum_{j=1}^{n} t_j x_j\right|^q P(t_1) dt_1 \cdots P(t_n) dt_n = \|\mathbf{x}\|_q^q = \nu_q \|\mathbf{x}\|_2^q.$$

We see that this expectation is independent of i. For convenience, let $\nu = \mathbb{E}(X_i^q) = \nu_q \|\mathbf{x}\|_2^q$ and

$$\xi_i := X_i^q - \nu. \quad (8.21)$$

Then ξ_i is of mean zero and all ξ_i are independent. To use Bernstein's inequality, we need to estimate $\mathbb{E}(\xi_i^k)$. By the definition,

$$\mathbb{E}(\xi_i^k) = \int_{-\infty}^{\infty} \cdots \int_{-\infty}^{\infty} \left(\left|\sum_{i=1}^{n} \phi_{ij} x_j\right|^q - \nu\right)^k P(t_1) \cdots P(t_n) dt_1 \cdots dt_n.$$

By using (8.19), we have

$$\mathbb{E}(\xi_i^k) = \int_{-\infty}^{\infty} M_p(u, \mathbf{x}) \left(|u|^q - \nu\right)^k du$$

$$= 2\int_0^{\nu^{1/q}} M_p(v; \mathbf{x})(|v|^q - \nu)^k dv + 2\int_{\nu^{1/q}}^{\infty} M_p(v; \mathbf{x})(|v|^q - \nu)^k dv.$$

We note that the first term on the right-hand side is negative when k is odd. Thus, we only need to estimate it when $k = 2\ell$ is even:

$$\int_0^{\nu^{1/q}} M_p(v;\mathbf{x})(v^q - \nu)^k dv = \nu^{k+1/q} \int_0^1 M_p(v\nu^{1/q};\mathbf{x})(v^q - 1)^k dv$$

$$\leq \|M_p(\cdot;\mathbf{x})\|_\infty \int_0^1 (1-u)^{2\ell} \frac{1}{q} u^{1/q-1} du$$

$$= \nu^{k+1/q} \|M_p(\cdot;\mathbf{x})\|_\infty \frac{1}{q} B(2\ell + 1, 1/q),$$

where $B(\cdot,\cdot)$ is the beta function defined by $B(m,n) = \int_0^1 x^m (1-x)^n dx$. It is easy to see the following result.

Lemma 8.22. *One can bound the beta function $B(2\ell + 1, 1/q)$ by*

$$B(2\ell + 1, 1/q) \leq q^{\ell+1} \ell!.$$

We leave this as Exercise 1, or see [6] for a proof. It follows from (8.20) and definition of ν that

$$\int_0^{\nu^{1/q}} M_p(v;\mathbf{x})(v^q - \nu)^{2\ell} dv \leq \nu^{k+1/q} \frac{(\Gamma((q+1)/2)/\sqrt{\pi})^{1/q}}{\sqrt{\pi}\nu^{1/q}} q^\ell \ell! = \ell! q^\ell \alpha_q \nu^{2\ell}, \quad (8.22)$$

where $\alpha_q = \frac{(\Gamma((q+1)/2)/\sqrt{\pi})^{1/q}}{\sqrt{\pi}}$. Note that $\Gamma(1/2) = \sqrt{\pi}$ and α_q is an increasing function of q. It thus is bounded below by $e^\gamma/(2\sqrt{\pi})$, with Euler's gamma number γ.

For the second term, we use the mean value theorem on $g(t) = t^q$ over the interval $[\nu^{1/q}, \infty)$ to get

$$v^q - \nu \leq (v - \nu^{1/q}) q \nu^{1-1/q}$$

and

$$\int_{\nu^{1/q}}^\infty M_p(v;\mathbf{x})(v^q - \nu)^k dv \leq q^k \nu^{k-k/q} \int_{\nu^{1/q}}^\infty M_p(v;\mathbf{x})(v - \nu^{1/q})^k dv$$

$$\leq q^k \nu^{k-k/q} \int_0^\infty M_p(v + \nu^{1/q};\mathbf{x}) v^k dv$$

$$\leq q^k \nu^{k-k/q} \int_0^\infty M_p(v;\mathbf{x}) v^k dv$$

$$= q^k \nu^{k-k/q} \frac{1}{2} \int_{-\infty}^\infty M_p(v;\mathbf{x}) |v|^k dv$$

$$= q^k \nu^{k-k/q} ((\Gamma((q+1)/2)/\sqrt{\pi})^{-1/q} \sqrt{\pi} \nu^{1/q})^k \int_{-\infty}^\infty e^{-t^2} t^k dt$$

$$= q^k \nu^k \alpha_q^{-k} \frac{(k-1)!!}{2^\ell} \leq q^k \nu^k \alpha_q^{-k} \ell!$$

by using the following estimate, whose proof is left as Exercise 9.

Lemma 8.23. *For any integer $k \geq 0$,*

$$\frac{1}{\sqrt{2\pi}} \int_{-\infty}^\infty e^{-t^2} t^k dt = \begin{cases} 0 & \text{if } k \text{ is odd}, \\ \frac{(2\ell)!}{2^{2\ell} \ell!} & \text{if } k = 2\ell. \end{cases}$$

Therefore, we have

$$\mathbb{E}(\xi_i^{2\ell}) \le \ell! q^\ell \nu^{2\ell} \alpha_q + q^{2\ell} \nu^{2\ell} \ell!! \alpha_q^{-2\ell} = \ell! q^\ell \nu^{2\ell} (\alpha_q + q^\ell \alpha_q^{-2\ell}). \qquad (8.23)$$

The result is similar for

$$\mathbb{E}(|\xi_i|^{2\ell-1}) \le q^{2\ell-1} \nu^{(2\ell-1)} \alpha_q^{-2\ell+1} \left(1 + \frac{(2\ell - 2)!!}{2^{\ell-1}} \right) \le \ell! q^{2\ell-1} \nu^{2\ell-1} \alpha_q^{-2\ell+1}. \qquad (8.24)$$

Let us summarize (8.23) and (8.24) in one formula: $\mathbb{E}(|\xi_i|^k) \le q^{k/2} \frac{([k/2]+1)!}{2} \nu^k \max\{\alpha_q, \alpha_q^{-k}\}$, where $[k/2]$ stands for the integer part of $k/2$. We are now ready to use Bernstein's inequality in Theorem 8.15 in the previous section. Rewrite the estimate as

$$\mathbb{E}(|\xi_i|^k) \le \frac{([k/2]+1)!}{2} \nu^2 H^{k-2},$$

with $H = \nu \sqrt{q} \max\{\alpha_q, \alpha_q^{-1}, 1\}$. Letting $t = \delta m \nu = \delta m \nu_q \|\mathbf{x}\|_2^q$, we use Theorem 8.15 to conclude that

$$P(|\|\Phi\mathbf{x}\|_q^q - m\|\mathbf{x}\|_q^q| < \delta m \|\mathbf{x}\|_q^q) \ge 1 - 2\exp\left\{ -\frac{\delta^2 m^2 \nu^2}{2(\delta m \nu^2 \sqrt{q} c_q + m\nu^2)} \right\}$$

$$= 1 - 2\exp\left\{ -\frac{\delta^2 m}{2(\delta \sqrt{q} c_q + 1)} \right\},$$

where $c_q = \max\{\alpha_q, \alpha_q^{-1}, 1\}$ is a positive constant independent of δ and $\|\mathbf{x}\|_2$. This completes the proof of the first part of Theorem 8.21.

We now work on a proof for the second part of Theorem 8.21. Let us consider a fixed index set $T \subseteq \{1, \dots, N\}$ of cardinality s and denote by $\mathbb{R}^T$ the space of vectors in $\mathbb{R}^n$ supported on T. Using Lemma 8.18, we can find a subset $\mathcal{U}_T$ of the unit sphere $\mathcal{S}_T$ of $\mathbb{R}^T$ such that

$$\min_{\mathbf{u} \in \mathcal{U}_T} \|\mathbf{x} - \mathbf{u}\|_2 \le \epsilon \quad \forall \mathbf{x} \in \mathcal{S}_T \quad \text{and} \quad \mathrm{card}(\mathcal{U}_T) \le \left(1 + \frac{2}{\epsilon}\right)^s.$$

Using the first part of Theorem 8.21, we obtain

$$P(|\|\Phi\mathbf{u}\|_q^q - m\nu_q \|\mathbf{u}\|_2^q| < \eta m\nu_q \|\mathbf{u}\|_2^q) \le \left(1 + \frac{2}{\epsilon}\right)^s 2\exp\left\{ -\frac{\eta^2 m}{2c} \right\} \qquad (8.25)$$

for all $\mathbf{u} \in \mathcal{U}_T$. For any $\mathbf{x} \in \mathcal{S}_T \backslash \mathcal{U}_T$, we can find $\mathbf{x}_0 \in \mathcal{U}_T$ such that $\|\mathbf{x} - \mathbf{x}_0\|_2 \le \epsilon$. Letting $\epsilon_1 = \|\mathbf{x} - \mathbf{x}_0\|_2$, we know that $(\mathbf{x} - \mathbf{x}_0)/\epsilon_1 \in \mathcal{S}_T$. Then we can find $\mathbf{x}_1 \in \mathcal{U}_T$ within ϵ. Continuing in this fashion, we obtain sequences $\{\epsilon_n, n \ge 1\}$ and $\{\mathbf{x}_n, n \ge 0\}$ such that $|\epsilon_n| \le \epsilon^n$ and $\|\mathbf{x} - \sum_{k=0}^n \epsilon_k \mathbf{x}_k\|_2 \le \epsilon^{N+1}$, where $\epsilon_0 = 1$. It follows that $\mathbf{x} = \sum_{n=0}^\infty \epsilon_n \mathbf{x}_n$. If any ϵ_n is zero, we just terminate the series. Now we calculate the probability of

$$\|\Phi\mathbf{x}\|_q^q = \left\|\sum_{n=0}^\infty \epsilon_n \Phi\mathbf{x}_n\right\|_q^q \le \sum_{n=0}^\infty |\epsilon_n|^q \|\Phi\mathbf{x}_n\|_q^q$$

$$\le \sum_{n=0}^\infty \epsilon^{nq} (1 + \eta) m\nu_q \|\mathbf{x}_n\|_2^q = \frac{1 + \eta}{1 - \epsilon^q} m\nu_q$$

$$= \left(1 + \frac{\eta + \epsilon^q}{1 - \epsilon^q}\right) m\nu_q,$$

since $\|\mathbf{x}_n\|_2 = 1$. Also, when $q \le 1$, we have

$$\|\Phi\mathbf{x}\|_q^q \ge \|\Phi\mathbf{x}_0\|_q^q - \left\|\sum_{n=1}^{\infty} \epsilon_n \mathbf{x}_n\right\|_q^q \ge (1-\eta)m\nu_q\|\mathbf{x}_0\|_2^q - \frac{(1+\eta)m\nu_q\epsilon^q}{1-\epsilon^q} = \left(1 - \frac{\eta+\epsilon^q}{1-\epsilon^q}\right)m\nu_q.$$

Thus, for any $\mathbf{x} \ne 0$ in $\mathbb{R}^T$, we have $\mathbf{x}/\|\mathbf{x}\|_2 \in \mathcal{S}_T$ and

$$\left(1 - \frac{\eta+\epsilon^q}{1-\epsilon^q}\right)m\nu_q \le \|\Phi\mathbf{x}/\|\mathbf{x}\|_2\|_q^q \le \left(1 + \frac{\eta+\epsilon^q}{1-\epsilon^q}\right)m\nu_q.$$

That is, with high probability, we have

$$(1-\delta)m\nu_q\|\mathbf{u}\|_2^q \le \|\Phi\mathbf{u}\|_q^q \le (1+\delta)m\nu_q\|\mathbf{u}\|_2^q \quad \forall \mathbf{u} \in \mathbb{R}^T,$$

where $\delta = \frac{\eta+\epsilon^q}{1-\epsilon^q} < 1$. That is, we have

$$P\left(\left|\|\Phi\mathbf{x}\|_q^q - m\nu_q\|\mathbf{x}\|_2^q\right| > \delta m\nu_q\|\mathbf{x}\|_2^q \; \forall \mathbf{x} \in \mathbb{R}^T\right) \le 2\left(1+\frac{2}{\epsilon}\right)^s \exp\left(-\frac{\eta^2 m}{c}\right).$$

We now observe that the set of all s-sparse vectors is the union of $\binom{n}{s} \le (eN/s)^s$ spaces $\mathbb{R}^T$ to deduce that

$$P\left(\left|\|\Phi\mathbf{x}\|_q^q - m\nu_q\|\mathbf{x}\|_2^q\right| > \delta m\nu_q\|\mathbf{x}\|_2^q \; \forall \mathbf{x} \in \mathbb{R}^n \text{ with } \|\mathbf{x}\|_0 \le s\right)$$

$$\le 2\binom{n}{s} \exp\left(-\frac{\eta^2 m}{c} + \frac{2s}{\epsilon}\right) \le 2\exp\left(-\frac{\eta^2 m}{c} + \frac{2s}{\delta} + s\ln\left(\frac{en}{s}\right)\right).$$

Here η, ϵ, δ are connected by a formula discussed before. That is,

$$\delta = \frac{\eta+\epsilon^q}{1-\epsilon^q} < 1.$$

For example, for any $\delta \ge 1/2$, we can choose $\epsilon = (1/4)^{1/q}$. Then $\eta = 3\delta/4 - 1/4 > 0$. If we impose that

$$\frac{2s}{\epsilon} + s\ln\left(\frac{eN}{s}\right) < \frac{1}{2}\frac{\eta^2 m}{c},$$

we shall have

$$P\left(\left|\|\Phi\mathbf{x}\|_q^q - m\nu_q\|\mathbf{x}\|_2^q\right| > \delta m\nu_q\|\mathbf{x}\|_2^q \; \forall \mathbf{x} \in \mathbb{R}^n \text{ with } \|\mathbf{x}\|_0 \le s\right) \le 2\exp\left(-\frac{\eta^2 m}{2c}\right).$$

Thus we have the second part of the result of Theorem 8.21 by choosing appropriate ϵ and η.

8.5 ▪ Sub-Gaussian Random Matrices (II)

In this section, we will show that a sensing matrix with each entry generated from a random variable subject to a sub-Gaussian distribution satisfies the RIP with high probability. Let us first recall the following definition.

Definition 8.24. *A random variable ξ is called sub-Gaussian if there exists a number $a \in [0, \infty)$ such that the inequality*

$$\mathbb{E}(\exp(\lambda\xi)) \le \exp\left(\frac{a^2\lambda^2}{2}\right)$$

holds for all $\lambda \in \mathbb{R}$. Let $\tau(\xi)$ be the smallest such number a so that the above inequality holds. We slightly abuse the notation by calling $\tau(\xi)$ the sub-Gaussian variance of ξ.

It is easy to verify that a Gaussian random variable ξ with mean zero and variance $\sigma^2 > 0$ satisfies

$$\mathbb{E}(\exp(\lambda\xi)) = \sum_{k=0}^{\infty} \frac{\lambda^k}{k!}\mathbb{E}(\xi^k) = \exp\left(\frac{\sigma^2\lambda^2}{2}\right)$$

for all λ, and hence ξ is a sub-Gaussian random variable.

One of the reasons to introduce sub-Gaussian random variables is the following.

Theorem 8.25. *Let $\xi_i = |X_i|^q - \nu$ be the random variable* (8.21) *defined in the previous section. Then ξ_i is a sub-Gaussian random variable.*

Proof. We use the estimates in the previous section, i.e., (8.23) and (8.24), to obtain

$$\mathbb{E}(\exp(\lambda\xi_i)) = \sum_{k=0}^{\infty} \frac{\lambda^k}{k!}\mathbb{E}(\xi_i^k) \le \sum_{k=0}^{\infty} \frac{\lambda^k}{k!}\sqrt{q}^k c_q^k \nu^k ([k/2])!$$

$$\le \sum_{k=0}^{\infty} \frac{\lambda^k}{2^k[k/2]!}\sqrt{q}^k c_q^k \nu^k \le \exp\left(\frac{\lambda^2 q c_q^2 \nu^2}{2}\right).$$

It follows from the definition of the sub-Gaussian random variable that $\tau(\xi_i) \le q c_q^2 \nu^2$. $\square$

Another interesting example of sub-Gaussian random variables is the random variable ξ with uniform distribution over $[-1, 1]$. That is, $p(x) = 1/2\chi_{[-1,1]}(x)$, with characteristic function $\chi_{[-1,1]}(x)$. It is easy to check that ξ is sub-Gaussian. Also, the Bernoulli random variable with probability $p(1) = 1/2$ at 1 and $p(-1) = 1/2$ at -1 is a sub-Gaussian random variable.

If ξ is a sub-Gaussian random variable, then there exists a positive constant $\tau > 0$ such that

$$P(|\xi| > x) \le 2\exp\left(-\frac{x^2}{2\tau^2}\right) \tag{8.26}$$

for all $x \in (-\infty, \infty)$. Indeed, by the Chebyshev–Markov inequality, we have

$$P(\xi \ge x) \le \exp(-\lambda x)\mathbb{E}(\exp(\lambda\xi)) \le \exp\left(\frac{1}{2}\lambda^2\tau^2 - \lambda x\right)$$

for all $\lambda > 0$ and $x > 0$. Minimizing the right-hand side for $\lambda > 0$ gives the inequality

$$P(\xi > x) \le \exp\left(-\frac{x^2}{2\tau^2}\right).$$

Similarly, we can show $P(\xi < -x) \le \exp(-\frac{x^2}{2\tau^2})$, and hence we conclude (8.26). In fact, when a random variable ξ with mean zero satisfies (8.26), ξ is a sub-Gaussian random variable. See Exercise 7.

Next let us consider a sequence of sub-Gaussian random variables $\xi_1, \ldots, \xi_m$.

Lemma 8.26. *Assume that $\xi_1, \ldots, \xi_m$ are independent sub-Gaussian random variables. Then the summation of all ξ_i is sub-Gaussian and its sub-Gaussian variance satisfies*

$$\tau^2\left(\sum_{i=1}^{m}\xi_i\right) \le \sum_{i=1}^{m}\tau^2(\xi_i). \tag{8.27}$$

Proof. For any $\lambda \in \mathbb{R}$, we have

$$\mathbb{E}\left(\exp\left(\lambda \sum_{i=1}^{m} \xi_i\right)\right) = \prod_{i=1}^{m} \mathbb{E}\exp(\lambda\xi_i)$$

$$\leq \prod_{i=1}^{m} \exp\left(\frac{\lambda^2\tau^2(\xi_i)}{2}\right) = \exp\left(\frac{\lambda^2}{2}\sum_{i=1}^{m}\tau^2(\xi_i)\right).$$

It follows from the definition of $\tau(\sum_{i=1}^{m}\xi_i)$ that we have (8.27).   □

We are now ready to prove a basic result similar to the Bernstein inequality for sub-Gaussian random variables.

Theorem 8.27. *Assume that $\xi_1, \ldots, \xi_m$ are independent sub-Gaussian random variables. Then the summation $S_m = \sum_{i=1}^{m}\xi_i$ satisfies*

$$P(|S_m| \geq x) \leq 2\exp\left(-\frac{x^2}{2\sum_{i=1}^{m}\tau^2(\xi_i)}\right). \tag{8.28}$$

In particular, if $\tau(\xi_i) \leq b$ for $i = 1, \ldots, m$, then

$$P(|S_m| \geq x) \leq 2\exp\left(-\frac{x^2}{2b^2m}\right). \tag{8.29}$$

Proof. By using (8.27), we know that S_m is a sub-Gaussian random variable with $\tau^2(S_m) \leq \sum_{i=1}^{m}\tau^2(\xi_i)$. By using (8.26) and (8.27), we conclude (8.28). The inequality in (8.29) follows immediately.   □

Instead of using the Bernstein inequality from the previous section, we should use Theorem 8.27 to conclude the following.

Theorem 8.28. *Suppose that $\Phi = [\phi_{ij}]_{1 \leq i \leq m, 1 \leq j \leq n}$ is a matrix with entries ϕ_{ij} being i.i.d. Gaussian random variables with mean zero and variance $1/\sqrt{m}$. Let $0 < q \leq 1$ and $\nu_q := 2^{q/2}\sigma^q\Gamma((q+1)/2)/\sqrt{\pi}$. Then for any $\delta > 0$, we have the probability*

$$P\left(\left|\|\Phi\mathbf{x}\|_q^q - m\nu_q\|\mathbf{x}\|_2^q\right| < \delta m\nu_q\|\mathbf{x}\|_2^q\right) \geq 1 - 2\exp\left(-\frac{\delta^2 m}{2qc_q^2}\right), \tag{8.30}$$

where c_q is a positive constant independent of δ and $\|\mathbf{x}\|_2$ for any $\mathbf{x} \in \mathbb{R}^n$. In particular, there exist two positive constants C_1 and C_2 such that when

$$m \geq C_1 s + qC_2 s\ln(N/s), \tag{8.31}$$

the matrix Φ satisfies the q-RIP

$$(1 - \delta_s)m\nu_q\|\mathbf{x}\|_2^q \leq \|\Phi\mathbf{x}\|_q^q \leq (1 + \delta_s)m\nu_q\|\mathbf{x}\|_2^q \qquad \forall\|\mathbf{x}\|_0 \leq s$$

with overwhelming probability.

Proof. We use Theorem 8.27 together with estimates (8.23) and (8.24) to get (8.30). Then we use the second part of the proof of Theorem 8.21 to obtain

$$P\left(\left|\|\Phi\mathbf{x}\|_q^q - m\nu_q\|\mathbf{x}\|_2^q\right| > \delta m\nu_q\|\mathbf{x}\|_2^q \ \forall \mathbf{x} \in \mathbb{R}^n \text{ with } \|\mathbf{x}\|_0 \leq s\right)$$

$$\leq 2\binom{n}{s}\left(1 + \frac{2}{\epsilon}\right)^s \exp\left(-\frac{\eta^2 m}{2qc_q^2}\right) \leq 2\exp\left(-\frac{\eta^2 m}{2qc_q^2} + s\ln\left(1 + \frac{2}{\epsilon}\right) + s\ln\left(\frac{en}{s}\right)\right).$$

Let us choose

$$s \ln\left(1 + \frac{2}{\epsilon}\right) + s \ln\left(\frac{eN}{s}\right) \leq s \ln\frac{3}{\epsilon} + s \ln\left(\frac{eN}{s}\right) \leq \frac{\eta^2 m}{4qc_q^2}.$$

Since $\delta = (\eta + \epsilon^q)/(1 - \epsilon^q) < 1$, we have $\epsilon^q = (\delta - \eta)/(1 + \delta)$. So we need

$$m \geq \frac{1}{\delta^2} 4qc_q^2 \left(s \ln 3 + \frac{s}{q} \ln(1 + \delta)/(\delta - \eta) + s \ln\left(\frac{en}{s}\right)\right)$$

$$= \frac{1}{\delta^2} \left(4c_q^2 s \ln(1 + \delta)/(\delta - \eta) + qs4c_q^2 \ln\left((e + 3)n/s\right)\right).$$

With $C_1 = \frac{4c_q^2}{\delta^2}(\ln(1 + \delta)/(\delta - \eta) + q \ln(e + 3))$ and $C_2 = \frac{4c_q^2}{\delta^2}$, we have (8.31). For such an m, we have

$$P\left(\left|\|\Phi x\|_q^q - m\nu_q\|x\|_2^q\right| > \delta m\nu_q\|x\|_2^q \ \forall x \in \mathbb{R}^n \text{ with } \|x\|_0 \leq s\right) \leq 2\exp\left(-\frac{\eta^2 m}{4qc_q^2}\right),$$

which has a small probability when m satisfies (8.31). That is, Φ satisfies the q-RIP with overwhelming probability. □

8.6 ▪ Pre-Gaussian Random Matrices (I)

Furthermore, let us show that a sensing matrix Φ with entries generated from a random variable subject to a pre-Gaussian distribution satisfies an RIP with high probability. Let us give another definition of pre-Gaussian random variables.

Definition 8.29. *A random variable ξ is pre-Gaussian if $\mathbb{E}(\xi) = 0$ and there exist constants $b > 0$ and $c > 0$ such that*

$$P(|\xi| > t) \leq b\exp(-ct) \qquad \forall t > 0.$$

There are many equivalent conditions. For example, consider the following lemma.

Lemma 8.30. *A random variable ξ with mean zero is a pre-Gaussian random variable if and only if there exists a constant $H > 0$ such that $\mathbb{E}(\exp(\lambda\xi)) < \infty$ for all $|\lambda| \leq H$.*

Proof. Indeed, suppose that ξ satisfies the above condition. Then we the use Chebyshev–Markov inequality to get

$$P(|\xi| \geq t) \leq \exp(-ct)\mathbb{E}(\exp(c|\xi|)),$$

where $c > 0$ is a constant satisfying $c < H$. Since $b = \exp(c|\xi|)$ is finite, we know that ξ is pre-Gaussian.

On the other hand, if ξ is pre-Gaussian, then letting $F(x)$ be the distribution function, we have

$$\mathbb{E}(\exp(\lambda|\xi|)) = 1 + \lambda \int_0^\infty (1 - F(x))\exp(\lambda x)dx \leq 1 + \lambda b \int_0^\infty \exp(-(c - a)x)dx$$

$$= 1 + \frac{b\lambda}{c - \lambda} < \infty$$

for $\lambda < c$. This shows that ξ satisfies the condition in this lemma. □

It is easy to give an example of a random variable that is pre-Gaussian, not sub-Gaussian. For example, let η be a Gaussian random variable in $N(0, \sigma^2)$. Then $\xi = \eta^2 - \sigma^2$ is pre-Gaussian. Indeed, $\mathbb{E}(\xi) = 0$ and

$$\mathbb{E}(\lambda(\eta^2 - \sigma^2)) = \exp(-\lambda\sigma^2)\mathbb{E}(\exp(\lambda\eta^2))$$
$$= \frac{\exp(-\lambda\sigma^2)}{\sqrt{1 - 2\sigma^2\lambda}} < \infty$$

for $|\lambda| < 1/(2\sigma^2)$. By the result in Lemma 8.30, ξ is pre-Gaussian. When $\lambda = 1/(2\sigma^2)$, $\mathbb{E}(\lambda(\eta^2 - \sigma^2)) = \infty$, and hence ξ cannot be a sub-Gaussian random variable.

Another equivalent definition of pre-Gaussian random variables is the following.

Theorem 8.31. *A random variable ξ with mean zero, i.e., $\mathbb{E}(\xi) = 0$, is pre-Gaussian if and only if there exists a constant $\theta > 0$ such that*

$$\theta(\xi) := \sup_{k \geq 1} \left[\frac{\mathbb{E}|\xi|^{2k}}{(2k)!} \right]^{1/2k} < \infty. \tag{8.32}$$

Proof. In fact, Theorem 8.31 can be stated with $\theta(\xi)$ replaced by

$$\theta'(\xi) := \sup_{k \geq 1} \left[\frac{\mathbb{E}|\xi|^k}{k!} \right]^{1/k}, \tag{8.33}$$

but the two statements are similar in view of the inequalities

$$\theta(\xi) \leq \theta'(\xi) \leq 2\theta(\xi).$$

The lower inequality is clear, while the upper inequality is a simple consequence of $\mathbb{E}(|\xi|^k) \leq \mathbb{E}(|\xi|^{2k})^{1/2}$. Assume that ξ satisfies the $\theta'(\xi)$ defined in (8.33). Let us show that there exists a positive number H such that $\mathbb{E}(\exp(\lambda\xi)) < \infty$ for all $|\lambda| \leq H$. Indeed, since $\mathbb{E}(\xi) = 0$,

$$\mathbb{E}(\exp(\lambda\xi)) \leq 1 + \sum_{m=2}^{\infty} \frac{|\lambda|^m \mathbb{E}(|\xi|^m)|}{m!} \leq 1 + \sum_{m=2}^{\infty} |\lambda\theta'(\xi)|^m$$
$$\leq 1 + \frac{(\lambda\theta'(\xi))^2}{1 - |\lambda\theta'(\xi)|} < \infty$$

for $|\lambda| < 1/\theta'(\xi)$. By Lemma 8.30, ξ is a pre-Gaussian random variable. $\quad\square$

The quantity $\theta(\xi)$ turns out to be more convenient and useful because of the following result.

Lemma 8.32. *If $\xi_1, \ldots, \xi_n$ are independent pre-Gaussian random variables, then*

$$\theta^2(x_1\xi_1 + \cdots + x_n\xi_n) \leq x_1^2 \theta^2(\xi_1) + \cdots + x_n^2 \theta^2(\xi_n).$$

We leave the proof to Exercise 9.

We are ready to discuss the RIP of general matrices with pre-Gaussian random variables. We shall assume that each entry ϕ_{ij} of a sensing matrix Φ of size $m \times N$ is a pre-Gaussian random variable satisfying

$$\mathbb{E}(|\phi_{ij}|) \geq \nu > 0 \quad \text{and} \quad \mathbb{E}(|\phi_{ij}|^2) \leq \sigma^2 < \infty$$

for all $1 \leq i \leq m, 1 \leq j \leq n$. We shall let

$$\|\mathbf{x}\|_i = \mathbb{E}\left(|\sum_{j=1}^{n} \phi_{i,j} x_j| \right), \quad i = 1, \ldots, m,$$

and

$$\|\mathbf{x}\| = \sum_{i=1}^{m} \|\mathbf{x}\|_i. \tag{8.34}$$

We are interested in the following modified restricted isometry property (MRIP): There exists a $\delta_s \in (0, 1)$ such that

$$(1 - \delta)\|\mathbf{x}\| \leq \|\Phi\mathbf{x}\|_1 \leq (1 + \delta)\|\mathbf{x}\| \quad \forall \mathbf{x} \in \mathbb{R}^n \text{ with } \|\mathbf{x}\|_0 \leq s. \tag{8.35}$$

In addition, we will show that when a sensing matrix Φ satisfies the above MRIP, the sparse solution can be found by using the ℓ_1 minimization.

We begin with a lemma.

Lemma 8.33. *Suppose that $\xi_1, \ldots, \xi_n$ are independent zero mean random variables satisfying*

$$\mathbb{E}(|\xi_j|) \geq \mu \quad \text{and} \quad \mathbb{E}(\xi_j^2) \leq \sigma^2 \quad \forall j = 1, \ldots, N.$$

Then letting $\|\mathbf{x}\|_\xi = \mathbb{E}(|\sum_{j=1}^{n} x_j \xi_j|)$,

$$\frac{\mu}{\sqrt{8}} \|\mathbf{x}\|_2 \leq \|\mathbf{x}\|_\xi \leq \sigma \|\mathbf{x}\|_2 \quad \forall \mathbf{x} \in \mathbb{R}^n.$$

Proof. For the upper estimate, the independence of the ξ_j's simply yields

$$\mathbb{E}\left| \sum_{j=1}^{n} x_j \xi_j \right| \leq \left(\mathbb{E}\left(\sum_{j=1}^{n} x_j \xi_j \right)^2 \right)^{1/2} = \left(\sum_{j=1}^{n} x_j^2\, \mathbb{E}(\xi_j)^2 \right)^{1/2} \leq \sigma \|\mathbf{x}\|_2.$$

As for the lower estimate, we use the symmetrization procedure (see Lemma 6.3 of [14]) to write

$$\mathbb{E}\left| \sum_{j=1}^{n} x_j \xi_j \right| \geq \frac{1}{2} \mathbb{E}\left| \sum_{j=1}^{n} \epsilon_j x_j \xi_j \right| = \frac{1}{2} \mathbb{E}_\xi \mathbb{E}_\epsilon \left| \sum_{j=1}^{n} \epsilon_j x_j \xi_j \right|, \tag{8.36}$$

where $(\epsilon_1, \ldots, \epsilon_n)$ is a Rademacher sequence independent of $(\xi_1, \ldots, \xi_n)$.

Next, using Khintchine's inequality with optimal constants due to Haagerup [10], we have

$$\mathbb{E}_\epsilon \left| \sum_{j=1}^{n} \epsilon_j x_j \xi_j \right| \geq \frac{1}{\sqrt{2}} \left(\sum_{j=1}^{n} x_j^2 \xi_j^2 \right)^{1/2} = \frac{\|\mathbf{x}\|_2}{\sqrt{2}} \left(\sum_{j=1}^{n} \frac{x_j^2}{\|\mathbf{x}\|_2^2} \xi_j^2 \right)^{1/2}.$$

Then, using the concavity of the function $t \mapsto t^{1/2}$, we derive

$$\mathbb{E}_\epsilon \left| \sum_{j=1}^{n} \epsilon_j x_j \xi_j \right| \geq \frac{\|\mathbf{x}\|_2}{\sqrt{2}} \sum_{j=1}^{n} \frac{x_j^2}{\|\mathbf{x}\|_2^2} |\xi_j|. \tag{8.37}$$

The desired estimate follows from (8.36), (8.37), and $\mathbb{E}(|\xi_j|) \geq \mu$.   □

In order to establish the MRIP (8.35) for all sparse vectors, we consider first individual vectors $\mathbf{x} \in \mathbb{R}^n$ and bound the tail probability

$$P\Big(\big|\|\Phi\mathbf{x}\|_1 - \|\!\|\mathbf{x}\|\!\|\big| > \delta \|\!\|\mathbf{x}\|\!\|\Big).$$

Theorem 8.34. *Suppose the entries of a matrix $\Phi \in \mathbb{R}^{m \times N}$ are independent pre-Gaussian random variables satisfying*

$$\mathbb{E}\big(|a_{i,j}|\big) \geq \mu \qquad and \qquad \mathbb{E}\big(|a_{i,j}|^{2k}\big) \leq (2k)!\,\theta^{2k}, \quad k \geq 1.$$

If $\|\!\| \cdot \|\!\|$ denotes the norm defined in (8.34) for the entries $\phi_{i,j}$, then

$$P\Big(\big|\|\Phi\mathbf{x}\|_1 - \|\!\|\mathbf{x}\|\!\|\big| > \delta \|\!\|\mathbf{x}\|\!\|\Big) \leq 2\exp\big(-\kappa\delta^2 m\big) \tag{8.38}$$

for any $\mathbf{x} \in \mathbb{R}^n$ and any $\delta \in (0,1)$, where the constant κ depends only on θ/μ.

Proof. Setting $Y_i := |(\Phi\mathbf{x})_i| - \|\!\|\mathbf{x}\|\!\|_i$, we observe that $Y_1, \ldots, Y_m$ are independent zero mean random variables and that

$$\|\Phi\mathbf{x}\|_1 - \|\!\|\mathbf{x}\|\!\| = \sum_{i=1}^m Y_i.$$

Then, since $\theta(a_{i,j}) \leq \theta$, Lemma 8.32 yields

$$\theta\big((\Phi\mathbf{x})_i\big) = \theta\left(\sum_{j=1}^n x_j a_{i,j}\right) \leq \theta\|\mathbf{x}\|_2, \qquad \text{hence} \qquad \theta'\big((\Phi\mathbf{x})_i\big) \leq 2\theta\|\mathbf{x}\|_2.$$

For an integer $k \geq 2$, it follows from the inequality $|Y_i| \leq \max\{|(\Phi\mathbf{x})_i|, \|\!\|\mathbf{x}\|\!\|_i\}$ that

$$\mathbb{E}\big(|Y_i|^k\big) \leq \max\{\mathbb{E}\big(|(\Phi\mathbf{x})_i|^k\big), \|\!\|\mathbf{x}\|\!\|_i^k\} \leq \max\{k!\,\big(\theta'\big((\Phi\mathbf{x})_i\big)\big)^k, \|\!\|\mathbf{x}\|\!\|_i^k\}$$
$$\leq \max\left\{\frac{k!}{2}2^{k/2}\big(\theta'\big((\Phi\mathbf{x})_i\big)\big)^k, \frac{k!}{2}\|\!\|\mathbf{x}\|\!\|_i^k\right\} \leq \frac{k!}{2}\max\{\sqrt{8}\,\theta\|\mathbf{x}\|_2, \|\!\|\mathbf{x}\|\!\|_i\}^k.$$

Since Lemma 8.33 implies $\mu\|\mathbf{x}\|_2/\sqrt{8} \leq \|\!\|\mathbf{x}\|\!\|_i \leq \sqrt{2}\,\theta\|\mathbf{x}\|_2$, we can apply Bernstein's inequality in Theorem 8.15 with

$$M = \max\{\sqrt{8}\,\theta\|\mathbf{x}\|_2, \sqrt{2}\,\theta\|\mathbf{x}\|_2\} = \sqrt{8}\,\theta\|\mathbf{x}\|_2, \quad v_i = M^2, \quad t = \delta\|\!\|\mathbf{x}\|\!\| \geq \delta\,m\,\mu\|\mathbf{x}\|_2/\sqrt{8}$$

to obtain

$$P\Big(\big|\|\Phi\mathbf{x}\|_1 - \|\!\|\mathbf{x}\|\!\|\big| > \delta m\|\!\|\mathbf{x}\|\!\|\Big) \leq 2\exp\left(-\frac{\delta^2 m^2 \mu^2 \|\!\|\mathbf{x}\|\!\|^2/8}{2\big(8\,m\,\theta^2\|\mathbf{x}\|_2^2 + \delta\,m\,\mu\,\theta\|\mathbf{x}\|_2^2\big)}\right)$$
$$= 2\exp\left(-\frac{\delta^2 m}{16\big(8(\theta/\mu)^2 + \delta(\theta/\mu)\big)}\right).$$

Since $\delta \leq 1$, the result follows with $\kappa := 1/\big(128(\theta/\mu)^2 + 16(\theta/\mu)\big)$. $\qquad\square$

We now continue to pass from the concentration inequality (8.38) for individual vectors to the MRIP (8.35) for all sparse vectors. In fact, we prove that (8.35) fails with an exponentially small probability.

Theorem 8.35. *Let $\Phi \in \mathbb{R}^{m \times n}$ be a random matrix, and let $\| \cdot \|$ be a norm on $\mathbb{R}^n$. Suppose that, for any $\mathbf{x} \in \mathbb{R}^n$ and any $\epsilon \in (0,1)$,*

$$P\left(\left| \|\Phi\mathbf{x}\|_1 - \|\mathbf{x}\| \right| > \delta \|\mathbf{x}\| \right) \le 2 \exp \left(-\kappa \delta^2 m \right). \tag{8.39}$$

Then there exist constants $c_1, c_2 > 0$ depending only on κ such that, for any $\delta \in (0,1)$,

$$P\left(\left| \|\Phi\mathbf{x}\|_1 - \|\mathbf{x}\| \right| > \delta \|\mathbf{x}\| \text{ for some } s\text{-sparse } \mathbf{x} \in \mathbb{R}^n \right) \le 2 \exp \left(-c_1 \delta^2 m \right)$$

provided that

$$m \ge \frac{c_2}{\delta^3} s \ln \left(\frac{eN}{s} \right).$$

Proof. We start by considering a fixed index set $S \subseteq \{1, \dots, N\}$ of cardinality s. Let $\mathcal{S}$ denote the unit sphere of the space $\mathbb{R}^S$ of vectors supported on S embedded with the norm $\|\mathbf{x}\|$. Using a proof similar to Lemma 8.18, we can find a subset $\mathcal{U}$ of $\mathcal{S}$ such that

$$\min_{\mathbf{u} \in \mathcal{U}} \|\mathbf{x} - \mathbf{u}\| \le \gamma := \frac{\delta}{3} \quad \forall \mathbf{x} \in \mathcal{S} \qquad \text{and} \qquad \text{card}(\mathcal{U}) \le \left(1 + \frac{2}{\gamma} \right)^s.$$

The concentration inequality (8.39), together with a union bound, gives

$$P(\left| \|\Phi\mathbf{u}\|_1 - \|\mathbf{u}\| \right| > \gamma \|\mathbf{u}\| \text{ for some } \mathbf{u} \in \mathcal{U}) \le \left(1 + \frac{2}{\gamma} \right)^s 2 \exp \left(-\kappa \gamma^2 m \right)$$

$$\le 2 \exp \left(-\kappa \gamma^2 m + \frac{2s}{\gamma} \right) = 2 \exp \left(-\frac{\kappa \delta^2 m}{9} + \frac{6s}{\delta} \right).$$

This means that the matrix Φ is drawn with high probability in such a way that

$$(1 - \gamma) \|\mathbf{u}\| \le \|\Phi\mathbf{u}\|_1 \le (1 + \gamma) \|\mathbf{u}\| \quad \forall \mathbf{u} \in \mathcal{U}. \tag{8.40}$$

Let $\tilde{\delta}$ be the smallest positive constant such that

$$\|\Phi\mathbf{x}\|_1 \le (1 + \tilde{\delta}) \|\mathbf{x}\| \qquad \forall \mathbf{x} \in \mathcal{S}. \tag{8.41}$$

Given $\mathbf{x} \in \mathcal{S}$, picking $\mathbf{u} \in \mathcal{U}$ with $\|\mathbf{x} - \mathbf{u}\| \le \gamma$, we derive

$$\|\Phi\mathbf{x}\|_1 \le \|\Phi\mathbf{u}\|_1 + \|\Phi(\mathbf{x} - \mathbf{u})\|_1 \le 1 + \gamma + (1 + \tilde{\delta}) \|\mathbf{x} - \mathbf{u}\| \le 1 + \gamma + (1 + \tilde{\delta})\gamma.$$

The minimality of $\tilde{\delta}$ implies that

$$1 + \tilde{\delta} \le 1 + \gamma + (1 + \tilde{\delta})\gamma \le 1 + 2\gamma + \tilde{\delta}/3, \qquad \text{so that} \quad \tilde{\delta} \le 3\gamma = \delta.$$

Substituting into (8.41), we obtain the upper estimate

$$\|\Phi\mathbf{x}\|_1 \le (1 + \delta) \|\mathbf{x}\| \qquad \forall \mathbf{x} \in \mathbb{R}^S. \tag{8.42}$$

Next, for the lower bound estimate, for $\mathbf{x} \in \mathcal{S}$ and $\mathbf{u} \in \mathcal{U}$ with $\|\mathbf{x} - \mathbf{u}\| \le \gamma$, we have

$$\|\Phi\mathbf{x}\|_1 \ge \|\Phi\mathbf{u}\|_1 - \|\Phi(\mathbf{x} - \mathbf{u})\|_1 \ge 1 - \gamma - (1 + \delta) \|\mathbf{x} - \mathbf{u}\| \ge 1 - \gamma - (1 + \delta)\gamma$$

$$\ge 1 - 3\gamma = 1 - \delta.$$

Thus, we obtain the lower estimate

$$\|\Phi\mathbf{x}\|_1 \ge (1 - \delta) \|\mathbf{x}\| \qquad \forall \mathbf{x} \in \mathbb{R}^S. \tag{8.43}$$

Since both upper and lower estimates (8.42) and (8.43) hold as soon as (8.40) holds, we obtain

$$P\Big(\big|\|\Phi\mathbf{x}\|_1 - \|\mathbf{x}\|\big| > \delta\|\mathbf{x}\|\ \text{ for some }\mathbf{x}\in\mathbb{R}^S\Big) \le 2\exp\Big(-\frac{\kappa\delta^2 m}{9} + \frac{6s}{\delta}\Big).$$

We now take into account that the set of s-sparse vectors is the union of $\binom{n}{s} \le (en/s)^s$ spaces $\mathbb{R}^S$ to deduce, using a union bound, that

$$P\Big(\big|\|\Phi\mathbf{x}\|_1 - \|\mathbf{x}\|\big| > \delta\|\mathbf{x}\|\ \text{ for some }s\text{-sparse }\mathbf{x}\in\mathbb{R}^n\Big)$$
$$\le \binom{n}{s} 2\exp\Big(-\frac{\kappa\delta^2 m}{9} + \frac{6s}{\delta}\Big) \le 2\exp\Big(-\frac{\kappa\delta^2 m}{9} + \frac{6s}{\delta} + s\ln\Big(\frac{en}{s}\Big)\Big)$$
$$\le 2\exp\Big(-\frac{\kappa\delta^2 m}{9} + \frac{7s}{\delta}\ln\Big(\frac{en}{s}\Big)\Big).$$

By imposing, say,

$$\frac{7s}{\delta}\ln\Big(\frac{eN}{s}\Big) \le \frac{\kappa\delta^2 m}{18}, \qquad \text{i.e.,} \qquad m \ge \frac{126}{\kappa\delta^3} s\ln\Big(\frac{eN}{s}\Big),$$

we ensure that

$$P\Big(\big|\|\Phi\mathbf{x}\|_1 - \|\mathbf{x}\|\big| > \delta\|\mathbf{x}\|\ \text{ for some }s\text{-sparse }\mathbf{x}\in\mathbb{R}^n\Big) \le 2\exp\Big(-\frac{\kappa\delta^2 m}{18}\Big).$$

This is the desired result, with $c_1 = \kappa/18$ and $c_2 = 126/\kappa$.   □

Furthermore, we verify that the MRIP (8.35) implies sparse recovery by ℓ_1 minimization.

Theorem 8.36. *Let $\|\cdot\|$ be a norm on $\mathbb{R}^n$ satisfying*

$$c\,\|\mathbf{x}\|_2 \le \|\mathbf{x}\| \le C\,\|\mathbf{x}\|_2 \qquad \forall \mathbf{x}\in\mathbb{R}^n.$$

If a matrix $\Phi\in\mathbb{R}^{m\times N}$ has an MRIP constant

$$\delta_{s+t} := \delta_{s+t}^{\|\cdot\|} < \frac{\sqrt{t/s} - C/c}{\sqrt{t/s} + C/c} \qquad \text{for some integer }t, \tag{8.44}$$

then any s-sparse vector $\mathbf{x}\in\mathbb{R}^n$ is exactly recovered as a solution of (1.20) with $\mathbf{y}=\Phi\mathbf{x}$.

Proof. As discussed in Chapter 7, it is necessary and sufficient to establish the NSP in the form

$$\|\mathbf{v}_S\|_1 < \|\mathbf{v}_{\overline{S}}\|_1 \quad \forall\mathbf{v}\in\ker\Phi\setminus\{0\}\ \text{and}\ \forall S\subseteq\{1,\dots,N\}\ \text{with card}(S)=s. \tag{8.45}$$

Given $\mathbf{v}\in\ker\Phi\setminus\{0\}$, we notice that it is enough to prove the latter for an index set S_0 of the s largest absolute value components of $\mathbf{v}$. We partition the complement of S_0 in $\{1,\dots,N\}$ as $\overline{S_0} = S_1 \cup S_2 \cup \cdots$, where

S_1 is an index set of the t largest absolute-value components of $\mathbf{v}$ in $\overline{S_0}$,

S_2 is an index set of the t next largest absolute-value components of $\mathbf{v}$ in $\overline{S_0}$, etc. Recall that $\delta_{s+t} := \delta_{s+t}^{\|\cdot\|}$. Using the MRIP (8.35), we have

$$\|\mathbf{v}_{S_0} + \mathbf{v}_{S_1}\| \le \frac{1}{1-\delta_{s+t}} \|\Phi(\mathbf{v}_{S_0} + \mathbf{v}_{S_1})\|_1 = \frac{1}{1-\delta_{s+t}} \left\| \Phi\left(-\sum_{k\ge 2} \mathbf{v}_{S_k}\right) \right\|_1$$

$$\le \frac{1}{1-\delta_{s+t}} \sum_{k\ge 2} \|\Phi\mathbf{v}_{S_k}\|_1 \le \frac{1+\delta_{s+t}}{1-\delta_{s+t}} \sum_{k\ge 2} \|\mathbf{v}_{S_k}\| \le C\frac{1+\delta_{s+t}}{1-\delta_{s+t}} \sum_{k\ge 2} \|\mathbf{v}_{S_k}\|_2.$$

For $k \ge 2$, the inequalities $|v_i| \le |v_j|$, $i \in S_k$, $j \in S_{k-1}$, averaged over j, raised to the power 2, and summed over i, yield

$$\|\mathbf{v}_{S_k}\|_2 \le \frac{1}{\sqrt{t}} \|\mathbf{v}_{S_{k-1}}\|_1.$$

We therefore have

$$\|\mathbf{v}_{S_0} + \mathbf{v}_{S_1}\| \le \frac{C}{\sqrt{t}} \frac{1+\delta_{s+t}}{1-\delta_{s+t}} \sum_{k\ge 2} \|\mathbf{v}_{S_{k-1}}\|_1 \le \frac{C}{\sqrt{t}} \frac{1+\delta_{s+t}}{1-\delta_{s+t}} \|\mathbf{v}_{\overline{S_0}}\|_1. \qquad (8.46)$$

Next, we observe that

$$\|\mathbf{v}_{S_0}\|_1 \le \sqrt{s} \|\mathbf{v}_{S_0}\|_2 \le \sqrt{s} \|\mathbf{v}_{S_0} + \mathbf{v}_{S_1}\|_2 \le \frac{\sqrt{s}}{c} \|\mathbf{v}_{S_0} + \mathbf{v}_{S_1}\|. \qquad (8.47)$$

Combining (8.46) and (8.47), we obtain

$$\|\mathbf{v}_{S_0}\|_1 \le \frac{C}{c} \sqrt{\frac{s}{t}} \frac{1+\delta_{s+t}}{1-\delta_{s+t}} \|\mathbf{v}_{\overline{S_0}}\|_1.$$

The NSP (8.45) follows from $\left(C\sqrt{s}\,(1+\delta_{s+t})\right)/\left(c\sqrt{t}\,(1-\delta_{s+t})\right) < 1$, which is just a rewriting of condition (8.44). The proof is now complete. $\qquad\square$

We finally combine all the results discussed above to prove that $m \times N$ pre-Gaussian random matrices allow for the reconstruction of all s-sparse vectors by ℓ_1 minimization with overwhelming probability provided that $m \ge c\,s\,\ln(eN/s)$. Note that the distributions of the entries of the matrix need not be related, so long as they obey the simple moment conditions stated in (8.48), which are automatically fulfilled when the entries are identically distributed.

Theorem 8.37. *Suppose the entries of a matrix $\Phi \in \mathbb{R}^{m\times N}$ are independent pre-Gaussian random variables satisfying*

$$\mathbb{E}(|a_{i,j}|) \ge \mu \qquad and \qquad \mathbb{E}(|a_{i,j}|^{2k}) \le (2k)!\,\theta^{2k}, \quad k \ge 1. \qquad (8.48)$$

Then, with probability at least

$$1 - 2\exp(-C_1 m),$$

any s-sparse vector $\mathbf{x} \in \mathbb{R}^n$ is exactly recovered as a solution of (1.20) with $\mathbf{b} = \Phi\mathbf{x}$, provided that

$$m \ge C_2\,s\,\ln(eN/s),$$

where the constants $C_1, C_2 > 0$ depend only on θ/μ.

Proof. Let $\nu_{i,j}$ denote the centered probability measure associated with the entry $a_{i,j}$, and let $\|\cdot\|$ be the norm defined in (8.34). According to Lemma 8.33, we have

$$m\frac{\mu}{\sqrt{8}} \|\mathbf{x}\|_2 \le \|\mathbf{x}\| \le m\sqrt{2}\,\theta\|\mathbf{x}\|_2 \qquad \forall \mathbf{x} \in \mathbb{R}^n.$$

Theorem 8.36 then guarantees s-sparse recovery by ℓ_1 minimization as soon as

$$\delta_{s+t}^{\|\cdot\|} < \frac{\sqrt{t/s} - 4\theta/\mu}{\sqrt{t/s} + 4\theta/\mu} \qquad \text{for some integer } t.$$

Let us choose an integer t such that $64\,(\theta/\mu)^2\,s < t \leq \left(64\,(\theta/\mu)^2 + 1\right)s$. Since then

$$\frac{\sqrt{t/s} - 4\theta/\mu}{\sqrt{t/s} + 4\theta/\mu} > \frac{8\theta/\mu - 4\theta/\mu}{8\theta/\mu + 4\theta/\mu} = \frac{1}{3},$$

s-sparse recovery by ℓ_1 minimization is guaranteed as soon as $\delta_{s+t}^{\|\cdot\|} \leq 1/3$. According to Theorems 8.34 and 8.35 with $\kappa := 1/\left(128(\theta/\mu)^2 + 16(\theta/\mu)\right)$ and $\delta = 1/3$, this is guaranteed with probability at least

$$1 - 2\exp\left(-\frac{c_1 m}{9}\right), \qquad c_1 = \frac{\kappa}{18},$$

provided that

$$m \geq 27\,c_2\,(s+t)\,\ln\left(\frac{eN}{s+t}\right), \qquad c_2 = \frac{126}{\kappa}.$$

This holds as soon as

$$m \geq 27\,c_2\,\left(64\,(\theta/\mu)^2 + 2\right)s\,\ln\left(\frac{eN}{s}\right).$$

The constants of the theorem are explicitly given by $C_1 = 1/\left(20736\,(\theta/\mu)^2 + 2592\,(\theta/\mu)\right)$ and $C_2 = \left(435456\,(\theta/\mu)^2 + 54432\,(\theta/\mu)\right)\left(64\,(\theta/\mu)^2 + 2\right)$. $\square$

8.7 ▪ Pre-Gaussian Random Matrices (II)

For a given random variable ξ we let P_ξ be the density function associated with ξ. For $\mathbf{x} = (x_1, \ldots, x_n)^\top$ a vector in $\mathbf{R}^n$, we define

$$\|\mathbf{x}\|_{\xi,q} := \left(\underbrace{\int_{-\infty}^{\infty} \cdots \int_{-\infty}^{\infty}}_{n} \left|\sum_{j=1}^{n} t_j x_j\right|^q P_\xi(t_1) \cdots P_\xi(t_n) dt_1 \cdots dt_n\right)^{1/q}, \tag{8.49}$$

where $0 < q < \infty$. We shall show that $\|\mathbf{x}\|_{\xi,q}$ behaves like a norm when $q \geq 1$, and $\|\mathbf{x}\|_{\xi,q}^q$ is a quasi-norm when $0 < q < 1$.

We will use the following well-known Khintchine inequality together with the exact constants (cf. Nazarov and Podkorytov [16]).

Lemma 8.38 (Khintchine). *Given $0 < p < 1$, for any integer N, we have*

$$A_q\left(\sum_{j=1}^{n} x_j^2\right)^{q/2} \leq \frac{1}{2^n} \sum_{\epsilon_1,\ldots,\epsilon_n = \pm 1} \left|\sum_{j=1}^{n} \epsilon_j x_j\right|^q \leq B_q\left(\sum_{j=1}^{n} x_j^2\right)^{q/2},$$

where

$$A_q := 2^{q/2-1}\min\left\{1, \frac{\Gamma((q+1)/2)}{\Gamma(3/2)}\right\}, \qquad B_q := 2^{q/2-1}\max\left\{1, \frac{\Gamma((q+1)/2)}{\Gamma(3/2)}\right\}.$$

Here and throughout this section, we shall call the quantity $\mathbb{E}(|\xi|^q)$ the qth moment of ξ for any $q \in (0, \infty)$. For simplicity, we shall denote it by σ_q. That is,

$$\sigma_q = \mathbb{E}(|\xi|^q).$$

Note that when q is an odd integer, the qth moment is different from the traditional moment, which is defined as $\mathbb{E}(\xi^q)$. Let us discuss a basic result on the quantity $\|\mathbf{x}\|_{\xi,q}$ defined in (8.49).

Lemma 8.39. *Let ξ be a random variable. Given $1 \leq q < \infty$, suppose that the qth moment of ξ is bounded. Then $\|\mathbf{x}\|_{\xi,q}$ defines a norm on $\mathbf{R}^n$. For $0 < q < 1$, $\|\mathbf{x}\|_{\xi,q}^q$ defines a quasi-norm if the second moment σ_2 of ξ is bounded.*

Proof. For any $q \in (0,\infty)$, it is readily seen that $\|\mathbf{x}\|_{\xi,q} = 0$ if and only if $\mathbf{x} = 0$ and that $\|\lambda\mathbf{x}\|_{\xi,q} = |\lambda|\|\mathbf{x}\|_{\xi,q}$ for all $\lambda \in \mathbb{R}$ and $\mathbf{x} \in \mathbf{R}^n$. Now suppose that $q \geq 1$. For $\mathbf{x}, \mathbf{y} \in \mathbf{R}^n$, using the triangle inequality for the L_p norm, we formally have

$$
\|\mathbf{x} + \mathbf{y}\|_{\xi,q}
$$

$$
= \left(\underbrace{\int_{-\infty}^{\infty} \cdots \int_{-\infty}^{\infty}}_{n} \left| \sum_{j=1}^{n} t_j(x_j + y_j) \right|^q P_\xi(t_1)\cdots P_\xi(t_n)dt_1\cdots dt_n \right)^{1/q},
$$

$$
\leq \left(\underbrace{\int_{-\infty}^{\infty} \cdots \int_{-\infty}^{\infty}}_{n} \left| \sum_{j=1}^{n} t_j x_j \right|^q P_\xi(t_1)\cdots P_\xi(t_n)dt_1\cdots dt_n \right)^{1/q}
$$

$$
+ \left(\underbrace{\int_{-\infty}^{\infty} \cdots \int_{-\infty}^{\infty}}_{n} \left| \sum_{j=1}^{n} t_j y_j \right|^q P_\xi(t_1)\cdots P_\xi(t_n)dt_1\cdots dt_n \right)^{1/q}
$$

$$
\leq \|\mathbf{x}\|_{\xi,q} + \|\mathbf{y}\|_{\xi,q}.
$$

We now show that $\|\mathbf{x}\|_{\xi,q}$ is well defined for each vector $\mathbf{x} \in \mathbf{R}^n$. When $q \geq 1$, letting σ_q be the qth moment of ξ, we have

$$
\|\mathbf{x}\|_{\xi,q} \leq \left(\underbrace{\int_{-\infty}^{\infty} \cdots \int_{-\infty}^{\infty}}_{n} \left(\sum_{j=1}^{n} |t_j||x_j| \right)^q P_\xi(t_1)\cdots P_\xi(t_n)dt_1\cdots dt_n \right)^{1/q}
$$

$$
= \|\mathbf{x}\|_1 \left(\underbrace{\int_{-\infty}^{\infty} \cdots \int_{-\infty}^{\infty}}_{n} \left(\sum_{j=1}^{n} |t_j|\frac{|x_j|}{\|\mathbf{x}\|_1} \right)^q P_\xi(t_1)\cdots P_\xi(t_n)dt_1\cdots dt_n \right)^{1/q}
$$

$$
\leq \|\mathbf{x}\|_1 \left(\sum_{j=1}^{n} \frac{|x_j|}{\|\mathbf{x}\|_1} \underbrace{\int_{-\infty}^{\infty} \cdots \int_{-\infty}^{\infty}}_{n} |t_j|^q P_\xi(t_1)\cdots P_\xi(t_n)dt_1\cdots dt_n \right)^{1/q}
$$

$$
= \|\mathbf{x}\|_1 \sigma_q^{1/q} < \infty
$$

for any given vector $\mathbf{x}$. This proves that $\|\mathbf{x}\|_{\xi,q}$ defines a norm on $\mathbf{R}^n$ for $q \geq 1$.

Similarly, we can prove that $\|\mathbf{x}\|_{\xi,q}^q$ defines a quasi-norm when $q < 1$. Indeed, we can first formally have $\|\mathbf{x}\|_{\xi,q}^q \le \|\mathbf{x}\|_{\xi,q}^q + \|\mathbf{y}\|_{\xi,q}^q$. Next, we use the Cauchy–Schwarz inequality twice to get

$$\|\mathbf{x}\|_{\xi,q}^q \le \underbrace{\int_{-\infty}^{\infty} \cdots \int_{-\infty}^{\infty}}_{n} \left(\sum_{j=1}^n |t_j|^2 \right)^{q/2} \|\mathbf{x}\|_2^q P_\xi(t_1) \cdots P_\xi(t_n) dt_1 \cdots dt_n,$$

$$\le \|\mathbf{x}\|_2^q \left(\underbrace{\int_{-\infty}^{\infty} \cdots \int_{-\infty}^{\infty}}_{n} \sum_{j=1}^n |t_j|^2 P_\xi(t_1) \cdots P_\xi(t_n) dt_1 \cdots dt_n \right)^{q/2}$$

$$\le \|\mathbf{x}\|_2^q (N\sigma_2)^{q/2} < \infty$$

for any $\mathbf{x} \in \mathbb{R}^n$. This completes the proof. □

Next we show that $\|\mathbf{x}\|_{\xi,q}$ is equivalent to the norm $\|\mathbf{x}\|_2$.

Theorem 8.40. *Suppose that P_ξ is a symmetric function, the second moment of ξ is bounded, and the qth moment is bounded when $q \ge 2$. Then there exist two constants c_q and C_q such that*

$$c_q \|\mathbf{x}\|_2^q \le \|\mathbf{x}\|_{\xi,q}^q \le C_q \|\mathbf{x}\|_2^q,$$

where c_q and C_q depend on the second and the qth moment of ξ as well as A_q and B_q in Theorem 8.38.

Proof. Let us estimate $\|\mathbf{x}\|_{\xi,q}^q$ in terms of $\|\mathbf{x}\|_2^q$. There are many ways to do this. We would like to have the constants in the estimate be independent of N. To this end, we assume that P_ξ is a symmetric function with respect to 0, and that the qth moment of ξ if $q \ge 1$ is bounded together with a bounded second moment. Under these assumptions, we decompose the integral into 2^n parts and rewrite them in the following fashion:

$$\|\mathbf{x}\|_{\xi,q}^q$$
$$= \sum_{\epsilon_1,\ldots,\epsilon_n=\pm 1} \epsilon_1 \cdots \epsilon_n \underbrace{\int_0^{\epsilon_1 \infty} \cdots \int_0^{\epsilon_n \infty}}_{n} \left| \sum_{j=1}^n t_j x_j \right|^q P_\xi(t_1) \cdots P_\xi(t_n) dt_1 \cdots dt_n$$

$$= \underbrace{\int_0^{\infty} \cdots \int_0^{\infty}}_{n} \sum_{\epsilon_1,\ldots,\epsilon_n=\pm 1} \left| \sum_{j=1}^n \epsilon_j t_j x_j \right|^q P_\xi(t_1) \cdots P_\xi(t_n) dt_1 \cdots dt_n.$$

We now call upon Khintchine's inequality to get

$$A_q 2^n \underbrace{\int_0^{\infty} \cdots \int_0^{\infty}}_{n} \left| \sum_{j=1}^n t_j^2 x_j^2 \right|^{q/2} P_\xi(t_1) \cdots P_\xi(t_n) dt_1 \cdots dt_n \tag{8.50}$$

$$\le \|\mathbf{x}\|_{\xi,q}^q \le B_q 2^n \underbrace{\int_0^{\infty} \cdots \int_0^{\infty}}_{n} \left| \sum_{j=1}^n t_j^2 x_j^2 \right|^{q/2} P_\xi(t_1) \cdots P_\xi(t_n) dt_1 \cdots dt_n.$$

Let us focus on bounding the right-hand side of (8.50) first. If $q < 2$, we use the increasing property of the L_q norm to derive

$$\|\mathbf{x}\|_{\xi,q}^q \leq B_q \left(2^n \underbrace{\int_0^\infty \cdots \int_0^\infty}_{n} \sum_{j=1}^n t_j^2 x_j^2 P_\xi(t_1) \cdots P_\xi(t_n) dt_1 \cdots dt_n \right)^{q/2}$$

$$= B_q \left(\sum_{j=1}^n x_j 2^n \underbrace{\int_0^\infty \cdots \int_0^\infty}_{n} t_j^2 P_\xi(t_1) \cdots P_\xi(t_n) dt_1 \cdots dt_n \right)^{q/2}$$

$$= B_q \|\mathbf{x}\|_2^q \sigma_2^{q/2},$$

where σ_2 is the second moment of ξ. If $q \geq 2$, then we can use the convexity of the function $t^{q/2}$ to derive

$$\|\mathbf{x}\|_{\xi,q}^q \leq B_q \|\mathbf{x}\|_2^q 2^n \underbrace{\int_0^\infty \cdots \int_0^\infty}_{n} \left(\sum_{j=1}^n \frac{x_j^2}{\|\mathbf{x}\|_2^2} t_j^2 \right)^{q/2} P_\xi(t_1) \cdots P_\xi(t_n) dt_1 \cdots dt_n$$

$$\leq B_q \|\mathbf{x}\|_2^q 2^n \underbrace{\int_0^\infty \cdots \int_0^\infty}_{n} \sum_{j=1}^n \frac{x_j^2}{\|\mathbf{x}\|_2^2} t_j^q P_\xi(t_1) \cdots P_\xi(t_n) dt_1 \cdots dt_n$$

$$= B_q \|\mathbf{x}\|_2^q \sigma_q,$$

where σ_q denotes the qth moment of ξ. Putting the two cases together, we obtain

$$\|\mathbf{x}\|_{\xi,q}^q \leq C_q \|\mathbf{x}\|_2^q,$$

where $C_q = B_q \sigma_q$ if $q \geq 2$ and $C_q = B_q \sigma_2^{q/2}$ if $q < 2$.

Next we bound the left-hand side of (8.50). We again separate into two cases. If $q \geq 2$, we use the increasing property of the L_q norm to derive

$$\|\mathbf{x}\|_{\xi,q}^q \geq A_q \left(2^n \underbrace{\int_0^\infty \cdots \int_0^\infty}_{n} \sum_{j=1}^n t_j^2 x_j^2 P_\xi(t_1) \cdots P_\xi(t_n) dt_1 \cdots dt_n \right)^{q/2}$$

$$= A_q \left(\sum_{j=1}^n x_j 2^n \underbrace{\int_0^\infty \cdots \int_0^\infty}_{n} t_j^2 P_\xi(t_1) \cdots P_\xi(t_n) dt_1 \cdots dt_n \right)^{q/2}$$

$$= A_q \|\mathbf{x}\|_2^q \sigma_2^{q/2}.$$

For the case $q < 2$, we use the concavity of the function $t^{q/2}$ to get

$$\|\mathbf{x}\|_{\xi,q}^q \geq A_q \|\mathbf{x}\|_2^q 2^n \underbrace{\int_0^\infty \cdots \int_0^\infty}_{n} \left(\sum_{j=1}^n \frac{x_j^2}{\|\mathbf{x}\|_2^2} t_j^2 \right)^{q/2} P_\xi(t_1) \cdots P_\xi(t_n) dt_1 \cdots dt_n$$

$$\geq A_q \|\mathbf{x}\|_2^q 2^n \underbrace{\int_0^\infty \cdots \int_0^\infty}_{n} \sum_{j=1}^n \frac{x_j^2}{\|\mathbf{x}\|_2^2} t_j^q P_\xi(t_1) \cdots P_\xi(t_n) dt_1 \cdots dt_n$$

$$= A_q \|\mathbf{x}\|_2^q \sigma_q.$$

Thus, $\|\mathbf{x}\|_{\xi,q}^q \geq A_q \|\mathbf{x}\|_2^q \sigma_q$. We can summarize the two cases to obtain

$$\|\mathbf{x}\|_{\xi,q}^q \geq c_q \|\mathbf{x}\|_2^q,$$

with $c_q = A_q \sigma_q$ if $q < 2$ and $c_q = A_q \sigma_2^{q/2}$ if $q \geq 2$. This completes the proof. $\square$

In particular, when $q = 2$, we know that $A_2 = B_2 = 1$. In this case, $C_2 = \sigma_2 = c_2$. We have

$$\|\mathbf{x}\|_{\xi,2}^2 = \sigma_2 \|\mathbf{x}\|_2^2.$$

Let us give a few examples to show the constants c_q and C_q.

Example 8.41. For a Gaussian random variable $\xi \in N(0, \sigma^2)$ with density function P_ξ, we can use a change of variables to get

$$\|\mathbf{x}\|_{\xi,q}^q = \|\mathbf{x}\|_2^q \underbrace{\int_{-\infty}^\infty \cdots \int_{-\infty}^\infty}_{n} |u_1|^q P_\xi(u_1) \cdots P_\xi(u_n) du_1 du_2 \cdots du_n$$

$$= \frac{2\|\mathbf{x}\|_2^q}{\sqrt{2\pi\sigma^2}} \int_0^\infty u^q \exp\left(-\frac{u^2}{2\sigma^2}\right) du$$

$$= \frac{2^{q/2}\sigma^q}{\sqrt{\pi}} \Gamma\left(\frac{q+1}{2}\right) \|\mathbf{x}\|_2^q.$$

That is, we have

$$c_q = C_q = (\sqrt{2}\sigma)^q \frac{\Gamma(\frac{q+1}{2})}{\sqrt{\pi}}.$$

Example 8.42. For a uniform random variable, we have

$$\|\mathbf{x}\|_{\xi,q}^q = \frac{1}{2^n} \underbrace{\int_{-1}^1 \cdots \int_{-1}^1}_{n} \left| \sum_{j=1}^n t_j x_j \right|^q dt_1 \cdots dt_n$$

$$= \frac{1}{2^n} \sum_{\substack{\epsilon_j = \pm 1 \\ j=1,\ldots,N}} \underbrace{\int_0^1 \cdots \int_0^1}_{n} \left| \sum_{j=1}^n \epsilon_j t_j x_j \right|^q dt_1 \cdots dt_n,$$

which can be estimated from below and above using Khintchine's inequality. In fact, we have

$$c_q \|\mathbf{x}\|_2^q \le \|\mathbf{x}\|_{\xi,q}^q \le C_q \|\mathbf{x}\|_2^q,$$

with $c_q = 2A_q/(1+q)$ if $q \le 2$, $c_q = A_q(2/3)^{q/2}$ if $q > 2$, $C_q = 2B_q/(1+q)$ if $q > 2$, and $C_q = B_q(2/3)^{q/2}$ if $q \le 2$, where A_q and B_q are two constants in Khintchine's inequality.

Example 8.43. For a Bernoulli random variable, we have

$$\|\mathbf{x}\|_{\xi,q}^q = \frac{1}{2^n} \sum_{\substack{\epsilon_j = \pm 1 \\ j=1,\dots,N}} \left| \sum_{j=1}^{n} \epsilon_j x_j \right|^q. \tag{8.51}$$

Khintchine's inequality in the previous section gives both bounds. That is,

$$A_q \|\mathbf{x}\|_2^q \le \|\mathbf{x}\|_{\xi,q}^q \le B_q \|\mathbf{x}\|_2^q,$$

with constants A_q and B_q given in Theorem 8.38.

Recall that $X_i := |\sum_{i=1}^{n} \phi_{ij} x_j|$ is a new random variable, and let

$$\nu_i = \mathbb{E}(X_i^q) \quad \text{and} \quad \Xi_i := X_i^q - \nu_i. \tag{8.52}$$

Then Ξ_i is of mean zero and

$$\|\Phi\mathbf{x}\|_q^q - m\|\mathbf{x}\|_{\xi,q}^q = \sum_{i=1}^{m}(X_i^q - \nu_i) = \sum_{i=1}^{m} \Xi_i \tag{8.53}$$

since

$$\mathbb{E}(X_i^q) = \int_{-\infty}^{\infty} \cdots \int_{-\infty}^{\infty} \left| \sum_{j=1}^{n} t_j x_j \right|^q P_\xi(t_1)dt_1 \cdots P_\xi(t_n)dt_n = \|\mathbf{x}\|_{\xi,q}^q$$

for $0 < q < \infty$. We see that this expectation is independent of i. We shall write $X = X_i$ and $\nu = \nu_i$ for all $i = 1, \dots, m$ for convenience.

Definition 8.44. *We say that the matrix* Φ *satisfies the modified restricted q-isometry property (MRqIP) (8.54) if there exists an integer $s > 0$ and a positive $\delta_s < 1$ such that*

$$(1 - \delta_s)m\|\mathbf{x}_T\|_q^q \le \|\Phi\mathbf{x}_T\|_q^q \le (1 + \delta_s)m\|\mathbf{x}_T\|_q^q \tag{8.54}$$

for any index set $T \subset \{1, 2, 3, \dots, n\}$ *with* $\#(T) \le s$.

When $q = 2$, we recall that $\|\mathbf{x}\|_q = \sigma\|\mathbf{x}\|_2$. The MRqIP in (8.54) is the standard RIP when $m\sigma^2 = 1$.

As before, we would like to have an exponential inequality like the one in (8.56) below.

We begin with the following inequality, which was proved in section 8.4. We now give another proof.

Lemma 8.45. *Let* Ξ_i *be i.i.d. random variables with mean zero. Suppose that there exists a constant $K \ge 0$ such that*

$$\mathbb{E}(|\Xi_i|^k) \le (k-1)!!(K\nu)^k \tag{8.55}$$

for all $k \geq 2$. Then for any $\epsilon \in (0, 1]$,

$$\mathbf{P}\left(\left|\|A\mathbf{x}\|_q^q - m\|\mathbf{x}\|_{\xi,q}^q\right| > \epsilon m \nu\right) \leq 2\exp\left(-\frac{\epsilon^2 m}{6K^2}\right). \tag{8.56}$$

Proof. Using the Chebyshev–Markov inequality, we have

$$P\left(\left|\sum_{i=1}^m \Xi_i\right| > umt\right) \leq \frac{\mathbb{E}(\exp(u\sum_{j=1}^m \Xi_j))}{\exp(umt)} \leq \prod_{j=1}^m \frac{\mathbb{E}(\exp(u\Xi_i))}{\exp(ut)}. \tag{8.57}$$

Since $\mathbb{E}(\Xi_i) = 0$, we have

$$\mathbb{E}(\exp(u\Xi))$$
$$= \mathbb{E}\left(\sum_{k=0}^\infty \frac{(u\Xi)^k}{k!}\right) = 1 + \sum_{k=2}^\infty \frac{\mathbb{E}((u\Xi)^k)}{k!} \leq 1 + \sum_{k=2}^\infty \frac{(k-1)!!}{k!}(uK\nu)^k$$
$$\leq 1 + \sum_{k=1}^\infty \frac{(uK\nu)^{2k}}{(2k)!!} + \sum_{k=1}^\infty \frac{(uK\nu)^{2k+1}}{(2k+1)!!}$$
$$\leq 1 + \sum_{k=1}^\infty \frac{(uK\nu)^{2k}}{2^k k!} + \frac{(uK\nu)^3}{3}\sum_{k=1}^\infty \frac{(uK\nu)^{2k-2}}{2^{k-1}(k-1)!}$$
$$= \exp\left(\frac{(uK\nu)^2}{2}\right) + \frac{(uK\nu)^3}{3}\exp\left(\frac{(uK\nu)^2}{2}\right).$$

It then follows that

$$\frac{\mathbb{E}(\exp(u\Xi_i))}{\exp(ut)} = \left(1 + \frac{(uK\nu)^3}{3}\right)\exp\left(\frac{(uK\nu)^2}{2} - ut\right).$$

By choosing $t = \epsilon\nu$ and $u = \epsilon/(K^2\nu)$, we can simplify the above to get

$$\frac{\mathbb{E}(\exp(u\Xi_i))}{\exp(ut)} \leq \left(1 + \frac{\epsilon^3}{3K^3}\right)\exp\left(-\frac{\epsilon^2}{2K^2}\right) \leq \exp\left(-\frac{\epsilon^2}{2K^2}\left(1 - \frac{2\epsilon}{3K}\right)\right).$$

Since $1 - \frac{2\epsilon}{3K} \geq \frac{1}{3}$, we summarize the above to get

$$\mathbf{P}\left(\sum_{i=1}^m \Xi_i > m\epsilon\nu\right) \leq \exp\left(-\frac{\epsilon^2 m}{6K^2}\right).$$

Replacing Ξ_i by $-\Xi_i$, we likewise obtain

$$\mathbf{P}\left(\sum_{i=1}^m \Xi_i < -m\epsilon\nu\right) = \mathbf{P}\left(\sum_{i=1}^m (-\Xi_i) > m\epsilon\nu\right) \leq \exp\left(-\frac{\epsilon^2 m}{6K^2}\right).$$

As a result, we derive

$$\mathbf{P}\left(\left|\sum_{i=1}^m \Xi_i\right| > m\epsilon\nu\right) \leq 2\exp\left(-\frac{\epsilon^2 m}{6K^2}\right).$$

The above arguments complete the proof.   □

We are now ready to prove the main result in this section.

Theorem 8.46. *Let ξ be a random variable with symmetric density function P_ξ. Suppose that $q \in (0, 1]$. Suppose that the lower and upper bounds c_q and C_q in Theorem 8.40 imply that there is a positive constant α such that*

$$\frac{C_{qj}}{c_q^j} \leq (j-1)!!\alpha^j \tag{8.58}$$

for all integers j with $j \geq 1$. Then

$$\mathbf{P}\left(\left|\|A\mathbf{x}\|_q^q - m\|\mathbf{x}\|_{\xi,q}^q\right| > m\epsilon\|\mathbf{x}\|_{\xi,q}^q\right) \leq 2\exp\left(-\frac{\epsilon^2 m}{6(1+\alpha)^2}\right).$$

Proof. By the definition,

$$\|\mathbf{x}\|_{\xi,q}^q = \int_{-\infty}^\infty M_\xi(u;\mathbf{x})|u|^q du.$$

For convenience, let $\nu = \|\mathbf{x}\|_{\xi,q}^q$. We use the basic properties of M_ξ to get

$$\mathbb{E}(|\Xi_i|^k) = \int_{-\infty}^\infty M_\xi(u,\mathbf{x})\left(|u|^q - \nu\right)^k du. \tag{8.59}$$

Let us assume that P_ξ is symmetric so that we can use Theorem 8.40. From (8.59) we use the estimates in Theorem 8.40 to get

$$\begin{aligned}
\mathbb{E}(|\Xi_i|^k) &= \sum_{j=0}^k \binom{k}{j}(-1)^j \int_{-\infty}^\infty M_\xi(u,\mathbf{x})|u|^{qj}\nu^{(k-j)}du \\
&\leq \sum_{j=0}^k \binom{k}{j}\|\mathbf{x}\|_{\xi,qj}^{qj}\|\mathbf{x}\|_{\xi,q}^{q(k-j)} \\
&\leq \sum_{j=0}^k \binom{k}{j}\frac{C_{qj}}{c_q^j}\|\mathbf{x}\|_{\xi,q}^{qk}.
\end{aligned}$$

The assumption in Theorem 8.46 implies that

$$\mathbb{E}(|\Xi|^k) \leq \sum_{j=0}^k \binom{k}{j}(j-1)!!\alpha^j\|\mathbf{x}\|_{\xi,q}^{qk} \leq (k-1)!!(1+\alpha)^k\|\mathbf{x}\|_{\xi,q}^{qk}.$$

Then Lemma 8.45 can be applied to conclude that

$$\mathbf{P}\left(\left|\sum_{i=1}^m \Xi_i\right| > m\epsilon\|\mathbf{x}\|_{\xi,q}^q\right) \leq 2\exp\left(-\frac{\epsilon^2 m}{6(1+\alpha)^2}\right). \qquad \square$$

Example 8.47. In particular, when ξ is a Gaussian random variable in $N(0,\sigma)$, it follows from Example 8.41 that

$$C_{qj} = 2^{qj/2}\sigma^{qj}\frac{\Gamma((qj+1)/2)}{\sqrt{\pi}} \leq 2^{qj/2}\sigma^{qj}(j-1)!!/2^{j/2}$$

and $c_q = 2^{q/2}\sigma^q\Gamma((q+1)/2)/\sqrt{\pi}$. It follows that

$$\frac{C_{qj}}{c_q^j} \leq (j-1)!!\left(\frac{\alpha}{\sqrt{2}}\right)^j$$

for $\alpha = \max_{p \in (0,1]} \left(\frac{\sqrt{\pi}}{\Gamma((p+1)/2)} \right) \le \sqrt{\pi}$. That is, we can apply Theorem 8.46 to Gaussian random variables to conclude the exponential inequality of the tail probability.

Example 8.48. In addition to Gaussian random variables, we also consider sub-Gaussian random variables, which have zero mean and

$$P(|\xi| \ge t) \le 2 \exp\left(-\frac{t^2}{B}\right)$$

for a constant $B > 0$. It is known that $\sigma_{2j} = \mathbb{E}(|\xi|^{2j}) \le (2j-1)!!b^{2j}$ for some positive number b and $\sigma_q = \mathbb{E}(|\xi|^q) < \infty$ (cf. [4]). Using the Cauchy–Schwarz inequality and induction, we can see that

$$\mathbb{E}(|\xi|^{2j+1}) \le (\mathbb{E}(|\xi|^{2j})\mathbb{E}(|\xi|^{2j+2}))^{1/2}$$
$$\le (2j-1)!!b^{2j+1}\sqrt{2j+1} \le (2j)!!b^{2j+1}.$$

Thus, for all $j \ge 1$, $\sigma_j = \mathbb{E}(|\xi|^j) \le (j-1)!!b^j$. Note that the uniform random variable is a typical example of sub-Gaussian random variables, and so is the Bernoulli random variable (cf. [4]).

When $q < 1/3$, we know that $c_q = \Phi_q \sigma_q = 2^{q/2-1}\sigma_q$ since $\Gamma((q+1)/2)/\Gamma(3/2) \ge 1$. When $qj \ge 2$, we have

$$C_{qj} \le B_{qj}\sigma_{qj} \le 2^{qj/2}\Gamma((qj+1)/2)/\sqrt{\pi}\sigma_{qj}$$
$$\le 2^{qj/2}[qj]!!/2^{([qj]+1)/2}([qj])!!b^{qj+1} \le ([qj]!!)^2 b^j,$$

where $[x]$ stands for the largest integer less than or equal to x. Note that $[qj]!! \le [j/2]!!$ and $([qj]!!)^2 \le (j-1)!!$ when $qj \ge 2$. Then we have

$$\frac{C_{qj}}{c_q^j} \le \left(\frac{2b}{\sigma_q}\right)^j (j-1)!!.$$

For $qj < 2$, we have

$$C_{qj} \le 2^{qj/2}\Gamma((qj+1)/2)/\sqrt{\pi}\sigma_2^{qj/2} \le 2^{qj/2}(b^2)^{qj/2} = 2^{qj/2}b^{qj}$$

and

$$\frac{C_{qj}}{c_q^j} \le \left(\frac{2b^q}{\sigma_q}\right)^j \le \left(\frac{2b}{\sigma_q}\right)^j (j-1)!!.$$

Then, when $q \le 1/3$, we can apply Theorem 8.46 to sub-Gaussian random variables to conclude the exponential inequality of the desired tail probability.

Example 8.49. Furthermore, we consider pre-Gaussian random variables. That is, a random variable ξ is called (strictly) pre-Gaussian if there exist positive numbers Λ and a such that

$$\mathbb{E}(\exp(\lambda\xi)) \le \exp\left\{\frac{\lambda^2 a^2}{2}\right\}$$

for all $\lambda \in (-\Lambda, \Lambda)$. It is known (cf. [4]) that a random variable ξ is pre-Gaussian if and only if there exists a positive constant b such that

$$\sigma_k = \mathbb{E}(|\xi|^k) \le k!b^k \quad \forall k \ge 1.$$

Our Theorem 8.46 can be applied to pre-Gaussian random variables when $q < 1/4$. Indeed, following the ideas in Example 8.48, we have, for $qj \geq 2$,

$$C_{qj} \leq B_{qj}\sigma_{qj} \leq 2^{qj/2}[qj]!!([qj]+1)!b^{qj+1} \leq 2^{qj/2}(j-1)!b^j,$$

where we have used the fact that when $q < 1/4$, $([qj]+1)! \leq (2[qj])!!$, and hence $[qj]!!([qj]+1)! \leq [qj]!!(2[qj])!! \leq (j-1)!!$. Therefore, we have

$$\frac{C_{qj}}{c_q^j} \leq \left(\frac{2b}{\sigma_q}\right)^j (j-1)!!.$$

When $qj < 2$, the estimate for $\frac{C_{qj}}{c_q^j}$ is the same as in the proof of Example 8.48. Then we apply Theorem 8.46 to get a desirable exponential inequality.

In this section, we provide a sufficient condition for every s-sparse vector $\mathbf{x} \in \mathbf{R}^n$ of linear system $\Phi\mathbf{x} = \mathbf{b}$ to be recovered by ℓ_q nonconvex minimization, where $\Phi = [\phi_{ij}]_{1 \leq i \leq m, 1 \leq j \leq n}$ is a rectangular matrix with ϕ_{ij} being i.i.d. copies of a random variable ξ. We shall mainly show that when the random variable ξ satisfies certain conditions, the matrix Φ satisfies the MRqIP with overwhelming probability provided that the number m of measurements is large enough.

Theorem 8.50. *Under the assumptions in Theorem 8.46, there exist constants k_1, k_2, k_3 dependent on $\delta \in (0,1)$ and $q \in (0,1)$ such that*

$$P\left(\left|\|A\mathbf{x}\|_q^q - m\|\mathbf{x}\|_{\xi,q}^q\right| > m\delta\|\mathbf{x}\|_{\xi,q}^q \; \forall\mathbf{x} \text{ with } \|\mathbf{x}\|_0 \leq s\right)$$
$$\leq 2\exp(-mk_1\delta)$$

provided that

$$m > k_2 s + k_3 s \log(eN/s). \tag{8.60}$$

Proof. We mainly use the ideas in the proof of Theorem 8.17 and adapt the proof when dealing with $\|\mathbf{x}\|_q$. In particular, we shall use Lemma 8.52 (see Exercise 8), whose proof is similar to that of Lemma 8.18. Let $S \subset \{1, 2, \ldots, N\}$ be a fixed index subset of cardinality $\#(S) = s$, and let $\mathbf{R}^S$ denote the space of vectors in $\mathbf{R}^n$ supported on S. According to Lemma 8.52, we can find a subset $\mathcal{U}$ of the unit sphere $\mathbf{B}_S$ of $\mathbf{R}^S$ related to the quasi-norm $\|\mathbf{x}\|_{\xi,q}^q$ such that

$$\min_{\mathbf{u} \in \mathcal{U}} \|\mathbf{z} - \mathbf{u}\|_{\xi,q}^q \leq \frac{\delta}{4} \quad \forall\mathbf{z} \in \mathbf{B}_S,$$

with $\#(\mathcal{U}) \leq (1 + 8/\delta)^{s/q}$. By using the assumption in Theorem 8.46, we have

$$P\left(\left|\|\Phi\mathbf{x}\|_q^q - m\|\mathbf{x}\|_{\xi,q}^q\right| > m\delta/2\|\mathbf{x}\|_{\xi,q}^q \; \forall\mathbf{x} \text{ in } \mathcal{U}\right)$$
$$\leq 2\#(\mathcal{U})\exp(-m\delta^2/(8(\delta K + 4K^2)))$$
$$\leq 2(1 + 8/\delta)^{s/q}\exp(-m\delta^2/(8(\delta K + 4K^2)))$$
$$\leq 2\exp\left(-\frac{m\delta^2}{8K(\delta + 4K)} + \frac{8s}{q\delta}\right).$$

That is, with overwhelming probability (if m satisfies the assumption (8.60)) with $k_2 = \frac{64K(\delta+4K)}{\delta^3}$),

$$\|\Phi\mathbf{x}\|_q^q - m\|\mathbf{x}\|_{\xi,q}^q < m\delta\|\mathbf{x}\|_{\xi,q}^q \; \forall\mathbf{x} \in \mathcal{U} \subset \mathbf{B}_S.$$

We claim that with the same overwhelming probability, the above equality holds for all $\mathbf{x} \in \mathbf{B}_S$. Indeed, following the ideas in [1], let $\widetilde{\delta}$ be the smallest number such that

$$\|\Phi\mathbf{x}\|_q^q < m(1+\widetilde{\delta})\|\mathbf{x}\|_{\xi,q}^q \tag{8.61}$$

for all $\mathbf{x} \in \mathbf{B}_S$. For each $\mathbf{x} \in \mathbf{B}_S$, let $\mathbf{u} \in \mathcal{U}$ such that $\|\mathbf{x}-\mathbf{u}\|_{\xi,q}^q \leq \delta/4$, and we have

$$\|\Phi\mathbf{x}\|_q^q \leq \|\Phi\mathbf{u}\|_q^q + \|\Phi(\mathbf{x}-\mathbf{u})\|_q^q \leq (1+\delta/2)m + m(1+\widetilde{\delta})\|\mathbf{x}-\mathbf{u}\|_{\xi,q}^q$$

$$\leq m(1 + \frac{\delta}{2} + \left(1+\widetilde{\delta}\right)\frac{\delta}{4}).$$

The minimality of $\widetilde{\delta}$ implies that

$$1+\widetilde{\delta} \leq (1+\delta/2)m + m(1+\widetilde{\delta})\|\mathbf{x}-\mathbf{u}\|_{\xi,q}^q \leq m\left(1+\frac{\delta}{2}+(1+\widetilde{\delta})\frac{\delta}{4}\right) \leq 1+\frac{3\delta}{4}+\frac{\widetilde{\delta}}{4},$$

with the assumption that $\delta \leq 1$. It follows that $\widetilde{\delta} \leq \delta$. By (8.61), this means that

$$\|\Phi\mathbf{x}\|_q^q < m(1+\delta)\|\mathbf{x}\|_{\xi,q}^q$$

for all $\mathbf{x} \in \mathbf{B}_S$ with the same overwhelming probability.

On the other hand, given $\mathbf{x} \in \mathbf{B}_S$, we can pick up a $\mathbf{u} \in \mathcal{U}$ such that $\|\mathbf{x}-\mathbf{u}\|_{\xi,q}^q \leq \delta/4$. Then

$$\|\Phi\mathbf{x}\|_q^q \geq \|\Phi\mathbf{u}\|_q^q - \|\Phi(\mathbf{x}-\mathbf{u})\|_q^q \geq \left(1-\frac{\delta}{2}\right)m - (1+\delta)m\|\mathbf{x}-\mathbf{u}\|_{\xi,q}^q$$

$$\geq \left(1-\frac{\delta}{2}-(1+\delta)\frac{\delta}{4}\right)m \geq (1-\delta)m.$$

Thus, we have obtained

$$(1-\delta)m\|\mathbf{x}\|_{\xi,q}^q \leq \|\Phi\mathbf{x}\|_q^q \leq (1+\delta)m\|\mathbf{x}\|_{\xi,q}^q$$

for all $\mathbf{x} \in \mathbf{B}_S$ with the same overwhelming probability. That is,

$$P\left(\left|\|\Phi\mathbf{x}\|_q^q - m\|\mathbf{x}\|_{\xi,q}^q\right| > m\delta\|\mathbf{x}\|_{\xi,q}^q \; \forall \mathbf{x} \text{ in } \mathbf{B}_S\right)$$

$$\leq 2\exp\left(-\frac{m\delta^2}{8K(\delta+4K)}+\frac{8s}{q\delta}\right).$$

Note that there are $\binom{N}{s}$ subsets $\mathbf{B}_S \subset \mathbb{R}^n$. Thus,

$$P\left(\left|\|\Phi\mathbf{x}\|_q^q - m\|\mathbf{x}\|_{\xi,q}^q\right| > m\delta\|\mathbf{x}\|_{\xi,q}^q \; \forall \mathbf{x} \text{ with } \|\mathbf{x}\|_0 \leq s\right)$$

$$\leq 2\binom{N}{s}\exp\left(-\frac{m\delta^2}{8K(\delta+4K)}+\frac{8s}{q\delta}\right).$$

Since $\binom{N}{s} \leq (eN/s)^s$, the term on the right-hand side yields

$$\exp\left(-\frac{m\delta^2}{8K(\delta+4K)}+\frac{8s}{q\delta}+s\log(eN/s)\right).$$

If

$$m > k_2 s + k_3 s \log(eN/s) \quad \text{with } k_2 = \frac{64K(\delta + 4K)}{q\delta^3}, \ k_3 = \frac{64K(\delta + 4K)}{\delta^2},$$

we have

$$P\left(\left|\|\Phi\mathbf{x}\|_q^q - m\|\mathbf{x}\|_{\xi,q}^q\right| > m\delta\|\mathbf{x}\|_{\xi,q}^q \ \forall \mathbf{x} \text{ with } \|\mathbf{x}\|_0 \leq s\right)$$
$$\leq 2\exp(-mk_1\delta^2),$$

with a positive constant k_1 dependent on δ and s. $\square$

We now work on sparse solution recovery by using the nonconvex minimization (1.27) in Chapter 1.

Theorem 8.51. *Suppose that the density function P_ξ is symmetric and ξ has all bounded moments. Let t be an integer > 0 such that the modified restricted q-isometry constant δ_t satisfies*

$$\delta_t \leq \frac{K_q - 1}{K_q + 1}, \ \text{with } K_q = 2\frac{c_q}{C_q}\left(\frac{t}{s}\right)^{1-q/2}. \tag{8.62}$$

Then any s-sparse vector $\mathbf{x}^*$ is exactly recovered as a solution of the minimization (1.27).

Proof. Let $\mathbf{x}$ be a solution of the minimization (1.27), and let $\mathbf{x}^*$ be the exact solution. We consider an arbitrary index set $S =: S_0$ with $|S| \leq s$. Let $\mathbf{v}$ be a vector in $\mathbb{R}^n$, which is an element of the null space of A. For instance, we will take $\mathbf{v} := \mathbf{x} - \mathbf{x}^*$. Fix $t > s$. We partition the complement of S in $\{1, \ldots, N\}$ as $\overline{S} = S_1 \cup S_2 \cup \cdots$, where

$$S_1 := \{\text{indices of the } t \text{ largest absolute value components of } \mathbf{v} \text{ in } \overline{S}\},$$
$$S_2 := \{\text{indices of the next } t \text{ largest absolute value components of } \mathbf{v} \text{ in } \overline{S}\},$$
$$\vdots$$

We first observe that by Theorem 8.40,

$$\|\mathbf{v}_{S_0}\|_{\xi,q}^q \leq \frac{1}{m(1-\delta_s)}\|A(\mathbf{v}_{S_0})\|_2^2$$
$$= \frac{1}{m(1-\delta_s)}\|A(-\mathbf{v}_{S_1} - \mathbf{v}_{S_2} - \cdots)\|$$
$$\leq \frac{1}{m(1-\delta_s)}\left(\|A(\mathbf{v}_{S_1})\|_q^q + \|A(\mathbf{v}_{S_2})\|_q^q + \cdots\right)$$
$$\leq \frac{1+\delta_t}{(1-\delta_s)}\left[\|\mathbf{v}_{S_1}\|_{\xi,q}^q + \|\mathbf{v}_{S_2}\|_{\xi,q}^q + \cdots\right]$$
$$\leq \frac{(1+\delta_t)}{(1-\delta_s)}C_q\left[\|\mathbf{v}_{S_1}\|_q^q + \|\mathbf{v}_{S_2}\|_q^q + \cdots\right].$$

It is somewhat classic to observe that

$$\|\mathbf{v}_{S_k}\|_2 \leq \frac{1}{t^{1/q-1/2}}\|\mathbf{v}_{S_{k-1}}\|_q$$

for all $k \geq 1$ due to the definition of S_k. Taking into account that

$$\|\mathbf{v}_{S_0}\|_q^q \leq s^{1-q/2}\|\mathbf{v}_{S_0}\|_2^q \leq \frac{s^{1-q/2}}{c_q}\|\mathbf{v}_{S_0}\|_{\xi,q}^q$$

by using Theorem 8.40, we have

$$\|\mathbf{v}_{S_0}\|_q^q \leq \frac{s^{1-q/2}}{c_q} \|\mathbf{v}_{S_0}\|_{\xi,q}^q \leq \frac{(1+\delta_t)}{(1-\delta_s)} \frac{C_q}{c_q} \frac{s^{1-q/2}}{t^{1-q/2}} \left[\|\mathbf{v}_{S_0}\|_q^q + \|\mathbf{v}_{S_1}\|_q^q + \cdots \right]$$

$$= \frac{(1+\delta_t)}{(1-\delta_s)} \frac{C_q}{c_q} \frac{s^{1-q/2}}{t^{1-q/2}} \left(\|\mathbf{v}_{S_0}\|_q^q + \|\mathbf{v}_{\overline{S}_0}\|_q^q \right),$$

where $\overline{S}_0$ denotes the complement of the set S_0 in $\{1, 2, \ldots, N\}$. Note that the assumption (8.62) implies that $\|\mathbf{v}_{S_0}\|_q^q < \|\mathbf{v}_{\overline{S}_0}\|_q^q$.

Now let $S = S_0$ be specified as the set of indices of the s largest absolute value components of $\mathbf{x}$, and let $\mathbf{v}$ be specified as $\mathbf{v} := \mathbf{x} - \mathbf{x}^*$. Because $\mathbf{x}^*$ is a minimizer of (1.27), we have

$$\|\mathbf{x}\|_q^q \geq \|\mathbf{x}^*\|_q^q, \qquad \text{i.e.,} \qquad \|\mathbf{x}_S\|_q^q + \|\mathbf{x}_{\overline{S}}\|_q^q \geq \|\mathbf{x}_S^*\|_q^q + \|\mathbf{x}_{\overline{S}}^*\|_q^q.$$

By the triangle inequality of the quasi-norm, we obtain

$$\|\mathbf{x}_S\|_q^q + \|\mathbf{x}_{\overline{S}}\|_q^q \geq \|\mathbf{x}_S\|_q^q - \|\mathbf{v}_S\|_q^q + \|\mathbf{v}_{\overline{S}}\|_q^q - \|\mathbf{x}_{\overline{S}}\|_q^q.$$

Rearranging the latter yields the inequality

$$\|\mathbf{v}_{\overline{S}}\|_q^q \leq 2 \|\mathbf{x}_{\overline{S}}\|_q^q + \|\mathbf{v}_S\|_q^q = 2 \|\mathbf{x}_{\overline{S}}\|_q^q + \|\mathbf{v}_S\|_q^q < 2 \|\mathbf{x}_{\overline{S}}\|_q^q + \|\mathbf{v}_{\overline{S}}\|_q^q. \qquad (8.63)$$

If the support of $\mathbf{x}$ is S, then $\mathbf{x}_{\overline{S}} = 0$, and the above equation is a contradiction. That is, $\mathbf{v}$ has to be a zero vector. $\square$

8.8 ▪ Remarks

Remark 8.8.1 (Fourier and Other Structured Random Matrices). *Often the design of the measurement matrix is subject to physical or other constraints of the application, or it is actually given to us without having the freedom to design anything, and therefore it is often not justifiable to assume that the sensing matrix follows a Gaussian or Bernoulli distribution. Moreover, Gaussian or other unstructured matrices have the disadvantage that no fast matrix multiplication is available, which may speed up recovery algorithms significantly, so that large-scale problems are not practicable with Gaussian or Bernoulli matrices. Even storing an unstructured matrix may be difficult. From a computational and an application-oriented viewpoint it is desirable to have measurement matrices with structure. Note that it is hard to rigorously prove good recovery conditions for deterministic matrices. For example, for a Vandermonde matrix as a sensing matrix, it is hard to know the exact restricted isometry constant (RIC), δ_s, as explained in Chapter 2. For another example, for a sensing matrix associated with the Poisson equation based on the finite element method, as in [12], it is difficult to estimate the RIC. To allow randomness to come into play, we should study some structured random matrices. For example, sensing matrices based on a randomly sampled row from a Fourier matrix have been studied extensively. See [18], [2], and [11].*

Remark 8.8.2. *The proof of Lemma 8.18 follows the same ideas as in [17]. One of the authors is grateful to Dr. Pisier for explaining the ideas to him during a class on the geometry of Banach spaces at Texas A&M University in 1989 and giving him a copy of his lecture notes before they were published [17]. The ideas of the covering set in [17] found their way to the study of the sparse solutions of underdetermined linear systems.*

Remark 8.8.3. *The advantage of taking the ℓ_1 norm rather than the ℓ_2 norm as the inner norm in the MRIP (8.35) is apparent. If we had used the ℓ_2 norm, we would have considered the random*

variable $Y_i = (\Phi \mathbf{x})_i^2 - \mathbb{E}((\Phi \mathbf{x})_i^2)$, *and we would have tried to bound the kth moment of* $(\Phi \mathbf{x})_i^2$
by $k! M^k$ *for some* $M > 0$ *in order to apply Bernstein's inequality. However, it proves difficult
to obtain more than* $\mathbb{E}((\Phi \mathbf{x})_i^{2k}) \leq (2k)! (\theta (\Phi \mathbf{x})_i)^{2k} \leq (2k)! (\theta \|\mathbf{x}\|_2)^{2k}$.

Remark 8.8.4. *The methods and results of sections 8.4 to 8.7 on sub-Gaussian and pre-Gaussian
random matrices are standard in the geometry of Banach spaces (see, e.g., [8], where the results
have been spelled out with the compressed sensing reader in mind).*

Remark 8.8.5. *The proof of Theorem 8.17 is based on [13], where the Bernstein inequality is
used. We could use other probability concentration inequalities to establish the result. See, e.g.,
[1].*

Remark 8.8.6. *The proof of Theorem 8.21 is based on the ideas in [6], although the results in
Theorem 8.21 are weaker. Stronger results are established in section 3 of [6].*

Remark 8.8.7. *Sections 8.4 and 8.5 on sub-Gaussian random matrices are based on [7], with
some simplification.*

8.9 ▪ Exercises

Exercise 1. Let $B(m, n)$ be the beta function. Show that $B(2\ell + 1, 1/q) \leq q^{\ell + 1} \ell!$.

Exercise 2. Show that

$$\frac{1}{\sqrt{2\pi}} \int_{-\infty}^{\infty} e^{-t^2} t^k dt = \begin{cases} 0 & \text{if } k \text{ is odd,} \\ \frac{(2\ell)!}{2^\ell \ell!} & \text{if } k = 2\ell. \end{cases}$$

Exercise 3. Show that $\Gamma((1 + q)/2)$ is an increasing function of $q \in (0, 1]$.

Exercise 4. Let $M_p(\cdot; \mathbf{x})$ be a function satisfying (8.19). Show that $M_p(\cdot; \mathbf{x})$ satisfies the following basic properties:

(1) $M_p(\cdot; \mathbf{x}) \geq 0$;

(2) $M_p(-u; \mathbf{x}) = M_p(u; \mathbf{x})$;

(3) $\int_{-\infty}^{\infty} M_p(u; \mathbf{x}) du = 1$; and

(4) $\lambda M_p(\lambda u; \lambda \mathbf{x}) = M_p(u; \mathbf{x})$.

Exercise 5. Show that the random variable ξ with uniform distribution over $[-1, 1]$ is sub-
Gaussian.

Exercise 6. Also show that the Bernoulli random variable with probability $p(1) = 1/2$ at 1 and
$p(-1) = 1/2$ at -1 is sub-Gaussian.

Exercise 7. Show that if a random variable ξ with mean zero satisfies

$$P(|\xi| > t) \leq 2 \exp\left(-\frac{t^2}{2b^2}\right)$$

for all $t > 0$ for a real number b, then ξ is sub-Gaussian.

Exercise 8. Prove the following lemma.

Lemma 8.52. *Let $\|\mathbf{x}\|$ be a quasi-norm for some $q \in (0,1]$, and define $\mathbf{S} = \{\mathbf{x} \in \mathbf{R}^s, \|\mathbf{x}\| = 1\}$ to be the unit sphere of $\mathbf{R}^s$ under the quasi-norm. Then for any real number $\delta \in (0,1)$, there exists a set $\mathcal{U}_q \subset \mathbf{S}$ with the following properties: For any $\mathbf{x} \in \mathbf{S}$,*

$$\min_{\mathbf{y} \in \mathcal{U}_q} \|\mathbf{x} - \mathbf{y}\|^q \leq \delta,$$

and the cardinality of $\mathcal{U}_q$ is less than or equal to $(1 + \frac{2}{\delta})^{s/q}$.

Exercise 9. Prove Lemma 8.32.

Exercise 10. Let $L_p(\Omega, P) = \{\xi \text{ measurable}, \mathbb{E}(|\xi|^p) < \infty\}$ equipped with $\|\xi\|_p = (\mathbb{E}(|\xi|^p))^{1/p}$ for $1 \leq p < \infty$. Show that $L_p(\Omega, P)$ is a Banach space.

Exercise 11. Show that for any $p > 0$, the absolute moment of a random variable X can be expressed as

$$\mathbb{E}(|\xi|^p) = p \int_0^\infty P(|\xi| > x) x^{p-1} dx.$$

Exercise 12. Because it has been useful in this chapter, prove Markov's inequality. That is, for any random variable ξ,

$$P(|\xi| > t) \leq \frac{\mathbb{E}(|\xi|)}{t} \quad \forall t > 0.$$

Exercise 13. Prove Lemma 8.7.

Exercise 14. Show that $\|\mathbf{x}\|_q^q$ is a quasi-norm when $0 \leq q \leq 1$ for $\mathbf{x} \in \mathbb{R}^n$.

Bibliography

[1] R. Baraniuk, M. Davenport, R. DeVore, and M. B. Wakin, A simple proof of the restricted isometry property for random matrices, Constr. Approx., 28 (2008), 253–263. (Cited on pp. 242, 272, 275)

[2] J. Bourgain, An improved estimate in the restricted isometry problem, in Geometric Aspects of Functional Analysis, Lecture Notes in Math. 2116, B. Klartag and E. Milman (Eds.). Springer International, 2014, 65–70. (Cited on p. 274)

[3] J. Bourgain, M. Meyer, V. Milman, and A. Pajor, On a geometric inequality, in Geometric Aspects of Functional Analysis, Lecture Notes in Math. 1317, Springer, Berlin, 1988, 271–282. (Cited on p. 235)

[4] V. V. Buldygin and Yu. V. Kozachenko, Metric Characterization of Random Variables and Random Processes, AMS Publication, Providence, RI, 2000. (Cited on pp. 237, 270)

[5] E. J. Candès and T. Tao, Decoding by linear programing, IEEE Trans. Inform. Theory, 51 (2005), 4203–4215. (Cited on p. 235)

[6] R. Chartrand and V. Staneva, Restricted isometry properties and nonconvex compressive sensing, Inverse Problems, 24(3) (2008), 035020. (Cited on pp. 247, 248, 250, 275)

[7] S. Foucart and M.-J. Lai, Sparse recovery with pre-Gaussian random matrices, Studia Math., 200 (2010), 91–102. (Cited on p. 275)

[8] A. A. Giannopoulos and V. D. Milman, Concentration property on probability spaces, Adv. Math., 156 (2000), 77–106. (Cited on p. 275)

[9] E. Gluskin and V. Milman, Geometric probability and random cotype 2, in Geometric Aspects of Functional Analysis, Lecture Notes in Math. 1850, Springer, Berlin, 2004, 123–138. (Cited on p. 235)

[10] U. Haagerup, The best constants in the Khintchine inequality, Studia Math., 70 (1982), 231–283. (Cited on p. 257)

[11] I. Haviv and O. Regev, The restricted isometry property for subsampled Fourier matrices, in Geometric Aspects of Functional Analysis, Springer, 2017, 163–179. (Cited on p. 274)

[12] H. Kang, M.-J. Lai, and X. Li, An economical representation of PDE solution by using compressive sensing approach. Comput.-Aided Des., 115 (2019), 78–86. (Cited on p. 274)

[13] M.-J. Lai, On sparse solution of underdetermined linear systems, J. Concr. Appl. Math., 8 (2010), 296–327. (Cited on p. 275)

[14] M. Ledoux and M. Talagrand, Probability in Banach Spaces, Springer-Verlag, 1991. (Cited on p. 257)

[15] S. Mendelson, A. Pajor, and N. Tomczak-Jaegermann, Uniform uncertainty principle for Bernoulli and sub-Gaussian ensembles, Constr. Approx., 28 (2008), 277–289. (Cited on p. 235)

[16] F. Nazarov and A. Podkorytov, Ball, Haagerup, and distribution functions, in Complex Analysis, Operators, and Related Topics, Oper. Theory Adv. Appl. 113, Birkhäuser, Basel, 2000, 247–267. (Cited on p. 262)

[17] G. Pisier, The Volume of Convex Bodies and Banach Space Geometry, Cambridge University Press, 1999. (Cited on p. 274)

[18] M. Rudelson and R. Vershynin, On sparse reconstruction from Fourier and Gaussian measurements, Comm. Pure Appl. Math., 61 (2008), 1025–1045. (Cited on p. 274)

Chapter 9

Matrix Completion and Recovery

One significant extension of the theory for sparse solutions of underdetermined linear systems is in the area of matrix recovery and matrix completion. The matrix completion problem was inspired by the Netflix problem in 2006. That is, in October 2006, Netflix challenged the tech community to predict some missing entries in a huge data set for a prize of $1,000,000 (cf. [2] and [30]). The data set is a movie-rating matrix for all movies and many customers. Clearly, not all movies were rated by all customers. The challenge is to fill in the entire matrix in a reasonable way such that the completed matrix matches closely with the entries that are open to the public as well as the hidden entries.

The problem was immediately attacked computationally by many researchers from mathematics, computer science, and engineering. Mathematically, the matrix completion problem can be explained as follows. We would like to recover a matrix $M \in \mathbb{R}^{m,n}$ of size $m \times n$ from a given sample set of partial entries $M_{ij}, (i,j) \in \Omega \subset \{1, \ldots, m\} \times \{1, \ldots, n\}$, by filling in the missing entries. The resulting matrix should have the lowest possible rank, taking into account that people have similar preferences when choosing a movie and many movies have similar themes, such as romance, comedy, tragedy, street fighting, war, and so on. In other words, we need to solve the rank minimization problem

$$\min_{X \in \mathbb{R}^{m \times n}} \quad \text{rank}\,(X): \quad \text{such that} \quad \mathcal{P}_\Omega(X) = \mathcal{P}_\Omega(M), \tag{9.1}$$

where $\mathcal{P}_\Omega$ is the projection which maps a matrix $X \in \mathbb{R}^{m \times n}$ into a vector $\mathbf{m}$ in $\mathbb{R}^{|\Omega|}$, where $\mathbf{m} = \mathcal{P}_\Omega(M)$ is the vector consisting of the entries of matrix M over indices in Ω.

So the matrix recovery problem is a generalization of the standard matrix completion problem.

Let us discuss the minimization (9.1) mathematically. For convenience, we shall consider $m = n$ in the rest of the chapter. First of all, does (9.1) have a solution if $\text{rank}(M) < n$? Letting $\mathcal{M}_r$ be the collection of all matrices of size $m \times m$ with rank $\leq r$, we need to make sure that the feasible set $\{X \in \mathcal{M}_r, \mathcal{P}_\Omega(X) = \mathcal{P}_\Omega(M)\}$ for the minimization (9.1) is nonempty. Certainly, it is dependent on the size $|\Omega|$, the distribution of the indices in Ω, and the values at Ω. If an entire column of the matrix M is missing, we will not be able to recover M uniquely by using the minimization (9.1). We will find some necessary and sufficient conditions on the patterns of indices in Ω to ensure that the feasible set is not empty. Even under the assumption that Ω is uniformly distributed over the index set $\{(i,j), i, j = 1, \ldots, n\}$, for an arbitrary given collection of real values for $\mathcal{P}_\Omega(M)$, we will still not be able to find a matrix M whose rank $r < n$, although we can find a matrix M of full rank satisfying (9.1).

In practice, the given entries $\mathcal{P}_\Omega(M)$ are from a matrix M of rank $r < n$. In this setting, we hope to use the minimization (9.1) to find M. However, the given entries may contain some noise, i.e., the given vector $\mathbf{m} = \mathcal{P}_\Omega(M) + \epsilon$ with noise vector $\epsilon \neq 0$. There may not be a matrix M of rank $r < n$ satisfying $\mathcal{P}_\Omega(M) = \mathbf{m}$. In other words, the feasible set may be empty. So the best way to approach the Netflix problem is to solve

$$\min_{X \in \mathbb{R}^{m \times m}} \operatorname{rank}(X) + \lambda \|\mathcal{P}_\Omega(X) - \mathbf{m}\|_2^2 \tag{9.2}$$

for a positive parameter $\lambda > 0$. However, $\operatorname{rank}(X)$ is not a good function to use for a minimization. One way to approximate $\operatorname{rank}(X)$ is to use the nuclear norm $\|X\|_*$ of matrix X, where

$$\|X\|_* = \sum_{i=1}^{n} \sigma_i(X)$$

and $\sigma_i(X)$ are the singular values of matrix X. Thus, one computational approach to approximating the minimization (9.1) is to solve

$$\min_{X \in \mathbb{R}^{n \times n}} \|X\|_* + \lambda \|\mathcal{P}_\Omega(X) - \mathcal{P}_\Omega(M)\|_2^2, \tag{9.3}$$

where $\lambda > 0$ is a parameter, as usual.

In addition, many computational approaches for sparse solutions of underdetermined linear systems discussed in previous chapters can be extended to this matrix completion setting. For example, the restricted isometry property (RIP) has a counterpart in the matrix recovery setting (cf., e.g., [7], [8]). So does the null space property (NSP). We shall mainly explain the existence and uniqueness of matrix completion and how many matrix completions we can find. In addition, we shall extend some computational approaches from previous chapters to study the matrix completion problem together with convergence analysis. These are nontrivial extensions which deserve some explanation. In addition, we shall present several well-known algorithms to complete a given partial matrix. They are the singular value thresholding (SVT) algorithm, the fixed point iterative algorithms, the orthogonal rank 1 pursuit algorithms, and the alternating projection algorithms. Many matrix completion algorithms, e.g., the algorithm in [27], are not included in this chapter due to limited space.

9.1 ▪ The Well-Posedness of Matrix Completion

Does problem (9.1) have a solution or a unique solution? If it has a solution, how many solutions are there? In this section, we mainly discuss some necessary and sufficient conditions on the given entries and locations of known entries to complete a rank r matrix. Let us start with two examples.

Example 9.1. We define M_Ω as the incomplete matrix with known entries

$$M_\Omega = \begin{bmatrix} \square & \square & -1 & \square \\ \square & \square & \square & 2 \\ 3 & 6 & -3 & 3 \\ 4 & \square & \square & \square \end{bmatrix},$$

where $\square$ stands for an unknown entry. This matrix has the following unique rank 1 completion:

$$M = \begin{bmatrix} 1 & 2 & -1 & 1 \\ 2 & 4 & -2 & 2 \\ 3 & 6 & -3 & 3 \\ 4 & 8 & -4 & 4 \end{bmatrix}.$$

We shall give a reason later.

Next, we give an example which has finitely many completions.

Example 9.2. Consider

$$M_\Omega = \begin{bmatrix} \square & 1 & 2 & 4 \\ 1 & \square & 3 & 5 \\ 1 & 3 & \square & 6 \\ 1 & 4 & 5 & \square \end{bmatrix}.$$

This matrix has exactly two rank 2 completions, which are

$$\begin{bmatrix} 1 & 1 & 2 & 4 \\ 1 & 2 & 3 & 5 \\ 1 & 3 & 4 & 6 \\ 1 & 4 & 5 & 7 \end{bmatrix} \quad \text{and} \quad \begin{bmatrix} -2/3 & 1 & 2 & 4 \\ 1 & 21/8 & 3 & 5 \\ 1 & 3 & 39/11 & 6 \\ 1 & 4 & 5 & 26/3 \end{bmatrix}.$$

This can be verified by using symbolic computer software, say Macaulay2.

To understand the general situation, let us introduce some necessary notation. We assume M is of size $n \times n$ for simplicity. We define the set Ω of indices (i, j) for the given known entries, where $\Omega \subset \{(i, j), 1 \leq i, j \leq n\}$. Let us say $|\Omega| = m < n^2$. Let the map $P_\Omega : M_{n \times n}(\mathbb{C}) \to \mathbb{C}^m$ denote the projection of a matrix onto the known entries, e.g., $P_\Omega : M \mapsto M_\Omega$, where $\mathbb{C}$ stands for the space of all complex values.

Let $\mathcal{M}_r = \{M \in M_{n \times n} \mid \text{rank}(M) = r\}$ be the space of $n \times n$ matrices of rank r over $\mathbb{C}$. Then the analytic closure of $\mathcal{M}_r$ is

$$\overline{\mathcal{M}_r} = \{M \in M_{n \times n} \mid \text{rank}(M) \leq r\}. \tag{9.4}$$

This is the set of $n \times n$ matrices with rank at most r.

A closed set $V \subset \mathbb{C}^n$ is called an algebraic variety if it is the zero set of a system of polynomial equations. The Zariski closure of a set is the smallest variety V which contains that set. Then $\overline{\mathcal{M}_r}$ is also the Zariski closure of $\mathcal{M}_r$, since it is the zero set of all $(r+1) \times (r+1)$ minors in $M_{n \times n}$. Moreover, $\overline{\mathcal{M}_r}$ is an irreducible variety, meaning it cannot be written as the union of two proper subvarieties. Given M_Ω, the partially complete matrix with known entries M_{ij} for $(i, j) \in \Omega$, let $\mathcal{A}_\Omega = \{X \mid X_{ij} = M_{ij}, (i, j) \in \Omega\} = P_\Omega^{-1}(M_\Omega)$ be the affine plane of all possible completions of M_Ω. Then the matrix completion problem amounts to finding a matrix M in the intersection $\mathcal{A}_\Omega \cap \overline{\mathcal{M}_r}$. In other words, X is a rank r completion of M_Ω if and only if $X_{ij} = M_{ij}$ and $\text{rank}(X) \leq r$, i.e., if and only if $X \in \mathcal{A}_\Omega \cap \overline{\mathcal{M}_r}$.

9.1.1 ▪ Number of Completions

It is possible for there to be finitely many ways to complete M_Ω to a rank r matrix. We present the following upper bound.

Theorem 9.3. *Suppose there are finitely many ways to complete an $n \times n$ partially complete matrix M_Ω to a rank r matrix. Then the number of ways to complete M_Ω to a rank r matrix will be less than or equal to $V_{n,r} := \prod_{i=0}^{n-r-1} \frac{\binom{n+i}{r}}{\binom{r+i}{r}}$.*

To prove this we need the notion of degree.

Definition 9.4. *Let $V \subset \mathbb{C}^m$ be an algebraic variety of dimension d. Then the degree of V is defined to be the number of points of intersection of V with d hyperplanes in general position.*

Example 9.5. The degree of the algebraic variety $\overline{\mathcal{M}_r}$ is

$$\prod_{i=0}^{n-r-1} \frac{\binom{n+i}{r}}{\binom{r+i}{r}}.$$

Proof. Refer to [12, Example 14.4.11]. □

It is known that the dimension of $\mathcal{M}_r$ is $2nr - r^2$ (cf. Proposition 12.2 in [13]). Since $\overline{\mathcal{M}_r}$ is the closure of $\mathcal{M}_r$ in the Zariski topology (cf. [37]), the dimension of $\overline{\mathcal{M}_r}$ is also $2nr - r^2$. Furthermore, it is clear that $\overline{\mathcal{M}_r}$ is an algebraic variety. In fact, $\overline{\mathcal{M}_r}$ is an irreducible variety, which is a standard result in algebraic geometry.

Lemma 9.6. $\overline{\mathcal{M}_r}$ *is an irreducible variety.*

Proof. Denote by $GL(n)$ the set of invertible $n \times n$ matrices. Consider the action of $GL(n) \times GL(n)$ on $M_n(R)$ given by $(G_1, G_2) \cdot M \mapsto G_1 M G_2^{-1}$ for all $G_1, G_2 \in GL(n)$. Fix a rank r matrix M. Then the variety $\mathcal{M}_r$ is the orbit of M. Hence, we have a surjective morphism, a regular algebraic map described by polynomials, from $GL(n) \times GL(n)$ onto $\mathcal{M}_r$. Since $GL(n) \times GL(n)$ is an irreducible variety, so is $\mathcal{M}_r$. Hence, the closure $\overline{\mathcal{M}_r}$ of the irreducible set $\mathcal{M}_r$ is also irreducible (cf. Example I.1.4 in [14]). □

Next we will use a generalized version of Bézout's theorem from Theorem 2.3 in [12].

Theorem 9.7. *Let $X_1, \ldots, X_s$ be irreducible varieties in $\mathbb{P}^n$, and let $Z_1, \ldots, Z_k$ be the irreducible components of $X_1 \cap \cdots \cap X_s$. Then*

$$\sum_{i=1}^{k} \deg(Z_i) \le \prod_{j=1}^{s} \deg(X_j).$$

We will now prove that $V_{n,r}$ is an upper bound to the number of possible rank r completions of M_Ω.

Proof of Theorem 9.3. Suppose there are exactly k possible rank r completions of M_Ω, $M_1, \ldots, M_k$. Then $\mathcal{A}_\Omega \cap \overline{\mathcal{M}_r} = \{M_1, \ldots, M_k\}$. Then from Theorem 9.7, we have

$$\sum_{i=1}^{k} \deg(M_i) \le \deg(\mathcal{A}_\Omega) \deg(\overline{\mathcal{M}_r}).$$

Note that the degree of a single point is equal to 1, so the left-hand side is equal to k. Also note that $\mathcal{A}_\Omega$ is a linear variety, so it is irreducible and has degree 1. So we have $k \le \deg(\overline{\mathcal{M}_r}) = V_{n,r}$. □

In general, $V_{n,r}$ is much larger than the actual number of possible rank r completions of M_Ω. For example, when $n = 4$ and $r = 2$, $V_{4,2} = 20$. However, from Example 9.2, $|\mathcal{A}_\Omega \cap \overline{\mathcal{M}_r}| = 2$. Some reasons for this could be as follows:

1. The intersections are at infinity in $\mathbb{P}^{n \times n}$.

2. The intersections have multiplicity.

3. Our hyperplanes are not in general position.

We leave the problem to the interested reader for further study to reduce the upper bound of possible completions.

9.1.2 • Necessary Conditions for rank r Matrix Completions

In general, the space of possible completions with rank at most r, $\mathcal{A}_\Omega \cap \overline{\mathcal{M}_r}$, can be high-dimensional, where the dimension of $\mathcal{A}_\Omega \cap \overline{\mathcal{M}_r}$ depends both on the positions of the known entries in Ω and on the values at known entries of M_Ω. It is easy to see that the dimension can depend on the known positions in Ω. For example, if M_Ω is missing the last column, then $\dim(\mathcal{A}_\Omega \cap \overline{\mathcal{M}_r}) \geq 1$, since for any rank r completion M, replacing the last column of M with any vector in the column space also gives us a valid rank r completion.

As an example of how $\dim(\mathcal{A}_\Omega \cap \overline{\mathcal{M}_r})$ depends on the known entries of M_Ω, consider the following three cases:

$$M_\Omega^1 = \begin{bmatrix} 1 & 0 \\ 1 & \square \end{bmatrix}, \quad M_\Omega^2 = \begin{bmatrix} 0 & 0 \\ 1 & \square \end{bmatrix}, \quad M_\Omega^3 = \begin{bmatrix} 0 & 1 \\ 1 & \square \end{bmatrix}.$$

Then M_Ω^1 has the unique rank 1 completion $M = \begin{bmatrix} 1 & 0 \\ 1 & 0 \end{bmatrix}$, so M_Ω^1 has a zero-dimensional space of possible rank 1 completions. For any x, M_Ω^2 has the rank 1 completion $M = \begin{bmatrix} 0 & 0 \\ 1 & x \end{bmatrix}$, so M_Ω^2 has a one-dimensional space of possible rank 1 completions. And, finally, M_Ω^3 has no possible rank 1 completion, so $\mathcal{A}_\Omega \cap \overline{\mathcal{M}_r} = \emptyset$. We will present necessary conditions for M_Ω to have finitely many rank r completions.

As a generalization to the first example above, where a missing column results in at least a one-dimensional space of possible completions, we have the following theorem as folklore.

Theorem 9.8. *A necessary condition for M_Ω to have finitely many completions in $\overline{\mathcal{M}_r}$ is that there must be at least r known entries in each row and column.*

Proof. Let Ω' be a set of known indices. Without loss of generality by permutation and transpose, suppose the last column of $M_{\Omega'}$ has k known entries, where k is strictly less than r. Suppose $M_{\Omega'}$ has the unique completion M. Let $\Omega \supset \Omega'$ such that M_Ω consists of all entries of M except for the last column. Then M_Ω also has the unique completion M. Therefore, it suffices to show that the incomplete matrix M_Ω never has a unique completion.

Again by permutation, let $M_\Omega = \begin{bmatrix} A & c \\ B & \square \end{bmatrix}$, where c is a $k \times 1$ block of known entries with $k < r$, and $\square$ is an $(n-k) \times 1$ block of unknown entries.

First, if $\text{rank}(\begin{bmatrix} A \\ B \end{bmatrix}) > r$, then any completion will have rank greater than r, so there will be no completions in $\mathcal{M}_r$.

Next, if $\text{rank}(\begin{bmatrix} A \\ B \end{bmatrix}) < r$, then any possible completion of M_Ω will have rank less than or equal to r, and so $\dim(\mathcal{A}_\Omega \cap \overline{\mathcal{M}_r}) = n - k > 0$. So M_Ω would not have a unique completion.

Suppose $\text{rank}(\begin{bmatrix} A \\ B \end{bmatrix}) = r$. Let $\text{rank}(A) = j$. Note that since A is a $k \times (n-1)$ matrix, then $j \leq \min(k, n-1) \leq k$.

Suppose c is not in the column space of A. In other words, suppose $\text{rank}(\begin{bmatrix} A & c \end{bmatrix}) = k+1$. Let M be any completion of M_Ω. Then, since c is linearly independent of the columns of A, the last column of M must be linearly independent of the columns of $\begin{bmatrix} A \\ B \end{bmatrix}$. Therefore, $\text{rank}(M) = r + 1$, and so there will be no completion of M_Ω in $\mathcal{M}_r$.

Suppose c is in the column space of A. In other words, suppose $\text{rank}(\begin{bmatrix} A & c \end{bmatrix}) = k$. Since $\text{rank}(\begin{bmatrix} A \\ B \end{bmatrix}) = r$, there exists an $r \times (n-1)$ submatrix $\begin{bmatrix} A' \\ B' \end{bmatrix}$ of rank r. We can choose A' such that it consists of j rows from A by first choosing j linearly independent rows from A, and then since the remaining rows of A are linearly dependent on the chosen rows, we can choose the remaining $r - j$ linearly independent rows from B.

Augmenting with the corresponding rows from $\begin{bmatrix} c \\ \Box \end{bmatrix}$, we get a matrix of the form $\begin{bmatrix} A' & c' \\ B' & \Box \end{bmatrix}$. Any completion $M' = \begin{bmatrix} A' & c' \\ B' & d' \end{bmatrix}$ of this submatrix will be full rank and therefore have rank r, since this is an $r \times n$ matrix. Moreover, there are $r - j$ degrees of freedom, which is greater than zero since $r > k \geq j$.

Every row b in B that is not in B' is contained in the row space of $\begin{bmatrix} A' \\ B' \end{bmatrix}$, and since $\begin{bmatrix} A' \\ B' \end{bmatrix}$ consists of r linearly independent rows, there exists a unique linear combination of rows that is equal to b. Therefore, there exists a unique d such that $\text{rank}(\begin{bmatrix} A' & c' \\ B & d \end{bmatrix}) = r$.

Moreover, note that $j = \text{rank}(A') \leq \text{rank}(\begin{bmatrix} A' & c' \end{bmatrix}) \leq \text{rank}(\begin{bmatrix} A & c \end{bmatrix}) = j$. Therefore, $\text{rank}(\begin{bmatrix} A' & c' \end{bmatrix}) = j$. So the row space of $\begin{bmatrix} A & c \end{bmatrix}$ is equal to the row space of $\begin{bmatrix} A' & c' \end{bmatrix}$. This implies that $\text{rank}(\begin{bmatrix} A & c \\ B & d \end{bmatrix}) = r$ since we are adding rows to $\begin{bmatrix} A' & c' \\ B & d \end{bmatrix}$ that are already in the row space.

Therefore, any completion M' of the submatrix $\begin{bmatrix} A' & c' \\ B' & \Box \end{bmatrix}$ extends uniquely to a rank r completion M of M_Ω. Therefore, $\dim(\mathcal{A}_\Omega \cap \overline{\mathcal{M}_r}) = r - j > 0$, and so M_Ω does not have a unique completion in $\mathcal{M}_r$.

This exhausts all possible cases for M_Ω, none of which have unique completions. Therefore, M_Ω does not have a unique completion in $\mathcal{M}_r$. $\Box$

Another point: to uniquely determine a complete rank r matrix, we need to know at least $2nr - r^2$ known entries. In other words, another necessary condition for M_Ω to have a zero-dimensional space of possible rank r completions is that we have at least $2nr - r^2$ known entries in M_Ω.

Theorem 9.9. *If* $\dim(\mathcal{A}_\Omega \cap \overline{\mathcal{M}_r}) = 0$, *then* M_Ω *must consist of at least* $2nr - r^2$ *known entries.*

Proof. Note that $\overline{\mathcal{M}_r}$ is an irreducible affine variety of dimension $2nr - r^2$. $\mathcal{A}_\Omega$ is also an algebraic variety which is the zero set of the system of equations $X_{ij} - M_{ij} = 0$ for all $(i, j) \in \Omega$, and the dimension of $\mathcal{A}_\Omega$ is equal to $n^2 - |\Omega|$.

Since both $\mathcal{A}_\Omega$ and $\overline{\mathcal{M}_r}$ are irreducible affine varieties, assuming that there is a nonempty intersection, we have that $\dim(\mathcal{A}_\Omega \cap \overline{\mathcal{M}_r}) \geq \dim(\mathcal{A}_\Omega) + \dim(\overline{\mathcal{M}_r}) - n^2 = 2nr - r^2 - |\Omega|$.

Therefore, if $\dim(\mathcal{A}_\Omega \cap \overline{\mathcal{M}_r}) = 0$, then we must have $|\Omega| \geq 2nr - r^2$. In other words, we must have at least $2nr - r^2$ known entries of M_Ω. $\Box$

Next we explain that if we have more than $2nr - e^2$ known entries, we may not have a unique completion. In particular, if the entries of M_Ω are randomly sampled, and the size m of Ω is larger than $2nr - r^2$, then we most likely will not be able to complete M_Ω to a rank r matrix.

Theorem 9.10. *If* $m = |\Omega| > 2nr - r^2$ *and we choose randomly the known entries for* M_Ω *over index set* Ω, *then the probability of being able to complete* M_Ω *to a rank* r *matrix is zero.*

To prove this no-chance result, let us recall the following result from Theorem 1.25 in section 6.3 of [33].

Lemma 9.11. *Let* $f : X \to Y$ *be a regular map between irreducible varieties. Suppose that* f *is surjective,* $f(X) = Y$, *and that* $\dim(X) = n$, $\dim(Y) = m$. *Then* $m \leq n$, *and*

1. $\dim(F) \geq n - m$ *for any* $y \in Y$ *and for any component* F *of the fiber* $f^{-1}(y)$;

2. *there exists a nonempty open subset* $U \subset Y$ *such that* $\dim(f^{-1}(y)) = n - m$ *for* $y \in U$.

Proof. Refer to Theorem 1.25 in section 6.3 of [33].   □

We are now ready to prove Theorem 9.10.

Proof of Theorem 9.10. We mainly use Lemma 9.11. Let $X = \overline{\mathcal{M}_r}$, which is an irreducible variety by Lemma 9.6. Let $Y = \mathcal{P}_\Omega(\overline{\mathcal{M}_r})$, which is also an irreducible variety because it is a continuous image of the irreducible variety $\overline{\mathcal{M}_r}$. Clearly, $\mathcal{P}_\Omega$ is a regular map. We have $\dim \mathcal{P}_\Omega(\overline{\mathcal{M}_r}) \leq \dim(\overline{\mathcal{M}_r}) = 2nr - r^2 < m$ by Lemma 9.11. Thus, $\mathcal{P}_\Omega(\overline{\mathcal{M}_r})$ is a proper lower-dimensional closed subset in $\mathbb{C}^m$. Almost none of the vectors in $\mathbb{C}^m$ belong to $\mathcal{P}_\Omega(\overline{\mathcal{M}_r})$. In other words, for almost all vectors $\mathbf{m} \in \mathbb{C}^m$, there is no matrix $X \in \overline{\mathcal{M}_r}$ such that $\mathcal{P}_\Omega(X) = \mathbf{m}$.   □

9.1.3 ▪ Sufficient Conditions for rank r Matrix Completion

From theorems in the previous subsection, we have that if $|\Omega| < 2nr - r^2$, then we cannot have a zero-dimensional space of possible rank r completions. Additionally, if $|\Omega| > 2nr - r^2$, then almost no choice of values for M_Ω will have any rank r completion. So is it possible for there to be an Ω with $|\Omega| = 2nr - r^2$ such that almost every choice of M_Ω has a unique rank r completion? The answer is yes. In this subsection, we discuss some sufficient conditions to ensure M_Ω has a unique matrix completion. Let us assume that $|\Omega| = 2nr - r^2$. Define $\chi_\Omega \subset \overline{\mathcal{M}_r}$ by

$$\chi_\Omega = \left\{ M \in \overline{\mathcal{M}_r} \mid \Phi_\Omega^{-1}(\Phi_\Omega(M)) \text{ is zero-dimensional} \right\}.$$

Note that $\overline{\mathcal{M}_r}$ is a singular variety whose singular points are exactly the matrices with rank $< r$ and $\mathcal{M}_r$ is the smooth part of $\overline{\mathcal{M}_r}$. So if $\mathcal{A}_\Omega \cap \overline{\mathcal{M}_r}$ has a finite number of points, then those points lie in the smooth part of $\overline{\mathcal{M}_r}$. When χ_Ω is not empty, the elements in χ_Ω must be isolated points. That is, χ_Ω has finitely many elements and we have many matrix completions. The number of completions was estimated in the previous subsection. We can further show the following result.

Theorem 9.12. *If χ_Ω is not empty, i.e., χ_Ω has finitely many completions of rank $\leq r$, then those completions must have rank $= r$.*

Proof. Without loss of generality, by permuting rows and columns, suppose the index $(1,1) \notin \Omega$. Let

$$M = \begin{bmatrix} m_{11} & m_{12} & \cdots \\ m_{21} & m_{22} & \\ \vdots & & \ddots \end{bmatrix} \in \chi_\Omega,$$

and suppose rank$(M) < r$. Then for any $x \in \mathbb{C}$,

$$X = \begin{bmatrix} x & m_{12} & \cdots \\ m_{21} & m_{22} & \\ \vdots & & \ddots \end{bmatrix}$$

has rank $\leq r$, so $X \in \overline{\mathcal{M}_r}$, and so $X \in \Phi^{-1}(\Phi(M))$. However, this implies that $\dim(\Phi^{-1}(\Phi(M))) \geq 1$, contradicting the supposition that $M \in \chi_\Omega$. Therefore, rank$(M) = r$.   □

Theorem 9.13 (Allen, 2021 [1]). *Suppose that M_Ω can be represented up to a permutation of rows and columns in the block form by $M_\Omega = \begin{bmatrix} A & B \\ C & \square \end{bmatrix}$, with block $\square$ for unknowns and A being $r \times r$. Then if A is invertible, then M_Ω has a unique rank r completion.*

Proof. Suppose A is invertible. Then for every column b_i of B there exists a unique $r \times 1$ vector x_i such that $Ax_i = b_i$. Moreover, since $\text{rank}(A) = r$, then $\text{rank}(\left[\begin{smallmatrix} A \\ C \end{smallmatrix}\right]) = r$, so the columns of $\left[\begin{smallmatrix} A \\ C \end{smallmatrix}\right]$ form a basis for the column space of any rank r completion M. Therefore, x_i is the unique vector such that $\left[\begin{smallmatrix} A \\ C \end{smallmatrix}\right]x_i = \left[\begin{smallmatrix} b_i \\ \square \end{smallmatrix}\right]$ for each of the $(i+r)$th columns of M_Ω. So a rank-r solution M exists and is unique, and the unknown block submatrix $\square = CA^{-1}B$. $\square$

Moreover, for this Ω, if the set of M_Ω does not have a unique rank r completion, we must have $\det(A) = 0$. Those given matrices M_Ω with $\det(A) = 0$ form a Zariski closed proper subset of $\mathbb{C}^m$. The Zariski closed set thus has measure zero. Therefore, almost every given matrix M_Ω has $\det(A) \neq 0$ and hence has a unique rank r completion. In other words, for almost all values in $\mathbb{C}^{2nr-r^2}$ for M_Ω, there is a unique completion.

Next we study a more general setting. For practical data sets, a more likely scenario would be the following: M_Ω takes the form

$$M_\Omega = \begin{bmatrix} A & B_\Omega \\ C_\Omega & D_\Omega \end{bmatrix},$$

where A is a $k \times k$ invertible matrix and $B_\Omega, C_\Omega,$ and D_Ω are partially known and unknown submatrices. Note that it is always possible to permute M_Ω into this for sufficiently small k.

How can we recover B_Ω, C_Ω, and D_Ω in this case? Note that the total number of equations in the system which defines $\mathcal{A}_\Omega \cap \overline{\mathcal{M}_r}$ that we need simultaneous solutions to is equal to $\binom{n}{r+1}^2$. Can we reduce the number of equations that we need to solve? The answer is yes. In fact, we only need to find simultaneous solutions to the system of $(r+1) \times (r+1)$ minors which contains A. Let us prove the following theorem.

Theorem 9.14 (Allen, 2021 [1]). *Let A be the top left $k \times k$ submatrix with variables in $M_{n \times n}$. Consider the variety V which is the zero set of all $(r+1) \times (r+1)$ minors containing A. Then $V = \overline{\mathcal{M}_r} \cup W$ for some W such that for all $M \in W$, the submatrix A of M is not invertible.*

To illustrate how this theorem reduces the number of simultaneous equations we need to solve, let us consider the case when $k = r$.

Example 9.15. If M_Ω contains an $r \times r$ known invertible submatrix, then $\mathcal{A}_\Omega \cap \overline{\mathcal{M}_r} = \mathcal{A}_\Omega \cap V$, where V is the zero set of all $(r+1) \times (r+1)$ determinants containing A. $\mathcal{A}_\Omega \cap V$ is the zero set of $(n-r)^2$ equations, which is a significant reduction from the $\binom{n}{r+1}^2$ equations which define $\mathcal{A}_\Omega \cap \overline{\mathcal{M}_r}$. Moreover, by the *Schur determinant identity*, i.e., Proposition 9.16 below, the equations of $\mathcal{A}_\Omega \cap V$ are easily written as the quadratic equations

$$c_i^\top A^{-1}b_j - d_{ij} = 0, \tag{9.5}$$

where $c_i^\top$ and b_j are the ith row and jth column of C_Ω and B_Ω, respectively, and d_{ij} is the (i,j)th entry of D_Ω.

Proposition 9.16. *Let $A, B, C,$ and D be block submatrices of M with A invertible. Then*

$$\det\left(\begin{bmatrix} A & B \\ C & D \end{bmatrix}\right) = \det(A)\det(D - CA^{-1}B).$$

In particular, when A is an $r \times r$ matrix, B and C are vectors, and D is a scalar, then

$$\det\left(\begin{bmatrix} A & B \\ C & D \end{bmatrix}\right) = (D - CA^{-1}B)\det(A).$$

Proof. Note that $\begin{bmatrix} A & B \\ C & D \end{bmatrix} = \begin{bmatrix} A & 0 \\ C & I \end{bmatrix} \begin{bmatrix} I & A^{-1}B \\ 0 & D - CA^{-1}B \end{bmatrix}$. Taking the determinant, we get the desired result. □

To prove Theorem 9.14, we can do induction on n, the size of our matrix, and on m (the size of Ω), as well as backwards induction on k. We need the following lemma, which gives an explicit description of W for $(r+1) \times n$ matrices.

Lemma 9.17. *Consider the space of all $(r + 1) \times n$ matrices. Let V be the zero set of all $(r + 1) \times (r + 1)$ minors containing the first k columns, with $k \leq r + 1$. Then $V = \overline{\mathcal{M}_r} \cup W$, where $W = \{M \mid$ the first k columns of M are linearly dependent$\}$.*

Proof. Note that $\overline{\mathcal{M}_r} \subset V$, since the set of equations which define $\overline{\mathcal{M}_r}$ contains the set of equations which define V. Also, $W \subset V$, since if $M \in W$, then the first k columns of M are linearly dependent, so every $(r + 1) \times (r + 1)$ minor containing the first k columns vanishes, so $M \in V$. Therefore, $\overline{\mathcal{M}_r} \cup W \subset V$.

For the opposite inclusion, we will induct on n, the number of columns in $M_{r+1 \times n}$, and use backwards induction on k. Consider $n = r + 1$. Then there is exactly one $(r + 1) \times (r + 1)$ minor. Note that in this case for all $k \leq r$, $W \subset \overline{\mathcal{M}_r}$, and so $V = \overline{\mathcal{M}_r} = \overline{\mathcal{M}_r} \cup W$. Fix n. For the case $k = r + 1$, both V and W are the zero set of the first $(r + 1) \times (r + 1)$ minor. In this case, $V = W$ and $\overline{\mathcal{M}_r} \subset W$. Therefore, $V = W = \overline{\mathcal{M}_r} \cup W$.

Now by induction we will assume that $V = \overline{\mathcal{M}_r} \cup W$ for all k for $(r+1) \times (n-1)$ matrices, and that $V = \overline{\mathcal{M}_r} \cup W$ for the first $k+1$ columns. Let $M \in V$. If $\text{rank}(M) \leq r$, then $M \in \overline{\mathcal{M}_r}$. Suppose $\text{rank}(M) = r+1$. Consider the submatrix M' obtained by deleting the $(k+1)$st column. Then since $M \in V$, then by our inductive hypothesis on n we have that the first k columns of M' are linearly dependent, or $\text{rank}(M') \leq r$. In the first case, we have that $M \in W$, so suppose $\text{rank}(M') \leq r$. Then since $\text{rank}(M) = r + 1$, we have that $\text{rank}(M') = r$, and the $(k + 1)$st column of M is linearly independent of the rest of the columns. In particular, it is linearly independent of the first k columns. Now note that M is in the zero set of all $(r + 1) \times (r + 1)$ minors containing the first $k + 1$ columns. By backwards induction on k, $M \in \overline{\mathcal{M}_r}$, or the first $k + 1$ columns are linearly dependent. However, $M \notin \overline{\mathcal{M}_r}$, so the first $k + 1$ columns must be linearly independent. We also have that the $(k + 1)$st column is linearly independent of the first k columns. Therefore, the first k columns are linearly dependent, so $M \in W$. Therefore, $M \in \overline{\mathcal{M}_r} \cup W$, and so $V \subset \overline{\mathcal{M}_r} \cup W$. So $V = \overline{\mathcal{M}_r} \cup W$. □

We will use this lemma as the base case to prove the main theorem in this subsection.

Proof of Theorem 9.14. We will induct on m. For the base case, let $m = r + 1$. Consider some $M \in V$, and suppose $\text{rank}(A) = k$. Then by the previous lemma, we have that $\text{rank}(M) \leq r$ or the rank of the first k columns is less than k. However, since $\text{rank}(A) = k$, the first k columns have rank k, and so we have $\text{rank}(M) \leq r$, so $M \in \overline{\mathcal{M}_r}$.

Now by induction we will assume that $V = \overline{\mathcal{M}_r} \cup W$ for $(m - 1) \times n$ matrices. Consider an $m \times n$ $M \in V$, and suppose $\text{rank}(A) = k$. Then we will show that $M \in \overline{\mathcal{M}_r}$.

Consider the submatrix M' which consists of the first $m-1$ rows of M. Now by the induction hypothesis, since $\text{rank}(A) = k$, we must have that $\text{rank}(M') \leq r$. If $\text{rank}(M') < r$, then by adding in the last row we will have $\text{rank}(M) \leq r$, which means that $M \in \overline{\mathcal{M}_r}$. So suppose $\text{rank}(M') = r$. Since $\text{rank}(A) = k$, then the first k rows are linearly independent. So without loss of generality, by permuting the rows, suppose that the first r rows are linearly independent and span the row space of M'. Now consider the submatrix M'' of M which consists of all but the second-to-last row. Then again by the induction hypothesis and since $\text{rank}(A) = k$, we must have that $\text{rank}(M'') \leq r$. However, since the first r rows of M'' span the row space, we have that the last row of M is contained in the span of the first r rows. Therefore, since $\text{rank}(M') = r$, and since the last row of M is contained in the row space of M', we must have $\text{rank}(M) = r$, so $M \in \overline{\mathcal{M}_r}$. □

Next we discuss the uniqueness of the matrix satisfying $\mathcal{P}_\Omega(M) = \mathbf{m}$. For each vector $\mathbf{m} \in \mathcal{R}_\Omega$, the preimage $\{Y \in \overline{\mathcal{M}_r} : \mathcal{P}_\Omega(Y) = \mathbf{m}\}$ may have more than one element. It is interesting to know how many elements are in the preimage of $\mathbf{m}$. Define the subset $\chi_\Omega \subset \overline{\mathcal{M}_r}$ by

$$\chi_\Omega = \left\{ X \in \overline{\mathcal{M}_r} \mid \mathcal{P}_\Omega^{-1}(\mathcal{P}_\Omega(X)) \text{ is zero-dimensional} \right\}.$$

Since we are working over Noetherian fields like $\mathbb{R}$ or $\mathbb{C}$, it is worthwhile to keep in mind that all zero-dimensional varieties over such fields will have only finitely many elements. Next we recall the following result from Proposition 11.12 in [13].

Lemma 9.18. *Let X be a quasi-projective variety and $\pi : X \to \mathbb{P}^m$ be a regular map; let Y be the closure of the image. For any $p \in X$, let $X_p = \pi^{-1}\pi(p) \subseteq X$ be the fiber of π through p, and let $\mu(p) = \dim_p(X_p)$ be the local dimension of X_p at p. Then $\mu(p)$ is an upper-semicontinuous function of p in the Zariski topology on X, i.e., for any m the locus of points $p \in X$ such that $\dim_p(X_p) > m$ is closed in X. Moreover, if $X_0 \subseteq X$ is any irreducible component, $Y_0 \subseteq Y$ is the closure of its image, and μ is the minimum value of $\mu(p)$ on X_0, then*

$$\dim(X_0) = \dim(Y_0) + \mu. \tag{9.6}$$

Proof. This is Proposition 11.12 in [13]. □

Because we have seen that $\dim(\mathcal{P}_\Omega(\overline{\mathcal{M}_r})) \leq \dim(\overline{\mathcal{M}_r})$, we can be more precise about these dimensions, as shown in the following lemma.

Lemma 9.19. *Assume $m > \dim(\overline{\mathcal{M}_r})$. Then χ_Ω is an open subset of $\overline{\mathcal{M}_r}$ and*

$$\dim(\overline{\mathcal{M}_r}) = \dim(\overline{\mathcal{P}_\Omega(\overline{\mathcal{M}_r})}) = \dim(\mathcal{P}_\Omega(\overline{\mathcal{M}_r}))$$

if and only if $\chi_\Omega \neq \emptyset$.

Proof. Assume $\dim(\overline{\mathcal{M}_r}) = \dim(\overline{\mathcal{P}_\Omega(\overline{\mathcal{M}_r})}) = \dim(\mathcal{P}_\Omega(\overline{\mathcal{M}_r}))$. Then, using Lemma 9.11, there exists a nonempty open subset $U \subset \mathcal{P}_\Omega(\overline{\mathcal{M}_r})$ such that $\dim(\mathcal{P}_\Omega^{-1}(y)) = 0$ for all $y \in U$. This implies that $\mathcal{P}_\Omega^{-1}(y) \in \chi_\Omega$. Hence $\chi_\Omega \neq \emptyset$.

We now prove the converse. Assume $\chi_\Omega \neq \emptyset$. We will apply Lemma 9.18 above by setting $X = \overline{\mathcal{M}_{r_g}}$, $Y = \mathcal{P}_\Omega(\overline{\mathcal{M}_{r_g}})$, and $\pi = \mathcal{P}_\Omega$. A couple of things to note here are that it does not matter whether we take the closure in $\mathbb{P}^m$ or in $\mathbb{C}^m$ since $\mathbb{C}^m$ is an open set in $\mathbb{P}^m$ and the Zariski topology of the affine space $\mathbb{C}^m$ is induced from the Zariski topology of $\mathbb{P}^m$. $\overline{\mathcal{M}_{r_g}}$ is an affine variety. Therefore, it is a quasi-projective variety.

By our assumption, χ_Ω is not empty. It follows that there is a point $p \in Y$ such that $\pi^{-1}(p)$ is zero-dimensional. Since zero is the least dimension possible, we have $\mu = 0$ in (9.6). Hence, using (9.6) above, we have $\dim(\overline{\mathcal{M}_r}) = \dim(\overline{\mathcal{P}_\Omega(\overline{\mathcal{M}_r})})$. But dimension does not change upon taking closure. So $\dim(\mathcal{P}_\Omega(\overline{\mathcal{M}_r})) = \dim(\overline{\mathcal{P}_\Omega(\overline{\mathcal{M}_r})})$. Also, using Lemma 9.20 below, $\chi_\Omega = \{x \in X : \dim(\phi^{-1}\phi(\mathbf{x})) < 1\}$ is an open subset of $\overline{\mathcal{M}_r}$. □

In the proof above, the following result was used.

Lemma 9.20. *Let* $\phi : X \to Y$ *be a morphism of affine varieties. Let* $\phi^{-1}\phi(\mathbf{x}) = Z_1 \cup \cdots \cup Z_j$ *be the irreducible components of* $\phi^{-1}\phi(\mathbf{x})$. *Let* $e(\mathbf{x})$ *be the maximum of the dimensions of the* Z_i's, $i = 1, \dots, j$. *Let* $S_n(\phi) := \{x \in X : e(\mathbf{x}) \geq n\}$. *Then, for any* $n \geq 1$, $S_n(\phi)$ *is a Zariski closed subset of* X. *Equivalently,* $\{x \in X : \dim(\phi^{-1}\phi(\mathbf{x})) < n\}$ *is an open subset of* X.

Proof. Refer to [29, I.8, Corollary 3]. □

We are now ready to prove another main result in this section.

Theorem 9.21. *Assume that there exists a finite* r-*feasible vector* $\mathbf{x} \in \mathbb{C}^m$ *over the given* Ω. *Then, with probability* 1, *the vector* $\mathbf{x}$ *is finite* r-*feasible. In other words, if we randomly choose a feasible vector* $\mathbf{x}$ *in the positions* Ω, *the matrix can be completed to a rank* r *matrix only in finitely many ways with probability* 1. *In addition, the number of ways to complete will be less than or equal to* $\prod_{i=0}^{n-r-1} \frac{\binom{n+i}{r}}{\binom{r+i}{r}}$.

Proof. We begin by noting that both $\overline{\mathcal{M}_r}$ and $\mathcal{P}_\Omega(\overline{\mathcal{M}_r})$ are irreducible varieties. So the closure $\overline{\mathcal{P}_\Omega(\overline{\mathcal{M}_r})}$ is also an irreducible variety. By the assumption and using Lemma 9.19, $\dim(\overline{\mathcal{M}_r}) = \dim(\overline{\mathcal{P}_\Omega(\overline{\mathcal{M}_r})})$. Hence, applying Lemma 9.11, there exists a nonempty open subset $U \subset \overline{\mathcal{P}_\Omega(\overline{\mathcal{M}_r})}$ such that $\mathcal{P}_\Omega^{-1}(y)$ is zero-dimensional for all $y \in U$. In other words, if we choose the m entries in positions Ω of a matrix from the open set U, then there are finitely many ways to complete the matrix. The result now follows by recalling that a Zariski open set in an irreducible variety is a dense set whose complement has Lebesgue measure zero.

When we fix m entries of a matrix M, the set of matrices of rank r which have those entries in the positions Ω are exactly the intersection points of the variety $\overline{\mathcal{M}_r}$ with m hyperplanes, namely, the hyperplanes defined by equations of the form $M_{ij} = constant$. Since $m > \dim(\overline{\mathcal{M}_r}) = 2nr - r^2$, the number of intersection points would be less than the degree of $\overline{\mathcal{M}_r}$ generically. Now using the exact formula for the degree from Example 9.5, the result follows. □

In this section, we presented our approach to studying the well-posedness of the matrix completion based on an elementary algebraic geometry (cf. [21]). In fact, more advanced algebraic geometry approaches (based on rigidity theory and combinatorics) have been used to study the well-posedness problem. We refer the reader to [34] and [17].

9.2 ▪ RIP and NSP for Matrix Recovery

In this section, we discuss using the nuclear norm minimization (9.3) to complete the matrix. We first extend the concepts of the null space property (NSP) and the restricted isometry property (RIP) from the compressive sensing setting to the matrix recovery setting. Consider the constraint minimization problem

$$\min_{X \in \mathbb{R}^{m \times n}} \|X\|_* : \mathcal{P}_\Omega(X) = \mathcal{P}_\Omega(M). \tag{9.7}$$

Let us discuss a necessary and sufficient condition that the minimization above will find the matrix M with given entries in Ω. To do so, we need some well-known results.

Theorem 9.22. *For any two matrices* A, B *which have the same size, say* $m \times n$ *with* $m \leq n$, *it holds that*

$$\sum_{i=1}^{k} |\sigma_i(A) - \sigma_i(B)| \leq \sum_{i=1}^{k} \sigma_i(A - B) \tag{9.8}$$

for all $k = 1, \dots, m$.

Proof. We leave the proof to the interested reader; see also [15], [16].   □

The following is an interesting case for $p \in (0, 1)$.

Lemma 9.23. *Let A and B be matrices of the same size, $m \times n$. Fix $p \in [0, 1]$. Let $\sigma_j(A)$ and $\sigma_j(B), j = 1, \ldots, n$, be singular values of A and B, respectively. For each k, we define*

$$\alpha = \max_{k=1,\ldots,n} \frac{\max_{j=1,\ldots,k} |\sigma_j(A - B)|}{\min_{j=1,\ldots,k} |\sigma_j(A) - \sigma_j(B)| \neq 0} \geq 1. \tag{9.9}$$

Then we have

$$\sum_{i=1}^{k} |\sigma_i^p(A) - \sigma_i^p(B)| \leq \alpha^{1-p} \sum_{i=1}^{k} \sigma_i^p(A - B) \tag{9.10}$$

for $k = 1, 2, \ldots, m$.

Proof. Let $\beta_k = \min_{j=1,\ldots,k} |\sigma_j(A) - \sigma_j(B)| \neq 0$ and $\gamma_k = \max_{j=1,\ldots,k} |\sigma_j(A - B)|$. Then for $p \in (0, 1]$, we have

$$\sum_{j=1}^{k} |\sigma_j^p(A) - \sigma_j^p(B)| \leq \sum_{j=1}^{k} |\sigma_j(A) - \sigma_j(B)|^p = \sum_{j=1}^{k} \beta_k^p \left| \frac{|\sigma_j(A) - \sigma_j(B)|}{\beta_k} \right|^p$$

$$\leq \beta_k^p \sum_{j=1}^{k} \left| \frac{|\sigma_j(A) - \sigma_j(B)|}{\beta_k} \right| \leq \beta_k^{p-1} \sum_{j=1}^{k} \sigma_j(A - B) = \beta_k^{p-1} \gamma_k \sum_{j=1}^{k} \frac{\sigma_j(A - B)}{\gamma_k}$$

$$\leq \beta_k^{p-1} \gamma_k \sum_{j=1}^{k} \left| \frac{\sigma_j(A - B)}{\gamma_k} \right|^p \leq \beta_k^{p-1} \gamma_k^{1-p} \sum_{j=1}^{k} \sigma_j^p(A - B) \leq \alpha^{1-p} \sum_{j=1}^{k} \sigma_j^p(A - B),$$

where we have used Theorem 9.22. This completes the proof.   □

Let $\mathcal{N}(\Omega) = \{X \in \mathbb{R}^{n,n}, \mathcal{P}_\Omega(X) = 0\}$ be the null space of matrices X of size $n \times n$. We can prove the following NSP.

Theorem 9.24. *The minimization problem (9.7) can recover all matrices M of rank$(M) \leq r$ if and only if for any matrix $W \in \mathcal{N}(\Omega)$,*

$$\sum_{i=1}^{r} \sigma_i(W) \leq \sum_{i=r+1}^{m} \sigma_i(W), \tag{9.11}$$

where $\sigma_i(W)$ stands for the ith singular value of W, $i = 1, \ldots, s$.

Proof. Let us prove the sufficiency first. Let X_Ω be a matrix of rank r whose values with indices in Ω are given. That is, $\mathcal{P}_\Omega(X_\Omega)$ is given. Since X_Ω is of rank $\leq r$, we have $\sigma_i(X_\Omega) = 0, i \geq r + 1$. Then for any $W \in \mathcal{N}(\Omega)$, using (9.8), we have

$$\|X_\Omega + W\|_* \geq \sum_{i=1}^{n} |\sigma_i(X_\Omega) - \sigma_i(W)| \geq \sum_{i=1}^{r} \sigma_i(X_\Omega) - \sigma_i(W) + \sum_{i=r+1}^{m} \sigma_i(W)$$

$$\geq \|X_\Omega\|_* - \sum_{i=1}^{r} \sigma_i(W) + \sum_{i=r+1}^{m} \sigma_i(W) \geq \|X_\Omega\|_*$$

by using (9.11). Hence, X_Ω is a minimizer.

Conversely, if (9.11) does not hold for some $W \in \mathcal{N}(\Omega)$, let W_r be the rank r approximation of W. That is, W_r is the matrix induced by setting all but the largest r singular values of W to zero. Let $X_\Omega = -W_r$. Clearly, X_Ω is of rank $\leq r$. But

$$\|X_\Omega + W\|_* = \sum_{i=r+1}^{m} \sigma_i(W) < \sum_{i=1}^{r} \sigma_i(W) = \|X_\Omega\|_*.$$

That is, for given $P_\Omega(X_\Omega)$, the minimization (9.7) cannot find X_Ω.   □

To introduce the RIP in the matrix version, we need to introduce the low rank matrix recovery problem. Clearly, we may replace $\mathcal{P}_\Omega$ by a set of linear mappings in $\mathcal{A}_\Omega$, i.e., each mapping $\mathbf{a}_\omega \in \mathcal{A}_\Omega$ is a linear or nonlinear functional of matrices $X \in \mathbb{R}^{m \times n}$ for $\omega \in \Omega$, where Ω is an index set. Let $|\Omega|$ be the number of total linear functionals in $\mathcal{A}_\Omega$. For simplicity, we use $\mathcal{A}_\Omega$ to denote the collection of some linear functionals. In the rest of this section, we shall use $\mathcal{A}_\Omega$ to replace $\mathcal{P}_\Omega$ in (9.1) and call the corresponding minimization problem the matrix recovery problem. So the matrix recovery problem is a generalization of the standard matrix completion problem.

The following definition is the natural extension of the RIP in the matrix recovery setting.

Definition 9.25. *Let* $m = |\Omega|$ *and* $\mathcal{P}_\Omega : \mathbb{R}^{n \times n} \to \mathbb{R}^m$ *be a set of* m *linear maps. Let* δ_r *be the smallest number such that*

$$(1 - \delta_r)\|X\|_F^2 \leq \|\mathcal{A}_\Omega(X)\|_2^2 \leq (1 + \delta_r)\|X\|_F^2$$

holds for all matrices X *of rank at most* r, *where* $\|\cdot\|_F$ *stands for the Frobenius norm for matrices. If* $\delta_r \in (0,1)$, *we say that the linear constraint set or linear map* $\mathcal{A}_\Omega$ *satisfies the RIP with constant* δ_r.

The next lemma follows immediately.

Lemma 9.26. *Suppose that the linear map* $\mathcal{A}_\Omega$ *of matrices possesses the RIP with* $\delta_{2r} < 1$ *for some integer* $r \geq 1$. *Then the minimization problem (9.3) has only one solution.*

Proof. On the contrary, let X_0 and X_1 be two minimizers. Then $Z = X_0 - X_1 \neq 0$, which is a matrix of rank at most $2r$. Since $\mathcal{A}(Z) = 0$, we have the contradiction $0 = \|\mathcal{A}(Z)\|^2 \geq (1 - \delta_{2r})\|Z\|_F^2 > 0$.   □

We now explain that there is an overwhelming probability that we can find a linear map $\mathcal{A}$ satisfying the RIP. We can always express any linear map $\mathcal{A}$ in its matrix representation

$$\mathcal{A}_\Omega(X) = \mathbf{A}\text{vec}(X),$$

where $\text{vec}(X) \in \mathbb{R}^{n^2}$ denotes the vectorized X with its columns stacked in order on top of one another and $\mathbf{A}$ is an $m \times n^2$ matrix. We can view $\mathcal{A}_\Omega$ as a random variable in the sense that the entries of $\mathbf{A}$ are i.i.d. random variables. For example, $\mathbf{A} = [A_{ij}]$ with $A_{ij} \sim N(0, 1/m)$.

Definition 9.27. *We say that* $\mathcal{A}_\Omega$ *is nearly isometrically distributed if*

(1) $\mathbb{E}(\|\mathcal{A}_\Omega(X)\|^2) = \|X\|_F^2$ *for all* $X \in \mathbb{R}^{n \times n}$;

(2) *for all $\epsilon \in (0,1)$,*

$$\mathbb{P}(|\|\mathcal{A}_\Omega(X)\|^2 - \|X\|_F^2| \geq \epsilon\|X\|_F^2) \leq 2\exp\left(-\frac{r^2}{2}\left(\frac{\epsilon^2}{2} - \frac{\epsilon^3}{3}\right)\right)$$

for matrix X of rank r; and

(3) *for all $t > 0$,*

$$\mathbb{P}\left(\|\mathcal{A}_\Omega\| \geq 1 + \sqrt{\frac{mn}{r}} + t\right) \leq \exp(-\gamma p t^2)$$

for some constant $\gamma > 0$.

Here $\mathbb{E}$ and $\mathbb{P}$ stand for the expectation and probability.

Unfortunately, for the matrix completion problem, $\mathcal{P}_\Omega$ does not induce a set of linear constraints which satisfy the matrix RIP. However, we can use the RIP to describe the exact matrix recovery problem. Many results from Chapter 7 can be generalized to the setting of matrix recovery theory. See such a general claim in [32]. Let us give an example. We begin with a technical lemma.

Lemma 9.28. *Let A and B be matrices of size $m \times n$. Then there exist matrices B_1 and B_2 such that*

(1) $B = B_1 + B_2$;

(2) $\text{rank}(B_1) \leq 2\,\text{rank}(A)$;

(3) $AB_2^\top = 0 = A^\top B_2$; and

(4) $\langle B_1, B_2 \rangle = \text{trace}(B_1^\top B_2) = 0$.

Proof. Let $A = U\Sigma V^\top$ be the singular value decomposition (SVD) with

$$\Sigma = \begin{bmatrix} \Sigma_A & 0 \\ 0 & 0 \end{bmatrix}.$$

Next let $\widetilde{B} = U^\top BV$. Partition $\widetilde{B}$ as

$$\widetilde{B} = \begin{bmatrix} B_{11} & B_{12} \\ B_{21} & B_{22} \end{bmatrix}$$

based on the partition of Σ_A. Now we choose

$$B_1 := U\begin{bmatrix} B_{11} & B_{12} \\ B_{21} & 0 \end{bmatrix}V^\top \quad \text{and} \quad B_2 := U\begin{bmatrix} 0 & 0 \\ 0 & B_{22} \end{bmatrix}V^\top.$$

The remaining part of the proof is to show that B_1 and B_2 satisfy the properties (1)–(4), which is left to the interested reader. $\square$

We are now ready to prove the following result.

Theorem 9.29 (Recht, Fazel, and Parrilo, 2010 [31]). *Suppose that $r \geq 1$ such that $\delta_{5r} < 1/10$. Then the minimization problem (9.7) has one and only one solution.*

Proof. Let X^* be a minimizer of (9.7). Let X_0 be another minimizer. Let $Z = X^* - X_0$. Applying Lemma 9.28 to the matrices X_0 and Z, there exist Z_1 and Z_2 satisfying the properties (1)–(4). Note that $\|Z_1\|_* \geq \|Z_2\|_*$.

Next we write $Z_2 = U\Sigma V^\top$ in SVD format and decompose its singular values $s_1, \ldots, s_m$ into blocks of size $3r$ in the following sense:

$$z_i = U_{T_i} \operatorname{diag}\left([s_j, j \in T_i]\right) V_{T_i}^\top,$$

where $T_i = \{3r(i-1)+1, \ldots, 3ri\}$. By construction, we have $\langle z_i, z_j \rangle = 0$ for $i \neq j$ and

$$s_k \leq \frac{1}{3r} \sum_{j \in I_i} s_j \quad \forall k \in I_{i+1}.$$

It follows that

$$\sum_{j \geq 2} \|z_j\|_F \leq \frac{1}{\sqrt{3r}} \sum_{j \geq 1} \|z_j\|_* = \frac{1}{\sqrt{3r}} \|Z_2\|_* \leq \frac{1}{\sqrt{3r}} \|Z_1\|_* \leq \frac{\sqrt{2r}}{\sqrt{3r}} \|Z_1\|_F.$$

Finally, note that the rank of $Z_1 + Z_2$ is at most $5r$, so we may put all this together as

$$0 = \|\mathcal{A}(Z)\| \geq \|\mathcal{A}(Z_1 + z_1)\| - \sum_{j \geq 2} \|\mathcal{A}(z_j)\|$$

$$\geq \sqrt{(1-\delta_{5r})} \|Z_1 + z_1\|_F - \sqrt{(1+\delta_{3r})} \sum_{j \geq 2} \|z_j\|_F$$

$$\geq \left(\sqrt{(1-\delta_{5r})} - \sqrt{2(1+\delta_{3r})/3}\right) \|Z_1\|_F.$$

Simple algebra shows that when $\delta_{5r} < 1/5$, $\sqrt{(1-\delta_{5r})} - \sqrt{2(1+\delta_{3r})/3} > 0$. It follows that $\|Z_1\|_F = 0$. Hence, $\|Z_2\|_F = 0$ and $Z = 0$, i.e., $X^* = X_0$. □

To uniquely recover all matrices of rank r or less by solving (9.7), it is sufficient for $\mathcal{A}_\Omega$ to satisfy $\delta_{5r} < 0.1$, as shown in [31], which has been improved to the RIP with $\delta_{4r} < \sqrt{2} - 1$ in [6] and to $\delta_{2r} < 0.307$, as well as expressions involving δ_{3r}, δ_{4r}, and δ_{5r}, in [28]. The algorithm SVP [24] provably achieves exact recovery if $\delta_{2r} < 1/3$. See [10] for more results on the uniqueness conditions for the matrix recovery problem.

Next, we present another RIP-based condition based on an extension of the ideas in [26] for sparse vector recovery.

Theorem 9.30 (RIP Condition for Exact Recovery by (9.3)). *Let $\mathbf{X}^0$ be a matrix with rank r or less. Minimization (9.3) exactly recovers $\mathbf{X}^0$ from measurements $\mathbf{b} = \mathcal{A}(\mathbf{X}^0)$ if $\mathcal{A}$ satisfies the RIP with $\delta_{2r} < 0.4931$.*

Proof. We establish the theorem by showing that (9.12) holds for any $H \in \operatorname{Null}(\mathcal{A}) \setminus \{0\}$:

$$\sum_{i=1}^{r} \sigma_i(H) < \sum_{i=r+1}^{m} \sigma_i(H). \tag{9.12}$$

Based on the SVD $H = \sum_{i=1}^{m} \sigma_i(H) \mathbf{U}_i \mathbf{V}_i^\top$, where $\sigma_i(H)$ is the ith largest singular value of H and $\mathbf{U}_i$ is the ith column of $\mathbf{U}$, and similar for $\mathbf{V}_i$. We decompose $H = H_0 + H_1 + H_2 + \cdots,$

where $H_0 = \sum_{i=1}^{r} \sigma_i(H)\mathbf{U}_i\mathbf{V}_i$, $H_1 = \sum_{i=r+1}^{2r} \sigma_i(H)\mathbf{U}_i\mathbf{V}_i$, $H_2 = \sum_{i=2r+1}^{3r} \sigma_i(H)\mathbf{U}_i\mathbf{V}_i, \ldots$. Following these definitions, condition (9.12) can be equivalently written as

$$\|H_0\|_* < \left\| \sum_{i \geq 1} H_i \right\|_*. \tag{9.13}$$

From $H \neq \mathbf{0}$ and the definition of H_0, we know that $H_0 \neq \mathbf{0}$ and thus $\mathcal{A}(H_0) \neq \mathbf{0}$ due to the RIP of $\mathcal{A}$. From $\mathcal{A}(H) = \mathbf{0}$ and $\mathcal{A}(H_0) \neq \mathbf{0}$, it follows that $\mathcal{A}(\sum_{i \geq 1} H_i) \neq \mathbf{0}$ and thus $\sum_{i \geq 1} H_i \neq \mathbf{0}$. Therefore, $\sum_{i \geq 1} \|H_i\|_* > 0$, and we can define $t := \|H_1\|_* / (\sum_{i \geq 1} \|H_i\|_*) > 0$ and $\rho := \|H_0\|_* / (\sum_{i \geq 1} \|H_i\|_*) > 0$.

Next, we present two inequalities without proofs (the interested reader can verify them by following the proofs of Lemmas 2.3 and 2.4 in [26]):

$$\frac{1 - \delta_{2r}}{r}(\rho^2 + t^2) \left(\sum_{i \geq 1} \|H_i\|_* \right)^2 \leq \|\mathcal{A}(H_0 + H_1)\|_2^2, \tag{9.14}$$

$$\frac{t(1-t) + \delta_{2r}(1 - 3t/4)^2}{r} \left(\sum_{i \geq 1} \|H_i\|_* \right)^2 \geq \left\| \mathcal{A}\left(\sum_{i \geq 1} H_i \right) \right\|_2^2. \tag{9.15}$$

Since $\mathcal{A}(H_0 + H_1) + \mathcal{A}\left(\sum_{i \geq 1} H_i \right) = \mathcal{A}(H) = \mathbf{0}$, the two right-hand sides of (9.14) equal each other. Hence,

$$\frac{1 - \delta_{2r}}{r}(\rho^2 + t^2) \left(\sum_{i \geq 1} \|H_i\|_* \right)^2 \leq \frac{t(1-t) + \delta_{2r}(1 - 3t/4)^2}{r} \left(\sum_{i \geq 1} \|H_i\|_* \right)^2,$$

and thus

$$\rho^2 \leq \frac{t(1-t) + \delta_{2r}(1 - 3t/4)^2 - (1 - \delta_{2r})t^2}{1 - \delta_{2r}},$$

or, after a simple calculation,

$$\rho \leq \sqrt{\frac{4(1 + 5\delta_{2r} - 4(\delta_{2r})^2)}{(1 - \delta_{2r})(32 - 25\delta_{2r})}} =: \theta_{2r}. \tag{9.16}$$

If $\delta_{2r} < (77 - \sqrt{1337})/82 \approx 0.4931$, then $\theta_{2r} < 1$ and thus $\rho < 1$. By definition, we get (9.13) and (9.12). $\square$

In fact, all of the RIP conditions for the sparse solution of underdetermined linear systems can be extended to the matrix recovery setting. The discussion above is just one such extension. We leave other extensions to the interested reader.

9.3 ▪ Schatten p Quasi-norm Minimization

Let us consider the recovery of low rank matrices by Schatten ℓ_p quasi-norm minimization:

$$\begin{array}{ll} \text{minimize} & \|X\|_p^p \\ \text{subject to} & \mathcal{A}(X) = y, \end{array} \tag{9.17}$$

where $\|X\|_p^p = \sum_{i=1}^k \sigma_i^p(X)$, $y = \mathcal{A}(M)$, and $\sigma_i(X)$ are singular values of matrix X of size $m \times n$. Note that when $p = 1$, $\|X\|_p^p = \|X\|_*$. We now discuss when a minimizer X^* of (9.17) is the minimal rank solution of (9.1). Some numerical results based on the minimization (9.17) can be found in [19].

To do so, we need a result on the Schatten p quasi-norm $\|X\|_p$.

Theorem 9.31 (Lai, Liu, Li, and Wang, 2020 [18]). *Let $p \in [0,1]$. For all real matrices A and B of size $m \times n$,*

$$\sum_{i=1}^k |\sigma_i^p(A) - \sigma_i^p(B)| \leq \sum_{i=1}^k \sigma_i^p(A - B) \tag{9.18}$$

for $1 \leq k \leq \ell = \min\{m,n\}$, where $\sigma_1(A) \geq \sigma_2(A) \geq \cdots \geq \sigma_\ell(A) \geq 0$, $\sigma_1(B) \geq \sigma_2(B) \geq \cdots \geq \sigma_\ell(B) \geq 0$, and $\sigma_1(A - B) \geq \sigma_2(A - B) \geq \cdots \geq \sigma_\ell(A - B) \geq 0$ are singular values of A, B, and $A - B$, respectively.

We shall provide a proof at the end of this section. Once we have the inequality (9.18), we are able to show the following result.

Theorem 9.32 (Lai, Liu, Li, and Wang, 2020 [18]). *Suppose that the matrix Φ whose matrix RIP constant $\delta_{2s} < 1$. Then there exists a number $p_0 \in (0,1)$ such that for any $p \leq p_0$, each minimizer X_p of the Schatten ℓ_p quasi-norm minimization (9.17) is the lowest rank solution of (9.1).*

First of all, we need the following lemma.

Lemma 9.33. *Suppose we have the inequality in (9.18). Let X^0 be the minimizer of (9.1) with rank s and X^* be a global minimizer of (9.17). Recall that both of them satisfy $\mathcal{A}(X^0) = \mathcal{A}(X^*)$. Let $H = X^0 - X^*$, which is in the null space of $\mathcal{A}$. Then*

$$\sum_{i=s+1}^m \sigma_i^p(H) \leq \sum_{i=1}^s \sigma_i^p(H). \tag{9.19}$$

Proof. We use (9.18) to get

$$\|X^0\|_p^p \geq \|X^*\|_p^p = \|X^0 - H\|_p^p = \sum_{i=1}^m \sigma_i^p(X^0 - H) \geq \sum_{i=1}^m |\sigma_i^p(X^0) - \sigma_i^p(H)|$$

$$\geq \sum_{i=1}^s (\sigma_i^p(X^0) - \sigma_i^p(H)) + \sum_{i=s+1}^m \sigma_i^p(H)$$

since $\sigma_i(X^0) = 0$ for $i \geq s + 1$. After rearranging the above inequality, since $\|X^0\|_p^p = \sum_{i=1}^s \sigma_i^p(X^0)$, we obtain the result. $\square$

Next we need an inequality which can be found in [5].

Lemma 9.34. *For any matrices X and Y with $\langle X, Y \rangle = \text{trace}(X^\top Y) = 0$ with $\text{rank}(X) \leq r$ and $\text{rank}(Y) \leq s$,*

$$\langle \mathcal{A}(X), \mathcal{A}(Y) \rangle \leq \delta_{r+s} \|X\|_F \|Y\|_F, \tag{9.20}$$

where $\|X\|_F$ stands for the Frobenius norm of X, and similar for $\|Y\|_F$.

Proof of Theorem 9.32. Let $H = X^0 - X^*$ and write $H = U\Sigma_H V^*$ in SVD format. We decompose $\Sigma_H = \Sigma_0 + \Sigma_1 + \Sigma_2 + \cdots$ such that Σ_0 contains the first s singular values of Σ_H, Σ_1 the second s singular values of Σ_H, and so on. Similarly, write $U = [U_0 \, U_1 \, U_2 \, \ldots]$ and $V = [V_0 \, V_1 \, V_2 \, \ldots]$ accordingly. Thus, $H = H_0 + H_1 + H_2 + \cdots$ with $H_i = U_i \Sigma_i V_i^\top$. Clearly, $\langle H_i, H_j \rangle = 0$ if $i \neq j$. Thus, we can use the definition of the matrix version of the RIP to obtain

$$
(1 - \delta_{2s})(\|H_0\|_F^2 + \|H_1\|_F^2)
$$

$$
\leq \|\mathcal{A}(H_0 + H_1)\|_2^2 = \left\| \mathcal{A}\left(H - \sum_{j \geq 2} H_j \right) \right\|_2^2
$$

$$
= \sum_{i,j \geq 2} \langle \mathcal{A}(H_j), \mathcal{A}(H_i) \rangle \leq \sum_{i \geq 2} (1 + \delta_{2s}) \|H_i\|_F^2 + \sum_{\substack{i,j \geq 2 \\ i \neq j}} \langle \mathcal{A}(H_i), \mathcal{A}(H_j) \rangle
$$

$$
\leq \sum_{i \geq 2} (1 + \delta_{2s}) \|H_i\|_F^2 + \delta_{2s} \sum_{\substack{i,j \geq 2 \\ i \neq j}} \|H_i\|_F \|H_j\|_F
$$

$$
\leq \sum_{i \geq 2} \|H_i\|_F^2 + \delta_{2s} \left(\sum_{j \geq 2} \|H_j\|_F \right)^2 \leq (1 + \delta_{2s}) \left(\sum_{j \geq 2} \|H_j\|_F \right)^2 .
$$

If $\|H_2\|_F = 0$, the right-hand side of the inequality above is zero and hence $H = 0$. Hence $X^* = X^0$. That is, the minimizer is the low rank solution of (9.1).

Otherwise, the above inequality can be rewritten as

$$
\|H_0\|_F \leq \left(\frac{1 + \delta_{2s}}{1 - \delta_{2s}} \right)^{1/2} \left(\sum_{j \geq 2} \|H_j\|_F \right) \tag{9.21}
$$

under the assumption that $\|H_2\|_F \neq 0$. In this situation, $\sigma_{2s+1}(H) \neq 0$. We have

$$
(2s)^{-1/p} \|H_0 + H_1 + H_2\|_p = (2s)^{-1/p} \sigma_{2s+1}(H) \left(\sum_{i=1}^{3s} \left(\frac{\sigma_i(H)}{\sigma_{2s+1}(H)} \right)^p \right)^{1/p}
$$

$$
\geq \sigma_{2s+1}(H)(2s)^{-1/p} \left(\sum_{i=1}^{2s+1} \left(\frac{\sigma_i(H)}{\sigma_{2s+1}(H)} \right)^p \right)^{1/p}
$$

$$
\geq \sigma_{2s+1}(H)(2s)^{-1/p}(2s + 1)^{1/p}
$$

$$
= \sigma_{2s+1}(H) \left(1 + \frac{1}{2s} \right)^{1/p} . \tag{9.22}
$$

On the other hand, we have

$$
\sum_{j \geq 2} \|H_j\|_F \leq \frac{n}{s} \|H_2\|_F \leq \frac{n}{s} \sqrt{s} \, \sigma_{2s+1}(H) = \frac{n}{\sqrt{s}} \sigma_{2s+1}(H). \tag{9.23}
$$

Now we claim that for any $\epsilon \in (0,1)$, there exists a real number $p_\epsilon \leq 1$ such that when $p \leq p_\epsilon$, we have

$$
(2s)^{1/p} \left(\frac{1 + \delta_{2s}}{1 - \delta_{2s}} \right)^{1/2} \sum_{j \geq 2} \|H_j\|_F \leq \epsilon \|H_0 + H_1 + H_2\|_p. \tag{9.24}
$$

For convenience, let

$$C_s = \left(\frac{1+\delta_{2s}}{1-\delta_{2s}}\right)^{1/2}.$$

We consider the following function $f(p, H)$, using (9.23) and (9.22):

$$f(p, H) = \left(\frac{1+\delta_{2s}}{1-\delta_{2s}}\right)^{1/2} \frac{(2s)^{1/p}\sum_{j\geq 2}\|H_j\|_F}{\|H_0 + H_1 + H_2\|_p}$$

$$\leq C_s \frac{(2s)^{1/p}\frac{n}{\sqrt{s}}\sigma_{2s+1}(H)}{\sigma_{2s+1}(H)(2s+1)^{1/p}} \leq \frac{C_s\frac{n}{\sqrt{s}}}{(1+\frac{1}{2s})^{1/p}} \leq \frac{C_s\frac{n}{\sqrt{s}}}{(1+\frac{1}{2s})^{1/p}}.$$

For $0 < \epsilon < 1$, since $(1+\frac{1}{2s})^{1/p} \to \infty$ when $p \to 0_+$, for any $H \in \mathbb{R}^{m\times n}$,

$$f(p, H) \leq \epsilon.$$

That is, we have (9.24).

Finally, we use the Hölder inequality, the inequality in (9.21), and then the inequality in (9.24) to get

$$\|H_0\|_p^p \leq s^{1-p/2}\|H_0\|_F^p \leq s^{1-p/2}\left(\epsilon(2s)^{-1/p}\|H_0 + H_1 + H_2\|_p\right)^p$$

$$\leq \epsilon^p s^{-p/2}2^{-1}\|H_0 + H_1 + H_2\|_p^p \leq \frac{\epsilon^p}{2s^{p/2}}\left(\|H_0\|_p^p + \sum_{i\geq s+1}\sigma_i^p(H)\right)$$

$$\leq \frac{\epsilon^p}{2s^{p/2}}2\|H_0\|_p^p = \frac{\epsilon^p}{s^{p/2}}\|H_0\|_p^p$$

by using Lemma 9.33.

Under the assumption that we used the ℓ_p minimization (9.17) with $p \leq p_\epsilon$ and $\epsilon < 1/2$, we get a contradiction if $\|H_0\|_F \neq 0$ since $t = \frac{\epsilon^p}{s^{p/2}} < 1$. Thus, $\|H_0\|_F = 0$. Therefore, $H = X^0 - X^* = 0$, and hence the minimizer of (9.17) is the minimal rank solution of (9.1). $\square$

Similarly, we shall discuss the noisy recovery case. Let X^* be the optimal solution of the problem

$$\min_{X\in\mathbb{R}^{m\times n}} \sum_{i=1}^{m}\sigma_i^p(X) \quad \text{such that} \quad \|\mathcal{A}(X) - \mathcal{A}(M)\|_2 \leq \eta, \tag{9.25}$$

where $\eta > 0$ is a noise level. We can establish the following result.

Theorem 9.35. *If $\delta_{2s} < 1$, there exists some $p_0 < 1$ such that for any $p \leq p_0$ we have*

$$\|X^* - X^0\|_p \leq \frac{2\cdot 3^{1/p}s^{1/p-1/2}\eta}{\sqrt{1-\delta_{2s}}(1-\frac{1}{2\sqrt{s}})}. \tag{9.26}$$

Proof. We leave the proof to the interested reader; see also [18]. $\square$

We now take some time to prove Theorem 9.31. First of all, we have the following proposition.

Proposition 9.36. *Suppose that the inequalities in (9.18) hold for any symmetric matrices A and B of the same size. Then the inequalities hold for any matrices A and B of size $m \times n$.*

Proof. For any matrix A and B of size $m \times n$, we consider the $(m + n) \times (m + n)$ symmetric matrices

$$\tilde{A} = \begin{bmatrix} 0 & A \\ A^\top & 0 \end{bmatrix} \quad \text{and} \quad \tilde{B} = \begin{bmatrix} 0 & B \\ B^\top & 0 \end{bmatrix}. \tag{9.27}$$

Then it is known that the eigenvalues of $\tilde{A}$ are $\pm\sigma_1(A), \ldots, \pm\sigma_\ell(A)$ together with $m + n - 2\ell$ zeros, where $\ell = \min\{m, n\}$. The case for $\tilde{B}$ is similar. For $k \leq \ell$, we use the assumption to obtain

$$\sum_{i=1}^{k} |\sigma_i^p(A) - \sigma_i^p(B)| = \sum_{i=1}^{k} |\sigma_i^p(\tilde{A}) - \sigma_i^p(\tilde{B})| \leq \sum_{i=1}^{k} \sigma_i^p(\tilde{A} - \tilde{B}) = \sum_{i=1}^{k} \sigma_i^p(A - B).$$

This completes the proof.   □

Let us focus on symmetric or Hermitian A and B.

Theorem 9.37. *When A and B are symmetric matrices of the same size, Theorem 9.31 is correct. That is, (9.18) holds for all symmetric matrices A and B of the same size.*

Proof of Theorem **9.31.** We simply use the results in the above, i.e., Proposition 9.36 and Theorem 9.37, to finish the proof of Theorem 9.31.   □

To prove Theorem 9.37, we begin with the definition of $|A|$ for any square matrix A: $|A| = \sqrt{A^\top A}$. If $A = U\Sigma_A V^\top$ in SVD format, we have $|A| = V\sqrt{\Sigma_A^\top \Sigma_A} V^\top$. Thus, $|A| = V|\Sigma_A|V^\top$. In particular, when A is symmetric, i.e., $A = C^\top \Lambda C$ for orthonormal matrix C and diagonal matrix Λ which contains all eigenvalues of A, we have $|A| = C^\top |\Lambda| C$. For any nonnegative function $f(t)$ defined on $[0, \infty)$ to $[0, \infty)$, we can define $f(|A|) = C^\top f(|\Lambda|)C$. We need to use the Bourin–Uchiyama triangle inequality [3, Corollary 2.6].

Lemma 9.38. *Let f be a nonnegative function defined on $[0, \infty)$ which is increasing and concave with $f(0) = 0$. For any matrices A and B of the same size, there exist unitary matrices U and V of appropriate size such that*

$$f(|A|) \preceq Uf(|A - B|)U^\top + Vf(|B|)V^\top. \tag{9.28}$$

In particular, for $f(t) = t^p$ for $p \in (0, 1)$ and $t \geq 0$, we have

$$|A|^p \preceq U|A - B|^p U^\top + V|B|^p V^\top. \tag{9.29}$$

That is, we have

$$|A|^p - V|B|^p V^\top \preceq U|A - B|^p U^\top \tag{9.30}$$

for any symmetric matrices A, B of size $m \times m$.

Proof. We refer the reader to [18] for an elementary proof.   □

We are now ready to establish the result in Theorem 9.37.

Proof of Theorem **9.37.** By Lemma 9.38, i.e., (9.30),

$$\sum_{i=1}^{k} \sigma_i(|A|^p - V|B|^p V^\top) \leq \sum_{i=1}^{k} \sigma_i(U|A - B|^p U^\top) = \sum_{i=1}^{k} \sigma_i(|A - B|^p) = \sum_{i=1}^{k} \sigma_i^p(A - B).$$

$$\tag{9.31}$$

On the other hand, when both A and B are symmetric, the left-hand side of the inequality (9.18) can be calculated as follows:

$$\sum_{i=1}^{k}|\sigma_i^p(A) - \sigma_i^p(B)| = \sum_{i=1}^{k}|\sigma_i^p(|A|) - \sigma_i^p(|B|)| = \sum_{i=1}^{k}|\sigma_i(|A|^p) - \sigma_i(|B|^p)|$$

$$= \sum_{i=1}^{k}|\sigma_i(|A|^p) - \sigma_i(V|B|^pV^\top)| \leq \sum_{i=1}^{k}\sigma_i(|A|^p - V|B|^pV^\top),$$

$$(9.32)$$

where we have used Theorem 9.22 in the last inequality. We now combine (9.31) and (9.32) to obtain the desired inequality (9.18).   □

9.4 ▪ Matrix Completion via the SVT Algorithm

Starting in this section, we discuss how to solve the low rank matrix completion problem. We begin with a well-known result on singular values of matrices.

Theorem 9.39. *For any two matrices A, B which have the same size, say $m \times n$ with $m \leq n$, it holds that*

$$\sum_{i=1}^{m}(\sigma_i(A) - \sigma_i(B))^2 \leq \sum_{i=1}^{m}\sigma_i^2(A - B). \qquad (9.33)$$

We refer the reader to [16] for a proof. The following singular value shrinkage operator is very useful. Write a matrix X in SVD format, i.e., $X = U\Sigma V^\top$. For each $\tau > 0$, we introduce the soft-thresholding operator $\mathcal{D}_\tau$ defined by

$$\mathcal{D}_\tau(X) = U(\Sigma - \tau I)_+ V^\top,$$

where $(\Sigma - \tau I)_+ = \text{diag}((\sigma_i - \tau)_+, i = 1, \ldots, \min\{m, n\})$, and $t_+ = \max\{0, t\}$. See [4] and [23] for other proofs of the following result.

Theorem 9.40. *For each $\tau > 0$ and matrix $X \in \mathbb{R}^{m \times n}$,*

$$\mathcal{D}_\tau(X) = \arg\min_Y \frac{1}{2}\|X - Y\|_F^2 + \tau\|Y\|_*.$$

Proof. Since the minimization functional $f(Y) = \tau\|Y\|_* + \frac{1}{2}\|X - Y\|_F^2$ is strictly convex, it is easy to see that there exists a unique minimizer. Let us prove that $\mathcal{D}_\tau(X)$ is a minimizer. First of all, let $\sigma_i(X), \sigma_i(Y), i = 1, \ldots, k = \min\{m, n\}$, be singular values of X and Y, respectively. In particular, let $X = U\Sigma(X)V^\top$ in SVD format. We have

$$\sum_{i=1}^{k}|\sigma_i(X) - \sigma_i(Y)|^2 \leq \|X - Y\|_F^2$$

by Theorem 9.39. Thus, we may consider the minimization problem

$$\min_{\mathbf{x} \in \mathbb{R}_+^k} \tau \sum_{i=1}^{m}\mathbf{x}_i + \frac{1}{2}\sum_{i=1}^{k}|\sigma_i(X) - \mathbf{x}_i|^2. \qquad (9.34)$$

The above minimization is the same as the one on the left-hand side of the inequality

$$\min_{Y \in \mathbb{R}^{m \times n}} \tau \sum_{i=1}^{k} \sigma_i(Y) + \frac{1}{2} \sum_{i=1}^{k} |\sigma_i(X) - \sigma_i(Y)|^2 \le \min_{Y \in \mathbb{R}^{m \times n}} \frac{1}{2} \|X - Y\|_F^2 + \tau \|Y\|_*. \quad (9.35)$$

The minimization on the left-hand side above or the minimization (9.34) gives a unique minimizer $\mathbf{x}^* = (\sigma_1^*, \ldots, \sigma_k^*)$ with $\sigma_i^* = \text{Shrinkage}(\sigma_i(X), \tau)$ for $i = 1, \ldots, k$. Recall that $X = U\Sigma(X)V^\top$ in SVD format. Letting $\Sigma(Y) = \text{diag}(\sigma_1^*, \sigma_2^*, \ldots, \sigma_k^*)$, we choose $\widehat{Y} = U(\sigma(Y))V^\top = U\mathcal{D}_\tau(\Sigma(X)V^\top$, which is $\mathcal{D}_\tau(X)$. It is easy to see that

$$\tau\|\widehat{Y}\|_* + \frac{1}{2}\|\widehat{Y} - X\|_F^2 = \tau \sum_{i=1}^{k} \sigma_i^* + \frac{1}{2} \sum_{i=1}^{k} |\sigma_i(X) - \sigma_i^*|^2 \le \min_{Y \in \mathbb{R}^{m \times n}} \frac{1}{2}\|X - Y\|_F^2 + \tau\|Y\|_*$$

by using (9.34). Thus, $\widehat{Y}$ is the unique minimizer. This completes the proof. □

To approximate the minimizer in (9.7), we consider the minimization

$$\min_{X \in \mathbb{R}^{n \times m}} \left\{ \alpha\|X\|_* + \frac{1}{2}\|X\|_F^2, \quad \mathcal{P}_\Omega(X) = \mathcal{P}_\Omega(M) \right\}. \quad (9.36)$$

That is, we add $\frac{1}{2}\|X\|_F^2$ to (9.7) to make the minimizer easy to compute. Indeed, note that the minimizing functional $\alpha\|X\|_* + \frac{1}{2}\|X\|_F^2$ is not only convex, but is also strongly convex. We can easily see that when α is large, the minimizer of this minimization problem is close to a minimizer of (9.3). We can solve it using the so-called Uzawa method. Letting

$$\mathcal{L}(x, \Lambda) = \alpha\|X\|_* + \frac{1}{2}\|X\|_F^2 + \langle \Lambda, \mathcal{P}_\Omega(X) - \mathcal{P}_\Omega(M) \rangle,$$

and starting with $\Lambda^0 = 0$, we solve

$$X^k = \arg \min_{X \in \mathbb{R}^{m \times n}} \mathcal{L}(X, \Lambda^{k-1}) \quad (9.37)$$

and

$$\Lambda^k = \Lambda^{k-1} + \rho(\mathcal{A}_\Omega(M) - \mathcal{A}_\Omega(X^k))$$

for $k = 1, 2, \ldots$, where $\rho > 0$ is a fixed step size. To solve (9.37), we rewrite the equation above as

$$\mathcal{L}(X, \Lambda^{k-1}) = \alpha\|X\|_* + \frac{1}{2}\|X - \mathcal{A}_\Omega^\top(\Lambda^{k-1})\|_F^2 - \frac{1}{2}\|\mathcal{A}_\Omega^\top(\Lambda^{k-1})\|_F^2 + \langle \Lambda^{k-1}, \mathcal{A}_\Omega(M) \rangle.$$

Since $\mathcal{A}_\Omega^\top(\Lambda) = \Lambda$, we use Theorem 9.40 to see that the minimizer of (9.37) is $X^k = \mathcal{D}_\alpha(\Lambda^{k-1})$. This leads to the following SVT algorithm.

ALGORITHM 9.1

Singular Value Thresholding (SVT) Algorithm [4]

1: **Input:** A partial matrix M_Ω, a total number of iterations, and a parameter $\rho > 0$.
2: Start with $\Lambda^0 = 0$.
3: Iteratively, we compute

$$\begin{cases} X^k = \mathcal{D}_\alpha(\Lambda^{k-1}), \\ \Lambda^k = \Lambda^{k-1} + \rho(\mathcal{A}_\Omega(M) - \mathcal{A}_\Omega(X^k)) \end{cases}$$

for $k = 1, \ldots$ until the total number of iterations.
4: **Output:** The last iterative matrix X^k.

Following the proof of the Uzawa method in Chapter 6, we can establish the convergence of this algorithm. The proof is left to the interested reader, or see [4].

Next we discuss when the minimizer of (9.36) is the solution of (9.7) and how large α must be so that the minimizer of (9.36) is the minimizer of (9.7). We have the following matrix NSP condition.

Theorem 9.41 (Lai and Yin, 2013 [20]). *Assume that $\|\mathbf{X}^0\|_2$ is a minimizer of (9.1) of rank r. The minimization (9.36) uniquely recovers all matrices $\mathbf{X}^0$ of rank r or less from measurements $\mathbf{b} = \mathcal{A}_\Omega(\mathbf{X}^0)$ if and only if*

$$\left(1 + \frac{\|\mathbf{X}^0\|_2}{\alpha}\right) \sum_{i=1}^{r} \sigma_i(H) \leq \sum_{i=r+1}^{m} \sigma_i(H) \tag{9.38}$$

holds for all matrices $H \in \mathrm{Null}(\mathcal{A}_\Omega)$.

Proof. Sufficiency. Pick any matrix $\mathbf{X}^0$ of rank r or less and let $\mathbf{b} = \mathcal{A}(\mathbf{X}^0)$. For any nonzero matrix $H \in \mathrm{Null}(\mathcal{A})$, we have $\mathcal{A}(\mathbf{X}^0 + H) = \mathcal{A}\mathbf{X}^0 = \mathbf{b}$. In particular, for the minimizer $\mathbf{X}^*$ of (9.36), consider $H = \mathbf{X}^* - \mathbf{X}^0$. By using (9.8) and (9.33), we have

$$\|\mathbf{X}^*\|_* + \frac{1}{2\alpha}\|\mathbf{X}^*\|_F^2$$

$$= \|\mathbf{X}^0 + H\|_* + \frac{1}{2\alpha}\|\mathbf{X}^0 + H\|_F^2 \geq \|s(\mathbf{X}^0) - s(H)\|_1 + \frac{1}{2\alpha}\|s(\mathbf{X}^0) - s(H)\|_2^2$$

$$= \sum_{i=1}^{r} |\sigma_i(\mathbf{X}^0) - \sigma_i(H)| + \sum_{i=r+1}^{m} \sigma_i(H)$$

$$+ \frac{1}{2\alpha}\sum_{i=1}^{r} \sigma_i(\mathbf{X}^0)^2 + \sigma_i(H)^2 - \frac{1}{\alpha}\sum_{i=1}^{r}\sigma_i(\mathbf{X}^0)\sigma_i(H) + \frac{1}{2\alpha}\sum_{i=r+1}^{m}\sigma_i(H)^2$$

$$\geq \sum_{i=1}^{r} \sigma_i(\mathbf{X}^0) + \frac{1}{2\alpha}\sum_{i=1}^{r}\sigma(\mathbf{X}^0)^2 - \sum_{i=1}^{r}\sigma_i(H)$$

$$+ \sum_{i=r+1}^{m}\sigma_i(H) - \frac{1}{\alpha}\sum_{i=1}^{r}\sigma_1(\mathbf{X}^0)\sigma_i(H) + \frac{1}{2\alpha}\sum_{i=1}^{m}\sigma(H)^2 \tag{9.39}$$

$$\geq \left[\|\mathbf{X}^0\|_* + \frac{1}{2\alpha}\|\mathbf{X}^0\|_F^2\right] + \left[\sum_{i=r+1}^{m}\sigma_i(H) - \left(1 + \frac{\|\mathbf{X}^0\|_2}{\alpha}\right)\sum_{i=1}^{r}\sigma_i(H)\right] + \frac{1}{2\alpha}\|H\|_F^2,$$

where $s(\mathbf{X}) = (\sigma_1(\mathbf{X}), \ldots, \sigma_m(\mathbf{X}))$ is the vector of all singular values of $\mathbf{X}$.

If $\|H\|_F > 0$, it follows that $\mathbf{X}^*$ is not a minimizer of (9.36). This is a contradiction. Thus, $H \equiv 0$ and $\mathbf{X}^0$ is the unique solution to minimization (9.36).

Necessity: For any nonzero $H \in \mathrm{Null}(\mathcal{A})$ obeying (9.38), let $H = \mathbf{U}\Sigma\mathbf{V}^\top$ be the SVD of H. Construct $\mathbf{X}^0 = -\mathbf{U}\Sigma_r\mathbf{V}^\top$, where Σ_r keeps only the largest r diagonal entries of Σ and sets the rest to 0. We use the argument above to get

$$\|\mathbf{X}^0 + tH\|_* + \frac{1}{2\alpha}\|\mathbf{X}^0 + tH\|_F^2$$

$$= \|\mathbf{X}^0\|_* + \frac{1}{2\alpha}\|\mathbf{X}^0\|_F^2 + \left[\sum_{i=r+1}^{m}\sigma_i(tH) - \left(1 + \frac{\|\mathbf{X}^0\|_2}{\alpha}\right)\sum_{i=1}^{r}\sigma_i(tH)\right] + \frac{1}{2\alpha}\|tH\|_F^2$$

for any $t > 0$ small enough. For $\mathbf{X}^0$ to be the unique solution to (9.3) given $\mathbf{b} = \mathcal{A}(\mathbf{X}^0)$, we must have

$$\left[\sum_{i=r+1}^{m} \sigma_i(tH) - \left(1 + \frac{\|\mathbf{X}^0\|_2}{\alpha}\right) \sum_{i=1}^{r} \sigma_i(tH) \right] + \frac{1}{2\alpha}\|tH\|_F^2 > 0$$

for all $t > 0$. Since the last term above has t^2, (9.38) is necessary. ☐

Next we study the condition for the augmented model (9.36).

Theorem 9.42 (RIP Condition for Exact Recovery). *Let $\mathbf{X}^0$ be a matrix with rank r or less. The augmented model (9.36) exactly recovers $\mathbf{X}^0$ from measurements $\mathbf{b} = \mathcal{A}(\mathbf{X}^0)$ if $\mathcal{A}$ satisfies the RIP with $\delta_{2r} < 0.4404$ and in (9.36) $\alpha \geq 10\|\mathbf{X}^0\|_2$.*

Proof. The proof of Theorem 9.30 establishes that any $H \in \text{Null}(\mathcal{A})$ satisfies $\|H_0\|_* \leq \theta_{2r}\|\sum_{i \geq 1} H_i\|_*$. Hence, (9.38) holds provided that $\left(1 + \frac{\|\mathbf{X}^0\|_2}{\alpha}\right)^{-1} \geq \theta_{2r}$. The rest of the proof is similar to that of Theorem 9.30. We leave it for the interested reader. ☐

Similar to that of Theorem 9.42, we present the stable recovery result as follows.

Theorem 9.43 (RIP Condition for Stable Recovery). *Let $\mathbf{X}^0 \in \mathbb{R}^{n_1 \times n_2}$ be an arbitrary matrix and $\sigma_i(\mathbf{X}^0)$ be its ith largest singular value. Let $\mathbf{b} := \mathcal{A}(\mathbf{X}^0) + \mathbf{n}$, where $\mathcal{A}$ is a linear operator and $\mathbf{n}$ is an arbitrary noise vector. If $\mathcal{A}$ satisfies the RIP with $\delta_{2r} \leq 0.3814$, then the solution $\mathbf{X}^*$ of (9.36) with any $\alpha \geq 10 \cdot \|\mathbf{X}^0\|_2$ satisfies the error bounds*

$$\|\mathbf{X}^* - \mathbf{X}^0\|_* \leq C_1 \cdot \sqrt{k}\|\mathbf{n}\|_2 + C_2 \cdot \hat{\sigma}(\mathbf{X}^0), \tag{9.40}$$

$$\|\mathbf{X}^* - \mathbf{X}^0\|_F \leq \bar{C}_1 \cdot \|\mathbf{n}\|_2 + (\bar{C}_2/\sqrt{r}) \cdot \hat{\sigma}(\mathbf{X}^0), \tag{9.41}$$

where $\hat{\sigma}(\mathbf{X}^0) := \sum_{i=r+1}^{\min\{n_1,n_2\}} \sigma_i(\mathbf{X}^0)$ is the best rank r approximation error of $\mathbf{X}^0$; $C_1, C_2, \bar{C}_1$, and $\bar{C}_2$ are given by formulas (6.61), (6.62), (6.64), and (6.65), in which θ_{2k} is replaced by θ_{2r} (given in (9.16)); and

$$C_3 := \frac{\alpha + \sigma_1(\mathbf{X}^0)}{\alpha - \sigma_{k+1}(\mathbf{X}^0)} \quad and \quad C_4 := \frac{2\alpha}{\alpha - \sigma_{r+1}(\mathbf{X}^0)}. \tag{9.42}$$

Proof. We leave the proof to Exercise 18; see also [20]. ☐

9.4.1 ▪ The Dual of the Minimization (9.36)

Similar to the study in section 6.5, we consider the dual version of the minimization in (9.36). Using the same arguments as in a previous section, we see that the minimization (9.36) can be converted to the following minimization problem: Letting $B = \mathcal{A}_\Omega(M) \in \mathbb{R}^{n \times n}$ and recalling that the inner product in the matrix setting is $\text{trace}(A, B)$ for matrices $A, B \in \mathbb{R}^{n \times n}$,

$$\min_{Y \in \mathbb{R}^{n \times n}} \left\{ -\text{trace}(B, Y) + \frac{\alpha}{2}\|\mathcal{A}_\Omega(Y) - \text{Proj}_{\{X:\|X\|_2 \leq 1\}}(\mathcal{A}_\Omega(Y))\|_F^2 \right\}, \tag{9.43}$$

where $\mathcal{A}_\Omega(Y) = \sum_{\omega \in \Omega} y_\omega \mathbf{A}_\omega$ and $\mathbf{A}_\omega$ is the zero matrix of size $n \times n$ with ω entry 1. It is the dual of the minimization (9.36). We leave the proof to Exercise 13.

Note that $Y - \text{Proj}_{\{X:\|X\|_2 \leq 1\}}(Y) = \mathcal{D}_1(Y)$ by using Theorem 9.40, where $\mathcal{D}_1$ is the matrix soft-thresholding operator $\mathcal{D}_1$.

Let $f(Y)$ be the minimizing functional in (9.43). It is easy to see that $f(Y)$ is convex (see Exercise 14). The set of minimizers of $f(Y)$ is not empty. Define $\mathcal{Y}^*$ to be the collection of all minimizers of (9.43). Clearly, $\mathcal{Y}^*$ is a convex set. The gradient of $f(Y)$ is

$$\nabla f(Y) = -B + \alpha \mathcal{A}_\Omega^\top (\mathcal{D}_1(\mathcal{A}_\Omega(Y))). \tag{9.44}$$

Applying the standard gradient descent method to (9.43), we have Algorithm 9.2.

ALGORITHM 9.2

Dual Gradient Descent Algorithm for Matrix Completion [20]

1: We start with an initial matrix Y_0.
2: For $k = 1, 2, \ldots$, do the iterations

$$Y_{k+1} = Y_k + h(B - \alpha \mathcal{A}_\Omega^*(\mathcal{D}_1(\mathcal{A}_\Omega(Y_k)))) \tag{9.45}$$

3: until the maximum number of iterations is reached.

We now claim that Algorithm 9.2 will converge in a linear fashion. But first, we need some preparatory results.

Lemma 9.44. *For any matrices* $A, B \in \mathbb{R}^{n \times n}$,

$$\|\mathcal{D}_1(A) - \mathcal{D}_1(B)\|_F \leq \|A - B\|_F. \tag{9.46}$$

Proof. We leave the proof to the interested reader; see also (2.16) in [9]. $\square$

Lemma 9.45. *For any matrices* $A, B \in \mathbb{R}^{n \times n}$,

$$\text{trace}(A - B, \mathcal{D}_1(A) - \mathcal{D}_1(B)) \geq \|\mathcal{D}_1(A) - \mathcal{D}_1(B)\|_F^2. \tag{9.47}$$

Proof. We leave the proof to the interested reader; see also (2.16) in [9]. $\square$

Letting $Y^* \in \mathcal{Y}^*$, since $\nabla f(Y^*) = 0$ or $B = \alpha \mathcal{A}_\Omega^\top (\mathcal{D}_1(\mathcal{A}_\Omega(Y^*)))$, we have

$$\begin{aligned}
\|Y_{k+1} - Y^*\|_F^2 &= \|Y_k - Y^*\|_F^2 - 2h\alpha \text{trace}(Y_k - Y^*, \mathcal{A}_\Omega^\top(\mathcal{D}_1(\mathcal{A}_\Omega(Y_k)) - \mathcal{D}_1(\mathcal{A}_\Omega(Y^*)))) \\
&\quad + h^2 \|\mathcal{D}_1(\mathcal{A}_\Omega(Y_k)) - \mathcal{D}_1(\mathcal{A}_\Omega(Y^*))\|^2 \\
&= \|Y_k - Y^*\|_F^2 - 2h\alpha \text{trace}(\mathcal{A}_\Omega(Y_k) - \mathcal{A}_\Omega(Y^*), \mathcal{D}_1(\mathcal{A}_\Omega(Y_k)) - \mathcal{D}_1(\mathcal{A}_\Omega(Y^*))) \\
&\quad + h^2 \|\mathcal{D}_1(\mathcal{A}_\Omega(Y_k)) - \mathcal{D}_1(\mathcal{A}_\Omega(Y^*))\|^2 \\
&\leq \|Y_k - Y^*\|_F^2 + (h^2 - 2h\alpha)\|\mathcal{D}_1(\mathcal{A}_\Omega(Y_k)) - \mathcal{D}_1(\mathcal{A}_\Omega(Y^*))\|^2,
\end{aligned}$$

where we have used Lemma 9.45. It thus follows that

$$\|Y_{k+1} - Y^*\|_F \leq \|Y_k - Y^*\|_F$$

if $h^2 - 2h\alpha < 0$ when $h > 0$ is sufficiently small. It follows immediately that $Y_k, k \geq 1$, are bounded and hence have a convergent subsequence. Hence, there exists a positive constant $C < \infty$ such that

$$\|Y_k\|_F \leq C \quad \forall k \geq 1 \tag{9.48}$$

as soon as $h < 2\alpha$.

However, this is not enough for us to establish the linear convergence. Let $B(0, 1) \subset \mathbb{R}^{n \times n}$ be the open unit ball in $\mathbb{R}^{n \times n}$. Fix any $\epsilon > 0$. We define $\mathcal{Y}_\epsilon^* = \mathcal{Y}^* + \epsilon B(0, 1) \subset \mathbb{R}^{n \times n}$, which is an open set. Let us state the main result in this section.

Theorem 9.46. *Fix an $\epsilon > 0$. Suppose that the linear mapping $\mathcal{A}_\Omega$ is normalized so that $\|\mathcal{A}_\Omega(B)\|_F \leq \|B\|_F$ for any square matrix B of size $n \times n$. Suppose that $h > 0$ small enough. Then*

$$\|Y_{k+1} - Y^*\|_F^2 \leq (1 - h(2\eta - h))\|Y_k - Y^*\|_F^2 \tag{9.49}$$

for all $Y_k, k \geq 1$, which are not in $\mathcal{Y}_\epsilon^*$, i.e., not within tolerance ϵ, where $Y^* \in \mathcal{Y}^*$, $h < 2\epsilon$, and $\eta > 0$ is to be defined in the proof.

Proof. Let Y^* be a minimizer of (9.43). Since $\nabla f(Y^*) = 0$, we have

$$\mathcal{A}_\Omega(\mathcal{D}_1(\mathcal{A}_\Omega(Y^*))) = B/\alpha = \mathcal{A}_\Omega(M)/\alpha \neq 0.$$

That is, $\mathcal{D}_1(\mathcal{A}_\Omega(Y^*)) \neq 0$. It follows that the leading singular value $\sigma_1(\mathcal{A}_\Omega(Y^*)) > 1$. If Y_k is such that $\mathcal{D}_1(\mathcal{A}_\Omega(Y_k)) - \mathcal{D}_1(\mathcal{A}_\Omega(Y^*)) = 0$, i.e., $\nabla f(Y_k) = 0$, then Y_k is a minimizer and the iteration (9.45) will not improve, and hence we are done.

Let us consider the situation where $Y_k \notin \mathcal{Y}^*$. $\|\mathcal{D}_1(Y_k) - \mathcal{D}_1(Y^*)\|_F \neq 0$. As mentioned above, we have $\|Y_k\|_F \leq C < \infty$ for all k for a positive constant C which is dependent on 2α. Since $\mathcal{Y}_\epsilon^* = \mathcal{Y}^* + \epsilon B(0,1) \subset \mathbb{R}^{n \times n}$ is an open set, the set $\{Y : \|Y\|_F \leq C\}\backslash\mathcal{Y}_\epsilon^*$ is a closed set. Next we let η be defined by

$$\eta = \min_{\{Y:\|Y\|_F \leq C\}\backslash\mathcal{Y}_\epsilon^*} \frac{\|\mathcal{D}_1(\mathcal{A}_\Omega(Y)) - \mathcal{D}_1(\mathcal{A}_\Omega(M/\alpha))\|_f^2}{\|\mathcal{A}_\Omega(Y) - \mathcal{A}_\Omega(Y^*)\|_F^2} \tag{9.50}$$

and claim that $\eta > 0$. Indeed, if $\eta = 0$, there exists a sequence $X_k \in \{Y : \|Y\|_F \leq C\}\backslash\mathcal{Y}_\epsilon^*$ such that

$$\frac{\|\mathcal{D}_1(\mathcal{A}_\Omega(X_k)) - \mathcal{D}_1(\mathcal{A}_\Omega(M/\alpha))\|_f^2}{\|\mathcal{A}_\Omega(X_k) - \mathcal{A}_\Omega(Y^*)\|_F^2} \leq \frac{1}{k}.$$

Letting X_* be the limit of X_k, which is inside $\{Y : \|Y\|_F \leq C\}\backslash\mathcal{Y}_\epsilon^*$ since it is closed, we have

$$\|\mathcal{D}_1(\mathcal{A}_\Omega(X_*)) - \mathcal{D}_1(\mathcal{A}_\Omega(Y^*))\|_F^2 = 0$$

and $X_* \in \mathcal{Y}^*$, which is a contradiction.

Thus, if $Y_k \notin \mathcal{Y}_\epsilon^*$, i.e., Y_k is not within the ϵ-neighborhood of the minimal set $\mathcal{Y}^*$, we have

$$\alpha\,\mathrm{trace}(\mathcal{A}_\Omega(Y_k) - \mathcal{A}_\Omega(Y^*), \mathcal{D}_1(\mathcal{A}_\Omega(Y_k)) - \mathcal{D}_1(\mathcal{A}_\Omega(Y^*))) \geq \eta\|\mathcal{A}_\Omega(Y_k) - \mathcal{A}_\Omega(Y^*)\|_F^2. \tag{9.51}$$

In this setting, we have

$$\begin{aligned}
\|Y_{k+1} - Y^*\|_F^2 &= \|Y_k - Y^*\|_F^2 - 2h\alpha\,\mathrm{trace}(\mathcal{A}_\Omega(Y_k - Y^*), \mathcal{D}_1(\mathcal{A}_\Omega(Y_k)) - \mathcal{D}_1(\mathcal{A}_\Omega(Y^*))) \\
&\quad + h^2\|\mathcal{D}_1(\mathcal{A}_\Omega(Y_k)) - \mathcal{D}_1(\mathcal{A}_\Omega(Y^*))\|_F^2 \\
&\leq \|Y_k - Y^*\|_F^2 - \|\mathcal{A}_\Omega(Y_k) - \mathcal{A}_\Omega(Y^*)\|_F^2 + \|\mathcal{A}_\Omega(Y_k) - \mathcal{A}_\Omega(Y^*)\|_F^2 \\
&\quad - 2h\eta\|\mathcal{A}_\Omega(Y_k) - \mathcal{A}_\Omega(Y^*)\|_F^2 + h^2\|\mathcal{A}_\Omega(Y_k) - \mathcal{A}_\Omega(Y^*)\|_F^2 \\
&\leq \|\mathcal{A}_\Omega(Y_k) - \mathcal{A}_\Omega(Y^*)\|_F^2(1 - 2h\eta + h^2).
\end{aligned}$$

Finally, we note that $\mathcal{A}_\Omega$ is linear and $\|\mathcal{A}_\Omega(Y)\|_F \leq \|Y\|_F$. Thus we have

$$\|Y_{k+1} - Y^*\|_F^2 \leq (1 - h(2\eta - h))\|Y_k - Y^*\|_F^2$$

if $h > 0$ is sufficiently small, say $h < 2\eta$. The discussion above finishes the proof of Theorem 9.46. □

9.5 ▪ Fixed Point Iterative Methods

An unconstrained version of the minimization (9.7) is (9.3):

$$\min_{X \in \mathbb{R}^{m \times n}} \|X\|_* + \frac{1}{2\lambda} \|\mathcal{A}_\Omega(X) - \mathcal{A}_\Omega(M)\|_F^2. \tag{9.52}$$

Since the objective function in (9.52) is convex, a minimizer X^* is the optimal solution to (9.52) if and only if

$$\mathbf{0} \in \lambda \partial \|X^*\|_* + g(X^*), \tag{9.53}$$

where $g(X) = \mathcal{A}_\Omega^\top(\mathcal{A}_\Omega(X) - \mathcal{A}_\Omega(M))$. For any $\alpha > 0$,

$$\mathbf{0} \in \alpha\lambda \partial \|X^*\|_* + \alpha g(X^*) = \alpha\lambda \partial \|X^*\|_* + X^* - (X^* - \alpha g(X^*)).$$

For simplicity, let us write

$$Y^* = X^* - \alpha g(X^*). \tag{9.54}$$

It follows that

$$\mathbf{0} \in \alpha\lambda \partial \|X^*\|_* + X^* - Y^*.$$

In other words, X^* is the optimal solution to

$$\min_{X \in \mathbb{R}^{n \times m}} \alpha \|X\|_* + \frac{1}{2\lambda} \|X - Y^*\|_F^2. \tag{9.55}$$

By Theorem 9.40, the solution X^* can be expressed in terms of the matrix shrinkage operator by $X^* = \mathcal{D}_{\alpha\lambda}(Y^*)$. Thus, Theorem 9.47 follows.

Theorem 9.47 (Ma, Goldfarb, and Chen, 2011 [23]). *X^* is an optimal solution to the minimization (9.52) if and only if $X^* = \mathcal{D}_{\alpha\lambda}(h(X^*))$, where $h(X^*) = X^* - \alpha g(X^*)$.*

Proof. The discussion above proved that an optimal solution X^* of (9.52) is a fixed point of the operator $T = \mathcal{D}_{\alpha\lambda}(h)$. On the other hand, if X^* is a fixed point, we know that X^* is a solution of (9.55), and then X^* will satisfy (9.53). Hence, it is an optimal solution of (9.52). ☐

The steps (9.54) and (9.55) trivially lead to the fixed point iterative algorithm below.

ALGORITHM 9.3
Fixed Point Iterative Algorithm [23]
  1: Starting with an initial guess X^1, we compute iteratively

$$\begin{cases} Y^k = X^k - \alpha g(X^k), \\ X^{k+1} = \mathcal{D}_{\alpha\lambda}(Y^k) \end{cases}$$

for $k \geq 1$ until a tolerance is achieved or a maximal number of iterations is reached.

We now discuss the convergence of Algorithm 9.3. Let us present two approaches to establishing the convergence. Define $f(X) = \frac{1}{2\lambda} \|\mathcal{A}_\Omega(X) - \mathbf{b}\|^2$ with $\mathbf{b} = \mathcal{A}_\Omega(M)$ and $G(X) = f(X) + \alpha \|X\|_*$. The following is a crucial preparatory result.

Lemma 9.48. *Let $X^k, k \geq 1$, be the sequence from Algorithm 9.3. For any $X \in \mathbb{R}^{m \times n}$,*

$$G(X) - G(X^{k+1}) \geq \frac{L}{2} \|X^k - X^{k+1}\|_F^2 + L\langle X^k - X, X^{k+1} - X^k \rangle,$$

where L is the Lipschitz constant of the differentiation of f.

Proof. By the convexity of function f and the convexity of the nuclear norm functional $\alpha \| \cdot \|_*$, we have

$$f(X) \geq f(X^k) + \langle \nabla f(X^k), X - X^k \rangle,$$
$$\alpha \|X\|_* \geq \alpha \|X^{k+1}\|_* + \langle \gamma_k, X - X^{k+1} \rangle_F,$$

where γ_k denotes a subdifferential of $\alpha \| \cdot \|_*$ at X^{k+1}. The summation of the two inequalities above yields

$$G(X) \geq f(X^k) + \langle \nabla f(X^k), X - X^k \rangle + \alpha \|X^{k+1}\|_* + \langle \gamma_k, X - X^{k+1} \rangle_F.$$

By using the convexity of f, we have

$$G(X^{k+1}) = \alpha \|X^{k+1}\|_* + f(X^{k+1})$$
$$\leq \alpha \|X^{k+1}\|_* + f(X^k) + \langle \nabla f(X^k), X^{k+1} - X^k \rangle + \frac{L}{2} \|X^{k+1} - X^k\|_F^2,$$

where L is the Lipschitz constant of the differentiation of f. It follows that

$$G(X) - G(X^{k+1}) \geq -\frac{L}{2} \|X^{k+1} - X^k\|_F^2 + \langle \nabla f(X^k), X - X^{k+1} \rangle + \langle \gamma_k, X - X^{k+1} \rangle.$$

Note that the optimality condition of X^{k+1} yields $\gamma_k \in -\nabla f(X^k) + L(X^k - X^{k+1})$. Using this relation in the above equation, we have

$$G(X) - G(X^{k+1}) \geq -\frac{L}{2} \|X^{k+1} - X^k\|_F^2 + L \langle X^k - X^{k+1}, X - X^{k+1} \rangle$$
$$= \frac{L}{2} \|X^{k+1} - X^k\|_F^2 + L \langle X^k - X^{k+1}, X - X^k \rangle.$$

This completes the proof. □

To show that Algorithm 9.3 is convergent, let us rewrite the iterations as

$$X^{k+1} = \mathcal{D}_{\alpha\lambda}(X^k - \alpha g(X^k)) = \mathcal{D}_{\alpha\lambda}((I - \alpha g)(X^k)),$$

which is a fixed point method. Next, we show that it is a convergent fixed point method.

Theorem 9.49. *The sequence $\{X^k, k \geq 1\}$ from Algorithm 9.3 has at least one subsequence which converges to a fixed point of operator $\mathcal{D}_{\alpha\lambda}$. Thus, the fixed point is an optimal solution of (9.52) by Theorem 9.47.*

Proof. It is clear that $X^k, k \geq 1$, are bounded. Therefore, there exists a convergent subsequence. Without loss of generality, we may assume that $X^k \to X^*$. We let $X = X^k$ in Lemma 9.48 to get

$$G(X^k) - G(X^{k+1}) \geq \frac{L}{2} \|X^k - X^{k+1}\|_F^2.$$

It follows that $G(X^1) \geq \frac{L}{2} \sum_{k=1}^{\infty} \|X^k - X^{k+1}\|_F^2$. Hence, $X^k - X^{k+1} = X^k - \mathcal{D}_{\alpha\lambda}(I - \alpha g)(X^k) \to 0$. That is, $X^* = \mathcal{D}_{\alpha\lambda}(I - \alpha g)(X^*)$ due to the continuity of the operator $\mathcal{D}_{\alpha\lambda}(I - \alpha g)$. □

Furthermore, we can establish the convergence rate of Algorithm 9.3.

Theorem 9.50. *Let* $\eta_k = G(X^k) - G(X^*)$, *where* X^* *is an optimal solution of* (9.52). *Then*

$$\eta_k \leq \frac{L\|X^1 - X^*\|_F^k}{2k}$$

for all $k \geq 1$, *where* L *is the Lipschitz constant of the differentiation of* f.

Proof. In Lemma 9.48, we choose $X = X^*$ (an optimal solution of (9.52)) in order to obtain

$$G(X^*) - G(X^{k+1}) \geq \frac{L}{2}\|X^k - X^{k+1}\|_F^2 + L\langle X^k - X^*, X^{k+1} - X^k\rangle$$

$$= \frac{L}{2}\|X^* - X^{k+1}\|_F^2 - \frac{L}{2}\|X^* - X^k\|_F^2.$$

That is, we can write

$$\eta_{k+1} \leq \frac{L}{2}\|X^* - X^k\|_F^2 - \frac{L}{2}\|X^* - X^{k+1}\|_F^2. \tag{9.56}$$

Note that $G(X^k)$ is monotonically decreasing. Indeed, letting $X = X^k$ in Lemma 9.48, we have $G(X^k) - G(X^{k+1}) \geq \frac{L}{2}\|X^k - X^{k+1}\|_F^2 \geq 0$. With this fact, we sum the inequalities in (9.56) from integers 1 to N to get

$$N\eta_N \leq \sum_{k=1}^{N} \eta_k \leq \frac{L}{2}\|X^* - X^1\|_F^2 - \frac{L}{2}\|X^* - X^{N+1}\|_F^2 \leq \frac{L}{2}\|X^* - X^1\|_F^2.$$

This completes the proof. $\square$

Furthermore, we can speed up the computation by using Nesterov's acceleration technique. That is, we have Algorithm 9.4.

ALGORITHM 9.4
Accelerated Fixed Point Iterative Algorithm [23]
1: Starting with any initial guess $X^0 = Y^0$ and $a_0 = 1$, we iteratively compute

$$\begin{cases} X^k = \mathcal{D}_{\alpha\lambda}((I - \alpha g)(Y^k)), \\ a_{k+1} = (1 + \sqrt{4a_k^2 + 1})/2, \\ Y^{k+1} = X^k + \frac{a_k - 1}{a_{k+1}}(X^k - X^{k-1}) \end{cases}$$

for $k \geq 0$.

As discussed in section 5.4, the updated rule on Y^{k+1} should be replaced by

$$Y^{k+1} = X^k + \frac{k}{k + \alpha}(X^k - X^{k-1}) \tag{9.57}$$

for a fixed $\alpha > 2$ to achieve a better convergence rate. That is, the resulting algorithm will lead to $o(1/k^2)$. See Exercise 19.

To establish the convergence of Algorithm 9.4, we use $\eta_k = G(X^k) - G(X^*)$ and show the following result.

Lemma 9.51. *For* $k \geq 1$,

$$\frac{2}{L}a_k^2\eta_k - \frac{2}{L}a_{k+1}^2\eta_{k+1} \geq \|\mathbf{u}_{k+1}\|^2 - \|\mathbf{u}_k\|^2, \tag{9.58}$$

where $\mathbf{u}_k = a_k X^k - (a_k - 1)X^{k-1} - X^*$.

Proof. We use Lemma 9.48. Choosing $X = X^k$, we have

$$\frac{2}{L}(\eta_k - \eta_{k+1}) \geq \|X^{k+1} - Y^{k+1}\|_F^2 + 2\langle X^{k+1} - Y^{k+1}, Y^{k+1} - X^k \rangle. \tag{9.59}$$

Choosing $X = X^*$, we have

$$-\frac{2}{L}\eta_{k+1} \geq \|X^{k+1} - Y^{k+1}\|_F^2 + 2\langle X^{k+1} - Y^{k+1}, Y^{k+1} - X^* \rangle. \tag{9.60}$$

Multiplying (9.59) by $a_{k+1} - 1$ and (9.60) by a_{k+1}, we add the two resulting inequalities to get

$$\frac{2}{L}((a_{k+1} - 1)\eta_k - a_{k+1}\eta_{k+1}) \geq a_{k+1}\|X^{k+1} - Y^{k+1}\|_F^2$$

$$+ 2\langle X^{k+1} - Y^{k+1}, a_{k+1}Y^{k+1} - (a_{k+1} - 1)X^k - X^* \rangle.$$

Multiplying the above inequality by a_{k+1} and using the relation $a_k^2 = a_{k+1}^2 - a_{k+1}$, we obtain

$$\frac{2}{L}(a_k^2\eta_k - a_{k+1}^2\eta_{k+1})$$

$$\geq \|a_{k+1}(X^{k+1} - Y^{k+1})\|_F^2 + 2a_{k+1}\langle X^{k+1} - Y^{k+1}, a_{k+1}Y^{k+1} - (a_{k+1} - 1)X^k - X^* \rangle$$

$$\geq \|a_{k+1}X^{k+1} - (a_{k+1} - 1)X^k - X^*\|_F^2 - \|a_{k+1}Y^{k+1} - (a_{k+1} - 1)Y^k - X^*\|_F^2$$

$$= \|\mathbf{u}_{k+1}\|_F^2 - \|\mathbf{u}_k\|_F^2,$$

where we have used the definition of Y^{k+1} and $\mathbf{u}_k$ to see that $a_{k+1}Y^{k+1} - (a_{k+1} - 1)Y^k - X^* = \mathbf{u}_k$. ☐

We are now ready to establish the following convergence rate of Algorithm 9.4.

Theorem 9.52. *Suppose that $X^k, k \geq 1$, are a sequence of matrices from Algorithm 9.4. Let $\eta_k = G(X^k) - G(X^*)$. Then*

$$\eta_k \leq \frac{C}{k^2}$$

for all $k \geq 1$, where $C > 0$ is a constant.

Proof. We use Lemma 9.51 and rewrite the inequality as

$$\frac{2a_k^2}{L}\eta_k + \|\mathbf{u}_k\|_F^2 \leq \frac{2a_{k+1}^2}{L}\eta_{k+1} + \|\mathbf{u}_{k+1}\|_F^2.$$

We repeat the above inequality to get

$$\frac{2a_{k+1}^2}{L}\eta_{k+1} \leq \frac{2a_{k+1}^2}{L}\eta_{k+1} + \|\mathbf{u}_k\|_F^2 \leq \cdots \leq \frac{2a_2^2}{L}\eta_2 + \|\mathbf{u}_1\|_F^2.$$

Letting C be the constant in the rightmost inequality above, we have

$$\eta_{k+1} \leq \frac{L}{2a_{k+1}^2}C.$$

It is easy to see that $a_{k+1} \geq a_k + 1/2 \geq \cdots \geq (k + 1)/2$. This shows that $\eta_k \leq C/k^2$ for another constant $C > 0$. ☐

Next we can use the theory of the nonexpansive operator to establish the convergence of Algorithm 9.3. Let us first show that $\mathcal{D}_\tau$ is nonexpansive as follows.

Lemma 9.53 (Ma, Goldfarb, and Chen, 2011 [23]). *The shrinkage operator $\mathcal{D}_\tau$ is nonexpansive. That is, for any $X, Y \in \mathbb{R}^{m\times n}$,*

$$\|\mathcal{D}_\tau(X) - \mathcal{D}_\tau(Y)\|_F \le \|X - Y\|_F.$$

Proof. Since $\mathcal{D}_\tau(X)$ is an optimal solution of (9.52) and the minimizing functional is convex, we know that

$$\mathbf{0} \in \tau\partial\|\mathcal{D}_\tau(X)\|_* + \mathcal{D}_\tau(X) - X.$$

That is, $X - \mathcal{D}_\tau(X) \in \tau\partial\|\mathcal{D}_\tau(X)\|_*$. Since $\|X\|_*$ is convex, by subdifferentiation, we have

$$\|Y\|_* - \|X\|_* \ge \langle \partial\|X\|_*, Y - X\rangle_F$$

for any matrices X and Y. It follows that

$$\|\mathcal{D}_\tau(X)\|_* - \|\mathcal{D}_\tau(Y)\|_* \ge \langle \partial\|\mathcal{D}_\tau(Y)\|_*, \mathcal{D}_\tau(X) - \mathcal{D}_\tau(Y)\rangle_F$$
$$= \left\langle \frac{1}{\tau}(Y - \mathcal{D}_\tau(Y)), \mathcal{D}_\tau(X) - \mathcal{D}_\tau(Y)\right\rangle_F.$$

Similarly, we have

$$\|\mathcal{D}_\tau(Y)\|_* - \|\mathcal{D}_\tau(X)\|_* \ge \left\langle \frac{1}{\tau}(X - \mathcal{D}_\tau(X)), \mathcal{D}_\tau(Y) - \mathcal{D}_\tau(X)\right\rangle_F.$$

Adding the above two inequalities, we have

$$0 \ge \frac{1}{\tau}\langle Y - \mathcal{D}_\tau(Y) - (X - \mathcal{D}_\tau(X)), \mathcal{D}_\tau(X) - \mathcal{D}_\tau(Y)\rangle_F.$$

It follows that

$$\langle X - Y, \mathcal{D}_\tau(X) - \mathcal{D}_\tau(Y)\rangle_F \ge \langle \mathcal{D}_\tau(X) - \mathcal{D}_\tau(Y), \mathcal{D}_\tau(X) - \mathcal{D}_\tau(Y)\rangle_F.$$

Using the Cauchy–Schwarz inequality on the term on the left-hand side of the above inequality yields the desired nonexpansiveness.   □

In addition, it is easy to see the following result.

Lemma 9.54. *Suppose that $\alpha \in (0, 2/\|A^\top A\|)$. Then $I - \alpha g$ is nonexpansive.*

Proof. For any two matrices X and $Y\mathbb{R}^{m\times n}$, we have

$$(I - \alpha g)(X) - (I - \alpha g)(Y) = X - Y - \alpha A^* A(X - Y) = (I - \alpha A^* A)(X - Y).$$

Since A^*A is nonnegative definite, as long as $\alpha \in (0, 2/\|A^*A\|)$, the norm

$$\|I - \alpha A^* A\| = \max_{\substack{\lambda \\ \text{eigenvalue of } A^*A}} \{|1 - \alpha\lambda|, 1\} \le 1.$$

That is, $I - \alpha g$ is nonexpansive.   □

Since the composition of two nonexpansive operators is also nonexpansive, letting $T = \mathcal{D}_{\alpha\lambda}(I - \alpha g)$, T is a nonexpansive operator. With additional properties of $\mathcal{D}_{\alpha\lambda}$ and $I - \alpha g$, we can show the convergence of Algorithm 9.3. Without these properties, we can formulate the fixed point algorithm $X^{k+1} = \mathcal{D}_{\alpha\lambda}(X^k - \alpha g(X^k))$ for $k \geq 1$ into a Mann iterative algorithm as follows.

ALGORITHM 9.5
An Averaged Fixed Point Algorithm

Input: Start with an initial guess X^0.
We iteratively compute

$$X^k = \frac{1}{2}X^{k-1} + \frac{1}{2}T(X^{k-1})$$

for $k = 1, 2, \ldots$ until a maximum number of iterations is reached.
Output: A completed matrix X^k.

We now show that the above Mann iterative algorithm converges.

Theorem 9.55. *The Mann iterative algorithm (Algorithm* 9.5*) converges.*

Proof. We first recall the following equality. For any $x, y, z \in K$ and a real number $\lambda > 0$, we have the identity

$$\|\lambda x + (1 - \lambda)y - z\|^2 = \lambda \|x - z\|^2 + (1 - \lambda)\|y - z\|^2 - \lambda(1 - \lambda)\|x - y\|^2$$

for any $x, y \in \mathbb{R}^n$. Let $\lambda = 1/2$ and $x = X^k$, $y = T(X^k)$, and $z = X^*$ be a fixed point or the solution to the problem (9.52). Then we have

$$\|X^{k+1} - X^*\|^2 = \frac{1}{2}\|X^k - X^*\|^2 + \frac{1}{2}\|T(X^k) - X^*\|^2 - \frac{1}{4}\|X^k - T(X^k)\|^2$$
$$\leq \|X^k - X^*\|^2 - \frac{1}{4}\|X^k - T(X^k)\|^2.$$

It follows that

$$\sum_{k=1}^{N} \frac{1}{4}\|X^k - T(X^k)\|^2 + \|X^{N+1} - X^*\|^2 \leq \|X^0 - X^*\|^2.$$

That is, $\|X^k - T(X^k)\| \to 0$. Let $\widehat{X}$ be the limit of any subsequence of $X^k, k \geq 1$. Then we have $\widehat{X} = T(\widehat{X})$.   □

Similarly, we can show the convergence of the following Mann iterative algorithm. The proof of the convergence is left to the interested reader.

ALGORITHM 9.6
An Averaged Fixed Point Algorithm

1: Starting with an initial guess X^1, we iteratively compute

$$X^{k+1} = \frac{1}{3}X^k + \frac{1}{3}T(X^k) + \frac{1}{3}T^2(X^k)$$

for $k = 1, 2, \ldots$.

9.6 ▪ Orthogonal Rank 1 Matrix Completion

In this section, we present a simple and efficient method of solving the low rank matrix completion problem by extending the orthogonal matching pursuit (OMP) procedure from the vector case to the matrix case. It is an iterative approach. In each iteration, a rank 1 basis matrix is generated by the left and right top singular vectors of the current approximation residual. We shall propose two algorithms in this section. In the first algorithm, we fully update the weights for all rank 1 matrices in the current basis set at the end of each iteration by performing an orthogonal projection of the observation matrix onto their spanning subspace. The most time-consuming step of the proposed algorithm is to calculate the top singular vector pair of a sparse matrix, which costs $O((n + m)|\Omega|)$ operations in each iteration. One drawback of the first algorithm is that it needs to store all rank 1 matrices in the current basis set for full weight updating, which contains $r|\Omega|$ elements in the rth iteration. This makes the storage complexity of the algorithm dependent on the number of iterations, which restricts the approximate rank of the target matrix for large-scale matrices. To relieve the storage complexity we propose an economic weight updating rule in our second algorithm. In this algorithm, we only track two matrices in each iteration. One is the current estimation of the target matrix and the other is the pursued rank 1 matrix. When restricted to the observations in Ω, each has $|\Omega|$ nonzero elements. So the storage requirement, i.e., $2|\Omega|$, stays the same for different iterations. To the best of the authors' knowledge, these proposed algorithms are the fastest among all matrix completion methods.

9.6.1 ▪ Orthogonal Rank 1 Matrix Pursuit

It is well known that any matrix $X \in \mathbb{R}^{n \times m}$ can be written as a linear combination of rank 1 matrices, i.e.,

$$X = M(\theta) = \sum_{i \in \mathcal{I}} \theta_i M_i, \tag{9.61}$$

where $\{M_i : i \in I\}$ is the set of all $n \times m$ rank 1 matrices with unit Frobenius norm. Clearly, θ is a finite-dimensional real vector. Such a representation can be obtained from the standard SVD decomposition of X. The original low rank matrix completion problem can be reformulated as

$$\begin{aligned} \min_{\theta} \quad & ||\theta||_0 \\ \text{subject to} \quad & P_\Omega(M(\theta)) = P_\Omega(Y), \end{aligned} \tag{9.62}$$

where $||\theta||_0$ denotes the cardinality of the set of nonzero elements of θ.

Let us reformulate the problem as

$$\begin{aligned} \min_{\theta} \quad & ||P_\Omega(M(\theta)) - P_\Omega(Y)||_F^2 \\ \text{subject to} \quad & ||\theta||_0 \le r. \end{aligned} \tag{9.63}$$

We solve it by an OMP-type greedy algorithm using rank 1 matrices as the basis. In particular, we find a suitable subset with overcomplete rank 1 matrix coordinates and learn the weight for each coordinate. This is achieved by executing two steps alternately: one is to pursue the basis matrices, and the other is to learn the weights of the basis matrices. Suppose that after the $(k - 1)$th iteration, the rank 1 basis matrices $M_1, \ldots, M_{k-1}$ and their current weight θ^{k-1} are already computed. In the kth iteration, we pursue a new rank 1 basis matrix M_k with unit Frobenius norm, which is mostly correlated with the current observed regression residual $R_k = P_\Omega(Y) - X_{k-1}$, where

$$X_{k-1} = (M(\theta^{k-1}))_\Omega = \sum_{i=1}^{k-1} \theta_i^{k-1}(M_i)_\Omega.$$

Therefore, M_k can be chosen to be an optimal solution of the problem

$$\max_{M}\{\langle M, R_k\rangle :\ \mathrm{rank}(M) = 1,\ \|M\|_F = 1\}. \tag{9.64}$$

Notice that each rank 1 matrix M with unit Frobenius norm can be written as the product of two unit vectors, namely, $M = \mathbf{u}\mathbf{v}^\top$ for some $\mathbf{u} \in \mathbb{R}^n$ and $\mathbf{v} \in \mathbb{R}^m$ with $\|\mathbf{u}\| = \|\mathbf{v}\| = 1$. We then see that problem (9.64) can be equivalently solved as

$$\max_{\mathbf{u},\mathbf{v}}\{\mathbf{u}^\top R_k\mathbf{v} :\ \|\mathbf{u}\| = \|\mathbf{v}\| = 1\}. \tag{9.65}$$

Clearly, the optimal solution $(\mathbf{u}_*, \mathbf{v}_*)$ of problem (9.65) is a pair of top left and right singular vectors of R_k. It can be efficiently computed by the power method. The new rank 1 basis matrix M_k is then readily available by setting $M_k = \mathbf{u}_*\mathbf{v}_*^\top$.

After finding the new rank 1 basis matrix M_k, we update the weights θ^k for all currently available basis matrices $\{M_1, \ldots, M_k\}$ by solving the least squares regression problem

$$\min_{\theta \in \mathbb{R}^k} \left\| \sum_{i=1}^{k} \theta_i M_i - Y \right\|_\Omega^2. \tag{9.66}$$

This can be done by reshaping the matrices $(Y)_\Omega$ and $(M_i)_\Omega$ into vectors $\dot{\mathbf{y}}$ and $\dot{\mathbf{m}}_i$, after which we can easily see that the optimal solution θ^k of (9.66) is given by

$$\theta^k = (\mathbf{M_k}^\top \mathbf{M_k})^{-1}\mathbf{M_k}^T \dot{\mathbf{y}}, \tag{9.67}$$

where $\mathbf{M_k} = [\dot{\mathbf{m}}_1, \ldots, \dot{\mathbf{m}}_k]$ is the matrix formed by all reshaped basis vectors. The row size of matrix $\mathbf{M}_k$ is the number of total observed entries.

We run the above two steps iteratively until some desired stopping condition is satisfied. We can terminate the method based on the rank of the approximation matrix or the approximation residual. That is, we can choose a preferred rank for the approximate solution matrix and run the method for that number of iterations. Alternatively, we can stop the method once the residual $\|R_k\|$ is less than a tolerance parameter ε. The main steps of the orthogonal rank 1 matrix pursuit (OR1MP) are given in Algorithm 9.7.

ALGORITHM 9.7
Orthogonal Rank 1 Matrix Pursuit Algorithm (OR1MP) [36]

1: **Input:** Y and a tolerance parameter ε.
2: **Initialize:** Set $X_0 = 0$, $\theta^0 = 0$ and $k = 1$.
3: **while** True **do**
4:      Step 1. Find a pair of top left and right singular vectors $(\mathbf{u}_k, \mathbf{v}_k)$ of the observed residual matrix $R_k = Y_\Omega - X_{k-1}$ and set $M_k = \mathbf{u}_k(\mathbf{v}_k)^\top$.
5:      Step 2. Compute the weight θ^k using the closed-form least squares solution

$$\theta^k = (M_k{}^\top M_k)^{-1}M_k{}^T \dot{\mathbf{y}}.$$

6:      Step 3. Set $X_k = \sum_{i=1}^{k} \theta_i^k (M_i)_\Omega$ and $k \leftarrow k + 1$
        until observed residual $\|R_k\|$ is smaller than ε.
7: **end while**
8: **Output:** Learned matrix $\hat{Y} = \sum_{i=1}^{k} \theta_i^k M_i$.

Our MATLAB implementation is given below.

```
%This is a demo program to demonstrate how to
% solve the image recovery problem by using low rank matrix approximation.
% It is written by Zheng Wang and modified by Ming-Jun Lai. If you find
% this program useful, please refer to our paper
% Wang, Z., Lai, M.-J., Lu, Z., Fan, W., Davulcu, H. and Ye, J.,
% Orthogonal Rank-One Matrix Pursuit for Low Rank Matrix Completion,
% SIAM Journal of Scientific Computing, vol. 37 (2015)  A488--A514.

image_name  = 'peppers512';
% read image
inData  = double(imread(image_name, 'bmp'));
mrate   = 0.8;  % missing ratio
amp     = 0; %0.01; % noise ratio
% image preprocessing
m = size(inData, 1);n = size(inData, 2);
% save the ground truth image, normalized into [0, 1], and plot the true
% image.
truthData = inData./256;
figure, image(inData), colormap gray(256)
missing_idx = rand(size(inData)) > mrate;
inData(missing_idx) = 0; % >254; % 0 is black, 255 is white
mask = inData(:,:) < 1;          % the mask matrix indicating the position
                                 % of the missing value: 1 means missing,
                                 % 0 means observed
figure, image(inData), colormap gray(256)
inData  = inData./256;           % normalization
noise   = randn(m, n);           % generate the Gaussian random noise
noise   = amp * norm(inData,'fro')/norm(noise,'fro')
inData  = inData + noise;

tr_idx     = find(~mask);            % observed position
tr_val_n   = inData(tr_idx);

% initial the parameters
r          = 50;
U          = []; V = []; rM = [];
estData    = 0;  i = 0;
val        = []; err_curve   = [];
datanorm   = norm(truthData, 'fro');
while (i < r) % & (gresnorm > epsilon )
% 1. find the top singular pair of the residual and update the residual
% each rank-one matrix u*v' can be viewed as a feature
    res = inData - estData;
    res(missing_idx) = 0;
    % top-1 singular vector decomposition
    [u, s, v] = svds(res, 1);
    % 2. update the weight w according to the basis uv'
U = [U u]; V = [V v];
```

```
    temp = u * v';
    mi = temp(tr_idx);
    rM = [rM mi];
    val = [val; mi'*tr_val_n];
    % solve the least squares problem for weight w
    [d1 d2] = size(rM);
    if (d1 > d2)
        Minv = inv(rM'*rM);
    else
        Minv = pinv(rM'*rM); % this is slow
    end
    w = Minv * val;
    estData = U * diag(w) * V'; %estData = Msup * Theta;
    i = i + 1;
end
fprintf( '\n Run for %d rounds! \n', i);
% calculate the approximation error in RMSE.
supp = length(truthData(:)); % which is 512 X 512
res = truthData - estData;
rmse = norm(res, 'fro') / sqrt(supp);
disp(['RMSE is ',num2str(rmse)])
figure,
image(estData*256), colormap gray(256)
```

We shall provide a few numerical experimental results in subsection 9.8.1.

9.6.2 ▪ Convergence Analysis

In this subsection, we will show that our Algorithm 9.7 converges at a linear rate. This main result is given in the following theorem. With this rate of convergence, we only need $O(\log(1/\epsilon))$ iterations to achieve an ϵ-accuracy solution.

Theorem 9.56. *The residual matrix from Algorithm* 9.7 *satisfies*

$$\|R_k\| \leq \left(\sqrt{1 - \frac{1}{\min(m,n)}} \right)^{k-1} \|Y\|_{\Omega} \quad \forall k \geq 1.$$

In order to prove Theorem 9.56, we need to establish some useful and preparatory properties. The first property says that R_{k+1} is perpendicular to all previously generated M_i for $i = 1, \ldots, k$.

Lemma 9.57. $\langle R_{k+1}, M_i \rangle = 0$ *for* $i = 1, \ldots, k$.

Proof. Recall that θ^k is the optimal solution of problem (9.66). By the first-order optimality condition, we have

$$\left\langle Y - \sum_{i=1}^{t} \theta_i^k M_i, M_i \right\rangle_{\Omega} = 0 \quad \text{for } i = 1, \ldots, k,$$

which together with $R_k = Y_{\Omega} - X_{k-1}$ and $X_k = \sum_{i=1}^{k} \theta_i^k (M_i)_{\Omega}$ implies that $\langle R_{k+1}, M_i \rangle = 0$ for $i = 1, \ldots, k$. □

The following property shows that as the number of rank 1 basis matrices M_i increases during our learning process, the residual $\|R_k\|$ does not increase.

Lemma 9.58. $\|R_{k+1}\| \leq \|R_k\|$ for all $k \geq 1$.

Proof. We observe that for all $k \geq 1$,

$$\|R_{k+1}\|^2 = \min_{\theta \in \mathbb{R}^k} \left\{ \left\| Y - \sum_{i=1}^{k} \theta_i M_i \right\|_\Omega^2 \right\}$$

$$\leq \min_{\theta \in \mathbb{R}^{k-1}} \left\{ \left\| Y - \sum_{i=1}^{k-1} \theta_i M_i \right\|_\Omega^2 \right\} = \|R_k\|^2,$$

and hence the conclusion holds. □

We next establish that $\{(M_i)_\Omega\}_{i=1}^k$ is linearly independent unless $\|R_k\| = 0$. It follows that formula (9.67) is well defined and hence θ^k is uniquely defined before the algorithm stops.

Lemma 9.59. Suppose that $R_k \neq 0$ for some $k \geq 1$. Then M_i has full column rank for all $i \leq k$.

Proof. Using Lemma 9.58 and the assumption $R_k \neq 0$ for some $k \geq 1$, we see that $R_i \neq 0$ for all $i \leq k$. We now prove this statement by induction on i. Indeed, since $R_1 \neq 0$, we clearly have $M_1 \neq 0$. Hence the conclusion holds for $i = 1$. We now assume that it holds for $i - 1 < k$ and need to show that it also holds for $i \leq k$. By the induction hypothesis, M_{i-1} has full column rank. Suppose for contradiction that M_i does not have full column rank. Then there exists $\alpha \in \mathbb{R}^{i-1}$ such that

$$(M_i)_\Omega = \sum_{j=1}^{i-1} \alpha_j (M_j)_\Omega,$$

which together with Lemma 9.57 implies that $\langle R_i, M_i \rangle = 0$. It follows that

$$\sigma_{\max}(R_i) = u_i^\top R_i v_i = \langle R_i, \mathbf{M}_i \rangle = 0,$$

and hence $R_i = 0$, which contradicts the fact that $R_j \neq 0$ for all $j \leq k$. Therefore, M_i has full column rank and the conclusion holds. □

We next build a relationship between two consecutive residuals $\|R_{k+1}\|$ and $\|R_k\|$. For convenience, define $\theta_k^{k-1} = 0$ and let

$$\theta^k = \theta^{k-1} + \eta^k.$$

In view of (9.66), we can observe that

$$\eta^k = \arg\min_\eta \left\| \sum_{i=1}^{k} \eta_i M_i - R_k \right\|_\Omega^2. \tag{9.68}$$

Let

$$L_k = \sum_{i=1}^{k} \eta_i^k (M_i)_\Omega. \tag{9.69}$$

By the definition of X_k, we can also observe that

$$X_k = X_{k-1} + L_k \quad \text{and} \quad R_{k+1} = R_k - L_k.$$

Lemma 9.60. $||R_{k+1}||^2 = ||R_k||^2 - ||L_k||^2$ and $||L_k||^2 \geq \langle M_k, R_k \rangle^2$, where L_k is defined in (9.69).

Proof. Since $L_k = \sum_{i \leq k} \eta_i^k (M_i)_\Omega$, it follows from Lemma 9.57 that $\langle R_{k+1}, L_k \rangle = 0$. We then have

$$
\begin{aligned}
||R_{k+1}||^2 &= ||R_k - L_k||^2 \\
&= ||R_k||^2 - 2\langle R_k, L_k \rangle + ||L_k||^2 \\
&= ||R_k||^2 - 2\langle R_{k+1} + L_k, L_k \rangle + ||L_k||^2 \qquad (9.70) \\
&= ||R_k||^2 - 2\langle L_k, L_k \rangle + ||L_k||^2 \\
&= ||R_k||^2 - ||L_k||^2.
\end{aligned}
$$

We next bound $||L_k||^2$ from below. If $R_k = 0$, $||L_k||^2 \geq \langle M_k, R_k \rangle^2$ clearly holds. We now suppose throughout the remaining proof that $R_k \neq 0$. It then follows from Lemma 9.59 that M_k has full column rank. Using this fact and (9.68), we have

$$\eta^k = \left(M_k^\top M_k \right)^{-1} M_k^\top \dot{\mathbf{r}}_k,$$

where $\dot{\mathbf{r}}_k$ is the reshaped residual vector of R_k. Invoking $L_k = \sum_{i \leq k} \eta_i^k (M_i)_\Omega$, we then obtain

$$||L_k||^2 = \dot{\mathbf{r}}_k^\top M_k (M_k^\top M_k)^{-1} M_k^\top \dot{\mathbf{r}}_k. \qquad (9.71)$$

Let $M_k = QU$ be the QR factorization of M_k, where $Q^\top Q = I$, U is a $k \times k$ nonsingular upper triangular matrix, and I is the identity matrix of size $k \times k$. We can observe that $(M_k)_k = \dot{\mathbf{m}}_k$, where $(M_k)_k$ denotes the kth column of the matrix M_k and $\dot{\mathbf{m}}_k$ is the reshaped vector of $(M_k)_\Omega$. Recall that $||M_k|| = ||\mathbf{u}_k \mathbf{v}_k^\top|| = 1$. Hence, $||(M_k)_k|| \leq 1$. Due to $Q^\top Q = I$, $M_k = QU$, and the definition of U, we have

$$0 < |U_{kk}| \leq ||U_k|| = ||(M_k)_k|| \leq 1.$$

In addition, by Lemma 9.57, we have

$$M_k^\top \dot{\mathbf{r}}_k = [0, \dots, 0, \langle M_k, R_k \rangle]^\top. \qquad (9.72)$$

Substituting $M_k = QU$ into (9.71), and using $Q^\top Q = I$ and (9.72), we obtain that

$$
\begin{aligned}
||L_k||^2 &= \dot{\mathbf{r}}_k^\top M_k (U^\top U)^{-1} M_k^\top \dot{\mathbf{r}}_k \\
&= [0, \dots, 0, \langle M_k, R_k \rangle] U^{-1} U^{-T} [0, \dots, 0, \langle M_k, R_k \rangle]^\top \\
&= \langle M_k, R_k \rangle^2 / (U_{kk})^2 \geq \langle M_k, R_k \rangle^2,
\end{aligned}
$$

where the last equality follows since U is upper triangular and the last inequality is due to $|U_{kk}| \leq 1$. □

We are now ready to prove Theorem 9.56.

Proof of Theorem 9.56. Using the definition of M_k, we have

$$\langle M_k, R_k \rangle = \langle \mathbf{u}^k (\mathbf{v}^k)^\top, R_k \rangle = \sigma_{\max}(R_k) \geq \sqrt{\frac{\sum_i \sigma_i^2(R_k)}{\mathrm{rank}(R_k)}} = \sqrt{\frac{||R_k||^2}{\mathrm{rank}(R_k)}} \geq \sqrt{\frac{||R_k||^2}{\min(m, n)}}.$$

Using this inequality and Lemma 9.60, we obtain

$$||R_{k+1}||^2 = ||R_k||^2 - ||L_k||^2 \le ||R_k||^2 - \langle M_k, R_k \rangle^2 \le \left(1 - \frac{1}{\min(m,n)}\right)||R_k||^2.$$

In view of this relation and the fact that $||R_1|| = ||Y||_\Omega^2$, we easily conclude that

$$||R_k|| \le \left(\sqrt{1 - \frac{1}{\min(m,n)}}\right)^{k-1} ||Y||_\Omega.$$

This completes the proof.   □

9.6.3 ▪ Economic Orthogonal Rank 1 Matrix Pursuit

The OR1MP algorithm has to track all pursued bases and save them in memory. It demands $O(r|\Omega|)$ storage complexity to save the bases. In large-scale problems, this is not negligible and restricts the rank of the approximate matrix. To adapt our algorithm for large-scale problems with arbitrary approximate rank, we simplify the orthogonal projection step. In this step, we only track the estimate matrix X_{k-1} and the rank 1 update matrix M_k. Then we calculate the best weights for these two matrices. In this new weight correction rule, we do not calculate the optimal least squares weights for all existing bases, but we solve the least squares problem

$$\alpha^k = \arg \min_{\alpha=\{\alpha_1,\alpha_2\}} ||\alpha_1 X_{k-1} + \alpha_2 M_k - Y||_\Omega^2. \tag{9.73}$$

This still fully corrects all weights of the existing bases, though the correction is suboptimal. The detailed procedure of this simplified method is given in Algorithm 9.8.

ALGORITHM 9.8
Economic Orthogonal Rank 1 Matrix Pursuit (EOR1MP) Algorithm [36]
1: **Input:** Y and a tolerance parameter ε.
2: **Initialize:** Set $X_0 = 0$, $\theta^0 = 0$, and $k = 1$.
3: **while** True **do**
4:     Step 1. Find a pair of top left and right singular vectors $(\mathbf{u}_k, \mathbf{v}_k)$ of the observed residual matrix $R_k = Y_\Omega - X_{k-1}$ and set $M_k = \mathbf{u}_k(\mathbf{v}_k)^\top$.
5:     Step 2. Compute the optimal weights α^k for X_{k-1} and M_k using least squares for

$$\arg \min_\alpha ||\alpha_1 X_{k-1} + \alpha_2 (M_k)_\Omega - Y_\Omega||^2.$$

6:     Step 3. Set $X_k = \alpha_1^k X_{k-1} + \alpha_2^k (M_k)_\Omega$ and $k \leftarrow k + 1$.
7:     until the observed residual $||R_k||$ is smaller than ε.
8: **end while**
9: **Output:** Learned matrix $\hat{Y} = X_k$.

This main result is given in the following theorem.

Theorem 9.61. *The EOR1MP algorithm, i.e., Algorithm 9.8, satisfies*

$$||R_k|| \le \left(\sqrt{1 - \frac{1}{\min(m,n)}}\right)^{k-1} ||Y||_\Omega \quad \forall k \ge 1.$$

The proof is similar to the one in the previous subsection. We leave the proof to Exercise 21; see also [36].

9.7 ▪ The Alternating Projection Algorithm

Suppose that the given entries $\mathcal{P}_\Omega(M)$ are from a matrix M of rank $r < n$. Many computational algorithms are designed to complete a rank r matrix by solving the least squares problem

$$\min\left\{\frac{1}{2}\|\mathcal{P}_\Omega(X) - \mathcal{P}_\Omega(M)\|_F^2 : X \in \mathcal{M}_r\right\}, \tag{9.74}$$

where $\mathcal{P}_\Omega(X)$ is the projection of X to the entries in Ω. Since $\mathcal{M}_r$ is not a linear space or a convex set or even a bounded set, the minimization (9.74) is nontrivial to solve. We shall introduce the so-called alternating projection method to complete a rank r matrix from given partial known entries. Then we shall explain its convergence.

9.7.1 ▪ Ideas for the Alternating Projection Algorithm

The alternating projection algorithm for matrix completion was first introduced in [22]. Let us describe the ideas for this algorithm as follows. Let $\mathcal{M}_r$ be the manifold in $\mathbb{R}^{n^2}$ consisting of $n \times n$ matrices of rank r, and denote by $P_{\mathcal{M}_r}$ the projection operator onto the manifold $\mathcal{M}_r$. Next consider the affine space $\mathcal{A}_\Omega$ defined by

$$\mathcal{A}_\Omega := \{X \mid \mathcal{P}_\Omega(X - M) = 0\}. \tag{9.75}$$

$\mathcal{A}_\Omega$ is a feasible set which consists of matrices which have exactly the same entries as M with indices in Ω. Although it is a convex set, $\mathcal{A}_\Omega$ is not a bounded set. Starting with an initial guess $X_1 \in \mathcal{A}_\Omega$, e.g., $X_1 = \mathcal{P}_\Omega(M)$ on $(i,j) \in \Omega$ and zero otherwise, or X_1 is another good initial guess, the alternating projection algorithm can be simply stated as follows.

ALGORITHM 9.9

Alternating Projection Algorithm for Matrix Completion

1: Given the rank r of the solution M and a tolerance ϵ, e.g., 1e-6,
2: initialize $X_1 = \mathcal{P}_\Omega(M)$ or any other good guess.
3: For $k = 1, 2, \ldots,$ do
4:   Step 1: $Y_k = \mathcal{P}_{\mathcal{M}_r}(X_k)$ and
5:   Step 2: $X_{k+1} = \mathcal{P}_{M_\Omega}(Y_k)$
6: until $\|X_{k+1} - X_k\|_F < \epsilon$.

In Algorithm 9.9, the projection $P_{\mathcal{M}_r}$ can be computed easily by using the SVD. Indeed, let us recall the following result.

Theorem 9.62. *Suppose that A is a matrix of size $n \times n$. Let $A = U\Sigma V^\top$ be the SVD of A, where U, V are unitary matrices and Σ is a diagonal matrix with entries $s_1 \geq s_2 \geq s_3 \cdots, s_n \geq 0$. Then for each $0 \leq r \leq n - 1$,*

$$\min_{\text{rank}(B) \leq r} \|A - B\|_2 = s_{r+1}. \tag{9.76}$$

Suppose $X_k = U\Sigma V^\top$; let Σ_r be the diagonal matrix with entries $s_1, \ldots, s_r$; and let $P_{\mathcal{M}_r}(X_k) = U\Sigma_r V^\top$ be a rank r matrix. $P_{\mathcal{A}_\Omega}$ is the projection onto $\mathcal{A}_\Omega$. The computation $P_{\mathcal{A}_\Omega}(Y_k)$ is obtained simply by setting the matrix entries of Y_k in positions Ω equal to the corresponding entries in M. Therefore, Algorithm 9.9 is simple and easy to compute. The purpose of this section is to show convergence under various conditions. Numerical results will be given in subsection 9.8.2.

9.7.2 ▪ Convergence of Algorithm 9.9

We start with some preliminary results.

Lemma 9.63. *Let L be a linear subspace of $\mathbb{R}^n$. Suppose P_L denotes the orthogonal projection onto L. Then for any $\mathbf{x} \in \mathbb{R}^n$,*

$$\|\mathbf{x}\| = \|P_L(\mathbf{x})\| \quad \text{if and only if} \quad \mathbf{x} \in L.$$

Equivalently,

$$\|P_L(\mathbf{x})\| < \|\mathbf{x}\| \quad \text{if and only if} \quad \mathbf{x} \notin L.$$

Proof. The "if" part is clear. So let us prove the "only if" part. Let $l_1, l_2, \ldots, l_k$ be an orthonormal basis of L. Extend it to an orthonormal basis $l_1, l_2, \ldots, l_n$ of $\mathbb{R}^n$. Then

$$\mathbf{x} = \sum_{i=1}^{n} \langle \mathbf{x}, l_i \rangle l_i$$

and

$$\|\mathbf{x}\|^2 = \sum_{i=1}^{n} \langle \mathbf{x}, l_i \rangle^2 = \|P_L(\mathbf{x})\|^2 + \sum_{i=k+1}^{n} \langle \mathbf{x}, l_i \rangle^2.$$

Now it follows that if $\|\mathbf{x}\| = \|P_L(\mathbf{x})\|$, then $\sum_{i=k+1}^{n} \langle \mathbf{x}, l_i \rangle^2 = 0$, which implies $\langle \mathbf{x}, l_i \rangle = 0$ for all $i \geq k+1$. Therefore, $\mathbf{x} = \sum_{i=1}^{k} \langle \mathbf{x}, l_i \rangle l_i \in L$. □

Lemma 9.64. *Let L_1 and L_2 be two linear subspaces of $\mathbb{R}^n$. Suppose P_{L_1} and P_{L_2} denote the orthogonal projections onto L_1 and L_2, respectively. Then $L_1 \cap L_2 = \{0\}$ if and only if*

$$\|P_{L_2} P_{L_1}\| < 1. \tag{9.77}$$

Proof. Assume $L_1 \cap L_2 = \{0\}$. Let $\mathbf{x} \neq 0 \in \mathbb{R}^n$. Then if $P_{L_1}(\mathbf{x}) = 0$, then $P_{L_2} P_{L_1}(\mathbf{x}) = 0 < \|\mathbf{x}\|$. Otherwise, $P_{L_1}(\mathbf{x}) \neq 0$. Since $L_1 \cap L_2 = \{0\}$, $P_{L_1}(\mathbf{x}) \notin L_2$. Therefore, using Lemma 9.63, we get

$$\|P_{L_2} P_{L_1}(\mathbf{x})\| < \|P_{L_1}(\mathbf{x})\| \leq \|P_{L_1}\| \|\mathbf{x}\| \leq \|\mathbf{x}\|.$$

Hence, we have

$$\|P_{L_2} P_{L_1}(\mathbf{x})\| < \|\mathbf{x}\|$$

for all nonzero $x \neq 0 \in \mathbb{R}^n$. So

$$\|P_{L_2} P_{L_1}\| < 1.$$

To prove the other direction, assume $\|P_{L_2} P_{L_1}\| < 1$. Assume, on the contrary, that $L_1 \cap L_2 \neq \{0\}$. Let $\mathbf{x} \neq 0 \in L_1 \cap L_2$ be a nonzero vector in the intersection. Then $P_{L_2} P_{L_1}(\mathbf{x}) = P_{L_2}(\mathbf{x}) = \mathbf{x}$, which implies that $\|P_{L_2} P_{L_1}(\mathbf{x})\| = \|\mathbf{x}\|$, contradicting the assumption. □

Lemma 9.65. *Let $M \in \mathcal{M}_r$. Then the projection operator $P_{\mathcal{M}_r}$ is well defined (single valued) in a neighborhood of M and is differentiable with gradient*

$$\nabla P_{\mathcal{M}_r}(M) = P_{T_{\mathcal{M}_r}(M)}, \tag{9.78}$$

where $T_{\mathcal{M}_r}(M)$ is the tangent space of $\mathcal{M}_r$ at M and $P_{T_{\mathcal{M}_r}(M)}$ is the projection operator onto the tangent space.

Proof. Since the projection $P_{\mathcal{M}_r}$ of a matrix X is obtained by hard thresholding the least $n - r$ singular values, we see that the projection is unique if $\sigma_r(M) \neq \sigma_{r+1}(M) \geq 0$. Now consider the neighborhood V of M given by

$$V := \left\{ X \in \mathbb{R}^{n \times n} : \quad \|X - M\|_F < \frac{\sigma_r(M)}{4} \right\}.$$

Then, by the perturbation bounds on singular values (cf. [25]), we have

$$|\sigma_r(X) - \sigma_r(M)| \leq \|X - M\|_F < \frac{\sigma_r(M)}{4}$$

and

$$|\sigma_{r+1}(X) - \sigma_{r+1}(M)| \leq \|X - M\|_F < \frac{\sigma_r(M)}{4}.$$

Hence, noting $\sigma_{r+1}(M) = 0$, we observe that

$$\sigma_{r+1}(X) < \frac{\sigma_r(M)}{4} < \frac{3\sigma_r(M)}{4} < \sigma_r(X).$$

In particular,

$$\sigma_r(X) \neq \sigma_{r+1}(X).$$

Therefore, $P_{\mathcal{M}_r}$ is single valued in the neighborhood V.

For the second part of the result, we refer the reader to Theorem 25 in [11], which is stated below. We have changed the notation for easy reading. In particular, note that although X has rank greater than r in [11], its easy to see that their proof is also valid when X has rank greater than or equal to r. Intuitively, it is easy to see that the gradient vector of the projection $P_{\mathcal{M}_r}$ of smooth manifold $\mathcal{M}_r$ at M will be the projection onto the tangent plane $T_{\mathcal{M}_r}$ at M in general. □

The following result was used in the proof above.

Theorem 9.66 (Feppon and Lermusiaux, 2017 [11]). *Consider $X \in \mathbb{R}^{n \times m}$ with rank greater than r and let $X = \sum_{i=1}^{r+k} \sigma_i u_i v_i^\top$ be its SVD, where the singular values are ordered decreasingly: $\sigma_1 \geq \sigma_2 \geq \cdots \geq \sigma_{r+k}$. Suppose the orthogonal projection $P_{\mathcal{M}_r}(X)$ of X onto $\mathcal{M}_r$ is uniquely defined, i.e., $\sigma_r(X) > \sigma_{r+1}(X)$. Then $P_{\mathcal{M}_r}$, the SVD truncation operator of order r, is differentiable at X and the differential in a direction Y is given by the formula*

$$\nabla_Y P_{\mathcal{M}_r}(X) = P_{T_{\mathcal{M}_r}(P_{\mathcal{M}_r}(X))}(Y)$$
$$+ \sum_{\substack{1 \leq i \leq r \\ 1 \leq j \leq k}} \left[\frac{\sigma_{r+j}}{\sigma_i - \sigma_{r+j}} \langle Y, \Phi_{i,r+j}^+ \rangle \Phi_{i,r+j}^+ - \frac{\sigma_{r+j}}{\sigma_i + \sigma_{r+j}} \langle Y, \Phi_{i,r+j}^- \rangle \Phi_{i,r+j}^- \right],$$

where

$$\Phi_{i,r+j}^\pm = \frac{1}{\sqrt{2}} (u_{r+j} v_i^\top \pm u_i v_{r+j}^\top)$$

are the principal directions corresponding to the principal curvature of the manifold of rank r matrices.

Proof. Refer to Theorem 25 in [11]. □

Next we need to know two tangent spaces $T_{\mathcal{M}_r}(M)$ and $T_{A_\Omega}(M)$.

Lemma 9.67. *The tangent space $T_{\mathcal{A}_\Omega}(M)$ at M can be given explicitly as*

$$T_{\mathcal{A}_\Omega}(M) = \{X \in \mathbb{R}^{n \times n} : P_\Omega(X) = 0\}. \tag{9.79}$$

Proof. Recall from (9.75) that

$$\mathcal{A}_\Omega := \{X : \quad P_\Omega(X - M) = 0\}.$$

Since $P_\Omega(X - M) = P_\Omega(X) - P_\Omega(M) = P_\Omega(X) - P_\Omega(P_\Omega(M)) = P_\Omega(X - P_\Omega(M))$, we have that the set $\mathcal{A}_\Omega$ is a translation of the linear space $\{X \in \mathbb{R}^{n \times n} \mid P_\Omega(X) = 0\}$ by M, i.e.,

$$\mathcal{A}_\Omega = \{X \in \mathbb{R}^{n \times n} \mid P_\Omega(X) = 0\} + M.$$

Hence we have that the tangent space of $\mathcal{A}_\Omega$ at M is equal to the tangent space of the vector space $\{X \in \mathbb{R}^{n \times n} \mid P_\Omega(X) = 0\}$ at the origin. But the tangent space of a vector space at any point is the vector space itself. Hence the result follows. □

Lemma 9.68. *The tangent space $T_{\mathcal{M}_r}(M)$ has an explicit description as follows:*

$$T_{\mathcal{M}_r}(M) = \{XM + MY \mid X \in \mathbb{R}^{n \times n} \text{ and } Y \in \mathbb{R}^{n \times n}\}. \tag{9.80}$$

Proof. First recall that the tangent space $T_{\mathcal{M}_r}(M)$ to a manifold $\mathcal{M}_r$ at a point M is the linear space spanned by all the tangent vectors at 0 to smooth curves $\gamma : \mathbb{R} \to \mathcal{M}_r$ such that $\gamma(0) = M$.

Now let $M \in \mathcal{M}_r$ be an $n \times n$ matrix of rank r. We can write $M = X_0 Y_0^\top$, where $X_0, Y_0 \in \mathbb{R}^{n \times r}$ and both X_0 and Y_0 have full column rank. This is possible because M has exactly rank r.

Let $\gamma(t) = X(t)Y(t)^\top$ be a smooth curve such that $X(0) = X_0$ and $Y(0) = Y_0$. Hence, $\gamma(0) = X_0 Y_0^\top = M$. Since X_0 and Y_0 have full column rank, X_0 and Y_0 have an $r \times r$ minor that does not vanish. Furthermore, there exists an open neighborhood of M in which all matrices have a nonvanishing minor of size $r \times r$. If we restrict the curve γ, we can assume that $X(t)$ and $Y(t)$ have full column rank. In other words, we can assume, without loss of generality, that $X(t)^\top X(t)$ and $Y(t)^\top Y(t)$ are invertible $r \times r$ matrices for all t.

By the product rule, we obtain

$$\begin{aligned}
\dot{\gamma}(0) &= \dot{X}(0)Y(0)^\top + X(0)\dot{Y}(0)^\top \\
&= \dot{X}(0)Y_0^\top + X_0\dot{Y}(0)^\top \\
&= \dot{X}(0)(X_0^\top X_0)^{-1}(X_0^\top X_0)Y_0^\top + X_0(Y_0^\top Y)(Y_0^\top Y_0)^{-1}\dot{Y}(0)^\top \\
&= \left(\dot{X}(0)(X_0^\top X_0)^{-1}X_0^\top\right)(X_0 Y_0^\top) + (X_0 Y_0^\top)\left(Y_0(Y_0^\top Y)^{-1}\dot{Y}(0)^\top\right) \\
&= \left(\dot{X}(0)(X_0^\top X_0)^{-1}X_0^\top\right)M + M\left(Y_0(Y_0^\top Y_0)^{-1}\dot{Y}(0)^\top\right) \\
&\in \{XM + MY \mid X \in \mathbb{R}^{n \times n} \text{ and } Y \in \mathbb{R}^{n \times n}\}.
\end{aligned}$$

Now to prove the reverse inclusion, let $AM + MB \in \{XM + MY \mid X \in \mathbb{R}^{n \times n} \text{ and } Y \in \mathbb{R}^{n \times n}\}$. Consider the smooth curve $\gamma(t) = X(t)Y(t)^\top$ defined by

$$X(t) = t(AX_0) + X_0$$

and

$$Y(t) = t\left((Y_0 B)^\top\right) + Y_0.$$

An easy computation shows that $\gamma(0) = M$ and $\dot{\gamma}(0) = AM + MB$. Hence we get the equality

$$T_{\mathcal{M}_r}(M) = \{XM + MY \mid X \in \mathbb{R}^{n \times n} \text{ and } Y \in \mathbb{R}^{n \times n}\}.$$

This completes the proof. □

We are now ready to establish the convergence of Algorithm 9.9 under a sufficient condition. Mainly we follow the arguments in [21].

Theorem 9.69. *Assume that $T_{\mathcal{A}_\Omega}(M) \cap T_{\mathcal{M}_r}(M) = \{0\}$. Then Algorithm 9.9 converges to M locally at a linear rate, i.e., there exists a neighborhood V around M such that if $X_1 \in V$, then there exists a positive constant $c < 1$ such that*

$$\|X_k - M\| < c^k \|X_0 - M\|, \tag{9.81}$$

where X_k is the kth iteration in Algorithm 9.9.

Proof. For notational convenience, let

$$f(X) := P_{\mathcal{A}_\Omega}(P_{\mathcal{M}_r}(X)).$$

Note that $\mathcal{A}_\Omega$ is an affine space, and the gradient $\nabla P_{\mathcal{A}_\Omega}$ of the projection $P_{\mathcal{A}_\Omega}$ is the projection onto the tangent space of the affine space $\mathcal{A}_\Omega$. By Lemma 9.65 and the chain rule, we have

$$(\nabla f)(X) = \mathbf{P}_{T_{\mathcal{A}_\Omega}(M)}(\mathbf{P}_{T_{\mathcal{M}_r}(M)}(X))$$

since $T_{\mathcal{A}_\Omega}(M) = T_{\mathcal{A}_\Omega}(X)$ for all X.

Now from the definition of differentiability of f at M, we have

$$\lim_{X \to M} \frac{\|f(X) - f(M) - \nabla f(M) \cdot (X - M)\|}{\|X - M\|} = 0.$$

Hence, there exists an open ball V, say $V = B_{r_0}(M)$ centered at M of radius r_0 around M, such that, for all $X \in V$,

$$\frac{\|f(X) - f(M) - \nabla f(M) \cdot (X - M)\|}{\|X - M\|} < \epsilon,$$

where $\epsilon = \frac{1 - \|\nabla f\|}{2} > 0$. Using our hypothesis and Lemma 9.64, we have

$$\|\nabla f(M)\| = \|\mathbf{P}_{T_{\mathcal{A}_\Omega}(M)}\mathbf{P}_{T_{\mathcal{M}_r}(M)}\| < 1.$$

Therefore, for all $X \in V$, we use $M = f(M)$ to obtain

$$\begin{aligned}
\|f(X) - M\| &= \|f(X) - f(M)\| \\
&\leq \|f(X) - f(M) - \nabla f(M) \cdot (X - M)\| + \|\nabla f(M) \cdot (X - M)\| \\
&< \epsilon\|X - M\| + \|\nabla f(M)\|\|(X - M)\| \\
&= (\epsilon + \|\nabla f(M)\|)\|X - M\| \\
&\leq \frac{1 + \|\nabla f(M)\|}{2}\|X - M\|.
\end{aligned}$$

where $\frac{1 + \|\nabla f(M)\|}{2} < 1$ since $\|\nabla f(M)\| < 1$, as discussed above.

Setting $c = \frac{1 + \|\nabla f(M)\|}{2} < 1$, we can rewrite the above inequality as follows:

$$\|f(X) - M\| < c\|X - M\| \quad \forall X \in V. \tag{9.82}$$

Hence, if $X_k \in V = B_{r_0}(M)$, we use $X_{k+1} = f(X_k)$ to get

$$\|X_{k+1} - M\| = \|f(X_k) - M\| < c\|X_k - M\| \leq r_0,$$

which implies $X_{k+1} \in V = B_{r_0}(M)$. So, if the initial guess $X_0 \in V$, we have, by induction,

$$X_k \in V \quad \forall k$$

and

$$\|X_k - M\| \leq c^k \|X_0 - M\|.$$

We have thus completed the proof.   □

9.8 ▪ Computational Results

In this section, we shall present some numerical results to demonstrate the effectiveness of various matrix completion algorithms. Examples of recovering images from partial given pixel values, enhancing images by removing unwanted wires/letters, and compressing images will be given.

9.8.1 ▪ Numerical Experiments Based on Algorithm 9.7

In the following, we show some numerical experiments to recover an image from a matrix with missing entries, say 50% to 90%. Two examples will be provided to demonstrate the efficiency and effectiveness of Algorithm 9.7.

Example 9.70. We first use rank $r = 50$ to recover the entire matrix from a given matrix as in Figure 9.1 with missing entries, say 50% to 90% without noise, to reveal the true image, which is a standard testing image. The pixel values of the image are in $[0, 1]$ and the standard root mean square error (RMSE) is calculated and shows how accurate the recovered images are (see Figures 9.2, 9.3, 9.4, 9.5, and 9.6). From the RMSEs shown in these figures, we can see that if more entries are known, the recovered images are better.

Example 9.71. Next we use ranks $r = 100$ and $r = 125$ to recover the entire matrix from a given matrix with missing entries, say 50% without noise, to reveal the true image, which is a standard testing image, as shown in Figure 9.1. The pixel values of the image are in $[0, 1]$, and the standard RMSE is calculated and shows how accurate the recovered images are for various ranks (see Figures 9.7 and 9.8). The higher the rank, the better the recovery.

Figure 9.1. *An original image for testing.*

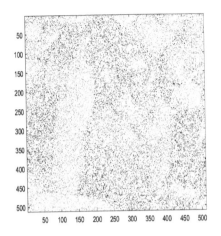 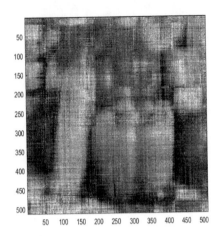

Figure 9.2. *A matrix with 10% known entries (left) and a recovered image by Algorithm 9.7 (right).*

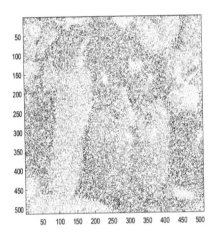 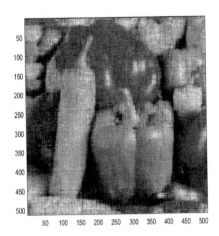

Figure 9.3. *A matrix with 20% known entries (left) and a recovered image (right) by Algorithm 9.7 with RMSE = 0.0889.*

9.8.2 ▪ Numerical Results from Algorithm 9.9

We have implemented Algorithm 9.9 in MATLAB. One suggestion is to use the completed matrix X_1 from the OR1MP algorithm, i.e., Algorithm 9.7, for an initial guess for Algorithm 9.9. This initial guess will make the convergence faster.

Example 9.72. One application is to improve the quality of the completed matrix from Algorithm 9.7. That is, we use the completed matrix from Algorithm 9.7 with rank $r = 50$ as an initial matrix for Algorithm 9.9 and apply the alternating projection iteration $r = 100$ times to obtain a new completed matrix with a better RMSE, as shown in Figure 9.9.

If we use rank $r = 100$, the completed matrices from Algorithms 9.7 and 9.9 are shown in Figure 9.10. From these experiments, we can see that Algorithm 9.9 can improve the quality of the recovered image.

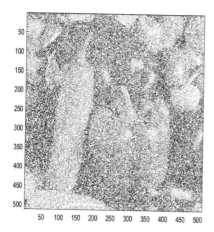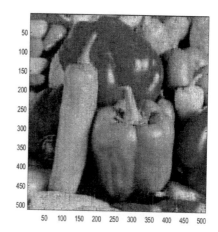

Figure 9.4. *A matrix with 30% known entries (left) and a recovered image (right) by Algorithm 9.7 with RMSE = 0.0666.*

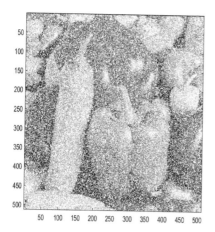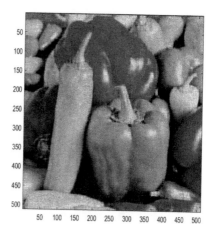

Figure 9.5. *A matrix with 40% known entries (left) and a recovered image (right) by Algorithm 9.7 with RMSE = 0.0552.*

Example 9.73. Let us present another example to show the quality of the completed matrix from Algorithm 9.9. That is, we use the completed matrix from Algorithm 9.7 with rank $r = 50$ as an initial matrix for Algorithm 9.9 and apply the alternating projection iteration $r = 100$ times to obtain a new completed matrix. In addition, we present the completed matrix from the standard SVT algorithm, Algorithm 9.1. Images with RMSE are shown in Figures 9.11 and 9.12.

Note that the computation of Algorithms 9.7 and 9.9 is much faster than that of Algorithm 9.1. These experiments further show that Algorithm 9.9 can improve the quality of the recovered image.

In addition, let us present an illustrative example for image in-painting.

Example 9.74. In this example, we use Algorithm 9.9 for image in-painting. See Figure 9.13 for an original image which has wires in the sky (the image on the left) and the image obtained

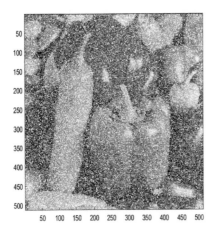 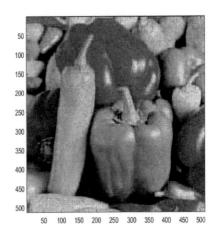

Figure 9.6. *A matrix with 50% known entries (left) and a recovered image (right) by Algorithm 9.7 with RMSE = 0.0483.*

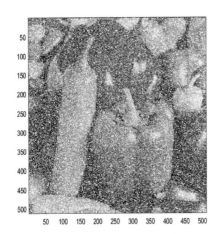 

Figure 9.7. *A matrix with 50% known entries (left) and a recovered image (right) by Algorithm 9.7 using 75 iterations to get a recovered image of RMSE = 0.0431.*

after Algorithm 9.9 (the image on the right). More precisely, we applied Algorithm 9.9 to two locations, as shown in Figure 9.14. The locations of wires are treated as unknown entries, and Algorithm 9.9 automatically fills in the entries with appropriate values. Because there are three color components for the image, we have to apply Algorithm 9.9 three times. We treated the black wires as missing and the rest as known. Algorithm 9.9 fills in the missing wires and produces a good image without wires in the sky.

Finally, we present an example of low rank image completion with an image missing a big block by using Theorem 9.13.

Example 9.75. We use an image of size $2976 \times 2976 \times 3$, as shown (left) in Figure 9.15 with a rank 186 approximation (on the right).

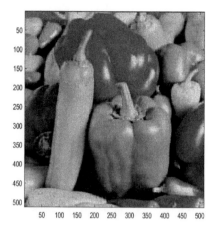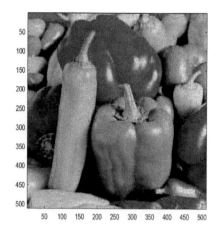

Figure 9.8. *From a given matrix with 50% known entries, two recovered images by Algorithm 9.7 using 100 and 125 iterations with RMSE* = 0.0413 *and* 0.0401, *respectively.*

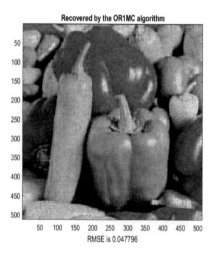

Figure 9.9. *A recovered matrix with 50% known entries by Algorithm 9.7 using 50 iterations (left) and a recovered image by Algorithm 9.9 (right) with rank* $r = 50$ *and 100 iterations.*

For each color, we take only the top and left bands with width 186 as the given known matrix, as shown on the far left and third left of Figure 9.16 and the far left of Figure 9.17. A huge block of each of three RGB images is missing. From the given images with missing entries, we cannot see what these images should be. The ratio of the known entries and the entire matrix is 0.1211, for a missing rate of 0.8789. Then we apply the method in Theorem 9.13 to recover the entire matrix of three colors to form an approximation (shown on the right of Figure 9.17) of the original image. The relative maximum norm of the approximation versus the original is 0.1803 and the RMSE is 1.7105. The total computational time is just a few minutes on a laptop computer.

Note that from these images with missing entries, we do not see anything. However, the matrix completion method explained in Theorem 9.13 can still recover the original image.

Figure 9.10. *A recovered matrix with 50% known entries by Algorithm 9.7 using 100 iterations (left) and a recovered image by Algorithm 9.9 (right) with rank $r = 100$ and 10 iterations.*

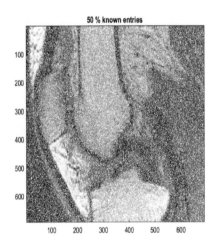
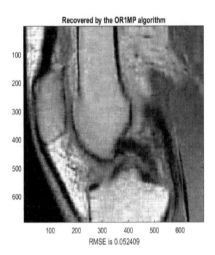

Figure 9.11. *An image with 50% known entries (left) and its completed image by Algorithm 9.7 using 50 iterations (right).*

9.9 ▪ Exercises

Exercise 1. Show that if A and B have singular values $s_1 \geq \cdots \geq s_m$ and $\tau_1 \geq \cdots \geq \tau_m$ and if $s_i \leq \tau_i$ for all $i = 1, \ldots, m$, then $\|A\| \leq \|B\|$ for every unitarily invariant norm $\| \cdot \|$.

Exercise 2. Let $\| \cdots \|$ be a unitarily invariant norm. Show that $\|AB\| \leq \|A\| \|B\|_2$ and $\|AB\| \leq \|A\|_2 \|B\|$.

Exercise 3. Define $\|A\|_k = (\sum_{i=1}^{k} \sigma_i(A)^2)^{1/2}$, where $\sigma_1(A) \geq \sigma_2(A) \geq \cdots \geq \sigma_m(A)$ are singular values of A, where A is a matrix of size $m \times m$. Show that $\|A\|_k$ is a norm for any integer $k \geq 1$.

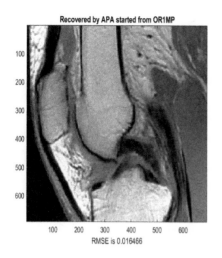
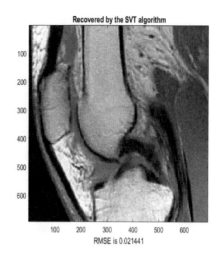

Figure 9.12. *A recovered image by Algorithm* 9.9 *based on the initial image from Algorithm* 9.7 *after* 50 *iterations (left) and a recovered image by the SVT algorithm, Algorithm* 9.1 *(right), with* 400 *iterations.*

Figure 9.13. *An image with wires and the image after Algorithm* 9.9.

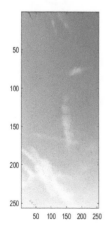

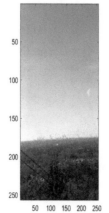

Figure 9.14. *Two locations where we have applied Algorithm* 9.9.

Figure 9.15. *An original image (left) agrees with the image (right) of rank* 186.

Figure 9.16. *The red image with known entries (far left) and the completed image (second from left), and the green image with known entries (third from left) and the completed image (far right).*

Figure 9.17. *The red image with known entries (far left) and the completed image (second from left), and the completed color image (based on three-color completed images) to approximate the original image (far right).*

Exercise 4. Show that $\|A\|_p = (\sum_{i=1}^{m}(\sigma_i(A))^p)^{1/p}$ is a norm for $p \geq 1$.

Exercise 5. Show that the following inequalities will hold for any matrix A of size $m \times n$ with rank at most r:

$$\|A\|_2 \leq \|A\|_F \leq \|A\|_* \leq \|A\|_p$$

for $p \geq 1$ and

$$\|A\|_* \leq \sqrt{r}\|A\|_F \leq r\|A\|_2,$$

where $\|A\|_2 = s_1(A)$ is the operator norm of A in the Euclidean space $\mathbb{R}^n$.

Exercise 6. In general, we can define the matrix dual norm as follows: For any matrix A of size $m \times n$ and matrix norm $\| \cdot \|$, let

$$\|A\|_{dual} := \sup\{\langle A, B \rangle : \quad B \text{ matrix of size } m \times n, \|B\| \leq 1\}.$$

This defines the dual norm to the original norm $\| \cdot \|$. Show that the dual norm of the Frobenius norm is the same as the Frobenius norm.

Exercise 7. Prove the following lemma.

Lemma 9.76. *The matrix dual norm of the operator norm* $\| \cdot \|_2$ *of matrices is the nuclear norm* $\| \cdot \|_*$.

Exercise 8. Show that $\|X\|_*$ is a convex function of matrix X.

Exercise 9. Prove the following result.

Lemma 9.77. *Let A and B be matrices of the same dimensions. If $AB^\top = 0$ and $A^\top B = 0$, then $\|A + B\|_* = \|A\|_* + \|B\|_*$.*

Furthermore, show that if the row and column spaces of A and B are orthogonal, then $\|A + B\|_* = \|A\|_* + \|B\|_*$.

Exercise 10. Suppose that we have $\mathcal{A}\mathcal{A}^* = I$, where I stands for the identity matrix. Show that for any $\alpha > 0$,

$$(I + \alpha \mathcal{A}^* \mathcal{A})^{-1} = I - \frac{\alpha}{1 + \alpha} \mathcal{A}^* \mathcal{A}.$$

Exercise 11. Let $B_\delta = \{Y, \|\mathcal{A}(Y) - \mathbf{b}\|_2 \leq \delta\}$ be the ball of radius δ centered at $\mathbf{b} = \mathcal{A}(M)$. Prove the following lemma.

Lemma 9.78. *For any matrix Y, let*

$$\mathcal{P}_\delta(Y) = \arg\min_{Z}\{\|Z - Y\|_F^2, Z \in B_\delta\}.$$

Then

$$\mathcal{P}_\delta(Y) = Y + \frac{\eta}{1 + \eta}\mathcal{A}^*(\mathbf{b} - \mathcal{A}(Y)),$$

where

$$\eta = \min\{\|\mathcal{A}(Y) - \mathbf{b}\|/\delta - 1, 0\}.$$

In particular, when $\delta = 0$, $\mathcal{P}_0(Y) = Y + \mathcal{A}^*(\mathbf{b} - \mathcal{A}(Y))$.

Exercise 12. Use the proof of the Uzawa method in Chapter 6 for sparse solutions to establish the convergence of Algorithm 9.1.

Exercise 13. Derive the dual problem of (9.36), and show that the dual is (9.43).

Exercise 14. Let $f(Y)$ be the minimization functional in (9.43). Show that $f(Y)$ is convex.

Exercise 15. Show that $\lambda_i(A^\top A) = \lambda_i(AA^\top)$ for any square matrix A.

Exercise 16. Establish the inequality in (9.39).

Exercise 17. Use a gradient descent method to solve (9.43) and show that the algorithm is convergent.

Exercise 18. Prove Theorem 9.43.

Exercise 19. Use the Attouch–Peypouqut acceleration technique discussed in Chapter 5 to speed up the computational Algorithm 9.4.

Exercise 20. Prove that Algorithm 9.6 converges.

Exercise 21. Prove Theorem 9.61.

Exercise 22. Write a MATLAB code for EOR1MP, i.e., Algorithm 9.8.

Exercise 23. Extend the result of Theorem 9.10 to the setting of the low rank matrix recovery problem. That is, prove the following theorem.

Theorem 9.79. *Let $\mathcal{A}_\Omega$ be a set of linear functionals over $\mathcal{M}_r$ and $\mathbf{a}_\Omega \in \mathbb{R}^m$ be a vector, where $m = |\Omega|$. Suppose that $m > 2nr - r^2$. Show that the chance of finding $M \in \mathcal{M}_r$ such that $\mathcal{A}_\Omega(M) = \mathbf{a}_\Omega$ is zero.*

Exercise 24. Implement Algorithm 9.9 in MATLAB and test your code.

Exercise 25. Consider a 2×2 rank 1 matrix M. Suppose that only M_{11} is missing. Show that M can be completed uniquely if $M_{22} \neq 0$.

Exercise 26. Consider the following 3×3 rank 1 matrix M:

$$\begin{bmatrix} \square & \square & a_{13} \\ \square & \square & a_{23} \\ a_{31} & a_{32} & a_{33} \end{bmatrix}.$$

Show that M has a unique rank 1 matrix completion if and only if $a_{33} \neq 0$.

Exercise 27. Consider the following rank 1 3×3 matrix M:

$$\begin{bmatrix} \square & \square & a_{13} \\ \square & a_{22} & \square \\ a_{31} & a_{32} & a_{33} \end{bmatrix}.$$

Show that M has a unique rank 1 matrix completion if and only if $a_{32} \neq 0$ and $a_{33} \neq 0$.

Exercise 28. Consider the following 3×3 rank 1 matrix M:

$$\begin{bmatrix} \square & \square & a_{13} \\ a_{21} & a_{22} & \square \\ a_{31} & a_{32} & \square \end{bmatrix}.$$

Show that M has many rank 1 matrix completions.

Exercise 29. Prove the following result.

Lemma 9.80 (Thompson, 1976 [35]). *For any real square matrices A and B of the same size, there exist orthonormal matrices U and V such that*

$$|A + B| \preceq U|A|U^T + V|B|V^\top, \tag{9.83}$$

namely, $U|A|U^T + V|B|V^\top - |A + B|$ is positive semidefinite.

Exercise 30. Prove the following result: If A and B are symmetric positive definite matrices and $\lambda_i(A) \leq \lambda_i(B)$ for all i, then there exists a matrix W such that $A \preceq WBW^\top$. (*Hint*: See Thompson, 1976 [35].)

Bibliography

[1] K. Allen, A Geometric Approach to Low-Rank Matrix and Tensor Completion, Dissertation, University of Georgia, Athens, GA. 2021. (Cited on pp. 285, 286)

[2] J. Bennett and S. Lanning, The Netflix Prize, in Proceedings of KDD Cup and Workshop, August 12, 2007. (Cited on p. 279)

[3] J.-C. Bourin and M. Uchiyama, A matrix subadditivity inequality for $f(A + B)$ and $f(A) + f(B)$. Linear Algebra Appl., 423 (2007), 512–518. (Cited on p. 298)

[4] J. Cai, E. Candés, and Z. Shen, A singular value thresholding algorithm for matrix completion, SIAM J. Optim., 20 (2010), 1956–1982. (Cited on pp. 299, 300, 301)

[5] E. J. Candès and Y. Plan, Tight oracle inequalities for low-rank matrix recovery from a minimal number of noisy random measurements, IEEE Trans. Inform. Theory, 57 (2011), 2342– 2359. (Cited on p. 295)

[6] E. Candés and Y. Plan, Matrix completion with noise, Proc. IEEE, 98 (2010), 925–936. (Cited on p. 293)

[7] E. Candés and B. Recht, Exact matrix completion via convex optimization, Found. Comput. Math., 9 (2009), 717–772. (Cited on p. 280)

[8] E. J. Candés and T. Tao, The power of convex relaxation: Near-optimal matrix completion, IEEE Trans. Inform. Theory, 56 (2010), 2053–2080. (Cited on p. 280)

[9] P. L. Combettes and V. R. Wajs, Signal recovery by proximal forward-backward splitting, Multiscale Model. Simul., 4 (2005), 1168–1200. (Cited on p. 303)

[10] Y. Eldar, D. Needell, and Y. Plan, Uniqueness conditions for low-rank matrix recovery, Appl. Comput. Harmon. Analy., 33 (2012), 309–314. (Cited on p. 293)

[11] F. Feppon and P. J. Lermusiaux, A Geometric Approach to Dynamical Model-Order Reduction, preprint, arXiv:1705.08521, 2017. (Cited on p. 320)

[12] W. Fulton, Intersection Theory, Springer-Verlag, 1984. (Cited on p. 282)

[13] J. Harris, Algebraic Geometry: A First Course, Grad. Texts in Math., Springer, 1992. (Cited on pp. 282, 288)

[14] R. Hartshorne, Algebraic Geometry, Grad. Texts in Math. 52, Springer, New York, 1977. (Cited on p. 282)

[15] R. Horn and C. Johnson, Matrix Analysis, Cambridge University Press, 1985. (Cited on p. 290)

[16] R. Horn and C. Johnson, Topics in Matrix Analysis, Cambridge University Press, 1991. (Cited on pp. 290, 299)

[17] F. J. Király, L. Theran, and R. Tomioka, The algebraic combinatorial approach for low-rank matrix completion, J. Mach. Learn. Res., 16(41) (2015), 1391–1436. (Cited on p. 289)

[18] M.-J. Lai, S. Li, L. Liu, and H. Wang, On the Schatten p-quasi-norm minimization for low-rank matrix recovery, J. Appl. Comput. Harmon. Anal., 51 (2021), 157–170. (Cited on pp. 295, 297, 298)

[19] M.-J. Lai, Y. Y. Xu, and W. T. Yin, Improved iteratively reweighted least squares for unconstrained smoothed ℓ_q minimization, SIAM J. Numer. Anal., 51 (2013), 927–957. (Cited on p. 295)

[20] M.-J. Lai and W. T. Yin, Augmented ℓ_1 and nuclear-norm models with a globally linearly convergent algorithm, SIAM J. Imaging Sci., 6 (2013), 1059–1091. (Cited on pp. 301, 302, 303)

[21] M.-J. Lai and A. Varghese, On Convergence of the Alternating Projection Method for Matrix Completion and Sparse Recovery Problems, preprint, arXiv:1711.02151v1 [math.OC], 2017. (Cited on pp. 289, 322)

[22] A. Lewis and J. Malick, Alternating projections on manifolds. Mathematics of Operations Research, 33(1) (2008), 216–234. (Cited on p. 318)

[23] S. Ma, D. Goldfarb, and L. Chen, Fixed point and Bregman iterative methods for matrix rank minimization, Math. Program. Ser. A., 128 (2011), 321–353. (Cited on pp. 299, 305, 307, 309)

[24] R. Meka, P. Jain, and I. S. Dhillon, Guaranteed Rank Minimization via Singular Value Projection, preprint, arXiv:0909.5457, 2009. (Cited on p. 293)

[25] L. Mirsky, Symmetric gage functions and unitarily invariant norms, Quart. J. Math., 11 (1960), 50–59. (Cited on p. 320)

[26] Q. Mo and S. Li, New bounds on the restricted isometry constant δ_{2k}, Appl. Comput. Harmon. Anal., 31 (2011), 460–468. (Cited on pp. 293, 294)

[27] K. Mohan and M. Fazel, Iterative reweighted least squares for matrix rank minimization, in Proceedings of the Allerton Conference on Controls and Communications, 2010. (Cited on p. 280)

[28] K. Mohan and M. Fazel, New restricted isometry results for noisy low-rank recovery, in Proceedings of the 2010 IEEE International Symposium on Information Theory (ISIT), 2010, 1573–1577. (Cited on p. 293)

[29] D. Mumford, The Red Book of Varieties and Schemes, Lecture Notes in Math. 1358, Springer, Berlin, 1988. (Cited on p. 289)

[30] Netflix, The Netflix Prize, http://www.netflixprize.com/, 2006. (Cited on p. 279)

[31] B. Recht, M. Fazel, and P. Parrilo, Guaranteed minimum-rank solutions of linear matrix equations via nuclear norm minimization, SIAM Rev., 52 (2010), 471–501. (Cited on pp. 292, 293)

[32] S. Oymak, K. Mohan, M. Fazel, and B. Hassibi, A Simplified Approach to Recovery Conditions for Low Rank Matrices, preprint, arXiv:1103.1178v3, 2011. (Cited on p. 292)

[33] I. R. Shafarevich, Basic Algebraic Geometry 1, Springer-Verlag, Berlin, Heidelberg, 2013. (Cited on pp. 284, 285)

[34] A. Singer and M. Cucuringu, Uniqueness of low-rank matrix completion by rigidity theory, SIAM J. Matrix Anal. Appl., 31(4) (2010), 1621–1641. (Cited on p. 289)

[35] R. C. Thompson, Convex and concave functions of singular values of matrix sums, Pacific J. Math., 66(1) (1976), 285–290. (Cited on p. 333)

[36] Z. Wang, M.-J. Lai, Z. Lu, W. Fan, H. Davulcu, and J. Ye, Orthogonal rank-one matrix pursuit for low rank matrix completion, SIAM J. Sci. Comput., 37 (2015), A488–A514. (Cited on pp. 312, 317)

[37] O. Zariski, On the purity of branch locus of algebraic functions, Proc. Natl. Acad. Sci. USA, 44 (1958), 791–796. (Cited on p. 282)

Chapter 10

Graph Clustering

Given a graph $G = (V, E)$ with the vector set V and the edge set E, let A be the adjacency matrix associated with G with entries $A_{ij} = 1$ if the vertex v_i and vector v_j have an edge in E and $A_{ij} = 0$ otherwise. More generally, we may consider a weighted adjacency matrix with $A_{ij} = w_{ij} \neq 0$ if $\{v_i, v_j\} \in E$ and $A_{ij} = 0$ if $\{v_i, v_j\} \notin E$. We are interested in determining the clusters of all vertices in V. That is, in the clean graph setting, all vertices of a cluster C_i are connected by edges in E and not connected to vertices of other clusters. If a graph G is clean, the vertices of V can be split into clusters with index sets $C_1, \ldots, C_k$ and edge set $E = E_1 \cup E_2 \cup \cdots \cup E_k$, with E_i being the set of edges among the vertices in C_i. Thus, G can be decomposed into several subgraphs such that $G = G_1 \cup G_2 \cup \cdots \cup G_k$ with disconnected subgraphs $G_1, \ldots, G_k$, where each G_i is associated with a cluster C_i, $i = 1, \ldots, k$. In practice, G is not a clean graph in general, i.e., G contains some noise in the sense that G can be decomposed into several clusters C_i, $i = 1, \ldots, k$, associated with subgraphs $G_i = (C_i, E_i)$ with $E_1 \cup \cdots \cup E_k \subset E$; $E_1 \cup \cdots \cup E_k \neq E$; and $\#(E_1 \cup \cdots \cup E_k) < \#(E)$ and $\approx \#(E)$, where $\#(E)$ stands for the cardinality of E. That is, in the noisy setting, all vertices of a cluster C_i are connected by edges in E, while some of the vertices in C_i are also connected to vertices of other clusters, but there are a lot of edges within the cluster C_i and a few edges outside of C_i. The characterization of more edges in-cluster and less edges out-of-cluster will be made more precise later in this chapter. Indeed, we shall use the stochastic block model and other models to describe the edges within each cluster and among other clusters. Noisy graphs make the clustering problem difficult yet interesting to solve due to its real-world applications.

Mathematically, for a clean graph, if we number the vertices according to the indices in C_1 followed by C_2, and so on, the new adjacent matrix A is a block diagonal matrix. For a noisy graph, the adjacent matrix A can be permuted to be a nearly block diagonal matrix. Thus, the clustering problem is a matrix reduction problem for this special symmetric and nonnegative adjacent matrix. For another example, the graph cluster problem is also equivalent to the communities detection problem since each community is a cluster of people of interest. If any person belongs to only one community, the communities detection problem can be a clean list of communities. If a few people belong to multiple communities, the communities detection problem will be noisy. That is, the adjacency matrix is noisy.

So far the most accurate way to solve a clean graph clustering problem is to exhaust all the permutations of columns and rows of the adjacency matrix A. This will need a nonpolynomial time computation as the size of A increases. A traditional computational approach called the spectral clustering method (cf. [21] or [22]) examines the eigenvectors of $\mathcal{L}$, the graph Laplacian (to be explained later) associated with the adjacency matrix A. Indeed, the number of eigenvectors associated with eigenvalue 0 of $\mathcal{L}$ is the number of clusters of G. When A is of size $n \times n$, determining the eigenvectors of a matrix with size $n \times n$ requires $\mathcal{O}(n^3)$ operations, which is prohibitively slow for truly large graphs. If the number of blocks is known, say k, then we still need $O(kn^2)$ to find the eigenvectors.

In this chapter, we present an approach based on sparse solution of underdetermined linear systems to attack the graph clustering problem. One of our ideas is to recast the graph clustering problem by finding a sparse solution of a linear system. Let $\mathbf{1}_{C_1}$ be the indicator vector of C_1 in $\mathbb{R}^n$, i.e., the entries of $\mathbf{1}_{C_1}$ are 1 for indices in C_1 and zero otherwise. Without loss of generality, we may assume that C_1 contains the first index. That is, if $C_1 = \{i_1, i_2, \ldots, i_{n_1}\} \subset \{1, 2, \ldots, n\}$ for some integer $n_1 > 1$, we assume that $i_1 = 1 \in C_1$. As mentioned above, $\mathbf{1}_{C_1}$ is an eigenvector of the graph Laplacian $\mathcal{L}$ associated with a zero eigenvalue. That is, $\mathcal{L}\mathbf{1}_{C_1} = 0$. If the size n_1 of cluster C_1 is small compared to n, then the indicator vector $\mathbf{1}_{C_1}$ is a sparse vector. Our approach is to solve the problem

$$\arg\min \|\mathcal{L}\mathbf{x}\|_2 \quad \text{subject to } x_1 = 1 \text{ and } \|\mathbf{x}\|_0 \leq n_1. \tag{10.1}$$

In this way, we can directly determine an approximation to the indicator vector of the cluster containing the first vertex. Once C_1 is found, we can reduce the graph to a new graph of smaller size and then repeat the above computation. In this way, we can determine all the clusters.

The computational approach for finding one cluster is called single cluster pursuit (SCP), and the approach for finding multiple clusters is called iterative single cluster pursuit (ISCP). Similarly, we can replace $x_1 = 1$ in (10.1) above by $x_j = 1$ to find a cluster containing vertex j of interest. Also, if we know multiple vectors in a cluster C_1, we can add all known constraints in to adjust the computation in (10.1). This computational procedure is called semisupervised cluster pursuit (SSCP). If we can decide on a metric to measure the connectivity of two vertices to get a good adjacency matrix, we can cluster the vertices based on a training set. Then we will have a lot of seeds to use to classify the testing and the unknown vertices. Designing a good metric can be done using the deep learning technique. We will not address that in this chapter.

In section 10.7 we will explain the cost of the computation of ISCP Algorithm 10.3, which is

$$O(n \log^2(n) \log(\log(n))),$$

i.e., the algorithm is very efficient. This chapter is devoted to the analysis and convergence of these graph clustering algorithms.

10.1 ▪ Preliminaries on Graph Theory

10.1.1 ▪ Some Necessary Notions and Definitions

Formally, by a graph G we mean a set of vertices V together with a subset $E \subset \{\{u, v\} : u, v \in V\}$ of edges. Since we are only concerned with finite graphs, we shall always identify the vertex set V with a finite set of consecutive natural numbers: $V = \{1, 2, \ldots, n\}$. In addition, a weighted graph is a graph G with the additional information of a weight $w_{ij} \in \mathbb{R}_+$ for every edge $\{i, j\} \in E$. Note that we only consider nonnegative weights for convenience in this chapter.

The degree of any vertex $i \in G$ is defined to be the total number of edges connected to i by edges in E, i.e.,

$$d_i = \#\{\{i, j\} \in E : 1 \leq j \leq n\}.$$

For any subset $S \subset V$ we define $\text{vol}(S) = \sum_{i \in S} d_i$ as the volume of S.

A subgraph G' of G is a subset of vertices $V' \subset V$ together with a subset of edges $E' \subset E \cap V' \times V'$. Given any subset $S \subset V$, we denote by G_S the subgraph with vertex set S and edge set all edges $\{i, j\}$ with $i, j \in S$.

A path in G is a set of "linked" edges $\{\{i_1, i_2\}, \{i_2, i_3\}, \ldots, \{i_{k-1}, i_k\}\}$, and we say that G is connected if there is a path linking any two vertices $i, j \in V$, and *disconnected otherwise*. If G is disconnected, any subgraph $G_S \subset G$ which is connected and maximal with respect to the property of being connected is called a connected component. If G is connected, we define the diameter of G to be the length of (i.e., the number of edges in) the longest path. Given any $i \in V$ and any nonnegative integer ℓ, we define the ball $B_\ell(i) \subset V$ to be the set of all vertices connected to i by a path of length ℓ or shorter. A good reference on elementary graph theory (and its applications) is [3].

10.1.2 ▪ Graph Laplacians

With any graph G with $|V| = n$ we associate a symmetric, $n \times n$, nonnegative *adjacency matrix* A. The various graph Laplacians of G are defined as follows.

Definition 10.1. *Let A denote the adjacency matrix of a graph G, and let D denote the matrix* $\mathrm{diag}(d_1, \ldots, d_n)$, *where d_i is the degree of the ith vertex. The graph Laplacian of G is $\mathcal{L} = D - A$. We define the normalized symmetric graph Laplacian of G as $\mathcal{L}_s := I_n - D^{-1/2}AD^{-1/2}$ and the normalized random walk graph Laplacian as $\mathcal{L}_{rw} := I_n - D^{-1}A$, where I_n is the identity matrix of size $n \times n$. In addition, the signless Laplacian is defined as $\mathcal{L}^+ = I_n + D^{-1}A$.*

We first list a few basic properties of graph Laplacians.

Theorem 10.2. *Suppose that $\mathcal{L} = \mathcal{L}_s$ or $\mathcal{L}_{rw}$. We have the following properties:*

1. *The eigenvalues of $\mathcal{L}$ are real and nonnegative.*

2. $\lambda_{n-1} \leq 2$.

3. *Let $\lambda_1 \leq \lambda_2 \leq \cdots \leq \lambda_n$ denote the eigenvalues of $\mathcal{L}$ in ascending order. Let k denote the number of connected components of G. Then $\lambda_i = 0$ for $i \leq k$, and $\lambda_i > 0$ for $i > k$.*

Proof. For $\mathcal{L} = \mathcal{L}_s$, items 1 to 3 follow from Lemma 1.7 in [7]. If $\mathcal{L} = \mathcal{L}_{rw}$, observe that $\mathcal{L}_{rw} = D^{-1/2}\mathcal{L}_s D^{1/2}$, and so the eigenvalues of $\mathcal{L}_{rw}$ and $\mathcal{L}_s$ coincide. Hence the above properties hold for L_{rw} as well. We leave the details to Exercise 2. □

Let $C_1, \ldots, C_k$ be the clusters of G. It is easy to see that $\mathcal{L}_{rw}\mathbf{1}_{C_i} = 0$ for $i = 1, \ldots, k$. Thus we have the following result.

Theorem 10.3. *The indicator vectors of the connected components of G, $\mathbf{1}_{C_1}, \ldots, \mathbf{1}_{C_k}$, form a basis for the zero eigenspace (i.e., the kernel) of $\mathcal{L}_{rw}$.*

Proof. We leave the proof to Exercise 3. □

The above results lead to the following well-known algorithm.

ALGORITHM 10.1
Spectral Clustering Algorithm [21]
 1: **Input:** The adjacency matrix A of a graph G.
 2: **Initialization:** Compute $\mathcal{L} = I_n - D^{-1}A$.
 3: **Find Eigenvectors.**

4: Compute the first k eigenvectors $\mathbf{v}_1, \mathbf{v}_2, \ldots, \mathbf{v}_k$ associated with the zero eigenvalue of $\mathcal{L}$.
5: **Perturbed Sparse Recovery**
6: Let $U \in \mathbb{R}^{n \times k}$ be the matrix containing the vectors $\mathbf{v}_1, \ldots, \mathbf{v}_k$ as columns.
7: For $i = 1, 2, \ldots, n$, let $\mathbf{r}_i \in \mathbb{R}^k$ be the vector corresponding to the ith row of U.
8: Cluster the vectors $\mathbf{r}_i$ for $i = 1, 2, \ldots, n$ in $\mathbb{R}^k$ into clusters $\mathbf{R}_1, \ldots, \mathbf{R}_k$ using any linear time clustering algorithm (e.g., the k-means algorithm; see [17]).

Output: Clusters $C_1, \ldots, C_k$ with $C_i = \{j \text{ if } \mathbf{r}_j \in \mathbf{R}_i\}$.

For convenience, we shall use $\mathcal{L}$ for $\mathcal{L}_{rw}$ in the rest of this chapter unless specified. We shall refer to the ith column of $\mathcal{L}$ as ℓ_i. We can easily check that

$$(\ell_i)_k = \delta_{ik} - \frac{A_{ik}}{d_k}.$$

Finally, we shall denote by $\mathcal{L}_{-i}$ the submatrix of L obtained by dropping the column ℓ_i.

10.1.3 ▪ Random Graph Theory

The stochastic block model (SBM) is a widely used mathematical model of random graphs for clustering. It was first introduced in a mathematical sociology study in [15] and is a generative model for random graphs with certain edge densities within and between underlying communities. Edges within communities are more dense than edges between communities. Let us begin with a simple random graph model, which is called the Erdős–Rényi model $\mathcal{G}(n_0, p)$ (cf. [12]).

Definition 10.4. *We say H is drawn from the Erdős–Rényi model $\mathcal{G}(n_0, p)$ if H has n_0 vertices and for all $i, j \in [1, \ldots, n_0]$ the edge $\{i, j\}$ is attached independently with probability p.*

It is known that an Erdős–Rényi random graph $G(n, p)$ is almost surely connected if $p = \frac{c \log n}{n}$ for $c \geq 1$ and is not only almost surely connected, but the degrees of almost all vertices are asymptotically equal, if $p \sim \frac{\omega(n) \log n}{n}$, where $\omega(n) \to \infty$ (cf. [12], [13], and [8]). We are now ready to define the SBM as follows.

Definition 10.5. *Given n, suppose that $V := [1, 2, \ldots, n]$ is partitioned into k subsets $C_1, \ldots, C_k$ of equal size n_0. We say G is drawn from the SBM $\mathcal{G}(n, k, p, q)$ if the following conditions hold:*

- *A partition $V = C_1 \cup C_2 \cup \cdots \cup C_k$ is fixed before any edges are assigned.*

- *Each cluster C_k is an Erdős–Rényi random graph in $G(n_k, p_k)$ with $p_k \in (0, 1]$. That is, each (unordered) pair of vertices $(u, v) \in V \times V$ is connected independently with probability $P_{u,v} = p_i$ if $u \in C_i, v \in C_i$. We define $P = \mathrm{diag}(p_1, \ldots, p_k)$ to be the diagonal matrix.*

- *Each (unordered) pair of vertices $(u, v) \in V \times V$ is connected independently with probability $Q_{u,v} = q_{ij}$ if $u \in C_i$, $v \in C_j$, and $i \neq j$, where $q_{i,j} \in [0, 1]$ is specified by a symmetric $k \times k$ matrix Q.*

We use $\mathrm{SSBM}(n, k, p, q)$ to denote the symmetric stochastic block model, where n is the total number of vertices; k is the number of communities; the matrix $P = \mathrm{diag}(p, \ldots, p)$, where p is the probability of having an edge within communities; and the off-diagonal entries of Q are $q_{i,j} = q$, where q is the probability of having an edge between two different communities. We are interested in the case where $q < p$, e.g., $p = a \log(n)/n$ and $q = b \log(n)/n$ with $a > b$. See [15], [2], and [1] for more in depth treatments of SBM.

Returning to the SBM, for any vertex i in community C_a, we define its *in-cluster degree* as $d_i^{in} = \#\{\{i,j\} \in E : j \in C_a\}$ and its *out-of-cluster degree* as $d_i^{out} := \#\{\{i,j\} \in E : j \notin C_a\}$. Similarly, the adjacency matrix A is written as $A = A^{in} + A^{out}$. We can easily see that

$$\mathbb{E}[d_i^{in}] = p(n_0 - 1) \quad \text{and} \quad \mathbb{E}[d_i^{out}] = q(n - n_0). \tag{10.2}$$

By definition, $d_i = d_i^{in} + d_i^{out}$. In fact, d_i^{in} is the degree of i considered as a vertex in the subgraph G_{C_a}. An important fact about degrees in Erdős–Rényi random graphs is that they concentrate around their means.

Lemma 10.6. *For any integer $a \in [1, \ldots, k]$, and for any real number $\alpha \in (0,1)$ with probability $1 - n_0 e^{-\alpha^2 n_0 p/2}$, we have*

$$(1 - \alpha)n_0 p \leq |d_i^{in}| \leq (1 + \alpha)n_0 p \quad \forall i \in C_a.$$

Proof. Each d_i^{in} is a binomial random variable, and so by the Chernoff inequality,

$$\mathbb{P}\left[|d_i^{in} - n_0 p| \geq \alpha n_0 p\right] \leq e^{-\alpha^2 n_0 p/2}.$$

Hence, we have

$$\mathbb{P}\left[\max_{i \in [n_0]} |d_i^{in} - n_0 p| \geq \alpha n_0 p\right]$$
$$= \mathbb{P}\left[|d_1^{in} - n_0 p| \geq \alpha n_0 p \text{ or } |d_2^{in} - n_0 p| \geq \alpha n_0 p \text{ or } \ldots \text{ or } |d_{n_0}^{in} - n_0 p| \geq \alpha n_0 p\right]$$
$$\leq \sum_{i=1}^{n_0} \mathbb{P}\left[|d_i^{in} - n_0 p| \geq \alpha n_0 p\right] \leq n_0 e^{-\alpha^2 n_0 p/2}.$$

Since $\mathbb{P}\left[|d_i^{in} - n_0 p| \leq \alpha n_0 p\right] = 1 - \mathbb{P}\left[|d_i^{in} - n_0 p| \geq \alpha n_0 p\right]$, the result of this lemma follows. $\square$

Returning to the SBM, we have the following result.

Lemma 10.7. *Suppose that G is drawn at random from $\mathcal{G}(n, k, p, q)$. For any $\alpha \in (0,1)$, as long as $p > \frac{4\ln(n)}{\alpha^2 n_0}$, we have*

$$(1 - \alpha)n_0 p \leq d_i^{in} \leq (1 + \alpha)n_0 p \quad \forall i \in [n]$$

with probability $1 - 1/n$.

Proof. The proof is similar to that of Lemma 10.6. Indeed,

$$\mathbb{P}\left[\max_{i \in [n]} |d_i^{in} - n_0 p| \geq \alpha n_0 p\right]$$
$$= \mathbb{P}\left[\max_{i \in C_1} |d_i^{in} - n_0 p| \geq \alpha n_0 p \text{ or } \max_{i \in C_2} |d_i^{in} - n_0 p| \geq \alpha n_0 p \text{ or } \ldots \max_{i \in C_k} |d_i^{in} - n_0 p| \geq \alpha n_0 p\right]$$
$$\leq \sum_{j=1}^{k} \mathbb{P}\left[\max_{i \in C_j} |d_i^{in} - n_0 p| \geq \alpha n_0 p\right] \leq \sum_{j=1}^{k} \left(n_0 e^{-\alpha^2 n_0 p/2}\right) \quad \text{by Lemma 10.6}$$
$$= k n_0 e^{-\alpha^2 n_0 p/2} = n e^{-\alpha^2 n_0 p/2}.$$

Clearly, $\mathbb{P}\left[\max_{i\in[n]}|d_i^{in}-n_0p|\leq\alpha n_0p\right]=1-\mathbb{P}\left[\max_{i\in[n]}|d_i^{in}-n_0p|\geq\alpha n_0p\right]$. As long as $p>\frac{4\ln(n)}{\alpha^2 n_0}=\frac{2\ln((n^2)}{\alpha^2 n_0}$, we have

$$1-n_0 k e^{-\alpha^2 n_0 p/2}=1-ne^{-\alpha^2 n_0 p/2}$$
$$\geq 1-ne^{-\alpha^2 n_0\left(2\ln((n^2)/(\alpha^2 n_0))\right)/2}\geq 1-n\left(e^{\ln(n^{-2})}\right)=1-1/n.$$

This completes the proof.   □

The following is an easy corollary.

Corollary 10.8. *Suppose that G is drawn at random from $\mathcal{G}(n,k,p,q)$ and that $p\geq 4k(\ln(n))^2/n$. Then*

$$(1-\alpha)n_0 p\leq d_i^{in}\leq(1+\alpha)n_0 p\quad\forall i\in\{1,\ldots,n\}$$

with probability $1-1/n$ for $\alpha=o(1)$ with respect to n.

Proof. Apply Lemma 10.7 with $\alpha=1/\sqrt{\ln(n)}$ by observing that for this value of α,

$$\frac{4\ln(n)}{\alpha^2 n_0}=\frac{4(\ln(n))^2}{(n/k)}=\frac{4k(\ln(n))^2}{n},$$

which completes the proof.   □

We remark that we can choose $\alpha=1/\sqrt{\log(\log(n))}$ in the proof above so that the probability $p=\omega\log(n)/n$ with $\omega=4k\log(\log(n))$. The second remarkable property of the Erdős–Rényi model is that the eigenvalues of $\mathcal{L}$ also concentrate around their means.

Theorem 10.9. *Let $\mathcal{L}$ be the Laplacian of a random graph drawn from $\mathcal{G}(n_0,p)$ with $p\gg(\ln(n_0))^2/n_0$. Then almost surely*

$$\max_{i\neq 0}|1-\lambda_i|\leq\frac{o(1)}{\sqrt{pn_0}}+\frac{g(n_0)\log^2(n_0)}{pn_0},$$

where $g(n_0)$ is a function tending to infinity arbitrarily slowly and $o(1)$ is a term which goes to zero when $n\to\infty$.

Proof. We leave the proof to the interested reader; see also Theorem 3.6 in [9].   □

Remark 10.1.1. *For our purposes in this chapter, it will be enough to note that Theorem 10.9 gives*

$$\lambda_2\geq 1-\frac{4}{\sqrt{pn_0}}-o\left(\frac{1}{\sqrt{n_0}}\right)\quad and\quad \lambda_{n_0}\leq 1+\frac{4}{\sqrt{pn_0}}+o\left(\frac{1}{\sqrt{n_0}}\right)$$

almost surely.

Theorem 10.10. *Let $C_1,\ldots,C_k$ denote the connected components of a graph G. Then $\mathbf{1}_{C_1},\ldots,\mathbf{1}_{C_k}$ form a basis for the kernel of $\mathcal{L}=D-A$.*

Proof. We start with $k=1$, in which case the graph is connected. Assume that $\mathbf{x}$ is an eigenvector with eigenvalue 0. Then we have

$$0=\mathbf{x}^\top\mathcal{L}\mathbf{x}=\sum_{i,j=1}^n a_{ij}(x_i-x_j)^2.\tag{10.3}$$

Since the a_{ij} are all nonnegative, this sum can only vanish if all the terms $a_{ij}(x_i - x_j)^2$ vanish. Thus if two vertices v_i and v_j are connected (i.e., $a_{ij} > 0$), then x_i needs to equal x_j.

Hence, for $k = 1$, we thus have only a constant vector, e.g., $\mathbf{1}$, as the eigenvector with eigenvalue 0, which is obviously the indicator vector of the connected component.

Now consider the case for which G has k connected components. Without loss of generality, we assume that the vertices are ordered according to the connected components they belong to. In this case, the adjacency matrix A has a block diagonal form, and the same is true for the graph Laplacian matrix $\mathcal{L}$:

$$\mathcal{L} = \begin{pmatrix} L_1 & & & \\ & L_2 & & \\ & & \ddots & \\ & & & L_k \end{pmatrix}.$$

Since $\mathcal{L}$ is block diagonal, we know that the spectrum of $\mathcal{L}$ is given by the union of the spectra of L_i, and the corresponding eigenvectors of $\mathcal{L}$ consist of the constant $\mathbf{1}$ vector of each L_i, filled with zeros at the positions of the other blocks. This completes the proof. □

10.2 ▪ RIP of Graph Laplacians

We first consider the RIP for the graph Laplacian matrix of a connected Erdős–Rényi graph drawn from $\mathcal{G}(n_0, p)$. We then extend the study to obtain the RIP for the Laplacian matrix of graphs drawn from the SBM $\mathcal{G}(n, k, p, 0)$, because these can be thought of as a disjoint union of k Erdős–Rényi graphs. Finally, we extend to graphs drawn from $\mathcal{G}(n, k, p, q)$ for $0 < q \ll p$ using a perturbation argument. The goal of this section is to show that certain graph Laplacian matrices satisfy an RIP property.

10.2.1 ▪ RIP for the Laplacian of Graphs in $\mathcal{G}(n_0, p)$

Lemma 10.11. *Let $\mathcal{L}$ be the Laplacian of a connected graph G with n vertices. For $S \subset [1, \dots, n]$ with $|S| = s < n$, let $\mathcal{L}_S$ be the Laplacian of subgraph G_S over the vertex set S. Then the smallest singular value $\sigma_{\min}(\mathcal{L}_S) \geq \sqrt{1 - s/n}\lambda_2$ and the restricted isometry constant (RIC)*

$$\delta_s = \max\left\{1 - \left(1 - \frac{s}{n}\right)\lambda_2^2, \lambda_n^2 - 1\right\},$$

where λ_i denotes the ith eigenvalue of $\mathcal{L}$, ordered from smallest to largest.

Proof. Let $\mathbf{w}_1, \dots, \mathbf{w}_n$ be an eigenbasis for $\mathcal{L}$ with eigenvalues $\lambda_1 \leq \cdots \leq \lambda_n$, where $\lambda_1 = 0$, $\lambda_2 > 0$, and $\mathbf{w}_i$ are normalized vectors. For any $\mathbf{v}$ supported on S, write $\mathbf{v} := \sum_{i=1}^n \alpha_i \mathbf{w}_i$. Note that $\mathbf{w}_1 = \frac{1}{\sqrt{n_0}}\mathbf{1}$ since $\ker(\mathcal{L}) = \mathrm{span}(\mathbf{w}_1)$. Then

$$\|\mathcal{L}\mathbf{v}\|_2^2 = \left\|\mathcal{L}\left(\sum_{i=1}^n \alpha_i \mathbf{w}_i\right)\right\|_2^2 = \left\|\left(\sum_{i=2}^n \alpha_i \mathcal{L}(\mathbf{w}_i)\right)\right\|_2^2 = \sum_{i=2}^n \alpha_i^2 \lambda_i^2 \geq \left(\sum_{i=2}^n \alpha_i^2\right)\lambda_2^2.$$

Because $\sum_{i=1}^n \alpha_i^2 = \|\mathbf{v}\|_2^2 = 1$, we have that $\sum_{i=2}^n \alpha_i^2 = 1 - \alpha_1^2$. Thus $\|\mathcal{L}\mathbf{v}\|_2^2 \geq \lambda_2^2(1 - \alpha_1^2)$, and so clearly this quantity is minimized by making α_1 as large as possible. Observe that

$$\alpha_1 = \left(\frac{1}{\sqrt{n_0}}\mathbf{1}_{C_1}\right)\cdot \mathbf{v} \leq \frac{1}{\sqrt{n}}\|\mathbf{v}\|_1 \leq \frac{1}{\sqrt{n}}(\sqrt{s}\|\mathbf{v}\|_2) = \frac{\sqrt{s}}{\sqrt{n}}.$$

We remark that this bound on α_1 is sharp and is achieved by taking $\mathbf{v} = \frac{1}{\sqrt{s}}\mathbf{1}_S$. Hence

$$\min_{\substack{\text{supp}(\mathbf{v}) \subset S \\ \|\mathbf{v}\|_2 = 1}} \|\mathcal{L}\mathbf{v}\|_2^2 \geq \left(1 - \left(\sqrt{\frac{s}{n}}\right)^2\right)\lambda_2^2 = \frac{n-s}{n}\lambda_2^2.$$

On the other hand,

$$\max_{\substack{\text{supp}(\mathbf{v}) \subset S \\ \|\mathbf{v}\|_2 = 1}} \|\mathcal{L}\mathbf{v}\|_2^2 \leq \max_{\|\mathbf{v}\|_2 = 1} \|\mathcal{L}\mathbf{v}\|_2^2 = \lambda_n^2.$$

It is known that $\lambda_n \leq 2$. Hence for any $S \subset [n]$ with $|S| = s$, and any $\mathbf{v}$ with $\text{support}(\mathbf{v}) \subset S$ and $\|\mathbf{v}\|_2 = 1$, we have

$$\left(1 - \frac{s}{n}\right)\lambda_2^2 \leq \|\mathcal{L}\mathbf{v}\|_2^2 \leq \lambda_n^2,$$

which completes the proof by using the arguments in the proof of Lemma 2.36 in Chapter 2. $\quad\square$

Theorem 10.12 (RIP for Laplacian of Erdős–Rényi Graphs). *Suppose that we are given a graph $G \in \mathcal{G}(n, p)$ with Laplacian $\mathcal{L} := \mathcal{L}_G$ and suppose that $p \gg (\ln(n))^2/n$. If $s = \gamma n$ with $\gamma \in (0, 1)$, then the RIC*

$$\delta_s(\mathcal{L}) \leq \gamma + (1 - \gamma)\frac{8}{\sqrt{pn}} + o\left(\frac{1}{\sqrt{n}}\right)$$

almost surely.

Proof. By Theorem 10.9 and Remark 10.1.1,

$$\lambda_2 \geq 1 - \frac{4}{\sqrt{pn}} - o\left(\frac{1}{\sqrt{n}}\right) \quad \text{and} \quad \lambda_n \leq 1 + \frac{4}{\sqrt{np}} + o\left(\frac{1}{\sqrt{n}}\right)$$

almost surely. Combining this with Lemma 10.11, we get that

$$\begin{aligned}
\delta_s \quad &\leq 1 - (1 - s/n)\left(1 - \frac{4}{\sqrt{pn}} - o\left(\frac{1}{\sqrt{n}}\right)\right)^2 \\
&= 1 - (1 - \gamma)\left(1 - \frac{8}{\sqrt{pn}} + o\left(\frac{1}{\sqrt{n}}\right)\right) \\
&= 1 - \left(1 - \gamma - (1 - \gamma)\frac{8}{\sqrt{pn}} + o\left(\frac{1}{\sqrt{n}}\right)\right) = \gamma + (1 - \gamma)\frac{8}{\sqrt{pn}} + o\left(\frac{1}{\sqrt{n}}\right).
\end{aligned}$$

Similarly, $\lambda_n^2 - 1 \leq \frac{8}{\sqrt{pn}} + o(\frac{1}{\sqrt{n}})$. When n is large enough, we have $\lambda_n^2 - 1 \leq \gamma$. We have thus completed the proof. $\quad\square$

We conclude this subsection with a lower bound for $\|\mathcal{L}\|_{2,s}$ as in (10.4) below, where $\mathcal{L}$ is the Laplacian of a connected graph G.

Lemma 10.13. *If $\mathcal{L}$ is the Laplacian of a connected graph G with n vertices, then $\|\mathcal{L}\|_{2,s} \geq \lambda_{s-1}$, where*

$$\|\mathcal{L}\|_{2,s} := \max_{\substack{S \subset [n] \\ |S| = s}} \|\mathcal{L}_S\|_2 = \max_{\substack{S \subset [n] \\ |S| = s}} \sigma_{\max}(\mathcal{L}_S). \tag{10.4}$$

Proof. We shall apply interlacing eigenvalues as in Theorem 10.14 (see below). For a symmetric nonnegative definite matrix $\mathcal{L}$, the eigenvalues of $\mathcal{L}$ are the singular values. We translate the notation of this theorem into the current situation: for $p = n$ and $q = s$, $\beta_i = \sigma_i(\mathcal{L}_S)$ and $\alpha_i = \lambda_i(\mathcal{L})$. Clearly, $\min(m,n) = \min(n,n) = n$ and $\min(p,q) = \min(n,s) = s$. From Theorem 10.14, we get

$$\sigma_{\max}(\mathcal{L}_S) \geq \lambda_{s-1}.$$

And so $\max_{\substack{S \subset [1,\ldots,n] \\ |S|=s}} \sigma_{\max}(L_S) \geq \lambda_{s-1}$. $\quad\square$

In the proof above, we have used the following classic interpolation theorem for singular values.

Theorem 10.14. *Let A be an $m \times n$ matrix with singular values $\alpha_1 \leq \alpha_2 \leq \cdots \leq \alpha_{\min(m,n)}$. Suppose that B is a $p \times q$ submatrix of A, with singular values $\beta_1 \leq \beta_2 \leq \cdots \leq \beta_{\min(p,q)}$. Then*

$$\alpha_{\min(m,n)-i} \geq \beta_{\min(p,q)-i} \quad \text{for } i = 0, 1, \ldots, \min(p,q) - 1$$

and

$$\beta_i \geq \alpha_{\min(m,n)-\min(p,q)-(m-p)-(n-q)-1+i}$$

for integer i satisfying $\min(p,q) + (m-p) + (n-q) + 2 - \min(m,n) \leq i \leq \min(m,n)$.

Proof. This is Theorem 1 in [24]. Note that they use the opposite notational convention: $\alpha_1 \geq \alpha_2 \geq \cdots \geq \alpha_{\min(m,n)}$ and $\beta_1 \geq \beta_2 \geq \cdots \geq \beta_{\min(p,q)}$. $\quad\square$

10.2.2 ▪ RIP for the Laplacian of Graphs from $\mathcal{G}(n, k, p, 0)$

Lemma 10.15. *Suppose that a graph G has k connected components $C_1, \ldots, C_k$, all of size n_0 (e.g., $G \in \mathcal{G}(n, k, p, 0)$). Let $G_{C_1}, \ldots, G_{C_k}$ denote the subgraphs on these components and let $\mathcal{L}^i = \mathcal{L}_{G_{C_i}}$ denote their Laplacians. Then for any $s < n_0$, $\delta_s(\mathcal{L}_G) = \max_i \delta_s(\mathcal{L}^i)$.*

Proof. Suppose $S \subset C_i$ for some i. For simplicity we assume $i = 1$, but the other cases are identical. In this case, $\mathcal{L}_S = \begin{bmatrix} \mathcal{L}^1 \\ 0 \end{bmatrix}$, where $\mathcal{L}^1$ denotes the Laplacian of G_{C_1} and 0 here is the zero matrix of the appropriate size. If $\text{support}(\mathbf{v}) \subset S$, then

$$\|\mathcal{L}_S \mathbf{v}\|_2^2 = \|\mathcal{L}^1 \mathbf{v}|_s\|_2^2$$

and so

$$(1 - \delta_s(\mathcal{L}^1))\|\mathbf{v}\|_2^2 \leq \|\mathcal{L}_S \mathbf{v}\|_2^2 \leq (1 + \delta_s(\mathcal{L}^1))\|\mathbf{v}\|_2^2.$$

It follows that, for all index sets S contained in a single component (i.e., $S \subset C_i$ for some i), we have

$$(1 - \max_i (\delta_s(\mathcal{L}^i)))\|\mathbf{v}\|_2^2 \leq \|\mathcal{L}_S \mathbf{v}\|_2^2 \leq (1 + \max_i (\delta_s(\mathcal{L}^i)))\|\mathbf{v}\|_2^2.$$

Now suppose that $S \not\subset C_i$. Write $S = \cup_i S_i$, where $S_i := S \cap C_i$. Given any $\mathbf{v}$ with $\text{support}(\mathbf{v}) = S$, we can write $\mathbf{v} = \sum_i \mathbf{v}_i$ with $\text{support}(\mathbf{v}_i) \subset S_i$. Then

$$\|\mathcal{L}_S \mathbf{v}\|_2^2 = \left\| \sum_i \mathcal{L}^i \mathbf{v}_i \right\|_2^2,$$

where $\mathcal{L}^i$ is the Laplacian of the subgraph G_{C_i}. Crucially, observe that all the terms $\mathcal{L}^i \mathbf{v}_i$ have disjoint support. Hence,

$$
\|\mathcal{L}_S \mathbf{v}\|_2^2 = \left\| \sum_i \mathcal{L}^i \mathbf{v}_i \right\|_2^2 = \sum_i \|\mathcal{L}_{S_i} \mathbf{v}_i\|_2^2 \geq \sum_i \left(1 - \delta_{s_i}(\mathcal{L}^i)\right) \|\mathbf{v}_i\|_2^2
$$

$$
\geq \min_i \left(1 - \delta_{s_i}(\mathcal{L}^i)\right) \sum_i \|\mathbf{v}_i\|_2^2 = \min_i \left(1 - \delta_{s_i}(\mathcal{L}^i)\right)
$$

$$
= 1 - \max_i \delta_{s_i}(\mathcal{L}^i) \geq 1 - \max_i \delta_s(\mathcal{L}^i),
$$

since $s_i \leq s$ for all i and δ_t is nondecreasing in t. An identical argument yields that

$$
\|\mathcal{L}_S \mathbf{v}\|_2^2 \leq 1 + \max_i \delta_{s_i}(\mathcal{L}^i),
$$

and so we have

$$
(1 - \max_i \left(\delta_s(\mathcal{L}^i)\right)) \|\mathbf{v}\|_2^2 \leq \|\mathcal{L}_S \mathbf{v}\|_2^2 \leq (1 + \max_i \left(\delta_s(\mathcal{L}^i)\right)) \|\mathbf{v}\|_2^2.
$$

This completes the proof.   □

Theorem 10.16 (RIP for Laplacian of Graphs from the SBM). *Suppose that we are given a graph $G \in \mathcal{G}(n, k, p, 0)$ with block sizes $n_0 = n/k$ and Laplacian $\mathcal{L}$. Suppose further that $p \geq (\ln(n_0))^2/n_0$ and that k is $O(1)$ with respect to n. If $s = \gamma n_0$ with $\gamma \in (0, 1)$, then*

$$
\delta_s(\mathcal{L}) \leq \gamma + (1 - \gamma)\frac{8}{\sqrt{pn_0}} + o\left(\frac{1}{\sqrt{n_0}}\right) \tag{10.5}
$$

almost surely.

Proof. Because $q = 0$, G will have k connected components with probability 1. Note that each subgraph G_i is an i.i.d. Erdős–Rényi graph, drawn from $\mathcal{G}(n_0, p)$. Let $\mathcal{L}^i$ denote the Laplacian of G_{C_i}. By Theorem 10.12,

$$
\delta_s(\mathcal{L}^i) \leq \gamma + (1 - \gamma)\frac{8}{\sqrt{pn_0}} + o\left(\frac{1}{\sqrt{n_0}}\right) \tag{10.6}
$$

almost surely. That is, there exists a function $f(n_0)$ going to zero as $n_0 \to \infty$ such that (10.6) holds with probability $1 - f(n_0)$. Because the G_{C_i} are i.i.d.,

$$
\max_{1 \leq i \leq k} \delta_s(\mathcal{L}^i) \leq \gamma + (1 - \gamma)\frac{8}{\sqrt{pn_0}} + o\left(\frac{1}{\sqrt{n_0}}\right) \tag{10.7}
$$

with probability $(1 - f(n_0))^k$. As long as k is bounded with respect to n_0 (which assuming k is $O(1)$ with respect to n will guarantee), $(1 - f(n_0))^k \to 1$ as $n_0 \to \infty$. Hence (10.7) holds almost surely.   □

Finally we conclude this subsection with a lower bound for $\|\mathcal{L}\|_{2,s}$ which is useful for estimates.

Lemma 10.17. *Suppose that a graph G has k connected components $C_1, \ldots, C_k$, all of size n_0. Let $G_{C_1}, \ldots, G_{C_k}$ denote the subgraphs on these components, and let $\mathcal{L}^i$ denote their Laplacians. Then $\|\mathcal{L}\|_{2,s} \geq \max_{i \in [k]} \lambda_{s-1}(\mathcal{L}^i)$.*

Proof. We leave the proof to the interested reader.   □

10.2.3 ▪ RIP for the Laplacian of Graphs from $\mathcal{G}(n, k, p, q)$ with $q > 0$

Next we study the RIP for the Laplacian of graphs from $\mathcal{G}(n, k, p, q)$. We need a preparatory result.

Theorem 10.18. *Let $\mathcal{L}$ be the Laplacian of a graph drawn at random from $\mathcal{G}(n, k, p, q)$. Suppose further that $\max_{1 \le i \le n} r_i := \max_{1 \le i \le n} d_i^{out}/d_i^{in} \le r$. Suppose that $r \ll 1$. Then $\mathcal{L} = \mathcal{L}^{in} + E$ with $E = E^1 + E^2$ such that $\text{support}(E^1) = \text{support}(A^{in})$ and $\text{support}(E^2) = \text{support}(A^{out})$ and*

1. $\|E^1\|_\infty, \|E^2\|_\infty \le r$;

2. $\|E^1\|_1, \|E^2\|_1 \le \beta^2 r$;

3. $\|E^1\|_2, \|E^2\|_2 \le \beta r$;

4. *for any $s < n$, $\|E^1\|_{2,s}, \|E^2\|_{2,s} \le \beta r$ and $\|E\|_{2,s} \le 2\beta r$*

for a positive constant $\beta > 0$ independent of r.

To prove the above results, we need the following lemma.

Lemma 10.19. *Suppose that $\mathcal{L}$ is the Laplacian of a graph G drawn from $\mathcal{G}(n, k, p, q)$, and let $\mathcal{L}^{in}$ denote the Laplacian of the underlying (disconnected) graph G^{in}, thought of as drawn from $\mathcal{G}(n, k, p, 0)$. Then $\mathcal{L} = \mathcal{L}^{in} + E$, where*

$$
E_{ij} = \begin{cases} \frac{r_i A_{ij}^{in}}{d_i^{in}} + \mathcal{O}(r^3) & \text{if } i, j \in C_a, \\ -\frac{1 - r_i}{d_i^{in}} A_{ij}^{out} + \mathcal{O}(r^3) & \text{if } i \in C_a \text{ and } j \in C_b \text{ for } a \ne b. \end{cases}
$$

Proof. Write $\mathcal{L} = [L_{ij}]$ with $L_{ij} = \delta_{ij} - \frac{A_{ij}}{d_i}$. Observe that $\frac{1}{d_i} = \frac{1}{d_i^{in} + d_i^{out}} = \frac{1}{d_i^{in}} \frac{1}{1 + \frac{d_i^{out}}{d_i^{in}}} = \frac{1}{d_i^{in}} \frac{1}{1 + r_i}$. Because $r_i \ll 1$ for all i, by Taylor's theorem $\frac{1}{d_i} = \frac{1}{d_i^{in}} (1 - r_i + \mathcal{O}(r_i^2))$. Direct computation reveals that, for $i, j \in C_a$,

$$
L_{ij} = \delta_{ij} - (1 - r_i) \frac{A_{ij}^{in}}{d_i^{in}} + \mathcal{O}\left(\frac{r_i^2}{d_i^{in}}\right) = \left(\delta_{ij} - \frac{A_{ij}^{in}}{d_i^{in}}\right) + \frac{r_i A_{ij}^{in}}{d_i^{in}} + \mathcal{O}(r^3) = L_{ij}^{in} + \frac{r_i A_{ij}^{in}}{d_i^{in}} + \mathcal{O}(r^3)
$$

(10.8)

since $1/d_i^{in} \le r_i = \mathcal{O}(r)$. The calculations for $i \in C_a, j \in C_b$ with $a \ne b$ are similar. □

Recall that $E = E^1 + E^2$, where E^1 will contain the "in-cluster" perturbation and E^2 will contain the "out-of-cluster" perturbation. Thus,

$$
E_{ij}^1 = \begin{cases} \frac{r_i A_{ij}^{in}}{d_i^{in}} + \mathcal{O}(r^3) & \text{if } i, j \in C_a, \\ 0 & \text{otherwise,} \end{cases}
$$

$$
E_{ij}^2 = \begin{cases} -\frac{1 - r_i}{d_i^{in}} A_{ij}^{out} + \mathcal{O}(r^3) & \text{if } i \in C_a \text{ and } j \in C_b \text{ for } a \ne b, \\ 0 & \text{otherwise.} \end{cases}
$$

We are now ready to prove Theorem 10.18.

Proof of Theorem 10.18. By definition, $\|\cdot\|_\infty$ is the maximum absolute row sum of the matrix in question. For any i, let C_a be the cluster to which it belongs. Then

$$\sum_j |E^1_{ij}| = \sum_{j \in C_a} \left(\frac{r_i A^{in}_{ij}}{d^{in}_i} + \mathcal{O}(r^3) \right) = \frac{r_i}{d^{in}_i} \sum_{j \in C_a} A^{in}_{ij} + \mathcal{O}(r^2) = \frac{r_i}{d^{in}_i} \left(d^{in}_i \right) + \mathcal{O}(r^2) = r_i + \mathcal{O}(r^2).$$

Hence, we have

$$\|E^1\|_\infty = \max_i \left(\sum_j |E^1_{ij}| \right) = \max_i(r_i) + \mathcal{O}(r^2) = r + \mathcal{O}(r^2).$$

Similarly,

$$\sum_j |E^2_{ij}| = \sum_{j \notin C_a} \left(\frac{(1 - r_i) A^{out}_{ij}}{d^{in}_i} + \mathcal{O}(r^3) \right) \le \frac{1}{d^{in}_i} \sum_{j \notin C_a} A^{out}_{ij} + \mathcal{O}(r^2)$$

$$= \frac{1}{d^{in}_i} \left(d^{out}_i \right) + \mathcal{O}(r^2) = r_i + \mathcal{O}(r^2),$$

and so again

$$\|E^2\|_\infty = \max_i \left(\sum_j |E^2_{ij}| \right) = \max_i(r_i) + \mathcal{O}(r^2) = r + \mathcal{O}(r^2).$$

In a similar fashion, because $\|\cdot\|_1$ is the maximum absolute column sum, for any j in cluster C_a, we have

$$\sum_{1 \le i \le n} |E^1_{ij}| = \sum_{i \in C_a} \left(\frac{r_i A^{in}_{ij}}{d^{in}_i} + \mathcal{O}(r^3) \right)$$

$$\le \frac{r}{d^{in}_{min}} \sum_{i \in C_a} A^{in}_{ij} + \mathcal{O}(r^2) \le \frac{r}{d^{in}_{min}} (d^{in}_{max}) + \mathcal{O}(r^2) = \frac{r(1 + \alpha)}{(1 - \alpha)} + \mathcal{O}(r^2)$$

and

$$\|E^1\|_1 = \max_j \left(\sum_i |E^1_{ij}| \right) \le \frac{r(1 + \alpha)}{(1 - \alpha)} + \mathcal{O}(r^2) = \beta^2 r + \mathcal{O}(r^2).$$

Because

$$\sum_{1 \le i \le n} |E^2_{ij}| = \sum_{i \notin C_a} \left(\frac{(1 - r_i) A^{out}_{ij}}{d^{in}_i} + \mathcal{O}(r^3) \right) \le \frac{1}{d^{in}_{min}} \sum_{i \notin C_a} A^{out}_{ij} + \mathcal{O}(r^2)$$

$$= \frac{(1 + \alpha)}{(1 - \alpha) d^{in}_{max}} (d^{out}_j) + \mathcal{O}(r^2) \le \frac{(1 + \alpha)}{(1 - \alpha)} \frac{d^{out}_j}{d^{in}_j} = \frac{(1 + \alpha)}{(1 - \alpha)} r_j,$$

we have

$$\|E^2\|_1 = \max_j \left(\sum_i |E^2_{ij}| \right) \le \frac{(1 + \alpha)}{(1 - \alpha)} \max_j(r_j) + \mathcal{O}(r^2) = \beta^2 r + \mathcal{O}(r^2).$$

Now, by the Riesz–Thorin interpolation theorem,

$$\|E^1\|_2 \leq \|E^1\|_1^{1/2}\|E^1\|_\infty^{1/2} \leq \left(\beta^2 r\right)^{1/2}(r)^{1/2} = \beta r,$$

and similarly we have $\|E^2\|_2 \leq \beta r$.

Item 4 in the statement of Theorem 10.18 follows from the simple fact that for any matrix B, $\|B\|_{2,s} \leq \|B\|_2$, and the bound for $\|E\|_{2,s}$ follows from the fact that $E = E^1 + E^2$ and the triangle inequality. □

We are now ready to state and prove the following preparatory result, which will be useful later.

Theorem 10.20. *Suppose that $G \in \mathcal{G}(n, k, p, q)$ with block sizes $n_0 = n/k$ and Laplacian $\mathcal{L} = \mathcal{L}_G$. Let $s = \gamma n_0$ for $\gamma \in (0, 1)$. If $p \geq (\log(n_0))^2/n_0$ and k is $O(1)$ with respect to n, then*

$$\delta_s(L) \leq \gamma + D_1 r + \frac{D_2}{\sqrt{n_0}} + D_3\frac{r}{\sqrt{n_0}}\mathcal{O}(r^2) + o\left(\frac{1}{\sqrt{n_0}}\right) \tag{10.9}$$

almost surely, where D_1, D_2, and D_3 are $\mathcal{O}(1)$ with respect to n_0

Proof. As in the previous section, we write $\mathcal{L} = \mathcal{L}^{in} + E$, where $\mathcal{L}^{in}$ is the Laplacian of the subgraph G^{in} thought of as drawn from $\mathcal{G}(n, k, p, 0)$. We define $\delta_s := \delta_s(\mathcal{L}^{in})$ and $\hat{\delta}_s := \delta_s(\mathcal{L})$. By Theorem 2.44,

$$\hat{\delta}_s \leq (1 + \delta_s)(1 + \epsilon_s^s)^2 - 1. \tag{10.10}$$

As defined in section 2.6, $\epsilon_{\mathcal{L}}^s := \|E\|_{2,s}/\|\mathcal{L}^{in}\|_{2,s}$. Theorem 10.18 gives us that $\|E\|_{2,s} \leq 2\beta r$, while Lemma 10.17 gives $\|\mathcal{L}^{in}\|_{2,s} \geq \max_{1 \leq a \leq k} \lambda_{s-1}(L^a)$. By an argument similar to that of Theorem 10.16, we can apply Lemma 10.9 (and see also Remark 10.1.1) to get

$$\max_{1 \leq a \leq k} \lambda_{s-1}(\mathcal{L}^a) \geq 1 - \frac{4}{\sqrt{pn_0}} - o\left(\frac{1}{\sqrt{n_0}}\right)$$

almost surely, assuming that k is bounded with respect to n_0. For convenience, choose n_0 large enough such that $4/\sqrt{pn_0} + o(\frac{1}{\sqrt{n_0}}) \leq 1/4$. Under this assumption,

$$\epsilon_{\mathcal{L}^{in}}^s = \frac{\|E\|_{2,s}}{\|\mathcal{L}^{in}\|_{2,s}} \leq \frac{2\beta r}{1 - 1/4} = \frac{8\beta r}{3}.$$

Combining this with (10.10) and Theorem 10.16 gives

$$\hat{\delta}_s \leq \left(1 + \gamma + (1 - \gamma)\frac{8}{\sqrt{pn_0}} + o\left(\frac{1}{\sqrt{n_0}}\right)\right)\left(1 + \frac{8\beta r}{3}\right)^2 - 1$$

$$= \left(1 + \gamma + (1 - \gamma)\frac{8}{\sqrt{pn_0}} + o\left(\frac{1}{\sqrt{n_0}}\right)\right)\left(1 + \frac{16\beta r}{3} + \frac{64\beta^2 r^2}{9}\right) - 1$$

$$= \gamma + r\left(\frac{16\beta(1 + \gamma)}{3}\right) + \frac{1}{\sqrt{n_0}}\left(8(1 - \gamma)p^{-1/2}\right) + \frac{r}{\sqrt{n_0}}\left(\frac{128(1 - \gamma)p^{-1/2}\beta}{3}\right)$$

$$+ \mathcal{O}(r^2) + o\left(\frac{1}{\sqrt{n_0}}\right),$$

and letting $D_1 = 16\beta(1+\gamma)/3$, $D_2 = 8(1-\gamma)p^{-1/2}$, $D_3 = 128(1-\gamma)p^{-1/2}\beta^2/3$ gives (10.9). Finally, because $\beta \to 1$ as $n \to \infty$, we indeed have that D_1, D_2, and D_3 are $\mathcal{O}(1)$ with respect to n_0. □

10.3 ▪ Mutual Coherence Properties of Graph Laplacians

In this section we study the mutual coherence properties of Laplacians of random graphs. Mainly, we compute the expected value of coherence $\mu(\mathcal{L}_G)$. We begin with a basic result for graphs from $\mathcal{G}(n, k, p, 0)$. Let A be the adjacency matrix of a graph G drawn at random from $\mathcal{G}(n, k, p, 0)$ and define $\chi_{i,j} = \sum_k A_{ik} A_{kj}$ for $i, j = 1, \ldots, n$. Recall that the coherence of A is

$$\mu = \max_{\substack{i,j=1,\ldots,n \\ i \neq j}} |\chi_{i,j}| = \max_{\substack{i,j=1,\ldots,n \\ i \neq j}} \chi_{i,j}. \tag{10.11}$$

Theorem 10.21. *Suppose that there exists an $\alpha \in (0,1)$ such that $(1-\alpha)pn_0 \leq d_i \leq (1+\alpha)pn_0$. Then for any $i, j \in C_1$ with $i \neq j$,*

$$(1-\alpha)n_0 p^2 - \sqrt{\frac{(1+\alpha)n_0 p \ln(1/\delta)}{2}} \leq \chi_{i,j} \leq (1+\alpha)n_0 p^2 + \sqrt{\frac{(1+\alpha)n_0 p \ln(1/\delta)}{2}} \tag{10.12}$$

with probability $1 - \delta$.

Proof. Let $B_1(i)$ denote the ball of radius 1 centered at i. That is, $B_1(i) := \{k \in [1, \ldots, n] : \{i, k\} \in E\}$. By definition, $\#B_1(i) = d_i$. Observe that

$$\chi_{i,j} = \sum_k A_{ik} A_{kj} = \sum_{k \in B_1(i)} A_{kj}.$$

Each A_{ik} is an i.i.d. Bernoulli random variable with parameter p. Applying the Chernoff bound to this sum, we get

$$\mathbb{P}\left[\sum_{k \in B_1(i)} A_{kj} \leq pd_i - \sqrt{\frac{d_i \ln(1/\delta)}{2}} \right] \leq \exp\left(-\frac{2\left(\sqrt{\frac{d_i \ln(1/\delta)}{2}}\right)^2}{d_i} \right) = \delta$$

and $(1 - \alpha)pn_0 \leq d_i \leq (1+\alpha)pn_0$, so

$$\mathbb{P}\left[\chi_{i,j} \leq (1-\alpha)n_0 p^2 - \sqrt{\frac{(1+\alpha)n_0 p \ln(1/\delta)}{2}} \right] \leq \mathbb{P}\left[\chi_{i,j} \leq pd_i - \sqrt{\frac{d_i \ln(1/\delta)}{2}} \right] \leq \delta.$$

The upper bound on $\chi_{i,j}$ is proved analogously. □

Let $\mathcal{L}_G$ be the Laplacian of a graph G drawn at random from $\mathcal{G}(n, k, p, 0)$. Write $\mathcal{L} = [\ell_1, \ldots, \ell_n]$. We now provide a bound on the inner product $|\langle \ell_i, \ell_j \rangle|$.

Theorem 10.22. *As in Theorem 10.21, suppose that there exists an $\alpha \in (0,1)$ such that $(1 - \alpha)pn_0 \leq d_i \leq (1+\alpha)pn_0$. Define $\beta^2 = (1 - \alpha)/(1 + \alpha)$ and assume that $\beta^2 \geq p$. Then for any $i, j \in C_a, a = 1, \ldots, k$ with $i \neq j$,*

$$|\langle \ell_i, \ell_j \rangle| \geq \frac{1}{n_0}\left(\frac{\beta^2}{(1+\alpha)} \right) - \frac{1}{n_0^{3/2}} \sqrt{\frac{-\ln(\delta)}{2(1+\alpha)^3 p^3}}$$

with probability $1 - \delta$.

Proof. Recall that

$$|\langle \ell_i, \ell_j \rangle| = \left| -\left(\frac{1}{d_i} + \frac{1}{d_j} \right) A_{ij} + \sum_{k=1}^{n} \frac{A_{ik} A_{jk}}{d_k^2} \right|.$$

There are two possibilities: either i and j are connected ($\{i,j\} \in E$) or they are not ($\{i,j\} \notin E$). We treat these cases separately. Suppose first that $\{i,j\} \notin E$. In this case, $A_{ij} = 0$ and so

$$|\langle \ell_i, \ell_j \rangle| = \sum_{k=1}^{n} \frac{A_{ik} A_{jk}}{d_k^2} \geq \frac{1}{d_{\max}^2} \chi_{ij}$$

by assumption, where $d_{\max} \leq (1+\alpha)pn_0$, and so by Theorem 10.21,

$$|\langle \ell_i, \ell_j \rangle| \geq \frac{1}{(1+\alpha)^2 p^2 n_0^2} \left((1-\alpha)n_0 p^2 - \sqrt{-\frac{(1+\alpha)n_0 p \ln(\delta)}{2}} \right)$$

$$= \frac{1}{n_0} \left(\frac{\beta^2}{(1+\alpha)} \right) - \frac{1}{n_0^{3/2}} \sqrt{\frac{-\ln(\delta)}{2(1+\alpha)^3 p^3}}. \tag{10.13}$$

Alternatively, suppose that $\{i,j\} \in E$. Then $A_{ij} = 1$ and

$$|\langle \ell_i, \ell_j \rangle| \geq \frac{2}{d_{\max}} - \frac{1}{d_{\min}^2} \chi_{ij} \geq \frac{2}{(1+\alpha)pn_0} - \frac{1}{(1-\alpha)^2 p^2 n_0^2} \chi_{ij}$$

$$\geq \frac{2}{(1+\alpha)pn_0} - \frac{1}{(1-\alpha)^2 p^2 n_0^2} \left((1+\alpha)n_0 p^2 + \sqrt{-\frac{(1+\alpha)n_0 p \ln(\delta)}{2}} \right)$$

$$= \frac{1}{n_0} \left(\frac{2}{(1+\alpha)p} - \frac{1}{(1+\alpha)\beta^2} \right) - \frac{1}{n_0^{3/2}} \sqrt{\frac{-\ln(\delta)}{2(1-\alpha)^3 p^3 \beta}} \tag{10.14}$$

by Theorem 10.21. Because $1 \geq \beta^2 > p$,

$$\frac{2}{(1+\alpha)p} - \frac{1}{(1+\alpha)\beta^2} > \frac{2}{(1+\alpha)\beta^2} - \frac{1}{(1+\alpha)\beta^2} = \frac{1}{(1+\alpha)\beta^2} \geq \frac{\beta^2}{(1+\alpha)}.$$

That is, the leading term in (10.14) is strictly larger than the leading term in (10.13). It follows that in the case where $\{i,j\} \in E$ we again have that

$$|\langle \ell_i, \ell_j \rangle| \geq \frac{1}{n_0} \left(\frac{\beta^2}{(1+\alpha)} \right) - \frac{1}{n_0^{3/2}} \sqrt{\frac{-\ln(\delta)}{2(1+\alpha)^3 p^3}}$$

for large enough n_0.   ☐

Next we consider the case where intercluster edges are present, i.e., where $q > 0$.

Theorem 10.23. *Let $\mathcal{L}$ be the Laplacian of G drawn from $\mathcal{G}(n,k,p,q), q > 0$. Assume further that*

1. *$p \geq 4k(\ln(n))^2/n$ and*

2. *$\max_i d_i^{out}/d_i^{in} \leq r$ with $r = r_0/\sqrt{n_0}$, where $r_0 = \mathcal{O}(1)$ with respect to n_0.*

Then with probability at least $1 - 1/n_0$,

$$|\langle \ell_i, \ell_j \rangle| \geq \frac{1}{n_0} \left(\frac{\beta^2}{1+\alpha} \right) - o\left(\frac{1}{n_0} \right)$$

for $i,j \in C_1$.

Proof. We leave the proof to the interested reader; see also the proof of Theorem 10.30 below. □

Remark 10.3.1. *Since the coherence μ of $\mathcal{L}$ is defined as $\mu = \max_{i \neq j} |\langle \ell_i, \ell_j \rangle|$, Theorem 10.23 also implies that*

$$\mu \geq \frac{1}{n_0} \left(\frac{\beta^2}{1 + \alpha} \right) - o \left(\frac{1}{n_0} \right).$$

10.4 ▪ Single Clustering Pursuit (SCP)

In this section, we consider graphs from $\mathcal{G}(n, k, p, 0)$. This is the case where there are no inter-cluster edges and it is easy to deal with. We may safely assume that, with high probability, each G_{C_i} is connected, as long as $p > \ln(n_0)/n_0$.

Lemma 10.24. *Let $\mathcal{L} = \mathcal{L}_G$ be the Laplacian of graph G with k connected components $C_1, \ldots, C_k$. Assume, without loss of generality, that vertex $1 \in C_1$. Define $\mathbf{x}^{\#}$ as*

$$\mathbf{x}^{\#} := \arg \min \{ \|\mathbf{x}\|_0 : \mathbf{x} \in \mathbb{R}^n, \ \mathcal{L}\mathbf{x} = \mathbf{0}, \ and \ x_1 = 1 \}. \tag{10.15}$$

Then $\mathbf{x}^{\#} = \mathbf{1}_{C_1}$

Proof. From Theorem 10.3, $\ker(\mathcal{L}) = \operatorname{span}\{\mathbf{1}_{C_1}, \ldots, \mathbf{1}_{C_k}\}$, and hence any $\mathbf{x}$ with $\mathcal{L}\mathbf{x} = \mathbf{0}$ can be written as $\mathbf{x} = \sum_{i=1}^k c_i \mathbf{1}_{C_i}$. If $x_1 = 1$ (recall that we are assuming that $1 \in C_1$), then $c_1 = 1$ and so $\mathbf{x} = \mathbf{1}_{C_1} + \sum_{i=2}^k c_i \mathbf{1}_{C_i}$. Recall that all of the $\mathbf{1}_{C_i}$ have disjoint support, and so clearly the sparsest solution $\mathbf{x}$ such that $\mathcal{L}\mathbf{x} = \mathbf{0}$ is that $\mathbf{x}$ has $c_2 = c_3 = \cdots = c_k = 0$. Hence, indeed, $\mathbf{x}^{\#} = \mathbf{1}_{C_1}$. □

We can rephrase this slightly as a standard compressed sensing problem. Recall that ℓ_i is the ith column of $\mathcal{L}$ and $\mathcal{L}_{-i} := [\ell_1, \ldots, \ell_{i-1}, \ell_{i+1}, \ldots, \ell_n]$, in which case

$$x_i = 1 \quad \Leftrightarrow \quad \mathcal{L}_{-i} \mathbf{x}_{-i} = -\ell_i,$$

where if $\mathbf{x} = [x_1, x_2, \ldots, x_n]^\top \in \mathbb{R}^n$, $\mathbf{x}_{-i} = [x_1, \ldots, x_{i-1}, x_{i+1}, \ldots, x_n]^\top \in \mathbb{R}^{n-1}$. Thus problem (10.15) becomes

$$\mathbf{x}^{\#} := \arg \min \{ \|\mathbf{x}\|_0 : \mathbf{x} \in \mathbb{R}^{n-1}, \ \mathcal{L}_{-1} \mathbf{x} = -\ell_1 \}. \tag{10.16}$$

We now show that problem (10.16) can be efficiently solved using the OMP Algorithm 3.2 discussed in Chapter 3. We shall use the study to establish the following result.

Theorem 10.25. *Suppose that G is a graph from $\mathcal{G}(n, k, p, 0)$ with k connected components $C_1, \ldots, C_k \subset V$, and let $n_0 := |C_1|$. The OMP Algorithm 3.2 in Chapter 3 applied to the sparse recovery problem (10.16) will return $\mathbf{x}^{\#} = \mathbf{1}_{C_1 \setminus \{1\}}$ after $n_0 - 1$ iterations. C_1 can then be recovered as $\{1\} \cup \operatorname{support}(\mathbf{x}^{\#})$.*

Remark 10.4.1. *Note that we place absolutely no constraints on the parameters n, k, and p, so this theorem holds for both dense and sparse graphs. In particular, if a symmetric matrix A is reducible, we can use this approach to find a permutation matrix to convert A into a block diagonal matrix.*

We shall appeal to the exact recovery condition, Theorem 3.1, to establish Theorem 10.25. Let us begin with the following lemma.

Lemma 10.26. *If $C_1 \subset V$ is a connected component, then $\mathcal{L}_{C_1 \setminus \{1\}}$ is injective, where $\mathcal{L}_{C_1 \setminus \{1\}}$ is the submatrix from $\mathcal{L}$ with column indices in $C_1 \setminus \{1\}$.*

Proof. First, observe that $\mathcal{L}_{C_1 \setminus \{1\}} = \begin{bmatrix} \mathcal{L}^1_{-1} \\ 0_{(n-n_0) \times (n_0-1)} \end{bmatrix}$, where $\mathcal{L}^1$ is the Laplacian of G_{C_1}. It suffices to show that $\mathcal{L}^1_{-1}$ is injective. We may assume that G_{C_1} is connected, so $\ker(\mathcal{L}^1) = \text{span}\{\mathbf{1}\}$, by Theorem 10.2. Suppose there exists a $\mathbf{u} \in \mathbb{R}^{n_0-1}$, $\mathbf{u} \neq 0$, such that $\mathcal{L}^1_{-1}\mathbf{u} = \mathbf{0}$. Then $[0, \mathbf{u}^\top]^\top$ is in $\ker(\mathcal{L}^1) = \text{span}\{\mathbf{1}\}$, a contradiction. $\square$

Lemma 10.27. *With notation as above (i.e., C_1 is the connected component of G containing the first vertex), we have that*

$$\|\mathcal{L}^\dagger_{C_1 \setminus \{1\}} \mathcal{L}_{C_1^c}\|_1 = 0, \tag{10.17}$$

where the pseudo-inverse of $\mathcal{L}_{C_1 \setminus \{1\}}$ is given by

$$\mathcal{L}^\dagger_{C_1 \setminus \{1\}} = \left(\mathcal{L}^\top_{C_1 \setminus \{1\}} \mathcal{L}_{C_1 \setminus \{1\}} \right)^{-1} \mathcal{L}^\top_{C_1 \setminus \{1\}}.$$

Proof. By Lemma 10.26, $\mathcal{L}_{C_1 \setminus \{1\}}$ is injective. The pseudo-inverse is well defined. As observed in the proof of Lemma 10.26, $\mathcal{L}_{C_1 \setminus \{1\}} = \begin{bmatrix} \mathcal{L}^1_{-1} \\ 0_{(n-n_1) \times (n_1-1)} \end{bmatrix}$. Similarly, $\mathcal{L}_{C_1^c} = \begin{bmatrix} 0_{n_1 \times (n-n_1)} \\ \tilde{\mathcal{L}} \end{bmatrix}$, where $\tilde{\mathcal{L}}$ denotes the Laplacian of $G_{C_1^c}$. To show (10.17), it will suffice to show that $\mathcal{L}^\top_{C_1 \setminus \{1\}} \mathcal{L}_{C_1^c} = 0$, but this follows easily because

$$\begin{bmatrix} (\mathcal{L}^1_{-1})^\top & 0_{(n_0-1) \times (n-n_0)} \end{bmatrix} \begin{bmatrix} 0_{n_0 \times (n_0-1)} \\ \tilde{\mathcal{L}} \end{bmatrix} = (\mathcal{L}^1_{-1})^\top 0_{n_0 \times (n_0-1)} + 0_{n_0 \times (n-n_0)} \tilde{\mathcal{L}} = 0.$$

This completes the proof. $\square$

We are now ready to establish Theorem 10.25.

Proof of Theorem 10.25. This follows easily from Theorem 3.1 in Chapter 3 by using Lemma 10.27 above, i.e., (10.17) above. $\square$

Let us remark that many other algorithms discussed in previous chapters will also return $\mathbf{x}^\# = \mathbf{1}_{C_1}$ by solving (10.16) or (10.18) below for some graphs. We leave the study to the interested reader.

Theorem 10.28. *If $p \geq \sqrt{\ln(n_0 - 1)}/\sqrt{2n_0}$, then with high probability $\mathcal{L}_{-1}(\mathcal{L}^\top_{-1}\ell_1) = C_1 \setminus \{1\}$.*

Proof. If $p \geq \sqrt{\ln(n_0 - 1)}/\sqrt{2n_0}$, then G_{C_1} has diameter at most 2, with high probability. It follows that $(\mathcal{L}^\top_{-1}\ell_1)_i = |\langle \ell_1, \ell_i \rangle| > 0$ with high probability for $i \in C_1 \setminus \{1\}$, and, by the same argument used to prove Theorem 10.27, $(\mathcal{L}^\top_{-1}\ell_1)_i = 0$ for $i \notin C_1 \setminus \{1\}$. The result follows. $\square$

If $p < \sqrt{\ln(n_0 - 1)}/\sqrt{2n_0}$ and so $\text{diam}(G_{C_1}) = d > 2$, it is possible to show that the iterative hard thresholding algorithm discussed in Chapter 4 will recover C_1 in at most $d - 1$ steps. We leave it to the interested reader. See Exercise 25.

10.5 ▪ SCP for General Graphs

We now turn our attention to the problem in the previous section for noisy graphs subject to $\mathcal{G}(n, k, p, q)$ with $0 < q \ll p$ in the sense that $p = \omega(n)\ln(n)/n$ and $q = \beta \ln(n)/n$ with $\beta \geq 1$ and $\omega(n) \to \infty$. Because q is much smaller than p, there are much fewer intercluster edges than intracluster edges, and so we may regard A as a small perturbation of the adjacency matrix A^{in} of the subgraph G^{in}, obtained from G by removing all intercommunity edges. Note that G^{in} has k connected clusters, $C_1, \ldots, C_k$, and can be thought of as drawn from $\mathcal{G}(n, k, p, 0)$. That is, $A = A^{in} + A^{out}$, where A^{out} corresponds to intercommunity edges:

$$A_{ij}^{in} = \begin{cases} 1 & \text{if } \{i, j\} \in E \text{ and } i, j \in C_a \text{ for an } a \in \{1, \ldots, k\}, \\ 0 & \text{otherwise}, \end{cases}$$

$$A_{ij}^{out} = \begin{cases} 1 & \text{if } \{i, j\} \in E \text{ and } i \in C_a \text{ and } j \in C_b \text{ for some } a \neq b, \\ 0 & \text{otherwise}. \end{cases}$$

Because $p \gg q$, A^{out} will be significantly sparser than A^{in}. Recall that, as in section 10.1.3, for any $i \in C_a$, d_i^{in} denotes the in-cluster degree (equivalently the degree of i in G_{C_a} or G^{in}) and d_i^{out} denotes the out-of-cluster degree. In terms of A^{in} and A^{out}, $d_i^0 = \sum_j A_{ij}^{in}$ and $d_i^{out} = \sum_j A_j^{out}$. Further, define $r_i = d_i^{out}/d_i^{in}$. We can assume that $d_i^{out} \ll d_i^{in}$ or, equivalently, $r_i \ll 1$, for all i.

Let $\mathcal{L}^{in}$ denote the Laplacian of the underlying graph G^{in}. It is tempting to assume that $\mathcal{L} = \mathcal{L}^{in} + \mathcal{L}^{out}$, where $\mathcal{L}^{out}$ is a Laplacian associated with A^{out}, but unfortunately this is not the case. In any case, $\mathcal{L}$ is a small perturbation of $\mathcal{L}^{in}$. That is, $\mathcal{L} = \mathcal{L}^{in} + E$, where the perturbation matrix $\|E\|_2 \ll \|\mathcal{L}^{in}\|_2$.

Denoting the ith column of $\mathcal{L}^{in}$ as ℓ_i^{in} and the ith column of E as $\mathbf{e}_i$, observe that

$$\ell_1 = \ell_1^{in} + \mathbf{e}_1 = -\mathcal{L}_{-1}^{in} \mathbf{1}_{C_1 \setminus \{1\}} + \mathbf{e}_1 \Rightarrow -\ell_1 = \mathcal{L}_{-1}^{in} \mathbf{1}_{C_1 \setminus \{1\}} - \mathbf{e}_1.$$

Define $\mathbf{x}^\#$ as

$$\mathbf{x}^\# := \arg\min\{\|\mathbf{x}\|_0 : \mathbf{x} \in \mathbb{R}^{n-1}, \|\mathcal{L}_{-1}\mathbf{x} - (-\ell_1)\|_2 \leq \eta\}, \tag{10.18}$$

where $\mathcal{L} = \mathcal{L}^{in} + E$, $\mathbf{y} = -\ell_1$, and $\mathbf{x}^* = \mathbf{1}_{C_1 \setminus \{1\}}$. We shall show that, provided E is small enough, $\text{supp}(\mathbf{x}^\#) = \text{supp}(\mathbf{1}_{C_1 \setminus \{1\}}) = C_1 \setminus \{1\}$. Hence, solving problem (10.18) is equivalent to finding the cluster C_1.

Unfortunately, a straightforward application of Algorithm 3.2 to (10.18) does not work well. A greedy step from Algorithm 3.2 will pick up several wrong indices due to the noise in E. We develop a new approach which consists of two stages. The first stage is to find a superset Ω containing a cluster C_a of interest for $a \in [k]$. The second stage is to find the indices in Ω, but not in C_a, by a compressive sensing approach. We shall explain one approach for finding Ω in this section. More approaches to find a subset Ω will be discussed later.

Let us first focus on the first stage. Recall that for any vector $\mathbf{v} = (v_1, \ldots, v_n)^\top \in \mathbb{R}^n$,

$$\mathcal{H}_s(\mathbf{v}) = \{(i_1, \ldots, i_s) : v_{i_1}, \ldots, v_{i_s} \text{ are the largest entries in absolute value in } \mathbf{v}\}$$

denotes the thresholding operator on $\mathbb{R}^n$. The main result in this section is as follows.

Theorem 10.29. *Let $\mathcal{L}$ be the Laplacian of a graph $G \in \mathcal{G}(n, k, p, q)$, and suppose there exists a real number $r = r_0/\sqrt{n_0}$ such that*

$$\max_{i \in [1, \ldots, n]} d_i^{out}/d_i^{in} \leq r, \tag{10.19}$$

where r_0 is a constant independent of n_0. Define $\Omega := \mathcal{H}_{\lceil 10(n_0-1)/9 \rceil}(L_{-1}^\top \ell_1)$. For n_0 large enough, we have that $C_1 \setminus \{1\} \subset \Omega$ with probability at least $1 - 1/n_0$.

Proof. Suppose otherwise. Then there exists an $i^* \in C_1 \setminus \{1\}$ not in Ω. Let $\Lambda = \Omega \cap C_1^c$. Since we assumed that $C_1 \setminus \{1\} \not\subset \Omega$, we have that $|\Lambda| \geq n_0/9 - 1/9$. Moreover, by definition of i^*, we have that $|\langle \ell_1, \ell_{i^*} \rangle| \leq |\langle \ell_1, \ell_j \rangle|$ for all $j \in \Omega$, and, in particular,

$$|\langle \ell_1, \ell_{i^*} \rangle| \leq |\langle \ell_1, \ell_j \rangle| \quad \forall j \in \Lambda.$$

Summing over Λ, we get

$$\#(\Lambda)|\langle \ell_1, \ell_{i^*} \rangle| \leq \sum_{j \in \Lambda} |\langle \ell_1, \ell_j \rangle| = \|L_\Lambda^T \ell_1\|_1. \tag{10.20}$$

We shall show that (10.20) cannot hold for n_0 large enough. From Theorem 10.18 we have that $\mathcal{L} = \mathcal{L}^{in} + E^1 + E^2$ with $\|E^j\|_\infty \leq r$ for $j = 1, 2$. Moreover, by construction, $\mathcal{L}_\Lambda^{in} = \begin{bmatrix} \mathbf{0}_{n_0 \times |\Lambda|} \\ \tilde{\mathcal{L}} \end{bmatrix}$, where $\tilde{\mathcal{L}}$ denotes the Laplacian of the subgraph $G_{C_1^c}$. Similarly, $\ell_1 = \ell_1^{in} + \mathbf{e}_1$ and we may write $\ell_1^{in} = \begin{bmatrix} \ell_1^1 \\ \mathbf{0}_{n-n_0} \end{bmatrix}$, where $\ell_1^1 \in \mathbb{R}^{n_0}$ is the first column of $\mathcal{L}^1$, the Laplacian of the subgraph G_{C_1}. Thus,

$$\|\mathcal{L}_\Lambda^T \ell_1\|_1 \leq \|(\mathcal{L}_\Lambda^{in})^T \ell_1^{in}\|_1 + \|(\mathcal{L}_\Lambda^{in})^T \mathbf{e}_1\|_1 + \|E^T \ell^1\|_1$$

$$= \left\| \begin{bmatrix} \mathbf{0}_{|\Lambda| \times n_0} & \tilde{L}^T \end{bmatrix} \begin{bmatrix} \ell_1^1 \\ \mathbf{0}_{n-n_0} \end{bmatrix} \right\| + \|(\mathcal{L}_\Lambda^{in})^T \mathbf{e}_1\|_1 + \|E^T \ell^1\|_1$$

$$\leq 0 + \|(\mathcal{L}_\Lambda^{in})^T\|_1 \|\mathbf{e}_1\|_1 + \|E^T\|_1 \|\ell^1\|_1$$

$$= \|\mathcal{L}_\Lambda^{in}\|_\infty \|\mathbf{e}_1\|_1 + \|E\|_\infty \|\ell^1\|_1 \leq (2)(r) + (2r)(2) = 6r = \frac{6r_0}{\sqrt{n_0}}.$$

On the other hand, by Theorem 10.30 below we have that

$$|\langle \ell_1, \ell_{i^*} \rangle| \geq \frac{1}{n_0}\left(\frac{\beta^2}{1+\alpha}\right) - o(1/n_0)$$

with probability at least $1 - 1/n_0$, and so we bound the left-hand side of (10.20) as

$$\#(\Lambda)|\langle \ell_1, \ell_{i^*} \rangle| \geq \frac{n_0}{9}\left(\frac{1}{n_0}\left(\frac{\beta^2}{1+\alpha}\right) - o(1/n_0)\right) = \frac{\beta^2}{9(1+\alpha)} - o(1).$$

Thus, if inequality (10.20) were true, it would imply that

$$\frac{\beta^2}{9(1+\alpha)} - o(1) \leq \frac{6r_0}{\sqrt{n_0}},$$

which cannot be true for n_0 large enough, because α is $o(1)$ with respect to n_0 and β^2 is $O(1)$, so $\beta^2/(9(1+\alpha)) - o(1) \to 1/9$ while $6r_0/\sqrt{n_0} \to 0$. Hence we may always take n_0 large enough so that, with probability at least $1 - 1/n_0$, inequality (10.20) cannot hold. Thus no such i^* which is in $C_1 \setminus \{1\}$ but not in Ω can exist, and so Theorem 10.29 is proved. $\square$

A key ingredient to the proof of Theorem 10.29 is the following novel lower bound on the inner product $|\langle \ell_i, \ell_j \rangle|$ when i, j are in the same cluster.

Theorem 10.30. *Let $\mathcal{L}$ be the Laplacian of G drawn from $\mathcal{G}(n, k, p, q)$, and suppose that $i, j \in C_1$. Assume further that $r = r_0/\sqrt{n_0}$, where r_0 is a constant, independent of n_0. Then with probability at least $1 - 1/n_0$,*

$$|\langle \ell_i, \ell_j \rangle| \geq \frac{1}{n_0}\left(\frac{\beta^2}{1+\alpha}\right) - o\left(\frac{1}{n_0}\right).$$

Proof. As before, let $\mathbf{e}_i^1$ (resp., $\mathbf{e}_j^1$) denote the ith (resp., jth) column of E^1, while $\mathbf{e}_i^2$ (resp., $\mathbf{e}_j^2$) denotes the ith (resp., jth) column of E^2. Then

$$
\begin{aligned}
\langle \ell_i, \ell_j \rangle &= \langle \ell_i^{in} + \mathbf{e}_i^1 + \mathbf{e}_i^2, \ell_j^{in} + \mathbf{e}_j^1 + \mathbf{e}_j^2 \rangle \\
&= \langle \ell_i^{in}, \ell_j^{in} \rangle + \langle \ell_i^{in}, \mathbf{e}_j^1 \rangle + \langle \ell_i^{in}, \mathbf{e}_j^2 \rangle + \langle \mathbf{e}_i^1, \ell_j^{in} \rangle + \langle \mathbf{e}_i^1, \mathbf{e}_j^1 \rangle \\
&\quad + \langle \mathbf{e}_i^1, \mathbf{e}_j^2 \rangle + \langle \mathbf{e}_i^2, \ell_j^{in} \rangle + \langle \mathbf{e}_i^2, \mathbf{e}_j^1 \rangle + \langle \mathbf{e}_i^2, \mathbf{e}_j^2 \rangle.
\end{aligned}
$$

By construction, $\mathbf{e}_i^1$ and $\mathbf{e}_i^2$ have disjoint support (and similarly for $\mathbf{e}_j^1$ and $\mathbf{e}_j^2$), as do ℓ_i^{in} and $\mathbf{e}_j^2$ (and similarly ℓ_j^{in} and $\mathbf{e}_i^2$). Hence,

$$
\langle \ell_i^{in}, \mathbf{e}_j^2 \rangle = 0, \ \langle \mathbf{e}_i^1, \mathbf{e}_j^2 \rangle = 0, \ \langle \mathbf{e}_i^2, \ell_j^{in} \rangle = 0, \ \langle \mathbf{e}_i^2, \mathbf{e}_j^1 \rangle = 0
$$

and so

$$
|\langle \ell_i, \ell_j \rangle| \geq |\langle \ell_i^{in}, \ell_j^{in} \rangle| - |\langle \ell_i^{in}, \mathbf{e}_j^1 \rangle| - |\langle \mathbf{e}_i^1, \ell_j^{in} \rangle| - |\langle \mathbf{e}_i^1, \mathbf{e}_j^1 \rangle| - |\langle \mathbf{e}_i^2, \mathbf{e}_j^2 \rangle|. \tag{10.21}
$$

Now ℓ_i^{in} and ℓ_j^{in} can be thought of as columns in the Laplacian of a graph drawn at random from $\mathcal{G}(n, k, p, 0)$. By Corollary 10.8 we have that $(1 - \alpha)pn_0 \leq d_i^{in} \leq (1 + \alpha)pn_0$ for all i, with probability $1 - 1/n$. Moreover, α is $o(1)$ and so $\beta^2 \to 1$ as $n_0 \to \infty$; thus $\beta^2 \geq p$ for large enough n_0 and we may apply Theorem 10.22 to get

$$
|\langle \ell_i^{in}, \ell_j^{in} \rangle| \geq \frac{1}{n_0} \left(\frac{\beta^2}{(1 + \alpha)} \right) - \frac{1}{n_0^{3/2}} \sqrt{\frac{-\ln(\delta)}{2(1 + \alpha)^3 p^3}}
$$

with probability $1 - \delta - 1/n$. Taking $\delta = e^{-p^3 \sqrt{n_0}}$ (and noting that $e^{-p^3 \sqrt{n_0}} < 1/\sqrt{n_0}$ for large enough n_0),

$$
\begin{aligned}
\frac{1}{n_0} \left(\frac{\beta^2}{(1 + \alpha)} \right) - \frac{1}{n_0^{3/2}} \sqrt{\frac{-\ln(\delta)}{2(1 + \alpha)^3 p^3}} &= \frac{1}{n_0} \left(\frac{\beta^2}{(1 + \alpha)} \right) - \frac{1}{n_0^{3/2}} \sqrt{\frac{n_0^{1/2}}{2(1 + \alpha)^3}} \\
&= \frac{1}{n_0} \left(\frac{\beta^2}{1 + \alpha} \right) - \frac{1}{n_0^{5/4}} \frac{1}{\sqrt{2}(1 + \alpha)^{3/2}} = \frac{1}{n_0} \left(\frac{\beta^2}{1 + \alpha} \right) - o(1/n_0)
\end{aligned}
$$

with probability at least $1 - 1/n_0 - 1/n = 1 - (k + 1)/n$. Next, we consider the term $\langle \ell_i^{in}, \mathbf{e}_j^1 \rangle$:

$$
\begin{aligned}
|\langle \ell_i^{in}, \mathbf{e}_j^1 \rangle| &= \left| \sum_k \left(\delta_{ik} - \frac{A_{ik}^{in}}{d_k^{in}} \right) \left(\frac{r_k A_{kj}^{in}}{d_k^{in}} + \mathcal{O}(r^3) \right) \right| = \left| \frac{r_i A_{ij}^{in}}{d_i^{in}} - \sum_k \frac{r_k A_{ik}^{in} A_{kj}^{in}}{(d_k^{in})^2} + \mathcal{O}(r^3) \right| \\
&\leq \frac{r}{d_{\min}^{in}} + \frac{r}{(d_{\min}^{in})^2} \sum_k A_{ik}^{in} A_{kj}^{in} + \mathcal{O}(r^3) \leq \frac{r}{d_{\min}^{in}} + \frac{r}{(d_{\min}^{in})^2} \min\{d_i^{in}, d_j^{in}\} + \mathcal{O}(r^3) \\
&\leq \frac{r}{d_{\min}^{in}} \left(1 + \frac{d_{\max}^{in}}{d_{\min}^{in}} \right) + \mathcal{O}(r^3).
\end{aligned}
$$

Now by assumption $d_{\min}^{in} = (1 - \alpha)pn_0$ and $d_{\max}^{in} = (1 + \alpha)pn_0$, and so

$$
|\langle \ell_i^{in}, \mathbf{e}_j^1 \rangle| \leq \frac{r}{(1 - \alpha)pn_0} (1 + \beta^2) + \mathcal{O}(r^3).
$$

Clearly the same bound holds for $|\langle \mathbf{e}_i^1, \ell_j^{in} \rangle|$. Finally, consider $|\langle \mathbf{e}_i^1, \mathbf{e}_j^1 \rangle|$. A similar calculation to the one above reveals that

$$
|\langle \mathbf{e}_i^1, \mathbf{e}_j^1 \rangle| \leq \frac{r^2}{n_0} \left(\frac{\beta^4}{(1 - \alpha)p} \right) + \mathcal{O}(r^4) \quad \text{as well as} \quad |\langle \mathbf{e}_i^2, \mathbf{e}_j^2 \rangle| \leq \frac{r}{(1 - \alpha)pn_0} + \mathcal{O}(r^3).
$$

Putting all these bounds back into (10.21), we get

$$
\begin{aligned}
|\langle \ell_i, \ell_j \rangle| \\
\geq \frac{1}{n_0}\left(\frac{\beta^2}{1+\alpha}\right) - o\left(\frac{1}{n_0}\right) - \left(\frac{2r(1+\beta^2)}{(1-\alpha)pn_0} + \mathcal{O}(r^3) + \frac{r^2}{n_0}\left(\frac{\beta^4}{(1-\alpha)p}\right)\right. \\
\left. + \mathcal{O}(r^4) + \frac{r}{(1-\alpha)pn_0} + \mathcal{O}(r^3)\right) \\
= \frac{1}{n_0}\left(\frac{\beta^2}{1+\alpha}\right) - \frac{r}{n_0}\left(\frac{2(1+\beta^2)}{(1-\alpha)p} + \frac{r\beta^4}{(1-\alpha)p} + \frac{1}{(1-\alpha)p}\right) + \mathcal{O}(r^3).
\end{aligned}
$$

Using the assumption that $r = r_0/\sqrt{n_0}$, we see that $r/n_0 = r_0/n_0^{3/2}$, which is of order $o(1/n_0)$. Similarly, $r^3 = r_0^3/n_0^{3/2}$, which is also of order $o(1/n_0)$. Thus,

$$
|\langle \ell_i, \ell_j \rangle| \geq \frac{1}{n_0}\left(\frac{\beta^2}{1+\alpha}\right) - o(1/n_0),
$$

as claimed.   □

With the results in Theorem 10.29 in mind, we derive a new algorithm to handle the graphs from $\mathcal{G}(n, k, p, q)$ for $p \gg q > 0$.

ALGORITHM 10.2
Single Clustering Pursuit Algorithm [18]
1: **Input:** The adjacency matrix A of a graph G, and in estimate of the size of clusters n_0.
2: (1) **Initialization** Compute $\mathcal{L} = I - D^{-1}A$.
3: (2) **Thresholding**
4: Let $\Omega = \mathcal{H}_{\lceil 10(n_0-1)/9 \rceil}(L_{-1}^T \ell_1)$.
5: (3) **Perturbed Sparse Recovery**
6: $y = \sum_{i \in \Omega} \ell_i + \ell_1$.
7: Solve $z^\# = \arg\min\{\|\mathcal{L}_\Omega z - y\|_2$ subject to $\|z\|_0 \leq n_0/9\}$ using OMP Algorithm 3.2.
8: $\Lambda^\# = \mathrm{support}(z^\#)$.
9: **Output:** $C_1^\# = \{1\} \cup (\Omega \setminus \Lambda^\#)$.

Now let us show the success of Algorithm 10.2 with high probability.

Theorem 10.31. *Let G be a graph drawn at random from the $\mathcal{G}(n, k, p, q)$ model. Suppose that $k = O(1)$ with respect to n and either*

1. $q = \frac{Q}{n}$ and $p = \frac{P \ln(n)}{\sqrt{n}\ln(\ln(n))}$ with Q and P being constants and $\frac{Q}{P} < \frac{1}{6}$, or

2. $q = \frac{Q\ln(n)}{n}$ with $Q \to \infty$ as $n \to \infty$ and $p = \frac{P\ln(n)}{\sqrt{n}}$ with $P \to \infty$ as $n \to \infty$ and $\frac{Q}{P} < \frac{1}{6}$
   for large enough n.

Then Algorithm 10.2 will recover C_1 (i.e., $C_1^\# = C_1$) almost surely.

Proof. For notational convenience, let $s := (n_0 - 1)/9$. In both cases, $p \geq 4k(\ln(n))^2/n$, and so by Corollary 10.8, $d_{\min}^{in} \geq (1-\alpha)n_0 p$ for $\alpha = o(1)$ with respect to n. In case 1, by (i) of Theorem 3.4 in [14], $d_{\max}^{out}$ is proportional to $\ln(n)/\ln(\ln(n))$. Hence

$$
\max_i \frac{d_i^{out}}{d_i^{in}} \leq \frac{d_{\max}^{out}}{d_{\min}^{in}} \leq \frac{\sqrt{k}}{P(1-\alpha)\sqrt{n_0}} := \frac{r_0}{\sqrt{n_0}}
$$

with $r_0 \leq 1/6$ for large enough n. In case 2, by (ii) of Theorem 3.4 in [14], $d_{\max}^{out}$ is proportional to $Q \ln(n)$ and so

$$\max_i \frac{d_i^{out}}{d_i^{in}} \leq \frac{d_{\max}^{out}}{d_{\min}^{in}} \leq \frac{Q\sqrt{n}}{P(1-\alpha)n_0} = \frac{Q\sqrt{k}}{P(1-\alpha)\sqrt{n_0}} = \frac{QO(1)}{P\sqrt{n_0}} := \frac{r_0}{\sqrt{n_0}},$$

and again $r_0 \leq 1/6$ for large enough n. Applying Theorem 10.29, $C_1 \setminus \{1\} \subset \Omega$ almost surely. As before, we define $\Lambda := \Omega \cap C_1^c$, and we know that $\mathbf{1}_\Lambda$ is a solution to

$$\mathbf{z}^\# := \arg\min\{\|L_\Omega^{in}\mathbf{z} - \mathbf{y}^{in}\|_2 : \text{subject to } \|\mathbf{z}\|_0 \leq n_0/9\}. \tag{10.22}$$

Because $\delta_{2s}(L_\Omega^{in}) < 1$, it is in fact the unique solution. Step 3 in Algorithm 10.2 is to solve the sparse approximation problem

$$\mathbf{1}_\Lambda := \arg\min\{\|L_\Omega\mathbf{z} - \mathbf{y}\|_2 : \text{subject to } \|\mathbf{z}\|_0 \leq n_0/9\}, \tag{10.23}$$

which we can identify with a (totally) perturbed version of (10.22), where $L_\Omega = L_\Omega^{in} + E$ and

$$\begin{aligned}
\mathbf{y} &:= \sum_{i \in \Omega} \ell_i + \ell_1 = (L_\Omega^{in} + E_\Omega)\mathbf{1}_\Omega + \ell_1^{in} + \mathbf{e}_1 \\
&= L_\Omega^{in}\mathbf{1}_\Lambda + (L_\Omega^{in}\mathbf{1}_{C_1\setminus\{1\}} + \ell_1^{in}) + (E_\Omega\mathbf{1}_\Omega + \mathbf{e}_1) = L_\Omega^{in}\mathbf{1}_\Lambda + \mathbf{0} + (E_\Omega\mathbf{1}_\Omega + \mathbf{e}_1) \\
&=: \mathbf{y}^0 + \mathbf{e}.
\end{aligned}$$

Thus, we appeal to the results of the RIP on totally perturbed compressed sensing. Define $\delta_t := \delta_t(\mathcal{L}^{in})$ and $\hat{\delta}_t := \delta_t(\mathcal{L})$. We shall use Theorem 10.32 below and will need to estimate $\epsilon_{\mathcal{L}^{in}}^s$, $\epsilon_{\mathbf{y}}$, $\hat{\rho}$, and $\hat{\tau}$. Let us bound on these quantities. Observe that in both cases, $p \gg (\ln(n_0))^2/n_0$ and, by assumption, k is $O(1)$ with respect to n; hence we may appeal to Theorem 2.44 in Chapter 2. As shown in the proof of Theorem 2.44, $\epsilon_{s,L^{in}} = 8\beta r/3 = 8\beta r_0/(3\sqrt{n_0})$ for n_0 large enough. Applying Theorem 2.44 with $t = 3s = n_0/3$, we get

$$\begin{aligned}
\hat{\delta}_{3s} &\leq \frac{1}{3} + D_1 r + \frac{D_2}{\sqrt{n_0}} + \frac{D_3 r}{\sqrt{n_0}} + \mathcal{O}(r^2) + o\left(\frac{1}{\sqrt{n_0}}\right) \\
&= \frac{1}{3} + \frac{(D_1 r_0 + D_2)}{\sqrt{n_0}} + \frac{D_3 r_0}{n_0} + \mathcal{O}\left(\frac{1}{n_0}\right) + o\left(\frac{1}{\sqrt{n_0}}\right) = \frac{1}{3} + \frac{(D_1 r_0 + D_2)}{\sqrt{n_0}} + o\left(\frac{1}{\sqrt{n_0}}\right).
\end{aligned}$$

Hence for large enough n_0, we certainly have $\hat{\delta}_{3s} \leq 0.4859$, as required. In fact, let us take n_0 large enough such that $\hat{\delta}_{3s} \leq 1/3 + 1/18 = 7/18 \approx 0.39$. Choosing n_0 larger if necessary, we shall also assume that $\epsilon_{s,L^{in}} < 1/3$. In this case, a straightforward but tedious calculation will reveal that $\hat{\rho} \leq 0.86$ and similarly $\hat{\tau} \leq 0.98 < 1$ (see the statement of Theorem 10.32 for the definitions of $\hat{\rho}$ and $\hat{\tau}$). We now turn our attention to $\epsilon_{\mathbf{y}}$:

$$\|\mathbf{y}^{in}\|_2 = \|L_{\Omega_1}^{in}\mathbf{1}_\Lambda\|_2 \geq \sqrt{1-\delta_s}\|\mathbf{1}_\Lambda\|_2 = \sqrt{1-\delta_s}\sqrt{s},$$

while

$$\|\mathbf{e}\|_2 = \|E\mathbf{1}_\Omega + \mathbf{e}_1\|_2 \leq \|E\|_2\|\mathbf{1}_{\Omega \cup \{1\}}\|_2 \leq 2\beta r\sqrt{10s}.$$

So we have

$$\epsilon_{\mathbf{y}} := \frac{\|\mathbf{e}\|_2}{\|\mathbf{y}^{in}\|_2} = \frac{2\sqrt{10}\beta r}{\sqrt{1-\delta_s}} = \frac{2\sqrt{10}\beta r_0}{\sqrt{n_0(1-\delta_s)}}.$$

Appealing to Theorem 10.16 with $s = n_0/9$, we obtain

$$\delta_s \leq 1/9 + \frac{64p^{-1/2}}{9\sqrt{n_0}} + o\left(\frac{1}{\sqrt{n_0}}\right).$$

Again, we will assume that n_0 is large enough such that $\delta_s \leq 2/9 \approx 0.22$. Under this assumption,

$$\epsilon_{\mathbf{y}} \leq \frac{2\sqrt{10}}{\sqrt{7/9}} \frac{\beta r_0}{\sqrt{n_0}} \leq 8 \frac{\beta r_0}{\sqrt{n_0}}$$

Finally, we use Theorem 10.32 again, this time with $t = s = n_0/9$, to get

$$\hat{\delta}_s \leq 1/9 + \frac{D_1 r_0 + D_2}{\sqrt{n_0}} + o\left(\frac{1}{\sqrt{n_0}}\right),$$

and we take n_0 large enough such that $\hat{\delta}_s \leq 2/9$ (this is possible since D_1 and D_2 are $\mathcal{O}(1)$ with respect to n_0). Now Theorem 10.32 below guarantees that

$$\frac{\|\mathbf{x}^m - \mathbf{1}_\Lambda\|_2}{\|\mathbf{1}_\Lambda\|_2} \leq \left(\hat{\tau} \frac{\sqrt{1+\hat{\delta}_s}}{1 - \epsilon^s_{L^{in}}} + 1\right)(\epsilon^s_{L^0} + \epsilon_{\mathbf{y}}),$$

where $m := \ln(\epsilon^s_{L^0} + \epsilon_{\mathbf{y}})/\ln(\hat{\rho}) = \mathcal{O}(\ln(n_0))$. Substituting in for the various constants gives us

$$\frac{\|\mathbf{x}^m - \mathbf{1}_\Lambda\|_2}{\|\mathbf{1}_\Lambda\|_2} \leq \left((1) \frac{\sqrt{1+2/9}}{1 - 1/3}\right) \left(\frac{8\beta r_0}{3\sqrt{n_0}} + \frac{8\beta r_0}{\sqrt{n_0}}\right) \leq 18 \frac{\beta r_0}{\sqrt{n_0}}.$$

We have

$$\|\mathbf{x}^m - \mathbf{1}_\Lambda\|_2 \leq 18 \frac{\beta r_0}{\sqrt{n_0}} \|\mathbf{1}_\Lambda\|_2 = 18 \frac{\beta r_0}{\sqrt{n_0}} \left(\frac{\sqrt{n_0}}{3}\right) = 6\beta r_0.$$

As in Lemma 10.33 below, as long as $\|\mathbf{x}^m - \mathbf{1}_\Lambda\|_2 < 1$, we have that $\text{support}(\mathbf{x}^m) = \text{support}(\mathbf{1}_\Lambda)$. Thus, if $\beta r_0 < 1/6$, we have that $C_1^\# = C_1$, and the result follows. □

Theorem 10.32 (Li, 2016 [20]). *Let $\mathbf{x}^*$, $\mathbf{y}$ $\hat{\mathbf{y}}$, Φ, and $\hat{\Phi}$ be as above and suppose that $\|\mathbf{x}^*\|_0 \leq s$. For any $t \in [n]$, let $\delta_t := \delta_t(\Phi)$. Define the constants*

$$\epsilon_{\mathbf{y}} := \|\mathbf{e}\|_2/\|\hat{\mathbf{y}}\|_2 \quad \text{and} \quad \epsilon^s_\Phi = \|M\|^s_2/\|\hat{\Phi}\|^s_2,$$

where for any matrix B, $\|B\|_2^{(s)} := \max\{\|B_S\|_2 : S \subset [n] \text{ and } |S| = s\}$. Define further

$$\rho = \frac{\sqrt{2\delta_{3s}^2(1 + \delta_{3s}^2)}}{1 - \delta_{3s}^2} \quad \text{and} \quad \tau = \frac{(\sqrt{2}+2)\delta_{3s}}{\sqrt{1 - \delta_{3s}^2}}(1 - \delta_{3s})(1 - \rho) + \frac{2\sqrt{2}+1}{(1 - \delta_{3s})(1 - \rho)}.$$

Assume $\delta_{3s} \leq 0.4859$, and let $\mathbf{x}^m$ be the output of Algorithm 10.6 below applied to problem (10.23) after m iterations. Then

$$\frac{\|\mathbf{x}^* - \mathbf{x}^m\|_2}{\|\mathbf{x}^*\|_2} \leq \rho^m + \tau \frac{\sqrt{1+\delta_s}}{1 - \epsilon^s_\Phi}(\epsilon^s_\Phi + \epsilon_{\mathbf{y}}).$$

Proof. This is Corollary 1 in [20]. Note that our convention on hats is different from theirs since our Φ is their $\hat{\Phi}$, so our ρ is their $\hat{\rho}$, and so on. □

Finally, the following result was used in the discussion above.

Lemma 10.33. *Suppose that $\mathbf{x}^* \in \mathbb{R}^n$ is a binary vector with $\|\mathbf{x}^*\|_0 = t$ and $\mathbf{x} \in \mathbb{R}^n$ is any other vector that also has $\|\mathbf{x}\|_0 = t$. If $\|\mathbf{x}^* - \mathbf{x}\|_2 < 1$; then $\text{support}(\mathbf{x}) = \text{support}(\mathbf{x}^*)$.*

Proof. Suppose otherwise. Then there exists an $i \in \text{support}(\mathbf{x}^*) \setminus \text{support}(\mathbf{x})$. Clearly,

$$\|\mathbf{x}^* - \mathbf{x}\|_2 \geq |x_i^* - x_i| = |1 - 0| = 1,$$

which contradicts the hypothesis. □

10.6 ▪ Semisupervised Clustering Pursuit

In this section, we assume that we know a few indices in a cluster C_1, say. We shall adapt the
SCP algorithm in the previous sections to this setting. In addition, we shall introduce a new
concept to establish the convergence of our approach. We begin with the main problem in this
section.

Problem 10.6.1 (Semisupervised Clustering Problem). *Given a graph $G = (V, E)$, with edges
possibly weighted, and a small subset Γ of vertices, find a $C \subset V$ such that*

1. *$\Gamma \subset C$ and*

2. *there are many edges between vertices in C but few edges between C and its complement.*

Suppose we have, for every n, a probabilistic model of graphs on n vertices $\mathcal{G}^n$ (such as the
stochastic block model introduced in section 10.1) containing an embedded cluster C^n. Let $\mathcal{A}$ be
any algorithm for Problem 10.6.1 with output $C^\#$. We say that algorithm $\mathcal{A}$ is an *almost exact
extraction* if

$$\mathbb{P}\left[\frac{|C^\# \triangle C^n|}{|C^n|} \leq o(1)\right] = 1 - o(1),$$

where for any two sets A and B, $A \triangle B$ denotes their symmetric difference:

$$A \triangle B = (A \setminus (A \cap B)) \cup (B \setminus (A \cap B)). \tag{10.24}$$

Note that the requirement $|C^n \setminus C^\#| = o(1)$ is not good enough to measure the accuracy of the
single cluster extraction because the output $C^\# = V$ will have $|C^n \setminus C^\#| = 0$.

As explained in the previous section, we propose a two-stage approach to solving Problem 10.6.1. Our main algorithm is Algorithm 10.3. In the first stage (see Algorithm 10.4) we
determine a superset $\Omega \supset C_1$, while in the second stage (Algorithm 10.5) we extract C_1 from
Ω by solving a sparse solution problem to find a vector supported on the *complement* of C_1 in
Ω. Note that the labeled data (Γ) is only used in the first stage. Moreover, these two stages
are independent of each other, so Algorithm 10.5 could be combined with any other reasonable
approach in this book to finding such an Ω containing C_1 or finding the complement of $C_1 \subset \Omega$.

ALGORITHM 10.3
Semisupervised Single Cluster Pursuit (SSCP) Algorithm [18]
1: **Input:** Adjacency matrix A, parameters $\epsilon, \rho \in (0, 1)$, $\Gamma \subset C$, and $n_0 = |C|$.
2: **Step 1** Perform the semisupervised single cluster thresholding algorithm, Algorithm 10.4,
   with input $(A, \epsilon, \Gamma, n_0)$ to obtain Ω.
3: **Step 2** Perform the cluster pursuit algorithm, Algorithm 10.5, with input $(A, \epsilon, \Omega, n_0)$ to
   obtain $C^\#$.
4: **Output:** $C^\#$.

ALGORITHM 10.4
Semisupervised Single Cluster Thresholding Algorithm
1: **Input:** Adjacency matrix A, a thresholding parameter $\epsilon \in (0, 1)$, $\Gamma \subset C$, and $n_0 = |C|$.
2: Compute $\mathbf{b} = \sum_{i \in \Gamma} \ell_i^+$.
3: Let $\mathbf{v} = (\mathcal{L}_{\Gamma^c}^+)^\top \mathbf{b}$.
4: Define $\widetilde{\Omega} = \mathcal{H}_{(1+\epsilon)n_0}(\mathbf{v})$.
5: **Output:** $\Omega = \widetilde{\Omega} \cup \Gamma$.

ALGORITHM 10.5

Cluster Pursuit Algorithm

1: **Input:** Adjacency matrix A, a rejection parameter $\rho \in (0,1)$, Ω, and $n_0 = |C|$.
2: Compute $\mathcal{L} = I - D^{-1}A$.
3: Compute $\mathbf{y} = \sum_{i \in \Omega} l_i$.
4: Compute $\epsilon = (|\Omega| - n_0)/n_0$.
5: Let $\mathbf{x}^m$ be the solution to

$$\operatorname{argmin}\{\|\mathcal{L}_\Omega \mathbf{x} - \mathbf{y}\|_2 : \|\mathbf{x}\|_0 \le \epsilon n_0\} \tag{10.25}$$

obtained after $m = O(\log(n))$ iterations of Algorithm 10.6.
6: Let $W = \{i : x_i^m > \rho\}$.
7: **Output:** $C^\# = \Omega \setminus W$.

ALGORITHM 10.6

Subspace Pursuit Algorithm [11]

1: **Input:** $\mathbf{y}$, Φ, and an integer $s \ge 1$.
2: **Initialization:**
3: (1) $T^0 = \mathcal{L}_s(\Phi^\top \mathbf{y})$.
4: (2) $\mathbf{x}^0 = \operatorname{argmin}\{\|\mathbf{y} - \Phi_{T^0}\mathbf{x}\|_2 : \operatorname{support}(\mathbf{x}) \subset T^0\}$.
5: (3) $\mathbf{r}^0 = \mathbf{y} - \Phi_{T^0}\mathbf{x}^0$.
6: **Iteration:**
7: **for** $k = 1 : m$ **do**
8:     (1) $\hat{T}^k = T^{k-1} \cup \mathcal{L}_s\left(\Phi^\top \mathbf{r}^{k-1}\right)$.
9:     (2) $\mathbf{u} = \operatorname{argmin}\{\|\mathbf{y} - \Phi_{\hat{T}^k}\mathbf{x}\|_2 : \mathbf{x} \in \mathbb{R}^N \text{ and } \operatorname{support}(\mathbf{x}) \subset \hat{T}^k\}$.
10:     (3) $T^k = \mathcal{L}_s(\mathbf{u})$ and $\mathbf{x}^k = \mathcal{H}_s(\mathbf{u})$.
11:     (4) $\mathbf{r}^k = \mathbf{y} - \Phi_{T^k}\mathbf{x}^k$.
12: **end for**
13: **Output:** T^k.

Note that in Algorithm 10.5, we found that defining W as the support of $\mathbf{x}^m$, i.e., $W = \{i : x_i^m > 0\}$, because we can assume $\mathbf{x}^m$ to be nonnegative), was a bit unstable to noise, hence the introduction of the rejection parameter ρ. The lower ρ is, the more vertices in Ω are "rejected," i.e., not included in $C^\#$. We found that by varying ρ from small to large, we can trade off decreasing specificity of Algorithm 10.3 (i.e., its robustness against incorrectly assigning a vertex to $C^\#$) for increasing sensitivity. This could prove useful in certain applications.

In this section, we prove that Algorithm 10.3 can almost exactly extract a cluster for graphs subject to the symmetric SBM. Our main result is as follows.

Theorem 10.34. *Suppose that Algorithm 10.3 is run on a graph $G \sim \text{SSBM}(n, k, p, q)$ with seed vertices $\Gamma \subset C_1$, where $|\Gamma| = gn_0$ for any fixed constant $g \in (0,1]$. Let $C_1^\#$ denote the output. If k is constant with respect to n, $p = \omega \ln(n_0)/n_0$ with ω any function of n such that $\omega \to \infty$ as $n \to \infty$, and $q = b \ln(n)/n$ for b constant, then*

$$\mathbb{P}\left[\frac{\left|C_1 \triangle C_1^\#\right|}{|C_1|} \le o(1)\right] = 1 - o(1).$$

We shall prove Theorem 10.34 by first showing that the thresholding step in Algorithm 10.4 returns an Ω containing a fraction $1 + o(1)$ of the vertices of C_1 with probability $1 - o(1)$. The analysis is based on the signless graph Laplacian instead of the traditional graph Laplacian used in the previous section. We then show that Algorithm 10.5, given such an Ω, will output a cluster $C_1^{\#}$ such that $C_1^{\#} \triangle C_1 = o(1)|C_1|$, again with probability $1 - o(1)$. We divide this section into three subsections.

10.6.1 ▪ Concentration in Erdős–Rényi Graphs

First of all, two concentration phenomena are key to our proof. The first is that the maximum and minimum degree of an Erdős–Rényi graph are within a small deviation of their expected value, almost surely. The second is that the second eigenvalue of the Laplacian of an Erdős–Rényi graph is within an $o(1)$ term of its expected value, again almost surely. We have already explained these concepts in section 10.1. For convenience, let us state these results more precisely so we can use them to establish Theorem 10.34. We shall use $\mathcal{G}(n, p)$ for the Erdős–Rényi model.

Theorem 10.35 (Bollobas, 1982 and 1998 [4, 6]). *Let $G \sim \mathcal{G}(n, p)$ with $q = (c+o(1)) \log(n)/n$. Then there exist functions $\eta_\delta(c)$ (resp., $\eta_\triangle(c)$) satisfying $-1 < \eta_\delta(c) < 0$ and $\lim_{c \to \infty} \eta_\delta(c) = 0$ (resp., $0 < \eta_\triangle(c) < 1$ and $\lim_{c \to \infty} \eta_\triangle(c) = 0$) such that*

$$d_{\max}(G) = (1 + \eta_\triangle(c)) c \log n + o(1) \text{ and } d_{\min}(G) = (1 + \eta_\delta(c)) c \log n + o(1)$$

almost surely.

Proof. We leave the proof to the interested reader; see also [4, 6]. □

Theorem 10.36 (Frieze and Karonski, 2016 [14]). *If $G \sim \mathcal{G}(n, p)$ with $p = \omega \log(n)/n$ where $\omega \to \infty$, then*

$$d_{\min}(G) = (1 - o(1))np \text{ and } d_{\max}(G) = (1 + o(1))np$$

almost surely.

Proof. We leave the proof to the interested reader; see also [14], Theorem 3.4 (ii). □

Theorem 10.37 (Chung and Radcliffe, 2011 [10]). *Suppose that $G \sim \mathcal{G}(n, p)$ with $p = \omega \log(n)$. Then almost surely*

1. $\lambda_{\max}(A) \leq (1 + o(1)) \omega \log(n)$,

2. $\lambda_i(A) \leq o(\omega \log n)$ *for* $\lambda_i < \lambda_{\max}$, *and*

3. $|\lambda_i(\mathcal{L}) - 1| \leq \sqrt{\frac{6 \log(2n)}{\omega \log(n_0)}} = o(1)$ *for all* $i > 1$.

Proof. See Theorems 3 and 4 in [10]. Note that, in their notation, $m = w_{\min} = pn = \omega \log n$. Their results refer to the symmetric normalized Laplacian, $\mathcal{L}^n := I - D^{-1/2} A D^{-1/2}$, but we can easily show that $\mathcal{L}^n$ and $\mathcal{L}$ have the same spectrum. □

Let G^{in} and G^{out} be as before. If $G \sim \text{SSBM}(n, k, p, q)$, then G^{in} consists of k disjoint graphs, each drawn identically and independently from the Erdős–Rényi model $\mathcal{G}(n_0, p)$. The graph G^{out} is not an Erdős–Rényi graph since there is 0 probability of it containing an edge between two vertices in the same cluster (because we have removed them). However, we can profitably think of G^{out} as a subgraph of some $\widetilde{G^{out}} \sim \mathcal{G}(n, q)$. In particular, any upper bounds on the degrees of vertices in $\widetilde{G^{out}}$ are automatically bounds on the degrees in G^{out}. Thus, we have the following corollaries of Theorems 10.36 and 10.35 pertaining to the concentration of degree for the SBM.

Corollary 10.38. *If $G \sim \mathrm{SSBM}(n, k, p, q)$ with $q = b \log(n)/n$, then $d_{\max}^{out}(G) \leq (1 + \eta_\triangle(b))b \log n + o(1)$ almost surely.*

Proof. Consider G^{out} as a subgraph of $\widetilde{G^{out}} \sim \mathcal{G}(n, q)$. By Theorem 10.35, $d_{\max}(\widetilde{G^{out}}) \leq (1 + \eta_\triangle(b))b \log n + o(1)$ almost surely, and the corollary follows. □

Corollary 10.39. *If $G \sim \mathrm{SSBM}(n, k, p, q)$ with $p = \omega \log(n_0)/n_0$, where $\omega \to \infty$, then $d_{\min}^{in}(G) = (1 - o(1))\omega \log(n_0)$ and $d_{\max}^{in}(G) = (1 + o(1))\omega \log(n_0)$ almost surely.*

Proof. If $i \in C_a$, then $d_i^{in} = d_i[G_a]$, where $G_a = G_{C_a} \sim \mathcal{G}(n_0, p)$. Clearly,

$$d_{\max}^{in}(G) = \max_i d_i^{in} = \max_a d_{\max}(G_a).$$

By Theorem 10.36, $d_{\max}(G_a) \leq (1 + o(1))n_0 p = (1 + o(1))\omega \log(n_0)$ almost surely. Note that the $d_{\max}(G_a)$ are i.i.d. random variables, and, since we are taking a maximum over $k = \mathcal{O}(1)$ of them, it follows that $\max_a d_{\max}(G_a) \leq (1 + o(1))\omega \log(n_0)$ almost surely too. Moreover, since $n_0 = n/k$, $o_{n_0}(1) = o_n(1)$. The proof for $d_{\min}^{in}(G)$ is similar. □

Corollary 10.40. *$G \sim \mathrm{SSBM}(n, k, p, q)$ with $p = \omega \log(n_0)/n_0$, where $\omega \to \infty$, $q = b \log(n)/n$, and $k = \mathcal{O}(1)$. Define $r := d_{\max}^{out}/d_{\min}^{in}$. Then $\max_i r_i \leq r$ and $r = o(1)$ almost surely.*

Proof. First, it is clear that

$$r_i = \frac{d_i^{out}}{d_i^{in}} \leq \frac{d_{\max}^{out}}{d_{\min}^{in}} = r.$$

From Corollaries 10.38 and 10.39, we have

$$\frac{d_{\max}^{out}}{d_{\min}^{in}} \leq \frac{(1 + \eta_\triangle(b))b \log n + o(1)}{(1 - o(1))n_0 p} = \frac{(1 + \eta_\triangle(b))b \log n + o(1)}{(1 - o(1))\omega \log(n_0)}$$
$$= \frac{(1 + \eta_\triangle(b))b \log n + o(1)}{(1 - o(1))\omega (\log(n) - log(k))} = o(1) \text{ as } \omega \to \infty,$$

because $n_0 = \frac{n}{k}$. This completes the proof. □

10.6.2 ▪ Reliably Finding Supersets

Let $U = C_1 \setminus (C_1 \cap \Omega)$ denote the "missed" indices, and let $W = \Omega \setminus (C_1 \cap \Omega)$ denote the "bad" indices (i.e., vertices in Ω that are not in C_1). Let $|U| = un_0$, in which case $|W| = (\epsilon + u)n_0$, since by construction $|\Omega| = (1 + \epsilon)n_0$. We shall prove that $u = o(1)$ in the following theorem.

Theorem 10.41 (Lai and Mckenzie, 2020 [18]). *Let $G \sim \mathrm{SSBM}(n, k, p, q)$, where $p = \omega \log(n_0)/n_0$, $\omega \to \infty$, and $q = b \log(n)/n$. Let $\Gamma \subset C_1$ with $|\Gamma| = gn_0$ for some constant $g \in (0, 1)$. For any $\epsilon > 0$, if Ω is the output of Algorithm 10.4, with inputs ϵ, Γ, and n_0, then $|C_1 \setminus (C_1 \cap \Omega)| = o(n_0)$.*

Proof. As in line 3 of Algorithm 10.4, define $\mathbf{v} := (\mathcal{L}_{\Gamma^c}^+)^\top \mathbf{b}$, where $\mathbf{b} = \sum_{i \in \Gamma} \ell_i^+$ and $\mathcal{L}_{\Gamma^c}^+$ is the signless graph Laplacian of Γ^c. Observe that

$$\left((\mathcal{L}_{\Gamma^c}^+)^\top \ell_j^+ \right)_i = \langle \ell_i^+, \ell_j^+ \rangle = \left(\frac{1}{d_i} + \frac{1}{d_j} \right) A_{ij} + \sum_{k=1}^n \frac{A_{ik} A_{kj}}{d_k^2}. \tag{10.26}$$

By the definition of the thresholding operator $\mathcal{H}_{(1+\epsilon)n_0}(\cdot)$, we must have $v_i \leq v_j$ for every $i \in U$ and $j \in W$. Summing over W,

$$(\epsilon + u)n_0 v_i \leq \sum_{j \in W} v_j,$$

and then summing over U,

$$(\epsilon + u)n_0 \sum_{i \in U} v_i \leq u n_0 \sum_{j \in W} v_j,$$

it follows that

$$\sum_{i \in U} v_i \leq \frac{u}{\epsilon + u} \sum_{j \in W} v_j \leq \sum_{j \in W} v_j. \tag{10.27}$$

Looking ahead, we shall show that $u = o(1)$ by showing that inequality (10.27) cannot hold if this is not the case. Now

$$\sum_{i \in U} v_i = \sum_{i \in U} \left((L_{\Gamma^c}^+)^\top \mathbf{b} \right)_i = \sum_{i \in U} \left(\sum_{j \in \Gamma} (L_{\Gamma^c}^+)^\top \ell_j^+ \right)_i = \sum_{i \in U} \sum_{j \in \Gamma} \langle \ell_i^+, \ell_j^+ \rangle.$$

From (10.26) we deduce that $\langle \ell_i^+, \ell_j^+ \rangle \geq \sum_{k=1}^n \frac{A_{ik} A_{kj}}{d_k^2}$. Moreover,

$$\sum_{k=1}^n \frac{A_{ik} A_{kj}}{d_k^2} \geq \frac{1}{d_{\max}^2} \sum_{k=1}^n A_{ik} A_{kj} \geq \frac{1}{d_{\max}^2} \sum_{k \in C_1} A_{ik} A_{kj},$$

and so

$$\sum_{i \in U} v_i \geq \frac{1}{d_{\max}^2} \sum_{i \in U} \sum_{j \in \Gamma} \sum_{k \in C_1} A_{ik} A_{kj}. \tag{10.28}$$

The triple sum above is precisely the number of length-2 paths from U to Γ contained in the Erdős–Rényi graph $G_{C_1} \sim \mathcal{G}(n, p)$. In [9] a neat formula for this quantity is given (they call it $e_2(U, \Gamma)$):

$$e_2(U, \Gamma) \geq p^2 n_0 |U||\Gamma| - o(p^2 n_0^3) \text{ almost surely}$$
$$= \left(\frac{\omega^2 \log^2(n_0)}{n_0^2} \right) n_0 (u n_0)(g n_0) - o\left(\frac{\omega^2 \log^2(n_0)}{n_0^2} n_0^3 \right)$$
$$= u g \omega^2 \log^2(n_0) n_0 - o(\omega^2 \log^2(n_0) n_0).$$

(Note that by Theorem 10.37, parts 1 and 2, G_{C_1} satisfies their conditions almost surely).

By Corollaries 10.38 and 10.39 above, $d_{\max} \leq d_{\max}^{in} + d_{\max}^{out} \leq (1 + o(1))\omega \log(n_0) + (1 + \eta_\triangle(b))b \log n + o(1) = (1 + o(1))\omega \log(n_0)$ as $\omega \to \infty$. Putting this all together, we get that

$$\sum_{i \in U} v_i \geq u g n_0 - o(1). \tag{10.29}$$

We now consider the right-hand side of inequality (10.27). We first rewrite the sum as an inner product of $\mathbf{v}$ with an indicator vector:

$$\sum_{j \in W} v_j = \sum_{j \in W} 1 v_j = \langle \mathbf{1}_W, \mathbf{v} \rangle.$$

In a similar vein, rewrite $\mathbf{b} = \sum_{i \in \Gamma} \ell_i^+ = \mathcal{L}_{\Gamma^c}^+ \mathbf{1}_\Gamma$. Now recall that $\mathbf{v} = \left(\mathcal{L}_{\Gamma^c}^+ \right)^\top \mathbf{b} = \left(\mathcal{L}_{\Gamma^c}^+ \right)^\top \mathcal{L}_{\Gamma^c}^+ \mathbf{1}_\Gamma$. It follows that

$$\sum_{j \in W} v_j = \langle \mathbf{1}_W, \mathbf{v} \rangle = \langle \mathbf{1}_W, \left(\mathcal{L}_{\Gamma^c}^+ \right)^\top \mathcal{L}_{\Gamma^c}^+ \mathbf{1}_\Gamma \rangle = \langle \mathcal{L}_{\Gamma^c}^+ \mathbf{1}_W, \mathcal{L}_{\Gamma^c}^+ \mathbf{1}_\Gamma \rangle.$$

Split L^+ into four submatrices as follows:

$$L^1 \in \mathbb{R}^{n_0 \times n_0} : L_{ij}^1 = (\mathcal{L}_{\Gamma^c}^+)_{ij} \quad \text{for } i, j \in C_1,$$
$$L^2 \in \mathbb{R}^{n_0 \times (n-n_0)} : L_{ij}^2 = (\mathcal{L}_{\Gamma^c}^+)_{ij} \quad \text{for } i \in C_1, j \notin C_1,$$
$$L^3 \in \mathbb{R}^{(n-n_0) \times n_0} : L_{ij}^3 = (\mathcal{L}_{\Gamma^c}^+)_{ij} \quad \text{for } i \notin C_1, j \in C_1,$$
$$L^4 \in \mathbb{R}^{(n-n_0) \times (n-n_0)} : L_{ij}^4 = (\mathcal{L}_{\Gamma^c}^+)_{ij} \quad \text{for } i, j \in C_1^c.$$

If we imagine the vertices to be ordered such that $C = \{1, \ldots, n_0\}$ and $C_1^c = \{n_0 + 1, \ldots, n\}$, then this decomposition looks like

$$\mathcal{L}_{\Gamma^c}^+ = \begin{bmatrix} L^1 & L^2 \\ L^3 & L^4 \end{bmatrix}.$$

Because $W \subset C^c$ and $\Gamma \subset C$,

$$\mathcal{L}_{\Gamma^c}^+ \mathbf{1}_\Gamma = \begin{bmatrix} L^1 \mathbf{1}_\Gamma \\ L^3 \mathbf{1}_\Gamma \end{bmatrix} \quad \text{and} \quad \mathcal{L}_{\Gamma^c}^+ \mathbf{1}_W = \begin{bmatrix} L^2 \mathbf{1}_W \\ L^4 \mathbf{1}_W \end{bmatrix}.$$

Hence

$$\langle \mathcal{L}_{\Gamma^c}^+ \mathbf{1}_W, \mathcal{L}_{\Gamma^c}^+ \mathbf{1}_\Gamma \rangle = \langle L^2 \mathbf{1}_W, L^1 \mathbf{1}_\Gamma \rangle + \langle L^4 \mathbf{1}_W, L^3 \mathbf{1}_\Gamma \rangle.$$

In Lemma 10.42 below, we provide bounds on $\|L^i\|_1$ and $\|L^i\|_\infty$ for $i = 1, \ldots, 4$. We shall use these bounds here to finish the proof. Indeed,

$$\langle L^2 \mathbf{1}_W, L^1 \mathbf{1}_\Gamma \rangle \leq \|L^2 \mathbf{1}_W\|_\infty \|L^1 \mathbf{1}_\Gamma\|_1 \leq \|L^2\|_\infty \|\mathbf{1}_W\|_\infty \|L^1\|_1 \|\mathbf{1}_\Gamma\|_1 \leq (r)(1)(2)|\Gamma|,$$
$$\langle L^4 \mathbf{1}_W, L^3 \mathbf{1}_\Gamma \rangle \leq \|L^4 \mathbf{1}_W\|_\infty \|L^3 \mathbf{1}_\Gamma\|_1 \leq \|L^4\|_\infty \|\mathbf{1}_W\|_\infty \|L^3\|_1 \|\mathbf{1}_\Gamma\|_1 \leq (2 + o(1))(1)(r)|\Gamma|,$$

and so

$$\sum_{j \in W} v_j \leq 4r|\Gamma| = g(4r + o(r))n_0.$$

Returning again to (10.27), and using the fact that $r = o(1)$ (from Corollary 10.40),

$$ugn_0 - o(1) \leq \sum_{i \in U} v_i \leq \sum_{j \in W} v_j \leq g(4r + o(r))n_0.$$

That is, $u \leq 4r + o(r) + o(1/n_0) = o(1)$ almost surely. This completes the proof. $\quad\square$

Lemma 10.42. *Let L^1, L^2, L^3, and L^4 be as in the above proof. Then $\|L^2\|_\infty, \|L^3\|_1 \leq r$, $\|L^1\|_1 \leq 2$, and $\|L^4\|_\infty \leq 2 + o(1)$.*

Proof. Recall that, for any matrix B, it is known that $\|B\|_1 = \max_i \sum_j |B_{ij}|$ and $\|B\|_\infty = \max_j \sum_i |B_{ij}|$. Now

$$\|L^2\|_\infty = \max_{j \in C_1^c} \sum_{i \in C_1} |L_{ij}^+| = \max_{j \in C_1^c} \sum_{i \in C_1} \frac{A_{ij}^{out}}{d_i} \le \frac{1}{d_{\min}} \max_{j \in C_1^c} d_j^{out} \le \frac{d_{\max}^{out}}{d_{\min}} \le r,$$

and the proof for $\|L^3\|_1$ is very similar. Additionally,

$$\|L^1\|_1 = \max_{i \in C_1} \sum_{j \in C_1} |L_{ij}^+| = \max_{i \in C_1} \left(1 + \sum_{j \in C_1} \frac{A_{ij}^{in}}{d_i}\right) \le \max_{i \in C_1} \left(1 + \frac{1}{d_i} \sum_{j \in C_1} A_{ij}^{in}\right) = \max_{i \in C_1}(2) = 2,$$

while

$$\|L^4\|_\infty = \max_{j \in C_1^c} \sum_{i \in C_1^c} |L_{ij}^+| = \max_{j \in C_1^c} \left(1 + \sum_{i \in C_1^c} \frac{A_{ij}^{in}}{d_i}\right) \le \max_{j \in C_1^c} \left(1 + \frac{1}{d_{\min}^{in}} \sum_{i \in C_1^c} A_{ij}^{in}\right) \le 1 + \frac{d_{\max}^{in}}{d_{\min}^{in}},$$

and, by Corollary 10.39, $d_{\max}^{in}/d_{\min}^{in} = (1 + o(1))/(1 - o(1)) = 1 + o(1)$. $\quad\square$

10.6.3 ▪ Extracting C_1 from Ω

As mentioned at the beginning of this section, it is not the case that $\mathcal{L} = \mathcal{L}^{in} + \mathcal{L}^{out}$. Instead, we write $\mathcal{L} = \mathcal{L}^{in} + E$, where E can be thought of as a perturbation, or error term.

Theorem 10.43. *Suppose that $G \sim \text{SSBM}(n, k, p, q)$, with $p = \omega \log(n_0)/n_0$, $\omega \to \infty$ as $n \to \infty$, $q = b \log(n)/n$, and $k = O(1)$ as $n \to \infty$. Then $\|E\|_2 \le 2r$, where $r = \max_i r_i = \max_i \frac{d_i^{out}}{d_i^{in}}$.*

Proof. Observe that

$$L_{ij} := \delta_{ij} - \frac{1}{d_i} A_{ij} = \delta_{ij} - \frac{1}{d_i^{in} + d_i^{out}} \left(A_{ij}^{in} + A_{ij}^{out}\right).$$

We shall use the following easily verifiable identity as a one-dimensional version of the Sherman–Morrison formula or the Woodbury formula in linear algebra:

$$\frac{1}{d_i^{in} + d_i^{out}} = \frac{1}{d_i^{in}} - \frac{1}{d_i^{in}} \left(\frac{r_i}{r_i + 1}\right).$$

Thus,

$$L_{ij} = \left(\delta_{ij} - \frac{1}{d_i^{in}} A_{ij}^{in}\right) - \frac{1}{d_i^{in}} A_{ij}^{out} + \frac{1}{d_i^{in}} \left(\frac{r_i}{r_i + 1}\right) A_{ij}$$

$$= L_{ij}^{in} - \frac{1}{d_i^{in}} A_{ij}^{out} + \frac{1}{d_i^{in}} \left(\frac{r_i}{r_i + 1}\right) \left(A_{ij}^{in} + A_{ij}^{out}\right)$$

$$= L_{ij}^{in} - \frac{1}{d_i^{in}} \left(\frac{1}{r_i + 1}\right) A_{ij}^{out} + \frac{1}{d_i^{in}} \left(\frac{r_i}{r_i + 1}\right) A_{ij}^{in}.$$

That is, writing $E = [E_{ij}]_{1 \le i,j \le n}$, $E_{ij} = -\frac{1}{d_i^{in}} \left(\frac{1}{r_i+1}\right) A_{ij}^{out} + \frac{1}{d_i^{in}} \left(\frac{r_i}{r_i+1}\right) A_{ij}^{in}$. To bound the spectral norm, we use Gershgorin's disk theorem, noting that $E_{ii} = 0$ for all i:

$$\|E\|_2 = \max_i \{|\lambda_i| : \lambda_i \text{ eigenvalue of } E\} \leq \max_i \left\{ \sum_j |E_{ij}| \right\}$$

$$= \max_i \left\{ \frac{1}{d_i^{in}} \left(\frac{1}{r_i + 1} \right) \sum_j A_{ij}^{out} + \frac{1}{d_i^{in}} \left(\frac{r_i}{r_i + 1} \right) \sum_j A_{ij}^{in} \right\}$$

$$= \max_i \left\{ \frac{1}{d_i^{in}} \left(\frac{1}{r_i + 1} \right) (d_i^{out}) + \frac{1}{d_i^{in}} \left(\frac{r_i}{r_i + 1} \right) (d_i^{in}) \right\}$$

$$= \max_i \left\{ \left(\frac{r_i}{r_i + 1} \right) + \left(\frac{r_i}{r_i + 1} \right) \right\} \leq 2r.$$

This completes the proof. $\square$

Recall that Algorithm 10.5 works by running Algorithm 10.6 for m iterations on the sparse solution problem

$$\operatorname{argmin}\{\|L_\Omega \mathbf{x} - \mathbf{y}\|_2 : \|\mathbf{x}\|_0 \leq \delta n_0\}$$

to obtain $\mathbf{x}^m$, and then obtaining an approximation to $W = \Omega \setminus (C_1 \cap \Omega)$ by considering the support of $\mathbf{x}^m$. We now show, using the theory of totally perturbed compressive sensing (particularly Theorems 2.44 and 10.32), that this is a provably good approximation.

Recall that G^{in} has connected components $C_1, \ldots, C_k$. From Theorem 10.10, it follows that $\mathcal{L}^{in} \mathbf{1}_{C_1} = 0$. If $C_1 \subset \Omega$, then $\mathcal{L}_\Omega^{in} \mathbf{1}_{C_1} = 0$ too. From this we deduce that

$$\mathcal{L}_\Omega^{in} \mathbf{1}_\Omega = \mathcal{L}_\Omega^{in} \left(\mathbf{1}_{C_1} + \mathbf{1}_{\Omega \setminus C_1} \right) = 0 + \mathcal{L}_\Omega^{in} \mathbf{1}_{\Omega \setminus C_1} \Rightarrow \mathcal{L}_\Omega^{in} \mathbf{1}_{\Omega \setminus C_1} = \mathcal{L}_\Omega^{in} \mathbf{1}_\Omega.$$

If C_1 is not completely contained in Ω, we can show the following result.

Lemma 10.44. *Suppose that* $|C_1 \setminus (\Omega \cap C_1)| = o(n_0)$. *Then*

$$\mathcal{L}_\Omega^{in} \mathbf{1}_{\Omega \setminus (\Omega \cap C_1)} = \mathcal{L}_\Omega^{in} \mathbf{1}_\Omega + \mathbf{e}_1,$$

where $\|\mathbf{e}\|_2 = o(\sqrt{n_0})$.

Proof. Let $U := C_1 \setminus (\Omega \cap C_1)$ and $W := \Omega \setminus (\Omega \cap C_1)$. Then

$$\mathcal{L}^{in} \mathbf{1}_\Omega + \mathcal{L}^{in} \mathbf{1}_U = \mathcal{L}^{in} \left(\mathbf{1}_{C_1 \cap \Omega} + \mathbf{1}_W \right) + \mathcal{L}^{in} \mathbf{1}_U = \mathcal{L}^{in} \left(\mathbf{1}_{C_1 \cap \Omega} + \mathbf{1}_U \right) + \mathcal{L}^{in} \mathbf{1}_W = 0 + \mathcal{L}^{in} \mathbf{1}_W \tag{10.30}$$

because $\mathbf{1}_{C_1 \cap \Omega} + \mathbf{1}_U = \mathbf{1}_{C_1}$. Letting $\mathbf{e}_1 = \mathcal{L}^{in} \mathbf{1}_U$, we have the result since $\|\mathbf{e}_1\|_2 \leq \|\mathcal{L}^{in}\|_2 \|\mathbf{1}_U\|_2 = (2)(\sqrt{|U|}) = 2o(\sqrt{n_0})$. $\square$

Ideally, in setting up the sparse recovery problem at the heart of Algorithm 10.5 (i.e., (10.25)), we would define $\mathbf{y} := L_\Omega^{in} \mathbf{1}_\Omega = \sum_{i \in \Omega} \ell_i^{in}$, but of course we do not have access to $\mathcal{L}^{in}$, only $\mathcal{L} = \mathcal{L}^{in} + E$. In the next lemma we prove that this introduces an error term with ℓ_2 norm of order $o(\sqrt{n_0})$.

Lemma 10.45. *Let* $\mathbf{y} := \sum_{i \in \Omega} \ell_i$ *and* $\mathbf{y}^{in} = \sum_{i \in \Omega} \ell_i^{in}$. *Then* $\mathbf{y} = \mathbf{y}^{in} + \mathbf{e}_2$ *with* $\|\mathbf{e}_2\|_2 = o(\sqrt{n_0})$.

Proof. Clearly $\mathbf{e}_2 := \mathbf{y} - \mathbf{y}^{in} = \mathcal{L}\mathbf{1}_\Omega - \mathcal{L}^{in}\mathbf{1}_\Omega = E\mathbf{1}_\Omega$. By Theorem 10.43, $\|E\|_2 \le o(1)$ and so

$$\|\mathbf{e}_2\|_2 \le \|E\|_2\|\mathbf{1}_\Omega\|_2 \le o(1)\left(\sqrt{(1+\epsilon)n_0}\right) = o(\sqrt{n_0}).$$

This completes the proof. $\qquad\square$

The net result of Lemmas 10.44 and 10.45 is that

$$\mathcal{L}_\Omega^{in}\mathbf{1}_{\Omega\backslash C_1} = \mathbf{y} + \mathbf{e},$$

where $\mathbf{y} := \sum_{i\in\Omega} \ell_i^{in}$ and $\mathbf{e} = \mathbf{e}_1 + \mathbf{e}_2$ with $\|\mathbf{e}\|_2 = o(\sqrt{n_0})$. In the notation of Theorems 2.44 and 10.32, we think of $\mathcal{L}_\Omega$ as Φ, the noisy measurement matrix, and $\mathcal{L}_\Omega^{in}$ as $\hat{\Phi}$. Similarly, we think of $\mathbf{y}^{in}$ as $\hat{\mathbf{y}}$, and the $\mathbf{y}$ defined above as the noisy signal.

We now give the following bound on the RIC of $\mathcal{L}_\Omega$.

Theorem 10.46. *Let $G \sim SSBM(n, k, p, q)$ with $p = \omega \ln(n_0)/n_0$ and $q = b\ln(n)/n$, where $\omega \to \infty$ as $n \to \infty$. Suppose further that $k = O(1)$. For any $t = \gamma n_0$ with $\gamma \in (0, 1)$, $\delta_t(\mathcal{L}_G) \le \gamma + o(1)$ almost surely.*

Proof. Recall that if $G \sim SSBM(n, k, p, q)$ and $\mathcal{L}_G$ denotes the Laplacian of G, then we may write $\mathcal{L}_G = \mathcal{L}^{in} + E$, where L_G^{in} is the Laplacian of the in-cluster subgraph $G^{in} \sim SSBM(n, k, p, 0)$ and $\|E\|_2 \le o(1)$ by Theorem 10.43. By Lemma 10.64, $\hat{\delta}_t := \delta_t(\mathcal{L}^{in}) \le t/n_0 + o(1)$ almost surely. Observe that, for any matrix B,

$$\|B\|_{2,t} := \max_{\substack{T\subset[n]\\|T|=t}} \|B_T\|_2 = \max_{\substack{T\subset[n]\\|T|=t}} \sigma_{\max}(B_T),$$

where $\sigma_{\max}(B_T)$ denotes the maximum singular value of B_T. By the interlacing property of singular values (cf. [24]), $\lambda_{t-1}(B) \le \sigma_{\max}(B_T) \le \lambda_t(B) \le \lambda_{\max}(B)$, and so $\|E\|_{2,t} \le \|E\|_2 \le o(1)$ almost surely by Theorem 10.43. Similarly, $\|\mathcal{L}^{in}\|_{2,t} \ge \lambda_{t-1}(\mathcal{L}^{in})$. The eigenvalues of $\mathcal{L}^{in}$ are the eigenvalues of the $\mathcal{L}_a$, counted with multiplicity. In particular, as long as $t > k + 1$, where t is proportional to n while k is fixed, $\lambda_{t-1}(\mathcal{L}^{in}) \ge \min_a \lambda_2(\mathcal{L}_a)$. By Theorem 10.37, $\lambda_2(\mathcal{L}_a) \ge 1 - o(1)$ almost surely, and since $k = O(1)$, we may apply the union bound to obtain $\|\mathcal{L}^{in}\|_{2,t} \ge 1 - o(1)$ almost surely. Hence

$$\epsilon_\Phi^t := \frac{\|E\|_{2,t}}{\|\mathcal{L}^{in}\|_{2,t}} \le \frac{o(1)}{1 - o(1)} = o(1) \text{ almost surely.} \qquad (10.31)$$

Applying Theorem 2.44,

$$\delta_t(\mathcal{L}) \le \left(1 + \frac{t}{n_0} + o(1)\right)(1 + o(1))^2 - 1 = \left(1 + \frac{t}{n_0} + o(1)\right)(1 + o(1)) - 1$$

$$= \left(1 + \frac{t}{n_0} + o(1)\right) - 1 = \frac{t}{n_0} + o(1) = \gamma + o(1) \quad \text{if } t = \gamma n_0. \qquad \square$$

Finally, we compute the various constants necessary to apply Theorem 10.32.

Lemma 10.47. *Let $G \sim SSBM(n, k, p, q)$ with $p = \omega \ln(n)/n$ and $q = b\ln(n)/n$, where $\omega \to \infty$. Suppose further that $k = O(1)$. For any $s = \gamma n_0$ with $\gamma \in (0, 1)$, we have that $\epsilon_\Phi^s, \epsilon_\Phi, \epsilon_\mathbf{y} = o(1)$ almost surely (these quantities are all defined in Theorem 10.32).*

Proof. That $\epsilon_\Phi^t = o(1)$ was shown in the proof of Theorem 10.46 (see equation (10.31)). Here, $\epsilon_\mathbf{y} := \|\mathbf{e}\|_2/\|\mathbf{y}^{in}\|_2$, where $\|\mathbf{e}\|_2 = o(n_0)$ by Lemmas 10.44 and 10.45. Rearranging equation

(10.30), we get that $y^{in} = \mathcal{L}^{in}1_\Omega = \mathcal{L}^{in}(1_U - 1_W)$, where $U := C_1 \setminus (\Omega \cap C_1)$ and $W := \Omega \setminus (\Omega \cap C_1)$. As in the proof of Theorem 10.48 below, $|U| = o(n_0)$ and $|W| = \epsilon n_0 + o(n_0)$; hence $\|1_U - 1_W\|_0 = o(n_0) + \epsilon n_0 + o(n_0) \le 2\epsilon n_0$ for n_0 large enough. It follows that

$$\|y^{in}\|_2^2 = \|\mathcal{L}^{in}(1_U - 1_W)\|_2^2 \ge (1 - \delta_{2\epsilon n_0})\|1_U - 1_W\|_2^2 \ge (2\epsilon + o(1))(\epsilon n_0 + o(n_0)),$$

where the bound on $\delta_{2\epsilon n_0} = \delta_{2\epsilon n_0}(\mathcal{L}^{in})$ comes from Lemma 10.64 below (see Exercise 23). Since ϵ is fixed, we obtain

$$\epsilon_y = \frac{o(n_0)}{2\epsilon^2 n_0 + o(n_0)} = o(1).$$

Note that the sparsity input for Algorithm 10.6, namely s, is set equal to ϵn_0. Because $\epsilon < 0.15$ by assumption, it follows that $\delta_{3\epsilon n_0} < 0.45 + o(1)$. When n is large enough, we may assume that $\delta_{3\epsilon n_0} \le 0.45$. We have thus completed the proof. $\quad\square$

Putting all of the above together, we can show that Algorithm 10.5 succeeds in the sense that if $C_1^\#$ is the output and C_1 is the true cluster, then $|C_1 \triangle C_1^\#| = o(n_0)$.

Theorem 10.48 (Lai and Mckenzie, 2020 [18]). *Let $G \sim \text{SSBM}(n, k, p, q)$ with $k = \mathcal{O}(1)$, $p = \omega \log(n_0)/n_0$, and $q = b \log(n)/n$, where $\omega \to \infty$. If $C_1^\#$ is the output of Algorithm 10.5 with inputs $\epsilon > 0$, Ω, and n_0, where $|\Omega| = (1 + \epsilon)n_0$ and $|C_1 \setminus (\Omega \cap C_1)| = o(n_0)$, then $|C_1^\# \triangle C_1| = o(n_0)$.*

Proof. Let x^m be the output of Algorithm 10.6 run for $m = \mathcal{O}(\log(n))$ iterations on the problem

$$\text{argmin}\{\|\mathcal{L}_\Omega x - y\|_2 : \|x\|_0 \le \delta n_0\}. \tag{10.32}$$

By Theorem 10.32, we have that

$$\frac{\|1_{\Omega \setminus (\Omega \cap C_1)} - x^m\|_2}{\|1_{\Omega \setminus (\Omega \cap C_1)}\|_2} \le \left(\hat{\tau}\frac{\sqrt{1 + \hat{\delta}_s}}{1 - \epsilon_\Phi^s} + 1\right)(\epsilon_\Phi^s + \epsilon_y) = o(1),$$

where the second equality follows from Lemma 10.47. Let $U = C_1 \setminus (\Omega \cap C_1)$ (think of this as the indices missed by the thresholding step). By assumption, $|U| = o(n_0)$. It follows that $|\Omega \setminus (\Omega \cap C_1)| = |\Omega| - |\Omega \cap C_1| = (1 + \epsilon)n_0 - (n_0 - |U|) = \epsilon n_0 + o(n_0)$. Hence $\|1_{\Omega \setminus (\Omega \cap C_1)}\|_2 = \sqrt{\epsilon n_0} + o(\sqrt{n_0})$ and thus

$$\|1_{\Omega \setminus (\Omega \cap C_1)} - x^m\|_2 \le o(\sqrt{n_0}).$$

From the following lemma, it follows that $|\text{supp}(x^m) \triangle (\Omega \setminus (\Omega \cap C_1))| \le o(n_0)$, and, consequently, since $C_1^\# = \Omega \setminus \text{supp}(x^m)$, we have that $|C_1^\# \triangle (\Omega \cap C_1)| \le o(n_0)$. Accounting for U, we have that

$$|C^* \triangle C| = |C^* \triangle (\Omega \cap C)| + |C^* \triangle U| \le o(n_0) + |U| = o(n_0),$$

which is what we need to conclude the proof. $\quad\square$

The following result was used in the proof above.

Lemma 10.49. *Let $T \subset [n]$ and $v \in \mathbb{R}^n$. If $\|1_T - v\|_2 \le D$ and $|T| = |\text{support}(v)| = m$, then $|T \triangle \text{supp}(v)| \le 2D^2$.*

Proof. Recall that $T \triangle \text{supp}(\mathbf{v}) = (T \setminus (T \cap \text{supp}(\mathbf{v}))) \cup (\text{supp}(\mathbf{v}) \setminus (T \cap \text{supp}(\mathbf{v})))$ is the union of two disjoint sets. Because $|T| = |\text{supp}(v)| = m$, we see that

$$T \setminus (T \cap \text{supp}(\mathbf{v})) = m - |T \cap \text{supp}(\mathbf{v})| = \text{supp}(\mathbf{v}) \setminus (T \cap \text{supp}(\mathbf{v})).$$

It follows that

$$|T \triangle \text{supp}(\mathbf{v})| = 2 \left(m - |T \cap \text{supp}(\mathbf{v})| \right) = 2 \left(T \setminus (T \cap \text{supp}(\mathbf{v})) \right).$$

But $T \setminus (T \cap \text{supp}(\mathbf{v}))$ cannot be too large because

$$D \geq \|\mathbf{1}_T - \mathbf{v}\|_2 \geq \| (\mathbf{1}_T - \mathbf{v}) \, |_{T \setminus (T \cap \text{supp}(\mathbf{v}))} \|_2 = \|\mathbf{1}_{T \setminus (T \cap \text{supp}(\mathbf{v}))} \|_2 = \sqrt{|T \setminus (T \cap \text{supp}(\mathbf{v}))|}.$$

Thus $|T \setminus (T \cap \text{supp}(\mathbf{v}))| \leq D^2$, and the result follows. $\square$

10.7 ▪ Computational Complexity

In this section, we bound the operation count needed by Algorithm 10.3. For convenience, we focus on graph $G \sim \text{SSBM}(n, k, p, q)$ with parameters as in Theorem 10.34.

Theorem 10.50. *Suppose Algorithm 10.3 runs on $G \sim \text{SSBM}(n, k, p, q)$ with parameters exactly as in Theorem 10.34. If we assume in addition that $\omega = O(\log(n))$, then Algorithm 10.3 requires $O(n \log^3(n))$ operations.*

Proof. There are three main steps in Algorithm 10.3, namely,

1. computing $\mathcal{L}$ and $\mathcal{L}^+$,

2. the thresholding step of Algorithm 10.2, and

3. solving the sparse recovery problem at the heart of Algorithm 10.5 using Algorithm 10.6.

We shall bound the complexity of each of these individually. We assume that A is stored as a sparse matrix.

Computing $\mathcal{L}$ and $\mathcal{L}^+$: Computing each d_i requires $d_i \leq d_{\max}$ additions. This is done n times to compute D, requiring $O(d_{\max} n)$ operations. Since D is diagonal, the cost of computing $D^{-1}A$ is equal to the number of nonzero entries in A, which is bounded by $d_{\max} n$. By Corollaries 10.38 and 10.39, $d_{\max} \leq d_{\max}^{in} + d_{\max}^{out} = (1 + o(1))\omega \log(n_0) + 2b \log(n) + o(1) = O(\omega \log(n))$. We conclude that the time required to compute $\mathcal{L}$ and $\mathcal{L}^+$ is $O(\omega \log(n)n)$.

Thresholding: Sorting the entries of a vector $\mathbf{v}$ in decreasing order, and then selecting the $(1 + \epsilon)n_0$-largest of them, as in line 4 of Algorithm 10.4, takes at most $O(n \log(n))$ operations (cf. [23]). Hence the computational cost of Algorithm 10.4 is dominated by the cost of the matrix-vector multiplication in line 3, namely, $\left(\mathcal{L}_{\Gamma^c}^+ \right)^\top \mathbf{b}$. Each row of $\left(\mathcal{L}_{\Gamma^c}^+ \right)^\top$ contains at most $d_{\max} + 1 \leq O(\omega \log(n))$ nonzero entries; hence this step requires at most $O(\omega \log(n)n)$ computations.

Running Algorithm 10.6: The computational cost of solving the perturbed sparse recovery problem (10.25) using Algorithm 10.6 is equal to the number of iterations, m, times the cost of each iteration. The cost of each iteration is determined by calculating the cost of each step in the iterative part of Algorithm 10.6:

1. Computing $\mathcal{L}_s(L_\Omega^\top r^{k-1})$ is dominated by the cost of the matrix-vector multiplication $L_\Omega^\top r^{k-1}$. Each row of L_Ω has at most d_{max} nonzero entries; hence the cost of this step is $O(\omega \log(n) n)$.

2. Solving the least squares problem in step 2 is the most computationally expensive step. We recommend using an iterative method, such as conjugate gradient (available in our implementation); we use MATLAB's backslash operation. Fortunately, as pointed out in [23], the matrix in question, $\mathcal{L}_\Omega|_{\hat{T}^k} = \mathcal{L}_{\hat{T}^k}$, is extremely well conditioned. This is because $|\hat{T}^k| = 2s$ and by assumption $\delta_{2s}(\mathcal{L}) \leq \delta_{3s}(\mathcal{L})$. As in the proof of Theorem 10.48, we may assume that $\delta_{3s}(\mathcal{L}) \leq 0.45$ for large enough n. By [23], specifically Proposition 3.1 and the discussion of section 5, this implies that the condition number is small:

$$\kappa(\mathcal{L}_{\hat{T}^k}^\top \mathcal{L}^{\hat{T}_k}) := \frac{\lambda_{max}(\mathcal{L}_{\hat{T}^k}^\top \mathcal{L}_{\hat{T}^k})}{\lambda_{min}(\mathcal{L}_{\hat{T}^k}^\top \mathcal{L}_{\hat{T}^k})} \leq \frac{1 + \delta_{2s}}{1 - \delta_{2s}} \leq 2.64.$$

It follows that we only require a constant number of iterations of the conjugate gradient to approximate the solution to the least squares problem, $\mathbf{u}$, to within an acceptable tolerance. The cost of each iteration of the conjugate gradient is equal to the cost of a matrix-vector multiplication by $\mathcal{L}_{\hat{T}_k}$ or $\mathcal{L}_{\hat{T}_k}^\top$, which is $O(\omega \log(n) n)$.

3. The cost of sorting and thresholding in step 3 is $\mathcal{O}(n \log(n))$.

4. Finally, the cost of computing the new residual r^k in step 4 is dominated by the matrix-vector multiplication $\mathcal{L}_{\hat{T}^k}^\top r^k$ and hence is $O(\omega \log(n) n)$.

We conclude that the cost of a single iteration of Algorithm 10.6 is $O(\omega \log(n) n)$. By the proof of Theorem 10.48, it suffices to take $m = O(\log(n))$; hence the cost of running Algorithm 10.6 is $O(\omega \log^2(n) n)$.

It follows that the computational cost of Algorithm 10.3 is dominated by the cost of the steps of Algorithm 10.6, which is $O(\omega \log^2(n) n)$. Specializing to the case where $\omega = O(\log(n))$, we see that the total computational cost is $O(\omega \log^2(n) n) = O(\log^3(n) n)$. $\square$

Since we can use $\omega = O(\log(\log(n+1)))$ in all our previous discussion, the cost of computation of Algorithm 10.3 is $O(n \log^2(n) \log \log(n+1))$. Thus, our Algorithm 10.3 is very efficient for graph clustering/communities detection.

10.8 ▪ Numerical Examples

In this section, we present a few examples to show the performance of the SCP and SSCP algorithms developed in previous sections, i.e., Algorithms 10.2 and 10.3. First of all, we explain a few methods of generating adjacency matrices. Generating a good adjacency matrix for any real-life problem is an art. For social network problems, one easy way is to use counting. That is, count the number of connections between two people if the vertices of a graph are people. In Example 10.51, the friendship between two people is represented simply as yes or no. Thus, the adjacency matrix A consists of 1 or 0. For Example 10.52, the similarity between customers is measured by the number of common products shopped for by these two customers. Then A will be a weighted adjacency matrix. For another type of application, separating points in Euclidean space $\mathbb{R}^d$ with $d \geq 2$ as in Example 10.54, we use the standard distance function $\|\mathbf{x} - \mathbf{y}\|$. Certainly, there are various ways to use the distance function, for example, weighted or powered, i.e., p-power. In Examples 10.55 and 10.56, we use an exponential distance function. Also, for points in high-dimensional space, we can use the angle between two vectors, i.e., the cosine angle

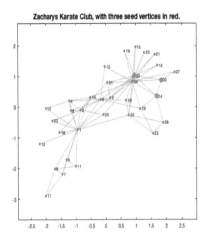
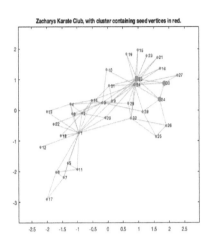

Figure 10.1. *Zachary's Karate Club data set (left), where red dots are seeds and two subclubs (members in red and blue) are found (right).*

$\langle \mathbf{x}, \mathbf{y} \rangle / (\|\mathbf{x}\| \|\mathbf{y}\|)$, to measure the closeness between two vertices $\mathbf{x}, \mathbf{y}$. Any detailed discussion is beyond the scope of this book.

Our first example is the classic Zachary's Karate Club problem.

Example 10.51 (Zachary's Karate Club). Zachary has a karate club which is too crowded. He needs to split it in two due to overcrowding. He has a mathematician friend who suggests he distribute a questionnaire to each member of the club: which friends do they want to attend the club with? These questionnaires generate a matrix A of all members with entries $A_{ij} = 1$ if the ith person is a friend of the jth person and $A_{ij} = 0$ otherwise. That is, A is an adjacency matrix. We now use our SSCP algorithm to permute A into a block diagonal matrix. See Figure 10.1 with 3 seeds. From the resulting matrix, Zachary is able to break the club into two subclubs with a good estimate of their sizes, and he can find the right-size rooms near the center of the members of each subclub to operate his business. In Figure 10.1, we show the friendship relations of all the club members. Instead of showing the adjacency matrix, we use MATLAB command graph.m, i.e., $G = \text{graph}(A)$, which converts an adjacency matrix into a set of space points in $\mathbb{R}^2$ and then plots G to show the connectivities among the members of the karate club. We apply Algorithm 10.2 to find two clusters based on three seeds (red dots shown in Figure 10.1).

Example 10.52 (Amazon's Shopping Recommendation). Amazon has a large data set on customers and products, i.e., who bought what. For simplicity, let us say Amazon has a one-month shopping record for a city, say 1000 customers who have bought 10,000 different products. Note that each customer has bought at least one product and definitely no customer has bought all 10,000 products. That is, we have a matrix A of size $1000 \times 10{,}000$ with entries 1 or 0, with $A_{ij} = 1$ indicating the ith customer bought the jth product and $A_{ij} = 0$ otherwise. A is a very sparse matrix. Amazon wants to recommend its products to its customers. Mathematically, this problem can be converted to a graph clustering problem by grouping customers into several groups where each group has a similar shopping tastes. Based on the list of products the group has bought, Amazon can send its promotion emails to each member in the group for the remaining products that the member has not bought yet. We can group customers using the following steps. Let $B = AA^\top$. Each entry B_{ij} is the similarity between the ith customer and the jth customer. We now use a graph clustering algorithm, say Algorithm 10.3, to convert B into a

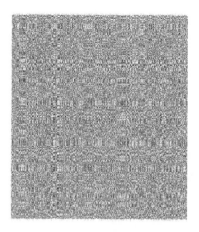 

Figure 10.2. *From the given matrix B (left), a new matrix with block structures is found by Algorithm* 10.2.

block diagonal matrix. Each block submatrix is a group with similar shopping tastes. Amazon can use this data to target their promotion. In this computation, we need to normalize each row in A. In Figure 10.2, we show the adjacency matrix B and the four groups that are found. Each group has similar tastes among its members. Note that the computation in Figure 10.2 is based on an artificial data set.

Example 10.53. We generate a nearly block diagonal matrix A which is an adjacency matrix generated from the stochastic block model (SBM) $\mathcal{G}(2400, 400, 0.8, 0.02)$. We first present the original adjacency matrix A. Next we permute A randomly and use it as the input matrix to our Algorithm 10.2. Our algorithm produces a set of clusterings whose rearranged indices are shown to be very similar to the original A.

We can see that the matrix on the right-hand side of Figure 10.4 is very close to the one in Figure 10.3.

Example 10.54 (Data Separation). We are given points in $\mathbb{R}^2$ or $\mathbb{R}^3$ as shown in Figure 10.5. We would like to separate the data into subgroups.

We use the distance between points and use $A_{ij} = \exp(-\text{dist}(\mathbf{x}_i, \mathbf{x}_j)/\alpha)$ to generate an adjacency matrix for a parameter $\alpha > 0$. We use Algorithm 10.3 with seeds as well as the standard spectral clustering algorithm to find the features of the given data sets. Due to the seeds, our SSCP algorithm is able to accurately separate each data set into subgroups, better than using the spectral clustering algorithm; see Figures 10.6 and 10.7.

Example 10.55 (Classification of MNIST Data). We need to train a computer to recognize a digit from any given image of size 28×28, as shown in Figure 10.8. The training set is of size 70,000, with more than 50,000 labeled.

We use the following procedure, presented in [16], to build an adjacency matrix of these digit images.

- Fix parameters r and K. Note that K, the number of neighbors, has no relation to k, the number of clusters.

- For all $i \in [n]$, define $\sigma_i := \|\mathbf{x}_i - \mathbf{x}_{[r,i]}\|$, where $\mathbf{x}_{[r,i]}$ denotes the rth closest point in $\mathcal{X}$ to $\mathbf{x}_i$. (If there is a tie, break it arbitrarily.) Let $\text{NN}(\mathbf{x}_i, K) \subset \mathcal{X}$ denote the set of the K closest points in $\mathcal{X}$ to $\mathbf{x}_i$. Again, we may break ties arbitrarily if they occur.

Figure 10.3. *A random matrix subject to the SBM* $\mathcal{G}(2400, 400, 0.8, 0.02)$.

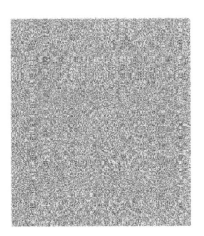 

Figure 10.4. *From the given matrix (left), a new matrix with block structures is found by Algorithm* 10.2 *and is very close to the one in Figure* 10.3.

- Define $\tilde{A}$ as

$$\tilde{A}_{ij} = \begin{cases} \exp\left(-\|\mathbf{x}_i - \mathbf{x}_j\|^2/\sigma_i\sigma_j\right) & \text{if } \mathbf{x}_j \in NN(\mathbf{x}_i, K), \\ 0 & \text{otherwise.} \end{cases}$$

- Observe that $\tilde{A}$ is not necessarily symmetric, since it may occur that $\mathbf{x}_j \in NN(\mathbf{x}_i, K)$ while $\mathbf{x}_i \notin NN(\mathbf{x}_j, K)$. So symmetrize $\tilde{A}$ to obtain A, the adjacency matrix of G. We considered two symmetrizations:

$$A_{\text{mult}} := \tilde{A}^\top \tilde{A} \quad \text{and} \quad A_{\text{max}}, \text{ where } (A_{\text{max}})_{ij} = \max\left\{\tilde{A}_{ij}, \tilde{A}_{ji}\right\}.$$

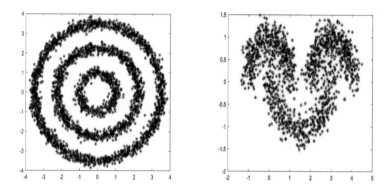

Figure 10.5. *Circular data set (left) and smile data set (right).*

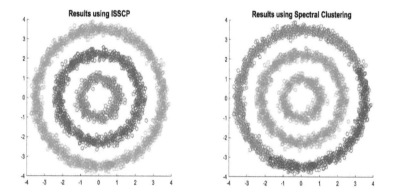

Figure 10.6. *Circular data separation by our Algorithm* 10.2 *(left) and by the spectral clustering algorithm (right).*

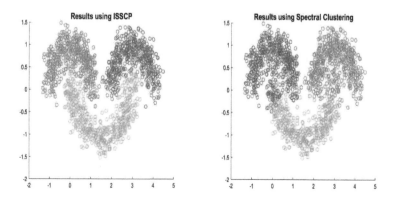

Figure 10.7. *Smile data separation by our Algorithm* 10.3 *(left) and by the spectral clustering algorithm (right).*

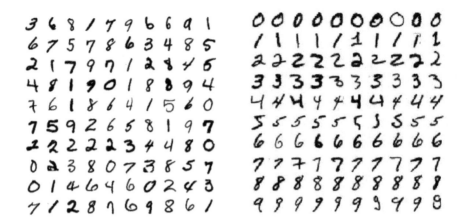

Figure 10.8. *A random sample of* 100 *images from* MNIST *data and a sorted data image by Algorithm* 10.3.

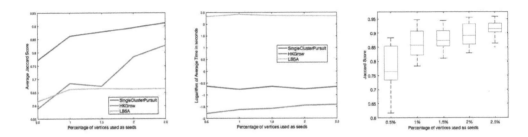

Figure 10.9. *Cluster extraction for* MNIST *with* $A_{\max}$ *as the adjacency matrix. Results are averaged over all* 10 *digits. From left to right: The Jaccard index as a function of the amount of labeled data, the run time as a function of the amount of labeled data, and a box plot of the Jaccard indices for Algorithm* 10.2.

Next let us recall the *Jaccard index* defined as

$$\mathrm{Jac}(C_a, C_a^{\#}) := \frac{|C_a \cap C_a^{\#}|}{|C_a \cup C_a^{\#}|}.$$

Note that the Jaccard index is symmetric and penalizes inaccurate algorithms as well as algorithms that attempt to "trivially" solve the cluster extraction problem by returning an overly large set of vertices containing Γ.

To present the performance of our SSCP algorithm, we also present the performance of two other representative algorithms called LIBA (cf. https://github.com/PanShi2016/LBSA) and HK-grow (cf. https://www.cs.purdue.edu/homes/dgleich/codes/hkgrow/).

From Figure 10.9, we can see that Algorithm 10.3 works well because the Jaccard index is highest.

Finally, let us look at an example of face classification.

Example 10.56. We are given a set of faces of 10 people, where each person has 10 faces with each face different from the others the same person. Thus, we have 100 thumbnail images of size 56×46. We use the exponential distance function defined in the previous example to find a good

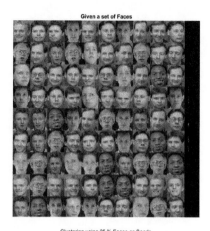
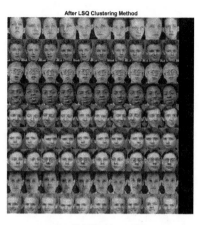

Figure 10.10. *AT&T face data clustering. (From Yale Face Database B.)*

adjacency matrix A with appropriate adjustment of some parameters. We then use a modified version of Algorithm 10.3 called LSQ (cf. [19]) to cluster the faces. With four seeds per person, we are able to cluster all faces. See Figure 10.10. With fewer seeds, we may not be able to find all corrected faces.

Similarly, we tested a database called YaleB faces.[1] There are 28 different people with 500 or more different lighting settings and face positions in the data set. For convenience, we chose 20 for each person. We ran our clustering code to sort these faces correctly, as shown in Figure 10.11 based on four seeds per person. Figure 10.12 shows the results.

10.9 ▪ Exercises

Exercise 1. Let A be an adjacency matrix. Let $A^2 = A A$. Show that the vertices i and j are connected if the entry A_{ij}^2 of A^2 is not zero. Furthermore, show that if A is fully connected, all entries of matrix A^p are not zero when $p \geq 1$ is large enough.

Exercise 2. Write the details of the proof of Theorem 10.2.

Exercise 3. Prove Theorem 10.3.

Exercise 4. Prove Theorem 10.9.

Exercise 5. Assume that a graph G satisfies the following four assumptions:

(A1) Its vertices $V = C_1 \cup \cdots \cup C_k$, where the C_a are disjoint, planted clusters and k is $O(1)$ as $n \to \infty$.

(A2) For all $a \in [k]$ we have that the second eigenvalue $\lambda_2(L_{G_{C_a}}) \geq 1 - \epsilon_1$ and $\lambda_{n_a}(L_{G_{C_a}}) \leq 1 + \epsilon_1$ almost surely.

---
[1]Publicly available at http://vision.ucsd.edu/content/yale-face-database.

Figure 10.11. *YaleB faces clustering (random arrangement). (From Yale Face Database B.)*

(A3) Letting $r_i := d_i^{out}/d_i^{in}$, we have $r_i \le \epsilon_2$ for all $i \in [n]$ almost surely.

(A4) If $d_{av}^{in} := \mathbb{E}[d_i^{in}]$, then $d_{\max}^{in} \le (1 + \epsilon_3)d_{av}^{in}$ and $d_{\min}^{in} \ge (1 - \epsilon_3)d_{av}^{in}$ almost surely.

Prove the following result.

Lemma 10.57. *Let $G \in \mathcal{G}_n$ satisfy Assumptions* (A1)–(A4)*. If $N_{G_{C_1}} := D_{G_{C_1}}^{-1/2} A_{G_{C_1}} D_{G_{C_1}}^{-1/2}$ and $U, \Gamma \subset C_1$, then*

$$\left| \langle D_{G_{C_1}}^{1/2} \mathbf{1}_U, N_{G_{C_1}}^t D_{G_{C_1}}^{1/2} \mathbf{1}_\Gamma \rangle - \frac{vol^{in}(U)vol^{in}(\Gamma)}{vol^{in}(G_{C_1})} \right| \le \epsilon_1^t \sqrt{vol^{in}(U)vol^{in}(\Gamma)}$$

(*Hint*: See [18].)

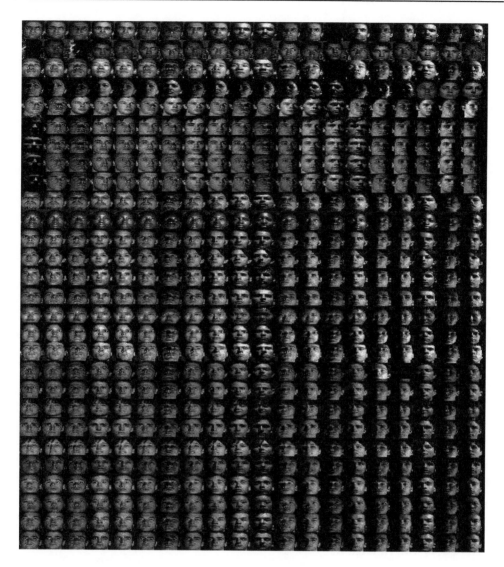

Figure 10.12. *YaleB Faces Clustering, assorted by a modified version of Algorithm* 10.3 *(cf.* [19]*). (From Yale Face Database B.)*

Exercise 6. Prove the following result.

Lemma 10.58. *Suppose that* $\mathcal{G}_n$ *satisfies assumptions* (A3) *and* (A4) *in the previous problem. For any* $S \subset V$, *recall that* $vol^{in}(S) = \sum_{i \in S} d_i^{in}$. *Then for any* $G \in \mathcal{G}_n$ *we have that*

(1) $(1 - \epsilon_3)|S|d_{av}^{in} \le vol^{in}(S) \le (1 + \epsilon_3)|S|d_{av}^{in}$ *and* (2) $vol^{in}(S) \le vol(S) \le (1 + \epsilon_2)vol^{in}(S)$.

(*Hint*: See [18].)

Exercise 7. Prove the following result.

Theorem 10.59. *Let* $G \sim ER(n, q)$, *with* $q = (\beta + o(1)) \log(n)/n$. *There exists a function* $\eta(\beta)$

satisfying $0 < \eta(\beta) < 1$ *and* $\lim_{\beta\to\infty} \eta(\beta) = 0$ *such that*

$$d_{\max}(G) = (1 + \eta(\beta))\beta \log n + o(1) \le 2\beta \log(n) + o(1) \text{ almost surely.}$$

(*Hint*: See [4, 5, 6].)

Exercise 8. Prove the following result.

Theorem 10.60. *If* $G \sim ER(n_a, p)$ *with* $p_a = \omega \log(n)/n_a$ *where* $\omega \to \infty$, *then* $d_{\min}(G) = (1 - o(1))\omega \log(n)$ *and* $d_{\max}(G) = (1 + o(1))\omega \log(n)$ *almost surely.*

(*Hint*: See [14, Theorem 3.4 (ii)].)

Exercise 9. Prove the following result.

Theorem 10.61. *Suppose that* $G \sim ER(n_a, p)$ *with* $p = \omega \log(n)/n_a$ *where* $\omega \to \infty$. *Then we have almost surely* $|\lambda_i(L) - 1| = O(\omega^{-1/2}) = o(1)$ *for all* $i > 1$.

(*Hint*: See [18].)

Exercise 10. Show that a graph G from the SBM satisfies the four assumptions mentioned in Exercise 5. (*Hint*: See [18].)

Exercise 11. Prove the following result.

$$|\lambda_i(L^{\text{sym}}) - 1| \le \sqrt{\frac{6 \log(2n_a)}{\omega \log(n)}}.$$

(*Hint*: See Theorem 4 in [10].)

Exercise 12. Implement Algorithm 3.2 to solve the minimization (10.15). Use adjacency matrices from $\mathcal{G}(n, k, p, 0)$ to see if your implementation is correct.

Exercise 13. Prove Theorem 10.10.

Exercise 14. Prove Lemma 10.17.

Exercise 15. Prove Theorem 10.23.

Exercise 16. Show that the triple summation

$$\sum_{i\in U}\sum_{j\in\Gamma}\sum_{k\in C} A_{ik}A_{kj} \tag{10.33}$$

is precisely the number of length-2 paths from U to $\Gamma \subset C$.

Exercise 17. Further show that

$$\sum_{i\in U}\sum_{j\in\Gamma}\sum_{k\in C} A_{ik}A_{kj} \ge p^2 n_0 |U||\Gamma| - o(p^2 n_0^3) \tag{10.34}$$

almost surely, where $\Gamma \subset C$.

Exercise 18. Prove Theorem 10.59.

Exercise 19. Prove Theorem 10.60.

Exercise 20. Show Theorem 10.61.

Exercise 21. Prove the following.

Lemma 10.62. *Let G be any connected graph on n_0 vertices, and let $t < n_0$. Then*

$$\delta_t(L) \leq \max\left\{1 - \lambda_2^2\left(\frac{d_{\min}}{d_{\max}} - \frac{d_{\max}}{d_{\min}}\frac{t}{n_0}\right), 1 - \lambda_{n_0}^2\right\}.$$

(*Hint:* See [18] for a proof.)

Exercise 22. Prove the following.

Lemma 10.63. *Suppose that $G \sim ER(n_0, p)$ with $p = \omega \ln(n_0)/n_0$ for some $\omega \to \infty$. Then $\delta_t(\mathcal{L}) \leq t/n_0 + o(1)$ almost surely.*

(*Hint:* See [18] for a proof.)

Exercise 23. Prove the following.

Lemma 10.64. *Suppose that $G \sim SSBM(n, k, p, 0)$ with $k = O(1)$. Then $\delta_t(\mathcal{L}) \leq \frac{t}{n_0} + o(1)$ almost surely.*

(*Hint:* See [18] for a proof.)

Exercise 24. Give a complete proof of Theorem 10.32.

Exercise 25. Show that the iterative hard thresholding algorithm (see Chapter 4) will recover C_1 in at most $d - 1$ steps if $p < \sqrt{\ln(n_0 - 1)}/\sqrt{2n_0}$ and so $\text{diam}(G_{C_1}) = d > 2$.

Bibliography

[1] E. Abbe, Community Detection and Stochastic Block Models: Recent Developments, preprint, arXiv:1703.10146, 2017. (Cited on p. 340)

[2] E. Abbe and Colin Sandon, Community detection in general stochastic block models: Fundamental limits and efficient algorithms for recovery, in IEEE 56th Annual Symposium on Foundations of Computer Science (FOCS), 2015, 670–688. (Cited on p. 340)

[3] J. M. Aldous and R. J. Wilson, Graphs and Applications: An Introductory Approach, Springer-Verlag, Berlin, 2000. (Cited on p. 339)

[4] B. Bollobas, Vertices of given degree in a random graph, J. Graph Theory, 6 (1982), 147–155. (Cited on pp. 362, 380)

[5] B. Bollobas, Random Graphs, Cambridge University Press, 2001. (Cited on p. 380)

[6] B. Bollobas, Modern Graph Theory, Springer-Verlag, 1998, 215–252. (Cited on pp. 362, 380)

[7] F. Chung, Spectral Graph Theory, AMS, Providence, RI, 1997. (Cited on p. 339)

[8] F. Chung and L. Lu, Complex Graphs and Networks, American Mathematical Society, Providence, RI, RI, 2006. (Cited on p. 340)

[9] F. Chung, L. Lu, and Van Vu, The spectra of random graphs with given expected degrees, Proc. Nat. Acad. Sci. USA, 100(11) (2003), 6313–6318. (Cited on pp. 342, 364)

[10] F. Chung and M. Radcliffe, On the spectra of general random graphs, Electron. J. Combin., 18 (2011), paper 215. (Cited on pp. 362, 380)

[11] W. Dai and O. Milenkovic, Subspace pursuit for compressive sensing signal reconstruction, IEEE
      Trans. Inform. Theory, 55(5) (2009), 2230–2249. (Cited on p. 361)

[12] P. Erdős and A. Rényi, On random graphs, I, Publ. Math. Debrecen, 6 (1959), 290–297. (Cited on
      p. 340)

[13] P. Erdős and A. Rényi, On the evolution of random graphs, Magyar Tud. Akad. Mat. Kutató Int. Közl.,
      5 (1960), 17–61. (Cited on p. 340)

[14] A. Frieze and M. Karonski, Introduction to Random Graphs, Cambridge University Press, 2016.
      (Cited on pp. 357, 358, 362, 380)

[15] P. W. Holland, K. B. Laskey, and S. Leinhardt, Stochastic block models: First steps, Social Networks,
      5(2) (1983), 109–137. (Cited on p. 340)

[16] M. Jacobs, E. Merkurjev, and S. Esedoglu, Auction dynamics: A volume constrained MBO scheme,
      J. Comput. Phys., 354 (2018), 288–310. (Cited on p. 373)

[17] A. K. Jain, Data clustering: 50 years beyond K-means, Pattern Recogn. Lett., 31(8) (2010), 651–666.
      (Cited on p. 340)

[18] M.-J. Lai and D. Mckenzie, Compressive sensing approach to cut improvement and local clustering,
      SIAM J. Math. Data Sci., 2 (2020), 368–395. (Cited on pp. 357, 360, 363, 369, 378, 379, 380, 381)

[19] M.-J. Lai and Z. Shen, An Effective Approach to Semi-supervised Cluster Extraction, submitted,
      2020. (Cited on pp. 377, 379)

[20] H. Li, Improved analysis of SP and CoSaMP under total perturbations, EURASIP J. Adv. Signal Proc.,
      2016(1) (2016), 112. (Cited on p. 359)

[21] U. Von Luxburg, A tutorial on spectral clustering, Stat. Comput., 17 (2007), 395–416. (Cited on
      pp. 337, 339)

[22] M. C. V. Nascimento and A. DeCarvalho, Spectral methods for graph clustering—A survey. European
      J. Oper. Res., 211(2) (2011), 221–231. (Cited on p. 337)

[23] D. Needell and J. A. Tropp, CoSaMP: Iterative signal recovery from incomplete and inaccurate sam-
      ples, Appl. Comput. Harmon. Anal., 26(3) (2009), 301–321. (Cited on pp. 370, 371)

[24] R. C. Thompson, Principal submatrices IX: Interlacing inequalities for singular values of submatrices,
      Linear Algebra Appl., 5(1) (1972), 1–12. (Cited on pp. 345, 368)

Chapter 11

Sparse Phase Retrieval

In this chapter we study phase retrieval for signals including sparse signals from their noisy measurements. The study of phase retrieval arises in many applied science applications, such as X-ray diffraction, crystallography, electron microscopy, and optical imaging. See [11] and [28] for X-ray diffraction. See [29] and [30] for crystallography. See [33], [16], and [15] for optical imaging applications. In particular, see [22] for an explanation of image recovery from its phaseless measurements. Mathematically, the phase retrieval problem can be described simply as follows. Given measurement values

$$b_j := |\langle \mathbf{a}_j, \mathbf{x_b} \rangle|^2 + e_j, \quad j = 1, \ldots, m, \tag{11.1}$$

where $\mathbf{a}_j \in \mathbb{R}^n$ are known measurement vectors and $e_j \in \mathbb{R}, j = 1, \ldots, m$, are unknown noises, our goal is to recover $\mathbf{x_b}$. Similarly, we can formulate the same problem in the complex variable setting. Typically, given the magnitude of the discrete Fourier transform of an unknown signal $\mathbf{x_b}$, we would like to find $\mathbf{x_b}$, mainly the phase of $\mathbf{x_b}$.

Clearly, if $\mathbf{x_b}$ is a solution, $\pm\mathbf{x_b}$ is also a solution for any $+$ and $-$ in the real or complex variable setting. Thus, the solution is always not unique. Also, in the complex variable setting, we will recover $\mathbf{x_b}$ up to a unimodular constant $c = e^{\mathbf{i}\theta}$ for a $\theta \in [0, 2\pi)$, where $\mathbf{i} = \sqrt{-1}$. A surprising fact is that we can find $\mathbf{x_b}$ from (11.1) up to a unimodular constant if $m \gg n$, e.g., $m = n(n+1)/2$ for $\mathbf{x_b} \in \mathbb{R}^n$ or $\mathbb{C}^n$. Another basic fact is that for any given $b_j, j = 1, \ldots, m$, the chance of finding $\mathbf{x_b} \in \mathbb{C}^n$ satisfying (11.1) is zero even with no noise (see Theorem 11.3). Our study assumes that there is a vector $\mathbf{x_b}$ which produces the given measurement values with some noise. Note that the targeted signal $\mathbf{x_b} \in \mathbb{C}^n$ may be a sparse vector. We hope to use few measurements to recover $\mathbf{x_b}$ and, in particular, to retrieve sparse solution vectors.

We shall address how many measurements are needed, i.e., how large m must be to retrieve a signal, and how to design these measurement vectors. Certainly, we have to choose measurement vectors $\mathbf{a}_j, j = 1, \ldots, m$, which span the whole Euclidean space $\mathbb{R}^n$ or $\mathbb{C}^n$. Because $m > n$, the measurement vectors, $\Phi = \{\mathbf{a}_1, \ldots, \mathbf{a}_m\}$, will form a frame. But this is not enough. They must be in generic position, e.g., any n measurements from Φ span $\mathbb{R}^n$ or $\mathbb{C}^n$. In the real variable setting, $m \geq 2n - 1$ is a necessary condition. However, in the complex variable setting, things are still unclear. One theory is known: $m \geq 4n - 4$ for n in the form $n = 2^k + 1, k \geq 1$, is necessary. We shall present another theory to provide a set of m ($\geq 3n - 2$) measurement vectors, which enables us to retrieve almost all signals.

Furthermore, the phase retrieval problem (11.1) can be viewed as solving a system of quadratic equations. The problem in (11.1) is a special case of the more general system of quadratic equations for $\mathbf{x_b} \in \mathbb{C}^n$:

$$|\langle \mathbf{a}_j, \mathbf{x_b} \rangle|^2 + \mathbf{c}_j^* \mathbf{x_b} + \mathbf{x_b^*} \mathbf{d}_j = b_j, \quad j = 1, \ldots, m, \tag{11.2}$$

where $\mathbf{a}_j, b_j, \mathbf{c}_j, \mathbf{d}_j, j = 1, \ldots, m$, are given. More generally, we may solve for $\mathbf{x_b} \in \mathbb{C}^n$:

$$|\langle \mathbf{a}_j, \mathbf{x_b} \rangle|^p = b_j, \quad j = 1, \ldots, m, \tag{11.3}$$

where $b_j > 0, j = 1, \ldots, m$, and $p \geq 3$ is an integer. The same questions can be raised: How large should m be in these cases and how can we find $\mathbf{x_b}$ from the given measurements in (11.2) or (11.3)?

In addition to using frame theory to reconstruct the unknown signal, we shall develop some numerical approaches to solve (11.1) and (11.2). One simple computational approach to retrieving $\mathbf{x_b}$ from (11.1) is to solve the least squares minimization

$$\min_{\mathbf{x} \in \mathbb{R}^n \text{ or } \mathbb{C}^n} \frac{1}{m} \sum_{j=1}^m (|\langle \mathbf{a}_j, \mathbf{x} \rangle|^2 - b_j)^2. \tag{11.4}$$

Although it is not a convex minimization problem, there are many computational algorithms available to solve (11.4), e.g., a gradient descent method (called the Wirtinger flow method due to the complex variable setting) and its variants, the Gauss–Newton method and its variants, and a DC (difference of convex functions) based method to be discussed in this chapter. The Wirtinger flow method and the Gauss–Newton method require $m = O(n \log(n))$ measurements. Our numerical experiments show that when $m \geq 3n$ for $n = 128$–1280, all three methods recover $\mathbf{x_b}$ from the measurements very well in the real and complex variable settings. When $m = 3n$ for $n = 128$–1280, the Gauss–Newton and DC-based methods work well. When $m = 2n$, only the DC-based method works well in the real variable setting.

Next we consider how to find the sparse solution $\mathbf{x_b}$ satisfying the given measurements (11.1). The following minimization is called the *compressive phase retrieval problem* [27], which is a pioneering work employing the ℓ_1 minimization approach to retrieve $\mathbf{x_b}$:

$$\min \|\mathbf{x}\|_1 \quad \text{subject to} \quad \big\| |A\mathbf{x}| - |\mathbf{b}| \big\|_2 \leq \epsilon, \tag{11.5}$$

where $A := [\mathbf{a}_1, \ldots, \mathbf{a}_m]^\top$, $|A\mathbf{x}| := [|\langle \mathbf{a}_1, \mathbf{x} \rangle|, \ldots, |\langle \mathbf{a}_m, \mathbf{x} \rangle|]^\top$, and $\mathbf{b} = (b_1, \ldots, b_m)^\top$ with b_i given in (11.1). Based on our studies in this book, we shall consider

$$\min_{\mathbf{x} \in \mathbb{R}^n \text{ or } \mathbb{C}^n} \lambda \|\mathbf{x}\|_1 + \frac{1}{m} \sum_{j=1}^m (|\langle \mathbf{a}_j, \mathbf{x} \rangle|^2 - b_j)^2, \tag{11.6}$$

where the signal $\mathbf{x_b}$ is approximately s-sparse, as well as many other approaches, as discussed before.

In addition, we can convert the phase retrieval problem into a matrix recovery problem (cf. [10]). As explained in Chapter 9, we can apply some computational methods to the problem

$$\min\{\text{rank}(X) : \text{tr}(A_j X) \approx b_j, j = 1, \ldots, m, X \succeq 0\}, \tag{11.7}$$

where $X = \mathbf{x}\mathbf{x}^\top$ and $A_j = \mathbf{a}_j \mathbf{a}_j^\top, j = 1, \ldots, m$. For example, an alternating projection method can be used. The details are left to the interested reader and the exercises at the end of this chapter.

11.1 ▪ Existence and Number of Solutions

In this section, we shall discuss the existence of solutions for phase retrieval and estimate the number of distinct solutions. To begin, we first recall (11.7), which shows the connection between phase retrieval and low rank matrix recovery.

Let $X = \mathbf{x}\mathbf{x}^\top$ and $A_j = \mathbf{a}_j\mathbf{a}_j^\top$, $j = 1, \ldots, m$. Then the constraints in (1.61) can be rewritten as

$$b_j = \text{tr}(A_j X), \quad j = 1, \ldots, m, \tag{11.8}$$

where $\text{tr}(\cdot)$ is the trace operator.

Note that the scaling of $\mathbf{x}$ by a unimodular constant c would not change X. Indeed, $(c\mathbf{x})(c\mathbf{x})^\top = |c|^2\mathbf{x}\mathbf{x}^\top = \mathbf{x}\mathbf{x}^\top = X$. Conversely, given a positive semidefinite matrix X with rank 1, there exists a vector $\mathbf{x}$ such that $X = \mathbf{x}\mathbf{x}^\top$. So the phase retrieval problem can be recast as a matrix recovery problem: Find $X \in \mathcal{M}_1$ satisfying the linear measurements $\text{tr}(A_j X) = b_j, j = 1, \ldots, m$, where $\mathcal{M}_r = \{X \in \mathbb{R}^{n \times n} : \text{rank}(X) = r\}$ is the manifold of all matrices of rank r and $r > 0$ is a fixed integer. Mathematically, we aim to solve the low rank matrix recovery problem

$$\min \ \text{rank}(X) \quad \text{subject to} \quad \text{tr}(A_j X) = b_j, \ j = 1, \ldots, m, X \in \mathcal{M}_1, \quad \text{and} \quad X \succeq 0. \tag{11.9}$$

As we will show in Theorem 11.3 below, for given $b_j \geq 0$, $j = 1, \ldots, m$, there may not exist a matrix $X \in \mathcal{M}_1$ satisfying the constraint conditions exactly unless b_j are exactly the measurement values from such a matrix X. Thus, to find the solution X, we reformulate the above problem as

$$\min \ \sum_{j=1}^m |\text{tr}(A_j X) - b_j|^2 \quad \text{subject to} \quad X \in \mathcal{M}_r \quad \text{and} \quad X \succeq 0. \tag{11.10}$$

Since $\mathcal{M}_r$ is a closed set, the above least squares problem will have a bounded solution if the following coercive condition holds:

$$\sum_{j=1}^m |\text{tr}(A_j X) - b_j|^2 \to \infty \quad \text{when} \ \|X\|_F \to \infty. \tag{11.11}$$

In the case that the above coercive condition does not hold, we must use other conditions to ensure that the minimizer of (11.10) is bounded. For example, if there is a matrix X_0 which is orthogonal to A_j in the sense that $\text{tr}(A_j X_0) = 0$ for all $j = 1, \ldots, m$, then the coercive condition will not hold since we can let $X = \ell X_0$ with $\ell \to \infty$.

We are now ready to discuss the existence of solutions for the phase retrieval problem. Recall that $\mathcal{M}_r$ is the set of $n \times n$ matrices with rank r, and let $\overline{\mathcal{M}}_r$ be the set of all matrices with rank no more than r. It is known that the dimension of $\mathcal{M}_r$ is $2nr - r^2$ (cf. Proposition 12.2 in [20]). Since $\overline{\mathcal{M}}_r$ is the closure of $\mathcal{M}_r$ in the Zariski topology (discussed in Chapter 9, or see [36]), the dimension of $\overline{\mathcal{M}}_r$ is also $2nr - r^2$. For convenience, let us recall some basic concepts in algebraic geometry. Let $\mathbb{F}$ be a field (specifically $\mathbb{F} = \mathbb{R}$ or $\mathbb{F} = \mathbb{C}$). A subset of $\mathbb{F}^d$ defined by the vanishing of finitely many polynomials in $\mathbb{F}[x_1, \ldots, x_d]$ is called an affine variety. If these polynomials are homogeneous, then their vanishing defines a subset of the projective space $\mathbb{P}(\mathbb{F}^d)$, which is called a projective variety. The Zariski topology on $\mathbb{F}^d$ (or $\mathbb{P}(\mathbb{F}^d)$) is defined by declaring affine (or projective) varieties to be closed subsets. Note that a Zariski closed set is also closed in the Euclidean topology. The complement of a variety is a Zariski open set. A nonempty Zariski open set is open and dense in the Euclidean topology. We say that a generic point of $\mathbb{F}^d$ (or $\mathbb{P}(\mathbb{F}^d)$) has a certain property if there is a nonempty Zariski open set of points having this property.

Furthermore, it is clear that $\overline{\mathcal{M}_r}$ is an algebraic variety. In fact, $\overline{\mathcal{M}_r}$ is an irreducible variety, which is a standard result in algebraic geometry.

Lemma 11.1. $\overline{\mathcal{M}_r}$ *is an irreducible variety.*

Proof. We refer the reader to the proof of Lemma 9.6 in Chapter 9. $\quad\square$

Define a map
$$\mathcal{A} : \mathcal{M}_1 \to \mathbb{R}^m$$
by projecting any matrix $X \in \mathcal{M}_1$ to $(b_1, \ldots, b_m)^\top \in \mathbb{R}^m$ in the sense that
$$\mathcal{A}(X) = (\mathrm{tr}(A_1 X), \ldots, \mathrm{tr}(A_m X))^\top.$$

Given the map $\mathcal{A}$, we define the range $\mathcal{R}_+ = \{\mathcal{A}(X) : X \in \overline{\mathcal{M}_1}, X \succeq 0\}$ and the range $\mathcal{R} = \{\mathcal{A}(X) : X \in \overline{\mathcal{M}_1}\}$. It is clear that the dimension of $\mathcal{R}_+$ is less than or equal to the dimension of $\mathcal{R}$. Since each entry $\mathrm{tr}(A_j X)$ of the map $\mathcal{A}$ is a linear polynomial on the entries of X, the map $\mathcal{A}$ is regular. We expect that $\dim(\mathcal{R})$ is less than or equal to the dimension of $\mathcal{M}_1$, which is equal to $2n - 1$. If $m > 2n - 1$, then $\mathcal{R}$ cannot occupy the whole space $\mathbb{R}^m$. The Lebesgue measure of the range $\mathcal{R}$ is zero, and hence a randomly chosen vector $b = (b_1, \ldots, b_m)^\top \in \mathbb{R}^m$, e.g., $b \in \mathbb{R}_+^m$, will not be in $\mathcal{R}$ with probability one, and hence not in $\mathcal{R}_+$. Thus, there will not be a solution $X \in \mathcal{M}_1$ such that $\mathcal{A}(X) = b$.

Certainly, this intuition should be made more precise. To this end, we first recall the following result from Theorem 1.25 in section 6.3 of [32].

Lemma 11.2. *Let* $f : X \to Y$ *be a regular map and* X, Y *be irreducible varieties with* $\dim(X) = n$ *and* $\dim(Y) = m$. *If* f *is surjective, then* $m \leq n$. *Furthermore, it holds that*

(a) *for any* $y \in Y$ *and for any component* F *of the fiber* $f^{-1}(y)$, $\dim(F) \geq n - m$, *and*

(b) *there exists a nonempty open subset* $U \subset Y$ *such that* $\dim(f^{-1}(y)) = n - m$ *for* $y \in U$.

We are now ready to prove the following result.

Theorem 11.3 (Huang, Lai, Varghese, and Xu, 2020 [21]). *If we randomly choose a vector* $b = (b_1, \ldots, b_m)^\top \in \mathbb{R}_+^m$ *with* $m > 2n - 1$, *the probability of finding a solution* X *satisfying the minimization (11.9) is zero. In other words, for almost all vectors* $b = (b_1, \ldots, b_m)^\top \in \mathbb{R}_+^m$, *the solution of (11.9) is a matrix with rank greater than or equal to 2.*

Proof. Let $X = \overline{\mathcal{M}_1}$ and $Y = \{\mathcal{A}(M), M \in \overline{\mathcal{M}_1}\}$. From Lemma 11.1, we know that X is an irreducible variety. Since Y is the continuous image of the irreducible variety $\overline{\mathcal{M}_1}$, it is also an irreducible variety. Note that $\mathcal{A}$ is a regular map. By Lemma 11.2, we have $\dim(Y) \leq \dim(\overline{\mathcal{M}_1}) = 2n - 1 < m$. Thus, Y is a proper lower-dimensional closed subset in $\mathbb{R}^m$. Almost all points in $\mathbb{R}^m$ do not belong to Y. In other words, for almost all points $b = (b_1, \ldots, b_n) \in \mathbb{R}^m$, there is no matrix $M \in \overline{\mathcal{M}_1}$ such that $\mathcal{A}(M) = b$, and hence no matrix $M \in \overline{\mathcal{M}_1}$ with $M \succeq 0$ such that $\mathcal{A}(M) = b$. $\quad\square$

Next we define the subset $\chi_b \subset \overline{\mathcal{M}_1}$ by
$$\chi_b = \left\{M \in \overline{\mathcal{M}_1} : \mathcal{A}(M) = b \text{ and } \mathcal{A}^{-1}(\mathcal{A}(M)) \text{ is zero-dimensional}\right\}.$$

Here, a set S is said to be zero-dimensional if the dimension of real points is zero for $S \subset \mathbb{R}^n$ or the dimension of complex points is zero for $S \subset \mathbb{C}^n$. Because we are dealing with fields like $\mathbb{R}$

or $\mathbb{C}$, if the fiber is zero-dimensional, then it has only a finite number of real or complex points. For those $\mathbf{b} \in \mathbb{R}_+^m$ with $\chi_{\mathbf{b}} \neq \emptyset$, we are interested in the upper bound of the number of solutions which satisfy the minimization (11.9). To determine this bound, we need more results from algebraic geometry. Mainly, we need Proposition 11.12 in [20], which was used in Chapter 9. For convenience, let us recall it here.

Lemma 11.4 (Proposition 11.12 in [20]). *Let X be a quasi-projective variety and $\pi : X \to \mathbb{R}^m$ be a regular map; let Y be the closure of the image. For any $p \in X$, let $X_p = \pi^{-1}\pi(p) \subseteq X$ be the fiber of π through p, and let $\mu(p) = \dim_p(X_p)$ be the local dimension of X_p at p. Then $\mu(p)$ is an upper-semicontinuous function of p in the Zariski topology on X, i.e., for any m the locus of points $p \in X$ such that $\dim_p(X_p) > m$ is closed in X. Moreover, if $X_0 \subseteq X$ is any irreducible component, $Y_0 \subseteq Y$ is the closure of its image, and μ is the minimum value of $\mu(p)$ on X_0, then*

$$\dim(X_0) = \dim(Y_0) + \mu. \tag{11.12}$$

As we have shown in the proof of Theorem 11.3, we have $\dim(\mathcal{R}) \le \dim(\overline{\mathcal{M}_r})$. Next we give a more precise characterization of these dimensions.

Lemma 11.5. *Assume $m > \dim(\overline{\mathcal{M}_r})$. Then $\dim(\overline{\mathcal{M}_r}) = \dim(\mathcal{R})$ if and only if $\chi_{\mathbf{b}} \neq \emptyset$ for some $\mathbf{b} \in \mathcal{R}$.*

Proof. Assume $\dim(\overline{\mathcal{M}_r}) = \dim(\mathcal{R})$. From Lemma 11.2, there exists a nonempty open subset $U \subset \mathcal{R}$ such that $\dim(\mathcal{A}^{-1}(\mathbf{b})) = 0$ for all $\mathbf{b} \in U$. This implies that $\chi_{\mathbf{b}}$ has finitely many points. Hence $\chi_{\mathbf{b}} \neq \emptyset$.

We now prove the converse. Assume $\chi_{\mathbf{b}} \neq \emptyset$. We will apply Lemma 11.4 by setting $X = \overline{\mathcal{M}_r}$, $Y = \mathcal{A}(\overline{\mathcal{M}_r})$, and $\pi = \mathcal{A}$. (To apply this lemma, note that it does not matter whether we take the closure in $\mathbb{P}^m$ or in $\mathbb{C}^m$ since $\mathbb{C}^m$ is an open set in $\mathbb{P}^m$ and the Zariski topology of the affine space $\mathbb{C}^m$ is induced from the Zariski topology of $\mathbb{P}^m$. $\overline{\mathcal{M}_r}$ is an affine variety. In particular, it is a quasi-projective variety.)

By our assumption, $\chi_{\mathbf{b}}$ is not empty. It follows that there is a point $p \in Y$ such that $\pi^{-1}(p)$ is zero-dimensional. Since zero is the least dimension possible, we have $\mu = 0$. Hence, using (11.12) above, we have $\dim(\overline{\mathcal{M}_1}) = \dim(\mathcal{R})$. $\square$

To conclude this section, we need the following definition.

Definition 11.6. *The* degree *of an affine or projective variety with dimension k is the number of intersection points with k hyperplanes in the general position.*

It has been shown [17, Example 14.4.11] that the degree of the algebraic variety $\overline{\mathcal{M}_r}$ is

$$\prod_{i=0}^{n-r-1} \frac{\binom{n+i}{r}}{\binom{r+i}{r}}.$$

In particular, the degree of $\mathcal{M}_1$ is

$$\prod_{i=0}^{n-2} \frac{n+i}{1+i} = \binom{2n-2}{n-1}. \tag{11.13}$$

We are now ready to prove another main result in this section.

Theorem 11.7. *Suppose a vector* $\mathbf{b} \in \mathbb{R}_+^m$ *lies in the range* $\mathcal{R}_+$. *Assume that* $\chi_{\mathbf{b}} \neq \emptyset$. *Then the number of distinct solutions in* $\chi_{\mathbf{b}}$ *is less than or equal to* $\binom{2n-2}{n-1}$.

Proof. For any fixed $\mathbf{b}$, the matrices M which satisfy $\mathcal{A}(M) = \mathbf{b}$ and $\mathrm{rank}(M) = 1$ are exactly the intersection points of the variety $\overline{\mathcal{M}_1}$ with m hyperplanes, namely, the hyperplanes defined by the equations $\langle A_i, M \rangle = b_i, i = 1, \ldots, m$. Since $m > \dim(\overline{\mathcal{M}_r}) = 2n - 1$, the number of intersection points would be less than the degree of $\overline{\mathcal{M}_1}$ generically. So the number of positive semidefinite matrices M which satisfy $\mathcal{A}(M) = \mathbf{b}$ and $\mathrm{rank}(M) = 1$ would be no more than the degree of $\overline{\mathcal{M}_1}$. Finally, using the exact formula for the degree from (11.13), the result follows. □

11.2 ▪ Number of Measurement Vectors

In this section, we assume that the measurement values are indeed obtained from a fixed unknown signal in $\mathbb{R}^n$ or in $\mathbb{C}^n$ and study the minimal number of measurements to recover the signal. In the real variable setting, $m = 2n-1$ measurements will be enough to recover the unknown signal up to a sign $+$ or $-$. However, in the complex setting, the minimal number m is not completely known. One theory requires $m \geq 4n - 4$. For $n = 4$, $m = 11 < 4n - 4$ will be enough. Certainly, not just any $4n - 4$ vectors can work. They must not be a Zariski closed set. Finding such a set is not easy. Another theory provides a design of $m = 3n - 2$ measurement vectors to recover almost all nonzero unknown signals $\mathbf{z}$ up to a unimodular constant c with $|c| = 1$. We shall explain these results in the next three subsections.

11.2.1 ▪ Uniqueness of Phase Retrieval in the Real Variable Setting

Let $\mathbf{a}_i \in \mathbb{R}^n$ or $\mathbb{C}^n$ for $i = 1, \ldots, m$ be given measurement vectors. For convenience, we let $A_m = \{\mathbf{a}_1, \ldots, \mathbf{a}_m\}$. We say $\mathbf{a}_i, i = 1, \ldots, m$ with $m \geq n$, are generic in the sense of linear independence if any n vectors chosen from $\mathbf{a}_i, i = 1, \ldots, m$, are linearly independent. In the literature, such a set of points is also said to be in the general position. Also, letting $\Phi_m = [\mathbf{a}_1, \ldots, \mathbf{a}_m]$ be a matrix of size $n \times m$, the vectors $\mathbf{a}_1, \ldots, \mathbf{a}_m$ in A_m are generic if and only if the matrix Φ_m is of completely full rank. Typically, a Vandermonde matrix is of completely full rank: For any set of distinct real numbers $x_1, \ldots, x_m$, let

$$\Phi_m = \begin{bmatrix} 1 & 1 & \cdots & 1 \\ x_1 & x_2 & \cdots & x_m \\ x_1^2 & x_2^2 & \cdots & x_m^2 \\ \vdots & \vdots & \ddots & \vdots \\ x_1^{n-1} & x_2^{n-1} & \cdots & x_m^{n-1} \end{bmatrix}. \tag{11.14}$$

Letting $\mathbf{a}_i = [1, x_i, x_i^2, \ldots, x_i^{n-1}]^\top, i = 1, \ldots, m$, these vectors are generic.

Theorem 11.8. *Suppose that* $\mathbf{a}_i, i = 1, \ldots, m$, *are generic and* $m \geq 2n - 1$. *Given measurements* $y_i = |\langle \mathbf{a}_i, \mathbf{x} \rangle|, i = 1, \ldots, m$, *of an unknown vector* $\mathbf{x} \in \mathbb{R}^n$, *there exists a unique solution* $\mathbf{x}$ *up to a sign* $\pm$ *such that* $|\langle \mathbf{a}_i, \mathbf{x} \rangle| = y_i$ *for all* $i = 1, \ldots, m$.

Proof. Suppose there are two vectors $\mathbf{x}, \mathbf{z} \in \mathbb{R}^n$ such that

$$|\langle \mathbf{a}_i, \mathbf{x} \rangle| = |\langle \mathbf{a}_i, \mathbf{z} \rangle|, \quad i = 1, \ldots, m. \tag{11.15}$$

Then there exist signs $\theta_i = \pm$ such that

$$\langle \mathbf{a}_i, \mathbf{x} \rangle = \theta_i \langle \mathbf{a}_i, \mathbf{z} \rangle, \quad i = 1, \ldots, m. \tag{11.16}$$

Define two sets as follows:

$$\Lambda_+ = \{i, \theta_i = 1\} \quad \text{and} \quad \Lambda_- = \{i, \theta_i = -1\}.$$

Since $m \geq 2n - 1$, either the cardinality of Λ_+ is greater than or equal to n or the cardinality of Λ_- is greater than or equal to n. Without loss of generality, we may assume that the cardinality $|\Lambda_+| \geq n$. From (11.16), we have

$$\langle \mathbf{x} - \mathbf{z}, \mathbf{a}_i \rangle = 0, \quad i \in \Lambda_+.$$

Due to the linear independence of $\mathbf{a}_i, i \in \Lambda_+$, we conclude that $\mathbf{x} - \mathbf{z} = 0$. $\quad\square$

For convenience, for the measurement vector set A_m, let $\mathcal{A}_m(\mathbf{x}) = [|\langle \mathbf{x}, \mathbf{a}_1 \rangle|, \ldots, |\langle \mathbf{x}, \mathbf{a}_m \rangle|]^\top \in \mathbb{R}_+^m$ be a map from $\mathbb{R}^n$ to $\mathbb{R}_+^m$. Clearly, $\mathcal{A}(-\mathbf{x}) = \mathcal{A}(\mathbf{x})$. Let $\mathbb{R}_\pm^n$ be the vector space of all $\{\{\pm\mathbf{x}\}, \mathbf{x} \in \mathbb{R}^n\}$. We say that $\mathcal{A}_m$ is an injective map if $\mathcal{A}_m(\mathbf{x}) = \mathcal{A}_m(\mathbf{y})$ implies $\mathbf{x} = \pm\mathbf{y}$, i.e., either $\mathbf{x} = \mathbf{y}$ or $\mathbf{x} = -\mathbf{y}$. Next we present a characterization of the injectivity of $\mathcal{A}_m$. We start with the following complement property (cf. [4]).

Definition 11.9. *We say $A_m = \{\mathbf{a}_i, i = 1, \ldots, m\} \subset R^n$ satisfies the complement property if for every $S \subset A$ either S or S^c spans $\mathbb{R}^n$.*

Theorem 11.10 (Bandeira, Cahill, Mixon, and Nelson, 2013 [7]). *In the real setting, $\mathcal{A}_m$ is injective if and only if A_m has the complement property.*

Proof. Suppose that $\mathcal{A}_m$ does not have the complement property. There exists a partition $S \subset A_m$ such that neither S nor S^c spans $\mathbb{R}^n$. That is, there exist nonzero vectors $\mathbf{u}, \mathbf{v} \in \mathbb{R}^n$ such that $\langle \mathbf{u}, \mathbf{a}_i \rangle = 0$ for all $\mathbf{a}_i \in S$ and $\langle \mathbf{v}, \mathbf{a}_j \rangle = 0$ for all $\mathbf{a}_j \in S^c$. Then $|\langle \mathbf{u} \pm \mathbf{v}, \mathbf{a}_i \rangle|^2 = |\langle \mathbf{u}, \mathbf{a}_i \rangle|^2 + |\langle \mathbf{v}, \mathbf{a}_i \rangle|^2$ for all $i = 1, \ldots, m$. That is, $|\langle \mathbf{u}+\mathbf{v}, \mathbf{a}_i \rangle|^2 = |\langle \mathbf{u}-\mathbf{v}, \mathbf{a}_i \rangle|^2$ or $\mathcal{A}_m(\mathbf{u}+\mathbf{v}) = \mathcal{A}_m(\mathbf{u} - \mathbf{v})$. Since $\mathbf{u}, \mathbf{v}$ are not zero, $\mathbf{u} + \mathbf{v} \neq \pm(\mathbf{u} - \mathbf{v})$. That is, $\mathcal{A}_m$ is not injective.

On the other hand, if $\mathcal{A}_m$ is not injective, there exist two nonzero vectors $\mathbf{x}, \mathbf{y} \in \mathbb{R}^n$ such that $\mathbf{x} \neq \pm\mathbf{y}$ while $\mathcal{A}_m(\mathbf{x}) = \mathcal{A}_m(\mathbf{y})$. Let

$$S = \{\mathbf{a}_i, \langle \mathbf{x}, \mathbf{a}_i \rangle = \langle \mathbf{y}, \mathbf{a}_i \rangle\} \subset A$$

and

$$S^c = \{\mathbf{a}_i, \langle \mathbf{x}, \mathbf{a}_i \rangle = -\langle \mathbf{y}, \mathbf{a}_i \rangle\} \subset A.$$

Then $\mathbf{u} = \mathbf{x} - \mathbf{y} \neq 0$ is orthogonal to $\text{span}(S)$ and $\mathbf{v} = \mathbf{x} + \mathbf{y} \neq 0$ is orthogonal to $\text{span}(S^c)$. Hence, neither S nor S^c spans $\mathbb{R}^n$. $\quad\square$

It is easy to see that when $A_m = \{\mathbf{a}_i, i = 1, \ldots, m\}$ is generic with $m \geq 2n - 1$, A_m has the complement property. However, when A has the complement property with $m \geq 2n - 1$, A_m may not be generic. Indeed, taking A_0 to be a set of m measurement vectors which is generic with $m \geq 2n - 1$ and letting $A_m = \{A_0, \mathbf{a}_1 + \mathbf{a}_2\}$, it is easy to see that A_m is not generic when $n \geq 2$. But this A_m has a complement property. Indeed, splitting A_m into two sets $A_m = S_1 \cup S_2$ with the cardinality $\#(S_1)$ greater than or equal to the cardinality $\#(S_2)$, we know that $\#(S_1) \geq n$. If $\#(S_1) \geq n+1$, we know that $\text{span}\{S_1\}$ will be $\mathbb{R}^n$ even if S_1 contains $\mathbf{a}_1 + \mathbf{a}_2$. If $\#(S_1) = n$, and if S_1 does not contain $\mathbf{a}_1 + \mathbf{a}_2$, S_1 spans $\mathbb{R}^n$. If $\#(S_1) = n$, but S_1 contains $\mathbf{a}_1 + \mathbf{a}_2$, then $\#(S_2) \geq n$ and S_2 spans $\mathbb{R}^n$. Hence, the generic property is stronger than the complement property. The set with the complement property may contain redundant vectors, such as $\mathbf{a}_1 + \mathbf{a}_2$ above.

11.2.2 ▪ Number of Measurement Vectors in the Complex Setting

Let us continue the study of the minimal number of measurements. For given measurement vectors $\mathbf{a}_i \in \mathbb{C}^n$ for $i = 1, \dots, m$ with $m \geq n$, recall the mapping

$$\mathcal{A}_m(\mathbf{z}) = [|\langle \mathbf{z}, \mathbf{a}_1 \rangle|, |\langle \mathbf{z}, \mathbf{a}_2 \rangle|, \dots, |\langle \mathbf{z}, \mathbf{a}_m \rangle|]^\top \in \mathbb{R}_+^m \qquad (11.17)$$

from $\mathbb{C}^n$ to the cone $\mathbb{R}_+^m$. We say that $\mathcal{A}_m$ is injective if for any two vectors $\mathbf{z}, \mathbf{y} \in \mathbb{C}^n$ with $\mathcal{A}_m(\mathbf{z}) = \mathcal{A}_m(\mathbf{y})$, $\mathbf{z} = c\mathbf{y}$ for a unimodular constant c. We begin with the following theorem.

Theorem 11.11. *Let $A_m = \{\mathbf{a}_1, \dots, \mathbf{a}_m\}$ be a given set of measurement vectors. If $\mathcal{A}_m$ is injective, then A_m satisfies the complement property.*

Proof. A proof can be given by showing that if A_m does not satisfy the complement property then A_m cannot be injective. We leave the details to Exercise 1. □

However, the complement property does not imply the injectivity property, as pointed out in [7]. The following characterization of the injectivity of $\mathcal{A}_m$ can be given.

Theorem 11.12 (Bandeira, Cahill, Mixon, and Nelson, 2013 [7]). *The map $\mathcal{A}_m$ is not injective if and only if there is a nonzero Hermitian matrix Q of size $n \times n$ with entries in $\mathbb{C}$ for which*

$$\mathrm{rank}(Q) \leq 2 \quad and \quad \mathbf{a}_i^* Q \mathbf{a}_i = 0 \quad \forall i = 1, \dots, m. \qquad (11.18)$$

Proof. If $\mathcal{A}_m$ is not injective, then there exist $\mathbf{x}, \mathbf{y} \in \mathbb{C}$ with $\mathbf{x} \neq \mathbf{y} \bmod \mathbb{T}$ such that $\mathcal{A}_m(\mathbf{x}) = \mathcal{A}_m(\mathbf{y})$, where $\mathbb{T} = \{e^{i\theta}, \theta \in [0, 2\pi]\}$. That is, we have $\mathbf{a}_i^* \mathbf{x} \mathbf{x}^* \mathbf{a}_i = \mathbf{a}_i^* \mathbf{y} \mathbf{y}^* \mathbf{a}_i$, and hence $Q = \mathbf{x}\mathbf{x}* -\mathbf{y}\mathbf{y}*$ is a Hermitian matrix with rank ≤ 2 satisfying the condition in (11.18).

On the other hand, if there is a matrix Q of rank 1 satisfying (11.18), we have $Q = \mathbf{x}\mathbf{x}^*$ for some nonzero vector $\mathbf{x} \in \mathbb{C}^n$, and hence $\mathcal{A}_m(\mathbf{x}) = 0$ by using (11.18). Clearly, the zero vector satisfies $\mathcal{A}_m(0) = 0 = \mathcal{A}_m(\mathbf{x})$. That is, $\mathcal{A}_m$ is not injective. If Q is of rank 2, then by the spectral theorem, there are orthonormal vectors $\mathbf{u}_1, \mathbf{u}_2 \in \mathbb{C}^n$ and nonzero eigenvalues $\lambda_1 \geq \lambda_2$ such that $Q = \lambda_1 \mathbf{u}_1 \mathbf{u}_1^* + \lambda_2 \mathbf{u}_2 \mathbf{u}_2^*$. The second condition in (11.18) yields

$$0 = \mathbf{a}_i^* Q \mathbf{a}_i = \lambda_1 |\langle \mathbf{u}_1, \mathbf{a}_i \rangle|^2 + \lambda_2 |\langle \mathbf{u}_2, \mathbf{a}_i \rangle|^2, \ i = 1, \dots, m. \qquad (11.19)$$

If λ_1 and λ_2 have the same sign, then $|\langle \mathbf{u}_1, \mathbf{a}_i \rangle|^2 = 0 = |\langle \mathbf{u}_2, \mathbf{a}_i \rangle|^2$ for all $i = 1, \dots, m$. Since $\mathbf{u}_1 \neq \mathbf{u}_2 \bmod \mathbb{T}$, $\mathcal{A}_m$ is not injective. If $\lambda_1 > 0$ and $\lambda_2 < 0$, we choose $\mathbf{x} = \sqrt{\lambda_1} \mathbf{u}_1$ and $\mathbf{y} = \sqrt{\lambda_2} \mathbf{u}_2$. Equation (11.19) implies that $\mathcal{A}_m(\mathbf{x}) = \mathcal{A}_m(\mathbf{y})$, while $\mathbf{x} \neq \mathbf{y} \bmod \mathbb{T}$. So $\mathcal{A}_m$ is not injective. □

Writing $\mathbf{a}_j = \mathbf{u}_j + i\mathbf{v}_j$ for $\mathbf{u}_j, \mathbf{v}_j \in \mathbb{R}^n$ and $Q = X + iY$ with $X, Y \in \mathbb{R}^{n \times n}$, the conditions in (11.18) can be rewritten in terms of $U = [\mathbf{u}_1, \dots, \mathbf{u}_m]$, $V = [\mathbf{v}_1, \dots, \mathbf{v}_m]$, X, and Y as

$$\mathrm{rank}(X + iY) \leq 2 \quad and \quad \mathbf{u}_j^\top X \mathbf{u}_j + \mathbf{v}_j^\top X \mathbf{v}_j - 2\mathbf{u}_j^\top Y \mathbf{v}_j = 0 \quad \forall j = 1, \dots, m, \quad (11.20)$$

where X is a symmetric matrix and Y is a skew-symmetric matrix. Let us define a set $B_{m,n}$ to be all quadruples of matrices (U, V, X, Y) which satisfy (11.20). We note quickly here that $[X, Y]$ and $[U, V]$ are homogeneous. In other words, they are invariant under scaling U and V by a nonzero scalar, and also X and Y by a nonzero scalar. Let $\mathbb{R}_{symm}$ be the space of all symmetric matrices of size $n \times n$ and $\mathbb{R}_{skew}$ the space of all skew-symmetric matrices of size $n \times n$. In addition, let $\mathbb{R}_{n \times m}$ be the space of all matrices of size $n \times m$. We define a projection $\mathcal{P}_{m,n}$ which maps any point

$$(\mathbb{R}_{n \times m} \times \mathbb{R}_{n \times m}) \times (\mathbb{R}_{symm} \times \mathbb{R}_{skew})$$

into $(\mathbb{R}_{n\times m} \times \mathbb{R}_{n\times m})$. Let us consider a subset of the above space:

$$\mathcal{J}_{m,n} = \{(U, V, X, Y) \in (\mathbb{R}_{n\times m} \times \mathbb{R}_{n\times m}) \times (\mathbb{R}_{symm} \times \mathbb{R}_{skew}),$$
$$X + iY \neq 0, \operatorname{rank}(X + iY) \leq 2,$$
$$\mathbf{u}_j^\top X \mathbf{u}_j + \mathbf{v}_j^\top X \mathbf{v}_j - 2\mathbf{u}_j^\top Y \mathbf{v}_j = 0 \quad \forall j = 1, \ldots, m\},$$

and $\mathcal{I} = \{(U, V) = \mathcal{P}_{m,n}(U, V, X, Y) \; \forall (U, V, X, Y) \in \mathcal{J}_{m,n}\}$. In terms of this new notation, the statement of Theorem 11.12 can be given as follows.

Theorem 11.13. *Fix a set of measurement vectors, and let U_m and V_m be the two matrices defined by the measurement vectors. Then the map $\mathcal{A}_m$ is injective if and only if $[U_m, V_m]$ does not belong to the image $\mathcal{I}$ of the projection $\mathcal{P}_{m,n}$ of the set $\mathcal{J}_{m,n}$.*

Proof. We leave the proof to the interested reader; see also [13]. □

Clearly, the dimension of $(\mathbb{R}_{n\times m} \times \mathbb{R}_{n\times m})$ is $2mn$. Because of the homogeneity, the space where the image $\mathcal{I}$ belongs has dimension $2mn - 1$. In order to have a nontrivial solution $[U_m, V_m]$, we hope that the dimension of the image $\mathcal{I}$ is strictly less than $2mn - 1$. That is, the dimension of $\mathcal{I}$ should be less than or equal to $2mn - 2$. Because the projection $\mathcal{P}_{m,n}$ will not increase the dimension, we want the dimension of the preimage of $\mathcal{P}_{m,n}$, i.e., $\mathcal{J}_{m,n}$, to be less than or equal to $2mn - 2$. Next let us compute the dimension of the set $\mathcal{J}_{m,n}$.

Theorem 11.14. *The dimension of the set $\mathcal{J}_{m,n}$ is less than or equal to $2mn - m + 4n - 6$.*

Proof. First we note that the set of matrices of rank 2 is an irreducible (affine) variety of dimension $4n - 4$. Because of the homogeneity of matrices $Q = X + iY$ in the set $\mathcal{J}_{m,n}$, the dimension should be $4n - 4 - 1 = 4n - 5$. For each $Q \in \mathcal{J}_{m,n}$, Q defines the nonzero polynomial equation

$$(\mathbf{u}_j - i\mathbf{v}_j)^\top Q(\mathbf{u}_j + i\mathbf{v}_j) = 0$$

on columns of U and V. Clearly, for each nonzero matrix Q, the polynomial on the left-hand side above is not identically zero. Indeed, let

$$q(\mathbf{u}, \mathbf{v}) = (\mathbf{u} - i\mathbf{v})^\top Q(\mathbf{u} + i\mathbf{v}). \tag{11.21}$$

Then q will not be equivalent to zero. We leave this to Exercise 3.

For each pair of columns $(\mathbf{u}_j, \mathbf{v}_j)$, this polynomial defines a hypersurface of dimension $2n - 1$ in $(\mathbb{R}^n)^2$. There are m copies of this hypersurface, and hence the dimension of the set in $\mathcal{J}$ with a fixed Q is less than or equal to $m(2n - 1)$. Again because of the homogeneity, the dimension of the set is less than or equal to $m(2n - 1) - 1$. Hence the dimension of the set $\mathcal{J}_{m,n}$ is less than or equal to $4n - 5 + m(2n - 1) - 1 = 2nm - m + 4n - 6$. □

To have $2nm - m + 4n - 6 \leq 2nm - 2$ it follows that $m \geq 4n - 4$. The discussion above forms a proof of the following result.

Theorem 11.15 (Conca, Edidin, Hering, and Vinzant, 2015 [13]). *If $m = 4n - 4$, then for a generic set $A_m = [\mathbf{a}_1, \ldots, \mathbf{a}_m]$ in the sense that $\mathbf{a}_i = \mathbf{u}_i + i\mathbf{v}_i, i = 1, \ldots, m$, with $U = [\mathbf{u}_1, \ldots, \mathbf{u}_m], V = [\mathbf{v}_1, \ldots, \mathbf{v}_m]$, and $[U, V]$ in the Zariski open set of the real vector space $\mathbb{R}_{n\times m} \times \mathbb{R}_{n\times m}$, the map $\mathcal{A}_m$ is injective.*

Proof. In the real vector space $(\mathbb{R}_{n\times m} \times \mathbb{R}_{n\times m})$, there is some nonzero polynomial that vanishes on all of the pairs (U, V) for which $\mathcal{A}_{U+iV} := \mathcal{A}_m$, with measurement vectors $\mathbf{a}_i = \mathbf{u}_i + i\mathbf{v}_i$

for $i = 1, \ldots, m$, is not injective. The complement of the zero set of this polynomial is a Zariski open subset of $(\mathbb{R}_{n \times m} \times \mathbb{R}_{n \times m})$, and for any pair (U, V) in this open set, $\mathcal{A}_{U + iV}$ is injective. Such a pair of vectors forms a set of measurement vectors which enable us to recover the phase of each signal uniquely up to a unimodular constant. $\quad \square$

Clearly, finding such a set of measurement vectors which form a generic set in the above Zariski open subset is nontrivial. A randomly chosen set of m ($\gg 4n - 4$) vectors may have a very good chance to have at least $4n - 4$ vectors in the Zariski open set such that $\mathcal{A}_m$ is injective.

11.2.3 ▪ A Design for Measurement Vectors

In addition to randomly selecting measurement vectors to retrieve the phase of a signal, which is very convenient for engineers to do, it is better to have a smart choice of measurement vectors so that we can find the phase more easily than by using random choice. In this subsection, we provide a good design for measurement vectors to retrieve the phase of a signal from its magnitude. First of all, if we are allowed to choose $n(n + 1)/2$ measurement vectors for the real case and n^2 for the complex case, we can easily retrieve the phase of a signal (see, e.g., [5]). To avoid a lot of measurements, the goal of this subsection is to study how to choose m measurement vectors with $m < n^2/2$. We start with n vectors

$$\mathbf{a}_1 = \mathbf{e}_1, \ldots, \mathbf{a}_n = \mathbf{e}_n,$$

where $\mathbf{e}_i$ is the standard vector in $\mathbb{R}^n$ with all zero entries except for the ith entry, which is 1. Then we choose additional vectors

$$\mathbf{a}_{n+2k-1} = \mathbf{e}_1 + \mathbf{e}_k, \mathbf{a}_{n+2k} = \mathbf{e}_1 + i\mathbf{e}_k$$

for $k = 2, \ldots, n$. We now show that such a choice can find the phase of almost all signals $\mathbf{z}$ from these magnitude measurements and that these vectors form an injective set for almost all $\mathbf{z} \in \mathbb{C}^n$.

Example 11.16. Consider $n = 2$ first. Then the set

$$\mathbf{e}_1, \mathbf{e}_2, \mathbf{e}_1 + \mathbf{e}_2, \mathbf{e}_1 + i\mathbf{e}_2 \tag{11.22}$$

is a generic set in the sense of linear independence. That is, any two members of the set are linearly independent. We now explain how to find the phase of a signal $\mathbf{z}$ from the measurements of these vectors. Indeed, writing $\mathbf{z} = \begin{bmatrix} x_1 + y_1 i \\ x_2 + y_2 i \end{bmatrix}$, we use the measurement vectors above to obtain the following magnitudes:

$$r_1^2 = x_1^2 + y_1^2 = |\langle \mathbf{z}, \mathbf{e}_1 \rangle|^2, \tag{11.23}$$

$$r_2^2 = x_2^2 + y_2^2 = |\langle \mathbf{z}, \mathbf{e}_2 \rangle|^2, \tag{11.24}$$

$$r_{1,1}^2 = (x_1 + x_2)^2 + (y_1 + y_2)^2 = |\langle \mathbf{z}, \mathbf{e}_1 + \mathbf{e}_2 \rangle|^2, \tag{11.25}$$

$$r_{1,2}^2 = (x_1 - y_2)^2 + (y_1 + x_2)^2 = |\langle \mathbf{z}, \mathbf{e}_1 + i\mathbf{e}_2 \rangle|^2. \tag{11.26}$$

It is not easy to solve the above system of quadratic equations directly. Let us introduce a creative way to do it. Without loss of generality, we may assume that $\mathbf{z}_1 \neq 0$. We write $\mathbf{z}_1 = x_1 + y_1 i = r_1 \exp(i\theta_1)$. Then we compute $\hat{\mathbf{z}} := \mathbf{z} \exp(-i\theta_1) = r_1$ instead. For simplicity, we still use $\mathbf{z}$ for $\hat{\mathbf{z}}$. In this case, we have $x_1 = |r_1|$ and $y_1 = 0$. Then we solve the equations in (11.23) to (11.26) to get

$$x_2 = \frac{r_{1,1}^2 - r_1^2 - r_2^2}{2|r_1|} \quad \text{and} \quad y_2 = \frac{r_{1,2}^2 - r_1^2 - r_2^2}{-2|r_1|}. \tag{11.27}$$

In this way, we obtain $\mathbf{z} = \hat{z}e^{\theta_1}$. Next we claim that the set in (11.22) is injective. Otherwise, there is another solution $\mathbf{w}$ which has the same measurement as $\mathbf{z}$, i.e., the same r_1, r_2, r_{11}, r_{12}. We can use the same technique to solve $\hat{\mathbf{w}} = \exp(\mathbf{i}\theta_2)\mathbf{w}$ so that $\hat{\mathbf{w}} = (u_1 + \mathbf{i}0, u_2 + \mathbf{i}v_2)$. Then $u_1 = x_1, u_2 = x_2$, and $v_2 = y_2$. Therefore, $\mathbf{w} = \exp(\mathbf{i}(-\theta_2))\hat{\mathbf{z}}\exp(\mathbf{i}\theta_1)\mathbf{z}$.

We now explain how to find the phase of a signal $\mathbf{z}$ from $\mathbb{C}^n$ for $n > 2$. Let

$$
\mathbf{z} = \begin{bmatrix} x_1 + y_1\mathbf{i} \\ x_2 + y_2\mathbf{i} \\ \vdots \\ x_n + y_n\mathbf{i} \end{bmatrix},
$$

for which we get the following observations:

$$
\begin{aligned}
r_k^2 &= x_k^2 + y_k^2, \\
r_{n+2k-1}^2 &= (x_1 + x_k)^2 + (y_1 + y_k)^2, \\
r_{n+2k}^2 &= (x_1 - y_k)^2 + (y_1 + x_k)^2
\end{aligned}
$$

for $k = 1, 2, \ldots, n$. Without loss of generality, we may assume that $r_1 \neq 0$. Let us use $z_1 = x_1 + y_1\mathbf{i}$ with $x_1 = |r_1|$ and $y_1 = 0$. Then we get the equations

$$
x_k = \frac{r_{n+2k-1}^2 - r_1^2 - r_k^2}{2|r_1|} \quad \text{and} \quad y_k = \frac{r_{n+2k}^2 - r_1^2 - r_k^2}{-2|r_1|}
$$

for $k = 2, \ldots, n$. Therefore, $\mathbf{z} = \mathbf{z}e^{\theta\mathbf{i}}$ for some θ.

This motivates us to prove the following general theorem.

Theorem 11.17 (Lai and Lee, 2020 [24]). *Suppose that $\mathbf{a}_i, i = 1, \ldots, n$, are orthonormal vectors in $\mathbb{C}^n$. In addition, we choose $\mathbf{a}_1 + \mathbf{a}_k, \mathbf{a}_1 + \mathbf{a}_k\mathbf{i}$ for $k = 2, \ldots, n$. Then we can recover all signals z by using these $3n - 2$ measurements generated by the $\mathbf{a}_i$'s, except for those signals z which have $\langle \mathbf{z}, \mathbf{a}_1 \rangle = 0$. Furthermore, for those signals z with $\langle \mathbf{z}, \mathbf{a}_1 \rangle \neq 0$, the signal z is uniquely recovered from the set of measurement vectors up to a factor of c with $|c| = 1$.*

Proof. We expand the signal $\mathbf{z} = \sum_{i=1}^n z_i \mathbf{a}_i$ and determine the coefficient vector $(z_1, \ldots, z_n)$. The proof of the recovery is similar to the discussion before the statement of this theorem. The details are omitted. We now show that the mapping

$$
\mathcal{A}_m(\mathbf{z}) = (|\langle \mathbf{z}, \mathbf{a}_1 \rangle|^2, \ldots, |\langle \mathbf{z}, \mathbf{a}_m \rangle|^2) \tag{11.28}
$$

is injective for all signals $\mathbf{z}$ such that $\langle \mathbf{z}, \mathbf{a}_1 \rangle \neq 0$. Suppose that there are two vectors $\mathbf{z}, \mathbf{y} \in \mathbb{C}^n$ satisfying $\mathcal{A}_m(\mathbf{z}) = \mathcal{A}_m(\mathbf{y})$. By the assumption that $|\langle \mathbf{z}, \mathbf{a}_1 \rangle|^2 \neq 0$, we also have $|\langle \mathbf{y}, \mathbf{a}_1 \rangle|^2 \neq 0$. Using the constructive procedure above, we can completely determine each entry of $\mathbf{z}$. Similarly, we can determine each entry of $\mathbf{y} = (y_1, \ldots, y_n)^\top$ starting from $y_1 = r_1 + \mathbf{i}0$. Note that corresponding entries of $\mathbf{z}$ and $\mathbf{y}$ are the same for all $i = 1, \ldots, n$. Hence, $\mathbf{z}$ and $\mathbf{y}$ are equal up to a constant c with $|c| = 1$. $\square$

To retrieve all signals, we can apply the above-designed vectors in the following way. We can first design n orthogonal vectors $\mathbf{a}_1, \ldots, \mathbf{a}_n$ and measure the unknown vector $\mathbf{z}$ once. If one of the measurements $|\langle \mathbf{z}, \mathbf{a}_i \rangle|, i = 1, \ldots, n$, is not zero, say $|\langle \mathbf{z}, \mathbf{a}_k \rangle| \neq 0$ for some k, then we use the new vectors $\mathbf{a}_k + \mathbf{a}_j, \mathbf{a}_k + \mathbf{i}\mathbf{a}_j, j = 1, 2, \ldots, k-1, k+1, \ldots, n$, to find the measurement values. Then we compute the unknown $\mathbf{z}$ up to a unimodular constant using the method in Theorem 11.17. These lead to a two-stage method of solving the phase retrieval problem.

Let us continue to discuss how to uniquely determine any signal $\mathbf{z}$ up to a unimodular constant. Instead of the two-stage method, we need to have more measurement vectors. For example, in $\mathbb{C}^3$, we can use nine measurement vectors to recover any signal in $\mathbb{C}^3$.

Example 11.18. Let us consider $\mathbb{C}^3$. We shall use an orthonormal basis $\mathbf{a}_1, \mathbf{a}_2, \mathbf{a}_3$ together with $\mathbf{a}_1 + \mathbf{a}_2, \mathbf{a}_1 + \mathbf{a}_3, \mathbf{a}_2 + \mathbf{a}_3$, as well as $\mathbf{a}_1 + i\mathbf{a}_2, \mathbf{a}_1 + i\mathbf{a}_3$, and $\mathbf{a}_2 + i\mathbf{a}_3$, to recover the phase of any signal $\mathbf{z} \in \mathbb{C}^3$. Write $\mathbf{z} = z_1\mathbf{a}_1 + z_2\mathbf{a}_2 + z_3\mathbf{a}_3$. Let us find $(z_1, z_2, z_3) \in \mathbb{C}^3$. Suppose that $z_1 = 0$. We claim that we can recover the remaining component of z_2, z_3 uniquely up to a unimodular constant. Indeed, we can view $\mathbf{z} \in \mathbb{C}^2$ in this case, and we use $\mathbf{a}_2, \mathbf{a}_3, \mathbf{a}_2 + \mathbf{a}_3, \mathbf{a}_2 + i\mathbf{a}_3$ to recover $\mathbf{z}$ by using the method in Example 11.16. Now let us consider the case $z_1 \neq 0$. We again write $\mathbf{z} = x_1 + i0$ with $x_1 = r_1$. Then we find $x_2 = (r_{12}^2 - r_1^2 - r_2^2)/(2r_1)$ and $x_3 = (r_{13}^2 - r_1^2 - r_3^2)/(2r_1)$. Hence,

$$y_2 = \pm\sqrt{r_2^2 - x_2^2} \quad \text{and} \quad y_3 = \pm\sqrt{r_3^2 - x_3^2}. \tag{11.29}$$

If both y_2 and y_3 are zero, we are done. Otherwise, we continue to decide the sign of y_2 and y_3. To do so, we use the three remaining measurements. First, we use the measurement from $\mathbf{a}_2 + \mathbf{a}_3$ to find

$$2y_2y_3 = r_{23}^2 - r_2^2 - r_3^2 - 2x_2x_3. \tag{11.30}$$

The sign of $r_{23}^2 - r_2^2 - r_3^2 - 2x_2x_3$ determines whether y_2 and y_3 have the same sign if $r_{23}^2 - r_2^2 - r_3^2 - 2x_2x_3 \neq 0$. Next, letting $\hat{r}_{12} = |\langle \mathbf{z}, \mathbf{a}_1 + i\mathbf{a}_2\rangle|$, we have

$$-2x_1y_2 = \hat{r}_{12}^2 - r_1^2 - r_2^2 \quad \text{and hence} \quad y_2 = (\hat{r}_{12}^2 - r_1^2 - r_2^2)/(2r_1). \tag{11.31}$$

We then use (11.30) to determine y_3. If $r_{23}^2 - r_2^2 - r_3^2 - 2x_2x_3 = 0$, $y_2y_3 = 0$. If $y_3 = 0$, we can determine y_2 as above. If $y_2 = 0$, we use the measurement $\hat{r}_{13} = |\langle \mathbf{z}, \mathbf{a}_1 + i\mathbf{a}_3\rangle|$ to get

$$-2x_1y_3 = \hat{r}_{13}^2 - r_1^2 - r_3^2, \quad \text{and hence} \quad y_4 = (\hat{r}_{13}^2 - r_1^2 - r_3^2)/(2r_1). \tag{11.32}$$

These completely determine all components of the signal $\mathbf{z}$. Finally, we verify that this set of measurement vectors is injective. If there is another signal $\mathbf{y}$ with the same measurement values, we will compute $\mathbf{y}$ in the same way as above. These lead to all components of signal $\mathbf{y}$ having the same values up to the unimodular factor $c_{\mathbf{z}}$ for $\mathbf{z}$ and $c_{\mathbf{y}}$ for $\mathbf{y}$ when we rotate z_1 and y_1. Therefore, this set of measurement vectors will uniquely determine the signal.

We conclude this section with the remaining question: how many measurement vectors do we need to uniquely determine any signal $\mathbf{z} \in \mathbb{C}^n$?

11.3 ▪ Frame for Phase Retrieval

In this section, we consider phaseless retrieval in the real variable setting with randomly chosen vectors of measurement. We study how to use frames to recover the signal from its measurement values. That is, we shall explain how to recover a signal $\mathbf{x}$ from linear measurements $\langle \mathbf{x}, \mathbf{a}_i \rangle$, $i = 1, \ldots, m$ with $m > n$. Suppose that the measurement vectors $\mathcal{D} = \{\mathbf{a}_1, \ldots, \mathbf{a}_m\}$ span the Euclidean space $\mathbb{R}^n$.

Definition 11.19. *The set $\mathcal{D}$ is a frame if and only if there are two positive constants $0 < A \leq B < \infty$ (called frame bounds) so that*

$$A\|\mathbf{x}\|^2 \leq \frac{1}{m}\sum_{k=1}^{m}|\langle \mathbf{x}, \mathbf{a}_k\rangle|^2 \leq B\|\mathbf{x}\|^2 \quad \forall \mathbf{x} \in \mathbb{R}^n. \tag{11.33}$$

When we can choose $A = B$, the frame is said to be tight. *For $A = B = 1$ the frame is called a* Parseval frame.

Typically, an orthonormal basis $\mathcal{D}$ is a frame. The union of two or more orthogonal bases forms a frame as well. The following is an famous example called the Mercedes-Benz frame.

Example 11.20. Consider $\mathbb{R}^2$ and let $a_1 = (1,0)$, $a_2 = (\cos(120°), \sin(120°))$, and $a_3 = (\cos(240°), \sin(240°))$. Then for any $x \in \mathbb{R}^2$, we have

$$x = \frac{2}{3} \sum_{i=1}^{3} \langle x, a_i \rangle a_i. \tag{11.34}$$

The example above is not a coincidence. The following is a classic result (cf. [14]).

Theorem 11.21. *The significance of frames is that one can recover x using $\langle x, a_k \rangle$, $k = 1, \ldots, m$, if $\mathcal{D} = \{a_1, \ldots, a_m\}$ is a frame.*

Proof. Indeed, we can construct a dual frame $\widetilde{a_k}$, $k = 1, \ldots, m$, so that x can be recovered by using the formula

$$x = \sum_{k=1}^{m} \langle x, a_k \rangle \widetilde{a_k} \quad \forall x \in \mathbb{R}^n. \tag{11.35}$$

For example, when a frame is tight, i.e., $A = B$, the dual frame is simply $\widetilde{a_k} = a_k/A$. We leave the proof to Exercise 24.

In general, consider a frame operator Φ associated with $\mathcal{D}$ defined by $\Phi = [a_1, a_2, \ldots, a_m]^\top / \sqrt{m}$ with the vectors $a_j \in \mathcal{D}$ as its columns. Then the largest A and smallest B in (11.33) are called the lower frame bound and upper frame bound of Φ, and they are given by

$$A = \lambda_{\min}(\Phi^*\Phi) = \sigma_n^2(\Phi), \quad B = \lambda_{\max}(\Phi^*\Phi) = \sigma_1^2(\Phi), \tag{11.36}$$

where $\lambda_{\max}$, $\lambda_{\min}$ denote the largest and smallest eigenvalues, respectively, while σ_1, σ_n denote the first and nth singular values, respectively. We leave the proof of (11.36) to Exercise 23.

Then the dual frame is defined by $\widetilde{a_k} = (\Phi^\top \Phi)^{-1} a_k$. Indeed, it is easy to see that we have

$$\Phi^\top \Phi(x) = \sum_{k=1}^{m} \langle x, a_k \rangle a_k. \tag{11.37}$$

We will have (11.35) if we choose $\widetilde{a_k} = (\Phi^\top \Phi)^{-1} a_k$. □

To invert $\Phi^\top \Phi$ when n is large, we shall use an iterative method. Let $R = I_n - \frac{2}{A+B} \Phi^\top \Phi$, where I_n is the identity matrix. Since $A I_n \leq \Phi^\top \Phi \leq B I_n$, we have $\|R\| \leq \frac{B-A}{B+A} < 1$. Therefore,

$$(\Phi^\top \Phi)^{-1} = \frac{B+A}{2}(I_n - R)^{-1} = \frac{B+A}{2} \sum_{i \geq 0} R^i. \tag{11.38}$$

For simplicity, consider $y_0 = \sum_{k=1}^{m} \langle x, a_k \rangle a_k$ and $y = \sum_{i \geq 0} R^i y_0$. Then y is the limit of $y_\ell = \sum_{i=0}^{\ell} R^i y_0$, which satisfies the recurrence relation

$$y_\ell = y_0 + R y_{\ell-1} \quad \forall \ell \geq 1. \tag{11.39}$$

Using the operator R, it follows that

$$y_\ell = y_0 + y_{\ell-1} - \frac{2}{A+B} \sum_{k \geq 1} \langle y_{\ell-1}, a_k \rangle a_k. \tag{11.40}$$

Because $\|R\| \leq (B - A)/(A + B) < 1$, the iterative procedure converges very quickly. That is, $\mathbf{x} = (\Phi^\top \Phi)^{-1} \mathbf{y}_0 = \frac{A+B}{2} \lim_{\ell \to \infty} \mathbf{y}_\ell$, which can be computed using the above iterative algorithm. The frame gives an efficient way to recover $\mathbf{x}$ from the measurements $\langle \mathbf{x}, \mathbf{a}_k \rangle, k = 1, \dots, m$. We refer the reader to [25] and [19] for more frame properties, the construction of spline frames, and frame application.

Next we discuss how to recover $\mathbf{x}$ from the nonlinear measurements $\{|\langle \mathbf{x}, \mathbf{a}_j \rangle| : 1 \leq j \leq m\}$. Let us recall a concept. A set of vectors $\mathcal{D}$ of the n-dimensional Hilbert space is said to be *completely full rank* if any subset of n vectors is linearly independent. Let us define the equivalence relation $\sim$ on $\mathbb{R}^m$: $\mathbf{x} \sim \mathbf{y}$ if and only if $\mathbf{y} = c\mathbf{x}$ for some unimodular constant c, $|c| = 1$. Since we focus on the real vector space $\mathbb{R}^n$, we have $\mathbf{x} \sim \mathbf{y}$ if and only if $\mathbf{x} = \pm \mathbf{y}$. So we will be looking at reconstruction on $\widetilde{\mathbb{R}^n} = \mathbb{R}^n / \sim$, whose elements are given by equivalent classes $\hat{\mathbf{x}} = \{\mathbf{x}, -\mathbf{x}\}$ for $\mathbf{x} \in \mathbb{R}^n$. The analogous analysis map for phaseless reconstruction is the nonlinear map

$$\alpha_\Phi : \widetilde{\mathbb{R}^n} \to \mathbb{R}^m_+, \quad \alpha_\Phi(\hat{\mathbf{x}}) = [|\langle \mathbf{x}, \mathbf{a}_1 \rangle|, |\langle \mathbf{x}, \mathbf{a}_2 \rangle|, \dots, |\langle \mathbf{x}, \mathbf{a}_m \rangle|]^\top. \tag{11.41}$$

Note that α_Φ can also be viewed as a map from $\mathbb{R}^n$ to $\mathbb{R}^m_+$. It is not easy to find the inverse of α_Φ directly.

Let us study some properties of α_Φ. We say a frame Φ is *phase retrievable* if we can reconstruct $\hat{\mathbf{x}} \in \widetilde{\mathbb{R}^n}$ for all $\hat{\mathbf{x}}$. In other words, α_Φ is injective on $\widetilde{\mathbb{R}^n}$. The main objective of this section is to analyze the robustness and stability of the inversion map and to give performance bounds of any reconstruction algorithm in the real variable setting.

To explain the injectivity of the nonlinear map α_Φ, we need one more concept. In general, a subset Z of a topological space X is said to be *generic* if its open interior is dense in X. However, we shall use *generic* in the Zariski topology: A set $Z \subset X := \underbrace{\mathbb{R}^n \times \cdots \times \mathbb{R}^n}_{m}$ is said to be *generic* if Z is dense in X and its complement is a finite union of zero sets of polynomials in mn variables with real coefficients.

Theorem 11.22 (Balan and Wang, 2015 [6]). *Let Φ be a frame in $\mathbb{R}^n$ with m elements. Then the following hold true:*

1. *The frame Φ is phase retrievable in $\widetilde{\mathbb{R}^n}$, i.e., the nonlinear map α_Φ is injective on $\widetilde{\mathbb{R}^n}$ if and only if for any disjoint partition of the frame set $\Phi = \Phi_1 \cup \Phi_2$, either Φ_1 spans $\mathbb{R}^n$ or Φ_2 spans $\mathbb{R}^n$.*

2. *If Φ is phase retrievable in $\widetilde{\mathbb{R}^n}$, then $m \geq 2n - 1$. Furthermore, for a generic (completely full rank) Φ with $m \geq 2n - 1$, the map α_Φ is phase retrievable in $\widetilde{\mathbb{R}^n}$.*

3. *Suppose that $m = 2n - 1$. Then Φ is phase retrievable in $\widetilde{\mathbb{R}^n}$ if and only if Φ is of completely full rank.*

4. *Let*

$$a_0 := \min_{\|\mathbf{x}\| = \|\mathbf{y}\| = 1} \sum_{j=1}^{m} |\langle \mathbf{x}, \mathbf{a}_j \rangle|^2 |\langle \mathbf{y}, \mathbf{a}_j \rangle|^2 \geq 0 \tag{11.42}$$

so that

$$\sum_{k=1}^{m} |\langle \mathbf{x}, \mathbf{a}_k \rangle|^2 |\langle \mathbf{y}, \mathbf{a}_k \rangle|^2 \geq a_0 \|\mathbf{x}\|^2 \|\mathbf{y}\|^2. \tag{11.43}$$

Then Φ is phase retrievable on $\widetilde{\mathbb{R}^n}$ if and only if $a_0 > 0$.

5. *For any* $\mathbf{x} \in \mathbb{R}^n$, *define the matrix* $R(\mathbf{x})$ *by*

$$R(\mathbf{x}) := \sum_{j=1}^{m} |\langle \mathbf{x}, \mathbf{a}_j \rangle|^2 \mathbf{a}_j \mathbf{a}_j^*. \tag{11.44}$$

Let $\lambda_{\text{SPRgoal2}}(R(\mathbf{x}))$ *denote the smallest eigenvalue of* $R(\mathbf{x})$, *and let*

$$a_0 = \min_{\|\mathbf{x}\|=1} \lambda_{\text{SPRgoal2}}(R(\mathbf{x})).$$

Equivalently, let a_0 *be the largest constant so that* $R(\mathbf{x}) \geq a_0 \|\mathbf{x}\|^2 I$ *for all* $\mathbf{x} \in \mathbb{R}^n$, *where* I *is the identity matrix. Then* Φ *is phase retrievable on* $\mathbb{R}^n$ *if and only if* $a_0 > 0$. *Additionally, the constant* a_0 *introduced here is the same as the constant* a_0 *given by* (11.42).

Proof. Some of these properties have been shown in a section 11.5. We leave them to the interested reader; see also the proof of results (1)–(3) in [4] and (4)–(5) in [3]. □

We now continue our discussion of how to retrieve the signal from the frame discussed above. Let $P = [p_{ij}]_{1 \leq i \leq m, 1 \leq j \leq 2^m}$ be the matrix with entries $p_{ij} = \pm$, which is the matrix of sign patterns for all possible patterns $\{-1, 1\}^m$. For $b_k = |\langle \mathbf{x}, \mathbf{a}_k \rangle|, k = 1, \ldots, m$, there is a column $\mathbf{p}_j$ of P such that

$$\langle \mathbf{x}, \mathbf{a}_k \rangle = p_{kj} b_k, \quad k = 1, \ldots, m.$$

Using the reconstruction formula (11.35), we have

$$\mathbf{x} = \sum_{k=1}^{m} p_{kj} b_k (\Phi^\top \Phi)^{-1} \mathbf{a}_k.$$

The only problem is to find which j is appropriate. If we know more information, say the sparsity of $\mathbf{x}$, we may be able to find $\mathbf{x}$ easily. For example, when $\|\mathbf{x}\|_0 = s$ is small, we only need $m \geq 2s - 1$ according to the result in the next section. Then we simply compute all the possible solutions

$$\mathbf{x}^{(j)} = \sum_{k=1}^{m} p_{kj} b_k (\Phi^\top \Phi)^{-1} \mathbf{a}_k \quad \forall j = 1, \ldots, 2^m$$

and find the sparsest signals from these $\mathbf{x}^j, j = 1, \ldots, 2^m$. For example, when $s = 5$, we can use $m = 10$, and hence we have 1024 possible solutions. Clearly, we only need to focus on 512 possible solutions since the others are just the negative of the 512 solutions. It is feasible to find the sparse solution. Let us call this approach the brute force method. When s is large, say $s = 100$, the computation will be impossible. Therefore, we need to continue exploring how to find an efficient method of retrieving the signal from its phaseless measurements.

Finally, we present an example of finding the solution with the correct phase. For convenience, let us consider a complex signal $\mathbf{x} \in \mathbb{C}^n$. We use the Mercedes-Benz frame.

Lemma 11.23 (Alexeev, Bandeira, Fickus, and Mixon, 2014 [1]). *Take* $\xi := \exp(2i\pi/3)$. *Then for any* $\mathbf{a}, \mathbf{b} \in \mathbb{C}$,

$$\mathbf{a}^* \mathbf{b} = \frac{1}{3} \sum_{k=0}^{2} \xi^k |\mathbf{a} + \xi^{-k} \mathbf{b}|^2. \tag{11.45}$$

Proof. We leave the proof to the interested reader. See Exercise 5. □

We can now use this polarization identity, i.e., Lemma 11.23, to determine the relative phases

$$\langle \mathbf{x}, \mathbf{a}_i \rangle^* \langle \mathbf{x}, \mathbf{a}_j \rangle = \frac{1}{3} \sum_{k=0}^{2} \xi^k |\langle \mathbf{x}, \mathbf{a}_i + \xi^{-k} \mathbf{a}_j \rangle|^2$$

for all $i, j = 1, \ldots, n$. Let

$$\rho_{ij} = \left(\frac{\langle \mathbf{x}, \mathbf{a}_i \rangle}{|\langle \mathbf{x}, \mathbf{a}_i \rangle|} \right)^{-1} \frac{\langle \mathbf{x}, \mathbf{a}_j \rangle}{|\langle \mathbf{x}, \mathbf{a}_j \rangle|} = \frac{\langle \mathbf{x}, \mathbf{a}_i \rangle^* \langle \mathbf{x}, \mathbf{a}_j \rangle}{|\langle \mathbf{x}, \mathbf{a}_i \rangle^* \langle \mathbf{x}, \mathbf{a}_j \rangle|}$$

for $j = 1, \ldots, n$. With $c_j = \rho_{ij} |\langle \mathbf{x}, \mathbf{a}_j \rangle|$, $\mathbf{x}$ can be recovered using the formula

$$\sum_{j \geq 1} c_j \tilde{\mathbf{a}}_j = \left(\frac{\langle \mathbf{x}, \mathbf{a}_i \rangle}{|\langle \mathbf{x}, \mathbf{a}_i \rangle|} \right)^{-1} \sum_{j \geq 1} \langle \mathbf{x}, \mathbf{a}_j \rangle \tilde{\mathbf{a}}_j = \left(\frac{\langle \mathbf{x}, \mathbf{a}_i \rangle}{|\langle \mathbf{x}, \mathbf{a}_i \rangle|} \right)^{-1} \mathbf{x} \qquad (11.46)$$

for each i, where $\tilde{\mathbf{a}}_j, j = 1, \ldots, n$, is the dual frame. That is, the above computation finds $\mathbf{x}$ in its equivalent class. If $\mathbf{x}$ is a real signal, we can choose $i = 1$ and the coefficient $\left(\frac{\langle \mathbf{x}, \mathbf{a}_i \rangle}{|\langle \mathbf{x}, \mathbf{a}_i \rangle|} \right)^{-1}$ to be 1 or -1, and then $\mathbf{x}$ can be easily determined in practice, whether $\mathbf{x}$ is more reasonable than $-\mathbf{x}$ or not.

We shall develop several numerical computational approaches to find the exact signal in later sections.

11.4 ▪ Number of Measurements for Sparse Signals

In this section, we study the problem of the minimal number of measurements required for s-sparse phase retrieval in the real variable setting. Our first theorem completely settles the minimality question for s-sparse phase retrieval.

Theorem 11.24 (Wang and Xu, 2014 [35]). *Let $\Phi = \{\mathbf{a}_1, \ldots, \mathbf{a}_m\}$ be a set of vectors in $\mathbb{R}^n$. Assume that Φ is s-sparse phase retrievable on $\mathbb{R}^n$. Then $m \geq \min\{2s, 2n-1\}$. Furthermore, $m > n - s$.*

Proof. Assume that Φ has $m < 2s$ elements. We claim that Φ does not have the s-sparse phase retrieval property. To prove the claim, we will construct $\mathbf{x}, \mathbf{y} \in \mathbb{R}^n$ with sparsity s such that $|\langle \mathbf{x}, \mathbf{a}_j \rangle| = |\langle \mathbf{y}, \mathbf{a}_j \rangle|, j = 1, \ldots, m$, but $\mathbf{x} \neq \pm \mathbf{y}$.

To do so, we divide Φ into two groups: $\Phi_1 = [\mathbf{a}_1, \ldots, \mathbf{a}_s]$ and $\Phi_2 = [\mathbf{a}_{s+1}, \ldots, \mathbf{a}_m]$. Consider the subspace

$$W = \{(x_1, \ldots, x_s, x_{s+1}, 0, \ldots, 0)^\top \in \mathbb{R}^n : x_i \in \mathbb{R}, i = 1, \ldots, s+1\}.$$

There exists a nonzero vector $\mathbf{u} \in W$ such that $\Phi_1^\top \mathbf{u} = 0$. Since $m < 2s$, there are only $m - s < 2s - s = s$ columns in Φ_2. It follows that the set

$$\{\mathbf{v} \in W : \Phi_2^\top \mathbf{v} = 0\}$$

has dimension at least 2. Hence, there exist linearly independent vectors $\alpha, \beta \in W$ so that $\mathbf{v} = x\alpha + y\beta \in W$ satisfies $\Phi_2^\top \mathbf{v} = 0$ for any $x, y \in \mathbb{R}$. Write $\alpha = (\alpha_1, \ldots, \alpha_n)^\top$ and $\beta = (\beta_1, \ldots, \beta_n)^\top$. Without loss of generality, we may assume that (α_1, α_2) and (β_1, β_2) are linearly independent. For any vector $\mathbf{u} = (u_1, \ldots, u_n)^\top \in W \subset \mathbb{R}^n$ such that $\Phi_1^\top \mathbf{u} = 0$, we can find x_0, y_0 such that $u_1 = x_0 \alpha_1 + y_0 \beta_1, -u_2 = x_0 \alpha_2 + y_0 \beta_2$. Now we let $\mathbf{w} = x_0 \alpha + y_0 \beta$ and

$$\mathbf{x} = \mathbf{u} + \mathbf{w}, \quad \mathbf{y} = \mathbf{u} - \mathbf{w},$$

which are sparse vectors with sparsity $= s$. Moreover,

$$\langle \mathbf{x}, \mathbf{a}_j \rangle = \langle \mathbf{u}, \mathbf{a}_j \rangle + \langle \mathbf{w}, \mathbf{a}_j \rangle = \begin{cases} \langle \mathbf{u}, \mathbf{a}_j \rangle, j > s, \\ \langle \mathbf{w}, \mathbf{a}_j \rangle, j \le s, \end{cases}$$

$$\langle \mathbf{y}, \mathbf{a}_j \rangle = \langle \mathbf{u}, \mathbf{a}_j \rangle - \langle \mathbf{w}, \mathbf{a}_j \rangle = \begin{cases} \langle \mathbf{u}, \mathbf{a}_j \rangle, j > s, \\ -\langle \mathbf{w}, \mathbf{a}_j \rangle, j \le s. \end{cases}$$

It follows that $|\langle \mathbf{x}, \mathbf{a}_j \rangle| = |\langle \mathbf{y}, \mathbf{a}_j \rangle|$ for $j = 1, \ldots, m$ but $\mathbf{x} \ne \pm\mathbf{y}$.

Furthermore, we claim that $m > n - s$. Otherwise, we have $m \le n - s$. Since a sparse signal $\mathbf{x}$ has a sparsity s, without loss of generality, we may assume that $\mathbf{x} \in \mathbb{R}^s$. If the measurement vectors are given in $(\mathbb{R}^s)^\perp$, it is easy to see that there are many nonzero sparse signals $\mathbf{x} \in \mathbb{R}^s$ which have the same measurement values. This contradicts the phase retrievability. $\qquad\square$

Theorem 11.25 (Wang and Xu, 2014 [35]). *Suppose that the set* $\Phi = [\mathbf{a}_1, \ldots, \mathbf{a}_m]$, *with* m *generically chosen vectors in* $\mathbb{R}^n$, *i.e., the matrix* Φ *is of completely full rank. If* $m \ge \min\{2s, 2n - 1\}$, *the s-sparse vector* $\mathbf{x}$ *is phase retrievable.*

Proof. Since the matrix Φ is of completely full rank, we have $m \ge n$. Let $I, J \subset \{1, \ldots, m\}$ with $\#(I) = \#(J) = s$. The goal is to prove that if $\mathbf{x}, \mathbf{y} \in \mathbb{R}^n$, with sparsity s and supp$(\mathbf{x}) \subset I$ and supp$(\mathbf{y}) \subset J$ satisfying

$$|\langle \mathbf{a}_j, \mathbf{x} \rangle|^2 = |\langle \mathbf{a}_j, \mathbf{y} \rangle|^2, \quad j = 1, \ldots, m,$$

then $\mathbf{x} = \pm\mathbf{y}$. Each of the above equations can be rewritten as

$$\langle \mathbf{a}_j, \mathbf{x} - \mathbf{y} \rangle \langle \mathbf{a}_j, \mathbf{x} + \mathbf{y} \rangle = 0.$$

It follows that either $\langle \mathbf{a}_j, \mathbf{x} + \mathbf{y} \rangle = 0$ or $\langle \mathbf{a}_j, \mathbf{x} - \mathbf{y} \rangle = 0$. The index set $[1, \ldots, m]$ is divided into two subsets,

$$\langle \mathbf{a}_j, \mathbf{x} + \mathbf{y} \rangle = 0, \, j \in [1, \ldots, k] \quad \text{and} \quad \langle \mathbf{a}_j, \mathbf{x} - \mathbf{y} \rangle = 0, \, j \in [k + 1, \ldots, m],$$

without loss of generality. Let $L = I \cap J$ and $\ell = \#(L)$. For convenience, we write

$$\mathbf{x} = \mathbf{u}_x + \mathbf{v}_x, \, \text{supp}(\mathbf{u}_x) \subset L, \, \text{supp}(\mathbf{v}_x) \subset I \setminus L,$$
$$\mathbf{y} = \mathbf{u}_y + \mathbf{v}_y, \, \text{supp}(\mathbf{u}_y) \subset L, \, \text{supp}(\mathbf{v}_y) \subset J \setminus L.$$

Without loss of generality, we may assume that $L = [1, \ldots, \ell]$, $I \setminus L = [\ell + 1, \ldots, s]$, and $J \setminus L = [s + 1, \ldots, 2s - \ell]$. Let

$$\mathbf{w}_- := \mathbf{u}_x - \mathbf{u}_y, \quad \mathbf{w}_+ = \mathbf{u}_x + \mathbf{u}_y, \quad \text{and} \quad \mathbf{z} = [\mathbf{v}_x \, \mathbf{v}_y \, \mathbf{w}_- \, \mathbf{w}_+]^\top.$$

Then the equations $\langle \mathbf{a}_j, \mathbf{x} - \mathbf{y} \rangle = 0$ and $\langle \mathbf{a}_j, \mathbf{x} + \mathbf{y} \rangle = 0$ can be rewritten as

$$\langle \mathbf{a}_j, \mathbf{x} - \mathbf{y} \rangle = \langle \mathbf{a}_j, \mathbf{v}_x \rangle - \langle \mathbf{a}_j, \mathbf{v}_y \rangle + \langle \mathbf{a}_j, \mathbf{w}_- \rangle = 0,$$
$$\langle \mathbf{a}_j, \mathbf{x} + \mathbf{y} \rangle = \langle \mathbf{a}_j, \mathbf{v}_x \rangle + \langle \mathbf{a}_j, \mathbf{v}_y \rangle + \langle \mathbf{a}_j, \mathbf{w}_+ \rangle = 0.$$

Set $A := \Phi^\top$. The above equations can be rewritten in matrix form as

$$\begin{bmatrix} A_{[1,\ldots,k],I \setminus L} & -A_{[1,\ldots,k],J \setminus L} & A_{[1,\ldots,k],L} & 0 \\ A_{[k+1,\ldots,m],I \setminus L} & A_{[k+1,\ldots,m],J \setminus L} & 0 & A_{[k+1,\ldots,m],L} \end{bmatrix} \mathbf{z} = \mathbf{0}. \qquad (11.47)$$

We claim that the linear equations above force $\mathbf{v}_x = 0, \mathbf{v}_y = 0$, and either $\mathbf{w}_- = 0$ or $\mathbf{w}_+ = 0$. Indeed, we first consider the case $k \geq 2s - \ell$. The coefficient matrix of the first k equations in (11.47) is

$$\left[A_{[1,\ldots,k],I\backslash L} - A_{[1,\ldots,k],J\backslash L} \; A_{[1,\ldots,k],L} \right],$$

which is of size $k \times (2s - \ell)$. The elements in Φ are generically chosen, and hence the coefficient matrix above has full rank. It follows that $\mathbf{v}_x = 0, \mathbf{v}_y = 0$, and $\mathbf{w}_- = 0$. That is, $\mathbf{x} = \mathbf{y}$.

We next consider the case with $m - k \geq 2s - \ell$. In the same fashion, we use the second block of the linear system in (11.47) to conclude that $\mathbf{x} = -\mathbf{y}$.

Finally, we consider the case when $k < 2s - \ell$ and $m - k < 2k - \ell$ and hence $k > \ell$. Similarly, we have $\ell < 2s - k$ and hence $2\ell < 2s$. In this case, we note that Φ is of full rank since all columns in Φ are in generic position. We claim that the matrix in (11.47) is of rank $2s$. Indeed, the first and third columns of the matrix in (11.47) are of full column rank. Similarly, the second and fourth columns of the matrix in (11.47) are full columns. It follows that $\mathbf{z} \equiv 0$. Hence, $\mathbf{x} = 0 = \mathbf{y}$. This completes the proof. $\quad\square$

11.5 ▪ The Minimizing Functional

One popular approach to retrieving the sparse or nonsparse signal is to use the least squares method. That is, let us study the minimization

$$\min_{\mathbf{z} \in \mathbb{C}^n} \frac{1}{2m} \sum_{i=1}^m \left(b_i - |\langle \mathbf{a}_j, \mathbf{z} \rangle|^2 \right)^2. \tag{11.48}$$

Note that the minimizing functional above is not convex, and hence a simple application of the standard gradient descent method does not give a good algorithm. Also, the minimizing functional $f(\mathbf{z})$ is not differentiable when $\mathbf{z}$ is a complex signal. Fortunately, these difficulties can be overcome, which will be discussed in the next few sections. In addition, three different algorithms will be developed based on the minimization (11.48) in later sections.

In this section, we mainly study some properties of the minimization functional

$$f(\mathbf{z}) = \frac{1}{2m} \sum_{i=1}^m \left(b_i - |\langle \mathbf{a}_j, \mathbf{z} \rangle|^2 \right)^2. \tag{11.49}$$

Because the minimizing function $f(\mathbf{z})$ is not a convex function, it is better to understand the shape or geometric properties of the functional f well enough. We start with the landscape of the expectation of $f(\mathbf{z})$ and give the characterization of the local/global minimizers. To this end, let us begin with some elementary lemmas. Their proofs are left to the interested reader, or see Exercise 10.

Lemma 11.26. *Suppose that the random vectors* $\mathbf{a}_k, k = 1, \ldots, m$, *follow the Gaussian model. Then for any fixed vector* $\mathbf{x} \in \mathbb{C}^n$,

$$\mathbb{E}\left(\frac{1}{m} \sum_{k=1}^m |\mathbf{a}_k^* \mathbf{x}|^2 \mathbf{a}_k \mathbf{a}_k^* \right) = \mathbf{x}\mathbf{x}^* + \|\mathbf{x}\|^2 I, \tag{11.50}$$

where I *is the identity matrix of size* $n \times n$ *and the expectation* $\mathbb{E}(f(\mathbf{z})) := \mathbb{E}_{\mathbf{a}_1,\ldots,\mathbf{a}_m}(f(\mathbf{z}))$. *Furthermore,*

$$\mathbb{E}\left(\frac{1}{m} \sum_{k=1}^m (\mathbf{a}_k^* \mathbf{x})^2 \mathbf{a}_k \mathbf{a}_k^* \right) = 2\mathbf{x}\mathbf{x}^\top. \tag{11.51}$$

With (11.50) and (11.51), we can prove the following.

Lemma 11.27. *For f in* (11.49), *we have*

$$\mathbb{E}(f(\mathbf{z})) = \|\mathbf{x}\|^4 + \|\mathbf{z}\|^4 - \|\mathbf{z}\|^2\|\mathbf{x}\|^2 - |\mathbf{x}^*\mathbf{z}|^2,$$

$$\nabla\mathbb{E}(f(\mathbf{z})) = \mathbb{E}(\nabla f(\mathbf{z})) = \left[\begin{array}{c} \nabla_{\mathbf{z}}\mathbb{E}(f(\mathbf{z})) \\ \nabla_{\bar{\mathbf{z}}}\mathbb{E}(f(\mathbf{z})) \end{array}\right] = \left[\begin{array}{c} ((2\|\mathbf{z}\|^2 - \|\mathbf{x}\|^2)I - \mathbf{x}\mathbf{x}^*)\mathbf{z} \\ ((2\|\mathbf{z}\|^2 - \|\mathbf{x}\|^2)I - \mathbf{x}\mathbf{x}^*)\bar{\mathbf{z}} \end{array}\right],$$

$$\nabla^2\mathbb{E}(f(\mathbf{z})) = \mathbb{E}(\nabla^2 f(\mathbf{z}))$$
$$= \left[\begin{array}{cc} 2\mathbf{z}\mathbf{z}^* - \mathbf{x}\mathbf{x}^* + (2\|\mathbf{z}\|^2 - \|\mathbf{x}\|^2)I & 2\mathbf{z}\mathbf{z}^* \\ 2\bar{\mathbf{z}}\mathbf{z}^* & 2\bar{\mathbf{z}}\mathbf{z}^\top - \bar{\mathbf{x}}\mathbf{x}^\top + (2\|\mathbf{z}\|^2 - \|\mathbf{x}\|^2)I \end{array}\right],$$

where I is the identity matrix of size $n \times n$.

Proof. We leave the proof to the interested reader. □

We now have the following characterization results on the landscape of $\mathbb{E}(f(\mathbf{z}))$.

Theorem 11.28 (Sun, Qu, and Wright, 2017 [34]). *Let $\mathbf{x}$ be the signal to be retrieved. Suppose that $\mathbf{x} \neq 0$. Then the critical points of $\mathbb{E}(f(\mathbf{z}))$ are* (1) $\mathbf{z} = 0$; (2) $\{\mathbf{z} : \mathbf{z}^*\mathbf{x} = 0, \|\mathbf{z}\| = \|\mathbf{x}\| = 1/\sqrt{2}\}$, *which are the maximizers;* (3) *the set of local and global minimizers; and* (4) *the set of saddle points.*

Proof. From Lemma 11.27, it follows that $\mathbf{z} = 0$ is a critical point and the Hessian is

$$\nabla^2\mathbb{E}(f(0)) = \left[\begin{array}{cc} -\mathbf{x}\mathbf{x}^* - \|\mathbf{x}\|^2 I & 0 \\ 0 & -\bar{\mathbf{x}}\mathbf{x}^\top - \|\mathbf{x}\|^2 I \end{array}\right],$$

which is negative definite. Hence, $\mathbf{z} = 0$ is a maximizer.

In the region $\{\mathbf{z} : 0 < \|\mathbf{z}\|^2 < \frac{1}{2}\|\mathbf{x}\|^2\}$, we have

$$\left[\begin{array}{c} \mathbf{z} \\ \bar{\mathbf{z}} \end{array}\right]^* \nabla\mathbb{E}(f(\mathbf{z})) = 2(2\|\mathbf{z}\|^2 - \|\mathbf{x}\|^2)\|\mathbf{z}\|^2 - 2|\mathbf{x}^*\mathbf{z}|^2 < 0.$$

So there is no critical point in this region.

When $\|\mathbf{z}\|^2 = \frac{1}{2}\|\mathbf{x}\|^2$, the gradient is $\nabla_{\mathbf{z}}\mathbb{E}(f(\mathbf{z})) = -\mathbf{x}\mathbf{x}^*\mathbf{z}$. Let $\mathcal{S} = \{\mathbf{z} \in \mathbb{C}^n, \mathbf{x}^*\mathbf{z} = 0, \|\mathbf{z}\| = \|\mathbf{x}\|/\sqrt{2}\}$. For any $\mathbf{x} \in \mathcal{S}$, we have $\nabla_{\mathbf{z}}\mathbb{E}(f(\mathbf{z})) = 0$. In this case, we can see that any point in $\mathcal{S}$ is a saddle point since the Hessian is not positive definite.

In the region $\{\mathbf{z}, \|\mathbf{x}\|^2/2 < \|\mathbf{z}\|^2 < \|\mathbf{x}\|^2\}$, any potential critical point must satisfy

$$(2\|\mathbf{z}\| - \|\mathbf{x}\|)\mathbf{z} - \mathbf{x}\mathbf{x}^*\mathbf{z} = 0.$$

That is, $2\|\mathbf{z}\| - \|\mathbf{x}\|$ is an eigenvalue of the nonnegative definite Hermitian matrix $\mathbf{x}\mathbf{x}^*$. A direct calculation shows that the eigenvalue of $\mathbf{x}\mathbf{x}^*$ is $\|\mathbf{x}\|$. That is, $2\|\mathbf{z}\| - \|\mathbf{x}\| = \|\mathbf{x}\|$ or $\|\mathbf{z}\| = \|\mathbf{x}\|$, which is a contradiction. In other words, there is no critical point in the region $\{\mathbf{z}, \|\mathbf{x}\|^2/2 < \|\mathbf{z}\|^2 < \|\mathbf{x}\|^2\}$.

When $\|\mathbf{z}\|^2 = \|\mathbf{x}\|^2$, critical points must satisfy

$$\|\mathbf{x}\|^2\mathbf{z} - \mathbf{x}\mathbf{x}^*\mathbf{z} = 0.$$

As discussed above, $\mathbf{z}$ is an eigenvalue of the Hermitian matrix $\mathbf{x}\mathbf{x}^*$, which is of rank 1. Thus, $\mathbf{z} = \alpha\mathbf{x}$ for scalar α. Since $\|\mathbf{z}\| = \|\mathbf{x}\|$, $\mathbf{z} = \mathbf{x}\exp(\theta)$ for any θ. Since $f(\mathbf{z}) = 0$, we see that $\mathbf{z} = \mathbf{x}\exp(\theta)$ is a global minimizer.

Finally, we consider the region $\{\mathbf{z}, \|\mathbf{z}\| > \|\mathbf{x}\|\}$:

$$\begin{bmatrix} \mathbf{z} \\ \bar{\mathbf{z}} \end{bmatrix}^* \nabla\mathbb{E}(f(\mathbf{z})) = 2(2\|\mathbf{z}\|^2 - \|\mathbf{x}\|^2)\|\mathbf{z}\|^2 - 2|\mathbf{x}^*\mathbf{z}|^2 > 0.$$

So there is no critical point in this region. This completes the proof. $\square$

With the results above, we can explain the geometry of the landscape of the minimizing function.

Theorem 11.29. *There exist positive constants C and c such that when $m \geq Cn\log^3(n)$, $f(\mathbf{z})$ has no spurious local minimizers and only local/global minimizers, which are exactly the targeted set $\mathcal{X}$ with probability $1 - cm^{-1}$. More precisely, let $\theta = \theta(\mathbf{z})$ be the phase of $\mathbf{z}$ and define*

$$R_1 = \left\{ \mathbf{z} : \begin{bmatrix} \mathbf{x}e^{i\theta} \\ \bar{\mathbf{x}}e^{-i\theta} \end{bmatrix}^* \nabla^2 f(\mathbf{z}) \begin{bmatrix} \mathbf{x}e^{i\theta} \\ \bar{\mathbf{x}}e^{-i\theta} \end{bmatrix} \leq -\frac{1}{100}\|\mathbf{x}\|^2\|\mathbf{z}\|^2 - \frac{1}{50}\|\mathbf{x}\|^4 \right\}, \quad (11.52)$$

$$R_3 = \left\{ \mathbf{z} : \text{dist}(\mathbf{z}, \mathcal{X}) \leq \frac{1}{\sqrt{7}}\|\mathbf{x}\| \right\}, \quad (11.53)$$

$$R_2 = (R_1 \cup R_3)^c. \quad (11.54)$$

Then when $\mathbf{z} \in R_1$, $f(\mathbf{z})$ has a negative curvature; when $\mathbf{z} \in R_3$, f is restricted strongly convex; and when $\mathbf{z} \in R_2$, $\|\nabla f(\mathbf{z})\| \geq \|\mathbf{x}\|^2\|\mathbf{z}\|/1000 > 0$ with high probability.

Proof. These statements are proved sequentially in Propositions 2.3, 2.5, 2.6, 2.4, and 2.7 in [34]. We leave these to the interested reader. $\square$

Next we give a deterministic description of the landscape of the minimizing function f. For convenience, let $A_\ell = \mathbf{a}_\ell\bar{\mathbf{a}}_\ell^\top$ be the Hermitian matrix of rank 1 for $\ell = 1, \dots, m$. We first need a definition.

Definition 11.30. *We say $\mathbf{a}_j, j = 1, \dots, m$, are a_0-generic if they satisfy*

$$\|(\mathbf{a}_{j_1}^*\mathbf{y}, \dots, \mathbf{a}_{j_n}^*\mathbf{y})\| \geq a_0\|\mathbf{y}\| \quad \forall \mathbf{y} \in \mathbb{C}^n \quad (11.55)$$

for a positive $a_0 \in (0,1)$ for any $1 \leq j_1 < j_2 < \cdots < j_n \leq m$.

Theorem 11.31. *Consider the real variable setting. Let $H_f(\mathbf{x})$ be the Hessian of the minimizing function $f(\mathbf{z})$ and let $\mathbf{x}^*$ be a global minimizer of (11.48). Suppose that $\mathbf{a}_j, j = 1, \dots, m$, are in a_0-generic position (see (11.55) above). Then $H_f(\mathbf{x}^\star)$ is positive definite.*

Proof. Recall that $A_\ell = \mathbf{a}_\ell\bar{\mathbf{a}}_\ell^\top$ for $\ell = 1, \dots, m$. It is easy to see that $\nabla f(\mathbf{x}) = 2\sum_{\ell=1}^m (\mathbf{x}^\top A_\ell\mathbf{x} - b_\ell)A_\ell\mathbf{x}$ and the entries h_{ij} of the Hessian $H_f(\mathbf{x})$ are

$$h_{ij} = \frac{\partial}{\partial x_i}\frac{\partial}{\partial x_j}f(\mathbf{x}) = 2\sum_{\ell=1}^m (\mathbf{x}^\top A_\ell\mathbf{x} - b_\ell)a_{ij}(\ell) + 4\sum_{p=1}^n a_{i,p}(\ell)x_p \sum_{q=1}^n a_{j,q}(\ell)x_q, \quad (11.56)$$

where $A_\ell = [a_{ij}(\ell)]_{ij=1}^n$. Because we have $(\mathbf{x}^*)^\top A_\ell\mathbf{x}^* = b_\ell, \ell = 1, \dots, m$, the first summation term of h_{ij} above is zero at $\mathbf{x}^*$. Letting $M(\mathbf{y}) = \mathbf{y}^\top H_f(\mathbf{x}^*)\mathbf{y}$ be a quadratic function of $\mathbf{y}$, we have

$$M(\mathbf{y}) = 4\sum_{\ell=1}^{m}(\mathbf{y}^\top A_\ell \mathbf{x}^*(\mathbf{x}^*)^\top A_\ell \mathbf{y}) = 4\sum_{\ell=1}^{m}|\mathbf{y}^\top A_\ell \mathbf{x}^*|^2$$

$$= 4\sum_{\ell=1}^{m}|\mathbf{y}^\top \mathbf{a}_\ell|^2|\bar{\mathbf{a}}_\ell^\top \mathbf{x}^*|^2 \geq 4a_0^4\|\mathbf{x}^*\|^2\|\mathbf{y}\|^2$$

using the definition of a_0 as in section 11.3. It follows that $H_f(\mathbf{x}^*)$ is positive definite. $\quad\square$

Next let us show that the global minimizer is $\mathbf{x}^*$ in the complex setting. In this case, the Hessian $H_F(\mathbf{x}^*)$ is no longer positive definite. Instead, it is nonnegative definite. To this end, let us fix some notation. Write $\mathbf{a}_\ell = a_\ell + ic_\ell$ for $\ell = 1, \ldots, m$. For $\mathbf{z} = \mathbf{x} + i\mathbf{y}$ with $\mathbf{x}, \mathbf{y} \in \mathbb{R}^n$, we have $\mathbf{a}_\ell^\top \mathbf{x}^* = b_\ell$ for the global minimizer $\mathbf{x}^*$. Writing $f_\ell(\mathbf{x}, \mathbf{y}) = |\mathbf{a}_\ell^\top \mathbf{z}|^2 - b_\ell = (a_\ell^\top \mathbf{x} - c_\ell^\top \mathbf{y})^2 + (c_\ell^\top \mathbf{x} + a_\ell^\top \mathbf{y})^2 - b_\ell$, we consider

$$f(\mathbf{x}, \mathbf{y}) = \frac{1}{m}\sum_{\ell=1}^{m} f_\ell^2. \tag{11.57}$$

The gradient of f can be easily found as follows: $\nabla f = [\nabla_{\mathbf{x}} f, \nabla_{\mathbf{y}} f]$ with

$$\nabla_{\mathbf{x}} f(\mathbf{x}, \mathbf{y}) = \frac{1}{m}\sum_{\ell=1}^{m}\nabla_{\mathbf{x}} f_\ell^2 = \frac{4}{m}\sum_{\ell=1}^{m} f_\ell(\mathbf{x}, \mathbf{y})[(a_\ell^\top \mathbf{x} - c_\ell^\top \mathbf{y})a_\ell + (c_\ell^\top \mathbf{x} + a_\ell^\top \mathbf{y})c_\ell] \tag{11.58}$$

and

$$\nabla_{\mathbf{y}} f(\mathbf{x}, \mathbf{y}) = \frac{1}{m}\sum_{\ell=1}^{m}\nabla_{\mathbf{y}} f_\ell^2 = \frac{4}{m}\sum_{\ell=1}^{m} f_\ell(\mathbf{x}, \mathbf{y})[(a_\ell^\top \mathbf{x} - c_\ell^\top \mathbf{y})(-c_\ell) + (c_\ell^\top \mathbf{x} + a_\ell^\top \mathbf{y})a_\ell]. \tag{11.59}$$

The Hessian of f is more complicated:

$$H_f(\mathbf{x}, \mathbf{y}) = \begin{bmatrix} \nabla_{\mathbf{x}}\nabla_{\mathbf{x}} f(\mathbf{x}, \mathbf{y}) & \nabla_{\mathbf{x}}\nabla_{\mathbf{y}} f(\mathbf{x}, \mathbf{y}) \\ \nabla_{\mathbf{y}}\nabla_{\mathbf{x}} f(\mathbf{x}, \mathbf{y}) & \nabla_{\mathbf{y}}\nabla_{\mathbf{y}} f(\mathbf{x}, \mathbf{y}) \end{bmatrix}, \tag{11.60}$$

with $\nabla_{\mathbf{x}}\nabla_{\mathbf{x}} f(\mathbf{x}, \mathbf{y}), \ldots, \nabla_{\mathbf{y}}\nabla_{\mathbf{y}} f(\mathbf{x}, \mathbf{y})$ given below:

$$\nabla_{\mathbf{x}}\nabla_{\mathbf{x}} f(\mathbf{x}, \mathbf{y}) = \frac{4}{m}\sum_{\ell=1}^{m} f_\ell(\mathbf{x}, \mathbf{y})[a_\ell a_\ell^\top + c_\ell c_\ell^\top]$$

$$+\frac{8}{m}\sum_{\ell=1}^{m}[(a_\ell^\top \mathbf{x} - c_\ell^\top \mathbf{y})a_\ell + (c_\ell^\top \mathbf{x} + a_\ell^\top \mathbf{y})c_\ell][(a_\ell^\top \mathbf{x} - c_\ell^\top \mathbf{y})a_\ell^\top + (c_\ell^\top \mathbf{x} + a_\ell^\top \mathbf{y})c_\ell^\top],$$

$$\nabla_{\mathbf{x}}\nabla_{\mathbf{y}} f(\mathbf{x}, \mathbf{y}) = \frac{4}{m}\sum_{\ell=1}^{m} f_\ell(\mathbf{x}, \mathbf{y})[a_\ell(-c_\ell)^\top + c_\ell a_\ell^\top]$$

$$+\frac{8}{m}\sum_{\ell=1}^{m}[(a_\ell^\top \mathbf{x} - c_\ell^\top \mathbf{y})a_\ell + (c_\ell^\top \mathbf{x} + a_\ell^\top \mathbf{y})c_\ell][(a_\ell^\top \mathbf{x} - c_\ell^\top \mathbf{y})(-c_\ell)^\top + (c_\ell^\top \mathbf{x} + a_\ell^\top \mathbf{y})a_\ell^\top],$$

$$\nabla_{\mathbf{y}}\nabla_{\mathbf{x}}f(\mathbf{x},\mathbf{y}) = \frac{4}{m}\sum_{\ell=1}^{m} f_\ell(\mathbf{x},\mathbf{y})[-c_\ell a_\ell^\top + a_\ell c_\ell^\top]$$

$$+\frac{8}{m}\sum_{\ell=1}^{m}[(a_\ell^\top\mathbf{x} - c_\ell^\top\mathbf{y})(-c_\ell) + (c_\ell^\top\mathbf{x} + a_\ell^\top\mathbf{y})a_\ell][(a_\ell^\top\mathbf{x} - c_\ell^\top\mathbf{y})a_\ell^\top + (c_\ell^\top\mathbf{x} + a_\ell^\top\mathbf{y})c_\ell^\top]$$

and

$$\nabla_{\mathbf{y}}\nabla_{\mathbf{y}}f(\mathbf{x},\mathbf{y}) = \frac{4}{m}\sum_{\ell=1}^{m} f_\ell(\mathbf{x},\mathbf{y})[c_\ell c_\ell^\top + a_\ell a_\ell^\top]$$

$$+\frac{8}{m}\sum_{\ell=1}^{m}[(a_\ell^\top\mathbf{x} - c_\ell^\top\mathbf{y})(-c_\ell) + (c_\ell^\top\mathbf{x} + a_\ell^\top\mathbf{y})a_\ell][(a_\ell^\top\mathbf{x} - c_\ell^\top\mathbf{y})(-c_\ell)^\top + (c_\ell^\top\mathbf{x} + a_\ell^\top\mathbf{y})a_\ell^\top].$$

We are now ready to prove the nonnegativity of the Hessian at a global minimizer $\mathbf{z}^*$.

Theorem 11.32. *At any global minimizer* $\mathbf{z}^* = (\mathbf{x}^*,\mathbf{y}^*)$, *we have the Hessian* $H_f(\mathbf{x}^*,\mathbf{y}^*) \geq 0$. *In fact,* $H_f(\mathbf{x}^*,\mathbf{y}^*) = 0$ *along the direction* $[-(\mathbf{y}^*)^\top,(\mathbf{x}^*)^\top]^\top$.

Proof. At $\mathbf{z}^* = \mathbf{x}^* + i\mathbf{y}^*$, we have

$$\nabla_{\mathbf{x}}\nabla_{\mathbf{x}}f(\mathbf{x}^*,\mathbf{y}^*)$$

$$= \frac{8}{m}\sum_{\ell=1}^{m}[(a_\ell^\top\mathbf{x}^* - c_\ell^\top\mathbf{y}^*)a_\ell + (c_\ell^\top\mathbf{x}^* + a_\ell^\top\mathbf{y}^*)c_\ell]\,[(a_\ell^\top\mathbf{x}^* - c_\ell^\top\mathbf{y}^*)a_\ell^\top + (c_\ell^\top\mathbf{x}^* + a_\ell^\top\mathbf{y}^*)c_\ell^\top],$$

$$\nabla_{\mathbf{x}}\nabla_{\mathbf{y}}f(\mathbf{x}^*,\mathbf{y}^*)$$

$$= \frac{8}{m}\sum_{\ell=1}^{m}[(a_\ell^\top\mathbf{x}^* - c_\ell^\top\mathbf{y}^*)a_\ell + (c_\ell^\top\mathbf{x}^* + a_\ell^\top\mathbf{y}^*)c_\ell]\,[(a_\ell^\top\mathbf{x}^* - c_\ell^\top\mathbf{y}^*)(-c_\ell)^\top + (c_\ell^\top\mathbf{x}^* + a_\ell^\top\mathbf{y}^*)a_\ell^\top],$$

and similar for the other two terms. It is easy to see that for any $\mathbf{w} = \mathbf{u} + i\mathbf{v}$ with $\mathbf{u},\mathbf{v}\in\mathbb{R}^n$, we have

$$[\mathbf{u}^\top\ \mathbf{v}^\top]^\top H_f(\mathbf{x}^*,\mathbf{y}^*)\begin{bmatrix}\mathbf{u}\\\mathbf{v}\end{bmatrix}$$

$$= \frac{8}{m}\sum_{\ell=1}^{m}[(a_\ell^\top\mathbf{x}^* - c_\ell^\top\mathbf{y}^*)a_\ell^\top\mathbf{u} + (c_\ell^\top\mathbf{x}^* + a_\ell^\top\mathbf{y}^*)c_\ell^\top\mathbf{u}]^2$$

$$+\frac{8}{m}\sum_{\ell=1}^{m}2[(a_\ell^\top\mathbf{x}^* - c_\ell^\top\mathbf{y}^*)a_\ell^\top\mathbf{u} + (c_\ell^\top\mathbf{x}^* + a_\ell^\top\mathbf{y}^*)c_\ell^\top\mathbf{u}]$$

$$\times[(a_\ell^\top\mathbf{x}^* - c_\ell^\top\mathbf{y}^*)(-c_\ell)^\top\mathbf{v} + (c_\ell^\top\mathbf{x}^* + a_\ell^\top\mathbf{y}^*)a_\ell^\top\mathbf{v}]$$

$$+\frac{8}{m}\sum_{\ell=1}^{m}[(a_\ell^\top\mathbf{x}^* - c_\ell^\top\mathbf{y}^*)(-c_\ell)^\top\mathbf{v} + (c_\ell^\top\mathbf{x}^* + a_\ell^\top\mathbf{y}^*)a_\ell^\top\mathbf{v}]^2$$

$$= \frac{8}{m}\sum_{\ell=1}^{m}[(a_\ell^\top\mathbf{x}^* - c_\ell^\top\mathbf{y}^*)a_\ell^\top\mathbf{u} + (c_\ell^\top\mathbf{x}^* + a_\ell^\top\mathbf{y}^*)c_\ell^\top\mathbf{u}$$

$$+(a_\ell^\top\mathbf{x}^* - c_\ell^\top\mathbf{y}^*)(-c_\ell)^\top\mathbf{v} + (c_\ell^\top\mathbf{x}^* + a_\ell^\top\mathbf{y}^*)a_\ell^\top\mathbf{v}]^2$$

$$\geq 0.$$

Furthermore, if we choose $\mathbf{u} = -\mathbf{y}^*$ and $\mathbf{v} = \mathbf{x}^*$, we can easily see that the Hessian H_f along this direction is zero:

$$[-(\mathbf{y}^*)^\top \ (\mathbf{x}^*)^\top]H_f(\mathbf{x}^*, \mathbf{y}^*)\begin{bmatrix} -\mathbf{y}^* \\ \mathbf{x}^* \end{bmatrix} = 0.$$

This completes the proof. □

11.6 ▪ The Wirtinger Flow Algorithm

For simplicity, let $f(\mathbf{z}) = (b - |\langle \mathbf{a}, \mathbf{z} \rangle|^2)^2$ be a function mapping from $\mathbb{C}$ to $\mathbb{R}$, where $b \in \mathbb{R}_+$ and $\mathbf{a} \in \mathbb{C}$. It is easy to see that $f(\mathbf{z}), \mathbf{z} \in \mathbb{C}$, is not holomorphic and hence not complex differentiable. However, $f(\mathbf{z})$ is differentiable in the following Wirtinger sense.

Definition 11.33. *Suppose a function $f(\mathbf{z}) = g(\mathbf{z}, \bar{\mathbf{z}})$ for another function g, where $\mathbf{z} \in \mathbb{C}$ and $\bar{\mathbf{z}}$ is the conjugate of $\mathbf{z}$. If the partial derivative of g with respect to $\mathbf{z}$ exists for any fixed $\bar{\mathbf{z}}$ and the partial derivative of g with respect to $\bar{\mathbf{z}}$ exists for any fixed $\mathbf{z}$, we say that $f(\mathbf{z})$ is differentiable in the Wirtinger sense.*

Suppose that f is Wirtinger differentiable. Let

$$\nabla f(\mathbf{z}) = (D_{\mathbf{z}}g(\mathbf{z}, \bar{\mathbf{z}}), D_{\bar{\mathbf{z}}}g(\mathbf{z}, \bar{\mathbf{z}})) \tag{11.61}$$

be the Wirtinger gradient of f. Similarly, we define the Hessian of f as

$$\nabla^2 f(\mathbf{z}) = \begin{bmatrix} D_{\mathbf{z}}(D_{\mathbf{z}}g) & D_{\mathbf{z}}(D_{\bar{\mathbf{z}}}g) \\ D_{\bar{\mathbf{z}}}(D_{\mathbf{z}}g) & D_{\bar{\mathbf{z}}}(D_{\bar{\mathbf{z}}}g) \end{bmatrix} \tag{11.62}$$

and the Taylor expansion of f in two variables $(\mathbf{z}, \bar{\mathbf{z}})$. See Exercises 11 and 12 at the end of this chapter. When $\mathbf{z} \in \mathbb{C}^n$ is a complex vector, we extend the Wirtinger derivative in the standard fashion. It is easy to check the following result.

Lemma 11.34 (Candés, Li, and Soltanolkotabi, 2015 [12]). *Our minimal function $f(\mathbf{z}) = \frac{1}{m}\sum_{i=1}^m (b_i - |\langle \mathbf{a}_i, \mathbf{z} \rangle|^2)^2$ in (11.48) is Wirtinger differentiable. Because f is real, $(D_{\bar{\mathbf{z}}}f)$ is the conjugate of $D_{\mathbf{z}}f$. In fact,*

$$(D_{\mathbf{z}}f)^* = \frac{2}{m}\sum_{i=1}^m (\bar{\mathbf{z}}^\top (\mathbf{a}_i \mathbf{a}_i^*)\mathbf{z} - b_i)(\mathbf{a}_i \mathbf{a}_i^*)\mathbf{z}. \tag{11.63}$$

Proof. We leave the proof to the interested reader. □

Instead of using ∇f, we simply use the gradient $D_{\mathbf{z}}f$ in the standard gradient descent method to solve (11.48). The Wirtinger flow algorithm for solving (11.48) can be given as follows.

ALGORITHM 11.1
The Wirtinger Flow Algorithm (cf. [12])

1: **Initialization:**   Start with an initial guess $\mathbf{z}^0 = \sqrt{\frac{1}{m}\sum_{k=1}^m |\mathbf{a}_k^*\mathbf{x_b}|^2}\bar{\mathbf{z}}_0$, where $\bar{\mathbf{z}}_0$ is the eigenvector associated with the leading eigenvalue λ of matrix $\mathbf{Y}$

$$\mathbf{Y} = \frac{1}{m}\sum_{i=1}^m b_i \mathbf{a}_i \mathbf{a}_i^\top, \tag{11.64}$$

which is normalized so that $\|\bar{\mathbf{z}}_0\| = 1$.

2: **Main Iterations:** For $k \geq 1$, we do

$$\mathbf{z}^k = \mathbf{z}^{k-1} - \frac{\tau_k}{\|\mathbf{z}_0\|} \sum_{i=1}^{m} (b_i - |\langle \mathbf{a}_i, \mathbf{z}^{k-1} \rangle|) \mathbf{a}_i (\mathbf{a}_i^* \mathbf{z}^{k-1}), \tag{11.65}$$

where τ_k is a variable step size for $k \geq 1$.

3: **Output:** $\mathbf{z}^k$ until a total number of iterations is achieved or the given tolerance is reached.

11.6.1 ▪ Initialization

Since a nonconvex minimization may have many local minimizers, it is extremely important to have an excellent starting point $\mathbf{z}_0$. A good initial guess provided in Algorithm 11.1 can be explained as follows. Recall that

$$\mathbf{Y} = \frac{1}{m} \sum_{i=1}^{m} b_i \mathbf{a}_i \mathbf{a}_i^\top, \tag{11.66}$$

and let $\mathbf{z}_0$ be an eigenvector associated with the largest eigenvalue λ of $\mathbf{Y}$. We normalize $\|\mathbf{z}_0\| = 1$. Let us show that this is indeed a good initial guess. To do so, we need a definition.

Definition 11.35. *Let* $\mathbf{x_b} \in \mathbb{C}^n$ *be any solution to the quadratic system* $b_i = |\langle \mathbf{x_b}, \mathbf{a}_i \rangle|^2, i = 1, \ldots, m$. *That is,* $\mathbf{x_b}$ *is the signal we wish to recover. For any* $\mathbf{z} \in \mathbb{C}^n$, *define*

$$\mathrm{dist}(\mathbf{z}, \mathbf{x_b}) = \inf_{\theta \in [0,1)} \|\mathbf{z} - \exp(\mathbf{i}2\pi\theta)\mathbf{x_b}\|. \tag{11.67}$$

Let us first state the following main result on the initialization.

Theorem 11.36 (Candés, Li, and Soltanolkotabi, 2015 [12]). *Suppose that a solution* $\mathbf{x_b}$ *is normalized such that* $\|\mathbf{x_b}\| = 1$. *Choose* $\tilde{\mathbf{z}}_0 = \sqrt{\frac{1}{m} \sum_{k=1}^{m} |\mathbf{a}_k^* \mathbf{x_b}|^2} \mathbf{z}_0$. *With high probability, we have*

$$\mathrm{dist}(\mathbf{z}_0, \mathbf{x_b}) \leq \frac{1}{8}. \tag{11.68}$$

To prove the above result, we need a preparatory lemma.

Lemma 11.37. *Recall that* $\mathbf{x_b}$ *is a solution which is independent of the sampling vectors* $\mathbf{a}_k$. *Furthermore, assume that the sampling vectors* $\mathbf{a}_k, k = 1, \ldots, m$, *are subject to the Gaussian distribution. Then*

$$\left\| I_n - \frac{1}{m} \sum_{i=1}^{m} \mathbf{a}_i \mathbf{a}_i^* \right\| \leq \delta \tag{11.69}$$

and

$$\|\mathbf{Y} - (\mathbf{x_b}\mathbf{x_b^*} + \|\mathbf{x_b}\|^2 I_n)\| \leq \delta \tag{11.70}$$

with high probability, where I_n *is the identity matrix of size* $n \times n$.

Proof. First of all, it is easy to see that $\mathbb{E}(\mathbf{a}_i \mathbf{a}_i^*) = I_n$. By the law of large numbers, we have

$$I_n - \frac{1}{m} \sum_{i=1}^{m} \mathbf{a}_i \mathbf{a}_i^* \to 0.$$

By the strong law of large numbers, we have (11.69).

Next we have

$$\mathbb{E}(\mathbf{Y}) = \mathbf{x_b}\mathbf{x_b^*} + \|\mathbf{x_b}\|^2 I_n,$$

whose proof is left to Exercise 10. We next use the strong law of large numbers to finish the proof. □

In the following, we fix $\delta > 0$ satisfying $2 + \delta - 2\sqrt{1 - 2\delta} < \frac{1}{8}$. Recall that $\mathbf{z_0}$ is an eigenvector corresponding to the leading eigenvalue λ_0 of $\mathbf{Y}$ with $\|\mathbf{z_0}\| = 1$. It is easy to see that

$$
\begin{aligned}
|\lambda_0 - (|\mathbf{z_0^*}\mathbf{x_b}|^2 + 1)| &= |\mathbf{z_0^*}\mathbf{Y}\mathbf{z_0} - \mathbf{z_0^*}(\mathbf{x_b}\mathbf{x_b^*} + I_n)\mathbf{z_0}| \\
&= |\mathbf{z_0^*}(\mathbf{Y} - (\mathbf{x_b}\mathbf{x_b^*} + I_n)\mathbf{z_0}|) \le \|\mathbf{Y} - (\mathbf{x_b}\mathbf{x_b^*} + \|\mathbf{x_b}\|^2 I_n)\| \le \delta.
\end{aligned}
$$

It follows that

$$|\mathbf{z_0^*}\mathbf{x_b}|^2 \ge \lambda_0 - 1 - \delta.$$

On the other hand, if $\mathbf{x_b}$ is normalized to have $\|\mathbf{x_b}\| = 1$, we have

$$\lambda_0 \ge \mathbf{x_b^*}\mathbf{Y}\mathbf{x_b} = \mathbf{x_b^*}(\mathbf{Y} - (I_n + \mathbf{x_b}\mathbf{x_b^*}))\mathbf{x_b} + 2 \ge 2 - \delta$$

by using (11.70). It then follows that

$$|\mathbf{z_0^*}\mathbf{x_b}|^2 \ge 1 - 2\delta.$$

Hence, we have

$$\text{dist}^2(\mathbf{z_0}, \mathbf{x_b}) = \inf_{\theta \in (0, 2\pi]} \|\mathbf{z_0}\|^2 - 2\exp(i\theta)\mathbf{z_0^*}\mathbf{x_b} + \|\mathbf{x_b}\|^2 = 2 - 2\sqrt{1 - 2\delta} < 1.$$

That is, $\text{dist}(\mathbf{z_0}, \mathbf{x_b}) < 2 - 2\sqrt{1 - 2\delta}$.

We are now ready to present a proof of Theorem 11.36.

Proof of Theorem 11.36. Recall that $\tilde{\mathbf{z}}_0 = \sqrt{\frac{1}{m}\sum_{k=1}^{m}|\mathbf{a_k^*}\mathbf{x_b}|^2}\mathbf{z_0}$. With high probability,

$$\left| \|\tilde{\mathbf{z}}_0\|^2 - 1 \right| = \left| \frac{1}{m}\sum_{k=1}^{m}\mathbf{x_b^*}\mathbf{a_k}\mathbf{a_k^*}\mathbf{x_b} - 1 \right| \le \delta$$

by using (11.69). It follows that $\left| \|\tilde{\mathbf{z}}_0\| - 1 \right| \le \delta$ and

$$\text{dist}(\tilde{\mathbf{z}}_0, \mathbf{x_b}) \le \|\mathbf{z_0} - \tilde{\mathbf{z}}_0\| + \text{dist}(\mathbf{z_0}, \mathbf{x_b}) = \left| \|\tilde{\mathbf{z}}_0\| - 1 \right| + \text{dist}(\mathbf{z_0}, \mathbf{x_b}) \le \delta + 2 - 2\sqrt{1 - 2\delta} < \frac{1}{8},$$

which completes the proof. □

11.6.2 ▪ Convergence of the Wirtinger Flow Algorithm

Finally, because the Hessian of f can be bounded below with high probability, we can show that the iterations $\mathbf{z}^k$ converge to a minimizer in a linear fashion. To state our main result in this section, we write $\mathbf{x_b}$ as the signal whose phase we need to recover, and let

$$P = \{\mathbf{x_b}\exp(i2\pi\theta), \theta \in [0, 1)\} \tag{11.71}$$

be the equivalence class of $\mathbf{x_b}$. For $\epsilon > 0$, let $\mathbf{E}(\epsilon) = \{\mathbf{z}, \text{dist}(\mathbf{z}, P) \le \epsilon\}$. We need a critical condition.

Definition 11.38 (Regularity Condition). *We say that the function f satisfies the regularity condition if for all vectors $\mathbf{z} \in \mathbf{E}(\epsilon)$, we have*

$$\mathrm{Re}(\langle ((D_{\mathbf{z}}f(\mathbf{z}))^*, \mathbf{z} - \exp(i2\pi\theta)\mathbf{x_b}\rangle) \geq \frac{1}{\alpha}\|\mathbf{z} - \exp(i2\pi\theta)\mathbf{x_b}\|^2 + \frac{1}{\beta}\|(D_{\mathbf{z}}f(\mathbf{z}))^*\|^2 \quad (11.72)$$

for two positive constants α, β.

Theorem 11.39 (Candés, Li, and Soltanolkotabi, 2015 [12]). *Suppose that $\mathbf{z}_0$ is sufficiently close to the equivalent class P of $\mathbf{x_b}$ to be retrieved, i.e., $\mathbf{z}_0 \in \mathbf{E}(1/8)$. Suppose that the Wirtinger gradient ∇f satisfies (11.72) for some α, β with high probability. Choosing step size $\tau_k = \nu \leq 1/n$ in (11.65), we have*

$$\mathrm{dist}(\mathbf{z}_k, \mathbf{x_b}) \leq C\left(1 - \frac{2\nu}{\alpha}\right)^{k/2} \quad (11.73)$$

for all $k \geq 1$, where $C > 0$ is a constant dependent on $\mathbf{x_b}$.

Proof. The proof is similar to the analysis of the gradient descent method for convex minimization. However, since our minimization is nonconvex, some justification is necessary. Mainly, we need to explain (11.72). We delay the explanation, which will hold with high probability to after the proof.

The rest is similar to the classic analysis of the linear convergence of the gradient descent method. By using (11.77) at the end of this subsection, we get

$$\|\mathbf{z}^{k+1} - \mathbf{x_b}\exp(i2\pi\theta)\|^2$$
$$= \|\mathbf{z}^k - \mathbf{x_b}\exp(i2\pi\theta)\|^2 - 2\nu\mathrm{Re}(\langle((D_{\mathbf{z}}f(\mathbf{z}^k))^*, \mathbf{z}^k - \mathbf{x}\exp(i2\pi\theta)\rangle) + \nu^2\|(D_{\mathbf{z}}f(\mathbf{z}^k))^*\|^2$$
$$\leq \|\mathbf{z}^k - \mathbf{x_b}\exp(i2\pi\theta)\|^2 - \frac{2\nu}{\alpha}\|\mathbf{z}^k - \mathbf{x_b}\exp(i2\pi\theta)\|^2 + \left(\nu^2 - \frac{2\nu}{\beta}\right)\|(D_{\mathbf{z}}f(\mathbf{z}^k))^*\|^2$$
$$\leq \left(1 - \frac{2\nu}{\alpha}\right)\|\mathbf{z}^k - \mathbf{x_b}\exp(i2\pi\theta)\|^2$$

since $\nu^2 - 2\nu/\beta < 0$ and $1 - 2\nu/\alpha > 0$ if $\nu > 0$ small enough. It follows that

$$\mathrm{dist}(\mathbf{z}^{k+1}, \mathbf{x_b})^2 \leq \left(1 - \frac{2\nu}{\alpha}\right)\mathrm{dist}(\mathbf{z}^k, \mathbf{x_b})^2.$$

Repeated application of the above inequality yields (11.73).   ☐

The remaining part of this subsection will establish the regularity condition (11.72). We need a series of lemmas.

Lemma 11.40. *Recall $(D_{\mathbf{z}}f)^*$ from Lemma 11.34. Let $\mathbf{z} \in \mathbb{C}^n$ be a fixed vector independent of the sampling vector. Then*

$$\mathbb{E}((D_{\mathbf{z}}f)^*) = (I_n - \mathbf{x_b}(\mathbf{x_b})^*)\mathbf{z} + 2(\|\mathbf{z}\|^2 - 1)\mathbf{z}. \quad (11.74)$$

Proof. We leave the proof to the interested reader; see also [12].   ☐

For convenience, we introduce

$$F(\mathbf{z}) = \frac{1}{2}\mathbf{z}^*(I - \mathbf{x_b}\mathbf{x_b}^*)\mathbf{z} + (\|\mathbf{z}\|^2 - 1)^2.$$

It is easy to see that $\nabla F(\mathbf{z}) = \mathbb{E}((D_{\mathbf{z}}f)^*)$ if $\mathbf{z} \in \mathbb{C}^n$ is a vector independent of the measurement vectors. Furthermore, letting $\mathbf{h} = \mathbf{z} - \mathbf{x_b}$, we have

$$\nabla F(\mathbf{z}) = (I_n - \mathbf{x_b}(\mathbf{x_b})^*)(\mathbf{x_b} + \mathbf{h}) + 2(\|\mathbf{x_b} + \mathbf{h}\|^2 - 1)(\mathbf{x_b} + \mathbf{h})$$
$$= (I_n - \mathbf{x_b}(\mathbf{x_b})^*)\mathbf{h} + 2(2\mathrm{Re}(\mathbf{x_b^*h}) + \|\mathbf{h}\|^2)(\mathbf{x_b} + \mathbf{h}).$$

It is easy to see that

$$\|\nabla F(\mathbf{z})\| \le \|I_n - \mathbf{x_b}(\mathbf{x_b})^*\|\|\mathbf{h}\| + 2(2\|\mathbf{h}\| + \|\mathbf{h}\|^2)(1 + \|\mathbf{h}\|) \le C\|\mathbf{h}\| \qquad (11.75)$$

for a positive constant C, when $\|\mathbf{h}\| \le 1/8$ and $\|\mathbf{x_b}\| = 1$. We now establish the regularity condition (11.72) in its expectation version. That is,

$$\mathrm{Re}(\langle \nabla F(\mathbf{z}), \mathbf{z} - \mathbf{x_b}\rangle)$$
$$= \mathrm{Re}(\langle (I_n - \mathbf{x_b}(\mathbf{x_b})^*)\mathbf{h} + 2(2\mathrm{Re}(\mathbf{x_b^*h}) + \|\mathbf{h}\|^2)(\mathbf{x_b} + \mathbf{h}), \mathbf{h}\rangle)$$
$$= \|\mathbf{h}\|^2 + 2\|\mathbf{h}\|^4 + 6\|\mathbf{h}\|^2\mathrm{Re}(\mathbf{x_b^*h}) + 3(\mathrm{Re}(\mathbf{x_b^*h}))^2 + (\mathrm{Im}(\mathbf{x_b^*h}))^2 \ge \frac{1}{4}\|\mathbf{h}\|^2,$$

where the last inequality holds because $\|\mathbf{h}\| \le 1/8$. Combining the above estimate with the one in (11.75), we have

$$\mathrm{Re}(\langle \nabla F(\mathbf{z}), \mathbf{z} - \mathbf{x_b}\rangle) \ge \frac{1}{4}\|\mathbf{h}\|^2 \ge \frac{1}{8}\|\mathbf{h}\|^2 + \frac{1}{8C}\|\nabla F(\mathbf{z})\|^2. \qquad (11.76)$$

Next we use the concentration inequality of random variable $(D_\mathbf{z}f)^*)$. That is, we have the following result.

Lemma 11.41 (Candés, Li, and Soltanolkotabi, 2015 [12]). *Suppose that $\mathbf{x_b}$ is a solution obeying $\|\mathbf{x_b}\| = 1$ which is independent of the sampling vectors $\mathbf{a}_k, k = 1, \dots, m$. Suppose that $\mathbf{a}_k$ are distributed according to the Gaussian model. Then, with high probability,*

$$\|(D_\mathbf{z}f(\mathbf{z}))^* - \nabla F(\mathbf{z})\| = \|(D_\mathbf{z}f(\mathbf{z}))^* - \mathbb{E}((D_\mathbf{z}f(\mathbf{z}))^*)\| \le \delta \mathrm{dist}(\mathbf{z}, \mathbf{x_b}).$$

Proof. We leave the proof to the interested reader; see also [12]. □

We are now ready to establish (11.72). Indeed, by Lemma 11.41 and (11.76),

$$\mathrm{Re}(\langle (D_\mathbf{z}f(\mathbf{z}))^*, \mathbf{z} - \exp(\mathbf{i}2\pi\theta)\mathbf{x_b}\rangle)$$
$$= \mathrm{Re}(\langle \nabla F(\mathbf{z}), \mathbf{z} - \exp(\mathbf{i}2\pi\theta)\mathbf{x_b}\rangle) + \mathrm{Re}(\langle (D_\mathbf{z}f(\mathbf{z}))^* - \nabla F(\mathbf{z}), \mathbf{z} - \exp(\mathbf{i}2\pi\theta)\mathbf{x_b}\rangle)$$
$$\ge \frac{1}{8}\|\mathbf{h}\|^2 + \frac{1}{8C}\|\nabla F(\mathbf{z})\|^2 - \|(D_\mathbf{z}f(\mathbf{z}))^* - \nabla F(\mathbf{z})\|\mathrm{dist}(\mathbf{z}, \mathbf{x_b})$$
$$\ge \frac{1}{8}\|\mathbf{h}\|^2 + \frac{1}{8C}\|\nabla F(\mathbf{z})\|^2 - \delta \mathrm{dist}(\mathbf{z}, \mathbf{x_b})^2$$
$$= \left(\frac{1}{8} - \delta\right)\mathrm{dist}(\mathbf{z}, \mathbf{x_b})^2 + \frac{1}{8C}\|\nabla F(\mathbf{z})\|^2$$

holds with high probability. Furthermore, by Lemma 11.41,

$$\|\nabla F(\mathbf{z})\|^2 \ge \|(D_\mathbf{z}f(\mathbf{z}))^*\|^2 - \|(D_\mathbf{z}f(\mathbf{z}))^* - \nabla F(\mathbf{z})\|^2 \ge \|(D_\mathbf{z}f(\mathbf{z}))^*\|^2 - \delta^2\mathrm{dist}(\mathbf{z}, \mathbf{x_b})^2$$

with high probability. Hence, we have

$$\mathrm{Re}(\langle (D_\mathbf{z}f(\mathbf{z}))^*, \mathbf{z} - \exp(\mathbf{i}2\pi\theta)\mathbf{x_b}\rangle)$$
$$\ge \left(\frac{1}{8} - \delta - \delta^2\right)\mathrm{dist}(\mathbf{z}, \mathbf{x_b})^2 + \frac{1}{8C}\|(D_\mathbf{z}f(\mathbf{z}))^*\|^2$$
$$=: \frac{1}{\alpha}\mathrm{dist}(\mathbf{z}, \mathbf{x_b})^2 + \frac{1}{\beta}\|(D_\mathbf{z}f(\mathbf{z}))^*\|^2, \qquad (11.77)$$

with $\alpha > 0$ and $\beta > 0$.

11.7 ▪ The Gauss–Newton Algorithm

In this section we derive a Gauss–Newton method to solve (11.48). Write $\mathbf{a}_j = \mathbf{a}_{j,R} + i\mathbf{a}_{j,I}$ and $f_j(\mathbf{x}) = \frac{1}{\sqrt{m}}((\langle \mathbf{a}_{j,R}, \mathbf{x} \rangle)^2 + (\langle \mathbf{a}_{j,I}, \mathbf{x} \rangle)^2 - b_j)$. Assuming $\mathbf{x}^{(k)}$ is given or found, the minimizing function can be approximated as

$$f(\mathbf{x}) = \frac{1}{2}\sum_{j=1}^{m}(f_j(\mathbf{x}))^2 \approx \frac{1}{2}\sum_{j=1}^{m}(f_j(\mathbf{x}^{(k)}) + \nabla_{\mathbf{x}}f_j(\mathbf{x}^{(k)})^{\top}(\mathbf{x} - \mathbf{x}^{(k)}))^2$$

$$= \frac{1}{2}\sum_{j=1}^{m}|f_j(\mathbf{x}^{(k)})|^2 + \sum_{j=1}^{m}f_j(\mathbf{x}^{(k)})\nabla_{\mathbf{x}}f_j(\mathbf{x}^{(k)})^{\top}(\mathbf{x} - \mathbf{x}^{(k)}) + \frac{1}{2}\sum_{j=1}^{m}|\nabla_{\mathbf{x}}f_j(\mathbf{x}^{(k)})^{\top}(\mathbf{x} - \mathbf{x}^{(k)})|^2.$$

The minimization (11.48) can be approximated by

$$\min_{\mathbf{x}\in\mathbb{R}^n} \sum_{j=1}^{m}f_j(\mathbf{x}^{(k)})\nabla_{\mathbf{x}}f_j(\mathbf{x}^{(k)})^{\top}(\mathbf{x} - \mathbf{x}^{(k)}) + \frac{1}{2}\sum_{j=1}^{m}|\nabla_{\mathbf{x}}f_j(\mathbf{x}^{(k)})^{\top}(\mathbf{x} - \mathbf{x}^{(k)})|^2, \qquad (11.78)$$

which is clearly a convex minimization problem. Since the minimization happens at the location where the gradient vector is zero, we can easily find the gradient vector, which is

$$\sum_{j=1}^{m}f_j(\mathbf{x}^{(k)})\nabla_{\mathbf{x}}f_j(\mathbf{x}^{(k)}) + \sum_{j=1}^{m}\nabla_{\mathbf{x}}f_j(\mathbf{x}^{(k)})\nabla_{\mathbf{x}}f_j(\mathbf{x}^{(k)})^{\top}(\mathbf{x} - \mathbf{x}^{(k)}) = 0. \qquad (11.79)$$

The classic Newton method for zeros of nonlinear equations is

$$\mathbf{x}^{(k+1)} = \mathbf{x}^{(k)} - \left(\sum_{j=1}^{m}\nabla_{\mathbf{x}}f_j(\mathbf{x}^{(k)})\nabla_{\mathbf{x}}f_j(\mathbf{x}^{(k)})^{\top}\right)^{-1}\sum_{j=1}^{m}f_j(\mathbf{x}^{(k)})\nabla_{\mathbf{x}}f_j(\mathbf{x}^{(k)}), \qquad (11.80)$$

which can simply be written as

$$\mathbf{x}^{(k+1)} = \mathbf{x}^{(k)} - \left(\sum_{j=1}^{m}\nabla f_j(\mathbf{x}^{(k)})\nabla f_j(\mathbf{x}^{(k)})^{\top}\right)^{-1}\nabla f(\mathbf{x}^{(k)}). \qquad (11.81)$$

Under the assumption that the matrix $\sum_{j=1}^{m}\nabla f_j(\mathbf{x}^{(k)})\nabla f_j(\mathbf{x}^{(k)})^{\top}$ is invertible, we obtain a Gauss–Newton method for solving the minimization (11.48). More precisely, the Gauss–Newton method is given in Algorithm 11.2.

ALGORITHM 11.2
The Gauss–Newton Algorithm (cf. [18])
   1: Start with an initial guess $\mathbf{x}^0$. For example, use Algorithm 11.4 below.
   2: For $k \geq 1$, we solve (11.81) in the real variable setting after a few iterations, say 10 to 15 iterations.
   3: Output the last iterative solution $\mathbf{x}^{k+1}$.

Note that when $\mathbf{x}$ is a complex-valued solution, the matrix $\sum_{j=1}^{m}\nabla f_j(\mathbf{x}^{(k)})\nabla f_j(\mathbf{x}^{(k)})^{\top}$ is not invertible. We use the pseudo-inverse to solve (11.79) instead, which leads to Algorithm 11.3.

ALGORITHM 11.3

The Complex-Valued Gauss–Newton Algorithm (cf. [18])

1: Start with an initial guess $\mathbf{z}^0$. For example, use Algorithm 11.4 below.
2: For $k \geq 1$, we solve (11.79) by using the pseudo-inverse method after a few iterations, say 10 to 15 iterations.
3: Output the last iterative solution $\mathbf{x}^{k+1}$.

Because the inverse of $(\sum_{j=1}^{m} \nabla_{\mathbf{x}} f_j(\mathbf{x}^{(k)}) \nabla_{\mathbf{x}} f_j(\mathbf{x}^{(k)})^{\top})^{-1}$ is not easy for large problems, we may use the Barzilai–Borwein technique to simplify the computation. The details will be explained in subsection 11.7.3.

11.7.1 ▪ Initialization

Next we explain an initialization for Algorithm 11.2 which was invented in [18]. It is given in Algorithm 11.4.

ALGORITHM 11.4

The Gauss–Newton Initialization Algorithm

1: **Input:** Observations $b_\ell \in \mathbb{R}, \ell = 1, \ldots, m$.
2: Set $\lambda^2 = \frac{1}{m} \sum_{j=1}^{m} b_j$.
3: Set $\mathbf{z}^{(1)}$ to be an eigenvector corresponding to the largest eigenvalue of matrix Y, where

$$Y = \frac{1}{m} \sum_{\ell=1}^{m} \left(\frac{1}{2} - \exp\left(-\frac{b_\ell}{\lambda^2} \right) \right) \mathbf{a}_\ell \mathbf{a}_\ell^* \tag{11.82}$$

for the complex case and

$$Y = \frac{1}{m} \sum_{\ell=1}^{m} \left(\frac{1}{\sqrt{3}} - \exp\left(-\frac{b_\ell}{\lambda^2} \right) \right) \mathbf{a}_\ell \mathbf{a}_\ell^{\top} \tag{11.83}$$

for the real case.
4: Then normalize the eigenvector such that $\|\mathbf{z}^{(1)}\|_2 = \lambda$.
5: **Output:** $\mathbf{z}^{(1)}$.

We now show that the initialization in Algorithm 11.4 is a good one.

Theorem 11.42 (Gao and Xu, 2017 [18]). *Suppose that* $\mathbf{a}_j \in \mathbb{C}^n$, $j = 1, \ldots, m$, *are Gaussian random measurements,* $\mathbf{x}_b \in \mathbb{C}^n$ *is the signal to be retrieved, and* $\mathbf{z}^{(1)}$ *is from Algorithm 11.4. For any* $\theta > 0$, *there exists a positive constant* C_θ *such that when* $m \geq C_\theta n$,

$$\text{dist}(\mathbf{z}^{(1)}, \mathbf{x}_b) \leq \sqrt{3\theta} \|\mathbf{x}_b\| \tag{11.84}$$

holds with high probability.

To prove the results above, we need some preparatory results. Let us begin with the following simple fact, which was used in the previous section. Since we need a slightly stronger version, let us explain it again. Since Gaussian random variables $\mathbf{a}_j$ are subject to $N(0, 1)$, we have

$$\mathbb{E}\left(\frac{1}{m} \sum_{j=1}^{m} \mathbf{a}_j \mathbf{a}_j^* \right) = I_{n \times n}, \tag{11.85}$$

where $I_{n \times n}$ is the identity matrix of size $n \times n$. Due to the Hoeffding inequality, we have that for any $\eta > 0$, there is an M dependent on n such that $m \geq M$, where

$$\left\| \frac{1}{m} \sum_{j=1}^{m} \mathbf{a}_j \mathbf{a}_j^* - I_{n \times n} \right\| \leq \frac{1}{4} \eta \tag{11.86}$$

and $M = Cn$ with $C \geq 1$. It follows that

$$|\lambda^2 - \|\mathbf{x_b}\|^2| \leq \frac{\eta}{4} \|\mathbf{x_b}\|^2, \tag{11.87}$$

where $\lambda^2 = \sum_{j=1}^{m} b_j / m$, as in Algorithm 11.4. Next we need the following lemma.

Lemma 11.43. *Suppose that* $\mathbf{a} = x + iy \in \mathbb{C}$ *with* x *and* y *i.i.d. random variables subject to* $\mathcal{N}(0, 1/2)$*. Show that the even moments of* $\mathbf{a}$ *satisfy*

$$\mathbb{E}(|\mathbf{a}|^{2k}) = k! \quad \forall k \geq 1,$$

where p is an integer.

Proof. Since x, y are i.i.d. random variables subject to $\mathcal{N}(0, 1/2)$, the even moment of the complex Gaussian random variable $\mathbf{a} = x + iy$ is

$$\mathbb{E}(|\mathbf{a}|^{2k}) = \mathbb{E}_{x,y}(|x|^2 + |y|^2)^k = \frac{1}{2^k} \mathbb{E}_{z \in \chi^2(2)}(z^k) = \frac{1}{2^k} 2^k k! = k!,$$

where $\chi^2(n)$ is the standard χ distribution of n Gaussian random variables. $\qquad \square$

With the above results, we can show that

$$\mathbb{E}\left(\frac{1}{m} \sum_{j=1}^{m} \exp\left(-\frac{|\mathbf{a}_j^* \mathbf{z}|^2}{\|\mathbf{z}\|^2} \right) \mathbf{a}_j \mathbf{a}_j^* \right) = \frac{1}{2} I_{n \times n} - \frac{\mathbf{z} \mathbf{z}^*}{4 \|\mathbf{z}\|^2}, \tag{11.88}$$

where the proof is left to Exercise 13. Again by the Hoeffding inequality, we have that

$$\left\| \frac{1}{m} \sum_{j=1}^{m} \exp\left(-\frac{|\mathbf{a}_j^* \mathbf{z}|^2}{\|\mathbf{z}\|^2} \mathbf{a}_j \mathbf{a}_j^* \right) - \frac{1}{2} I_{n \times n} - \frac{\mathbf{z} \mathbf{z}^*}{4 \|\mathbf{z}\|^2} \right\| \leq \frac{1}{8} \eta \tag{11.89}$$

holds with high probability when $m \geq Cn$. Instead of Y in (11.82), we let

$$Y_1 = \frac{1}{m} \sum_{j=1}^{m} \left(\frac{1}{2} - \exp\left(-\frac{|\mathbf{a}_j^* \mathbf{x_b}|^2}{\|\mathbf{x_b}\|^2} \right) \right) \mathbf{a}_j \mathbf{a}_j^*.$$

Based on the discussion above, it is easy to see that

$$\|Y_1 - \mathbb{E}(Y_1)\| \leq \frac{1}{2} \left\| \frac{1}{m} \sum_{j=1}^{m} \mathbf{a}_j \mathbf{a}_j^* - I_{n \times n} \right\|$$

$$+ \left\| \frac{1}{m} \sum_{j=1}^{m} \exp\left(-\frac{|\mathbf{a}_j^* \mathbf{x_b}|^2}{\|\mathbf{x_b}\|^2} \mathbf{a}_j \mathbf{a}_j^* \right) - \frac{1}{2} I_{n \times n} - \frac{\mathbf{x_b} \mathbf{x_b}^*}{4 \|\mathbf{x_b}\|^2} \right\|$$

$$\leq \frac{1}{4} \eta.$$

Finally, recalling Y defined in (11.82),

$$Y - Y_1 = \frac{1}{m} \sum_{j=1}^{m} \left[\exp\left(-\frac{|a_j^* x_b|^2}{\|z\|^2} \right) - \exp\left(-\frac{|a_j^* x_b|^2}{\lambda^2} \right) \right] a_j a_j^*.$$

To estimate the difference of $Y - Y_1$, we recall a standard inequality: $|1 - e^x| \leq |x|$ for all x and $e^{-x} x \leq 1$ for all $x \geq 0$. Since $\|a_j\| \leq 1$, it follows that

$$\|Y - Y_1\| \leq \frac{1}{m} \sum_{j=1}^{m} \exp\left(-\frac{|a_j^* x_b|^2}{\|x_b\|^2} \right) \left| 1 - \exp\left(|a_j^* x_b|^2 \left(\frac{1}{\|x_b\|^2} - \frac{1}{\lambda^2} \right) \right) \right|$$

$$\leq \frac{1}{m} \sum_{j=1}^{m} \exp\left(-\frac{|a_j^* x_b|^2}{\|x_b\|^2} \right) |a_j^* x_b|^2 \frac{|\lambda^2 - \|x_b\|^2|}{\|x_b\|^2 \lambda^2}$$

$$\leq \frac{1}{m} \sum_{j=1}^{m} \frac{1}{\lambda^2} |\lambda^2 - \|x_b\|^2| \leq \eta$$

by using (11.87), where we have used the fact that λ is around $1/4$ to be given below.

We are now ready to prove Theorem 11.42.

Proof of Theorem 11.42. Let z_0 be an eigenvector associated with the largest eigenvalue λ_{max} of matrix Y with unit norm. We first claim that λ_{max} is around $1/4$ with high probability. Indeed, we have

$$\mathbb{E}(Y_1) = \frac{x_b x_b^*}{4\|x_b\|^2}$$

and $\|Y - \frac{x_b x_b^*}{4\|x_b\|^2}\| \leq \|Y - Y_1\| + \|Y_1 - \mathbb{E}(Y_1)\| + \|\mathbb{E}(Y_1) - \frac{x_b x_b^*}{4\|x_b\|^2}\|$, which can be made very small when m is very large, say less than η. Since the largest eigenvalue of $\frac{x_b x_b^*}{4\|x_b\|^2}$ is $1/4$, we have

$$\left| \lambda_{max}(Y) - \frac{1}{4} \right| \leq \left\| Y - \frac{x_b x_b^*}{4\|x_b\|^2} \right\| \leq \eta.$$

It follows that

$$\eta \geq \left| z_0^* \left(Y - \frac{x_b x_b^*}{4\|x_b\|^2} \right) z_0 \right| = \left| \lambda_{max} - \frac{1}{4} + \frac{1}{4} - \frac{|x_b^* z_0|^2}{4\|x_b\|^2} \right|$$

$$\geq \left| \frac{1}{4} - \frac{|x_b^* z_0|^2}{4\|x_b\|^2} \right| - \left| \lambda_{max} - \frac{1}{4} \right| \geq \left| \frac{1}{4} - \frac{|x_b^* z_0|^2}{4\|x_b\|^2} \right| - \eta.$$

Hence, we have $\frac{|x_b^* z_0|^2}{\|x_b\|^2} \geq 1 - 8\eta$ or

$$|x_b^* z_0|^2 \geq (1 - 8\eta) \|x_b\|^2. \tag{11.90}$$

That is, setting $\hat{z}_0 = \lambda_{max} z_0$ and letting $z = x_b$ for convenience, we have

$$\text{dist}^2(\hat{z}_0, z) = \min\{\|z - \exp(i\theta)\lambda_{max} z_0\|^2\}$$

$$\leq \|z\|^2 + \lambda_{max}^2 - 2\lambda_{max}|z^* z_0|$$

$$\leq \|z\|^2 + (1 + \eta/4)\|z\|^2 - 2(1 - 8\eta)\|z\|^2 = 3\theta\|z\|^2$$

for an appropriate $\theta > 0$, where we have used (11.87) and (11.90).

Similarly, we can establish the estimate for the real variable setting. The details are left to Exercise 18. This completes the proof. □

11.7.2 ▪ Convergence Analysis

Let us present a convergence analysis of the Gauss–Newton Algorithm 11.2 as follows.

Theorem 11.44 (Gao and Xu, 2017 [18]). *Let $x_b \in \mathbb{R}^n$ be an arbitrary vector with $\|x_b\| = 1$ and $b_j = |\langle a_j, x_b \rangle|^2$, where $a_j \in \mathbb{R}^n$, $j = 1, \ldots, m$, are Gaussian random measurement vectors with $m \geq Cn \log n$. Suppose $0 < \delta \leq 1/81$ is a constant and $x^{(k)} \in \mathbb{R}^n$ satisfies $\mathrm{dist}(x^{(k)}, x_b) \leq \sqrt{\delta}$. Suppose that $x^{(k+1)}$ is from the Gauss–Newton algorithm, Algorithm 11.2. With high probability, we have*

$$\mathrm{dist}(x^{(k+1)}, x_b) \leq \beta \mathrm{dist}^2(x^{(k)}, x_b), \tag{11.91}$$

where

$$\beta = \frac{24(4 + \frac{\delta}{2})(1 + \sqrt{\delta})}{(16 - \delta)(1 - \sqrt{\delta})^2}. \tag{11.92}$$

To prove Theorem 11.44 we need some essential lemmas.

Lemma 11.45. *Suppose that a_j, $j = 1, 2, \ldots, m$, are Gaussian random measurement vectors and $m \geq Cn \log n$, where C is sufficiently large. Set*

$$S := \frac{1}{m} \sum_{j=1}^{m} |a_j^* e_1|^2 a_j a_j^*.$$

Then for any $\delta > 0$, $\|S - \mathbb{E}(S)\| \leq \frac{\delta}{4}$ holds with high probability.

Proof. We leave the proof to the interested reader; see also Lemma 7.4 in [12]. □

Recall that $S_k := \{\lambda x_b + (1 - \lambda) x^{(k)} : 0 \leq \lambda \leq 1\}$, $J(x) := (\nabla F_1(x), \ldots, \nabla F_m(x))$ and

$$J(x)^\top J(x) = \sum_{j=1}^{m} \nabla F_j(x) \nabla F_j(x)^\top = \frac{4}{m} \sum_{j=1}^{m} \langle a_j, x \rangle^2 \cdot a_j a_j^\top.$$

Lemma 11.46. *Suppose that $x^{(k)} \in \mathbb{R}^n$ and $x_b \in \mathbb{R}^n$ with $\|x_b\| = 1$ and $\|x^{(k)} - x_b\| \leq \sqrt{\delta}$, where $0 < \delta \leq 1/81$ is a constant. Suppose that the measurement vectors a_j, $j = 1, \ldots, m$, are Gaussian random measurements which are independent of $x^{(k)}$ and x_b. Then when $m \geq Cn \log n$,*

$$J(x)^\top J(x) = \frac{4}{m} \sum_{j=1}^{m} \langle a_j, x \rangle^2 a_j a_j^\top$$

is L_J-Lipschitz continuous on S_k with high probability, i.e., for any $x, y \in S_k$,

$$\|J(x)^\top J(x) - J(y)^\top J(y)\| \leq L_J \|x - y\|,$$

where $L_J = 8(4 + \frac{\delta}{2})(1 + \sqrt{\delta})$.

Proof. For $x, y \in S_k$, we can write x, y in the form

$$x = \lambda_1 x^{(k)} + (1 - \lambda_1) x_b, \quad \lambda_1 \in [0, 1],$$
$$y = \lambda_2 x^{(k)} + (1 - \lambda_2) x_b, \quad \lambda_2 \in [0, 1].$$

Since the measurement vectors $\mathbf{a}_j$, $j = 1, \ldots, m$, are rotationally invariant and independent of $\mathbf{x}^{(k)}$ and $\mathbf{x}_b$, without loss of generality, we may assume that $\mathbf{x}_b = \mathbf{e}_1$ and decompose $\mathbf{x}^{(k)}$ in the following format: $\mathbf{x}^{(k)} = \|\mathbf{x}^{(k)}\|(\alpha \mathbf{e}_1 + \sqrt{1 - \alpha^2}\mathbf{e}_2)$, where $\mathbf{e}_2$ is an orthogonal direction to $\mathbf{e}_1$ and $\alpha = \langle \mathbf{x}^{(k)}, \mathbf{x}_b \rangle / \|\mathbf{x}^{(k)}\|$. Because $\|\mathbf{x}^{(k)} - \mathbf{x}_b\| \leq \sqrt{\delta}$, we have $\langle \mathbf{x}^{(k)}, \mathbf{x}_b \rangle \geq 0$, i.e., $\alpha \geq 0$. It follows that

$$\|J(\mathbf{x})^\top J(\mathbf{x}) - J(\mathbf{y})^\top J(\mathbf{y})\|$$

$$= 4 \left\| \frac{1}{m} \sum_{j=1}^{m} (\langle \mathbf{a}_j, \mathbf{x} \rangle^2 - \langle \mathbf{a}_j, \mathbf{y} \rangle^2) \mathbf{a}_j \mathbf{a}_j^\top \right\| = 4 \left\| \frac{1}{m} \sum_{j=1}^{m} \langle \mathbf{a}_j, \mathbf{x} + \mathbf{y} \rangle \langle \mathbf{a}_j, \mathbf{x} - \mathbf{y} \rangle \mathbf{a}_j \mathbf{a}_j^\top \right\|$$

$$\leq 4 \|\mathbf{x} + \mathbf{y}\| \|\mathbf{x} - \mathbf{y}\| \left\| \frac{1}{m} \sum_{j=1}^{m} ((\mathbf{a}_j^\top \mathbf{e}_1)^2 + (\mathbf{a}_j^\top \mathbf{e}_2)^2) \mathbf{a}_j \mathbf{a}_j^\top \right\|, \tag{11.93}$$

where the last inequality is obtained by the Cauchy–Schwarz inequality.

Next, letting

$$S := \frac{1}{m} \sum_{j=1}^{m} ((\mathbf{a}_j^\top \mathbf{e}_1)^2 + (\mathbf{a}_j^\top \mathbf{e}_2)^2) \mathbf{a}_j \mathbf{a}_j^\top,$$

we know that $\mathbb{E}(S) = 2I_n + 2\mathbf{e}_1 \mathbf{e}_1^\top + 2\mathbf{e}_2 \mathbf{e}_2^\top$ and $\|\mathbb{E}(S)\| = 4$. According to Lemma 11.45, for $0 < \delta \leq 1/81$ and $m \geq Cn \log n$, $\|S - \mathbb{E}(S)\| \leq \frac{\delta}{2}$ holds with high probability, so

$$\|S\| \leq 4 + \frac{\delta}{2}. \tag{11.94}$$

On the other hand, since $\|\mathbf{x}^{(k)} - \mathbf{x}_b\| \leq \sqrt{\delta}$, we have

$$1 - \sqrt{\delta} \leq \|\mathbf{x}^{(k)}\| \leq 1 + \sqrt{\delta}. \tag{11.95}$$

Thus

$$\|\mathbf{x} + \mathbf{y}\| = \|(\lambda_1 + \lambda_2)\mathbf{x}^{(k)} + (2 - \lambda_1 - \lambda_2)\mathbf{x}_b\|$$
$$\leq (\lambda_1 + \lambda_2)\|\mathbf{x}^{(k)}\| + (2 - \lambda_1 - \lambda_2) \leq 2(1 + \sqrt{\delta}). \tag{11.96}$$

Putting (11.94) and (11.96) into (11.93), we obtain

$$\|J(\mathbf{x})^\top J(\mathbf{x}) - J(\mathbf{y})^\top J(\mathbf{y})\| \leq 8 \left(4 + \frac{\delta}{2} \right) (1 + \sqrt{\delta})\|\mathbf{x} - \mathbf{y}\|.$$

So we conclude that when $m \geq Cn \log n$, $J(\mathbf{x})^\top J(\mathbf{x})$ is Lipschitz continuous on the line S_k with high probability. □

Next we let

$$H(\mathbf{x}) := \nabla^2 f(\mathbf{x}) - J(\mathbf{x})^\top J(\mathbf{x}) = \frac{2}{m} \sum_{j=1}^{m} (\langle \mathbf{a}_j, \mathbf{x} \rangle^2 - b_j) \mathbf{a}_j \mathbf{a}_j^\top. \tag{11.97}$$

Corollary 11.47. *Under the same conditions as in Lemma 11.46, the Hessian matrix in (11.97) is Lipschitz continuous on S_k with Lipschitz constant $L_H = \frac{1}{2}L_J$ with high probability.*

Proof. For any $\mathbf{x}, \mathbf{y} \in S_k$, we have

$$H(\mathbf{x}) - H(\mathbf{y}) = \frac{2}{m} \sum_{j=1}^{m} (\langle \mathbf{a}_j, \mathbf{x} \rangle^2 - b_j - \langle \mathbf{a}_j, \mathbf{y} \rangle^2 + b_j) \mathbf{a}_j \mathbf{a}_j^\top$$

$$= \frac{2}{m} \sum_{j=1}^{m} (\langle \mathbf{a}_j, \mathbf{x} \rangle^2 - \langle \mathbf{a}_j, \mathbf{y} \rangle^2) \mathbf{a}_j \mathbf{a}_j^\top = \frac{1}{2} (J(\mathbf{x})^\top J(\mathbf{x}) - J(\mathbf{y})^\top J(\mathbf{y})).$$

Then by using the Lipschitz differentiability of $J(\mathbf{x})^\top J(\mathbf{x})$, it follows that

$$\|H(\mathbf{x}) - H(\mathbf{y})\| = \left\| \frac{1}{2} (J(\mathbf{x})^\top J(\mathbf{x}) - J(\mathbf{y})^\top J(\mathbf{y})) \right\| \leq \frac{1}{2} L_J \|\mathbf{x} - \mathbf{y}\|.$$

So $H(\mathbf{x})$ is Lipschitz continuous on S_k with constant $L_H = \frac{1}{2} L_J$.   □

Next we present an estimation of the largest eigenvalue of $(J(\mathbf{x}^{(k)})^\top J(\mathbf{x}^{(k)}))^{-1}$.

Lemma 11.48. *Suppose that $\|\mathbf{x}^{(k)} - \mathbf{x_b}\| \leq \sqrt{\delta}$, where $\mathbf{x}^{(k)}, \mathbf{x_b} \in \mathbb{R}^n$, with $\|\mathbf{x_b}\| = 1$ and $0 < \delta \leq 1/81$. Suppose that $\mathbf{a}_j$, $j = 1, \ldots, m$, are Gaussian random measurements which are independent of $\mathbf{x}^{(k)}$. If $m \geq Cn \log n$ for a sufficiently large constant C, then with high probability*

$$\|(J(\mathbf{x}^{(k)})^\top J(\mathbf{x}^{(k)}))^{-1}\| \leq \frac{4}{(16 - \delta)(1 - \sqrt{\delta})^2}.$$

Proof. Set

$$S := J(\mathbf{x}^{(k)})^\top J(\mathbf{x}^{(k)}) = \frac{4}{m} \sum_{j=1}^{m} \langle \mathbf{a}_j, \mathbf{x}^{(k)} \rangle^2 \mathbf{a}_j \mathbf{a}_j^\top.$$

After a simple calculation, we obtain $\mathbb{E}(S) = 4(\|\mathbf{x}^{(k)}\|^2 I_n + 2\mathbf{x}^{(k)}(\mathbf{x}^{(k)})^\top)$, and the minimum eigenvalue of $\mathbb{E}(S)$ is $\lambda_{\min}(\mathbb{E}(S)) = 4\|\mathbf{x}^{(k)}\|^2$. According to Lemma 11.45, for $0 < \delta \leq 1/81$ and $m \geq Cn \log n$,

$$\|S - \mathbb{E}(S)\| \leq \frac{\delta}{4} \|\mathbf{x}^{(k)}\|^2$$

holds with probability at least $1 - 5 \exp(-\gamma_\delta n) - 4/n^2$. Then, according to the Wely theorem, we have

$$|\lambda_{\min}(S) - \lambda_{\min}(\mathbb{E}(S))| \leq \|S - \mathbb{E}(S)\| \leq \frac{\delta}{4} \|\mathbf{x}^{(k)}\|^2,$$

which implies that

$$\lambda_{\min}(S) \geq \left(4 - \frac{\delta}{4} \right) \|\mathbf{x}^{(k)}\|^2 \geq \left(4 - \frac{\delta}{4} \right) (1 - \sqrt{\delta})^2,$$

where we have used (11.95) in the last inequality. Then, with high probability,

$$\lambda_{\max}(S^{-1}) = 1/\lambda_{\min}(S) \leq \frac{4}{(16 - \delta)(1 - \sqrt{\delta})^2}, \tag{11.98}$$

which establishes the result.   □

Finally, we are ready to present the proof of Theorem 11.44.

Proof of Theorem 11.44. Without loss of generality, we only consider the case where $\langle x^{(k)}, x_b \rangle \geq 0$, i.e., $\text{dist}(x^{(k)}, x_b) = \|x^{(k)} - x_b\|$. Then we just need to prove that when $\|x^{(k)} - x_b\| \leq \sqrt{\delta}$ and $m \geq Cn \log n$,

$$\text{dist}(x^{(k+1)}, x_b) = \|x^{(k+1)} - x_b\| \leq \beta \|x^{(k)} - x_b\|^2 = \beta \text{dist}^2(x^{(k)}, x_b)$$

holds with high probability. Since x_b is an exact solution to (11.48), we have $\nabla f(x_b) = H(x_b) = 0$. The definition of $x^{(k+1)}$ shows that

$$x^{(k+1)} - x_b = x^{(k)} - x_b - (J(x^{(k)})^\top J(x^{(k)}))^{-1} \nabla f(x^{(k)}) \tag{11.99}$$

$$= \left(J(x^{(k)})^\top J(x^{(k)}) \right)^{-1} \cdot \left((J(x^{(k)})^\top J(x^{(k)})) \cdot (x^{(k)} - x_b) - (\nabla f(x^{(k)}) - \nabla f(x_b)) \right).$$

Recall that $S_k := \{x^{(k)} + t(x_b - x^{(k)}) : 0 \leq t \leq 1\}$ and $x(t) = x^{(k)} + t(x_b - x^{(k)})$. Then we have

$$\nabla f(x^{(k)}) - \nabla f(x_b) = \nabla f(x(0)) - \nabla f(x(1)) = -\int_0^1 \frac{d(\nabla f(x(t)))}{dt} dt \tag{11.100}$$

$$= -\int_0^1 \nabla^2 f(x(t)) \cdot x'(t) dt = -\frac{1}{\|x^{(k)} - x_b\|} \int_{S_k} \nabla^2 f(x) \cdot (x_b - x^{(k)}) dx.$$

The integral in (11.100) is interpreted as elementwise. Combining (11.97) and $H(x_b) = 0$, we have

$$\left\| (J(x^{(k)})^\top J(x^{(k)})) \cdot (x^{(k)} - x_b) - (\nabla f(x^{(k)}) - \nabla f(x_b)) \right\|$$

$$= \frac{1}{\|x^{(k)} - x_b\|} \left\| \int_{S_k} (J(x^{(k)})^\top J(x^{(k)})) \cdot (x^{(k)} - x_b) - \nabla^2 f(x) \cdot (x^{(k)} - x_b)) dx \right\|$$

$$= \frac{1}{\|x^{(k)} - x_b\|} \left\| \int_{S_k} (J(x^{(k)})^\top J(x^{(k)}) - J(x)^\top J(x) - H(x)) \cdot (x^{(k)} - x_b) dx \right\|$$

$$\leq \frac{1}{\|x^{(k)} - x_b\|} \left(\left\| \int_{S_k} \left(J(x^{(k)})^\top J(x^{(k)}) - J(x)^\top J(x) \right) \cdot (x^{(k)} - x_b) dx \right\| \right.$$

$$\left. + \left\| \int_{S_k} (H(x) - H(x_b)) \cdot (x^{(k)} - x_b) dx \right\| \right). \tag{11.101}$$

Since $J(x)^\top J(x)$ and $H(x)$ are Lipschitz continuous on the line S_k with high probability, it follows from (11.101) that

$$\left\| (J(x^{(k)})^\top J(x^{(k)})) \cdot (x^{(k)} - x_b) - (\nabla f(x^{(k)}) - \nabla f(x_b)) \right\|$$

$$\leq \int_{S_k} \left\| J(x^{(k)})^\top J(x^{(k)}) - J(x)^\top J(x) \right\| dx + \int_{S_k} \|H(x) - H(x_b)\| dx$$

$$\leq \int_{S_k} L_J \|x^{(k)} - x\| dx + \int_{S_k} L_H \|x - x_b\| dx$$

$$= L_J \|x^{(k)} - x_b\| \int_0^1 t \cdot \|x^{(k)} - x_b\| dt + L_H \|x^{(k)} - x_b\| \int_0^1 (1-t) \cdot \|x^{(k)} - x_b\| dt$$

$$= \frac{L_J + L_H}{2} \cdot \left\| x^{(k)} - x_b \right\|^2 = 6 \left(4 + \frac{\delta}{2} \right) (1 + \sqrt{\delta}) \left\| x^{(k)} - x_b \right\|^2.$$

Thus, according to Lemma 11.48 and (11.99), when $m \geq Cn \log n$,

$$
\begin{aligned}
\|\mathbf{x}^{(k+1)} - \mathbf{x_b}\| &= \|(J(\mathbf{x}^{(k)})^\top J(\mathbf{x}^{(k)}))^{-1}\| \\
&\quad \cdot \|(J(\mathbf{x}^{(k)})^\top J(\mathbf{x}^{(k)})) \cdot (\mathbf{x}^{(k)} - \mathbf{x_b}) - (\nabla f(\mathbf{x}^{(k)}) - \nabla f(\mathbf{x_b}))\| \\
&\leq \frac{4}{(16 - \delta)(1 - \sqrt{\delta})^2} \cdot 6 \left(4 + \frac{\delta}{2}\right)(1 + \sqrt{\delta}) \left\|\mathbf{x}^{(k)} - \mathbf{x_b}\right\|^2 = \beta \left\|\mathbf{x}^{(k)} - \mathbf{x_b}\right\|^2 \quad (11.102)
\end{aligned}
$$

holds with high probability. Based on the discussion before, we have

$$
\|\mathbf{x}^{(k+1)} - \mathbf{x_b}\| \leq \beta \left\|\mathbf{x}^{(k)} - \mathbf{x_b}\right\|^2 \leq \sqrt{\delta}.
$$

Then we have $\langle \mathbf{x}^{(k+1)}, \mathbf{x_b} \rangle \geq 0$, i.e., $\mathrm{dist}(\mathbf{x}^{(k+1)}, \mathbf{x_b}) = \|\mathbf{x}^{(k+1)} - \mathbf{x_b}\|$. Then (11.102) implies the conclusion. $\square$

In Theorem 11.44 above, we require $0 < \delta \leq 1/81$ in order to guarantee $\beta\delta \leq \sqrt{\delta}$, i.e., when

$$
\mathrm{dist}(\mathbf{x}^{(k)}, \mathbf{x_b}) \leq \sqrt{\delta},
$$

the condition $\mathrm{dist}(\mathbf{x}^{(k+1)}, \mathbf{x_b}) \leq \beta\delta \leq \sqrt{\delta}$ still holds and we can use Theorem 11.44 at the $(k+1)$st iteration. Note that for any $0 < \delta \leq 1/81$ and $0 < \theta \leq \delta/8$, we have

$$
\mathrm{dist}(\mathbf{x}^{(0)}, \mathbf{x_b}) = \min\{\|\mathbf{x}^{(0)} - \mathbf{x_b}\|, \|\mathbf{x}^{(0)} + \mathbf{x_b}\|\} \leq 2\sqrt{2\theta} \leq \sqrt{\delta}
$$

with high probability. Combining this initialization result with Theorem 11.44, we have the following conclusion on the number of iterative steps.

Theorem 11.49. *Suppose that* $\mathbf{x_b} \in \mathbb{R}^n$ *with* $\|\mathbf{x_b}\| = 1$ *is an arbitrary vector and* $\mathbf{a}_j \in \mathbb{R}^n$, $j = 1, \ldots, M$, *are Gaussian random measurements. Suppose that* ϵ *is an arbitrary constant within the range* $(0, 1/2)$ *and* $\delta \in (0, 1/81]$ *is a fixed constant. Set* $\mathbf{b} = (\langle \mathbf{a}_j, \mathbf{x_b} \rangle)^2, j = 1, \ldots, m$, *and* $k_0 \geq \max\{0, \log_2 \log_2 \frac{1}{\epsilon} - \log_2 \log_2 \frac{1}{\beta\sqrt{\delta}}\}$, *and let* β *be as given in* (11.92). *If* $M \geq C \cdot \log_2 \log_2 \frac{1}{\epsilon} \cdot n \log n$. *Then with high probability, Algorithm 11.2 outputs* $\mathbf{x}^{(k)}$ *satisfying*

$$
\mathrm{dist}(\mathbf{x}^{(k)}, \mathbf{x_b}) < \epsilon
$$

for all $k \geq k_0$, *where* C *is a constant depending on* δ.

Proof. Suppose that $0 < \delta \leq 1/81$. According to the initialization discussed in the previous subsection, we have

$$
\mathrm{dist}(\mathbf{x}^{(0)}, \mathbf{x_b}) \leq \sqrt{\delta}
$$

with high probability. As mentioned above, we have $\beta\delta \leq \sqrt{\delta}$, where β is defined in Theorem 11.44. Iterating (11.91) in Theorem 11.44 k times with $k \geq k_0$ leads to

$$
\begin{aligned}
\mathrm{dist}(\mathbf{x}^{(k)}, \mathbf{x_b}) &\leq \beta \cdot \mathrm{dist}^2(\mathbf{x}^{(k-1)}, \mathbf{x_b}) \leq \beta^{2^k - 1} \mathrm{dist}^{2^k}(\mathbf{x}^{(0)}, \mathbf{x_b}) \\
&\leq \beta^{2^k - 1} \cdot (\sqrt{\delta})^{2^k} \leq (\beta \cdot \sqrt{\delta})^{2^k} \leq \epsilon,
\end{aligned}
$$

which holds with high probability. We have thus completed the proof. $\square$

We leave the study of the convergence of the Gauss–Newton algorithm for complex variables to the interested reader.

11.7.3 ▪ A Barzilai–Borwein Implementation of the Gauss–Newton Algorithm 11.2

Since the computation (11.81) is expensive, we will need a lot of time when the size of the problem is large although it has quadratic convergence. To speed things up, we use the Barzilai–Borwein method, which is an excellent approach for a large-scale minimization problem (cf. [8]). Letting

$$G(\mathbf{x}^{(k)}) := \sum_{j=1}^{m} \nabla_{\mathbf{x}} f_j(\mathbf{x}^{(k)}) \nabla_{\mathbf{x}} f_j(\mathbf{x}^{(k)})^{\top},$$

we compute

$$\beta_j = (\mathbf{x}^{(j)} - \mathbf{x}^{(j-1)})^{\top} (G(\mathbf{x}^{(j)}) - G(\mathbf{x}^{(j-1)})) / \|\mathbf{x}^{(j)} - \mathbf{x}^{(j-1)}\|^2. \tag{11.103}$$

The iteration step in Algorithm 11.2, i.e., the computation in (11.81), can be replaced by

$$\mathbf{x}^{(k+1)} = \mathbf{z}^{(k)} - \beta_k^{-1} \nabla f(\mathbf{x}^{(k)}) \tag{11.104}$$

to speed up the computation when the size of the phase retrieval problem is very large. Therefore, we have obtained Algorithm 11.5 to solve the phase retrieval problem.

ALGORITHM 11.5

The Barzilai–Borwein Version of Gauss–Newton Algorithm 11.2

1: Let $\mathbf{u}^{(1)} = \mathbf{z}^{(1)}$ be an initial guess from Algorithm 11.4.
2: For $k \geq 1$, we solve the minimization in (11.78) by computing β_k according to (11.103) and update $\mathbf{x}^{(k)}$ by using (11.104) until a maximum number of iterations is achieved.
3: Output the last iterate $\mathbf{x}^{(k+1)}$.

We have implemented Algorithm 11.5 and it works very well. In section 11.10, we shall present some numerical examples to show that the Gauss–Newton algorithm, Algorithm 11.2 (using Algorithm 11.5 instead), performs very well. It is better than the Wirtinger flow algorithm discussed in the previous section. However, it does not perform as well as the algorithm to be discussed in the next section.

11.8 ▪ The DC-Based Algorithm

In this section, we propose another method of solving (11.48). Note that

$$f(\mathbf{x}) = \sum_{i=1}^{m} (\mathbf{x}^{\top} A_i \mathbf{x})^2 - 2b_i \mathbf{x}^{\top} A_i \mathbf{x} + b_i^2 = F_1(\mathbf{x}) - F_2(\mathbf{x}),$$

where $F_1(\mathbf{x}) = \sum_{i=1}^{m} (\mathbf{x}^{\top} A_i \mathbf{x})^2$ and $F_2(\mathbf{x}) = \sum_{i=1}^{m} 2b_i \mathbf{x}^{\top} A_i \mathbf{x} - b_i^2$. It is easy to see that both F_1 and F_2 are convex functions when $\mathbf{x} \in \mathbb{R}^n$. When $\mathbf{x} \in \mathbb{C}^n$ and $\mathbf{a}_j \in \mathbb{C}^n$, $j = 1, \ldots, m$, we have to write $\mathbf{x} = \mathbf{x}_R + \sqrt{-1}\mathbf{x}_I$, and similarly for $\mathbf{a}_j$. Letting $\mathbf{c} = [\mathbf{x}_R^{\top} \mathbf{x}_I^{\top}]^{\top} \in \mathbb{R}^{2n}$, we view $F_1(\mathbf{x})$ as a function in $G_1(\mathbf{c}) = F_1(\mathbf{x}_R + \sqrt{-1}\mathbf{x}_I)$. Then $G_1(\mathbf{c})$ is a convex function of variable $\mathbf{c}$. Similarly, $G_2(\mathbf{c}) = F_2(\mathbf{x}_R + \sqrt{-1}\mathbf{x}_I)$ is a convex function of $\mathbf{c}$. For convenience, we simply discuss the case when $\mathbf{x}, \mathbf{a}_j, j = 1, \ldots, m$, are real. The complex variable setting can be treated in the same fashion.

It is easy to see that the minimization in (11.48) can happen when $\|\mathbf{x}\| \leq L_0\|\mathbf{b}\| + 1$ for a positive number $L_0 > 0$, where $\mathbf{b} = (b_1, b_2, \ldots, b_m)^{\top}$ is the vector of the given measurement values, since we can easily see that $f(\mathbf{x}) \to \infty$ when $\|\mathbf{x}\| \to 0$. That is, the minimization (11.48) is coercive.

Let us introduce a new computational method as follows. Starting from any iterative solution $\mathbf{x}^{(k)}$, we solve the convex minimization problem

$$\mathbf{x}^{(k+1)} = \arg\min_{\mathbf{x}\in\mathbb{R}^n} F_1(\mathbf{x}) - \nabla F_2(\mathbf{x}^{(k)})^\top (\mathbf{x} - \mathbf{x}^{(k)}) \tag{11.105}$$

for $k \geq 1$, where $\mathbf{x}^{(1)}$ is an initial guess discussed in section 11.7. Our goal in this section is to show that $\mathbf{x}^{(k)}$, $k \geq 1$, converge to a local minimizer. If the initial guess $\mathbf{x}^{(1)}$ is not too far away from the global minimizer, then the limit of $\mathbf{x}^{(k)}$ will be the global minimizer. We shall call our algorithm a DC-based algorithm. Although it is standard to solve a convex minimization problem with differentiable minimizing functional, we have to solve (11.105) by using an iterative method; for example, we can use a gradient descent method with various acceleration techniques such as Nesterov's, Barzilai–Borwein, and other techniques or the Newton method. Hence, there will be two iterative procedures. The iterative procedure for solving (11.105) is an inner iteration which will be discussed in the next section. In this section, we mainly discuss the outer iteration.

A standard result for any DC-based algorithm is as follows. We shall further present many more results on the convergence of the iterative solutions $\mathbf{x}^{(k)}$ due to the differentiability of F_1 and F_2. In particular, we shall use the Kurdyka–Lojasiewicz inequality to establish the convergence rate of $\mathbf{x}^{(k)}$.

Theorem 11.50. *Starting from any initial guess $\mathbf{x}^{(1)}$, let $\mathbf{x}^{(k+1)}$ be the solution in (11.105) for all $k \geq 1$. Then*

$$f(\mathbf{x}^{(k+1)}) \leq f(\mathbf{x}^{(k)}) \quad \forall k \geq 1 \tag{11.106}$$

and $\nabla F_1(\mathbf{x}^{(k+1)}) - \nabla F_2(\mathbf{x}^{(k)}) = 0$ for $k \geq 1$.

Proof. By the convexity of F_2, we have $F_2(\mathbf{x}^{(k+1)}) \geq F_2(\mathbf{x}^{(k)}) + \nabla F_2(\mathbf{x}^{(k)})^\top (\mathbf{x}^{(k+1)} - \mathbf{x}^{(k)})$. From (11.105), we see that

$$f(\mathbf{x}^{(k+1)}) = F_1(\mathbf{x}^{(k+1)}) - F_2(\mathbf{x}^{(k+1)}) \leq F_1(\mathbf{x}^{(k+1)}) - \nabla F_2(\mathbf{x}^{(k)})^\top (\mathbf{x}^{(k+1)} - \mathbf{x}^{(k)}) - F_2(\mathbf{x}^{(k)})$$
$$\leq F_1(\mathbf{x}^{(k)}) - F_2(\mathbf{x}^{(k)}) = f(\mathbf{x}^{(k)}).$$

The second property, $\nabla F_1(\mathbf{x}^{(k+1)}) - \nabla F_2(\mathbf{x}^{(k)}) = 0$, follows from the minimization (11.105) directly. □

In fact, we can prove more. Note that $F_2(\mathbf{x})$ is a strongly convex function. Let S be the matrix

$$S = \frac{1}{m} \sum_{\ell=1}^m b_\ell A_\ell.$$

Lemma 11.51. *Suppose that $\mathbf{a}_j$, $j = 1, \ldots, m$, satisfy*

$$\|(\mathbf{a}_{j_1}^*\mathbf{y}, \ldots, \mathbf{a}_{j_n}^*\mathbf{y})\| \geq a_0\|\mathbf{y}\| \quad \forall \mathbf{y} \in \mathbb{C}^n \tag{11.107}$$

for a positive $a_0 \in (0,1)$ for any $1 \leq j_1 < j_2 < \cdots < j_n \leq m$. Suppose that $b_j > 0$ for $j = 1, \ldots, m$. Then the Hessian of F_2 is strictly positive definite.

Proof. It is easy to see that the Hessian of F_2 is $2mS$. A simple calculation shows that

$$\mathbf{y}^* S \mathbf{y} \geq \frac{1}{m} \sum_{j=1}^m b_j \mathbf{y}^* \mathbf{a}_j \mathbf{a}_j^* \mathbf{y} \geq \frac{1}{m} \min_{j=1,\ldots,m} b_j \sum_{i=1}^n \|\mathbf{a}_{j_i}^* \mathbf{y}\|^2 \geq \frac{a_0}{m} \min_{j=1,\ldots,m} b_j \|\mathbf{y}\|^2.$$

That is, F_2 is strongly convex. □

One of our major theorems in this section is as follows.

Theorem 11.52 (Huang, Lai, Varghese, and Xu, 2020 [21]). *Starting from any initial guess* $\mathbf{x}^{(1)}$, *let* $\mathbf{x}^{(k+1)}$ *be the solution to (11.105) for all* $k \geq 1$. *Under the assumptions in Lemma 11.51,* $\mathbf{x}^{(k)}$, $k \geq 1$, *converge to a critical point of* f. *Furthermore, if we let* $\mathbf{x}^*$ *be the unique limit, then*

$$\|\mathbf{x}^{(k+1)} - \mathbf{x}^*\| \leq C\tau^k \tag{11.108}$$

for a positive constant C *and* $\tau \in (0,1)$.

Proof. Since F_2 is strongly convex, we have

$$F_2(\mathbf{x}^{(k+1)}) \geq F_2(\mathbf{x}^{(k)}) + \nabla F_2(\mathbf{x}^{(k)})^\top (\mathbf{x}^{(k+1)} - \mathbf{x}^{(k)}) + \frac{\ell}{2}\|\mathbf{x}^{(k+1)} - \mathbf{x}^{(k)}\|^2$$

for $\ell = \min_{j=1}^m b_j a_0$. Now we use the same idea as in the proof in Theorem 11.50 to get

$$f(\mathbf{x}^{(k+1)}) \leq f(\mathbf{x}^{(k)}) - \frac{\ell}{2}\|\mathbf{x}^{(k+1)} - \mathbf{x}^{(k)}\|^2. \tag{11.109}$$

That is, we have

$$\frac{\ell}{2}\|\mathbf{x}^{(k+1)} - \mathbf{x}^{(k)}\|^2 \leq f(\mathbf{x}^{(k)}) - f(\mathbf{x}^{(k+1)}). \tag{11.110}$$

It is easy to see that for all $k \geq 1$, $\mathbf{x}^{(k)} \in \{\mathbf{x} \in \mathbb{R}^n, f(\mathbf{x}) \leq f(\mathbf{x}^{(1)})\}$, which is a closed set. There exists a cluster point $\mathbf{x}^*$ and a subsequence $\mathbf{x}^{(k_i)}$ such that $\mathbf{x}^{(k_i)} \to \mathbf{x}^*$ and $f(\mathbf{x}^{(k_i)}) \to f(\mathbf{x}^*)$. In fact, we have $f(\mathbf{x}^{(k)}) \to f(\mathbf{x}^*)$ for all $k \geq 1$.

Note that $f(\mathbf{x})$ is a polynomial of variable $\mathbf{x}$. We claim that

$$C_1\|\mathbf{x}^{(k+1)} - \mathbf{x}^{(k)}\| \leq \sqrt{f(\mathbf{x}^{(k)}) - f(\mathbf{x}^*)} - \sqrt{f(\mathbf{x}^{(k+1)}) - f(\mathbf{x}^*)} \tag{11.111}$$

for a positive constant C_1. To establish this claim, we need to use the well-known Kurdyka–Lojasiewicz inequality, which is given after this proof. Considering $g(t) = \sqrt{t}$, which is concave over $[0,1]$, we have $g(t) - g(s) \geq g'(t)(t-s)$. Then there exists a positive constant $c_0 > 0$ and $\delta > 0$ such that

$$\|g'(f(\mathbf{x}) - f(\mathbf{x}^*))\nabla f(\mathbf{x}))\| \geq c_0 > 0 \tag{11.112}$$

for all $\mathbf{x}$ in the neighborhood $B(\mathbf{x}^*, \delta)$ of $\mathbf{x}^*$. Since $f(\mathbf{x}^{(j)}) - f(\mathbf{x}^*) \to 0$ as $j \to \infty$, let us say there is an integer j large enough such that $\delta \geq 2\sqrt{2(f(\mathbf{x}^{(j)}) - f(\mathbf{x}^*))/\ell}$. Also, $\mathbf{x}^{(k_i)} \to \mathbf{x}^*$ as $k_i \to \infty$. Without loss of generality, we may assume that $\mathbf{x}^{(1)} \in B(\mathbf{x}^*, \delta/2)$ and $\delta \geq 2\sqrt{2(f(\mathbf{x}^{(1)}) - f(\mathbf{x}^*))/\ell}$. Let us show that $\mathbf{x}^{(k)}$, $k \geq 1$, will be in the neighborhood $B(\mathbf{x}^*, \delta)$. We shall use induction to do so. By (11.110) we have

$$\|\mathbf{x}^{(2)} - \mathbf{x}^*\| \leq \|\mathbf{x}^{(2)} - \mathbf{x}^{(1)}\| + \|\mathbf{x}^{(1)} - \mathbf{x}^*\| \leq \sqrt{2(f(\mathbf{x}^{(1)}) - f(\mathbf{x}^*))/\ell} + \|\mathbf{x}^{(1)} - \mathbf{x}^*\| \leq \delta.$$

Assume that $\mathbf{x}^{(k)} \in B(\mathbf{x}^*, \delta)$ for $k \leq K$. Multiplying by $g'(f(\mathbf{x}^{(k)}) - f(\mathbf{x}^*))$ on both sides of (11.110), we have

$$\frac{1}{2\sqrt{f(\mathbf{x}^{(k)}) - f(\mathbf{x}^*)}}\frac{\ell}{2}\|\mathbf{x}^{(k+1)} - \mathbf{x}^{(k)}\|^2 \leq g'(f(\mathbf{x}^{(k)}) - f(\mathbf{x}^*))\left(f(\mathbf{x}^{(k)}) - f(\mathbf{x}^{(k+1)})\right)$$

$$\leq \sqrt{f(\mathbf{x}^{(k)}) - f(\mathbf{x}^*)} - \sqrt{f(\mathbf{x}^{(k+1)}) - f(\mathbf{x}^*)} \tag{11.113}$$

by using the concavity of g. We use the Kurdyka–Lojasiewicz inequality to obtain

$$|g'(f(\mathbf{x}^{(k)}) - f(\mathbf{x}^*))| \geq \frac{c_0}{\|\nabla F(\mathbf{x}^{(k)})\|} = \frac{c_0}{\|\nabla F_1(\mathbf{x}^{(k)}) - \nabla F_2(\mathbf{x}^{(k)})\|}$$

$$= \frac{c_0}{\|\nabla F_2(\mathbf{x}^{(k-1)}) - \nabla F_2(\mathbf{x}^{(k)})\|}.$$

By using one of the properties in Theorem 11.50, we have

$$\|\nabla f(\mathbf{x}^{(k)})\| = \|\nabla F_1(\mathbf{x}^{(k)}) - \nabla F_2(\mathbf{x}^{(k)})\| = \|\nabla F_1(\mathbf{x}^{(k-1)}) - \nabla F_1(\mathbf{x}^{(k)})\| \leq L_1 \|\mathbf{x}^{(k)} - \mathbf{x}^{(k-1)}\|$$

using the Lipschitz differentiability of F_2 with constant $L_1 > 0$ with L_1 a positive constant dependent on $\mathbf{a}_i, i = 1, \dots, m$. In other words, we use (11.113) to get

$$\sqrt{f(\mathbf{x}^{(k)}) - f(\mathbf{x}^*)} - \sqrt{f(\mathbf{x}^{(k+1)}) - f(\mathbf{x}^*)} \geq \frac{c_0}{L_1 \|\mathbf{x}^{(k)} - \mathbf{x}^{(k-1)}\|} \|\mathbf{x}^{(k)} - \mathbf{x}^{(k+1)}\|^2.$$

Multiplying by $\|\mathbf{x}^{(k)} - \mathbf{x}^{(k-1)}\|$ on both sides of the above inequality, taking the square root of both sides, and using the standard inequality $2ab \leq a^2 + b^2$ on the left-hand side, we have

$$\|\mathbf{x}^{(k)} - \mathbf{x}^{(k-1)}\| + \frac{L_1}{c_0} \left(\sqrt{f(\mathbf{x}^{(k)}) - f(\mathbf{x}^*)} - \sqrt{f(\mathbf{x}^{(k+1)}) - f(\mathbf{x}^*)} \right) \geq 2\|\mathbf{x}^{(k)} - \mathbf{x}^{(k+1)}\|$$

for all $2 \leq k \leq K$. It follows that

$$\frac{L_1}{c_0} \sqrt{f(\mathbf{x}^{(1)}) - f(\mathbf{x}^*)} \geq \sum_{j=1}^{K} \|\mathbf{x}^{(j+1)} - \mathbf{x}^{(j)}\| + \|\mathbf{x}^{(K+1)} - \mathbf{x}^{(K)}\|.$$

That is, we have

$$\|\mathbf{x}^{(K+1)} - \mathbf{x}^*\| \leq \|\mathbf{x}^{(K+1)} - \mathbf{x}^{(1)}\| + \|\mathbf{x}^{(1)} - \mathbf{x}^*\|$$

$$\leq \sum_{j=1}^{K} \|\mathbf{x}^{(j+1)} - \mathbf{x}^{(j)}\| + \|\mathbf{x}^{(1)} - \mathbf{x}^*\|$$

$$\leq \frac{L_1}{c_0} \sqrt{f(\mathbf{x}^{(1)}) - f(\mathbf{x}^*)} + \|\mathbf{x}^{(1)} - \mathbf{x}^*\| \leq \delta.$$

That is, $\mathbf{x}^{(K+1)} \in B(\mathbf{x}^*, \delta)$. This shows that all $\mathbf{x}^{(k)}$ are in $B(\mathbf{x}^*, \delta)$.

We are now ready to prove the claim in (11.111). Using one of the properties in Theorem 11.50, we have

$$\|\nabla f(\mathbf{x}^{(k)})\| = \|\nabla F_1(\mathbf{x}^{(k)}) - \nabla F_2(\mathbf{x}^{(k)})\| = \|\nabla F_1(\mathbf{x}^{(k)}) - \nabla F_1(\mathbf{x}^{(k+1)})\| \leq L \|\mathbf{x}^{(k)} - \mathbf{x}^{(k+1)}\|$$

by using the Lipschitz differentiability of F_1 with constant $L > 0$, with L a positive constant dependent on $L_0 \|\mathbf{b}\| + 1$ and $\mathbf{a}_i, i = 1, \dots, m$. Therefore, combining the above four inequalities, we have

$$\sqrt{f(\mathbf{x}^{(k)}) - f(\mathbf{x}^*)} - \sqrt{f(\mathbf{x}^{(k+1)}) - f(\mathbf{x}^*)}$$

$$\geq |g'(f(\mathbf{x}) - f(\mathbf{x}^*))| \frac{\ell}{2} \|\mathbf{x}^{(k+1)} - \mathbf{x}^{(k)}\|^2$$

$$\geq \frac{1}{\|\nabla F_1(\mathbf{x}^{(k)}) - \nabla F_1(\mathbf{x}^{(k+1)})\|} \frac{\ell}{2} \|\mathbf{x}^{(k+1)} - \mathbf{x}^{(k)}\|^2$$

$$\geq \frac{\ell}{2} \|\mathbf{x}^{(k+1)} - \mathbf{x}^{(k)}\|^2 \frac{c_0}{L \|\mathbf{x}^{(k)} - \mathbf{x}^{(k+1)}\|}$$

$$= \frac{\ell c_0}{2L} \|\mathbf{x}^{(k+1)} - \mathbf{x}^{(k)}\|,$$

which is the claim (11.111) with $C_1 = \ell c_0/(2L)$. By summing the inequality in (11.111) above, it follows that

$$\sum_{k \geq 1} \|\mathbf{x}^{(k+1)} - \mathbf{x}^{(k)}\| \leq \frac{1}{C_1}\sqrt{f(\mathbf{x}^{(1)}) - f(\mathbf{x}^*)}.$$

That is, $\mathbf{x}^{(k)}$ is a Cauchy sequence, and hence it is convergent. Let us say $\mathbf{x}^{(k)} \to \mathbf{x}^*$. By the second property in Theorem 11.50, we have $\nabla f(\mathbf{x}^*) = \lim_{k\to\infty} \nabla F_1(\mathbf{x}^{(k+1)}) - \nabla F_2(\mathbf{x}^{(k)}) = 0$ by using a result in [31]. □

In the proof above, we have used the following result.

Theorem 11.53 (Lojasiewicz [26]). *Let f be a real analytic function in a neighborhood of $0 \in \mathbb{R}^n$ with $f(0) = 0$ and $\mathbf{x} = 0$ being a critical point. Then there exists a number $\theta \in (0,1)$ such that*

$$|f(\mathbf{x})|^\theta \leq C_L\|\nabla f(\mathbf{x})\| \tag{11.114}$$

in a neighborhood $B(0,\delta)$ of $\mathbf{x} = 0$ for $\delta > 0$, where $C_L > 0$ is a constant independent of $\mathbf{x}$.

Proposition 11.54. *Suppose that $f : \mathbb{R}^n \mapsto \mathbb{R}$ is a continuously twice differentiable function whose Hessian $H(f)(\mathbf{x})$ is invertible at a critical point $\mathbf{x}^*$ of f. Then there exists a positive constant C, an exponent $\theta = 1/2$, and a positive r such that*

$$|f(\mathbf{x}) - f(\mathbf{x}^*)|^{1/2} \leq C\|\nabla f(\mathbf{x})\| \quad \forall \mathbf{x} \in B(\mathbf{x}^*, r), \tag{11.115}$$

where $B(\mathbf{x}*, r)$ is a ball at $\mathbf{x}^*$ with radius r.

Proof. Since f is continuously twice differentiable, using the Taylor formula for f and noting that $f(\mathbf{x}^*) = 0$, we have

$$|f(\mathbf{x}) - f(\mathbf{x}^*)| \leq c_1\|\mathbf{x} - \mathbf{x}^*\|^2 \quad \forall \mathbf{x} \in B(\mathbf{x}^*, r)$$

for some $r > 0$. On the other hand, we have $\|\nabla f(\mathbf{x})\| = \|\nabla f(\mathbf{x}) - \nabla f(\mathbf{x}^*)\| \geq c_2\|\mathbf{x} - \mathbf{x}^*\|$ because the Hessian is invertible. Thus, (11.115) follows with $\theta = 1/2$ and $C = \sqrt{c_1}/c_2$. □

The importance of the Lojasiewicz inequality is to establish the inequality in (11.115) when f may not have an invertible Hessian at the critical point $\mathbf{x}^*$. The proof is based on knowledge from algebraic geometry, mainly the curve-selecting lemma. See [23] for a more general setting.

A few obvious consequences are as follows:

(1) For any given initial point $\mathbf{x}^{(1)}$, let $D = f(\mathbf{x}^{(1)}) - f(\mathbf{x}^*) > 0$, where $\mathbf{x}^*$ is one of the global minimizers of (11.48). Then

$$f(\mathbf{x}^{(k)}) - f(\mathbf{x}^*) \leq D - \frac{\ell}{2}\sum_{j=1}^{k-1}\|\mathbf{x}^{(j+1)} - \mathbf{x}^{(j)}\|^2.$$

That is, $\mathbf{x}^{(k)}$ is closer to one of the global minimizers than the initial guess point is.

(2) Since our approach can find a critical point, if a global minimizer $\mathbf{x}^*$ is a local minimizer over a neighborhood $N(\mathbf{x}^*)$ and an initial vector $\mathbf{x}^{(1)}$ is in $N(\mathbf{x}^*)$, then our approach finds $\mathbf{x}^*$.

(3) Our analysis can be extended to deal with more general systems of polynomial equations:

$$\min \left\{ \frac{1}{m} \sum_{i=1}^{m} (g(\langle \mathbf{a}_i, \mathbf{x} \rangle) - b_i)^2 \right\}, \tag{11.116}$$

where $g(t) \geq 0$ is a continuously differentiable and strongly convex function, e.g., $g(t) = t^4$.

11.8.1 ▪ Computation of the Inner Minimization (11.105)

We now discuss how to compute the minimization in (11.105). For convenience, we rewrite the minimization in the form

$$\min_{\mathbf{x} \in \mathbb{R}^n} G(\mathbf{x}) \tag{11.117}$$

for a differentiable convex function $G(\mathbf{x}) := F_1(\mathbf{x}) - \langle \nabla F_2(\mathbf{x}^{(k)}), \mathbf{x} - \mathbf{x}^{(k)} \rangle$. The first approach is to use the gradient descent method:

$$\mathbf{z}^{(j+1)} = \mathbf{z}^{(j)} - h \nabla G(\mathbf{z}^{(j)}) \tag{11.118}$$

for $j = 1, \ldots,$ with $\mathbf{z}^{(1)} = \mathbf{x}^{(k)}$, where $h > 0$ is a fixed step size or a variable step size. It is well known that we need to choose $h = \nu/L$ for the Lipschitz differentiability constant L of $G(\mathbf{x})$ and the ν-strong convexity of G and then use the Nesterov acceleration technique as explained in Chapters 5, 6, and 9. Indeed, we have the following result.

Lemma 11.55 (The Nesterov Acceleration). *Let $f : \mathbb{R}^n \to \mathbb{R}$ be a ν-strong convex function, and let the gradient function have an L-Lipschitz constant. Starting at an arbitrary initial point $\mathbf{u}_1 = \mathbf{z}_1$, the Nesterov accelerated gradient descent*

$$\mathbf{z}^{j+1} := \mathbf{u}^{(j)} - \frac{\nu}{L} \nabla f(\mathbf{u}^{(j)}),$$

$$\mathbf{u}^{(j+1)} = \mathbf{z}^{(j+1)} - q(\mathbf{z}^{(j+1)} - \mathbf{z}^{(j)})$$

satisfies

$$f(\mathbf{z}^{j+1}) - f(\mathbf{z}^*) \leq \frac{\nu + L}{2} \|\mathbf{z}^{(1)} - \mathbf{z}^*\|^2 \exp\left(-\frac{j}{\sqrt{L/\nu}}\right), \tag{11.119}$$

where $\mathbf{z}^*$ is the optimal solution and $q = (\sqrt{L/\nu} - 1)/(\sqrt{L/\nu} + 1)$ is a constant.

The significance of the Nesterov acceleration above is to reduce the number of iterations in (11.118) significantly.

Since G is twice differentiable, we can certainly use the Newton method to solve (11.105) because it has quadratic convergence. However, we will not pursue it here because when the number of variables of $\mathbf{z}$ is large, the Newton method is extremely slow.

Instead, another method to choose a good h is to use the Barzilai–Borwein method, which is an excellent approach for a large-scale minimization problem (cf. [8]). The iteration of the Barzilai–Borwein method can be described as

$$\mathbf{z}^{(j+1)} = \mathbf{z}^{(j)} - \beta_j^{-1} \nabla G(\mathbf{z}^{(j)}), \tag{11.120}$$

where the step size is

$$\beta_j = (\mathbf{z}^{(j)} - \mathbf{z}^{(j-1)})^\top (\nabla G(\mathbf{z}^{(j)}) - \nabla G(\mathbf{z}^{(j-1)}))/\|\mathbf{z}^{(j)} - \mathbf{z}^{(j-1)}\|^2. \tag{11.121}$$

We shall use Algorithm 11.6 to solve the minimization (11.105).

ALGORITHM 11.6

The Barzilai–Borwein Algorithm for the Inner Minimization

1: Let $\mathbf{u}^{(1)} = \mathbf{z}^{(1)}$ be an initial guess.
2: For $j \geq 1$, we solve the minimization in (11.117) by computing β_k according to (11.121) and update

$$\mathbf{z}^{j+1} := \mathbf{u}^{(j)} - \beta_j^{-1} \nabla G(\mathbf{u}^{(j)}), \tag{11.122}$$

$$\mathbf{u}^{(j+1)} = \mathbf{z}^{(j+1)} - q(\mathbf{z}^{(j+1)} - \mathbf{z}^{(j)}) \tag{11.123}$$

3: until a maximum number of iterations is achieved, where q is an accelerated factor as in Nesterov's acceleration technique or the Attouch–Peypouquet acceleration.

Finally, we add one more speed-up technique. To retrieve a real value signal, if we know the signs of the measurement values, we can simply solve the system of linear equations by using the least squares method. That is, if we know $s_i = \pm$ for each measurement value $\mathbf{a}_i^\top \mathbf{x_b}$, $i = 1, \ldots, m$, we simply solve $\mathbf{x_b}$ from the linear system

$$\langle \mathbf{a}_i, \mathbf{x} \rangle = s_i \sqrt{b_i}, \quad i = 1, \ldots, m.$$

This holds because at the end of the computation of each inner iteration, since $\mathbf{u}^{(j+1)}$ is a good approximation of true signal $\mathbf{x_b}$, we obtain the signs $s_i = \text{sign}(\langle \mathbf{a}_i, \mathbf{u}^{(j+1)} \rangle)$ and use the least squares method to find a better approximation, say $\mathbf{v}^{(j+1)}$. We feed it back into the iteration to help the DC algorithm perform better.

11.9 ▪ Sparse Phase Retrieval

In previous sections, several computational algorithms have been developed for the phase retrieval problem in (11.48). We now extend the approaches to study sparse phase retrieval. Suppose that $\mathbf{x_b}$ is a sparse solution to the given measurements (11.1). Let us use the DC-based algorithm to explain how to do this. First, we consider

$$\min_{\mathbf{x} \in \mathbb{R}^n \text{ or } \mathbb{C}^n} \lambda \|\mathbf{x}\|_1 + \sum_{i=1}^{m} (|\langle \mathbf{a}_i, \mathbf{x} \rangle|^2 - b_i)^2 \tag{11.124}$$

by adding $\lambda \|\mathbf{x}\|_1$ to (11.48) as a standard approach in compressive sensing. As before, let $f(\mathbf{x}) = \sum_{i=1}^{m} (|\langle \mathbf{a}_i, \mathbf{x} \rangle|^2 - b_i)^2$ and let $F(\mathbf{x}) = \lambda \|\mathbf{x}\|_1 + f(\mathbf{x})$. It is easy to see that $F : \mathbb{C} \to \mathbb{R}_+$ is a continuous function which satisfies the coercive condition

$$F(x) \to \infty \quad \text{when } |x| \to \infty.$$

It follows that minimization problem (11.124) has a minimizer. We now discuss how to solve it numerically.

We approach it by using a method similar to one in the previous section. Indeed, for the case $\mathbf{x} \in \mathbb{R}^n$ and $\mathbf{a}_i \in \mathbb{R}^n$, we rewrite $F(\mathbf{x}) = \sum_{i=1}^{m} (f(\langle \mathbf{a}_i, \mathbf{x} \rangle) - b_i)^2$ as

$$F(\mathbf{x}) = F_1(\mathbf{x}) - F_2(\mathbf{x}) := \sum_{i=1}^{m} f^2(\langle \mathbf{a}_i, \mathbf{x} \rangle) + b_i^2 - \sum_{i=1}^{m} 2b_i f(\langle \mathbf{a}_i, \mathbf{x} \rangle).$$

The minimization (11.124) will be approximated by

$$\mathbf{x}^{(k+1)} := \arg \min \lambda \|\mathbf{x}\|_1 + F_1(\mathbf{x}) - \nabla F_2(\mathbf{x}^{(k)})^\top (\mathbf{x} - \mathbf{x}^{(k)}) \tag{11.125}$$

for any given $\mathbf{x}^{(k)}$. We call this algorithm the sparse DC-based method. When $\mathbf{x} \in \mathbb{C}^n$ and $\mathbf{a}_j \in \mathbb{C}^n$, $j = 1, \ldots, m$, we have to write $\mathbf{x} = \mathbf{x}_R + \sqrt{-1}\mathbf{x}_I$, and similarly for $\mathbf{a}_j$. Letting $\mathbf{c} = [\mathbf{x}_R^\top \mathbf{x}_I^\top]^\top \in \mathbb{R}^{2n}$, we view $F_1(\mathbf{x})$ as a function in $G_1(\mathbf{c}) = F_1(\mathbf{x}_R + \sqrt{-1}\mathbf{x}_I)$. Then $G_1(\mathbf{c})$ is a convex function of variable $\mathbf{c}$. Similarly, $G_2(\mathbf{c}) = F_2(\mathbf{x}_R + \sqrt{-1}\mathbf{x}_I)$ is a convex function of $\mathbf{c}$. We can formulate the same minimization problem as in (11.125). For convenience, we simply discuss the case when $\mathbf{x}$, $\mathbf{a}_j$, $j = 1, \ldots, m$, are real. The complex variable setting can be treated in the same fashion.

To solve (11.125), we use a proximal gradient method: For any given $\mathbf{y}^{(k)}$,

$$\mathbf{y}^{(k+1)} := \arg\min \lambda\|\mathbf{y}\|_1 + F_1(\mathbf{y}^{(k)}) + (\nabla F_1(\mathbf{y}^{(k)}) - \nabla F_2(\mathbf{y}^{(k)}))^\top (\mathbf{y} - \mathbf{y}^{(k)}) + \frac{L_1}{2}\|\mathbf{y} - \mathbf{y}^{(k)}\|^2$$
$$(11.126)$$

for $k \geq 1$, where L_1 is the Lipschitz differentiability of F_1. The above minimization can be easily solved using the shrinkage and thresholding technique as in [9]. The discussion above furnishes a computational method for the sparse phase retrieval problem (11.124). Let us point out that one significant difference from (11.126) is that we can find $\mathbf{y}^{(k+1)}$ by using a formula while the solution $\mathbf{x}^{(k+1)}$ in (11.105) has to be computed using an iterative method, as explained before. Thus sparse phase retrieval is more efficient in this sense.

Let us study the convergence of our sparse phase retrieval method. We again start with a standard result for a DC-based algorithm.

Theorem 11.56. *Assume that F_2 is a strongly convex function with parameter ℓ. Starting from any initial guess $\mathbf{y}^{(1)}$, let $\mathbf{y}^{(k+1)}$ be the solution in (11.126) for all $k \geq 1$. Then*

$$\lambda\|\mathbf{y}^{(k+1)}\|_1 + F(\mathbf{y}^{(k+1)}) \leq \lambda\|\mathbf{y}^{(k)}\|_1 + F(\mathbf{y}^{(k)}) - \frac{\ell}{2}\|\mathbf{y}^{(k+1)} - \mathbf{y}^{(k)}\|^2 \quad \forall k \geq 1 \quad (11.127)$$

and $\partial g(\mathbf{y}^{(k+1)}) + \nabla F_1(\mathbf{y}^{(k+1)}) - \nabla F_2(\mathbf{y}^{(k)}) + \frac{L_1}{2}(\mathbf{y}^{(k+1)} - \mathbf{y}^{(k)}) = 0$, where $g(\mathbf{x}) = \lambda\|\mathbf{x}\|_1$.

Proof. The Lipschitz differentiability of F_1 tells us that

$$F_1(\mathbf{y}^{(k+1)}) \leq F_1(\mathbf{y}^{(k)}) + \nabla F_2(\mathbf{y}^{(k)})^\top (\mathbf{y}^{(k+1)} - \mathbf{y}^{(k)}) + \frac{L_1}{2}\|\mathbf{y}^{(k+1)} - \mathbf{y}^{(k)}\|^2,$$

where L_1 is the Lipschitz differentiability of F_1. By the strong convexity of F_2, we have

$$F_2(\mathbf{y}^{(k+1)}) \geq F_2(\mathbf{y}^{(k)}) + \nabla F_2(\mathbf{y}^{(k)})^\top (\mathbf{y}^{(k+1)} - \mathbf{y}^{(k)}) + \frac{\ell}{2}\|\mathbf{y}^{(k+1)} - \mathbf{y}^{(k)}\|^2.$$

With the above two inequalities, we see that

$$\lambda\|\mathbf{y}^{(k+1)}\|_1 + F(\mathbf{x}^{(k+1)}) = \lambda\|\mathbf{y}^{(k+1)}\|_1 + F_1(\mathbf{y}^{(k+1)}) - F_2(\mathbf{y}^{(k+1)})$$

$$\leq \lambda\|\mathbf{y}^{(k+1)}\|_1 + F_1(\mathbf{y}^{(k)}) + \nabla F_1(\mathbf{y}^{(k)})^\top (\mathbf{y}^{(k+1)} - \mathbf{y}^{(k)}) + \frac{L}{2}\|\mathbf{y}^{(k+1)} - \mathbf{y}^{(k)}\|^2$$

$$- F_2(\mathbf{y}^{(k)}) - \nabla F_2(\mathbf{y}^{(k)})^\top (\mathbf{y}^{(k+1)} - \mathbf{y}^{(k)}) - \frac{\ell}{2}\|\mathbf{y}^{(k+1)} - \mathbf{y}^{(k)}\|^2$$

$$= F_1(\mathbf{y}^{(k)}) - F_2(\mathbf{y}^{(k)}) - \frac{\ell}{2}\|\mathbf{y}^{(k+1)} - \mathbf{y}^{(k)}\|^2$$

$$+ \lambda\|\mathbf{y}^{(k+1)}\|_1 + (\nabla F_1(\mathbf{y}^{(k)}) - \nabla F_2(\mathbf{y}^{(k)}))^\top (\mathbf{y}^{(k+1)} - \mathbf{y}^{(k)}) + \frac{L}{2}\|\mathbf{y}^{(k+1)} - \mathbf{y}^{(k)}\|^2$$

$$\leq F_1(\mathbf{y}^{(k)}) - F_2(\mathbf{y}^{(k)}) - \frac{\ell}{2}\|\mathbf{y}^{(k+1)} - \mathbf{y}^{(k)}\|^2 + \lambda\|\mathbf{y}^{(k)}\|_1$$

$$= \lambda\|\mathbf{y}^{(k)}\|_1 + F(\mathbf{y}^{(k)}) - \frac{\ell}{2}\|\mathbf{y}^{(k+1)} - \mathbf{y}^{(k)}\|^2,$$

where we have used the optimization condition in (11.126).

Letting $g(\mathbf{x}) = \lambda \|\mathbf{x}\|_1$, the second property, $\partial g(\mathbf{y}^{(k+1)}) + \nabla F_1(\mathbf{y}^{(k)}) - \nabla F_2(\mathbf{y}^{(k)}) + \frac{L_1}{2}(\mathbf{y}^{(k+1)} - \mathbf{y}^{(k)}) = 0$, follows from the minimization (11.126). □

Next we show that the sequence $\mathbf{y}^{(k)}, k \geq 1$, from (11.126) has a convergent subsequence which converges to a critical point $\mathbf{y}^*$.

Theorem 11.57. *Suppose that F_2 is strongly convex. Let $\mathbf{y}^{(k)}, k \geq 1$, be the sequence obtained from (11.126). Then it has a convergent subsequence which converges to a critical point $\mathbf{y}^*$ of F.*

Proof. Recall that $F(\mathbf{x}) = g(\mathbf{x}) + f(\mathbf{x})$. From Theorem 11.56, we have

$$\frac{\ell}{2}\|\mathbf{y}^{(k+1)} - \mathbf{y}^{(k)}\|^2 \leq F(\mathbf{y}^{(k)}) - F(\mathbf{y}^{(k+1)}). \tag{11.128}$$

That is, $F(\mathbf{y}^{(k)}), k \geq 1$, is a strictly decreasing sequence. Due to the coerciveness, we know that

$$\mathcal{R} := \{\mathbf{x} \in \mathbb{R}^n, F(\mathbf{y}) \leq F(\mathbf{y}^{(1)})\}$$

is a bounded set. It follows that the sequence $\{\mathbf{y}^{(k)}\}_{k=1}^{\infty} \subset \mathcal{R}$ is a bounded sequence and there exists a cluster point $\mathbf{y}^*$ and a subsequence $\mathbf{y}^{(k_i)}$ such that $\mathbf{y}^{(k_i)} \to \mathbf{y}^*$. Note that $\{F(\mathbf{y}^{(k)})\}_{k=1}^{\infty}$ is a bounded monotonic descending sequence, so $F(\mathbf{y}^{(k)}) \to F(\mathbf{y}^*)$ for all $k \geq 1$. Let us prove that $\|\nabla F(\mathbf{y}^*)\| = 0$ holds, that is, $\mathbf{y}^*$ is a critical point of F. Indeed, using one of the properties in Theorem 11.56, we have

$$\begin{aligned}
\|\partial F(\mathbf{y}^{(k)})\| &= \|\partial g(\mathbf{y}^{(k)}) + \nabla F_1(\mathbf{y}^{(k)}) - \nabla F_2(\mathbf{y}^{(k)})\| \\
&\leq \|\nabla F_1(\mathbf{y}^{(k)}) - \nabla F_1(\mathbf{y}^{(k-1)})\| + \|\nabla F_2(\mathbf{y}^{(k)}) - \nabla F_2(\mathbf{y}^{(k-1)})\| \\
&\quad + \frac{L}{2}\|\mathbf{y}^{(k)} - \mathbf{y}^{(k-1)}\|
\end{aligned}$$

by using the second conclusion of Theorem 11.56. Combining this with (11.128) and the Lipschitz differentiation of F_1 and F_2, we get that $\|\partial f(\mathbf{x}^{(k_i)})\| \to 0$. By a property of the subgradient of g and the continuity of the gradients of F_1 and F_2, we have $\|\partial F(\mathbf{x}^*)\| = 0$ when $\mathbf{y}^{(k_i)} \to \mathbf{y}^*$. Thus, $\mathbf{y}^* \in \text{domain}(\partial F)$, the set of all critical points of F. □

Finally, let us show that the convergence is linear. We begin with the following result.

Lemma 11.58. *Let $g(\mathbf{x}) = \lambda\|\mathbf{x}\|_1$ for $\lambda > 0$. Then for any $\mathbf{x}$, there exists a $\delta > 0$ such that for any $\mathbf{y} \in B(\mathbf{x}, \delta)$, the open ball of radius δ at $\mathbf{x}$, we have*

$$(\partial g(\mathbf{y}) - \partial g(\mathbf{x}))^\top (\mathbf{y} - \mathbf{x}) = 0. \tag{11.129}$$

Proof. For simplicity, consider $\mathbf{x} \in \mathbb{R}^1$. Then if $\mathbf{x} \neq 0$, we can find $\delta = |\mathbf{x}| > 0$ such that when $\mathbf{y} \in B(\mathbf{x}, \delta)$, we have $\partial g(\mathbf{y}) = \partial g(\mathbf{x})$, and hence we have (11.129). If $\mathbf{x} = 0$, for any $y \neq 0$, we choose $\partial g(0)$ according to $\mathbf{y}$, i.e., $\partial g(0) = 1$ if $\mathbf{y} > 0$ and $\partial g(0) = -1$ if $\mathbf{y} < 0$. Then we have (11.129). □

Lemma 11.59. *Let $F(\mathbf{x}) = g(\mathbf{x}) + f(\mathbf{x})$. Suppose that f is L-Lipschitz differentiable. Let $\mathbf{x}^*$ be a critical point of F, as explained in Theorem 11.57. Then there exists $\delta > 0$ such that for all $\mathbf{x} \in B(\mathbf{x}^*, \delta)$,*

$$|F(\mathbf{x}) - F(\mathbf{x}^*)| \leq C\|\mathbf{x} - \mathbf{x}^*\|^2. \tag{11.130}$$

Proof. At $\mathbf{x}^*$, we have $\partial F(\mathbf{x}^*) = \partial g(\mathbf{x}^*) + \nabla f(\mathbf{x}^*) = 0$. It follows that

$$F(\mathbf{x}) - F(\mathbf{x}^*) = g(\mathbf{x}) - g(\mathbf{x}^*) + f(\mathbf{x}) - f(\mathbf{x}^*)$$

$$\leq \partial g(\mathbf{x})^\top (\mathbf{x} - \mathbf{x}^*) + \nabla f(\mathbf{x}^*)(\mathbf{x} - \mathbf{x}^*) + \frac{1}{2}(\mathbf{x} - \mathbf{x}^*)^\top \nabla^2 f(\xi)(\mathbf{x} - \mathbf{x}^*)$$

$$= (\partial g(\mathbf{x}) - \partial g(\mathbf{x}^*))^\top (\mathbf{x} - \mathbf{x}^*) + \frac{1}{2}(\mathbf{x} - \mathbf{x}^*)^\top \nabla^2 f(\xi)(\mathbf{x} - \mathbf{x}^*)$$

$$\leq \frac{1}{2}(\mathbf{x} - \mathbf{x}^*)^\top \nabla^2 f(\xi)(\mathbf{x} - \mathbf{x}^*),$$

where ξ is a point between $\mathbf{x}$ and $\mathbf{x}^*$. That is, $|F(\mathbf{x}) - F(\mathbf{x}^*)| \leq C\|\mathbf{x} - \mathbf{x}^*\|^2$ for a positive constant C. $\square$

We are now ready to establish the following result on the rate of convergence.

Theorem 11.60. *Suppose that F_2 is strongly convex. Starting from any initial guess $\mathbf{x}^{(1)}$, let $\mathbf{x}^{(k+1)}$ be the solution to (11.105) for all $k \geq 1$. Without loss of generality, we assume that $\mathbf{x}^{(k)}, k \geq 1$, converge to a critical point $\mathbf{x}^*$ of F by using Theorem 11.57. Then for any $\epsilon > 0$, either $\mathbf{x}^{(k+1)} \in B(\mathbf{x}^*, \epsilon)$ or*

$$\|\mathbf{x}^{(k+1)} - \mathbf{x}^*\| \leq C_\epsilon \tau^k \tag{11.131}$$

for a positive constant C_ϵ dependent on ϵ and $\tau \in (0,1)$ independent of ϵ.

Proof. According to the results in Theorems 11.56 and 11.57, we have

$$C_0 \|\mathbf{x}^{(k+1)} - \mathbf{x}^{(k)}\|^2 \leq (F(\mathbf{x}^{(k)}) - F(\mathbf{x}^*)) - (F(\mathbf{x}^{(k+1)}) - F(\mathbf{x}^*)) \tag{11.132}$$

for a positive constant C_0. We now claim that

$$C_1 \|\mathbf{x}^{(k+1)} - \mathbf{x}^{(k)}\| \leq \sqrt{F(\mathbf{x}^{(k)}) - F(\mathbf{x}^*)} - \sqrt{F(\mathbf{x}^{(k+1)}) - F(\mathbf{x}^*)} \tag{11.133}$$

for a positive constant C_1. To establish this claim, we need to use the result in Lemma 11.59. Let us rewrite the inequality in Lemma 11.59 as

$$\frac{1}{\sqrt{F(\mathbf{x}^{(k+1)}) - F(\mathbf{x}^*)}} \geq \frac{C}{\|\mathbf{x}^{(k+1)} - \mathbf{x}^*\|}.$$

Multiplying the above inequality by the inequality in (11.133), we have

$$C_1 C \frac{\|\mathbf{x}^{(k+1)} - \mathbf{x}^{(k)}\|^2}{\|\mathbf{x}^{(k+1)} - \mathbf{x}^*\|} \leq \frac{1}{\sqrt{F(\mathbf{x}^{(k+1)}) - F(\mathbf{x}^*)}} \left[(F(\mathbf{x}^{(k)}) - F(\mathbf{x}^*)) - (F(\mathbf{x}^{(k+1)}) - F(\mathbf{x}^*)) \right].$$

$$\tag{11.134}$$

Considering $g(t) = \sqrt{t}$, which is concave over $[0,1]$, we know that $g(t) - g(s) \geq g'(t)(t - s)$. Thus, the right-hand side above is less than or equal to the right-hand side of (11.133). We now work on the left-hand side of the inequality above. Let us first note that F is strongly convex outside the ball $B(\mathbf{x}_b, \epsilon)$ (in the real variable setting). If $\mathbf{x}^{(k+1)}$ is within $B(\mathbf{x}_b, \epsilon)$, we do not need to iterate further when $\epsilon > 0$ is a tolerance. Otherwise, we use the strong convexity of F to get

$$C_\epsilon \|\mathbf{x}^{(k+1)} - \mathbf{x}^*\| \leq \|\nabla F(\mathbf{x}^{(k+1)}) - \nabla F(\mathbf{x}^*)\|$$

for a positive constant dependent on ϵ. The second property of Theorem 11.56 implies that

$$\partial g(\mathbf{x}^{(k+1)}) + \nabla F_1(\mathbf{x}^{(k+1)}) - \nabla F_2(\mathbf{x}^{(k)}) + \frac{L_1}{2}(\mathbf{x}^{(k+1)} - \mathbf{x}^{(k)}) = 0$$

and $\partial g(\mathbf{x}^*) + \nabla f(\mathbf{x}^*) = 0$. By using Lemma 11.58, it follows that

$$\nabla F(\mathbf{x}^{(k+1)}) - \nabla F(\mathbf{x}^*) = \nabla F_2(\mathbf{x}^{(k)}) - \nabla F_2(\mathbf{x}^{(k+1)}) - \frac{L_1}{2}(\mathbf{x}^{(k+1)} - \mathbf{x}^{(k)}).$$

In other words,

$$C_\epsilon \|\mathbf{x}^{(k+1)} - \mathbf{x}^*\| \le \|\nabla F_2(\mathbf{x}^{(k)}) - \nabla F_2(\mathbf{x}^{(k+1)})\| + \frac{L_1}{2}\|\mathbf{x}^{(k+1)} - \mathbf{x}^{(k)}\|.$$

Using the Lipschitz differentiability of F_2, we have

$$\|\mathbf{x}^{(k+1)} - \mathbf{x}^*\| \le \frac{L_2 + L_1}{C_\epsilon}\|\mathbf{x}^{(k+1)} - \mathbf{x}^{(k)}\|.$$

The left-hand side of the equation in (11.134) can be simplified to be

$$C_0 C \frac{L_1 + L_2}{C_\epsilon}\|\mathbf{x}^{(k+1)} - \mathbf{x}^{(k)}\|,$$

which is the desired term on the left-hand side of the inequality in (11.133). These establish the claim.

By summing the inequality in (11.133) above, it follows that

$$\sum_{k \ge 1} \|\mathbf{x}^{(k+1)} - \mathbf{x}^{(k)}\| \le \frac{1}{C_1}\sqrt{f(\mathbf{x}^{(1)}) - f(\mathbf{x}^*)}.$$

That is, $\mathbf{x}^{(k)}$ is a Cauchy sequence and hence is convergent.

The remaining part of the proof is to establish the convergence rate. The proof is similar to one in section 11.8. We leave the details to the interested reader. □

We now present some numerical results to demonstrate that the ℓ_1 DC-based Algorithm 11.7 below works well.

ALGORITHM 11.7
The ℓ_1 DC-Based Algorithm
  1: We use the same initialization as in section 11.7.
  2: **repeat** For $k \ge 1$, do the following:
  3:     Solve (11.126) using a thresholding technique to get $\mathbf{y}^{(k+1)}$
  4: **until** the maximal number of iterations is reached.

We have experimented with Algorithm 11.7 numerically for retrieving solutions in the real variable setting. We use $n = 100$ and $m = 1.1n, 1.2n, \ldots, 2.5n$. All measurement vectors $\mathbf{a}_j, j = 1, \ldots, m$, are Gaussian random vectors. So is $\mathbf{x_b}$. We used Algorithm 11.7 to recover $\mathbf{x_b}$ from $b_j = |\langle \mathbf{a}_j, \mathbf{x_b} \rangle|^2, j = 1, \ldots, m$, based on 5000 iterations when $n = 1000$. To recover a general solution $\mathbf{x_b}$, we used a small value, $\lambda = 1e - 5$. We repeated the experiment 100 times and summarized the frequency of retrievals in Table 11.1.

Next we explain how to use the ℓ_1 DC-based algorithm to retrieve sparse solutions. We should use m smaller than $2n$, as shown in Table 11.1; when $m \ge 2n$, we can retrieve any solution, no matter whether it is sparse or not, with a high success rate. However, when $m \approx 1.5n$, we cannot retrieve solutions. For retrieving sparse $\mathbf{x_b}$, the performance of Algorithm 11.7 is not very good. We improve Algorithm 11.7 by using an alternating projection method. That is, we project $\mathbf{y}^{(k+1)}$ from (11.126) to the set of all s-sparse vectors and use it as an initial guess for Algorithm 11.7. We repeat the projection and gradient iteration.

Table 11.1. *The numbers of successful retrieved solutions over* 100 *repeated runs based on numbers of measurements satisfying the listed relations* m/n, *where* $n = 100$ *is the size of the solution.*

m/n	1.5	1.6	1.7	1.8	1.9	2	2.1	2.2	2.3	2.4	2.5
DC alg.	0	0	2	8	28	57	72	91	93	93	99

Table 11.2. *The numbers of successes with sparsities* $s = 2, 4, 5, 10, 20$ *over* 1000 *repeated trials of Algorithm* 11.8.

m/n	0.5	0.6	0.7	0.8	0.9	1.0	1.1	1.2	1.3	1.4	1.5	1.6	1.7	1.8	1.9	2
2	422	517	566	619	662	710	745	829	846	837	833	849	862	870	877	946
4	187	333	429	474	590	673	693	747	778	811	819	832	840	862	873	919
5	71	116	264	383	452	594	618	674	726	771	802	821	837	858	869	894
10	0	0	0	52	151	247	416	385	537	482	590	680	701	737	796	812
20	0	0	0	0	0	55	156	180	227	271	368	422	451	527	574	599

ALGORITHM 11.8
ℓ_1 DC with Hard Thresholding Algorithm

1: We use the same initialization as in section 11.7.
2: **repeat** For $k \geq 1$, do the following:
3:      Solve (11.126) using the thresholding technique to get $y^{(k+1)}$.
4:      Project $y^{(k+1)}$ into the collection $\mathcal{R}_s$, an s-sparse set. That is, let $z^{(k+1)}$ solve the minimization problem

$$\sigma_s(x^k) = \min_{z \in \mathcal{R}_s} \|y^{(k+1)} - z\|_1. \tag{11.135}$$

5:      Let $y^{(k+1)} = z^{(k+1)}$ and repeat the computation
6: **until** the maximal number of iterations is reached.

Example 11.61. In this example, we show that our ℓ_1 DC-based algorithm with hard thresholding, Algorithm 11.8, works well. We choose $n = 128$ and the number of measurements $m = 0.5n, 0.6n, \ldots, 2n$. For each m, we test the performance of Algorithm 11.8 with sparsities $s = 2, 4, 5, 10, 20$. The experiments are implemented for 1000 repeated trials. The results on the number of successes are presented in Table 11.2. From Table 11.2, we can see that Algorithm 11.8 recovers sparse solutions with high probability.

11.10 · Numerical Experimental Results

In this section, we report some computational results based on the Wirtinger flow algorithm, the Gauss–Newton algorithm, and the DC-based algorithm. The significant result is to demonstrate that the DC-based algorithm retrieves real signals of size n from $m > 2n$ measurements with high probability around 80% over 1000 repeated runs. It is better than both the Gauss–Newton algorithm and the Wirtinger flow algorithm, which requires $m \geq 3n$ to be able to retrieve the real variable solution.

Example 11.62. In this example, we consider recovering a real signal x_b from the given measurements (11.1) using Gaussian random measurement vectors $a_j, j = 1, \ldots, m$. We fix $n = 128$ and consider the number of measurements m as around the twice the dimension of x_b, i.e., $m = kn/16$ for $k = 24, 25, \ldots, 35$. For the initialization, we first obtain an initial guess by the initialization Algorithm 11.4 and then improve the initial guess by applying the alternating

Table 11.3. *The number of successes over* 1000 *repeated trials versus the number of measurements* m/n *listed.*

m/n	1.5	1.5625	1.6250	1.6875	1.75	1.8125	1.875	1.9375	2	2.0625	2.125	2.1875
WF	0	0	0	0	0	1	11	10	24	27	41	64
GN	0	0	0	18	13	36	71	114	167	251	315	415
DC	50	78	119	182	266	318	406	542	600	681	744	807

Table 11.4. *The number of successes over* 1000 *repeated trials versus the number of measurements* m/n *listed, where* $n = 128$.

m/n	2.4375	2.5	2.5625	2.625	2.6875	2.75	2.8125	2.875	2.937	3
WF successes	168	220	254	352	372	459	513	612	641	706
GN successes	728	749	844	886	908	934	931	963	968	982
DC successes	944	952	975	982	984	989	994	993	995	998

projection method discussed in Algorithm 11.8. We say a trial is successful if the relative error is less than 10^{-5}. Table 11.3 gives the empirical success rate of recovering x_b for DC, Wirtinger flow (WF), and Gauss–Newton (GN) methods. From Table 11.3, we can see that the DC-based algorithm can recover the solutions with probability larger than 60% for $m \geq 2n$. According to the result in section 11.2, we need $m \geq 2n - 1$ measurements to guarantee the recovery of all real signals. Thus the DC-based algorithm almost reaches the theoretical low bound.

Example 11.63. We next repeat Example 11.62 using a few more measurements. The number of successes for the Wirtinger flow algorithm, the Gauss–Newton algorithm, and the DC-based algorithm is shown in Table 11.4. One can see that the performance of the DC-based algorithm is the best and can achieve a 95% success rate with $m = 2.5n$.

Example 11.64. This example shows the robustness of the DC-based algorithm. We repeat the computation in Example 11.62 for noisy measurements. There are two ways to generate the noisy measurements. One way is to add the noises η_j to b_j directly and obtain

$$\hat{b}_j = |\langle a_j, x_b \rangle|^2 + \eta_j, \quad j = 1, \dots, m. \tag{11.136}$$

Another way is

$$\tilde{b}_j = |\langle a_j, x_b \rangle + \delta_j|^2 + \eta_j, \quad j = 1, \dots, m, \tag{11.137}$$

where δ_j and η_j are noises. For noisy measurements (11.136), we assume that η_j are subject to uniform random distribution between $[-\beta, \beta]$ with mean zero, where $\beta = 10^{-1}, 10^{-3}$, and 10^{-5}. If the stopping tolerance ϵ satisfies $\epsilon \geq \beta$, then the Gauss–Newton method and the DC-based method produce the same empirical success rate as in Table 11.4. For noisy measurements (11.137), we assume that both ϵ_j and δ_j are subject to uniform distribution in $[-\beta, \beta]$ with mean zero. Similarly, if the stopping tolerance ϵ satisfies $\epsilon \geq \beta$, then both algorithms can recover the true solution.

Example 11.65. In this example, we use the DC-based algorithm and the Gauss–Newton method to recover the complex signals. We choose $n = 128$, and the number of measurements m is around $3n$, i.e., $m \approx 3n$. For Gaussian random measurements $a_j = a_{j,R} + i a_{j,I}, j = 1, \dots, m$, we aim to recover $z \in \mathbb{C}^n$ with $z = x + iy$ from $|\langle a_j, z \rangle|^2, j = 1, \dots, m$. The maximum iteration numbers for Wirtinger flow, Gauss–Newton, and DC are 3000, 100, and 1000, respectively. We

Table 11.5. *The number of successes over* 1000 *repeated trials based on* m/n *listed for the complex case.*

m/n	2.938	3	3.062	3.125	3.187	3.25	3.312	3.375	3.437	3.5	3.562	3.625	3.687	3.75
WF	0	0	0	0	0	0	0	0	0	56	192	204	322	401
GN	191	338	304	416	452	536	594	739	744	762	801	815	910	912
DC	422	563	537	565	623	730	829	887	881	894	954	946	981	986

Table 11.6. *The number of successes over* 1000 *repeated trials based on Algorithm* 11.9.

m/n	1.5	1.6	1.7	1.8	1.9	2	2.1	2.2	2.3	2.4	2.5
ℓ_1 DC alg.	0	0	206	317	352	557	724	913	938	947	994

say a trial is successful if the relative error is less than 10^{-5}. Table 11.5 gives the number of successes for Wirtinger flow, Gauss–Newton, and DC with 1000 repeated trials. From Table 11.5, we can see that the DC-based algorithm can recover the complex signals very well when $m \geq 3n$, which is slightly better than the Gauss–Newton algorithm and much better than the Wirtinger flow algorithm.

We next present some numerical experiments to demonstrate that the ℓ_1 DC-based algorithm works well. The procedure is presented in Algorithm 11.9, where a modified Attouch–Peypouquet technique [2] is used. We use the step size $\beta_k = k/(k + \alpha)$ for the first few k iterations, say $k \leq K$, and then a fixed step size β_K for the remaining iterations.

ALGORITHM 11.9
Accelerated ℓ_1 DC-Based Algorithm
1: We use the same initialization as in the previous examples.
2: For $k \geq 1$
3: Step 1. Solve (11.126) to get $\mathbf{y}^{(k+1)}$.
4: Step 2. Apply a modified Attouch–Peypouquet technique to get a new $\mathbf{y}^{(k+1)}$
5: until the maximal number of iterations is reached.
6: End **return** $\mathbf{y}^\top$.

Example 11.66. In this example, we test the performance of Algorithm 11.9 for recovering the real signals without sparsity. We choose $n = 128$ and the number of measurements $m = 1.1n, 1.2n, \ldots, 2.5n$. The target signal $\mathbf{x_b}$ and the measurement vectors $\mathbf{a}_j, j = 1, \ldots, m$, are Gaussian random vectors. We choose the parameter $\lambda = 10^{-5}$ in Algorithm 11.9. The number of successes is summarized in Table 11.6. The results show that the ℓ_1 DC-based algorithm can recover the real signals with high probability if $m \geq 2.2n$.

11.11 ▪ Exercises

Exercise 1. Prove Theorem 11.11.

Exercise 2. Prove Theorem 11.13.

Exercise 3. Show that each polynomial in (11.21) is not identically zero if Q is not a zero matrix.

Exercise 4. Let $\mathbf{e}_i, i = 1, \ldots, n \in \mathbb{R}^n$, be the standard unit vectors. Let $\mathbf{e}_i + \mathbf{e}_j, i, j = 1, \ldots, n$, and $i < j$. Show that if one uses these measurement vectors to find measurements (11.1) with $\epsilon_i = 0$ of a unknown vector $\mathbf{z}$, one can recover the phase of vector $\mathbf{z}$ up to a unimodular constant.

Exercise 5. Prove the polarization identity, i.e., Lemma 11.23.

Exercise 6. Prove Lemma 11.34.

Exercise 7. Prove the following lemma.

Lemma 11.67. *Suppose that* $\mathbf{a} = x + iy \in \mathbb{C}$ *with* x *and* y *i.i.d. random variables subject to* $N(0, 1/2)$. *Then the expectation of the even moments of* $\mathbf{a}$ *is*

$$\mathbb{E}(|\mathbf{a}|^{2k}) = k! \quad \forall k \geq 1.$$

Exercise 8. Suppose that a, b, c, d are i.i.d. Gaussian random variables subject to $N(0, 1)$. Compute the expectations $\mathbb{E}(abcd)$, $\mathbb{E}(a^2bc)$, $\mathbb{E}(a^2b^2)$, $\mathbb{E}(a^3b)$, and $\mathbb{E}(a^4)$.

Exercise 9. Suppose the sequence $\{\mathbf{a}_i, i = 1, \ldots, m\}$ follows the Gaussian model. Show that for any fixed vector $\mathbf{x} \in \mathbb{C}^n$,

$$\mathbb{E}\left(\frac{1}{m}\sum_{r=1}^{m}(\mathbf{a}_r^*\mathbf{x})^2\mathbf{a}_r\mathbf{a}_r^*\right) = 2\mathbf{x}\mathbf{x}^*. \tag{11.138}$$

Exercise 10. Suppose the sequence $\{\mathbf{a}_i, i = 1, \ldots, m\}$ follows the Gaussian model. Show that for any fixed vector $\mathbf{x} \in \mathbb{C}^n$,

$$\mathbb{E}\left(\frac{1}{m}\sum_{r=1}^{m}|\mathbf{a}_r^*\mathbf{x}|^2\mathbf{a}_r\mathbf{a}_r^*\right) = \mathbf{x}\mathbf{x}^* + \|\mathbf{x}\|^2 I_n. \tag{11.139}$$

Exercise 11. Consider any continuous function $f(\mathbf{z}) : \mathbb{C}^n \mapsto \mathbb{R}$ with continuous first- and second-order Wirtinger derivatives. For any $\delta \in \mathbb{C}^n$ and scalar t, show that

$$f(\mathbf{z} + t\delta) = f(\mathbf{z}) + t \int_0^1 \begin{bmatrix} \delta \\ \bar{\delta} \end{bmatrix}^* \nabla f(\mathbf{z} + ts\delta)ds, \tag{11.140}$$

$$f(\mathbf{z} + t\delta) = f(\mathbf{z}) + t \begin{bmatrix} \delta \\ \bar{\delta} \end{bmatrix}^* \nabla f(\mathbf{z}) + t^2 \int_0^1 (1 - s) \begin{bmatrix} \delta \\ \bar{\delta} \end{bmatrix}^* \nabla^2 f(\mathbf{z} + st\delta) \begin{bmatrix} \delta \\ \bar{\delta} \end{bmatrix} ds.$$

Exercise 12. Consider any continuous function $f(\mathbf{z}) : \mathbb{C}^n \mapsto \mathbb{R}$ with continuous first- and second-order Wirtinger derivatives. Suppose that $\nabla^2 f(\mathbf{z})$ is L-Lipschitz continuous. Then let

$$\widehat{f}(\mathbf{z}, \delta) = f(\mathbf{z}) + \begin{bmatrix} \delta \\ \bar{\delta} \end{bmatrix}^* \nabla f(\mathbf{z}) + \frac{1}{2} \begin{bmatrix} \delta \\ \bar{\delta} \end{bmatrix}^* \nabla^2 f(\mathbf{z}) \begin{bmatrix} \delta \\ \bar{\delta} \end{bmatrix}. \tag{11.141}$$

Show that around each point $\mathbf{z}$, we have the estimate

$$|f(\mathbf{z} + t\delta) - \widehat{f}(\mathbf{z}, \delta)| \leq \frac{1}{3}L\|\delta\|^2. \tag{11.142}$$

Exercise 13. Prove (11.88). *Hint:* Use Lemma 11.67.

Exercise 14. Prove Lemma 11.40.

Exercise 15. Prove Lemma 11.41.

Exercise 16. Let A be an $m \times n$ matrix with complex random variables in its entries with $m > n$. That is, $A = [a_{ij}]$ with $a_{ij} = x_{ij} + iy_{ij}$ and $x_{ij}, y_{ij} \in \mathcal{N}(0, 1/2)$. Show that

$$\sqrt{m} - \sqrt{n} \leq \mathbb{E}(\sigma_{max}(A)) \leq \sqrt{m} + \sqrt{n}. \tag{11.143}$$

Furthermore, for each $t > 0$, show that it holds with probability at least $1 - 2\exp(-t^2)$ that

$$\sqrt{m} - \sqrt{n} - t \leq \mathbb{E}(\sigma_{max}(A)) \leq \sqrt{m} + \sqrt{n} + t. \tag{11.144}$$

Exercise 17. Prove Lemma 11.45.

Exercise 18. Show that under the initialization Y in (11.83), letting $\mathbf{z}^{(1)}$ be an eigenvector corresponding to the largest eigenvalue of matrix Y and $\|\mathbf{z}^{(1)}\| = \lambda$,

$$\text{dist}(\mathbf{z}^{(1)}, \mathbf{x_b}) \leq C_\theta \|\mathbf{x_b}\| \tag{11.145}$$

holds with high probability for a positive constant C_θ.

Exercise 19 (Modified Phase Retrieval Problem).   We need to solve the modified phase retrieval problem

$$y_j = |\langle \mathbf{a}_j, \mathbf{z}\rangle|^2 + \langle \mathbf{b}_j, z\rangle, \quad j = 1, 2, \ldots, m, \tag{11.146}$$

where $\mathbf{z} \in \mathbb{C}^n$ is a targeted variable, $\mathbf{a}_j, \mathbf{b}_j \in \mathbb{C}^n$ are known sampling vectors, and $y_j \in \mathbb{R}$ for $j = 1, \ldots, m$. Show that if $m = 2n$ and $\mathbf{b}_j, \in \mathbb{C}^n, j = 1, \ldots, m$, are linearly independent in $\mathbb{R}^n$ in the sense that matrix $B = [\text{Re}(\mathbf{b}_j)^\top \; \text{Im}(\mathbf{b}_j)^\top]_{j=1,\ldots,2n}$ of size $2n \times 2n$ is of rank $2n$, then there exists a unique solution $\mathbf{z}$ satisfying the equations in (11.146).

Exercise 20 (Modified Phase Retrieval Problem).   Develop a numerical method, say the Wirtinger flow algorithm, to solve (11.146) by solving the minimization

$$\min_{\mathbf{z}} f(\mathbf{z}) = \frac{1}{2m} \sum_{j=1}^{m} (|\langle \mathbf{a}_j, \mathbf{z}\rangle|^2 + \langle \mathbf{b}_j, \mathbf{z}\rangle - y_j)^2, \quad j = 1, 2, \ldots, m, \tag{11.147}$$

where $\mathbf{b}_j \in \mathbb{C}^n, j = 1, \ldots, m \geq 2n$, may not be linearly independent in $\mathbb{R}^n$.

Exercise 21 (Generalized Phase Retrieval Problem).   Study the generalized phase retrieval problem

$$y_j = |\langle \mathbf{a}_j, \mathbf{z}\rangle|^4, \quad j = 1, 2, \ldots, m, \tag{11.148}$$

by finding sufficient conditions on m and $\mathbf{a}_j$ such that this problem has a unique solution in $\mathbb{R}^n$ up to $\pm$ and in $\mathbb{C}^n$ up to a unimodular constant.

Exercise 22 (Generalized Phase Retrieval Problem).   Develop a numerical approach for the generalized phase retrieval problem (11.148).

Exercise 23. Prove (11.36).

Exercise 24. Show that when a frame is tight, i.e., $A = B$, one has (11.35) with $\widetilde{\mathbf{a}_k} = \mathbf{a}_k/A$.

Bibliography

[1] B. Alexeev, A. Bandeira, M. Fickus, and D. Mixon, Phase retrieval with polarization, SIAM J. Imaging Sci., 7 (2014), 35–66. (Cited on p. 397)

[2] H. Attouch and J. Peypouquet, The rate of convergence of Nesterov's accelerated forward-backward method is actually faster than $1/k^2$, SIAM J. Optim., 26 (2016), 1824–1834. (Cited on p. 432)

[3] R. Balan, Reconstruction of signals from magnitudes of redundant representations, Found. Comput. Math., 16 (2016), 677–721. (Cited on p. 397)

[4] R. Balan, P. Casazza, and D. Edidin, On signal reconstruction without phase, Appl. Comput. Harmon. Anal., 20 (2006), 345–356. (Cited on pp. 389, 397)

[5] R. Balan, B. Bodmann, P. Casazza, and D. Edidin, Painless reconstruction from magnitudes of frame coefficients, J. Fourier Anal. Appl., 15(4) (2009), 488–501. (Cited on p. 392)

[6] B. Balan and Y. Wang, Invertibility and robustness of phaseless reconstruction, Appl. Comput. Harmon. Anal., 38 (2015), 469–488. (Cited on p. 396)

[7] A. S. Bandeira, J. Cahill, D. G. Mixon, and A. A. Nelson, Saving phase: Injectivity and stability for phase retrieval, Appl. Comput. Harmon. Anal., 37(1) (2014), 106–125. (Cited on pp. 389, 390)

[8] J. Barzilai and J. Borwein, Two-point step size gradient methods. IMA J. Numer. Anal., 8(1) (1988), 141–148. (Cited on pp. 419, 424)

[9] A. Beck and M. Teboulle, A fast iterative shrinkage-thresholding algorithm for linear inverse problems, SIAM J. Imaging Sci., 2 (2009), 183–202. (Cited on p. 426)

[10] E. J. Candés, Y. C. Eldar, T. Strohmer, and V. Voroninski, Phase retrieval via matrix completion, SIAM J. Imaging Sci., 6 (2013), 199–225. (Cited on p. 384)

[11] E. J. Candés, X. Li, and M. Soltanolkotabi, Phase retrieval from coded diffraction patterns, Appl. Comput. Harmon. Anal., 39 (2015), 277–299. (Cited on p. 383)

[12] E. J. Candés, X. Li, and M. Soltanolkotabi, Phase retrieval via Wirtinger flow: Theory and algorithms, IEEE Trans. Inform. Theory, 61 (2015), 1985–2007. (Cited on pp. 405, 406, 408, 409, 414)

[13] A. Conca, D. Edidin, M. Hering, and C. Vinzant, An algebraic characterization of injectivity in phase retrieval, Appl. Comput. Harmon. Anal., 38(2) (2015), 346–356. (Cited on p. 391)

[14] I. Daubechies, Ten Lectures on Wavelets, SIAM, 1992. (Cited on p. 395)

[15] C. Dainty and J. R. Fienup, Phase retrieval and image reconstruction for astronomy, in Image Recovery: Theory and Application, Academic Press, 1987, 231–275. (Cited on p. 383)

[16] J. R. Fienup, Phase retrieval algorithms: A comparison, Appl. Optics, 21 (1982), 2758–2769. (Cited on p. 383)

[17] W. Fulton, Intersection Theory, Springer-Verlag, 1984. (Cited on p. 387)

[18] B. Gao and Z. Xu, Phaseless recovery using the Gauss–Newton method. IEEE Trans. Signal Process., 65(22) (2017), 5885–5896. (Cited on pp. 410, 411, 414)

[19] W. H. Guo and M. J. Lai, Box spline wavelet frames for image edge analysis, SIAM J. Imaging Sci., 6 (2013), 1553–1578. (Cited on p. 396)

[20] J. Harris, Algebraic Geometry: A First Course, Grad. Texts in Math., Springer, 1992. (Cited on pp. 385, 387)

[21] M. Huang, M. J. Lai, A. Varghese, and Z. Xu, On DC Based Methods for Phase Retrieval, Approximation XVI, San Antonio, 2019, Springer Verlag, 2021. (Cited on pp. 386, 421)

[22] K. Jaganathan, Y. C. Eldar, and B. Hassibi, Phase Retrieval: An Overview of Recent Developments, preprint, arXiv:1510.07713v1, 2015. (Cited on p. 383)

[23] K. Kurdyka, On gradients of functions definable in o-minimal structure, Ann. Inst. Fourier, 48 (1998), 769–783. (Cited on p. 423)

[24] M.-J. Lai and J. Lee, Phase Retrieval and Beyond, private communication, 2020. (Cited on p. 393)

[25] M.-J. Lai and J. Stoeckler, Construction of multivariate compactly supported tight wavelet frames, Appl. Comput. Harmon. Anal., 21 (2006), 324–348. (Cited on p. 396)

[26] S. Lojasiewicz, Une propriété topologique des sous-ensembles analytiques réels, in Les Équations aux Dérivées Partielles, Éditions du centre National de la Recherche Scientifique, Paris, 1963, 87–89. (Cited on p. 423)

[27] M. Moravec, J. Romberg, and R. Baraniuk, Compressive phase retrieval, in Proceedings of SPIE, International Society for Optics and Photonics, 2007. (Cited on p. 384)

[28] J. Miao, T. Ishikawa, B. Johnson, E. H. Anderson, B. Lai, and K. O. Hodgson, High resolution 3D X-Ray diffraction microscopy, Phys. Rev. Lett., 89 (2002), 088303. (Cited on p. 383)

[29] R. P. Millane, Phase retrieval in crystallography and optics, J. Opt. Soc. Amer. A, 7 (1990), 394–411. (Cited on p. 383)

[30] R. W. Harrison, Phase problem in crystallography, J. Opt. Soc. Amer. A, 10 (1993), 1046–1055. (Cited on p. 383)

[31] R. T. Rockafellar, Convex Analysis, Princeton University Press, 1970. (Cited on p. 423)

[32] I. R. Shafarevich, Basic Algebraic Geometry 1, Springer-Verlag, Berlin, 2013. (Cited on p. 386)

[33] Y. Shechtman, Y. C. Eldar, O. Cohen, H. N. Chapman, J. Miao, and M. Segev, Phase retrieval with application to optical imaging: A contemporary overview, Signal Process. Mag., IEEE, 32(3) (2015), 87–109. (Cited on p. 383)

[34] J. Sun, Q. Qu, and J. Wright, A geometric analysis of phase retrieval, Found. Comput. Math., 18 (2017), 1131–1198. (Cited on pp. 401, 402)

[35] Y. Wang and Z. Xu, Phase retrieval for sparse signals, Appl. Comput. Harmon. Anal., 37 (2014), 531–544. (Cited on pp. 398, 399)

[36] O. Zariski, On the purity of branch locus of algebraic functions, Proc. Natl. Acad. USA, 44 (1958), 791–796. (Cited on p. 385)

Index